W9-CLX-019
LA MOSCOVIE
St. Michel Archangel
Pologne
Constantinople
Turquie
Arabie
LE SAARA ou DESERT
CONGO
OCEAN
Mer de
INDIEN
Ligne Equino
I. de Bourbon
I. de Ficos
I. de Tristan de
OCEAN

# THE QUEST FOR LONGITUDE

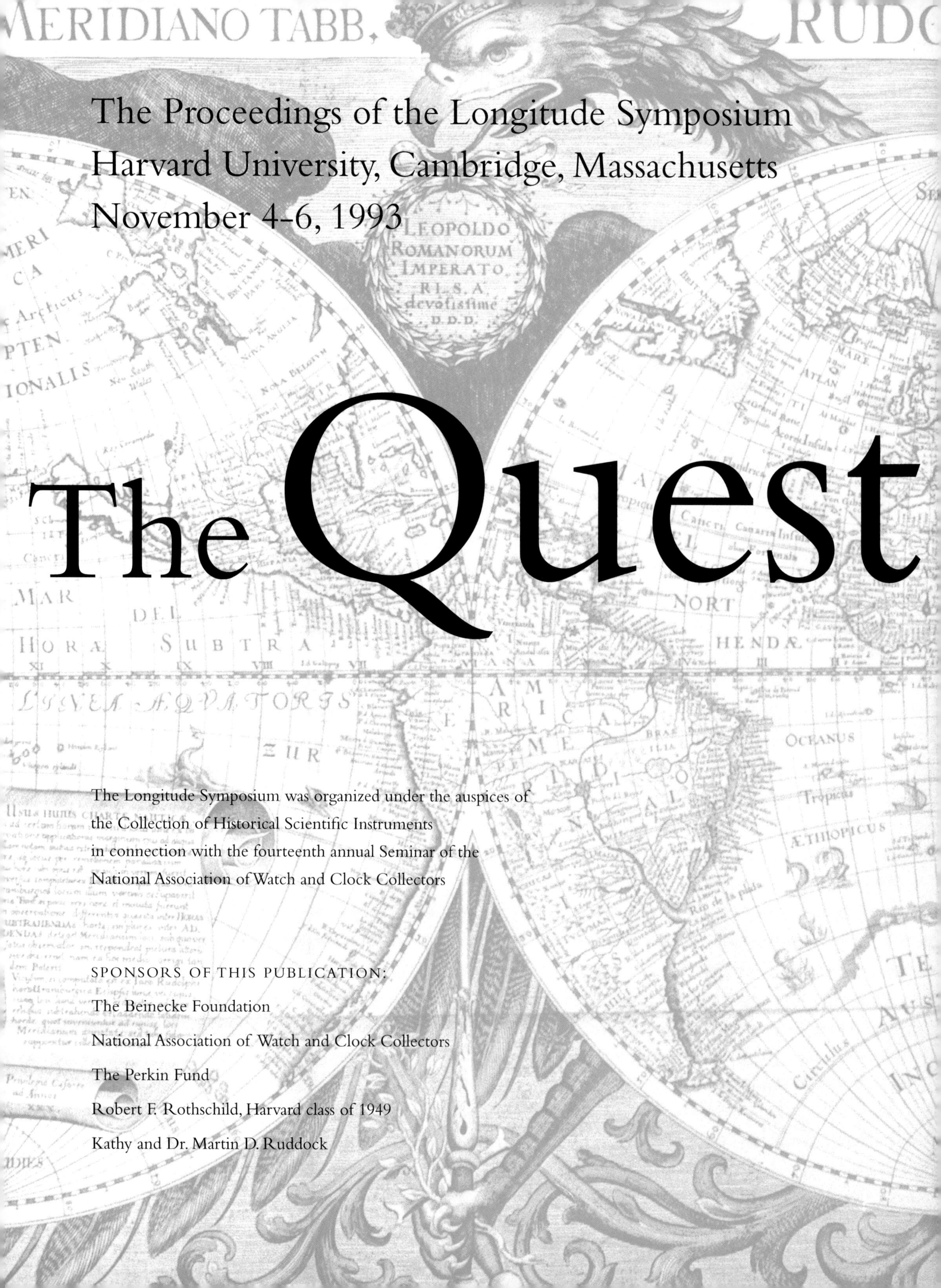

The Proceedings of the Longitude Symposium
Harvard University, Cambridge, Massachusetts
November 4–6, 1993

# The Quest

The Longitude Symposium was organized under the auspices of
the Collection of Historical Scientific Instruments
in connection with the fourteenth annual Seminar of the
National Association of Watch and Clock Collectors

SPONSORS OF THIS PUBLICATION:
The Beinecke Foundation
National Association of Watch and Clock Collectors
The Perkin Fund
Robert F. Rothschild, Harvard class of 1949
Kathy and Dr. Martin D. Ruddock

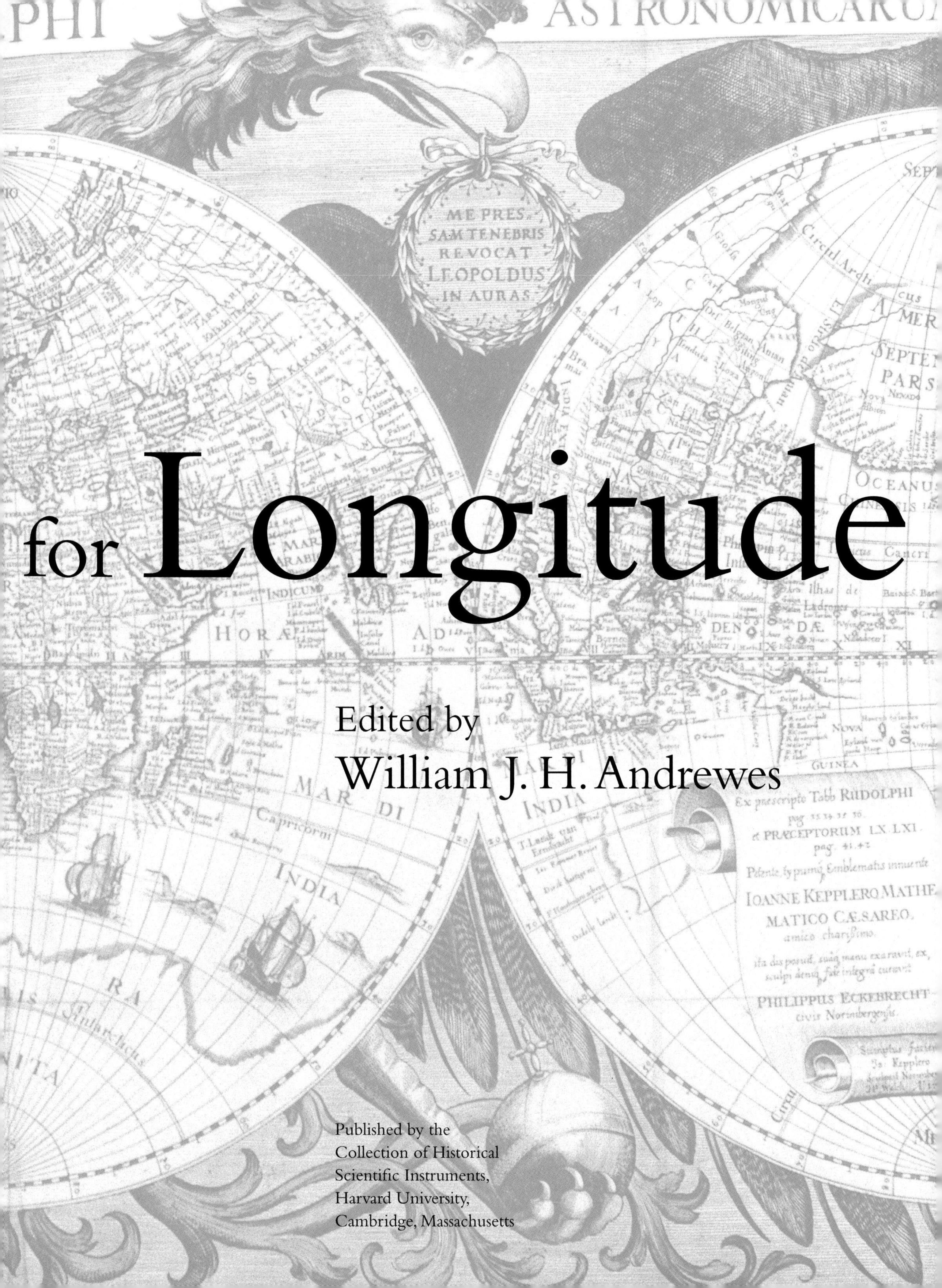


# for Longitude

Edited by
William J. H. Andrewes

Published by the
Collection of Historical
Scientific Instruments,
Harvard University,
Cambridge, Massachusetts

*The Quest for Longitude*

Produced and edited under the direction of William J. H. Andrewes
Editorial advisor: Bruce Chandler
Copy editor: Peggy Liversidge
Production assistant: Martha R. Richardson
Reader: Susan Rossi-Wilcox
Design: Pamela Geismar

Printing: Toppan Printing Company, Shenzhen Ltd.
Printed on a pH-neutral paper.

Publisher: Collection of Historical Scientific Instruments



Printed in China

*Library of Congress Cataloguing-in-Publication data*
Longitude Symposium (1993: Harvard University)
The quest for longitude: the proceedings of the Longitude Symposium, Harvard University, Cambridge, Massachusetts, November 4–6, 1993 / edited by William J. H. Andrewes.
p. cm.
"The Longitude Symposium was organized under the auspices of the Collection of Historical Scientific Instruments in connection with the fourteenth annual seminar of the National Association of Watch and Clock Collectors."
Includes bibliographical references and index.
ISBN 0-9644329-0-0
1. Longitude—Measurement—History—Congresses.
2. Harrison, John, 1693–1776—Congresses.
3. Horology—History—Congresses.
I. Andrewes, William J. H., 1950–. II. Harvard University. Collection of Historical Scientific Instruments. III. National Association of Watch and Clock Collectors. Seminar (14th: 1993: Cambridge, Mass.) IV. Title.
QB225.L66 1996 96-25845
527'.2—dc20 CIP

**Overleaf: World map, dated 1630, by Philipp Eckebrecht** **(see p. 55).**

*Dedicated to the memory of*

David Pingree Wheatland (1898–1993)

*whose years of devotion*
*to the Collection of Historical*
*Scientific Instruments*
*provided the foundation*
*for this work.*

# Contents

**John Harrison**



**Perfecting the Marine Timekeeper**



**The James Arthur Lecture**



**Appendices**

# Introduction

On November 4, 1993, 500 people from seventeen countries gathered in Cambridge, Massachusetts, to attend the Longitude Symposium, a conference about the history of finding longitude at sea. Shortly before Alistair Cooke rose to deliver the concluding address, the well-known physicist Philip Morrison made a brief announcement that gave a modern perspective to this historical problem. From his pocket he produced a hand-held instrument and, with the push of a button, informed us of our location on the Earth's surface:

Latitude: North 42° 22.546', Longitude: West 071° 06.904'.

With this simple act, he demonstrated how advances in science and technology had reduced the problem of finding longitude—once the most crucial challenge facing every seafaring nation—to a small "black box" that provides a precise answer almost instantaneously on a miniature screen. Like so many products of modern technology, the Global Positioning System receiver conceals the nature of the problem being solved: the theoretical complexity, the technical difficulties, and the untold thousands of hours of thought and labor expended to tell us, so simply, exactly where we are. Technology has become such an integral part of our lives that we tend to take it entirely for granted. It has raised our expectations, changed our priorities, and left us perilously ignorant of how helpless we would be without it. Traveling overseas, we now complain when delayed for an hour: we have forgotten that once there were problems finding continents.

Because the problem of finding longitude has been so totally eclipsed by technology and no longer presents an obstacle in the path of progress, it might appear to be an obscure subject for a conference. Yet the solution, and the search for that solution, were to have an immense impact upon the expansion of Western civilization. Finding longitude became the nucleus that, both directly and indirectly, influenced the growth of many disciplines, and it is therefore a topic of considerable historical interest.

With the advent of ocean voyages by European explorers in the late fifteenth century, the inability to establish a ship's longitude at sea became of increasing concern to every seafaring nation: unlike navigation in the well-traveled waters of the English and European coastal regions, ocean voyages took mariners out of sight of land for several weeks. When clouds obscured their celestial guides and storms swept their vessels off course, they lost track of where they were and in which direction their destination lay. While a ship's latitude could be quickly re-established when the skies cleared, there was no method for finding its longitude. To add to the hazards facing those who ventured on an ocean voyage, no one knew exactly where land was. Maps were full of errors until a practicable and useful method of finding longitude allowed accurate surveys of coastlines to be made.

Had power and profit not been found in exploration, colonization, and trade, finding longitude might never have been regarded as a serious problem. But immense wealth and opportunity were, of course, found in the Americas and beyond, and as shipping increased, so did the significance of the longitude problem. Before the end of the sixteenth century, the loss of lives, expensive ships, and valuable cargoes had gained the serious attention of government. In 1598, Philip III of Spain offered substantial financial incentives (a perpetual pension of 6,000 ducats as a starter) to the "discoverer of the longitude." A few decades later, the States General of the Dutch Republic followed suit with a prize rumored to be 30,000 florins. Other rewards have also been recorded. The largest and most famous came in 1714 from the British government, which offered three prizes corresponding to three levels of accuracy. The highest reward of £20,000 may be equated, by comparing the average pay for skilled labor then and now, to approximately $12 million.

The longitude problem was, in fact, a daunting, multifaceted scientific and technological challenge. The many methods proposed may be divided into three basic categories: terrestrial, celestial, and mechanical. The first includes those that required a precise knowledge of the Earth's magnetic fields and measurements of their varying strengths by the deviation of a compass needle either in a horizontal plane, which gives its variation from true north, or in a vertical plane, which gives its inclination or dip from the horizontal. In theory, a sufficient number of observations would allow the creation of maps that showed lines of either magnetic variation or magnetic inclination superimposed on the grid of the lines of latitude and longitude. Thus, by measuring the latitude and the magnetic deviation of a particular location and determining the place through the corresponding observations marked on the map, a navigator could, in theory, use this method to find his position. However, when investigations revealed that the Earth's magnetic fields were shifting unpredictably, this method was abandoned. Another terrestrial method used sound signals. Although this was also abandoned because it was totally impracticable at the time it was suggested, it is in certain respects a distant ancestor of the Global Positioning System used today.

Celestial methods relied upon some astronomical occurrence to determine the time at a place of known longitude, which could then be compared with the local time at the place of observation: with the Earth rotating 1° of longitude in every four minutes of time, the longitude difference can easily be calculated once the time difference between two places is known. The Moon, which travels approximately 12° across the heavens each day, served as the hand of the celestial clock, with its dial of stars, and therefore presented several possibilities for finding the time at a place of known longitude: by eclipses, by her occultation of stars, by meridian transits, or by measurements of her distance from stars; the last became known as the lunar-distance method. The frequency and regularity of the eclipses of Jupiter's satellites also provided a celestial method with considerable potential, and it became, in fact, the principal method adopted for determining longitude on land.

The third category, the mechanical, was theoretically the most straightforward solution to the problem, in that the time of a place of known longitude was provided directly by a timekeeper carried aboard the ship. The ship's local time, found by celestial observation, could then be compared to the time on the clock, and from the difference between these times, the longitude could easily be calculated.

The only proposals that eventually became practicable for use at sea were the lunar-distance and the timekeeper methods. Neither was even close to being viable when first proposed during the first half of the sixteenth century. On the contrary, both required a substantial increase in the understanding of the Earth and the heavens, as well as a significant refinement of technological knowledge and expertise. How to turn theory into practice continued to elude the greatest scientific minds for more than two centuries.

It is difficult to imagine how restricted we would be without the technologies that have become part of our daily existence. Although we do not possess any natural sense that is superior to those of our forebears, science and technology make us think differently about the world we live in. We have come to accept—and expect—things that they, with great imagination, could only dream of. Unaided, the human faculties are very limited in their ability to delve into the secrets of nature. When demand becomes apparent, however, human creativity and ingenuity seem to find a way to solve almost any problem. The limitations of our natural senses were overcome through the invention and development of instruments. Those that enhanced the ability to see and to measure revealed new horizons of the natural world, and as a result, knowledge of natural philosophy began to advance rapidly: the telescope opened up the celestial sphere; the microscope gave new insight into the structure of matter; improvements in angular-measuring instruments and timekeeping devices allowed the form and size of the natural world to be more clearly defined. It was with the creation and development of devices that allowed insight into worlds beyond the range of the human faculties that the scientific revolution commenced.

Of all the demands placed upon instruments, none have been more stringent than those placed upon the mechanical clock, the device that became of central importance to the longitude story. A timekeeper has to keep going 24 hours a day, 365 days a year, for several years without attention—a requirement far beyond that expected of almost any other mechanical device. Furthermore, our expectations are such that if it doesn't keep good time, it is of no use. We think of 99 per cent as being good enough for most things in life. But for the accuracy of a precision timekeeper, it is unacceptable. In the measure of the day, which has 86,400 seconds, 99 per cent allows for a variance of 864 seconds, almost fifteen minutes. Even 99.99 per cent, permitting a variance of 8.64 seconds per day, would not qualify for the lowest award (£10,000) offered by the English government for a solution to the longitude problem. To qualify for the highest prize of £20,000, the timekeeper had to perform within two seconds per day and be capable of maintaining that kind of rate under the grueling conditions of an ocean voyage. It is not surprising, therefore, that when the prize was first offered, Isaac Newton and other leading academics, believing that degree of precision to be unattainable, considered the timekeeper method to be impracticable. Yet by the time Captain Cook returned from his exploration of the South Seas in 1775, such a performance had been accomplished on a voyage of three years' duration.

When Cook embarked upon this voyage, his second, in 1772, both the lunar-distance method and the timekeeper method had been perfected to a sufficient degree of accuracy to determine longitude at sea. This voyage, however, would be the most arduous trial either method had ever been subjected to. Despite the achievement of an English clockmaker, John Harrison, in developing a successful marine timekeeper, the lunar-distance method was still preferred by academics and astronomers and was recognized at the outset of the voyage as the primary method that Cook should use for charting the exact locations of his discoveries. Accordingly, he and his astronomers had been trained to undertake all the necessary calculations and were equipped with the required lunar-distance tables and with the finest sextants for making angular-distance observations of the Moon and stars. But lunar-distance observations had to be timed precisely, and therefore an accurate timekeeper was also needed.

The marine chronometer was still in its very early stage of development, the first successful ocean trials having been made only in the previous decade. By November 1771, eight months before Cook set sail, less than half a dozen such timekeepers had been completed in England, and it is a clear indication of the importance of the voyage that by the time that he embarked in mid-July, he had four with him, two on each of his ships. Three of these turned out to be erratic, but the other—Larcum Kendall's K.1, a copy of Harrison's prize-winning watch, H.4—performed with such extraordinary accuracy and reliability that Cook and William Wales, the astronomer aboard his ship, *Resolution,* began to use it on a regular basis for determining longitude. Practice showed that a timekeeper represented a much easier method of finding longitude than using lunar-distance measurements.

Cook's praise for the timekeeper method helped to promote the recognition of its reliability and ease of use and contributed to the rapid growth of its popularity. For fifteen years following his second voyage, the marine timekeeper underwent an intense period of development and evolved into the device known as the marine chronometer. As further refinements in design and production improved performance, reduced price, and increased availability, the chronometer started to be employed as the principal method of finding longitude at sea, while the lunar-distance method was maintained primarily to check results at periodic intervals. By 1790, the chronometer's fundamental design was so sophisticated that it remained largely unaltered until developments with quartz crystal changed the principles upon which the mechanism worked. Despite the introduction of new materials and production techniques, the chronometers made during World War II retained all of the essential elements of the design perfected 150 years earlier. Very few devices remained so untouched by the remarkable advances in science and technology that accompanied the Industrial Revolution and two world wars.

As with so many aspects of history, there is much to be learned from the longitude story. Solutions to problems do not always come from expected sources, but from unknown individuals in remote areas who, being imbued with a passion and determination to succeed, can approach the problem without the restricted vision that traditional academic thought can sometimes impose on novel ideas. An incentive, usually of fame and fortune, that allows any promising individual regardless of background or education to participate will often provide the fastest and most cost-efficient results. When the £20,000 prize was offered as a stimulus to people from all nations and all walks of life to solve the longitude problem, the British Parliament also established a select group of scholars and high ranking officials, known as the Board of Longitude, to review and, when necessary, provide encouragement to promising individuals. The reward provided an important impetus for not just one, but two, practicable solutions during the next 50 years. This may seem like a long period of time, but it is only a fraction of the two and a half centuries of effort already devoted to the problem.

The person who developed the solution to what would ultimately become the principal method for finding longitude at sea had neither university education nor formal training in an associated field. John Harrison, the inventor of the first successful marine timekeeper, was trained as a carpenter and came from a small, remote village. Despite his humble origins, his remarkable insight into the laws of physics and the nature of materials, combined with endless perseverance and determination, enabled him to overcome problems that were thought by the greatest scientific minds of that time to be insurmountable.

Developments in one field often give rise to the evolution of specialized branches in another. Such was the case with the technologies developed for keeping time. The increase in the accuracy of these mechanisms led to the demand for improvements in the instruments used for finding time. Some of the most famous makers of precision instruments in the eighteenth century were closely associated with, or even employed by, some of the greatest clockmakers. When the new theory that made the lunar-distance method viable was advanced in the 1750s, it took little time to develop the octant into a more precise angular-measuring instrument that allowed this method to become practicable for finding longitude at sea: the sextant, like the marine chronometer, was of such good design that it too was never improved upon. The black plastic examples produced today retain all the principal features of their more elegant polished brass ancestors.

Although the longitude problem was, first and foremost, a scientific and technological one, it was also inextricably a part of the larger nexus of law, politics, and empire building, of social and cultural history, and of commerce and economics. Improvements in navigation allowed not only safer but also more direct (and hence faster) passage across the oceans, resulting in greater intercontinental trade and the creation of new markets. The ability to chart geographic discoveries with ever-greater precision opened to European colonization and development regions that hitherto had been limited to explorers and pioneers. Subsequent discoveries of new plant and animal life greatly expanded the field of natural history, and the revelation of large deposits of gold and other natural resources led to commercial exploitation. These developments in turn caused massive shifts in population, significantly expanding the influence of some cultures while suppressing or even eradicating others. The enormous sums that had been invested in maritime enterprise began to pay off once the longitude problem was solved: for the next 150 years, Great Britain was the world's dominant naval power.

The purpose of this volume is to provide a detailed and well-illustrated account of the longitude story that will serve as a reliable resource for those interested in various aspects of the subject. In addition to serving the needs of teachers, students, scholars, and collectors, it is hoped that this book will also appeal to those with a more general interest in history. Some readers may have been recently introduced to the longitude story by two engaging and highly successful books published in 1995, Umberto Eco's *The Island of the Day Before* and Dava Sobel's *Longitude*. The latter, related to this volume in that it shares the same parent, was the first offspring of the Longitude Symposium and, as such, is this book's sibling. Sobel was engaged to cover the symposium by *Harvard Magazine*. Her article, which appeared in the March/April 1994 issue, inspired George Gibson of Walker and Company to commission her to write *Longitude,* which provides a general and highly readable account of the longitude story, focused on John Harrison's achievements. It is hoped that readers who have been introduced to the subject by these works will find the many illustrations and the wider context of this volume useful and interesting.

The story of finding longitude has, of course, been told many times before, but usually from the point of view of one field of interest. However, the subject—even when limited to the period of 1500 to 1800—involves so many specialized disciplines that it would be almost impossible for one person to undertake a detailed account of the many facets of this story. In its analysis of the history of this subject from the point of view several disciplines, it is hoped that this volume will represent a landmark in collaborative research: the individuals who were chosen to represent a particular period of development or aspect of the problem at the Longitude Symposium are all leading authorities in their respective fields. I was fortunate to have had the opportunity of meeting most of them when I worked at The Time Museum in Rockford, Illinois, and earlier at the Old Royal Observatory at Greenwich.

To these friends and colleagues, who, having served as speakers at the symposium, now contribute as authors to this volume, I owe a special debt of gratitude. Their involvement has represented a significant commitment of time away from their regular work, especially for those who do not have the good fortune to have the ongoing support of an academic institution. Of all who have given so freely and generously of their time, I would like to express particular appreciation to two individuals whose presence added greatly to the conference and to this volume: Alistair Cooke, whose magnanimity in participating has drawn the attention of a much wider audience than would otherwise have been possible; and Jeremy Knowles, who, amid ever-consuming responsibilities as Dean of the Faculty of Arts and Sciences at Harvard, took the time to show his support for this endeavor.

The history of how the Longitude Symposium evolved is given in the first section of Appendix B, but it is appropriate to mention here two of the main elements that enabled the conference to be a success. The first was its organizational structure. The opportunity for arranging the conference came about as a result of my involvement with the National Association of Watch and Clock Collectors (NAWCC), the largest horological society in the world. The NAWCC has for some time held annual seminars on the history of time measurement, and it was as their fourteenth annual seminar that the Longitude Symposium was originally organized. Although the symposium grew to be larger than any seminar in this series, the NAWCC provided the ideal vehicle for the conference. Through this large, well-established organization, immediate support was forthcoming for the conference committee, a core of interest in the event was generated by the regular attendees of the seminars, and funding was granted by several of the NAWCC's chapters, as well as by collectors and dealers engaged in the field of horology. Through the encouragement of the symposium's assistant chairman, Robert Cheney, and others, the NAWCC's Council became one of the symposium's main sponsors. In summary, the NAWCC played a major role in the success of the Longitude Symposium.

The second element in the symposium's success was that it was organized at a major university that was equipped with all the resources required for an international conference on such a seemingly unusual subject. It is impossible to overestimate the important role that Harvard University has played, both directly and indirectly, in the creation of this volume. While the expenses of those hired to work on this project represent direct costs covered by our generous donors, the expenses involved in my own time have been covered through my appointment as Curator of the Collection of Historical Scientific Instruments, and I could not have undertaken this project without that support. My appointment was made possible through the generosity of David Pingree Wheatland (Harvard Class of 1922), who not only founded the Collection, but also served as its main source of support and growth for 45 years. Wheatland also contributed greatly to the Harvard Libraries, which, in addition to being one of the world's greatest repositories of books and manuscripts, must also rank as one of the most accessible—and therefore useful—research libraries for students and scholars. Sadly, Wheatland died on May 4, 1993, exactly six months before the Longitude Symposium took place. But through his devotion to his alma mater, he had established the firm foundation upon which the conference was organized. Hence, it is to him that this work is dedicated.

Securing funding for this volume has been of critical importance. Initial capital was generated by the Longitude Symposium, but it was evident that additional financing would be needed. Some was raised by inviting attendees of the symposium to subscribe to this volume by paying for their copy in advance; a list of these individuals is contained in Appendix A. We also sought major donations. The first to give to this project was Marty Ruddock, a long-time friend and dedicated horologist with a passion for precision timekeepers. Help was also forthcoming from the Perkin Fund and the NAWCC. Soon after the symposium, I approached James Baker, who, in his capacity as a trustee of the Perkin Fund, had helped to secure one of the major donations for the Longitude Symposium. The success of the conference encouraged the board of trustees to support this endeavor as well. The Perkin Fund's generous donation was matched, as it had been for the symposium, by the NAWCC, through the good offices of its president, Jim Coulson, and other colleagues on the council. My friends Robert and Maurine Rothschild, who have encouraged many activities in the history of science at Harvard, also came forward in support of this project. Finally, I would like to thank John R. and John B. Robinson for the assistance they provided to make the printing of this work possible. This publication would not have been possible without the generosity of these organizations and individuals.

The team engaged for the design and production of this volume has its origins in my former place of employment, The Time Museum. Seth Atwood, who created this outstanding collection of timekeeping devices, came up with the idea of publishing a series of scholarly works covering the entire history of time measurement that were intended not only to provide a permanent record of his collection, but also to make a valuable contribution to the available literature. Atwood recognized that no established publisher would consider such an undertaking without placing stringent restrictions on the finished product. To avoid problems, he decided to publish the work "in house." As curator, it was my duty, and privilege, to supervise the production team as well as the scholars contributing to each field.

After coming to Harvard in 1987, I engaged this team for the catalogues of the Collection of Historical Scientific Instruments and followed the same course of publishing in house. Bruce Chandler, who was general editor of The Time Museum catalogues, has continued in that capacity for the catalogues of Harvard's scientific instruments collections. In the preparation of this volume, to which he also contributed as an author, he has served as my mentor on the editorial board. Pamela Geismar undertook all of the design responsibilities of the team soon after I came to Harvard, when our former designer and her former employer, Lucy Sisman (who still serves in an advisory capacity), became occupied with other projects.

Of the other people mentioned in the credits, there are two who deserve special praise. Peggy Liversidge was originally hired as copy editor for this volume, but through her expertise and extraordinary diligence she has become an indispensable member of the editorial team, attending also to the bibliography and the index. More than anyone else involved, she has been responsible for clarifying many details; the majority of the papers—my own included—have benefited greatly from her astute suggestions. Martha Richardson, as mentioned in Appendix B, was hired shortly before the Longitude Symposium and became an invaluable support throughout the event. During the production of this volume, she has undertaken all of the accounting duties and helped to coordinate the production. Her unswerving dedication to the task at hand has maintained the Collection of Historical Scientific Instruments on an even keel and allowed me to devote my time to this project.

In order to use Harvard's outstanding library resources in the most effective manner, I have relied greatly upon the assistance of friends and colleagues, in particular Roger Stoddard, Curator of Rare Books in the Houghton Library, and his assistant, Mariana Oller, who has helped unremittingly to ferret out many rare works that needed to be checked. I am also grateful to all the students who have helped during the past two years to track down other details associated with this work. The many other individuals and institutions that have kindly provided information and illustrations are listed in the second part of Appendix A.

There are several people who are not mentioned anywhere in the book because their help has been inconspicuous: for sharing in the good times and for providing enduring encouragement during periods of doubt, I am most grateful to my parents, John and Pol Andrewes, my wife, Cathy Andrewes, and her parents, Jim and Marge Stewart. To Roderick and Marjorie Webster, who have encouraged so many people engaged in the history of scientific instruments, I also express sincere appreciation.

The collaboration in this volume of authors from a number of countries brought up some grammatical issues of which English to use: British English or American English. National preferences were found to run deep, so to avoid the dispute, the spelling and punctuation in papers by all the English and European authors are in British English, while those written by authors on this side of the Atlantic are in American English. Although we did not adhere completely to a single style manual for each, Judith Butcher's *Copy-Editing: The Cambridge Handbook for Editors, Authors and Publishers,* 3rd ed., has served as our guide in the former, while *The Chicago Manual of Style,* 13th ed., has been our primary reference for the latter. All general parts of the book, such as the bibliography and the index, adhere to American English. The bibliography, in addition to giving a complete list of the works that have been of direct use in the research for this book, provides a reference to the passages in the text where the material has been cited, so that readers may work back from the bibliography to the text to see where the material has been used and how it has been interpreted. As many references as possible have been checked with their original sources.

This work has grown into a much larger volume than originally anticipated, and, as a result, it has taken far longer than expected to bring it to fruition. The embarrassment of having to inform my colleagues and supporters of continued delays will, however, be of little consequence once the book is published. As the passion, the money, and the thousands of hours that have gone into its production are forgotten, the only thing that will remain visible, and therefore significant, will be its content. If this volume remains of interest and use as a source of information to future generations, then all the effort that so many have given to this project will have been worthwhile. For like the works herein described, this contribution will have withstood the test of time, the ultimate trial for any action of the human race.

*William J. H. Andrewes*
April 1996

# Opening Address at the Longitude Symposium

*Jeremy R. Knowles*

Dean of the Faculty of Arts and Sciences
Harvard University

It is inevitable that a symposium on longitude should bring such a variety of distinguished experts together, and it is entirely appropriate that it should take place at a major university, where we may presume that intellectual and disciplinary boundaries are readily crossed. For the solution to the longitude problem was not driven by narrow scientific curiosity or the challenge of a technological advance. The £20,000 reward that was offered by Queen Anne's government in 1714 was the product of economic and political forces, rather than the simple desire to know where one was, at sea.

Harrison's route to success was strewn with the rocks of human envy and competitiveness, and it shows a number of parallels with the way that scientific advances are made today. The unusual precision of Harrison's wooden-wheel clocks convinced Halley and Graham that he deserved support. Their encouragement soon resulted in Harrison's first marine timekeeper (H.1), which proved its value on a trial voyage. In today's terms, we should say that the applicant's preliminary results were shown to be convincing on peer review, he secured research support, and enjoyed early success. He applied to the Board of Longitude for support to continue his work, and H.2 followed in two years. But Harrison was not satisfied, and it took another nineteen years before H.3 was perfected. I fear that this agonizing period of development would, today, have meant loss of research funding long before H.3 was ready for trial. Then, in a manner that is not so uncommon, Harrison realized that he had solved the problem in related work—the ultimately successful H.4. Sadly, recognition came only slowly, as he had to wrestle with the competing lunar-distance method and—even with the support of George III—with a Board unwilling to grant that the problem had been solved. Then at the end, when poor Harrison, at 80, received the award, a mean-minded government chipped away at it by every conceivable device.

In welcoming you all to Harvard, let me acknowledge our institutional debt to David P. Wheatland (Harvard A.B., 1922), thanks to whose munificence we have one of the three largest and most glorious university collections of scientific instruments in the world. This proud legacy, and the high skill and enthusiasm of its imaginative curator, Will Andrewes, in establishing this meeting, makes me certain that the symposium will be successful. As David Landes has put it: "The search for longitude was perceived from the beginning as a project of intellectual and humanitarian concern, transcending national interests and boundaries." He could as well have been describing this meeting.

Welcome!

# La Salle: When Ignorance Was Death

## *Alistair Cooke*

**Alistair Cooke was born in Manchester, England, in 1908 and was educated at Cambridge, Yale, and Harvard. He first arrived in the United States in 1932 but went back to England two years later to become the BBC's film critic. He returned to the United States in 1937, when he started his lifetime career as a reporter on American life, serving first as a special correspondent for *The Times* and then for 27 years as chief American correspondent of *The [Manchester] Guardian.* He became an American citizen in 1941. His famous weekly radio broadcast, "Letter from America," which the BBC originally commissioned in 1946 to run for thirteen weeks, is now heard in 52 countries and has become the longest running program in the history of broadcasting.**

**Cooke began his career in television in the 1950s, when he served as master of ceremonies for the memorable "Omnibus" series, and he became known to millions as the regular host of PBS's "Masterpiece Theatre." His own television masterpiece is "America: A Personal History of the United States." Among his earlier books are *A Generation on Trial* (1950), *One Man's America* (1952), and *Talk about America* (1968). Cooke's personal recollections of six contemporary titans—Chaplin, Edward VIII, Mencken, Russell, Stevenson, and Bogart—comprise *Six Men* (1977).**

**Cooke has received honorary degrees from the universities of Cambridge, St. Andrews, Edinburgh, and Manchester, as well as Yale's rarely awarded Howland Medal. In 1973, Queen Elizabeth II made him an Honorary Knight Commander of the British Empire for "his outstanding contribution over many years to Anglo-American mutual understanding."**

# La Salle: When Ignorance Was Death

## *Alistair Cooke*

Mr. Andrewes, Distinguished Guests, Ladies and Gentlemen. I hope nobody here, from a younger generation, takes offense at being called either a lady or a gentleman. You'll have noticed immediately that I belong to a defunct generation, which was taught that the first requirement of social contact of any kind was civility. I'm not so sure these days. I helped a young woman on with her coat on a bus. She hissed: "I-don't-need-that!!"

Many of you must wonder what I'm doing here. When I mentioned this date to several friends, not one of them said, "Oh, really?" All of them said, in their separate ways, "A longitude conference? You're not serious." Well, it will be no news to you that your specialty is so special that the vast majority of the human race is unaware of its existence. Why did I accept Mr. Andrewes's invitation?

Because it presented a ludicrous challenge. In the past quarter century, I have found myself—as the French say—talking often before specialists, as a layman, about their specialty. The first invitation came from the Mayo Clinic—almost 30 years ago. Would I do the closing speech before the Annual Convocation of the Mayo Graduate School of Medicine? I jumped at it. The director, or president, of the clinic wrote to me to assure me that "although all of us in Rochester who are involved in this program are primarily in some branch of medicine, we do not necessarily expect an address related to medicine. Any topic of broad general interest would be suitable."

Little did he know that I had no intention of talking on a topic of broad general interest. I had been waiting 40 years to talk back to doctors! What came out, in Rochester, Minnesota, was a mildly scolding talk—lamenting the doctors' infatuation with the pomp of Latin and Greek names for things, befogging the layman by telling him he had something called a tibia and a femur and a clavicle—when, all the time, all he had was a shinbone, a thighbone, and a collarbone. I ended with a fatal passing thought. "I suppose," I said, "that an appetite for jargon might be imbibed with the mother's milk, a matter of genes. Maybe," I went on, "you can no more cure a naturally pompous person that you can reflower a virgin." Well, the Mayo Clinic

Bulletin (which reprinted the lecture) goes around the world, not least to Japan. And the first written response came from Tokyo University Department of Gynecology. It said: "But, Mr. Cooke, we do it!"

You'd think, after that, I'd be offered no more invitations from specialists, not from doctors, anyway. On the contrary, four other distinguished institutions rushed at me. (Doctors, you may have noticed, from their student days are notable for, among other things, a rather coarse taste in humor.)

If I knew little about their specialty, I discovered—from a distinguished American cabinet officer—a way of overcoming that liability. The only time I met Mr. Robert McNamara he was already the Secretary of Defense and in charge—poor man—of the Vietnam War. I wondered aloud how a CEO of a famous motor-car company had come to take on the first great land war that the United States had fought in Asia. How did he come to be an expert on Vietnam? "Very simple," he said, "I looked it up in the *Encyclopaedia Britannica.*" Ah, so!

So, when the Royal College of Surgeons asked me to do their annual—I don't know—er—John Hunter Lecture, I didn't flinch. I looked up surgery in the *Encyclopaedia Britannica.* Right? Not so right. It said: "Surgery is the treatment of malformations and diseases by manual operation." An alarming definition to me. The last time I'd been manually operated on for a muscle spasm in the lower back was by a chiropractor, whose manual operation succeeded in violently aggravating the existing malfunction. So much so that ever afterwards, I remained incurably prejudiced against "manual operators" and often lulled myself to sleep by reading aloud H. L. Mencken's beautiful tribute to chiropractors as social servants and instruments of God's will: "It eases and soothes me to see these preposterous quacks so prosperous, for they counteract the evil work of so called public health, which now seeks to make imbeciles immortal. If a man, being ill of appendicitis, resorts to a shaved and fumigated longshoreman and submits willingly to a treatment which involves playing on his vertebrae as on a concertina, then I'm willing to believe that that man is badly wanted in Heaven. And if that same man, having achieved lawfully a lovely baby, hires a blacksmith to cure its diphtheria by pulling its neck, then I am prepared to believe it is the Divine will that there shall be one less radio fan later on."

Of course, a mere horror at chiropractic wasn't enough to satisfy the surgeons, but I soon boned up—shall I say?—enough to be able to talk to them and describe, by way of recounting my own operations, a plausible voyage through the Islets of Langherhans and round McBurney's Point. They stayed with me long enough, anyway, until I told them something I doubt any other speaker had ever told them, either in public or in private. I repeated the confidential remark of a friend of mine who was a surgeon of great repute. "Gentlemen"—I said to these panting stags—"gentlemen—I hand on to you a watchword from a great surgeon. 'Do not,' he told me, 'ever trust a man who is eager to use the knife. Never forget—surgery is defeat!'" At that, the entire audience headed for the woods—or at least the adjoining drawing room, in which we held what was euphemistically called "a reception."

So—you must already be tapping your teeth with curiosity, if not impatience. What is he going to do with us? Will you honor us, Mr. Andrewes wrote, by addressing—on the last night—our symposium on—wait for it—longitude? Absolutely. No problem. I turned to the *Encyclopaedia Britannica.*

Longitude. It said—see Latitude.

I soldiered on. And came from *o* to *a*. And there are two paragraphs about your specialty. The first gives a concise, if dense, definition of longitude and in the second paragraph begins to get interesting with the sentence "As for the relationship between longitude and time, see Time"!

I decided at that point (in time, as we say in Washington) to throw myself on your mercy and allow me, at last, to justify my appearance here tonight by telling you my one longitudinal story. Rather, the story of one of the great figures in the settlement of North America—and perhaps the most notable tragic hero, or victim, of the contemporary ignorance about longitude.

The man I'm talking about is, to Canadians, one of the great heroes of the French conquests in North America and, like other French-Canadian heroes—Cadillac, Pontiac—is thought by Americans to be an automobile: La Salle. René Robert Cavelier, Sieur de La Salle. Born exactly 350 years ago this month in Rouen.

At the age of 23 he emigrated to Montreal and acquired a grant of land on the St. Lawrence River. But he had an itch for exploration and—with the rather amused approval of Frontenac, the governor of Canada—he very early on developed a grand ambition: to spread the fur trade through the wilderness to the West, and to build enough French forts down riverbanks of the mid-continent so as to hem in the English east of the Appalachians.

More than any other man, he opened up the Great Lakes to the trappers and furriers, and by his mid-thirties, he had imagined the farthest reach of his ambition: by building forts and cultivating friendly relations with the Indians (he learned a dozen Indian dialects), to protect the approaches to the Great River they had marveled about, the Mississippi; to slide down its legendary length, and at its mouth found a city—and announce the birth of an empire.

At the start, in Wisconsin, he lost a ship loaded with a fortune in furs, and with it many of his supplies and the money he expected from his backers. So he raced back to Ontario to talk his sponsors into providing more funds. By "raced," I mean he performed a canoe-and-carry trip of just over a thousand miles in 65 days. He got enough money and supplies to start again, the next winter. When he arrived at the southernmost fort at St. Joseph, Missouri, he found it abandoned and his men deserted. So he walked and canoed the thousand miles back again to Kingston, Ontario.

An indestructible optimist, as well as a sweet-talking client—and plainly a man of iron resolve—he began the long-dreamed-of voyage down the continent's central watershed. It was December 1681. He was 48. He had with him 23 Frenchmen and 31 Indians.

The weather was appalling. The rivers were frozen along a several-hundred-mile stretch. That did not deter him. He made sledges for the canoes, and the men pulled them across the Chicago River and down the Illinois, every man wearing out a pair of leather moccasins a day. But they came at last to running water. The snows melted, the downriver currents quickened, the spring began to blossom, and La Salle and his men now launched into the happiest, perhaps the only happy, passage of his explorer's life, a journey as effortless, compared with what had gone before, as a cruise on the *QE2:* they paddled down the 1,100 miles of the river and came to its mouth.

And there, on April 9, 1682, on a bluff overlooking the Gulf, La Salle raised the Cross of Christ, planted the banner of France, and annexed *"this country*...in the name of the Most High, Mighty, invincible and victorious Louis the Great, by the Grace of God, King of France and of Navarre." "This country" that he claimed he called Louisiana. It was not at all the compact little—or large!—southern state we know. La Salle understandably claimed the regions where his huge ordeal had started, up at the headwaters of, and down the whole watershed of, the Mississippi and its tributaries, comprising thirteen of today's central and southern states, from Montana to Louisiana: one-third of the land mass of today's United States.

That was a great day in the history of imperial France. In high spirits, La Salle made the 2,000-mile return journey confident of King Louis XIV's permission to go back and found a colonial capital at New Orleans, as a prelude to consolidating the whole of middle and south-central North America as France's North American empire. One thing of which La Salle was rightly proud: he had painstakingly made friends with all the Indian tribes (except one) along the way and could claim, in the face of the comparatively appalling Spanish experience, that friendly natives were the best kind of ally an empire builder could seek.

But once La Salle arrived home, his euphoria expired. There was, in Quebec, a new governor, who looked on La Salle's great expedition as a sort of buccaneering exploit and was not about to finance a return trip. So, this unyielding man sailed off home to France, personally appealed to the king in whose name he thought he had planted an imperial capital. The king, who had been primed by the new governor, at first told La Salle that his vast explorations had been a waste of time. But, after a month or more, and considering the unpleasant effect that a French empire in America would have on his enemy, the King of Spain, Louis relented and set up La Salle with four ships, soldiers, and colonists of both sexes to build that fort on the Gulf and, as King Louis nonchalantly put it, "proceed to control the continent."

Invigorated by this royal backing and the vision of a growing French empire, La Salle was back in Quebec, outfitted his expedition, and sailed south in 1684, more than two years after he had planted the cross and the flag.

I will not wring your emotions with an account of the atrocious sufferings and ill-luck of the nearly three years it took La Salle to descend again those 2,000 miles. Enough to say that he started with 400 men and finished with 45. Shipwreck, mutinies, illness, desertion of the supply ship. And, finally, the crowning error of missing the Passes through to the Mississippi. He landed, with those 45 men, on the coast of Texas—about 400 miles west of the river's mouth, of his own pronounced French capital of New Orleans. If we want to indulge a bout of historical sadism, we can imagine the wanderings of these men, the native Indians always pointing west, La Salle begging his men—like Columbus facing a similar mutiny—just another day, another score of miles, one more try. Well, the dreadful day came when twenty of his men agreed on a final try: to march across a stretch of rolling prairie, find again the great river, and know at last where they were. Well, over the prairie horizon was yet another prairie horizon. In the end, they stopped close by what today is Navasota, Texas, which has a bronze statue of La Salle, a few stores, an automobile sales lot. It was here that the last of the crew murdered La Salle and mooched off, leaving his naked body to the wind and the buzzards.

If he had been born 50 years later, if he had managed to triumph over the contemporary ignorance of longitude, we might this evening be meeting in the French capital of New England.

That is the only thing I know about longitude, and I have told it to you in the fervent hope that it's the only thing you don't know. I've known about La Salle, and grieved for him, for many years. (I wish we'd done better by him in my television history of "America.") But I learned only recently about John Harrison. And I'm eager to share the little I've learned. What strikes me, as a total stranger to his life and works, is that he possessed an eighteenth- and nineteenth-century gift which I believe in this century we have lost. A gift that every scientist and novelist and historian and many statesmen of the eighteenth and nineteenth century had: a gift of industry, of industriousness, of regarding 24 hours a day as little enough time in which to live a life and pursue your interests. I think of Darwin, going down to the seashore—any seashore at hand—for 41 years, with a broken teacup (he had no grant from a national science foundation) and scooping up sand and algae and brooding over them. And at the end of those 41 years, feeling confident enough to publish the *Origin of Species*.

But I find John Harrison to be a close competitor for tenacious scholarly industry. He hears, when he's 21, that the government is offering a prize (the stupendous amount of £20,000) for an accurate marine timekeeper. "Promptly," it says in my account, he settled to the problem and solved it—promptly?—in 45 years. And, promptly, the British government paid him his prize—fourteen years later, when he was 80! Well done! No wonder you're celebrating his birth tonight, promptly 300 years after the event.

But when you next offer the toast of John Harrison, spare a glass, and a tear, for the tragic hero of your specialty—René Robert Cavelier, Sieur de La Salle.

# Finding the Point at Sea

*David S. Landes*

David S. Landes is the Coolidge Professor of History and Professor of Economics in Harvard University. His books include *The Unbound Prometheus* (1959), *Bankers and Pashas* (1969), and *Revolution in Time* (1983). Professor Landes is a member of the National Academy of Sciences (U.S.A.), Accademia dei Lincei (Rome), American Academy of Arts and Sciences, American Philosophical Society, British Academy, Fondation Royaumont pour le Progrès des Sciences de l'Homme, Koninklijke Academie van Belgie, and Royal Historical Society (U.K.).

# Finding the Point at Sea

*David S. Landes*

Whenever anyone wants to demonstrate the dangers and costs of not being able to know where one is at sea (of being "at sea"), he cites George Anson's voyage around South America and into the Pacific. For want of a way to find the longitude, Anson's ships bounced back and forth east and west of the Juan Fernandez islands, those godforsaken dots in the Pacific where Daniel Defoe found his model for Robinson Crusoe, and were lucky to fetch up on land after loss of almost all hands. Of a thousand sailors who set forth with him, a hundred survived.

George Anson (1697-1762), of course, was not the only such wanderer. Too many were the ships that dashed aimlessly and fruitlessly about, too far this way, too near that, until scurvy and thirst killed off or incapacitated so many hands that the crew could no longer man the rigging and direct the vessel; and then the ship would float helpless with its population of skeletons and ghosts, another "flying Dutchman," to ground one day on reef or sand or ice and provide the stuff of legend.

I have my own such tale, more interesting in a way because it is not the story of missing tiny islands but of missing continents. Land masses may be big, but the sea is even bigger. In 1591 George Raymond, a veteran seafarer who had fought the Great Armada some three years earlier, led a fleet of three small ships (none over 350 tons) out of Plymouth harbor into the Channel to begin a voyage around Africa to the Indies, there to "annoy" the Spaniards and the "Portingalls (nowe our enemyes)" and perhaps sell English "comodities."[1] This was the first effort by the English to move into a route that had been thought of as Portuguese property, if only because the Portuguese knew it as no one else did or could. As Thomas Stevens, Jesuit missionary to India, wrote in 1579, "there is not a fowl that appeareth or sign in the air or in the sea that they have not written down."[2]

The voyage started easily enough. Raymond made out into the Atlantic, rounded Finisterre, and headed south, followed by a good northeast wind. All went well until the Canaries, when scurvy struck on all three vessels; two crewmen died of it before they reached the equator. Along the west coast of Africa, the weather was foul, the winds and currents contrary; Raymond did

1. Russell Miller, *The Seafarers: The East Indiamen* (Alexandria, Virginia: Time-Life Books, 1980), p. 12.

2. J. S. Furnivall, *Netherlands India: A Study of Plural Economy* (Cambridge: Cambridge University Press, 1939; reprinted 1967), p. 2.

not know, as the Portuguese had learned, that in sailing south to round Africa, the shortest distance is not the fastest: one wants to swing far westward, almost to the coast of South America, and then head east-southeast for the Cape of Good Hope. By the time they reached shelter and fresh water at Table Bay (South Africa), over three months had gone by. They stayed there a month, got to do some trading with the natives (two knives for an ox, one knife for a sheep or bullock), then sent one ship back home with a skeleton crew and proceeded with two, well manned. And so up the eastern side of Africa.

There, off Cape Correntes (Mozambique), they ran into a storm that destroyed the flagship; as the purser of the other vessel put it, "We saw a great sea breake over our admirall, the *Penelope*, and their light strooke out; and after that we never saw them any more."[3] So now one small ship continued up the African coast toward India, under the command of James Lancaster (fl. 1591-1618), a London merchant, also a veteran of the fight against the Armada. After much hardship and peril, including a lightning strike that melted the spikes in the masts, they came to harbor in Zanzibar. It was now the end of November, almost seven months since they had left Plymouth. There Lancaster found a friendly reception, in spite of Portuguese warnings against violent, cannibalistic Englishmen, and there he decided to winter, repair, and wait for favorable monsoon winds.

3. Miller, *The Seafarers* (*op. cit.*, note 1), pp. 12-13.

It was three months later, end of February, that Lancaster set sail for the southern tip of India. He never got there. Owing to navigational errors and unfavorable winds and currents, he found himself at the island of Gomes off the coast of Sumatra. It was now early June, over a year since they had left Plymouth. With summer monsoons coming, Lancaster decided to shelter at Penang. Meanwhile sickness was taking its toll: by August, the 97 men who had left the Cape were down to 34, and only two-thirds of these were fit to work.

Now it was time to make something of this voyage. Lancaster attacked four ships in the straits of Malacca, picking up a cargo of Portuguese pepper; and then attacked a large Portuguese vessel, homeward bound for Goa, and benevolently deposited the crew on a nearby shore. In early December, Lancaster found himself off the coast of Ceylon, ready to do some more "annoyinge" of the "Portingalls"; but his crew had had enough and insisted that he return to England. So February found them near the coast of Africa, and on April 3, 1593, two years after setting sail, they landed at St. Helena, where they got new provisions and water. The rest should have been easy, but they got stuck in the doldrums for over a month and drifted westward. Lancaster then tried to reach Trinidad but missed the island and got bounced around the Caribbean by a series of nasty storms. Then, off a small island near Puerto Rico, the ship's cables were somehow cut, and she drifted off with five men and a boy, never to be seen again. Lancaster, who had gone ashore, was rescued by a French ship and finally reached England on May 24, 1594, with a handful of men and the clothes on his back.

Financially the expedition had been a bust, but it had proved that Portuguese space could be penetrated. Meanwhile the capture off the Azores in 1592 of the great Portuguese treasure ship *Madre de Deus*, with a cargo worth half the entire English exchequer—the greatest haul since the ransom of the Inca Atahualpa—showed how big the stakes were. The English vessels that took her had been waiting there for Spanish treasure ships from the Caribbean, and here a Portuguese carrack bursting with booty fell into their hands.

Talk about good fortune! Nothing thereafter could keep the English from these eastern waters, but it would obviously be a lot easier, safer, and more profitable if they could find a way to find where they were—longitude and latitude, the point at sea.

★★★★★

Now that we can find the longitude, we are inclined to think of it as indispensable; this, I think, is especially true of those of us who are aficionados of time measurement and chronometry. Yet we surely know better. The great voyages of discovery, those feats of derring-do that opened the world, all took place before anyone could ascertain the longitude at sea. Indeed, some of these voyages (I am thinking here of the Norse voyages across the North Atlantic or the Polynesian jumps from island to island in the vast emptiness of the Pacific) were the work of sailors who had no artificial instruments of navigation at their command—not compass, nor means of calculating the altitude of the Sun, nor device to estimate the speed of the vessel.[4] We may well ask ourselves how this was possible.

One thing is clear: these successes were not a matter of luck or accident. There were such voyages, no doubt: ill-equipped and unprepared men who sailed away beyond the horizon in search of treasures and wonders. But they typically did not return and, like their goals, are remembered as myth or legend. If there is anything that characterizes the successful voyages of the pre-longitude period, it is their rationality. There was method to their madness.

We should not underestimate the cunning of these early oceanic sailors. They had learned from experience to recognize the clues to direction (stars, Sun, migratory birds, currents, prevailing winds) and location, especially nearness to land (birds, cumulus clouds, reflections on the sky, floating vegetation and debris, depth of water, smell and taste of water, character of bottom); and they had learned to limit their exposure as much as possible by keeping in touch with the land. Much could be accomplished by coasting (consider the early portolan journals and maps, which are essentially a guide to landmarks) and by short-distance island hopping.[5] This latter was the secret of the Norsemen, among others, and the topography of the North Atlantic lent itself well to this strategy: the Faroes, then Iceland, then Greenland, then Newfoundland. They also used ravens as Noah used the dove.

By the same token, the Spanish and Portuguese prepared unwittingly for the transoceanic era by the discovery of the Atlantic islands off the European and African coasts: the Azores, the Canaries, the Madeiras, the Cape Verde Islands. And once the Spanish got to the other side of the Atlantic, they found themselves in an island-sprinkled sea that allowed them to probe and probe without ever cutting the cord to nearby land. The same held for the waters of Southeast Asia and the Indian Ocean. These are splattered with islands and peninsulas, in such wise that the topography not only facilitated sequential advances but almost dictated routes and channels. When the Europeans got there, they found very active maritime traffic linking Africa to the Middle East and India, India to Indonesia and the South China Sea,

4. According to M. A. P. Meilink-Roelofsz, *Asian Trade and European Influence in the Indonesian Archipelago between 1500 and about 1630* (The Hague: M. Nijhoff, 1962), pp. 104-5, Dutch seamen remarked that native pilots whom they took on board were not familiar with the compass and showed no interest in it. When Steven van der Hagen (*ca.*1560–*ca.*1610) made his first voyage to Indonesia in 1603, part of his cargo consisted of some hundreds of compasses of all qualities. He found no takers and had to bring them back to Holland.

5. On the importance of coasting, *cf.* Luis de Albuquerque, "Portuguese Navigation: Its Historical Development," in Jay A. Levenson, ed., *Circa 1492: Art in the Age of Exploration* (New Haven: Yale University Press, 1991), pp. 35-9.

and China to Japan and the Moluccas and Philippines. They found ships that were in some ways superior to theirs, in some ways inferior. But above all, they found sailors who did not like the high seas. Thus Father Mendoza (1540–1617) in 1577: the Chinese, he said, "are afraid of the sea, as one might expect of people who are not habituated to plunging into it." And Rodrigo de Vivero, regarding voyages from Osaka to Nagasaki in the Inland Sea of Japan: "they sleep practically every night on land." And Father du Halde (1674–1743), in 1693, regarding Chinese navigators: "[g]ood coasting pilots, but pretty poor on the high seas."[6]

None of this navigation really depended on finding the longitude, though obviously it would have helped to be able to do so. It was chart- and map-dependent, and that was why the Portuguese were at such an advantage over the other Europeans who followed them into the Indian Ocean. They had the maps. They had put them together voyage by voyage, and they had paid dearly for them, with loss of ships and men. Much has been made of Prince Henry the Navigator (1394-1460) and his so-called school of Sagres, a fifteenth-century think tank for astronomers, navigators, and meteorologists. Not so, says Luis de Albuquerque. There was no science of meteorology as yet; and quoting Luciano Pereira da Silva (1864–1926): "the school of Sagres was the planks of the caravels."[7] They were sailing, to begin with, by spit and feel.

Naturally the Portuguese treated these maps as a state secret, while making it a policy to present these oceanic voyages as far more difficult than they actually were. When a Portuguese pilot boasted to King John II that any old ship in good shape could make the trip back from El Mina (west African coast) to Portugal, the king told him to shut up if he wanted to stay out of prison.[8] Not until the Dutchman Cornelis de Houtman (*ca.*1560–*ca.*1605) had been able to infiltrate and spend two years spying in Lisbon was any foreigner ready and able to follow effectively in their traces.[9]

There was still the task of getting to these places—the long runs crossing the Atlantic or getting from Europe to the Indian Ocean. How was one able to do that without knowing the longitude? The answer, of course (but "of course" begs the knowledge, ingenuity, courage, tenacity, and nose, taste, and feel required), lay in sailing the parallel.[10] Here is where science did come in. In the northern latitudes, the pole-star was the surest guide, its height diminishing as one approached the equator. South of the equator, the sailors found new constellations, but navigators came to depend increasingly on solar readings.

A new difficulty now reared its ugly head: the Sun does not describe a path parallel to the equator, and so its altitude has to be corrected for its declination, that is, its seasonally changing zenith above and below the equator. Here Iberian science, transmitted to the Portuguese by Jews and conversos, made all the difference. Thanks to the researches promoted by such visionaries as Henry the Navigator and the astronomical observations and the calculations of such savants as Abraham Zacut (1452-1515?), one now possessed tables of the Sun's declination;[11] and these in combination with instruments for measuring altitude (the mariner's astrolabe, the quadrant, the cross-staff) made it possible for an arithmetically intelligent seaman of the sixteenth century to calculate latitude within one or two degrees.[12] So one could know where one was on the north-south axis, and if one ran the appropriate parallel, one would eventually fetch up at a place known to be at that latitude.

6. Fernand Braudel, *Civilisation matérielle, économie et capitalisme, XV^e-XVIII^e siècle,* vol. 1, *Les structures du quotidien* (Paris: Armand Colin, 1979), pp. 360-1.

7. Albuquerque, "Portuguese Navigation" (*op. cit.,* note 5), p. 37.

8. Braudel, *Civilisation matérielle* (*op. cit.,* note 6), p. 361.

9. On Houtman's return to Amsterdam, a syndicate of merchants formed the Compagnie van Verre (the Company of Far Away), with capital of 290,000 florins, and sent Houtman off the following year at the head of four ships to reconnoiter the Indies. He returned in 1597 after reaching Java, and although the venture was not a commercial success, it showed that the Portuguese were not the only ones who could do it. Within less than five years, ten such companies had been formed. These sent out fourteen fleets, comprising 65 ships, of which 54 made it back safely. Furnivall, *Netherlands India* (*op. cit.,* note 2), p. 21.

10. On the psychological leap required to move from coastal sailing and island hopping to navigation on the high seas, *cf.* Braudel, *Civilisation matérielle* (*op. cit.,* note 6), pp. 359-61.

11. How good and usable these tables were, is a question. Albuquerque ("Portuguese Navigation" [*op. cit.,* note 5], pp. 38-9) points out that because of the difference between the Julian year and the true solar year, Zacut's tables had to be corrected on a year-to-year basis, a task "beyond the understanding of a pilot." At the end of the fifteenth and in the early sixteenth century, corrected yearly tables, some perhaps prepared by Zacut himself, were made available to navigators.

12. This accuracy may be an exaggeration, as I infer it from the reckoning made by the Portuguese of the latitude of the southern tip of Africa—a calculation made on land. (Vasco da Gama was actually within half a degree of true [Vincent Jones, *Sail the Indian Sea* (London/New York: Gordon and Cremonesi, 1978), p. 48].) At sea, it was much harder to get a good reading, especially of the Sun (as against the pole-star). But near and south of the equator, there was no pole-star to work with; hence the need to reckon by the Sun, a technique tried and worked out in the second half of the fifteenth century. Albuquerque, "Portuguese Navigation" (*op. cit.,* note 5), p. 38.

The operative words, of course, are "known to be"; without that knowledge, one was heading straight into the unknown. That was what Columbus did—more or less. Francis Maddison tells us that when Columbus thought he was some 34° from the equinoctial line, he was in fact under 20° north.[13] Columbus was apparently not getting good use from his quadrant. No matter; there were islands all around. Columbus was convinced that the Indies could be reached by going west; and like a darts player leaning as far forward as possible, the closer to get to the target, he jumped out into the ocean sea from the westernmost port in the Canaries. He did so not knowing how far he would have to go before reaching land, or, rather, thinking that the world was far smaller than it is and the Indies that much closer; and even so, he systematically underestimated his daily run with a view to keeping his men patient.[14]

So he was lucky; he made good use of the compass, and perhaps he just knew and rode his winds. He also knew his birds, and in one month's time he bumped into an island in the Caribbean.[15] Which one, no one can be sure of, although numerous scholars and yachtsmen have expended large sums in "proving" that it had to be this one or that. No matter. Once people knew that there was land out there in those latitudes, the voyage became almost routine for those who followed, even though they could not find the point at sea, or on land for that matter. Incidentally, Columbus followed a similar strategy for his return, made all the easier because he knew the latitudes of Europe. He sought out favorable winds and currents, which took him well north of his westward (outward) route, and these brought him smartly back, again without concern for longitude.

The discovery of the sea route to India is an example of a similar navigational strategy. Convinced that there had to be a water route across or around Africa—an article of faith—and that enormous wealth and spiritual merit were to be had by bypassing or outflanking the Muslims in what we now call the Middle East, the Portuguese began, already in the early fifteenth century, to work their way down the African coast. This was no easy task, partly because of psychological barriers (the bugaboo of Cape Bojador), even more because of geographical obstacles. The idea was, as always, to cling to the coast. The trouble was that, in these waters, winds and currents run contrary to southing vessels, so that ships took forever to advance, and this intermittently and along a shoreline that, south of the equator, offers little in the way of shelter or refreshment. Had it not been for the happy accident of the island of St. Helena, this must have become the scurviest reach on the high seas.

Still, they made it. It took three-quarters of a century, but in 1488, Bartolomeu Dias (fl. 1481-1500) finally rounded the tip of Africa and saw the coast turn north into the Indian Ocean; and when he got back to Lisbon, he brought the news that, yes, there was a southern end to Africa, and, yes, one could sail to the Indies. Most important, he told them the latitude of that critical turning point (35° S.), and that made it possible for his successor, Vasco da Gama (*ca.* 1460-1524), to introduce an innovation that transformed the whole problem.[16] What Gama did was follow the traditional route down to the Cape Verde Islands (16° N.) and then, instead of beating his way against the wind and currents, swing out west, far out to where he may have sighted the coast of Brazil. He then swung back east to connect with Africa one or two degrees north of Good Hope, and then sailed into

13. Francis Maddison, *Medieval Scientific Instruments and the Development of Navigational Instruments in the XVth and XVIth Centuries,* Agrupamento de Estudos de Cartografia Antiga, no. 30 (Coimbra, Portugal, 1969), p. 92.

14. Patricia Seed, in "Enacting Colonialism: The Politics and Ceremony of European Rule over the New World, 1492-1640" (typescript, 1994), chap. 4, suggests that the Portuguese were better informed about the size of the world and therefore understood that Columbus did not know what he was talking about. Hence their reluctance to finance the voyage. The Spanish, in their ignorance, were persuaded to go along. In the short run, the Spanish were wrong but right; in the long run, they were right but wrong. But that is another story.

15. See Francis Maddison, "Tradition and Innovation: Columbus' First Voyage and Portuguese Navigation in the Fifteenth Century," in Jay A. Levenson, ed., *Circa 1492: Art in the Age of Exploration* (New Haven: Yale University Press, 1991), p. 92. For a more favorable, if tentative, view of Columbus's navigational science, see Arne B. Molander, "Columbus and the Method of Lunar Distance," *Terrae Incognitae: The Journal for the History of Discoveries,* vol. 24 (1992), pp. 65-78; Molander is trying to test (show), among other things, whether Columbus tried to find longitude by means of lunar distances.

16. On Gama's navigation and its influence on his successor Cabral, see Samuel Eliot Morison, *The European Discovery of America: The Southern Voyages, 1492-1616* (Oxford: Oxford University Press, 1974). On the role of science in enabling the Portuguese to solve the special navigational difficulties of the South Atlantic, see Seed, "Enacting Colonialism" (*op. cit.*, note 14), chap. 4.

the Indian Ocean and north along the east African coast until he could pick up a pilot at Malindi to guide him on the regular route between Africa and the Malabar coast of India. Fortunately he found his man, because this was a tricky route, pocked with islands and coral atolls, and with the help of early monsoon winds, Gama made a reach of 2,300 miles (as the crow flies) in 27 days. Going back, the Portuguese no longer had their pilot (he had simply vanished in India), and the same stretch took three months. Thirty crewmen died of scurvy on this stage alone, and the survivors were barely strong enough to push the corpses over the side. "[A]nother fortnight," writes the chronicler, "there would have been no men at all to navigate the ships."[17] Gama eventually lost 116 of his original 170 men, more than two out of three.

In sum, one could do astonishing things without knowledge of longitude. It took courage, fortitude, and a certain foolhardiness; but it could be done, and with experience it could be routinized. Note, moreover, that in these early centuries of global opening, the great bulk of the oceanic travel was taking place in the Atlantic, where knowledge of the longitude, however desirable, was less critical than in the Pacific and Indian oceans. The VOC (Vereenigde Oost-Indische Compagnie, or Dutch East India Company), for example, sent to all Asia some 29 ships a year in the first half of the eighteenth century, 25 a year in the second half; whereas traffic to Surinam alone used 60 to 70 vessels a year. Similarly the English sent seven ships to China in 1721, eight in 1778, and 21 in 1779; whereas more than 300 ships a year were plying the Jamaica route from the 1730s.[18]

17. Richard Humble, *The Seafarers: The Explorers* (Alexandria, Virginia: Time-Life Books, 1979), p. 102.

18. Jean Mayer, *Les Européens et les autres: de Cortes à Washington* (Paris: Armand Colin, 1975), p. 245.

19. See Silvio A. Bedini, *The Pulse of Time: Galileo Galilei, the determination of longitude, and the pendulum clock,* Biblioteca di Nuncius, Studi e Testi, vol. 3 (Florence: Leo S. Olschki, 1991). See also W. F. J. Mörzer Bruyns's and Albert Van Helden's papers in this volume.

★ ★ ★ ★ ★

Whence, then, the pressure for a way to find the longitude? For pressure there was. Otherwise one could not understand the extent of interest in this question. Before the effort was rewarded with success, the better part of the scientific community, and not only in maritime nations, had been drawn into the campaign. Why this generalized curiosity?

At one level, the answer lies in money. The major sea powers, those with direct interest in transoceanic voyages, were all ready to offer princely rewards to anyone who would discover a means of finding the longitude; the great English prize of 1714, the one that actually led to a solution, was only the last of such. A hundred years earlier, the great Galileo (1564-1642) was already panting after the Spanish and Dutch prizes, to the point of promising results he was in no position to provide.[19] Money talks.

But to say this is only to push the question back a step: Why were rulers ready to offer so much for this particular discovery?

The answer is that ignorance cost, and the cost was rising. The world's naval and merchant shipping was increasing in size and number as international competition and a growing volume of trade brought new entrants into the market. Where once there had been only Spain and Portugal, blithely dividing the world between them with the blessing of the Holy Father (the Treaty of Tordesillas, 1494)—a new high in the annals of arrogance—there were now, from the seventeenth century onward, Holland, England, and France as well as such lesser parties as Denmark, Sweden, even Austria and Prussia, and the North American and West Indian colonies of England.

This competition, moreover, frequently took violent forms. For much of this critical period, for example, Spain was at war with the rebellious provinces of the northern Netherlands (what we know also as Holland), later established in confederation as the United Provinces, and with England; and from 1580 to 1640, the crowns of Spain and Portugal were joined in the person of the king of Spain, so that these enemies of Spain also took Portugal as a target. Much of this war was fought at sea, to the point where the oceans swarmed with privateers and ships of the line lying in wait for merchant and naval vessels (but especially merchant vessels, because they carried the rich cargoes and were most vulnerable) flying the wrong flag.

Now one may well ask how, in this vast emptiness, these slow-moving sailing vessels were able to find one another; and the answer is that the routes taken were adapted to the shortcomings of navigational technique and familiar to all. Ignorance of location, we have seen, made ships dependent on following conventional lanes, their lines traced and channeled by winds, currents, and topographical features. It is not an accident, for example, that English vessels waiting for Spanish ships coming home from the New World saw a Portuguese vessel from the Indies sail right into their hands: from west and south they came that way, looking for good winds and stopping for water and supplies before the final run to the European continent.[20]

An economist would say there was a trade-off: captains and navigators were afraid to stray and yet, by sticking to familiar paths, exposed themselves to predators. To minimize loss, ships often proceeded in convoys, accompanied by naval vessels; but that took time to organize and slowed the run to the speed of the slowest. Even in familiar home waters, knowledge of longitude could make a difference in safety and speed, in the ability to keep going or the necessity to wait before proceeding. Better to be safe than sorry; but caution took time, and whether in war or peace (and leaving piracy aside), commercial competition necessarily enhanced the value of time.

The earlier strategies of navigation took forever—two years, or even three, for the round trip between Europe and the Indies—partly because one was prisoner of favorable winds (there was a time for staying and a time for going), partly because it was not possible to take the shortest, direct course to a destination.[21] There were also delays en route: English seamen, for example, scorned the Spanish because they would stop at night rather than advance in the dark. The longer the trips took, moreover, the harder on the crew (and consequently the harder the recruitment of seamen),[22] the longer the immobilization of funds, and the slower the communication between merchant capitalists in Europe and agents in the field. Much of the history of the great chartered companies, the VOC and the English East India Company, is the tale of instructions unheeded and flouted, of proconsular *faits accomplis*, of involuntary commitments of common resources in the interest of rogue private entrepreneurs. That, in spite of all this free riding (privatizing the gains and socializing the overhead costs), these colonial ventures proved so profitable is testimony to the enormous discrepancies between values and prices in East and West; also to the ability of these new and more powerful intruders to take over the exactions of their predecessors (the local rulers) and extract surplus more efficiently.[23]

But just because an enterprise is making money does not mean it would not like to make more. Even good returns are no consolation for profits forgone or stolen, so that shipping and naval interests in Europe had every reason

20. *Cf.* Jones, *Sail the Indian Sea* (*op. cit.*, note 12), p. 36.

21. On the length of these journeys, see Miller, *The Seafarers* (*op. cit.*, note 1), p. 62. On the wind problem, see J. C. van Leur, *Indonesian Trade and Society: Essays in Asian Social and Economic History* (The Hague/Bandung: W. van Hoeve, 1955), p. 135: "Trade came and went with the alternating semi-annual winds, and the nature of the trade staying over was just as much determined by that periodicity." See also Robert Durand, *Histoire du Portugal* (Paris: Hatier, 1992), p. 94.

22. Life at sea was never agreeable, but given a choice, seamen opted for the easier, shorter voyages. Hence the need for crimpers (*zielverkoopers*) and press gangs. Miller, *The Seafarers* (*op. cit.*, note 1), pp. 62-4.

23. The VOC was able to distribute dividends of 300 per cent toward the beginning; and although these later fell to around 30 per cent toward the end of the seventeenth century, this still represented a splendid performance, the more so as much of the potential profit was siphoned off by private dealing and corruption. Mayer, *Les Européens* (*op. cit.*, note 18), p. 241.

to clamor for more efficient navigation. And this, by the seventeenth century, meant a sustained and organized effort to find a way to find the longitude.

★ ★ ★ ★ ★

So much for the demand side. On the supply side, scientific knowledge and capabilities were coming together to render possible what had long been visionary. The savants knew what had to be done. Ever since the work of Johann Werner of Nuremberg (*In hoc opere...,* 1514) and Peter Apian (*Cosmographicus Liber...*, 1524), for what I shall call the astronomical method, and of Gemma Frisius (*...de Principiis Astronomiae & Cosmographiae,...*, 1530), for the clock method, they understood the nature and solution of the problem.[24] They just lacked the means.

In effect, some way had to be found to learn the time in place and the simultaneous time at another place of known longitude, then convert the difference in time into the distance between. The key to the conundrum lay in this matter of simultaneity. How to know the time at some distant place at the same moment as the time *in situ?*

We all know the answers; they are two. One was to keep the time at the distant reference point—to keep it for weeks and even months over the course of a voyage—so as to be able to know it and compare it with local time at will. To do that required a timekeeper far more accurate and reliable than anyone could build in the sixteenth century.

The other was to observe and time some celestial event, then compare that time with the time of the same event as observed at the distant reference point. But without means of instantaneous communication with that distant place, how was one to know when said event was visible there? The answer lay in knowing (predicting) when said event would take place as seen from the point of reference. I shall omit discussion of the conditions such an event would have to meet, conditions of regularity (occasional events will not do), observability, and precision of prediction and observation that would leave little margin for errors of comparison. Suffice it to say that the best kind of event for the purpose, at least at sea, was the rapidly changing position of our Moon relative to fixed stars.[25]

Note that we are talking about finding the point at sea. This was not the only interest of measurements of longitude. The early sea captains also made efforts to ascertain longitude on land, where observation and measurement were easier (more stable instruments), using such events as eclipses and putting time of observation in place against that, say, in Madrid or Lisbon. Such comparisons *ex post,* months after the event, were obviously not intended to help the captain know where he was at the time. Rather they had a cartographic purpose, to make possible more accurate maps and, among other things, to enable the Spanish and Portuguese to sort out their respective shares of the world. These opportunistic measures, we now know, were grossly inaccurate, which may have suited statesmen better than astronomers and mapmakers.

Two methods, then: the horological and astronomical. For the first, one needed a better clock. The key advances here were, indirectly, the pendulum controller (invented by Huygens in 1656), the balance spring (again Huygens, 1675), better escapements, and temperature compensation (Harrison and

24. An excellent discussion of these early scientific proposals is to be found in Derek Howse, *Greenwich time and the discovery of the longitude* (Oxford: Oxford University Press, 1980), pp. 6-11. See also Derek Howse's paper and Appendix C in this volume. Note that Arne Molander ("Columbus" [*op. cit.,* note 15], pp. 65-6), following on Adm. Thomas Davies and quoting from Amerigo Vespucci's journal, makes a strong case for a lunar-distance calculation of longitude (however inaccurate) as early as 1499.

25. Vespucci, for example, was well aware of the importance of the rapidity of the Moon's movement across the sky. Molander, "Columbus" (*op. cit.,* note 15), p. 65.

Graham, in the 1720s). Through the work of clockmakers, in particular John Harrison, Pierre Le Roy, John Arnold, and Thomas Earnshaw, the marine chronometer reached a high state of perfection by the end of the eighteenth century. For the second, one needed better instruments of observation (telescope, octant, then sextant), an accumulation of lunar observations, mathematical advances (both for the prediction of lunar orbits [principally Euler] and for the comparisons of lunar distances [spherical trigonometry, logarithms]), and—yes, here too—a good watch that could keep fairly accurate time between the noon fixing and the lunar-stellar (nocturnal) observations. After some disappointing attempts with pendulum timekeepers in the late seventeenth century (no good on shipboard), the scientists gave up on clocks and devoted their efforts to the astronomical attack on the problem; and when they finally succeeded in the seventh decade of the eighteenth century, they had accomplished a feat that rested on an extraordinary array of talent and diversity of contributions. One can well understand the passion of their commitment to this achievement and the pleasure they derived from explaining it in handbooks of navigation and teaching it to generations of bright young cadets.

Yet this method, the lunar-distance method, was obsolete—I use the word advisedly—from the time it was introduced. It was complex and so time-consuming (four hours of calculation in the beginning) that the technique was confined to a small elite of trained specialists. Royal navies could prepare people to do it in their academies; the great chartered companies could find and hire the necessary expertise. But the vast bulk of the world's merchant shipping had to do without it. This they did the more willingly because from the end of the eighteenth century, they had at their disposal clocks that could keep and tell them the time at a place of known longitude; so all they had to do to find where they were, at any hour of day or night, sky clear or cloudy, was check their own time and subtract it from the time indicated by their chronometer(s).

These sea clocks were not cheap, but they were a bargain in performance and time saved. The scientists did not like them. They instinctively do not like magic boxes, which enable people to avoid and evade the necessity of thinking. (Compare the preference of mathematicians for traditional solutions as against computer-driven number crunching.) But in matters of technology, economy and practicality will win out over principle every time.

It is the invention of these clocks that we are celebrating here, and we do appropriate honor to the memory of the great John Harrison (1693-1776), autodidact, genius, and skilled craftsman who outdid the expectations of the greatest intellects of his day and showed that it could be done and how it could be done. It was there that his great achievement lay: he did not finish the task, but he announced its completion. His work was in one sense imitable, but also inimitable in practical terms. So that much remained to be done. As a result, the marine chronometer did not become a serious factor in navigation and trade for another 50 years. It took the simplifying improvements of such people as Earnshaw and Arnold[26]—I give the names in that order—combined with the division of labor implicit in batch production to make the diffusion of the chronometer possible.

In right good time. The nineteenth century saw a huge increase in the size of navies and merchant marine as well as of individual ships, thereby

26. See Jonathan Betts's paper in this volume.

enhancing considerably the pay-off (in cost savings) to superior navigation.[27] These gains, obviously, were related to savings in time—but, and this is a surprise, not so much as a result of superior speed (the nineteenth-century sailing packets were making their crossings in about the same time as the transoceanic vessels of the early eighteenth century) but as a result of faster turnarounds. Where the typical Atlantic sailing ship in the earlier period made one round trip a year, its nineteenth-century successors were doing two and then three.[28]

How was this possible? And where does the marine chronometer come in? Part of the answer lay in the facilities for loading and unloading, which were improved in an effort to deal with increased shipping volume. The ports were getting crowded. But part, I suggest, was the result of increased navigational autonomy: the ability to leave port in any season, to seek better winds outside the conventional routes and channels, to lose oneself in emptiness because one could never be lost.

Some last thoughts: the chronometer came in, and then only partially, on the eve of a major change in the technology of marine transport—the shift from sail to steam. The effect of the new mode of locomotion on navigation was contradictory: it made finding the longitude less important, but also more important. On the one hand, it reduced the risks of drift and loss of way due to storms; on the other, by increasing speed, it made accuracy of navigation that much more important. (The rule is, the faster one or it moves, the smaller the tolerance of error.)

But the greatest contribution of the new technique for finding longitude was in mapping, in locating and shaping more accurately the land points and masses on the Earth's surface, the better to find one's way. (Here its superiority to the astronomical method was overwhelming.) This was crucial, because chronometers are only as good and useful as the maps and charts that go with them, and it was these same chronometers that made the maps and charts reliable to their measure. This, moreover, was a reciprocal dynamic: once one had better charts and maps, the usefulness of the chronometer was correspondingly enhanced.

Today, of course, finding the point at sea is just that. Thanks to satellite signals and triangulation, a properly equipped ship can know its location within a matter of yards or even feet.[29] The old marine chronometers are no longer needed, although some ships still carry them out of habit or just in case. (It is not at all clear, by the way, that on these ships the clocks are carefully set to time on departure; or that they are regularly used and maintained; or that the crew even know where to find them. And why should they, when a good quartz wrist watch will do even better?)

There are, then, far more marine chronometers in private collections and museums than there are on the high seas. Science, which has a way of junking or erasing the things it no longer needs or uses, has long since left these clocks behind. But that is why we need history—to remind us of our debts and make us understand that knowledge and technique are not manna from heaven. We have to plant and cultivate them.

27. Among the factors making it possible to build bigger: the introduction of iron members—replacing the old wooden crutches, very hard to come by—to strengthen the hull. These came in well in advance of all-iron construction in the 1840s. On the shift to larger ships, see Braudel, *Civilisation matérielle (op. cit.,* note 6), p. 372.

28. On the economic consequences of changes in shipbuilding technology and the organization of trade, see Douglass C. North, "Sources of Productivity Change in Ocean Shipping, 1600-1850," *Journal of Political Economy,* vol. 76, no. 5 (September/October 1968), pp. 953-70. North stresses the gains in turnaround time but says nothing about the adoption of the chronometer.

29. This is the so-called Global Positioning System (GPS). It is now being developed and adapted for civilian use, not only by vessels at sea but by automobile drivers and even hikers. See "Navigation: Finding the future," editorial in Science and Technology section of *The Economist,* vol. 329, no. 7836 (November 6, 1993), p. 115.

## FURTHER READING

du Halde, Jean-Baptiste. 1735. *Description geographique, historique, chronologique, politique et physique de l'empire de la Chine et de la Tartarie chinoise.* 4 vols. Paris. The Hague, 1736. English translation: London, 1736, 1741. German translation: 1748.

Marcus, G. J. 1980. *The Conquest of the North Atlantic.* Ipswich, England: Boydell.

Taylor, E. G. R. 1971. *The Haven-Finding Art: A History of Navigation from Odysseus to Captain Cook.* New ed. London: Hollis & Carter for the Institute of Navigation.

Thrower, Norman J. W., ed. 1984. *Sir Francis Drake and the Famous Voyage, 1577-1580. Essays Commemorating the Quadricentennial of Drake's Circumnavigation of the Earth.* Berkeley: University of California Press.

Turner, Anthony. 1987. *Early Scientific Instruments: Europe 1400-1800.* London: Sotheby's Publications.

Longitude in
Context
A New and Correct CHART Shewing the VARIATIONS of the COMPASS in the WESTERN & SOUTHERN OCEANS as Observed in ye Year 1700 by his Maties Command by Edm. Halley.
NOVA BRITANNIA
CANADA
NEW-FRANCE
NOVA SCOTIA
NEW-ENGLAND
NEWYORK
PENSYLVANIA
MARYLAND
VIRGINIA
CAROLINA
the Banks of Newfoundland
West Variation
THE WESTERN OCEAN
AZORES
WESTERN ISLES
CANARY ISLES
CAPE DE VIRDE ISLES
Tropic of Cancer
Line of no Variation
East Variation
The Æquinoctiall
Tropick of Capricorn
East Vari-ation
West Varia
THE SOUTHERN OCEAN
East Variation
THE ICEY
DEGREES of Va
EAST WE
SPAIN
FRANCE
BAY OF BISCAY
BARBARY
GUINEA
AFRI
SOUTH AME-RICA
WILD BRASILE
BRASILE
PARAGUAY
PERU
CHILI
PATAGONIA
The Sea in these parts abounds with two sorts of Animalls

See Thrower Figure 9 for caption for this and previous page

# Mathematics

## *Bruce Chandler*

Bruce Chandler is Professor of Mathematics at The College of Staten Island of The City University of New York. He received his Ph.D. from the Courant Institute of New York University in 1963. His main research interests are in combinatorial group theory and the history of mathematics and scientific instruments. He is the author of *The History of Combinatorial Theory* (with Wilhelm Magnus, 1982) as well as a number of books and articles. He was the founding editor (with Harold Edwards) of *The Mathematical Intelligencer* (1978) and the General Editor of the catalogue project of The Time Museum, Rockford, Illinois. He is currently the General Editor of the *Catalogue of the Collection of Scientific Instruments,* Harvard University, and of the catalogue project of The Adler Planetarium, Chicago, Illinois.

# Longitude in the Context of Mathematics

*Bruce Chandler*

ἃ τῆς ἀνωτάτω καὶ καλλίστης ἐστὶ θεωρίας, ἐπιδεικνύντα διὰ τῶν μαθημάτων ταῖς ἀνθρωπίναις καταλήψεσι....

*It belongs to the highest and most admirable theoretical discipline to demonstrate these things through mathematics to human understanding....*[1]

Ptolemy, *The Geography, ca.* A.D. 150

By the end of the eighteenth century there were two practical solutions to the problem of determining longitude at sea.[2] Both solutions were based on a simple mathematical idea: if the local time at a place of known longitude, which can be obtained by an astronomical observation, is compared at the same instant with the local time at another location, a simple calculation using the difference in local times will give the longitude of the second location. This is because the Earth rotates about its axis—360° in longitude—once in 24 hours, or 15° every hour. Thus a difference of four minutes in time is equivalent to a change of 1° of longitude. The difficulty, of course, is how to know the local time of one location at another place on the Earth.

One of the practical solutions to this problem was achieved, after many years, by John Harrison (1693-1776): a clock accurate and stable enough on a long sea voyage to carry local time (such as the time at the port of embarkation, the longitude of which was known) to any other location. The second solution was to use the local times of an astronomical event that could be seen simultaneously at any two points on Earth to find the difference in longitude between these locations. Since the relative positions of the fixed stars and the Moon change rapidly, the distance between a star and the Moon was used. For this reason it was called the "lunar-distance" method.[3] To make this method practicable, what was needed was a table giving the local times of the distance between the Moon and certain stars at a location of known longitude. A prerequisite for computing these tables in advance was a theory of the motion of the Moon that, with adequate observational data, could be used to predict the position of the Moon relative to the fixed stars with sufficient accuracy. One of the great mathematical achievements of the eighteenth century was the development of such a theory.

1. *Claudii Ptolemaei Geographia,* ed. C. F. A. Nobbe, 3 vols. (Leipzig, 1898), vol. 1, p. 5. I would like to thank Gisela Striker for the translation. The English translation of *The Geography* (*Claudius Ptolemy: The Geography,* trans. and ed. Edward Luther Stevenson [New York: Dover, 1991]) should be used with caution.

2. For a history and explanation of various methods for finding longitude after the eighteenth century, see Edmond Guyot, *Histoire de la détermination des longitudes* (La Chaux-de-Fonds, Switzerland, 1955), especially chaps. 3-5.

3. See Derek Howse's paper in this volume for a detailed description of this method.

4. A. J. Turner, "France, Britain and the Resolution of the Longitude Problem in the 18th Century," *Vistas in Astronomy,* vol. 28, parts 1/2 (1985), p. 315.

5. See Christiaan Huygens, *Horologium* (The Hague, 1658) and *Horologium Oscillatorium Sive De Motu Pendulorum Ad Horlogia Aptato Demonstrationes Geometricae* (Paris, 1673).

6. See Isaac Newton, *Philosophiae Naturalis Principia Mathematica* (London, 1687).

7. Huygens, *Horologium Oscillatorium* (*op. cit.,* note 5). In this book Huygens developed cycloidal cheeks that, in theory, allowed the pendulum to be isochronous—that is, to perform each oscillation in the same period of time regardless of its amplitude of swing. He also developed here the first mathematical theory of the composite, or compound, pendulum and many new mathematical concepts, *e.g.,* the evolute of a curve and the center of oscillation of a figure.

The cycloid (see Figure 1) is a curve (AEBC) generated by a point (E) on a circle as the circle (ELK) rolls, without slipping, on a straight line (AC). The construction of the cycloid had in fact already been used by Charles Bouvelles (*ca.* 1470-*ca.* 1553), in 1503, in his attempt to square the circle; see *Dictionary of Scientific Biography,* ed. Charles Coulston Gillispie and Frederic L. Holmes, 18 vols. (New York: Charles Scribner's Sons, 1970–90), vol. 2, pp. 360-1, and note 25 in this paper. But it was not until the seventeenth century that the curve was studied in

At the "Longitude Zero" Symposium in 1984, one of the speakers asked: "why did it take so long for the problem of the longitude to be resolved?" The author went on to say that "there is little involved either in the chronometers of John Harrison or in the instruments and tables essential to the lunar-distance method which were beyond the limits of 17th century techniques if they had been seriously prosecuted."[4] I think, however, that this judgement seriously underrates the difficulties, technological and mathematical, involved in the solution of the problem and distorts the historical situation. A closer look at the history of mathematics is necessary in order to understand why it took "so long" to arrive at a solution to the longitude problem.

Two of the greatest achievements in applied mathematics in the second half of the seventeenth century were the application by Christiaan Huygens (1629-1695) of his mathematical theory of the motion of the pendulum to horology[5] and the application by Isaac Newton (1642-1727) of his mathematical discoveries to astronomy.[6] Both were essential to the solution to the problem of finding longitude at sea, but both were also inadequate for solving the problem.

Huygens, following his application of the pendulum to a clock in December 1656, developed a mathematical theory of the pendulum.[7] He continued, until the time of his death, to pursue these ideas for a clock for finding longitude at sea. He was unsuccessful.[8] Harrison's great genius was to understand the theory of precision timekeeping and to provide solutions to resolve the technical problems, such as the effect of temperature on the accuracy of a clock, that no scientist or clockmaker of the seventeenth century had been able to overcome.[9]

Newton's astronomical theory, though spectacularly successful in analyzing the motions of each of the planets around the Sun by proving that the force of attraction was inversely proportional to the distance from the Sun,[10] was not so successful in solving the "three body problem": the motion of the Moon revolving about the Earth, which, in turn, revolves around the Sun. This was central to making the predictions of the Moon's position that were to make the lunar-distance method possible. D. T. Whiteside sums up the problems with Newton's lunar theory:

> In the *Principia's* third book, it is true, he [Newton] put up a brave public show of deriving such principal periodic and secular inequalities as its annual variation, latitudinal wobble and mean nodal regress

detail. Galileo (1564-1642), at the end of the sixteenth century, called the curve a cycloid, and Marin Mersenne (1588-1648), around 1630, suggested that the cycloid could be used as a test curve for the new advanced mathematical techniques that were then being developed. Other seventeenth-century mathematicians who studied the cycloid were Pierre Fermat (1601-1665), Gilles Roberval (1602-1675), John Wallis (1616-1703), René Sluse (1622-1685), Blaise Pascal (1623-1662), Huygens, Christopher Wren (1632-1723), Newton, Gottfried Leibniz (1646-1716), Jakob Bernoulli (1654-1705), and Johann Bernoulli (1667-1748). Most of the work on the cycloid was purely mathematical except for Huygens's application of it to the clock and the proofs, given in 1696-7 by Jakob and Johann Bernoulli, Newton, and Leibniz, that the branchistochrone—the curve along which a body must descend so that it goes from one point to another, not directly below it, in the least possible time—was a cycloid. The cycloid was also called a trochoid or a roulette.

For the history and mathematical exposition of the cycloid, see Gino Loria, *Spezielle Algebraische und Transscendente Ebene Kurven. Theory und Geschichte,* trans. Fritz Schütte (Leipzig: B. G. Teubner, 1902), pp. 460-78; *Dictionary of Scientific Biography*, vol. 10, pp. 336-7, and vol. 2, p. 53; and D. J. Struik, *A Source Book in Mathematics, 1200-1800* (Cambridge: Harvard University Press, 1969), pp. 232-3 and 392-9.

8. See J. H. Leopold's paper in this volume.

9. See papers by William Andrewes and Anthony Randall in this volume for a history and description of Harrison's sea clocks.

10. Newton, *Philosophiae Naturalis* (*op. cit.,* note 6), p. 395.

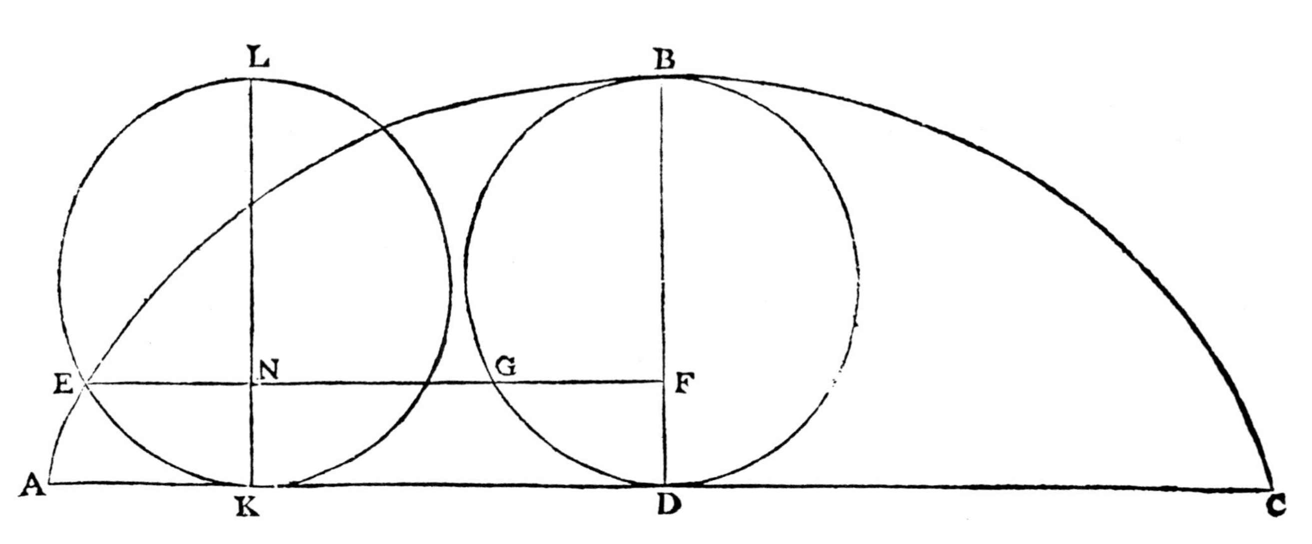


**Figure 1. The generation of a cycloid. From Christiaan Huygens, *Horologium Oscillatorium...* (Paris, 1673), proposition 14, p. 38. By permission of the Houghton Library, Harvard University. Ref. no. Typ 615.73.457F.**

reasonably well from the simplified hypothesis that the undeviated lunar orbit is a circle uniformly traversed round the earth at its centre, but the mass of his contemporary and subsequent worksheets where he sought doggedly to deepen and refine those basic findings tell a different tale of repeated false starts, the myopic pursuit of dead-end trails and a near-total lack of success....We may well understand why, when long afterwards John Machin [1680-1751] praised his lunar theory as 'all sagacity', he [Newton] sadly 'smiled & said his head never ached but with his studies on the moon'. A cogent explanation of the inequality was not in fact to be achieved, through the combined efforts of [Leonhard] Euler [1707-1783], [Jean Le Rond] D'Alembert [1717-1783] and [Alexis-Claude] Clairaut [1713-1765], for more than another sixty years, and then only by discarding Newton's ineffectual approach by geometrical limit-approximations in favour of more powerful analytical series methods.[11]

There was a quiet revolution that occurred in mathematics in the first half of the eighteenth century. The discarding of "geometrical limit-approximations in favour of more powerful analytical series methods" was only part of this revolution. Fundamental was the clarification and recognition of the concepts of function and differential equation.[12] As Clifford Truesdell, the

11. *The Mathematical Works of Isaac Newton,* ed. D. T. Whiteside (Cambridge: Cambridge University Press, 1974), vol. 6, p. 27.

12. See H. J. M. Bos, "Mathematics and rational mechanics," in G. S. Rousseau and Roy Porter, eds., *The Ferment of Knowledge: Studies in the Historiography of Eighteenth-Century Science* (Cambridge: Cambridge University Press, 1980), pp. 333-51.

**Figure 2. Portrait of Isaac Newton in 1703, by Charles Jervas (*ca.* 1675-1739). It was in this year that Newton was elected president of the Royal Society, a position that he held until his death in 1727. This portrait was presented to the Royal Society by Newton in 1717. Oil on canvas. 125.7 x 101.6 cm. (49.5 x 40 in.). By permission of the President and Council of the Royal Society, London.**

**Figure 3. Portrait of Leonhard Euler, painted in 1753 by Emanuel Handmann (1718-1791). Pastel on paper. 57 x 44 cm. (22.4 x 17.3 in.). Courtesy of the Öffentliche Kunstsammlung Basel, Kunstmuseum. Acc. no. 276.**

eminent historian of mathematics, points out, it was Euler, not Newton, who first published what are commonly referred to as "Newtonian equations":

> [Euler's] paper called "Discovery of a new principle of mechanics", published in 1752, presents the [differential] equations
>
> $$F_x = Ma_x \qquad F_y = Ma_y \qquad F_z = Ma_z$$
>
> where the mass $M$ may be either finite or infinitesimal, as the axioms which "include all the laws of mechanics." Later he called them "the first principles of mechanics". These are the famous "Newtonian equations", here proposed for the first time as general, explicit [differential] equations for mechanical problems of all kinds. The discovery of this principle seems so easy, from the Newtonian ideas, that it has never been attributed to anyone but Newton; such is the universal ignorance of the true history of mechanics.[13]

And as William Whewell (1794-1866), the influential nineteenth-century historian of science, put it when discussing lunar theory,

> Newton's successors, in the next generation, abandoned the hope of imitating him in this intense mental effort [using his geometrical

13. Clifford Truesdell, "A Program toward Rediscovering the Rational Mechanics of the Age of Reason," in *Essays in the History of Mathematics* (New York: Springer-Verlag, 1968), pp. 116-17.

408 PHILOSOPHIÆ NATURALIS

De Mundi Systemate

PROPOSITIO XXXI. PROBLEMA XII.

*Invenire motum horarium Nodorum Lunæ in Orbe Elliptico.*

Deſignet *Qpmaq* Ellipſin, axe majore *Qq*, minore *ab* deſcriptam, *QAq* Circulum circumſcriptum, *T* Terram in utriuſque centro communi, *S* Solem, *p* Lunam in Ellipſi motam, & *pm* arcum quem data temporis particula quam minima deſcribit, *N* & *n* Nodos linea *Nn* junctos, *pK* & *mk* perpendicula in axem *Qq* demiſſa & hinc inde producta, donec occurrant Circulo in *P* & *M*,

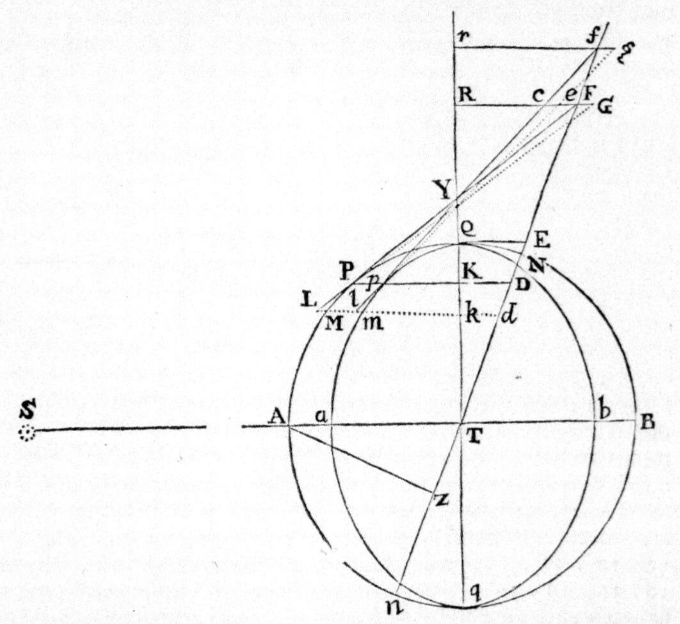


& lineæ Nodorum in *D* & *d*. Et ſi Luna, radio ad Terram ducto, aream deſcribat tempori proportionalem, erit motus Nodi in Ellipſi ut area *pDdm*.

Nam ſi *PF* tangat Circulum in *P*, & producta occurrat *TN* in *F*, & *pf* tangat Ellipſin in *p* & producta occurrat eidem *TN* in

**Figure 4. Newton's geometrical method for finding the hourly motion of the nodes of the Moon, assuming an elliptical orbit. From Isaac Newton, *Philosophiae Naturalis Principia Mathematica*, 2nd ed. (Cambridge, 1713), book 3, proposition 31, problem 12, p. 408. Courtesy of The Adler Planetarium and Astronomy Museum, Chicago.**

> methods]; they gave the subject over to the operation of algebraical reasoning, in which symbols think for us, without our dwelling constantly upon their meaning,....[14]

The differences between Newton's and later eighteenth-century mathematical techniques are illustrated in Figures 4 and 5.

It has often been said that the motivation for scientists and clockmakers to solve the longitude problem was the financial incentives given by the Longitude Act of 1714. For example, "both [the invention of the timekeeper with the requisite precision and reliability and the theory that could predict the position of the Moon against the fixed stars] were to be achieved...thanks largely to the incentives promised by the Longitude Act, just as Parliament had intended."[15] I cannot comment on Harrison's and other clockmakers' motivations, but it seems clear that the reasons for finding an adequate lunar theory, and hence the possibility of using the lunar-distance method, had very little to do with the longitude prize. The problem of developing a mathematical theory of the Moon's motions has been a fundamental problem in Western astronomy from before the time of Claudius Ptolemy (fl. A.D. 150) to the present.

14. William Whewell, *History of the Inductive Sciences from the Earliest to the Present Time*, 3rd ed. with additions, 2 vols. (New York: D. Appleton, 1858), vol. 1, p. 411.

15. Derek Howse, *Nevil Maskelyne: The Seaman's Astronomer* (Cambridge: Cambridge University Press, 1989), p. 14.

Figure 5. The differential equations of the motion of the Moon. From Pierre-Simon de Laplace, *Traité de Mécanique Céleste* (Paris, 1802), vol. 3, part 2, book 7, chap. 1, p. 181. Courtesy of the Adler Planetarium and Astronomy Museum, Chicago.

SECONDE PARTIE, LIVRE VII. 181

## CHAPITRE PREMIER.

### *Intégration des équations différentielles du mouvement lunaire.*

1. Reprenons les équations différentielles (*K*) du n°. 15 du second livre, et donnons-leur la forme suivante,

$$dt = \frac{dv}{hu^2 \cdot \sqrt{1 + \frac{2}{h^2} \cdot \int \left(\frac{dQ}{dv}\right) \cdot \frac{dv}{u^2}}}$$

$$\left.\begin{aligned} 0 &= \left(\frac{ddu}{dv^2} + u\right) \cdot \left\{1 + \frac{2}{h^2} \cdot \int \left(\frac{dQ}{dv}\right) \cdot \frac{dv}{u^2}\right\} + \frac{du}{h^2u^2 \cdot dv} \cdot \left(\frac{dQ}{dv}\right) \\ &\quad - \frac{1}{h^2} \cdot \left(\frac{dQ}{du}\right) - \frac{s}{h^2 u} \cdot \left(\frac{dQ}{ds}\right) \\ 0 &= \left(\frac{dds}{dv^2} + s\right) \cdot \left\{1 + \frac{2}{h^2} \cdot \int \left(\frac{dQ}{dv}\right) \cdot \frac{dv}{u^2}\right\} + \frac{1}{h^2u^2} \cdot \frac{ds}{dv} \cdot \left(\frac{dQ}{dv}\right) \\ &\quad - \frac{s}{h^2u} \cdot \left(\frac{dQ}{du}\right) - \frac{(1+ss)}{h^2u^2} \cdot \left(\frac{dQ}{ds}\right) \end{aligned}\right\} ; (L)$$

Dans ces équations, $t$ exprime le temps, et l'on a

$$Q = \frac{1}{r} - \frac{m' \cdot (xx' + yy' + zz')}{r'^3} + \frac{m'}{\sqrt{(x'-x)^2 + (y'-y)^2 + (z'-z)^2}}.$$

$M$, $m$ et $m'$ sont les masses de la terre, de la lune et du soleil; $x, y, z$ sont les coordonnées de la lune rapportées au centre de gravité de la terre, et à une écliptique fixe; $x', y', z'$ sont les coordonnées du soleil; $r$ et $r'$ sont les rayons vecteurs de la lune et du soleil; $s$ est la tangente de la latitude de la lune au-dessus du plan fixe; $\frac{1}{u}$ est la projection de son rayon vecteur sur le même plan; $v$ est l'angle fait par cette projection, et par l'axe des $x$; enfin $h^2$ est une constante arbitraire dépendante principalement de la distance de la lune à la terre.

The Longitude Act of 1765 awarded prizes to two scientists for their contributions to solving the longitude problem: Leonhard Euler for his lunar theory and Tobias Mayer (1723-1762) for his lunar tables, which are based on Euler's theory.[16] Yet it is clear that for both men the hope of being awarded the longitude prize was not their primary motivation. In the extensive correspondence between Euler and Mayer—31 letters dating from 1751 to 1755, dealing primarily with the problem of lunar theory[17]—the first mention of the longitude prize is in the 21st letter, dated June 11, 1754. Here Euler writes:

> In coming now to your new important discoveries, I first of all congratulate you wholeheartedly upon them, and wish that their importance would soon be known to everyone....In England...following the representations which I made on the importance of your tables, I have received such an answer from which I can with reason conclude that you would be regarded as a worthy competitor for the award established for the discovery of longitude.[18]

In all the letters there is no hint on Mayer's part that the longitude prize had anything to do with his working on these problems. On the contrary,

16. *Ibid.*, p. 79, and Eric Forbes, *The Euler-Mayer Correspondence (1751-1755)* (New York: American Elsevier, 1971), pp. 17-20. The awards given were £300 for Euler and £3,000 for Mayer's heirs.

17. Forbes (*ibid.*).

18. *Ibid.*, p. 86.

Mayer's concern was not to find longitude at sea; rather, as his biographer, Eric Forbes, says, "the primary motivation for Mayer's astronomical researches was his devotion to the aim of improving the state of cartography by providing accurate determination of terrestrial latitudes and longitudes."[19] There was no prize for this work! It was Euler who broaches the subject of the prize and, in the last letter, dated May 27, 1755, congratulates Mayer on "the fact that your tables found such great acclaim in England." He continues:

> Since you now have found means of improving them still more, and are in a position to be able to determine through them the longitude at sea, so much is certain; that until now no one has been able to make such a well-established claim on the fixed prize, and since in addition you have solved this problem in the most accurate manner, I heartily wish that this will be recognised in England in accordance with its worth and that the maximum prize will be offered to you;....[20]

But Euler's motivation also was surely not the longitude prize, for himself or for Mayer. This was incidental to his major concern of developing a lunar theory. One sees this clearly in 1749 when the St. Petersburg Academy of Sciences, "in imitation of the Paris and Berlin academies, had decided to stage a series of prize contests,"[21] the first of which was scheduled for 1751. They asked Euler, the most prominent living mathematician, to make suggestions for the topics to be considered for the prize. One of the four problems he proposed was "Whether all the inequalities which are observed in the motion of the Moon are in agreement with Newtonian theory or not? And what is the true theory whereby the place of the Moon at any time can be very exactly defined?"[22] When the Academy chose this as the prize question, Euler wrote to Teplov, a councillor of the imperial Russian chancellery:

> ...at present it is the subject which most interests the greatest *savants* of Europe. Their opinions are still divided over it, some maintaining that Newton's theory...is not sufficient to account for all the inequalities observed in the movement of the planets and above all in the Moon [Euler's view at the time], while others maintain the contrary....And whatever the outcome,...one can rightly say that never has any academy proposed so important a question, nor one whose explanation promised so much benefit for the advancement of the sciences.[23]

No talk of longitude or longitude prizes here. Euler then renounced entering the competition himself and proposed that he be one of the judges. The French mathematician Alexis-Claude Clairaut won the competition, and Euler wrote to him:

> ...the more I consider this happy discovery [Clairaut's prize essay], the more important it seems to me, and in my opinion it is the greatest discovery in the Theory of Astronomy....For it is very certain that it is only since this discovery that one can regard the law of attraction reciprocally proportional to the squares of the distances as solidly established; and on this depends the entire theory of astronomy.[24]

TOBIAS MAYER

*Gebohren zu Marbach im Wirtemberg d. 17. Febr. 1723. Gestorben zu Göttingen d. 20. Febr. 1762.*

**Figure 6. Portrait of Tobias Mayer, engraved by Westermayr (probably Konrad, 1765–1834) from a painting by Joel Paul Kaltenhofer (d. 1777). From Franz Xaver von Zach, *Allgemeine Geographische Ephemeriden Verfasset von einer Gesellschaft Gelehrten und herausgegeben*, vol. 3 (Weimar, 1799), frontispiece.**

19. *Ibid.*, p. ix.

20. *Ibid.*, p. 97.

21. Curtis A. Wilson, "Perturbations and Solar Tables from Lacaille to Delambre: the Rapprochement of Observation and Theory, Part I," *Archive for History of Exact Sciences* (New York), vol. 22, no. 1/2 (1980), p. 139.

22. *Ibid.*, p. 139.

23. *Ibid.*, pp. 140-1.

24. *Ibid.*, p. 143. The search for a lunar theory does not end in the eighteenth century. Some of the most important mathematics in the twentieth century was developed to solve this problem (*e.g.*, chaos theory, so fashionable today, was an outgrowth of the mathematical ideas introduced by Henri Poincaré (1854-1912) in a prize-winning essay in 1889-90 concerning this problem). For the history of the problem from the eighteenth century to the present, see June Barrow-Green, *Poincaré and the Three Body Problem* (Providence, Rhode Island: American Mathematical Society, 1996) and Jeremy Gray, "Poincaré, topological dynamics, and the stability of the solar system," in P. M. Harman and Alan E. Shapiro, eds., *The investigation of difficult things: Essays on Newton and the history of the exact sciences in honour of D. T. Whiteside* (Cambridge: Cambridge University Press, 1992), pp. 503-24.

**Figure 7. The three conic sections: a) parabola, b) ellipse, and c) hyperbola. From Apollonius of Perga, *Apollonii Pergaei Conicorum Libri Quatuor,...* (Pistoia, 1696), pp. 30, 34, 36.**

**a. A parabola is the section (DFE) made by a plane cutting through a cone (ABC) when the plane is parallel to the side (AC) of the cone.**

**b. An ellipse is the section (ELD) made by a plane cutting through the entire cone (ABC) at any angle.**

**c. A hyperbola is the formation of two sections (DEF and GHK) made by a plane cutting through a double cone (BAC and XAO) parallel to the axis of the cones.**

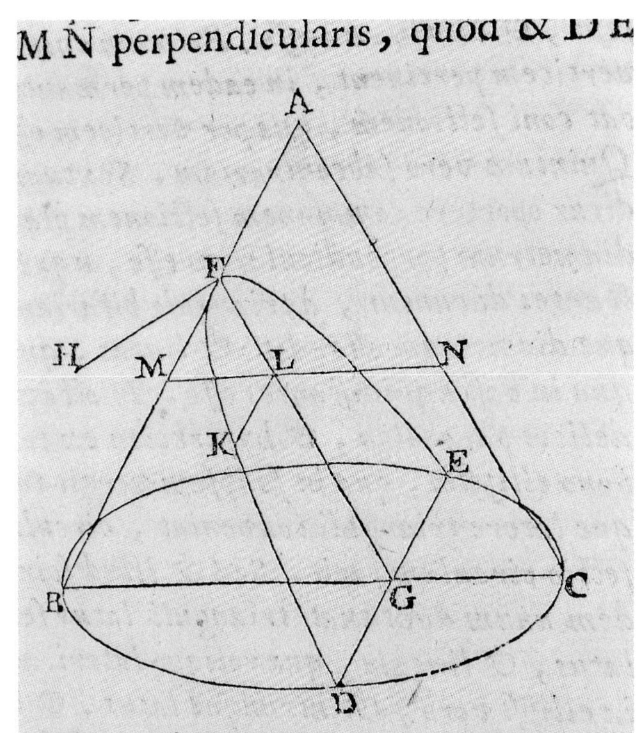


a

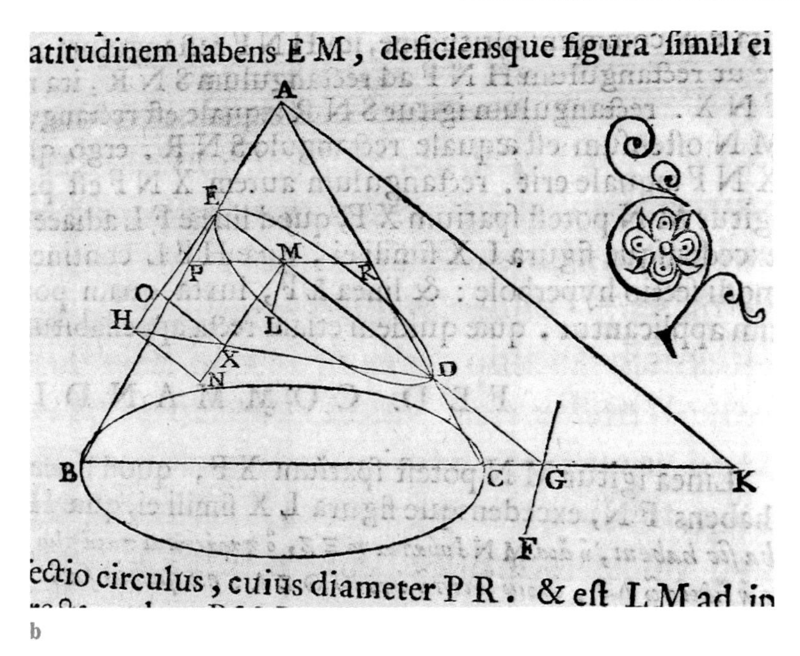


b

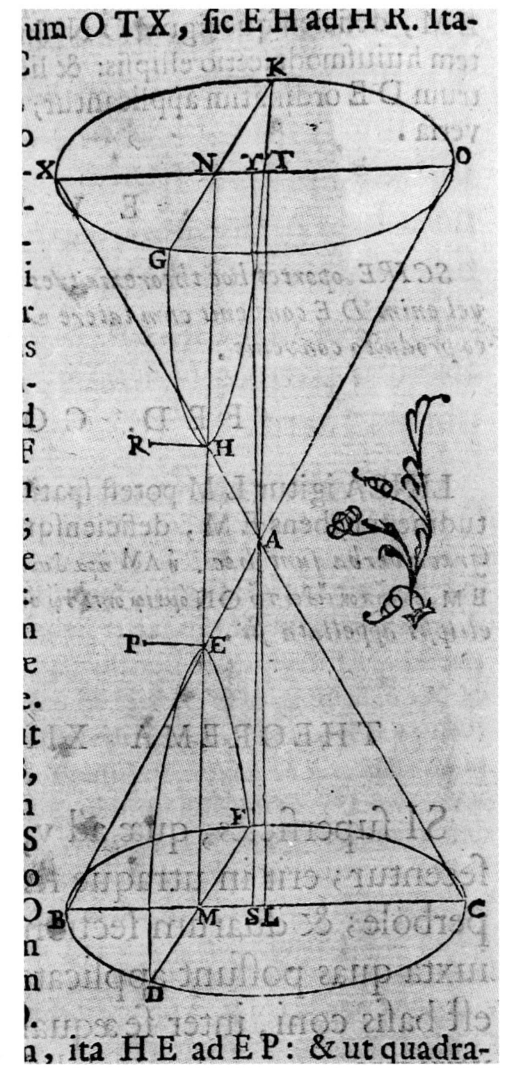


c

Though the motivations of an individual's scientific work can be diverse, I think it is very important to realize that the drive to solve problems that are internal to a science is fundamental. I will close with an example from the history of mathematics that illustrates this and also illuminates important aspects of the relation between mathematics and the finding of longitude at sea.

When discussing the major achievements of Greek mathematics, historians usually point to the discovery of the incommensurable (*i.e.*, in modern terms, the irrational number), the introduction of proof into mathematics, the axiomatic method, and the method of exhaustion—the precursor to the integral calculus. These advances changed mathematics and were crucial to the development of Western science. One achievement that is usually not mentioned but is as important as the others is the mathematical investigation of curves other than the straight line and the circle. These "higher" curves were introduced into mathematics to solve problems that arose because of the peculiar way the Greeks framed their mathematics—*i.e.*, geometrically. Three problems (sometimes referred to as the "three famous problems of antiquity") were recognized early as being central to the development of Greek mathematics: (1) duplication of the cube, (2) squaring of the circle, and (3) trisection of an angle.[25]

The Greek mathematicians, beginning in the fourth century B.C., introduced curves as auxiliaries to the solution of these problems.[26] By the end of the fourth century, the mathematical study of the curves themselves, especially conic sections (see Figure 7), had become the core of Greek higher mathematics. The works of the two greatest mathematicians of antiquity, Archimedes of Syracuse (*ca.* 287-212 B.C.) and Apollonius of Perga (*ca.* 262-190 B.C.), are mainly concerned with problems related to these curves. It is hard to exaggerate the importance of these developments to the history of Western mathematics, science, and technology. These curves became the curves of the universe in the work of Johannes Kepler (1571-1630), the curves of motion in the work of Galileo Galilei (1564-1642), and the main objects of study in the coordinate geometry of René Descartes (1596-1650). These developments, above all others, made the scientific revolution possible. Without Kepler's laws of elliptical orbits of the planets, Galileo's analysis of motion, and the unification of these ideas, using the calculus, in the work of Isaac Newton, there would have been no accurate lunar theory and no possibility of finding longitude at sea by astronomical means. Further, without the interest in the properties of curves in the seventeenth century, the work of Huygens in horology would not have been possible either. The cycloid, the curve of Huygens's isochronous pendulum, was discovered and studied by mathematicians early in that century, before any applications were evident.[27] It would appear that Harrison's clocks owe much to Huygens's discoveries. Finally, most fundamental to the finding of longitude at sea is the purely mathematical concept of longitude.[28]

25. See Thomas Heath, *A History of Greek Mathematics,* 2 vols. (Oxford: Oxford University Press, 1921), vol. 1, pp. 218-70, and Ivor Thomas, *Selections Illustrating the History of Greek Mathematics,* 2 vols. (Cambridge: Harvard University Press, 1957), vol. 1, pp. 256-363. Thomas gives the texts and translations of most of the important sources for the solutions to these problems in Greek mathematics.

26. The duplication problem—to find the side of a cube with volume twice that of a given cube—can be stated in modern algebraic notation as finding the solution to the cubic equation $x^3 = 2a^3$, where $a$ is the side of the given cube. The mathematician Hippocrates of Chios (fl. *ca.* 440 B.C.) reduced the problem to finding two mean proportionals—in modern notation, finding $x$ and $y$ such that $a/x = x/y = y/2a$. Menaechmus (fl. mid-4th century B.C.) solved the mean proportional problem using conic sections. In modern notation using coordinate geometry, $a/x = x/y$ gives $ay = x^2$ (the equation of a parabola) and $a/x = y/2a$ gives $2a^2 = xy$ (the equation of a hyperbola). The $x$-coordinate of the intersection of these two curves is the required solution. The squaring problem—to find the side of a square with area equal to that of a given circle—led to the introduction into Greek mathematics of the "method of exhaustion," the precursor of the seventeenth-century development of the calculus. The trisection problem—to find one-third of a given angle—also leads to the solution of a cubic equation, which was solved using conic sections and other curves. All three problems cannot be solved with a straight edge and compass, the tools postulated in Euclid's *Elements* for geometric constructions. The Greek mathematicians knew this but did not prove it. It was only in the nineteenth century that this was shown to be true.

27. See note 7.

28. It is often forgotten that a spherical Earth with a coordinate system of longitude and latitude circles is a mathematical abstraction constructed by the ancient Greek mathematicians. For the history of longitude and latitude in ancient Greece, see O. Neugebauer, *A History of Ancient Mathematical Astronomy,* part II (New York: Springer-Verlag), pp. 934-40.

# Navigation

## *W. F. J. Mörzer Bruyns*

**Willem Mörzer Bruyns is Senior Curator of Navigation at the Nederlands Scheepvaartmuseum in Amsterdam. Trained as a navigation officer at the Amsterdam Nautical College, he sailed for the Netherland Line (Amsterdam) as a junior officer before joining the museum in 1969 as Deputy Curator of Navigation.**

**From 1980 to 1989, Mörzer Bruyns was co–editor of the scholarly maritime history periodical *Tijdschrift voor zeegeschiedenis.* He served as Secretary General of the International Congress of Maritime Museums from 1984 to 1990 and as Chairman of the Dutch Museums Association from 1989 to 1993. In 1993 he was elected Fellow of the Royal Institute for Navigation in London.**

**Mörzer Bruyns has lectured on many occasions in the United States, Australia, and Europe. In addition to writing articles for various scholarly periodicals on the history of navigation and navigational instruments, he recorded the history of Dutch nineteenth-century Arctic exploration in *De eerste tocht van de Willem Barents naar de Noordelijke Ijszee 1878* (1985). More recently he published *The Cross–staff. History and Development of a Navigational Instrument* (1994).**

# Longitude in the Context of Navigation

*W. F. J. Mörzer Bruyns*

There can be no doubt that, in the entire history of seafaring, the problem of finding longitude at sea was the most difficult one. Although in theory the problem was solved early in the sixteenth century, the practical solution did not become available until the second half of the eighteenth century.

The Louvain professor Gemma Frisius (1508-1555) published a proposal in 1530 suggesting that the difference in longitude between two places could be established with an accurate clock measuring the difference in time between the places.[1] In order to measure this, an extremely accurate clock is required. It would have to 'keep the time' on the meridian of departure during an entire voyage. The timekeeper would have to be resistant to considerable changes in temperature and relative humidity and to the movement of a ship. Not until the mid-eighteenth century was such an instrument constructed with success by John Harrison (1693-1776).[2] This development, however, did not stop scholars, scientists, and inventors from searching for methods to solve the problem with or without a timekeeper. They were stimulated by monetary rewards offered by the Spanish kings Philip II in 1567 and Philip III in 1598, by the Dutch government around 1600,[3] and by the British Parliament in 1714.[4] The rewards attracted applicants from all over Europe, among them Galileo Galilei (1564-1642), who suggested a method that involved observing the eclipses of Jupiter's satellites.[5] The eclipsing of the satellites in the shadow of Jupiter, which can be seen to occur at precisely the same moment regardless of the location of the observer on Earth, indeed provided a practical method for finding longitude on land, but not on a moving ship. Much earlier than Galileo, the German astronomer Johann Werner (1468-1522) had, in 1514, proposed another solution for the longitude problem.[6] He suggested measuring the distance between the Moon and a fixed star. This method, later known as lunar distance, described and illustrated by Petrus Apianus (1495-1552) as early as 1524, had to wait until the mid-eighteenth century before sufficiently accurate instruments and tables were available for its practical use.[7]

1. *Gemma Phrysius de Principiis Astronomiae & Cosmographiae, Deque usu Globi ab eodem editi* (Louvain and Antwerp, 1530). The passage concerning finding longitude at sea with a clock is in part 2, *cap.* 18 ('De novo modo inveniendi longitudinē'), fols. $D2^v$-$D3^r$. (See Appendix C.)

2. For the life and work of John Harrison, see papers in this volume by William Andrewes, Martin Burgess, Andrew King, and Anthony Randall.

3. Derek Howse, *Greenwich time and the discovery of the longitude* (Oxford: Oxford University Press, 1980), pp. 10-12.

4. *Ibid.*, pp. 51-2.

5. For details of this method, see Albert Van Helden's paper in this volume.

6. Johann Werner, *In hoc opere haec cōtinentur Noua translatio primi libri geographiae Cl'. Ptolomaei:...* (Nuremberg, 1514). The passage concerning Werner's description of the lunar-distance method of finding longitude is in *cap.* 4, note 8, sig. $dv^v$. (See Appendix C.)

7. For details of this method, see Derek Howse's paper in this volume.

8. For details of Huygens's work, see John Leopold's paper in this volume.

**Figure 1.** ***Allegory on the Glory of Navigation.*** **Anonymous, Dutch School, *ca.* 1680. Ceiling painting, oil on canvas. 250 x 380 cm. (98.4 x 149.6 in.). Courtesy of the Rijksmuseum-Stichting, Amsterdam. Inv. no. A 2716.**

Many of the solutions for finding longitude concerned the variation of the compass needle, the angle between the direction of true and magnetic north. This angle varies annually with the change in location of the magnetic pole. Although there is indeed a relation between variation and longitude, it was not directly available for practical use, since it requires a vast amount of data on the degree of variation in combination with the position and knowledge of the annual change. Two sixteenth-century Dutch scholars, Petrus Plancius (1552-1622) and Simon Stevin (1548-1620), developed methods to find longitude from variation, but due to the lack of data, their methods were unsuccessful.

During the seventeenth century, timekeepers were developed and tested at sea. A well-known attempt at a 'sea clock', mounted in gimbals and with a pendulum, was made by the Dutchman Christiaan Huygens (1629-1695).[8] However, at sea, a pendulum is deranged by the movement of the ship, and its period of oscillation is influenced by gravity, which changes with latitude due to the fact that the Earth has not a perfect spherical shape.

An important stimulant to solving the problem of finding longitude at sea was the founding of Greenwich Observatory in 1675 by Charles II. It was initiated by the Royal Society and inspired by the Paris Observatory, founded eight years previously by the French king. John Flamsteed (1646-1719) was appointed Astronomer Royal with the instruction to '[rectify] the tables of the motions of the heavens, and the places of the fixed stars, so as to find the so-much-desired longitude of places for perfecting the art of navigation'.[9] In 1698, the astronomer Edmond Halley (*ca.*1656-1742), later to become the second Astronomer Royal, was given the command of the *Paramore* for the specific purpose of collecting data on the variation of the compass needle. This resulted in the production of a number of charts with isogonals, lines indicating the places on the Earth's surface with the same magnetic variation. The first chart, published in 1701, was of the Atlantic Ocean. A chart of the English Channel and a world chart followed a year later with additional observations drawn from journals and logs of navigators other than Halley. The world chart was also published in France and in the Netherlands. The chart of the English Channel was published in Amsterdam as early as 1705, with a printed explanation in Dutch and French. Around 1740, the Dutch East India Company adopted the Halley charts for its ships, specifically for estimating longitude from variation.[10]

A new quadrant, later known as an octant, was presented to the Royal Society in 1731 by the Englishman John Hadley (1682-1744). Hadley's quadrant was an angle-measuring instrument for observing the altitude of the Sun or a star above the horizon at sea. Soon it was also employed for finding longitude by lunar distances.[11] This method became available for practical use as the necessary tables were calculated and published. They were compiled by the German astronomer Tobias Mayer (1723-1762) and published in Göttingen in 1753. His contribution to the solving of the longitude problem was recognised by the British Board of Longitude, which in 1765 awarded £3,000 to his widow. Lunar tables appeared in the first edition of *The Nautical Almanac*, published in 1766 for the years 1767-9. One of the first seamen to use this almanac for lunar distances was James Cook (1728-1779), who in 1769 established the longitude of several places in New Zealand.

Because the angle between the Moon and a fixed star often exceeds 90°, the range of the octant soon proved too limited to take lunar distances. By 1757, the British naval officer John Campbell (1720-1790) commissioned the London instrument maker John Bird (1709-1776) to construct an octant with an arc extended to 120°, thus creating the sextant. Initially, like octants, Bird made his sextant frames of wood, but later used brass, which proved to be far more rigid, increasing the reliability of the instrument. After Bird, other instrument makers produced high-quality sextants: in England, these included Jesse Ramsden (1735-1800), George Adams (*ca.* 1704-1773), Benjamin Martin (*ca.* 1704-1782), and Edward Troughton (1753-1836); in France, Étienne Lenoir (1744-1827); and in the Netherlands, the Amsterdam firm of Van Keulen.

These instrument makers were using the circular dividing engine invented about 1770 by Jesse Ramsden. Although Bird's dividing work had been of a high quality, Ramsden's invention made far more accurate and compact instruments possible. The reflecting circle, or Borda's circle, came into use when the range of the sextant and even the quintant[12] proved too limited. This instrument was developed in 1770 by the French naval officer Jean-Charles de Borda (1733-1799), based on an earlier design by Tobias Mayer. This full-circle instrument measured angles up to 180°. Troughton in England and Lenoir in France manufactured excellent reflecting circles

9. Howse, *Greenwich time* (*op. cit.*, note 3), p. 28.

10. W. F. J. Mörzer Bruyns, 'Navigation on Dutch East India Company Ships Around the 1740s', *The Mariner's Mirror*, vol. 78, no. 2 (1992), pp. 143-54, and C. A. Davids, 'Finding longitude at sea by magnetic declination on Dutch East-Indiamen, 1596-1795', *The American Neptune,* vol. 50 (1990), pp. 281-90.

11. As an experiment by Capt. John Campbell of HMS *Chatham* in 1747. See Charles H. Cotter, *A history of the navigator's sextant* (Glasgow: Brown, Son & Ferguson, 1983), pp. 137-8.

12. An extended sextant with which angles up to 150° could be measured.

that, by the end of the century, were available for the lunar-distance method of finding longitude.

The method of finding longitude of Gemma Frisius, using a clock, became practical when better timekeepers were developed. Encouraged by the Act of Parliament of 1714 offering a reward of £20,000 for an invention that would solve the problem of finding longitude at sea, scholars, scientists, and instrument makers worked on inventions for the rest of the century. The English carpenter John Harrison qualified for this prize with his fourth marine timekeeper, completed in 1759. By order of the Board of Longitude, an exact copy of this timekeeper was completed in 1770 by the London watchmaker Larcum Kendall (1719-1790). After a trial at the Royal Observatory in Greenwich, Kendall's copy was, with great success, tried out at sea by James Cook during his second voyage (1772-5). About this time, the French horologists Pierre Le Roy (1717-1785) and Ferdinand Berthoud (1727-1807) were developing marine timekeepers in France.[13]

By the end of the eighteenth century the marine chronometer was being manufactured commercially, largely due to the work of the English chronometer makers John Arnold (1736-1799) and Thomas Earnshaw (1749-1829).[14] Their compact instruments were used on both English and Dutch ships to establish the longitude of geographically important places. By 1815, the introduction of chronometers at sea was well under way. Those who could not afford the relatively expensive chronometer could still find their longitude by the lunar-distance method. Next to finding longitude by chronometer, this method, at least in the British and Dutch royal navies, remained in use throughout the nineteenth century to establish the error and rate of chronometers.[15]

13. For details of Le Roy and Berthoud, see Catherine Cardinal's paper in this volume.

14. For details of Arnold and Earnshaw, see Jonathan Betts's paper in this volume.

15. Howse, *Greenwich time* (*op. cit.*, note 3), p. 79, and C.A. Davids, 'De zeevaartkunde en enkele maatschappelijke veranderingen in Nederland tussen 1850 en 1914', *Mededelingen van de Nederlandse Vereniging voor Zeegeschiedenis*, no. 40/41 (1980), pp. 51-83 [55]. See also Derek Howse's paper in this volume.

APPENDIX 1

# FURTHER READING

Davids, C. A. 1990. 'Finding longitude at sea by magnetic declination on Dutch East-Indiamen, 1596-1795'. *The American Neptune,* vol. 50, pp. 281-90.

Howse, Derek. 1980. *Greenwich time and the discovery of the longitude.* Oxford: Oxford University Press.

Mörzer Bruyns, W. F. J. 1992. 'Navigation on Dutch East India Company Ships Around the 1740s'. *The Mariner's Mirror,* vol. 78, no. 2, pp. 143-54.

# Cartography

## *Norman J. W. Thrower*

**Norman Thrower is Professor of Geography and was Director of the (Columbus) Quincentenary Programs at the University of California, Los Angeles (UCLA) from 1989 to 1993. He was appointed Clark Library Professor at UCLA in 1975 and served from 1981 to 1987 as the Director of the Clark Library. He is the recipient of many honors, including the Distinguished Mentor Award of the U.S. National Council for Geographical Education (1991) and the Cross 1st Class of the Orden del Mérito Civil by H. M. King Juan Carlos of Spain (1993). He was awarded a Guggenheim Fellowship (1963) and has served on the Educational Advisory Board of the Guggenheim Foundation since 1978. He was appointed to the Sir Francis Drake Commission in 1973 and served as its President from 1975 to 1981.**

**Thrower was born in England and emigrated to the United States in 1947. After receiving his B.A. Honors from the University of Virginia and his M.A. and Ph.D. degrees from the University of Wisconsin, he joined the faculty of UCLA in 1957. His professional interests include cadastral surveys, cultural cartography, navigation, and piracy. He has published extensively on geographical discoveries and cartography. His major works include *Original Survey and Land Subdivision* (1966), *Maps and Man* (1972), *The Three Voyages of Edmond Halley in the* Paramore, *1698-1701* (1981), *Sir Francis Drake and the Famous Voyage, 1577-1580* (1984), *Standing on the Shoulders of Giants* (1990), *A Buccaneer's Atlas* (ed. with Derek Howse, 1992), and *Maps and Civilization* (1996).**

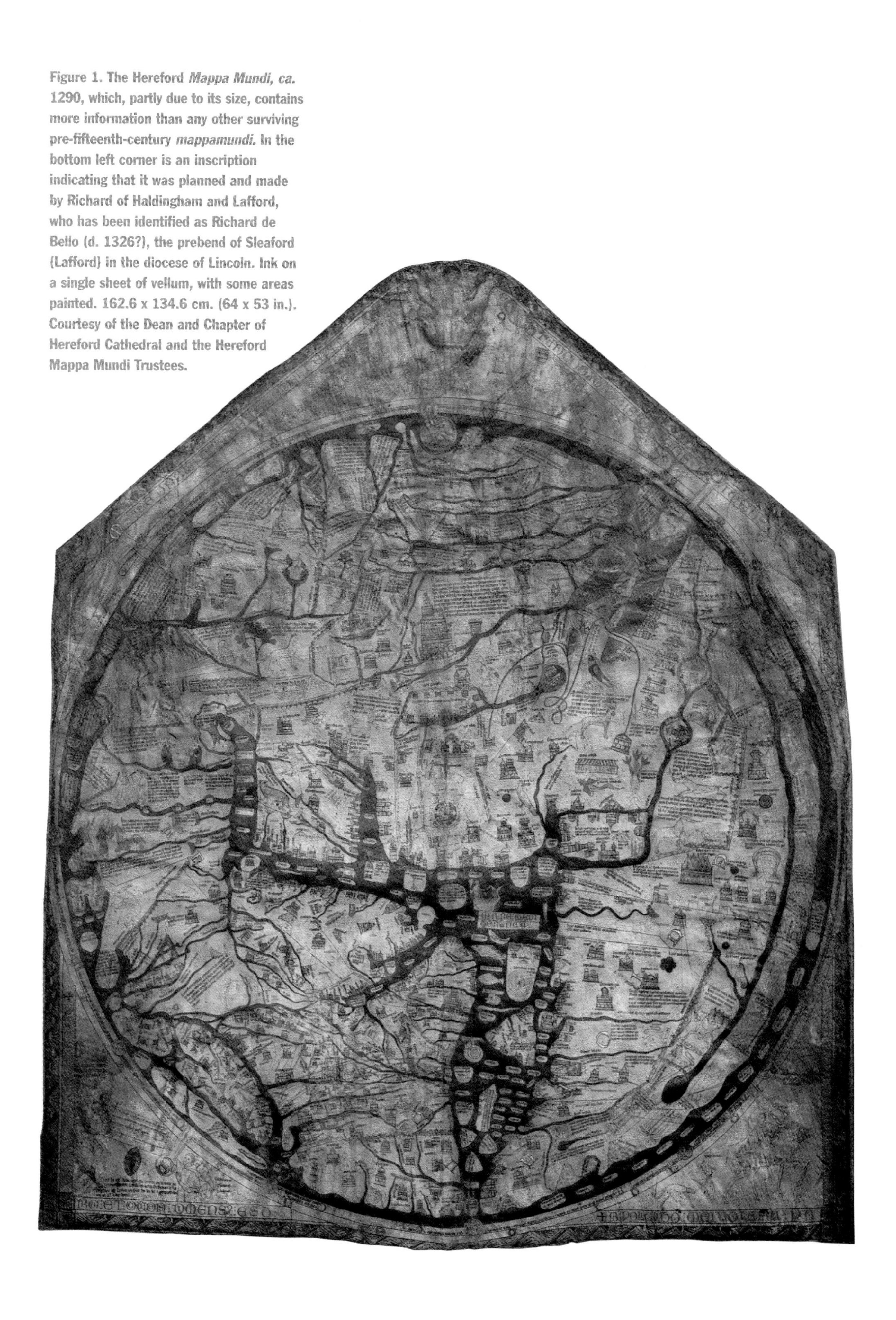

Figure 1. The Hereford *Mappa Mundi, ca.* 1290, which, partly due to its size, contains more information than any other surviving pre-fifteenth-century *mappamundi.* In the bottom left corner is an inscription indicating that it was planned and made by Richard of Haldingham and Lafford, who has been identified as Richard de Bello (d. 1326?), the prebend of Sleaford (Lafford) in the diocese of Lincoln. Ink on a single sheet of vellum, with some areas painted. 162.6 x 134.6 cm. (64 x 53 in.). Courtesy of the Dean and Chapter of Hereford Cathedral and the Hereford Mappa Mundi Trustees.

# Longitude in the Context of Cartography

*Norman J. W. Thrower*

There are two related ways in which longitude is crucially important in cartography: for the plotting of geographical locations in a global context and for the construction of map projections. Of course, it is possible to have projectionless maps, and, indeed, many examples of these exist.

The European maps of the Middle Ages, such as the Hereford world map, *ca.* 1290 (Figure 1), have no projection, although attempts have been made, recently, to fit projections to them after the fact.[1] Likewise the contemporaneous portolan or haven-finding charts of southern Europe (Figure 2) have no formal projection, although some scholars have seen them as related to an equirectangular (plane chart) base, or even as proto-Mercator projections.[2] At the same time that the Europeans were making such maps and charts, the Chinese developed rectangular grid systems, although these were not specifically related to latitude and longitude. The maps of preliter-

1. Waldo R. Tobler, "Medieval Distortions: The Projections of Ancient Maps," *Annals of the Association of American Geographers,* vol. 56 (1966), pp. 351-60.

2. See J. B. Harley and David Woodward, *The History of Cartography,* 2 vols. (Chicago: University of Chicago Press, 1987 and 1992), vol. 1, pp. 385-6, for a summary of opinions on this, with references.

**Figure 2. *Carte Pisane, ca.* 1275, the oldest extant example of a portolan chart. Although supposedly discovered in Pisa during the nineteenth century, it is generally thought to be of Genoese origin. Ink on vellum. 105 x 50 cm. (41.3 x 19.7 in.). Courtesy of the Bibliothèque Nationale, Paris. Ref. no. Rés. Ge. B 1118.**

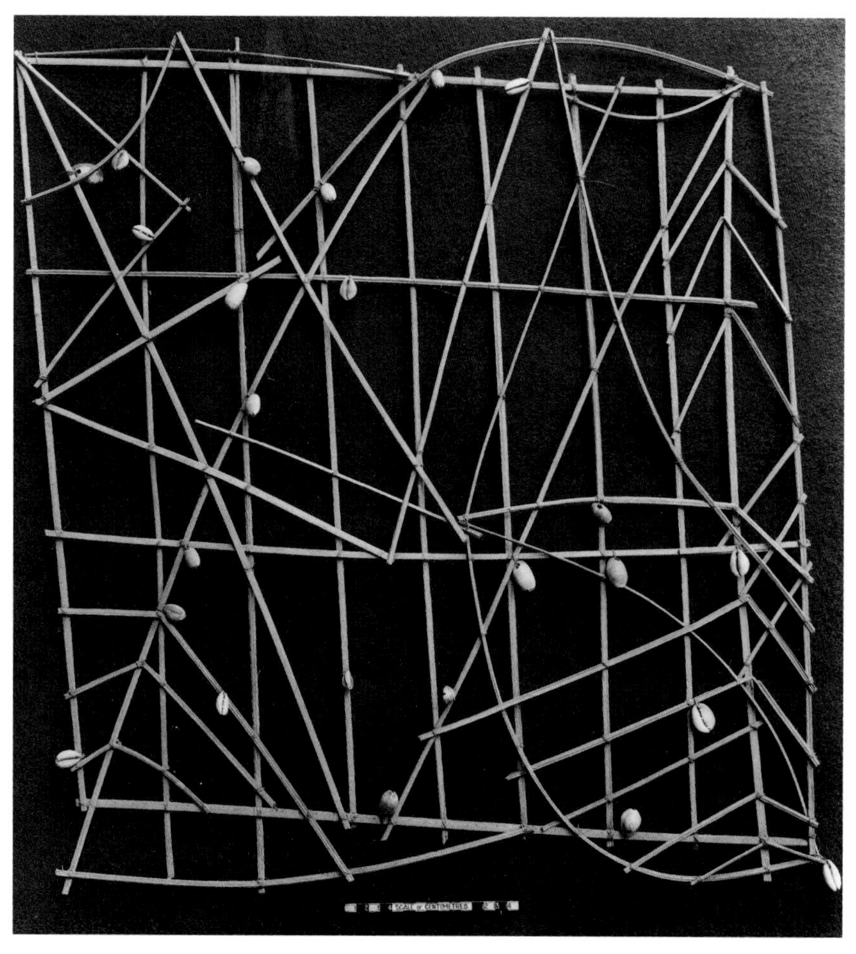

**Figure 3.** ***Rebbelib (rebbelith)*** **or sea chart used in the Marshall Islands in the Micronesian archipelago. The Micronesian navigators "feel" the wave patterns shown on such charts, breaking around low islands they cannot see. The center ribs of palm, bound together with lengths of palm fiber, indicate the altered direction taken by ocean swells when deflected by islands, which are represented by cowrie shells. Dimensions: approx. 67.5 x 71 cm. (26.6 x 28 in.). Courtesy of the Science Museum/Science and Society Picture Library, London. Inv. no. 1927-8.**

ate peoples also have no formal projection, though some, such as the Marshall Island stick charts, were useful for open-ocean navigation to the Micronesians (Figure 3). Even many modern maps are without projections, and, indeed, for very small areas it is unnecessary to relate the map data to global latitude and longitude; a simple Cartesian square grid system will suffice.[3] Similarly some real estate maps, fictional maps, TV maps, mental maps, and children's maps lack projections or have artificial reference grids. In fact, we can generalize and say that the smaller the area being mapped, the less important projection becomes, and, conversely, the larger the area, especially the whole globe, the more important is a systematic graticule[4] related to the Earth itself (*e.g.,* latitude and longitude).

As we know, longitude is much more difficult to determine than latitude, which has natural points of origin—the poles, 90° equidistant north and south from the equator, which is 0° latitude (Figure 4). Longitude has no such fixed line, so over the centuries and in different countries many prime meridians (0° longitude) have been used on maps. In order to have a spherical reference system, abscissae (*e.g.,* meridians or longitude) as well as ordinates (*e.g.,* parallels or latitude) are needed. For example, Eratosthenes

3. Cartesian coordinates (named after René Descartes, 1596-1650) are a system for locating points on a plane surface by their distance from two intersecting right-angled lines; a network of uniformly spaced vertical and horizontal lines locates points on a map, chart, etc.

4. A graticule is a network of parallels (lines of latitude) and meridians (lines of longitude) on a map, chart, or globe.

5. Ptolemy's work included *Syntaxis mathematica* (known in the Middle Ages as the *Almagest*) and the later *Geographike Hyphegesis* (generally known as *Geography* or *Geographia*). See O. A. W. Dilke, *Greek and Roman Maps* (Ithaca, New York: Cornell University Press, 1985) for further information on this subject, especially chap. 5, pp. 72-86.

6. Dilke (*ibid.*), especially chap. 2, pp. 21-38, and particularly note 42 with references.

(*ca.* 276?-*ca.* 195? B.C.), who measured the circumference of the Earth with great accuracy about 200 B.C., used a central, perpendicular meridian through Rhodes for his now-lost map of the *oikoumene,* the area of the Earth, encompassing 138° of longitude, that was believed to be inhabited, and the *diaphragma,* a line running east to west through Gibraltar, Athens, and Rhodes to the extreme north of the Indian subcontinent, parallel to the equator, for latitude. Hipparchus (*ca.* 180-125 B.C.) is credited with first devising a regular global system of latitude and longitude about 150 B.C. in which parallels and meridians are each divided into 360 degrees, each degree is divided into 60 minutes, and each minute is divided into 60 seconds (the sexagesimal system, a legacy of Babylonia). Hipparchus is also credited with inventing the orthographic and stereographic projections, which, however, were used by him and others for astronomical purposes and only much later for geographical maps and charts.

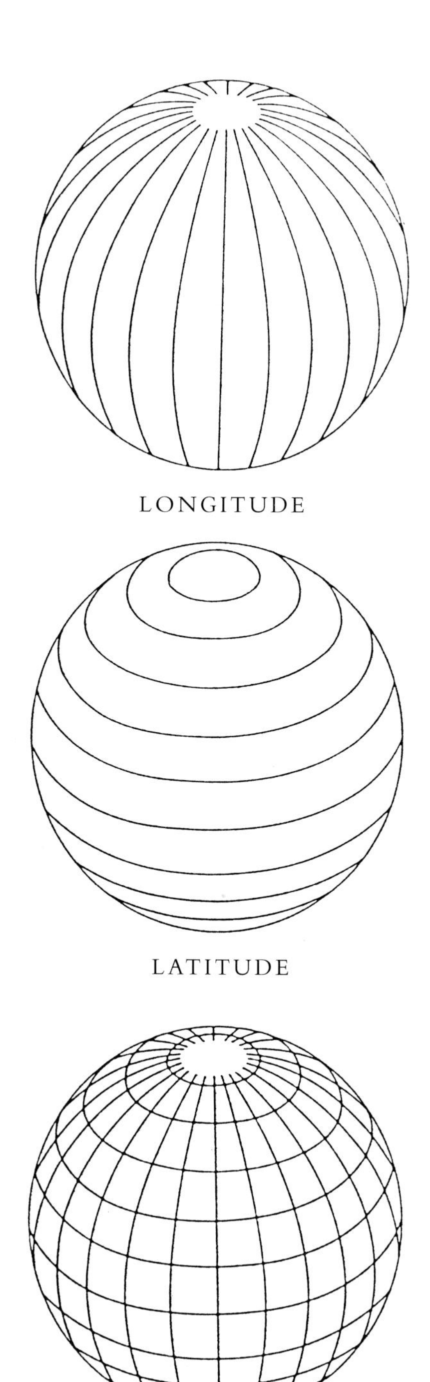


**Figure 4. The coordinate system of the globe. Drawing courtesy of the author.**

Another Greek, Claudius Ptolemy, or Claudius Ptolemaeus (fl. A.D. 150), devised two projections for the Earth, one pseudoconical and the other *chlamys*, or cloak-shaped (Appendix B, Figure 7). Following others, he used *klimata* (the length of the longest day of the year at a particular place) as well as degrees for latitude. For instance, the latitude of Rhodes could be expressed either as 36° N. or as 14.5 hours, the duration of the day at the summer solstice. For longitude, he at first proposed the meridian of Alexandria, numbering longitude east 90° and west 90° from that point, showing only half of the globe. This was adequate for mapping the inhabited world of the later Greeks and Romans. In later work, however, the Canary (Fortunate) Islands were used as the prime meridian, with longitude measured 180° east of this line, and it was this system that was adopted when Ptolemy's work was "rediscovered" after over a millennium, during the European Renaissance. Thus many maps of this period, reconstructed from his instructions, used the island of Ferro (Hierro) in the Canaries as the prime meridian, and longitude was numbered 180° eastward from a line passing through this point.

Following the translation of Ptolemy's *Geography* in the Latin West in the fifteenth century,[5] which coincided with the beginning of the great European overseas discoveries, it became the business of cartographers to attempt to map places, including those newly found, with increasing accuracy. Ptolemy's *Geography* listed some 8,000 places whose longitude had been determined mostly by dead reckoning (distance as a function of time traveled), but, after the rounding of the southern cape of Africa and the discovery of America by Europeans, this number increased greatly. Well into the sixteenth century, these determinations were still made by dead reckoning, although it had been known from antiquity that observations of simultaneous eclipses of the Sun and the Moon could be used to determine longitude; this method was at first rarely used.[6]

During the Renaissance there was also a flurry of interest in developing new projections of the expanding world on which map data could be plotted, and after the invention of printing in Europe in the fifteenth century, maps were more widely distributed. Figure 5 shows a few of the ingenious ways of representing the curving figure of the globe with lines of latitude and longitude devised during this period. Gerardus Mercator (1512–1594), in the famous projection of 1569 that bears his name, used a longitude numbering system with 0° in the Canary Islands, running eastward 360° around the world.

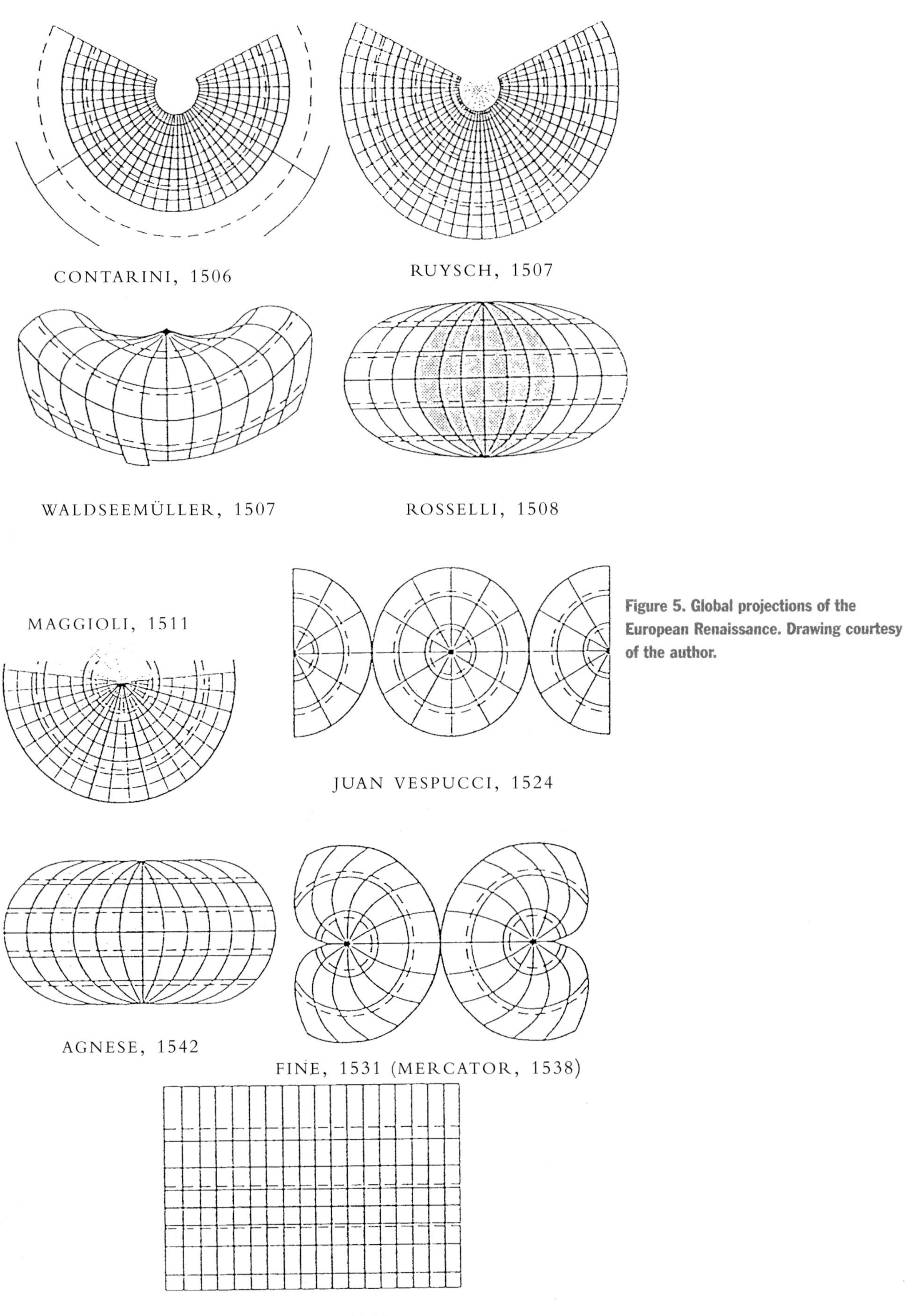


**Figure 5. Global projections of the European Renaissance. Drawing courtesy of the author.**

**Figure 6. World map, dated 1630, by Philipp Eckebrecht using difference in time to establish longitude, based on the 1627 Rudolphine Tables of Johannes Kepler. Dimensions of actual image: 38.6 x 68.6 cm. (15.2 x 27 in.). By permission of the Houghton Library, Harvard University. Ref. no. *fGC6.K4436.627tad.**

A world map that established longitude by difference in time through observation of celestial phenomena, drawn by Philipp Eckebrecht (1594-1667) of Nuremberg and dated 1630, was added to certain copies of the Rudolphine Tables of Johannes Kepler (1571-1630) (Figure 6).[7] Eckebrecht's map was based on lunar eclipses using star positions from these tables and employed a prime meridian running through the observatory of Tycho Brahe (1546-1601) at Uraniborg on the island of Ven (formerly known as Hven or Hveen), between Denmark and Sweden.

In spite of attempts on the part of cartographers to improve the delineation of the world's coastlines, many inaccuracies remained, owing especially to the lack of knowledge of longitude. Sometimes there was retrogression, as in the case of California appearing as an island on maps of the seventeenth century, whereas in the previous century it had been represented as part of the mainland. After the establishment of the Paris Observatory and the appointment of Giovanni Domenico (later known as Jean-Dominique) Cassini (1625-1712) as director in 1669, a planisphere (world map) was laid out on the floor of the observatory.[8] Only places the latitude and longitude of which had been determined astronomically were plotted, although there were errors in the delineation of coastlines on copies of this map (Figure 7). By this time, Cassini's tables for telescopic observations of the satellites of Jupiter had been used in determining longitude (following the method originated by Galileo) in various places in Europe, Africa, French Guiana, and southern and eastern Asia.[9] Such observations led to the immediate improvement of the map of France, as shown by an illustration published in 1693, of the coastline as determined by surveys by the scientists

7. Johannes Kepler, *Tabulae Rudolphinae, Quibus Astronomicae Scientiae,...* (Ulm, 1627). The Houghton Library at Harvard University has several copies of this work, but only one contains Eckebrecht's map, which is mounted on a stub bound in the book between the index (contents) and p. 1. Although the map is dated 1630, its dedication to Leopold I (1640-1705) would suggest that it was not printed before 1658, the date that Leopold was elected Holy Roman Emperor.

8. This planisphere, which was drawn in ink on the floor of the observatory, had completely disappeared by around 1700, although it had been retraced in 1690. An engraving of this map, entitled *Planisphere Terrestre,* was, however, published by Jean Baptiste Nolin in 1696. Very few copies are known to have survived. The history of the planisphere is described in Lloyd A. Brown's book, *Jean Domenique Cassini and his World Map* (Ann Arbor: University of Michigan Press, 1941).

9. See Albert Van Helden's paper in this volume; also Jean-Dominique Cassini, *Éclipses du premier satellite de Jupiter pendant l'année* (Paris, 1692).

of the Académie Royale des Sciences superimposed on the best previous delineation, the map by Sanson that was presented to the Dauphin in 1679.[10]

The English astronomer Edmond Halley (*ca.* 1656-1742), who visited Cassini at the Paris Observatory in 1681 and made some observations there, drew a map of the trade winds in 1686, shown in Figure 8, with 10° lines of latitude and 15° lines of longitude on a Mercator projection or, as Halley called it, the Nautical projection,[11] perhaps recalling the role of his fellow countryman Edward Wright (1561-1615) in its development and adoption. Below this map are modern delineations of the coastlines on the same projection. Later, when Halley was appointed captain of the *Paramore,* his instructions (which he wrote himself), dated October 15, 1698, called for him "to improve the knowlege of the Longitude and variations of the Compasse."[12] While in the New World Halley thus made telescopic observations and used Cassini's tables to correct his dead-reckoning estimates of longitude, which had been off by more than 400 miles during his three-month crossing of the Atlantic. He also took readings of the "variations of the Compasse" and in 1701 published a map of the Atlantic Ocean (Figure 9) and the following year a world map on which were plotted isogonic lines, which indicate places with the same amount of magnetic variation.

Magnetic variation, also known as magnetic declination,[13] is the difference between true north and magnetic north as measured by the angle that a compass needle diverges from the astronomical north-south line, or, to use Halley's words, "the deflection of the Magnetic Needle from the true Meridian."[14] This phenomenon was first proposed as a possible method of determining longitude at sea soon after Columbus observed the effect dur-

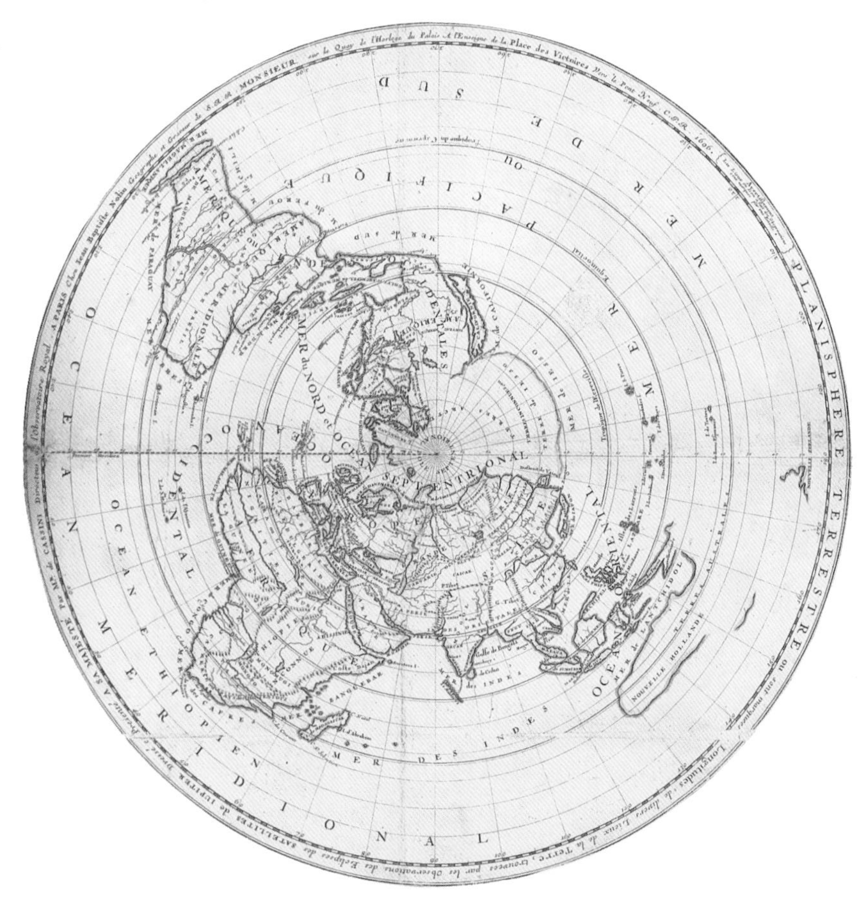


Figure 7. Cassini's *Planisphere Terrestre* (world map) published by Jean Baptiste Nolin in 1696. The original planisphere, based on longitude observations made by Jean-Dominique Cassini and others of the Académie Royale des Sciences, was drawn on the floor of the Paris Observatory. Copper plate engraving with color-wash outline. Diameter: 55.2 cm. (21.75 in.). Courtesy of the Clements Library, University of Michigan, Ann Arbor. Ref. no. ATLAS E10B.

10. This map is shown in Figure 9 of Albert Van Helden's paper in this volume. It was first published in *Recueil d'Observations Faites en Plusieurs Voyages par Ordre de Sa Majesté, pour Perfectionner l'Astronomie et la Geographie...* (Paris, 1693), on pp. 91-2 of the third section concerning observations made by Picard and de La Hire under the passage entitled "Pour la Carte de France corrigée sur les Observations de MM. Picard & de la Hire." The map is mounted on a stub bound into the volume following this passage. It was later published in *Memoires de l'Académie Royale des Sciences. Depuis 1666. jusqu'à 1699* (Paris, 1729), vol. 7, part 1, pp. 429-30, with the map bound in as a fold-out plate opposite p. 1.

11. "The Projection of this Chart is what is call'd Mercator's but from its particular Use in Navigation, ought rather to be named the Nautical; as being the only True and Sufficient Chart for the Sea." Edmond Halley, "The Description and Uses of a New and Correct Sea Chart of the whole World, shewing the Variations of the Compass." This explication appears beneath Halley's world chart of magnetic variation, which was published in 1702. A copy of this map may be found in the Royal Geographical Society, London.

12. Norman J. W. Thrower, ed., *The Three Voyages of Edmond Halley in the* Paramore, *1698-1701,* Hakluyt Society Publications, 2nd series, no. 156-7 (London, 1981), vol. 1, pp. 268-9. This is the substance of other orders and instructions concerning the first and second of Halley's voyages.

13. At sea, the term used for this phenomenon is magnetic variation, whereas on land, the term commonly employed by surveyors is declination.

14. Edmond Halley, "A Theory of the Variation of the Magnetical Compass," *Philosophical Transactions* of the Royal Society, London *(Phil. Trans.),* vol. 13, no. 148 (1683), p. 208.

15. Halley, "The Description and Uses" (*op. cit.,* note 11).

16. "There is yet a further difficultie, which is the change of the variation, one of the discoveries of this last Century; which shews, that it will require some Hundreds of years to establish a compleat doctrine of the Magnetical System." Halley, "A Theory of the Variation" (*op. cit.,* note 14), p. 220.

17. Thrower, *The Three Voyages* (*op. cit.,* note 12), p. 22. In a recent publication, Josef Konvitz (*Cartography in France, 1660-1846: Science, Engineering, and Statecraft* [Chicago: University of Chicago Press, 1987], p. 70), cites a letter written by Philippe Buache (1700-1773); paraphrased in part, this letter described "Halley's theory of

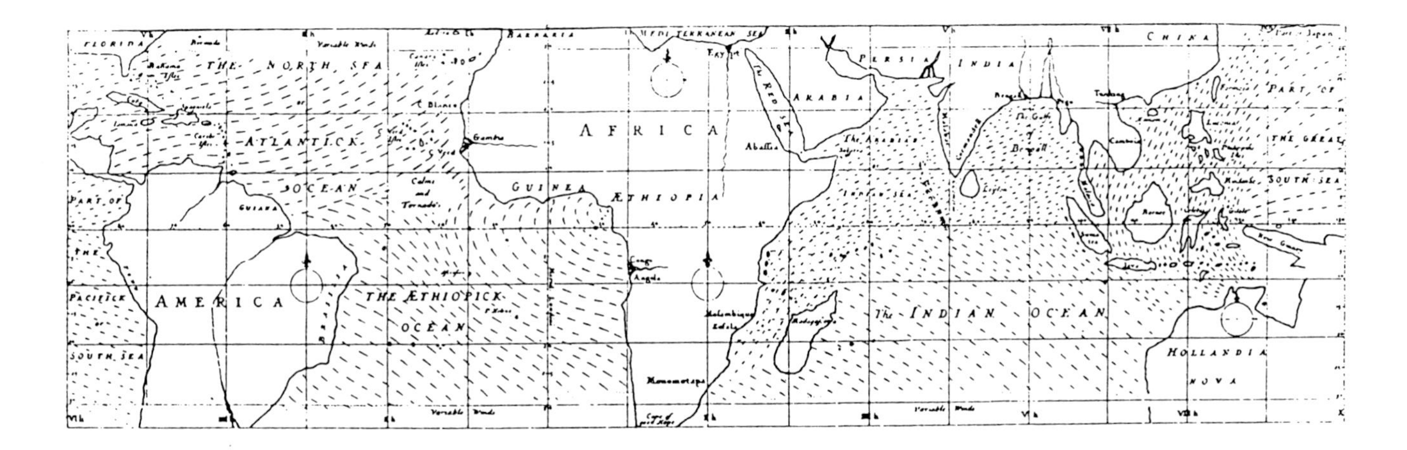


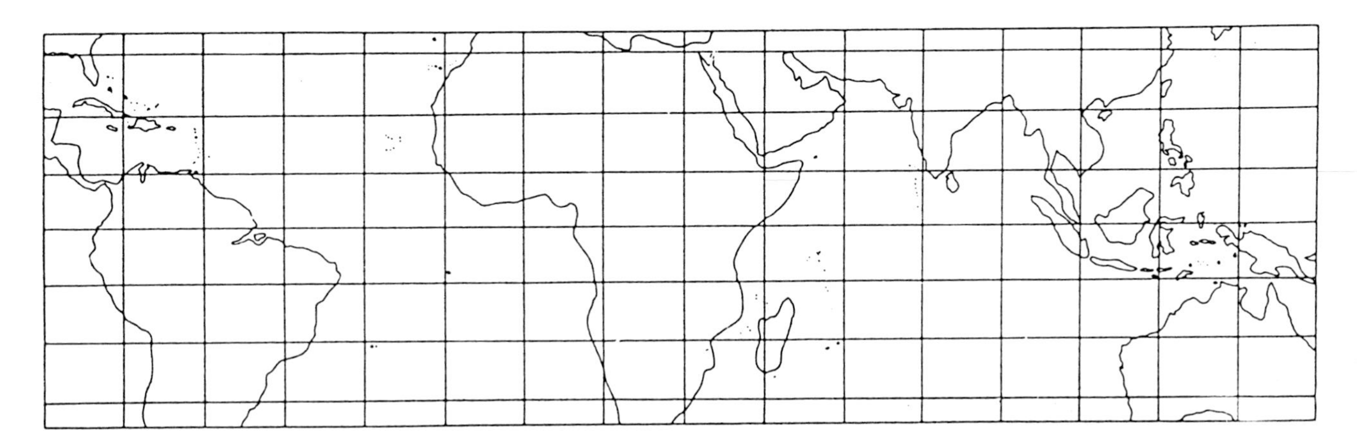

**Figure 8. Above: Chart of the trade winds by Edmond Halley. *Philosophical Transactions* of the Royal Society, London, vol. 16, no. 183 (1686), opposite p. 151. Below: A map of the world's coastlines from modern sources on the same projection (10° of latitude, 15° of longitude) as above. Drawing courtesy of the author.**

ing his famous voyage across the Atlantic in 1492–3. Halley's voyage and the subsequent publication of the Atlantic map thus aroused considerable interest, but, as Halley himself correctly pointed out in his explication of the world map of 1702, this method did not prove to be useful except where the isogonic lines are parallel to a coast, as in southern Africa at the time. In this explication he also stated that "there is a perpetual, tho' slow Change in the Variation almost every where, which will make it necessary in time to alter the whole System."[15] Halley thus reaffirmed the concept of secular or temporal variation, which he had observed as a schoolboy in 1672 and had written about in the *Philosophical Transactions* of 1683.[16] His ideas on this subject had been based on the discovery, announced by Henry Gellibrand (1597-1636) in England (and, independently, by others on the Continent) in the mid-1630s,[17] that "the variation is accompanied with a variation."[18] Halley's isogonic charts, the first printed and published maps of any type to employ isolines to indicate a specific value of a particular measurement,[19] continued to be revised until the early nineteenth century.

Another less well-known method relating to the Earth's magnetism for finding longitude at sea involved measuring the inclination, or "dip," of the compass needle from the horizontal. An instrument for measuring this phenomenon was devised by the English instrument maker Robert Norman (fl. 1560-1596), who in 1581 published a description of the effect in *The newe Attractive*.[20] It was not long before magnetic inclination was proposed for determining latitude, a method endorsed by several well-known individuals, including William Gilbert (1540-1603), Thomas Blundeville (fl. 1560-1602), Edward Wright, and Henry Briggs (1561-1630). In 1676, in his book *The*

magnetic variation as a constant that did not vary from year to year in the same place"! Buache apparently wanted the French to receive major credit for development of the isoline concept, in which he had a great interest.

18. Henry Gellibrand, *A Discourse Mathematical On The Variation Of The Magneticall Needle. Together with Its admirable Diminution lately discovered* (London, 1635), p. 7.

19. Two earlier isoline maps in manuscript are known: one, drawn in 1584 by Peter Bruinss, is a simple chart with lines passing through points having equal depth of water (isobaths); the other is a map of the same type drawn in 1697 by Pierre Ancelin. See Thrower, *The Three Voyages* (*op. cit.*, note 12), p. 57, note 3. It appears that Halley did not know of these: in the explication of his world map of 1702 he stated, "What is here properly New, is the Curve Lines drawn over the several Seas, to show the degrees of the variation of the Magnetical Needle, or Sea Compass." Halley, "The Description and Uses" (*op. cit.*, note 11).

20. Robert Norman, *The newe Attractive, Containyng a short discourse of the Magnes or Lodestone,...* (London, 1521).

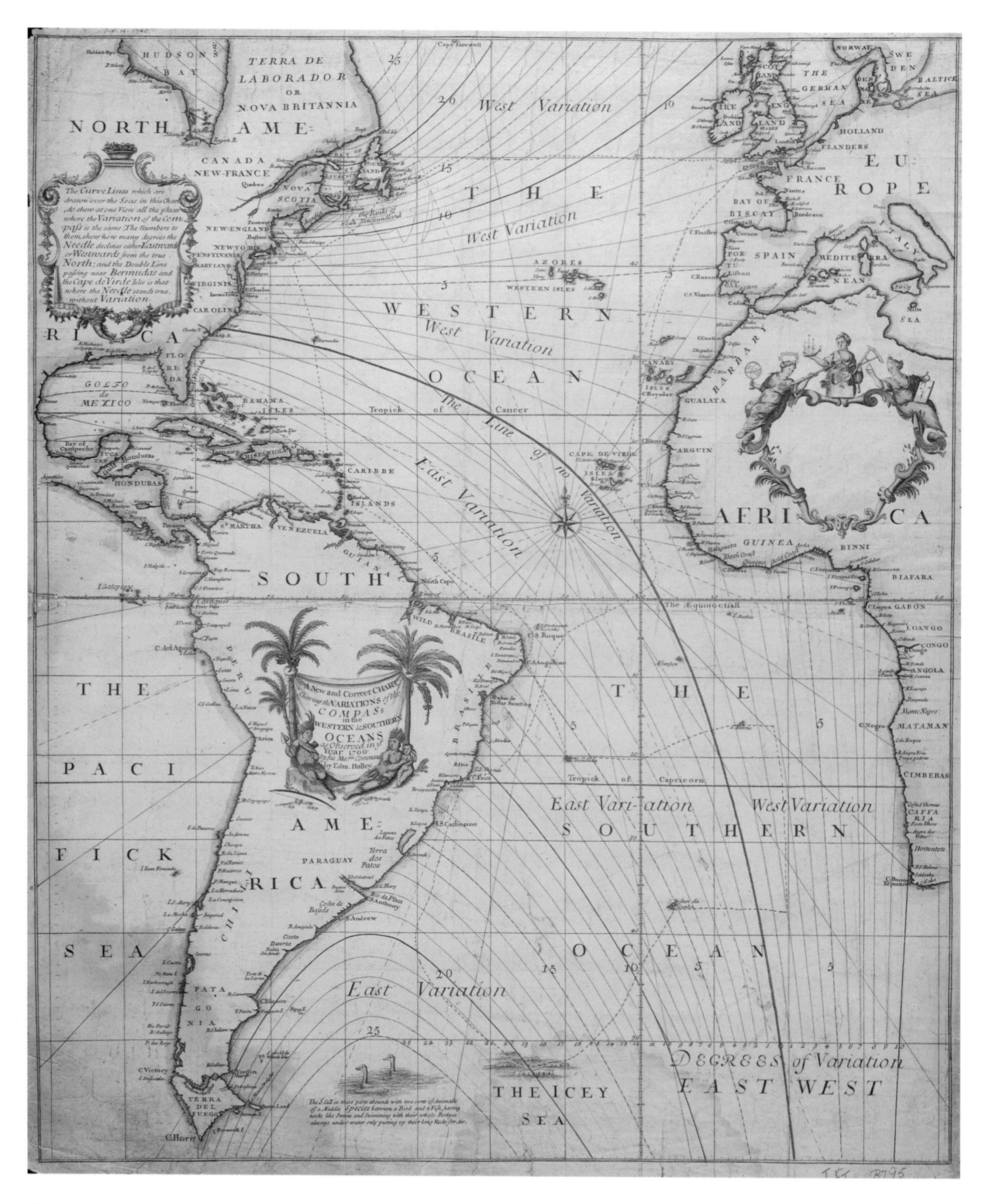


Figure 9. Isogonic chart of the Atlantic (1701) by Edmond Halley showing lines of magnetic variation. This is the "first state," or proof, of the earliest printed and published isoline map. Subsequent impressions of this map included a dedication to William III. 58 x 49 cm. (22.8 x 19.3 in.). By kind permission of the William Andrews Clark Memorial Library, University of California, Los Angeles. Ref. no. MAP G9101 C93 1700 H34. Photograph courtesy of Rand McNally.

*Longitude Found,*[21] the English teacher of mathematics and navigation Henry Bond (*ca.* 1600-1678) suggested its use for finding longitude. In theory, determining the latitude by observation and comparing that with the angle of dip known to coincide with that latitude could establish the position, and thereby the longitude, of the observer. A major problem with this method was accurate measurement of the dip. By 1768, however, Johan Carl Wilcke (1732-1796)[22] had acquired sufficient information to publish a world map of magnetic inclination (Figure 10).[23] This appears to be the first global chart of isoclines (lines passing through points having equal magnetic inclination, or dip), and shows these lines on a Mercator projection in relation to postulated north and south magnetic poles, whose locations were not known at the time. Nevertheless, as with magnetic variation, this method proved impractical for finding longitude.

Edmond Halley, in addition to his contributions to cartography, played an important role in the history of the longitude problem by the encouragement and support he offered to John Harrison (1693-1776) in the early part of the latter's career. About 1730, when Halley was the Astronomer Royal and thus also an *ex officio* member of the Board of Longitude, he was visited by the then-unknown John Harrison, who wished to obtain support for constructing his first marine timekeeper.[24] Despite the fact that Halley himself was committed to the lunar-distance solution to the problem,[25] he offered Harrison good advice, strongly recommending that he visit George Graham (*ca.* 1674-1751), the influential London clockmaker, before bringing his proposal to the Board. Graham, in turn, was to provide Harrison with the encouragement and money that enabled him to build his first marine timekeeper, H.1. When this timekeeper was completed in 1735, Halley was one of the members of the Royal Society who strongly recommended its

21. Henry Bond, *The Longitude Found: Or, A Treatise Shewing An Easie and Speedy way...to find the Longitude,...* (London, 1676).

22. Wilcke was appointed Thamian Lecturer in Experimental Physics at the Royal Academy of Sciences in Stockholm in 1759.

23. As he acknowledged in the legend of the map, Wilcke depended on observations of a number of eighteenth-century natural philosophers, including his fellow Swede, Capt. Carl Gustaf Ekeberg (1716-1784), who served with the Swedish East India Company and made observations of magnetic inclination during his voyages to China in 1766-7 and 1770-1. Ekeberg was provided with instruments for measuring the dip by the Royal Academy of Sciences. One of these dip needles is still in the collections of the Academy. The author and editor are grateful to Olov Amelin of the Royal Swedish Academy of Sciences for bringing this rare map to our attention and providing the photograph for this publication. For further information on the Academy and the instruments used, see Gunnar Pipping, *The Chamber of Physics: Instruments in the History of Sciences Collections of the Royal Swedish Academy of Sciences, Stockholm* (Stockholm: Almqvist & Wiksell, 1977).

24. For further details of Harrison's visit to London, see Andrew King's and William Andrewes's papers in this volume.

25. For example, see Edmond Halley, "A Proposal of a Method for finding the Longitude at Sea within a Degree, or twenty Leagues...," *Phil. Trans.*, vol. 37, no. 421 (1731), pp. 185-95.

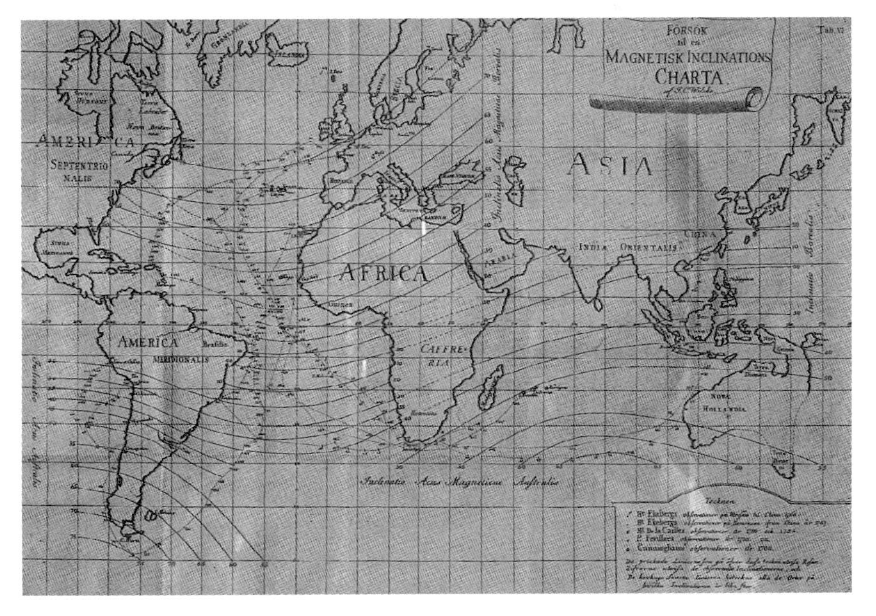


**Figure 10. World map, the first of its kind, showing isoclines, lines of magnetic inclination, superimposed on lines of latitude and longitude. This map was produced by the Swedish natural philosopher Johan Carl Wilcke in 1768. Magnetic inclination had been proposed as a method for finding longitude in the previous century, but the theory of using precise measurements of magnetic inclination and latitude to determine the position, and thus the longitude, of a ship did not prove successful in practice. 45 x 34 cm. (17.7 x 13.4 in.). Courtesy of the Royal Swedish Academy of Sciences, Stockholm.**

trial at sea. Following this trial, Halley, as a member of the Board of Longitude, assisted in obtaining financial support from the Board for construction of a second timekeeper, and in January 1741/2, shortly before he died, he recommended further financial assistance for John Harrison. In later years, Harrison sorely missed Halley's support, but, as we know, he eventually received the final part of the prize for the solution to the longitude problem in 1773, more than three decades after Halley's death in 1742.[26] Many of Harrison's problems in gaining the reward were due to the opposition of the Astronomer Royal Nevil Maskelyne (1732-1811) and others who favored the lunar-distance method for finding longitude, which became a viable solution to the longitude problem in the late 1750s. In contrast, Halley, who had also been committed to an astronomical solution of the problem, nevertheless had given encouragement and good advice to Harrison.

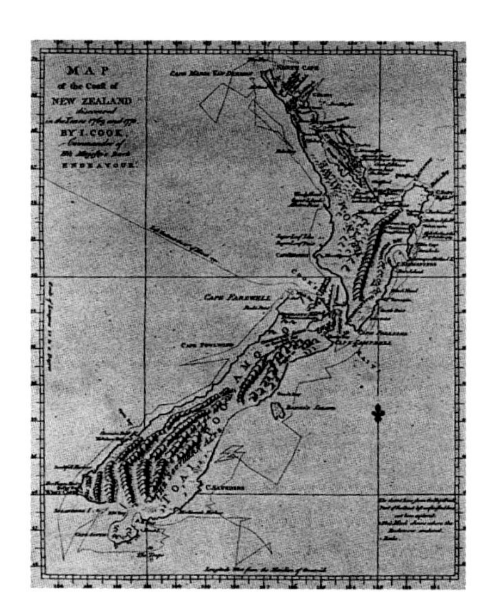


**Figure 11. Map of the coast of New Zealand, based on the surveys made in 1769 and 1770 by James Cook during his first Pacific voyage. This map, almost certainly drawn by Cook's assistant Isaac Smith (1752-1831), was engraved by B. Longmate and published in *A Journal of a Voyage to the South Seas, in his Majesty's Ship, The Endeavour. Faithfully transcribed from the Papers of the late Sydney Parkinson*, ed. Stanfield Parkinson (London, 1773), plate xxv, opposite p. 124.**

It was Maskelyne, nonetheless, who was responsible for the fact that James Cook (1728-1779) took several marine timekeepers with him on his second voyage to the Pacific between 1772 and 1775. On his first Pacific voyage, between 1768 and 1771, Cook had made a survey of New Zealand (Figure 11), based entirely on lunar-distance calculations derived from observations taken aboard ship at Mercury Bay (North Island). As expected, latitudes were excellent but longitudes were not generally accurate, with the result that New Zealand was placed too far to the east by an average of about 25' (about 20 miles at that latitude), with errors of up to 40' (32 miles).[27] At Maskelyne's suggestion, marine timekeepers (later known as chronometers) were put aboard the ships for Cook's second Pacific voyage: on the *Resolution,* the first copy (K.1) of Harrison's No. 4 by Larcum Kendall (1719-1790) and one made by John Arnold (1736-1799); on the *Adventure,* two other timekeepers by Arnold. The *Adventure* returned to England prematurely, but the *Resolution* completed a three-year voyage around the world, during which K.1 performed with remarkable reliability, recording the ship's longitude to a high degree of accuracy.[28]

Although Cook's celebrated voyages, along with those of George Vancouver (*ca.* 1757-1798), Louis-Antoine de Bougainville (1729-1811), and Jean-François de Galaup, Comte de La Pérouse (1741-*ca.* 1788), greatly contributed to the accuracy of the maps of the late eighteenth and early nineteenth centuries, there still remained places to be discovered and charted, as shown by an 1808 world map by the quasi-official cartographic firm of

26. For details of Harrison's prize-winning watch, see Anthony Randall's paper in this volume.

27. Derek Howse, *Nevil Maskelyne: The Seaman's Astronomer* (Cambridge: Cambridge University Press, 1989), pp. 122-3, and Derek Howse and Michael Sanderson, *The Sea Chart* (Newton Abbot, England: David and Charles, 1973), pp. 98-9.

28. This was Cook's greatest voyage, according to his biographer, John Beaglehole (1901-1971); see J. C. Beaglehole, *The Life of Captain James Cook R.N., F.R.S.,* Hakluyt Society Publications, extra series, vol. 37 (London, 1974). It was after this voyage that Cook was appointed to the rank of captain; he had been a lieutenant on the first and a commander on the second.

29. Although this map was reasonably up-to-date, some geographical discoveries that had been made by 1808 were not shown; *e.g.,* the entire coast of Australia was charted by Matthew Flinders (1774-1814) by 1803, but he was delayed in reporting his discoveries until 1810, being detained by the French on Mauritius.

30. Because the stereographic is "projected" from the back side of the globe, only one hemisphere can be shown at a time; by using two hemispheres, in this instance projected from the equator at exactly opposite points, the whole Earth is represented.

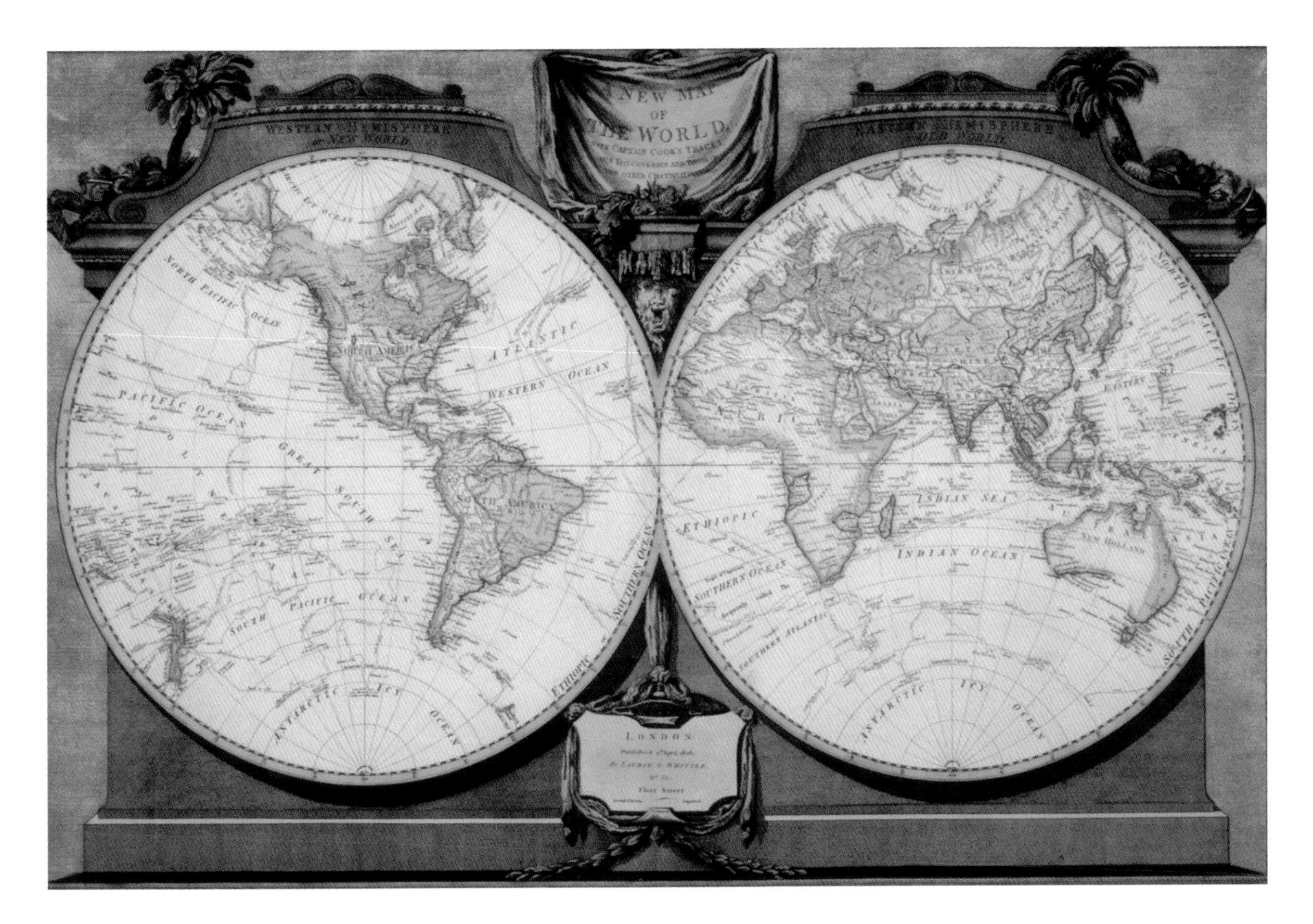


**Figure 12. "A New Map of the World" (1808) by Laurie and Whittle, London. Copper plate engraving with color wash. 32.4 x 48.3 cm. (12.75 x 19 in.). Courtesy of Historic Urban Plans, Ithaca, New York. Inv. no. EW HM 808.**

Laurie and Whittle of London (Figure 12).[29] Interestingly, this map is double-hemispheric, based on the stereographic projection invented by Hipparchus some 2,000 years earlier.[30] Like Mercator's 1569 projection, the stereographic projection is conformal (*i.e.,* shapes around a point are geometrically correct), and this one has a grid of 15° for longitude, which runs east and west of the prime meridian in the Canary Islands, and a grid of 10° for latitude.

Later, new projections were devised for the Earth grid, and at the International Meridian Conference held in Washington, D.C., in 1884, Greenwich was adopted as the universal prime meridian, with longitude counted in two directions east and west up to 180°. Meanwhile, use of the chronometer at sea, as well as telescopic observations of longitude ashore in different parts of the globe by workers from many nations, led to production of more accurate charts and maps. In this enterprise, the work of the hydrographic offices of the maritime powers was of the greatest importance during the nineteenth and twentieth centuries.

APPENDIX 1
SELECTED REFERENCES

Brown, Lloyd A. 1949. *The Story of Maps*. Boston: Little, Brown.

David, Andrew, ed. 1988 and 1992. *The Charts and Coastal Views of Captain Cook's Voyages*. 2 vols. Hakluyt Society Publications, extra series, no. 43. London.

Friis, Herman, ed. 1967. *The Pacific Basin: A History of its Geographical Exploration*. New York: American Geographical Society.

Howse, Derek, ed. 1990. *Background to Discovery: Pacific Exploration from Dampier to Cook*. Berkeley/Los Angeles: University of California Press.

Robinson, Arthur H., *et al.* 1978. *Elements of Cartography*. 4th ed. New York: John Wiley and Sons.

Snyder, John P. 1993. *Flattening the Earth: Two Thousand Years of Map Projections*. Chicago: University of Chicago Press.

Thrower, Norman J. W. 1972. *Maps and Man: An Examination of Cartography in Relation to Culture and Civilization*. Englewood Cliffs, New Jersey: Prentice-Hall.

———. 1996. *Maps and Civilization: Cartography in Culture and Society*. Chicago: University of Chicago Press.

# The History of Science

## *Michael S. Mahoney*

**Michael S. Mahoney is a professor of history of science and technology at Princeton University, where he earned his Ph.D. in 1967. He is the author of a variety of studies on mathematics from antiquity through the seventeenth century, including *The Mathematical Career of Pierre Fermat* (2nd ed., 1994) and "Christiaan Huygens: The measurement of time and longitude at sea" (1980). He has more recently turned to the history of computing, with articles on "The History of Computing in the History of Technology" (1988) and "The Roots of Software Engineering" (1990), and he is currently completing a book on the formation of theoretical computer science as a mathematical discipline.**

# Longitude in the Context of the History of Science

*Michael S. Mahoney*

Having determined that longitude is a matter of time, one can measure time in two ways: either by reading the celestial clock directly or by building a working model of it. Both approaches present practical difficulties. The dials of the celestial clock are all askew and it is often hard to read the hands, especially when the observer is moving about in several directions at the same time. The model—that is, a mechanical clock—has the advantage of being designed to be read easily and directly, once one figures out how to make it work as regularly and uninterruptedly as the original. As the speakers of the previous sessions have shown, practitioners over the period from the late sixteenth to the late eighteenth century ultimately found it preferable to build the model.

That is where the interest and the context of the history of science lie: in the model. Shaped in the seventeenth and eighteenth centuries by its application to the problem of longitude, the mechanical clock assumed its own metaphysical, epistemological, and social status. It came to embody, indeed to symbolize as ideal, Europeans' understanding of how the world works. As Johannes Kepler (1571-1630) put it in speaking of his new, physical astronomy in 1605:

> I mean to speak here of the celestial machine not on the model of a divine animal but on the model of a clock—if you think a clock to be animate, you attribute glory to the work of a craftsman. In [that machine] almost all the variety of motions [stems] from one most simple, physical magnetic force; just as in the clock all motions stem from a simple weight. And I mean to call this [form of] reasoning "physics [done] with numbers and geometry".[1]

The model would capture not only time, but the very forces determining the motions of which time is the measure.

Just how the celestial clock works became the driving question of astronomy in the seventeenth century, and the mechanical model became the driving force behind the answer. In the hands of Galileo (1564-1642), the pendulum became the tool for measuring and analyzing the fall of bodies

1. Kepler to Herwart von Hohenberg, February 10, 1605, in *Johannes Kepler Gesammelte Werke,* vol. 15, *Briefe 1604-1607*, ed. Max Caspar (Munich: C. H. Beck, 1951), p. 146.

into the laws of uniform and uniformly accelerated motion. His proposal to use a pendulum to regulate a mechanical clock remained just a sketch until translated into a first model in 1649 by his son, Vincenzio (d. 1649), and then into a working clock by Johann Philipp Treffler (1625-1697) in the late 1650s.[2] Following up on that proposal, initially with the aim of automating the counting of the swings of an astronomical pendulum, the young Christiaan Huygens (1629-1695) in 1656 devised a crutch that delicately balanced the pendulum's regulating period with the escapement's driving force to achieve a clock that told time well enough to make the variations of a simple pendulum significant. Huygens's analysis of the period of a pendulum and his subsequent discovery of the tautochronic properties of the cycloid led him in turn to the mechanical principle of simple harmonic oscillation underlying those properties and thus to the balance spring in 1675 and to the series of "perfect marine balances" on which he was still working at the time of his death in 1695. In the process, he created the mathematical theory of evolutes, thereby opening a path to the analysis of motion along a curve defined by a changing radius, and he determined the center of oscillation of a compound pendulum, thereby laying the foundations of the dynamics of rigid bodies.[3]

Inspired by Huygens's *Horologium Oscillatorium* of 1673, Isaac Newton (1642-1727) laid down in his *Principia* in 1687 the laws of motion that govern the celestial clock in particular and any system of mutually attracting or repelling bodies in general. The universal force of gravity, rather than the great magnetic force of the Sun, joins all the planets in a self-regulating mechanical system, in which both the planets and the hands of a clock move by the same rules. As Bruce Chandler emphasized in placing the problem of longitude in the astronomical and mathematical context, Newton's work did not end the story.[4] The mathematical complexity of a system involving more than two bodies eluded his understanding, as did the behavior of bodies moving through resisting media. At the turn of the eighteenth century, there was much more to understand about moons and pendulums before either the lunar method or the mechanical clock could meet the demands of the Longitude Act. Nonetheless, Newton and his predecessors had created a model of the world that set a new standard of scientific knowledge.

We are so imbued with both this model and the standard it set that we tend to overlook where it came from. As in the case of the steam engine in the nineteenth century, the physical device and the technical capacity to build it preceded and directed the science, and not the other way around. Ultimately, the science explained, or rather figured out, how the device worked "in principle" and only then could suggest how to improve its operation. The quotation marks are a reminder that the principles by which the clock functions were themselves being worked out in the process, especially the principles that explain the behavior of the materials from which the clock is constructed and the effects on it of the environment in which it is running. As Martin Burgess shows, only in trying to build a pendulum regulator does one discover the problems to be solved.[5] Nature clearly has more surprises than we can imagine *a priori* or than we can handle mathematically. Will Andrewes's opening quotation from Newton bears out the limitations of purely theoretical judgements on such matters.[6]

Four themes of the history of science seem particularly pertinent to the

2. Silvio A. Bedini, *The Pulse of Time: Galileo Galilei, the determination of longitude, and the pendulum clock,* Biblioteca di Nuncius, Studi e Testi, vol. 3. (Florence: Leo S. Olschki, 1991), chap. 3.

3. Michael S. Mahoney, "Christiaan Huygens: The measurement of time and longitude at sea," in H. J. M. Bos *et al.*, eds., *Studies on Christiaan Huygens* (Lisse, the Netherlands: Swets, 1980), pp. 234-70.

4. See Bruce Chandler's paper in this volume.

5. See Martin Burgess's paper in this volume.

6. See Will Andrewes's paper in this volume.

process of working out the material aspects of the problem of longitude in the eighteenth century: the extension of the mechanical model to the structure and behavior of matter, the empirical study of materials, precision measurement, and the role of craft skill in articulating the new science. It may help to look a bit more closely at each of these in turn.

In the *Principia*, Newton worked out the abstract machine that modeled the heavenly clockwork. In his *Opticks* of 1704, he explored a mechanical model of light that accounted for the laws of reflection and refraction and for the differential refrangibility that in turn explained the phenomena of colors. In a second set of "Queries" added to the text in 1713, he turned his attention to the chemical and electrical properties of bodies and wondered rhetorically whether they might not be explained in terms of small particles of matter attracting and repelling one another by central forces of a different sort from gravity. "And thus Nature will be very conformable to her self," he mused, "and very simple, performing all the great Motions of the heavenly Bodies by the Attraction of Gravity which intercedes those Bodies, and almost all the small ones of their Particles by some other attractive and repelling Powers which intercede the particles."[7]

As promising as that vision looks to us in hindsight, it proved elusive in the short run. Huygens had been baffled by the effect of heat and humidity on the materials of which his mechanical systems were constructed. He never had much interest in investigating such phenomena and preferred to seek mechanical configurations that overrode those effects or canceled them reciprocally. Newton, by contrast, was at home in the alchemist's laboratory and eager to probe beneath the surface of bodies. Nonetheless, as Newton himself recognized, it was one thing to imagine miniature constellations of atoms performing chemical or electrical motions, and quite another to identify precisely what motions were involved in what properties:

> For we must learn from the Phaenomena of Nature what Bodies attract one another, and what are the Laws and Properties of the Attraction, before we enquire the Cause by which the Attraction is perform'd. The Attractions of Gravity, Magnetism, and Electricity, reach to very sensible distances, and so have been observed by vulgar Eyes, and there may be others which reach to so small distances as hitherto escape Observation; and perhaps electrical Attraction may reach to such small distances, even without being excited by Friction.[8]

For much of the eighteenth century, the scientific investigation of materials focused on properties and behavior that could be observed empirically and controlled experimentally.

Through all this period of experimentation, the mechanical model remained the goal, which made its influence felt through a growing emphasis on precise measurement, that is, by the imposition of numerical scales on what seemed quantitative or quantifiable about the phenomena. Precision clocks, both in their building and their working, set the standard of what it meant to measure accurately. Hence their makers became the source of the craft knowledge and the skill to build the instruments of quantitative science, of clockwork science.

7. Isaac Newton, *Opticks, or a Treatise of the Reflections, Refractions, Inflections & Colours of Light*, 4th ed. (London, 1730; reprinted New York: Dover Publications, 1952), p. 396.

8. *Ibid*, p. 276.

London makers in particular played a direct role in the work of the Royal Society, to which several of them belonged and from which several, including John Harrison (1693-1776), received the institution's highest accolade, the Copley Medal. As Richard Sorrenson has recently shown, George Graham (*ca.* 1674-1751) designed the zenith sector that made Bradley's measurement of stellar aberration theoretically significant.[9] In another development that revealed the shortcomings of Newton's practical understanding, John Dollond (1706-1761) tied chromatic aberration to the material of the lens and was thereby able to produce achromatic lenses while advancing the physical theory of light. In transferring some of the maker's skill to a machine, the dividing engine of Jesse Ramsden (1735-1800) heralded a new industrial age in which precision would become a matter of routine, not only for scientific instruments but also for tools of production. Indeed, in the nineteenth century it would become increasingly difficult to distinguish between scientific instruments and tools of production, as science itself, both conceptually and institutionally, reflected its new industrial context.

At the same time, the roles of scientist and artisan diverged, as theory curiously emphasized its independence of craft skill. Christiaan Huygens never quite appreciated the contributions his clockmakers made to his inventions, evidently believing that his theories fully explained their working and his sketches sufficed to indicate their construction. His clockmakers knew better what it took to turn those sketches into working devices and thus how much Huygens's ideas depended on their craft knowledge.[10] More generally, Steven Shapin has spoken of the "invisible technician" behind the experiments of the seventeenth century. Indeed, the people who actually build—and, in building, often design—the instruments and machines of science largely remain unseen by the audience of science, including Nobel Prize committees.[11]

In treating instruments and machines as embodied knowledge, historians of science and technology have been working recently to bring craft skill back into view. For that, they are coming to recognize their own reliance on craftsmen and curators when reading the primary record that past artisans have left of their thinking, namely the products of their hands. As Andrew King points out, Harrison did not tell us how he designed his early clocks, only how he tested them.[12] To get to the reasoning behind the design we need the sort of artisanally informed reading that Anthony Randall presents of Harrison's watch in collaboration with Martin Burgess and Jonathan Betts.[13] As Henry Ford put it, "Machines are to a mechanic what books are to a writer. He gets ideas from them, and if he has any brains he will apply those ideas."[14] Historians with any brains will learn what mechanics have to teach them.

The determination of longitude lies at the intersection of a complex of both continuing and recent concerns of the history of science. It is a theme around which one can unforcedly organize many of the theoretical developments that constitute the monumental changes we characterize as the "Scientific Revolution." It presents one of the most solid links we have between technology and science during that period. Even more pointedly, it ties artisanal practice to theoretical inquiry. But perhaps more—even most—importantly, it offers insight into the elusive central question of the

9. Richard Sorrenson, "Scientific Instrument Makers at the Royal Society of London, 1720-1780" (Ph.D. diss., Princeton University, 1993).

10. J. H. Leopold, "Christiaan Huygens and his instrument makers," in H. J. M. Bos *et al.*, eds., *Studies on Christiaan Huygens* (Lisse, the Netherlands: Swets, 1980), pp. 221-33.

11. Steven Shapin, "The Invisible Technician," *American Scientist*, vol. 77, no. 6 (1989), pp. 554-63.

12. See Andrew King's paper in this volume.

13. See Anthony Randall's, Martin Burgess's, and Jonathan Betts's papers in this volume.

14. Henry Ford, *My Life and Work* (Garden City, New York: Doubleday, 1922), pp. 23-4.

development of modern science: Why did science first assume its characteristically modern shape in Europe in the seventeenth century? Why there? Why then?

The underlying theme of the papers presented to this symposium might be termed "the values of precision," which conveys by its ambiguity both the value of precision to science and technology and the social values that place a premium on precision itself.[15] That theme takes us back to the main point of David Landes's presentation.[16] Longitude became a problem not with the European voyages of discovery, but with the subsequent European drive to exploit those discoveries. Europeans were not the first to explore the world by sea, nor were they alone in their willingness to sail over open water. What seems peculiar about them was precisely the drive not only to discover but to control and command the world they encountered: to map it, to move freely over it, to know where everything is. You don't have to know your precise longitude to sail across oceans, but it is necessary if you want to do so efficiently and reliably; "ignorance is expensive." To this day, the sense of the world we convey to our children in school is the familiar map in the front of the room, the spherical Earth in Mercator projection, a view designed for navigation. So too, clockwork nature is a nature built for control and command. In relating the intricacies of pendulum regulators, Martin Burgess urges humility before the complexity of nature. Nonetheless, the pride of conquest showed through. It drove Europeans in the seventeenth century, and it continues to drive us today.

15. "The Values of Precision" was the theme of the Princeton Workshop in History of Science in 1991-2; the papers presented at the workshop appear in a volume by that title, edited by M. Norton Wise (Princeton, New Jersey: Princeton University Press, 1994).

16. See David Landes's paper in this volume.

# Early Attempts to Find Longitude

See Howse Figure 4 for caption
for this and previous page.

# The Longitude Problem: The Navigator's Story

*Alan Stimson*

**Alan Stimson was recently appointed Chief Executive of the Maritime Trust, having served as Head of Development at the National Maritime Museum, Greenwich, England. He joined the museum in 1963, after a ten-year career in the Merchant Navy, and was for many years Curator of Navigation. Stimson has made many contributions to the literature of navigation. His most recent publication is *The Mariner's Astrolabe* (1988).**

# The Longitude Problem: The Navigator's Story

*Alan Stimson*

In 1683 that most acute observer of professional competence, Samuel Pepys (1653-1714), recorded the following caustic remarks about the abilities of the fleet's navigators in his journal during a voyage to Tangier:

> It is most plain from the confusion all these people are to be in how to make good their reckonings (even each man's with itself) and the nonsensical arguments they would make use of to do it, and disorder they are in about it, that it is by God Almighty's providence and great chance and the wideness of the sea that there are not a great many more misfortunes and ill chances in navigation than there are.[1]

Figure 1 shows one of a series of Nicholas Pocock's watercolours. Although made a century later, I think this brilliantly illustrates the perpetual dilemma of the navigator in having to make a critical decision, often *in extremis* and based upon too little accurate information. The result could be catastrophic, as Pocock's other illustrations go on to show.

Progress toward accurate navigation has been a slow and difficult journey, with the seamen of classical and early medieval times relying on the Sun by day, the stars at night, and an accumulation of knowledge of capes and harbours, seasonal winds, and hidden dangers passed on by word of mouth from father to son and from master to apprentice. This knowledge was not committed to print until the late fifteenth century.[2]

The earliest aid to navigation was the sounding lead, originally a long pole that was developed into a lead weight at the end of a length of rope, which, when 'armed' with tallow, would indicate not only the depth of the water beneath the keel but also the nature of the seabed. In northern European waters, there was gradually a build-up of knowledge of the sea floor that, coupled with the ability to predict from the age of the Moon the times of high water at points around the coast, meant that its dangers could be avoided.

The invention of the magnetic compass was probably the single most important improvement in navigation technique, and yet there is still considerable argument as to the origin of the invention.[3] In Europe, a strong tra-

1. *The Tangier Papers of Samuel Pepys,* ed. E. Chappell (London: Navy Record Society, 1935), p. 129.

2. A Portolano (sailing directions) attributed to Alvise Ca'da Mosto (1432-1488), published in Venice by Bernardino Rizo (1490).

3. The invention of the compass occurred earlier in China. See Joseph Needham, *Science and Civilisation in China,* vol. 3, *Mathematics and the Sciences of the Heavens and the Earth* (Cambridge: Cambridge University Press, 1959), p. 576, and vol. 4, *Physics and Physical Technology* (1962), pp. 229-315 [especially 245-9, 279-88].

**Figure 1. 'The Consultation', a watercolour by Nicholas Pocock (1741-1821), used as an illustration for the poem 'The Shipwreck' (1762) by William Falconer (1732-1769). Courtesy of the National Maritime Museum, Greenwich.**

dition links its invention with Amalfi in southern Italy, sometime in the twelfth century, but the earliest written reference is made by an English monk, Alexander Neckham (1157-1217), in 1187.[4]

At this time, the magnetised iron needle was used only in overcast conditions when it was impossible to see the Sun or stars. At first the secret of the lodestone, a naturally occurring magnetic ore, was treated as a form of magician's trick. About 1240, the Dominican scholar Vincent of Beauvais (*ca.* 1190-*ca.* 1264) wrote that

> ...when clouds prevented sailors from seeing the sun they take a needle and rub its point on the magnetic stone. They then thrust it through a straw and place it in a bowl of water. The stone is then moved round and round the bowl faster and faster until the needle which follows it is whirling swiftly. At this point the stone is suddenly snatched away, and the needle turns its point towards the Stella Maris. From that position it does not move.[5]

At about the same time as the invention of the compass, the so-called portolan chart began to be used in the Mediterranean. Drawn on a whole skin of vellum and at first probably intended to illustrate the *portolani* or Italian sailing directions for the Mediterranean Sea, the earliest surviving examples date from the end of the thirteenth century. They all bear a striking similarity in style and content and are amazingly accurate in their representation of the coastline. This is not surprising when it is considered that the accumulated knowledge of several thousand years had gone into their compilation, coupled perhaps with the increasing accuracy of direction-finding provided by the magnetic compass.

Just when the compass needle was first pivoted on a brass spike instead of floated in water, and when the needle was first fixed to a compass card, are still open to debate, but the available evidence points to sometime in the late thirteenth century. The practice of a helmsman steering the ship by a compass seems to have begun sometime in the fourteenth century, and the first reference to a compass binnacle is found in an English ship inventory of

4. W. E. May, *A History of Marine Navigation* (Henley-on-Thames, England: G. T. Foulis, 1973), p. 45.

5. E. G. R. Taylor, *The Haven-Finding Art: A History of Navigation from Odysseus to Captain Cook,* new ed. (London: Hollis & Carter for the Institute of Navigation, 1971), p. 94.

1410-12. The binnacle protected the compass from the elements, and by placing a lamp in it, the helmsman was able to steer by the compass at all times, day or night.

How the ship was steered is beautifully recorded by a German monk, Felix Faber (1441[2?]-1502), on a pilgrimage to the Holy Land in 1483:

> ...they have as compass a Stella Maris near the mast and a second one on the topmost deck of a poop. And beside it all night long a lantern burns, and they never take their eyes off it, and there is always a man watching the star [the compass rose], and he sings out a sweet tune telling that all goes safely and with the same chant directs the man at the tiller how to turn the rudder. Nor does the helmsman dare to move the tiller in the slightest degree except at the orders of the one who watches the Stella Maris from which he sees whether the ship ought to go straight on, or curve or turn sideways.

Faber then goes on to describe how

> ...they have a chart on which is a scale of inches showing length and breadth, on which thousands of lines are drawn across the sea on which regions are marked by dots and numbers of miles. Over this chart they hang and can see where they are even when the stars are hidden.[6]

At last a form of seafaring routine is beginning to emerge with which a modern seaman can identify as being not too dissimilar from his own, and which over the next five centuries will come more clearly into focus.

The Portuguese, inspired by Prince Henry the Navigator (1394-1460), became the originators of the next phase of navigational improvements when they undertook extensive voyages of exploration down the West African coast. The Canary Islands were colonised by 1420, and the Azores too were visited and colonised by 1433. During this early period, the wind systems in the North Atlantic were still being thoroughly investigated, and as soon as the mariner was confident of being able to return to Portugal safely, a determined attempt was made to round Cape Bojador and push down the west coast of Africa.[7]

These voyages were made partly in an attempt to find a way around the Islamic empire at the eastern end of the Mediterranean to the rich trading goods and spices of the Orient and partly, it is said, to satisfy Prince Henry's interest in adventurous voyages of discovery. Under the direction of Prince Henry, a school of navigation was established at Sagres, where the best Mediterranean navigators and astronomical experts were gathered together. Here were developed simple astronomical instruments and methods that the uneducated seaman was taught to use and apply.

The new techniques meant that the 'dead reckoning', a progressively less accurate estimation of the ship's position, could be checked by observation of the height of the pole-star above the horizon with a simplified version of the astronomical quadrant. This was a wooden quarter circle with two sights fixed on one straight edge and a plumb-line suspended from its apex, so as to read over a scale of 90° marked upon its limb.

With this instrument, the seaman was at first taught to observe the pole-star at his port of departure and mark on the limb the place where the

6. *Ibid.*, pp. 144-5.

7. D. W. Waters, *The Art of Navigation in England in Elizabethan and Early Stuart Times* (London: Hollis & Carter, 1958), pp. 39-40.

plumb-line cut the scale. During the course of the voyage he marked the height of the star at successive headlands, river estuaries, and islands so that on his return the cartographers gradually built up a table of coastal latitudes expressed as 'altura', or heights of the Pole.

Because of its circumpolar motion, the navigator was warned to observe the pole-star only when the 'Guards' (pointing stars in the constellation of the Little Bear) were in a certain position. When he wished to visit a port or place of known altura, all he had to do was look for its name on the scale of his quadrant, and when the height of the Pole agreed with the mark, he was east or west of his destination.

Later he was taught to convert the number of degrees of altura into distance sailed by multiplying each degree by 16⅔ leagues (a league being reckoned as three miles) in order to find the distance sailed south or north and eventually, as his understanding increased, to convert the altitude of the pole-star into latitude.

By 1471 the equator, from where the pole-star was no longer visible, had been reached. What was the navigator to do? The problem of how to navigate in the Southern Hemisphere was resolved by the commission set up by King John II of Portugal in 1485. A 'Regiment of the Sun' was drawn up. This enabled the navigator to use the Sun as a means of latitude determination. The 'Regiment' gave the Sun's declination (its seasonal distance north or south of the equator) for each day of the year so that, by measuring the altitude of the Sun at noon and correctly applying its declination for that day, a navigator could find his latitude everywhere on the Earth's surface. Additional rules were necessary to explain how to work the calculation for the various cases of the observer being to the north or south of the Sun when it lay in either the Northern or Southern hemispheres and when the observer and the Sun were in different hemispheres. Mercifully, this was often expressed as a diagram, which most practical seamen found much easier to understand than the written word.[8]

The plumb-bob made the quadrant a difficult instrument to use at sea, and an unwanted visit ashore was often necessary to get a satisfactory result. The astronomer's astrolabe was eventually simplified for use at sea and soon proved itself to be a superior instrument. First recorded in 1481,[9] the mariner's astrolabe (Figures 2 and 3) was initially made of wood but was eventually cast in brass. It was a heavy circular instrument, from six to eight inches in diameter, suspended by a hinged suspension ring that allowed it to hang vertically. A sighting alidade was pivoted at its centre in such a way that pointers at each end could in turn read from a scale of degrees marked on the upper limb.[10]

In his *Breve compendio* of 1551, the Spaniard Martín Cortés (d. 1582) describes the use of this instrument. In the English translation of 1561, the passage reads as follows:

> To take the altitude of the Sunne, hange up the Astrolabie by the rynge: and set the Alhidada against the Sunne. And rayse it or put it downe in the quarter that is graduate, untyll the beames of the Sunne enter in by the lyttle hole of the tablet or raysed plate, and precysely by the other lyttle hole of the other tablet. Then looke uppon the lyne of confydence. And howe manye degrees it sheweth

8. D. W. Waters, *Science and Technique of Navigation in the Renaissance*, National Maritime Museum Monograph (London, 1976).

9. D. W. Waters, *The Sea- or Mariner's Astrolabe*, Agrupamento de Estudos de Cartografia Antiga, no. 15 (Coimbra, Portugal, 1966), pp. 8-9.

10. Alan Stimson, *The Mariner's Astrolabe: A survey of known, surviving sea astrolabes* (Utrecht: HES, 1988), p. 16.

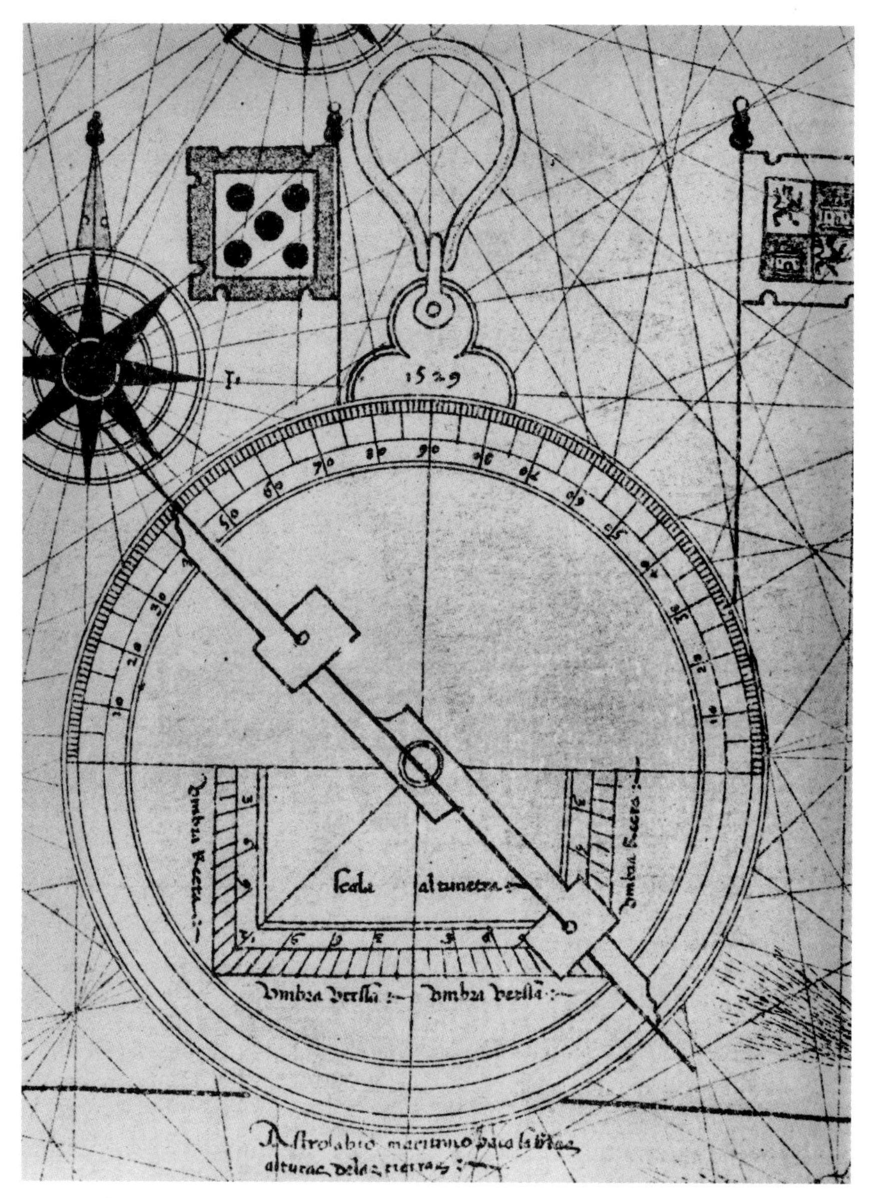


**Figure 2. An early illustration of a mariner's astrolabe by Diego Ribeira, 1529. The earliest one drawn by him is dated 1525 and is on the Castiglioni Planisphere, Archivo Marchesi, Castiglioni, Mantua. Courtesy of the National Maritime Museum, Greenwich.**

> in the quarter that is graduate (begynnynge from the Horizontall lyne) so many degrees of height hath the Sunne. In lyke maner shall you doe to take the altitude of any other Starre lookynge thorough the greate holes, because this may hardelye be seene by the lyttle holes.[11]

A further altitude instrument, the cross-staff, was introduced early in the sixteenth century, again a development of an astronomical instrument (see Figure 5). The astronomical cross-staff was invented by the Provençal Jew Levi ben Gerson (1288-1344) about 1330 to measure the diameter of the Sun or the distance between any two celestial objects.[12] Its introduction at sea closely followed the voyage in 1498 of Vasco da Gama (*ca.* 1460-1524), who, while crossing the Indian Ocean, had encountered the Arab *kamal*, a navigation instrument of similar principle. It is possible that the two events may have been connected.

The cross-staff consisted of a square-sectioned staff approximately 30 inches in length on which three transversals or crosses could be mounted, each appropriate to an altitude scale in degrees and minutes engraved upon the staff. In use, the observer selected the cross appropriate to the altitude

11. Martín Cortés, *Breve compendio de la sphera y de la arte de navegar...* (Seville, 1551). Translated by Richard Eden, *The Arte of Navigation,...* (London, 1561), fols. lxx$^{v}$-lxxi$^{r}$.

12. Anthony Turner, *Early Scientific Instruments: Europe 1400-1800* (London: Sotheby's Publications, 1987), p. 20. See also B. R. Goldstein, 'Preliminary Remarks on Levi Ben Gerson's Contributions to Astronomy', *The Israel Academy of Sciences and Humanities, Proceedings III* (1969) and p. 378 in Appendix C of this volume.

**Figure 3. The earliest illustration of a seaman using a mariner's astrolabe, 1545. From Pedro de Medina's *Arte de Navegar...* (Valladolid, 1545). Courtesy of the National Maritime Museum, Greenwich.**

and placed one end of the staff to his eye. He then slid the cross in, or out, until the top end of the cross just covered the Sun or star while at the same moment the lower end rested in line with the horizon. The altitude was read from where the cross cut the scale.

Although much more accurate than the astrolabe, its one major defect was ably expressed by William Bourne (d. 1583) in his *Regiment for the Sea* of 1574:

> ...but if it [the altitude] doe exceed 50 degrees then by the means of casting your eye upwardes and downwards so muche, you may soone commit error and then in like manner the degrees be so small marked, that if the Sunne doth pass 50 or 60 degrees in height, you must leave the cross staffe and use the Mariner's ring, called by them the Astrolaby which they ought to call the Astrolabe.[13]

13. William Bourne, *A Regiment for the Sea* (London, 1574).

The new Portuguese methods quickly spread to Spain, which, after Columbus's epic voyage of 1492, rapidly built up an empire in the New World across the Atlantic. The portolan chart of the Mediterranean was expanded to accommodate the new discoveries and progressed from being based on direction and estimated distance only to incorporating the latitude of places and a latitude scale. Some charts also carried longitude scales, but these were almost always in error because it was not yet possible to find longitude at sea.

By the middle of the sixteenth century, altitude observations and charts were beginning to be used by north European seamen, although not without a certain amount of resistance by the more conservative elements. This resistant attitude is succinctly described by William Bourne in the preface to his *Regiment*:

> ...I have known within this twenty years that them that were ancient masters of ships hath derided and mocked them that have been busy with their cards and plats [charts] and also the observation of the altitude of the Pole, saying they care not for their parchment for they could keep a better account on a board.
>
> ...and when they did take the latitude they would call them star-shooters, and would ask if they had hit it.[14]

Fortunately, this resistance to new methods did not pervade all English seamen, for by the 1540s William Hawkins of Plymouth (*ca.* 1490-1554) had made 'three long and famous voyages unto the coast of Brazil, a thing in those days very rare, especially for our nation'.[15] Not many years later Sir Francis Drake (*ca.* 1543-1596) circumnavigated the globe.

For the next two centuries very little progress in method or technique was made. Seamen could generally find their latitude to within about fifteen miles and found their destinations by practising the art of 'running down the latitude'—that is, sailing well to seaward of a desired landfall of known latitude, observing the Sun or pole-star until that latitude was reached, and then sailing east or west, until land was sighted ahead.

Distance sailed in English ships was estimated by log and line. As a flat piece of wood attached to a thin line was payed out over the stern of the ship, its length was timed against a sand-glass. The length of line run out in 28 seconds was proportional to the speed of the ship in one hour. Knots were made in the line to assist this process, but considerable differences of opinion existed as to how far apart the knots should be. The correct distance of 51 feet had been recommended by Richard Norwood (1590-1665) in 1637, but as John Hamilton Moore (1738-1807) explained in his *Practical Navigator* as late as 1782,

> ...experienced commanders find that the allowing of 50 feet a knot generally makes the ship ahead of the reckoning; and to avoid Danger, mostly divide the log-line of 7 or 7½ fathoms and of 6 feet each [42 to 45 feet]....[16]

Improved altitude instruments continued to be invented. The backstaff, a development of the cross-staff, was first described in 1595 by John Davis (1552-1605) in his *Seamans Secrets* and quickly became popular.[17] Routine timekeeping on a ship at sea was controlled by sand-glasses, which were reset at noon each day if the Sun was visible, a practice that continued into the nineteenth century.

During the seventeenth century, ships of many nations came to trade regularly across the Atlantic Ocean and to India and the Far East around the Cape of Good Hope. Inability to find a ship's longitude proved the greatest obstacle to the improvement of navigational practice and resulted in unduly prolonged voyages, if not in actual shipwreck. The commercial implications of the development of accurate navigational techniques could not fail to be

14. *Ibid.*, Preface.

15. R. Hakluyt, *Principal Navigations* (London, 1599).

16. John Hamilton Moore, *The Practical Navigator* (London, 1782).

17. John Davis, *The Seamans Secrets,...* (London, 1595). Additional information and illustrations are given in Appendix D.

Figure 4. Navigators and their instruments at the end of the seventeenth century. The title page from John Seller's *Practical Navigation or An Introduction to that whole Art* (London, 1672). Courtesy of the National Maritime Museum, Greenwich.

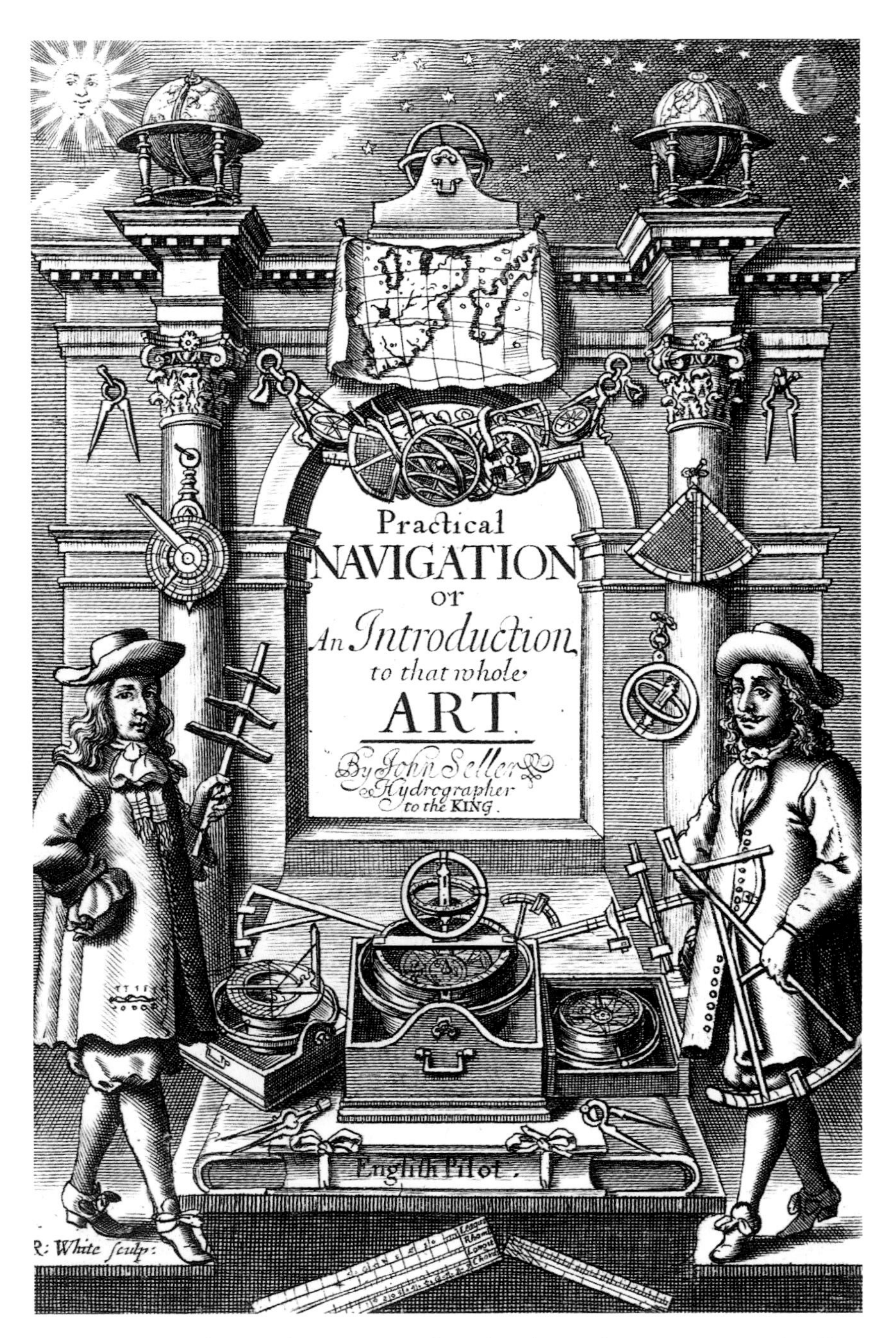


apparent, and, during the seventeenth and eighteenth centuries, much effort was expended in attempting to find a solution to the problem:

> They have one great imperfection yet in their arte, and hitherto by no man supplyed, and that is the wante of exact rules to knowe the longitude...without the which they can not truely geeve the place or situation of any Coaste, Harberough, Rode or Towne, ne yet in saylinge discerne how the place they sail unto beareth....[18]

18. Thomas Digges, 'Errors in the Arte of Navigation commonly practized', essay in 'The Addition', published in Thomas Digges's corrected and augmented edition of Leonard Digges, *A Prognostication everlastinge of righte good effecte,...* (London, 1576).

Thomas Digges (*ca.* 1546-1595), writing this in 1576 in his 'Errors in the Arte of Navigation', clearly defined the last major navigational problem to be conquered, little knowing that the solution still lay some 200 years in the future.

Longitude is basically distance east or west on the Earth's surface and can be expressed as the difference in time between two places. In order to find how far east or west his ship has sailed, a navigator must be able to find the

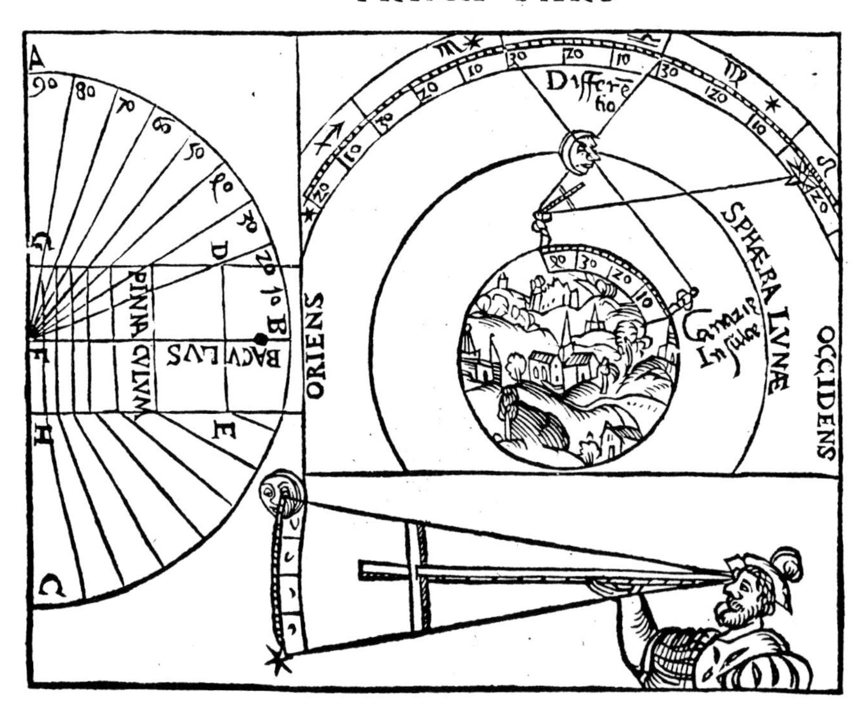


**Figure 5. Lunar-distance diagram from the second edition of Peter Apian's *Cosmographicus Liber...*, revised and corrected by Gemma Frisius (Antwerp, 1529), fol. XVI, sig. Div$^{v}$. The suggestion that a cross-staff should be used to measure the lunar distance has obvious shortcomings. By permission of the Houghton Library, Harvard University.**

time at his ship and at the same instant be able to calculate the time at his place of departure or the time at a standard meridian.

A theoretically correct answer to the problem had been put forward by Johann Werner (1468-1522) of Nuremberg as early as 1514.[19] Werner recommended using the Moon as a clock as it moved across the star background by measuring its distance from selected stars, or from the Sun, such distance being equivalent to the time at a standard meridian. This method was, however, impractical at that time, because of the inability to predict the Moon's position accurately, inexact plotting of star positions, and the lack of a sufficiently accurate instrument with which to take the 'lunar distances' between Moon and stars (Figure 5).

An alternative method was suggested in 1530 by Gemma Frisius (1508-1555), who put forward the idea of carrying a standard time in a ship with a clock that would continue to function accurately in seagoing conditions. The difference between standard time by the clock and local time, found by observation of the Sun, would give the longitude.[20] However, the technology of the sixteenth century was not sufficiently advanced for such a clock to be constructed.

While later speakers will address the matters of lunar distance and chronometers in some detail, I intend to concentrate on the third element of the equation: how an accurate instrument became freely available to the navigator. As world trade expanded, the need for a solution became more pressing. Prizes for a practical method were offered by Philip II of Spain in 1567 and by Philip III of Spain in 1598. Other offers of prizes were to follow.

In 1675, Charles II made the first official step in England toward a solution of the longitude problem by establishing a Royal Observatory at Greenwich, near London. The first Astronomer Royal, John Flamsteed (1646-1719), was directed to

> ...apply himself with the utmost care and diligence to rectifying the tables of the motions of the heavens and the places of the fixed stars

19. Johann Werner, *In hoc opere haec cōtinentur Noua translatio primi libri geographiae Cl'. Ptolomaei:...* (Nuremberg, 1514). The passage concerning Werner's description of the lunar-distance method of finding longitude is in *cap.* 4, note 8, sig. dv$^{v}$. (See Appendix C.)

20. *Gemma Phrysius de Principiis Astronomiae & Cosmographiae, Deque usu Globi ab eodem editi* (Louvain and Antwerp, 1530). The passage concerning finding longitude at sea with a clock is in part 2, *cap.* 18 ('De novo modo inveniendi longitudinē'), fols. D2$^{v}$-D3$^{r}$. (See Appendix C.)

21. Eric G. Forbes, *Greenwich Observatory, The Royal Observatory at Greenwich and Herstmonceux, 1675-1975*, vol. 1, *Origins and Early History (1675-1835)* (London: Taylor and Francis, 1975).

22. Also spelt Clowdisley Shovell, the way it appears he spelt his own name.

23. Act 12 Anne *cap.* 15 (1714): *An Act for Providing a Publick Reward for such Person or Persons as shall Discover the Longitude at Sea.*

24. By a strange coincidence a similar instrument was invented in Philadelphia by a window glazier and amateur scientist, Thomas Godfrey (1704-1749), at the same time that Hadley was developing his invention. See Silvio A. Bedini, *At the Sign of the Compass and Quadrant: The Life and Times of Anthony Lamb,* Transactions of the American Philosophical Society, vol. 74, part 1 (Philadelphia, 1984), pp. 37-44. For further information and other references, see Appendix D in this volume.

25. John Hadley, 'The Description of a new Instrument for taking Angles. By John Hadley, Esq; Vice-Pr. R. S. communicated to the Society on May 13. 1731', *Philosophical Transactions* of the Royal Society, London, vol. 37, no. 420 (1731), pp. 147-57.

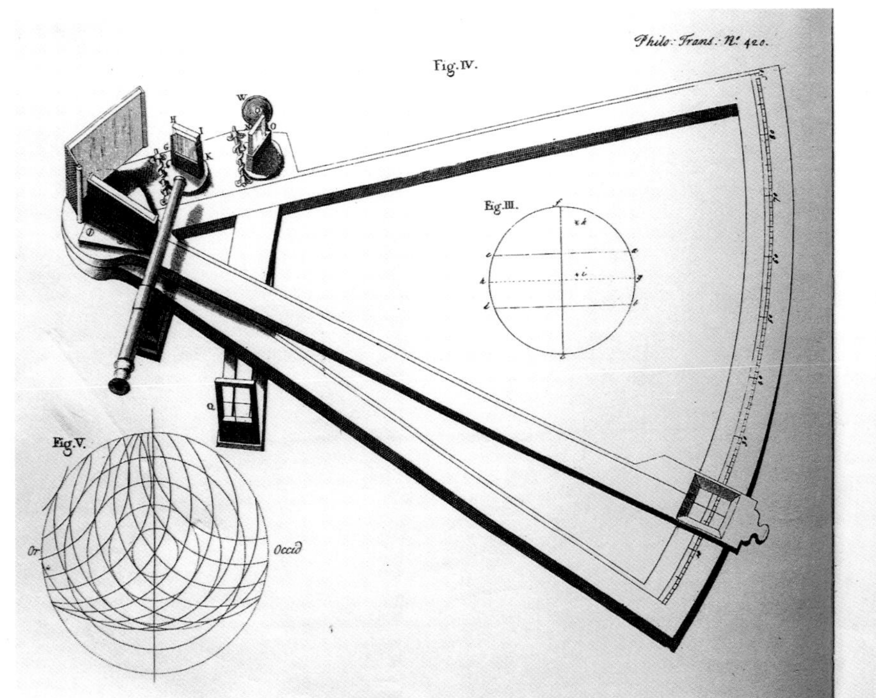


Numb. 420.

PHILOSOPHICAL TRANSACTIONS

For the Months of *August* and *September* 1731.

The CONTENTS.

I. *The Description of a new Instrument for taking* Angles. *By* John Hadley, *Esq; Vice-President R. S. communicated to the Society on* May 13. 1731.

II. *An Extract of a Dissertation* De Stylis Veterum, & diversis Chartarum generibus, [*by the Hon. Sir* John Clerk, *one of the Barons of the Exchequer in* Scotland, *and F. R. S.*] *By* Roger Gale, *Esq; Vice-President and Treasurer R. S.*

III. *Some Experiments concerning the* poisonous Quality *of the* Simple Water *distilled from the* LAURO-CERASUS, *or* common Laurel, *made upon Dogs, at* Toppingo-Hall *in* Essex, August *the* 24*th* 1731. *and others made before the* ROYAL SOCIETY *in their Repository. By* Cromwell Mortimer, *M. D. R. S. Secret.*

IV. *A*

Figure 6. John Hadley's octant, 1731. This diagram was published to illustrate Hadley's article in the *Philosophical Transactions* of the Royal Society, London, vol. 37, no. 420 (1731). Courtesy of the National Maritime Museum, Greenwich.

so as to find out the so much to be desired longitude of places for perfecting the art of navigation....[21]

A series of maritime disasters, culminating in the wrecking of part of a returning squadron of naval ships under the command of Sir Cloudesley Shovel (*ca.* 1650-1707)[22] on the Scilly Islands in 1707, eventually led to the British Parliament setting up a committee to examine the problem. This, in turn, resulted in an Act in 1714 that offered a prize of £20,000, a fortune in the eighteenth century, for any method of determining a ship's longitude to within half a degree. A panel of commissioners, known as the Board of Longitude, was constituted at the same time to examine and adjudicate on proposed solutions.[23]

In addition to the terms of the longitude award, the Act also contained a provision that authorised the Board to award sums of money up to £2,000 to projects that they considered worthy of encouragement although not fulfilling the terms of the Act. This was, in effect, a government fund to finance technological and scientific experimentation which had a direct influence on the productivity and excellence of scientific and mathematical instrument making in England during the second half of the eighteenth century.

The offer of this great sum of money concentrated the minds of scientists and instrument makers throughout Europe on finding a solution to the longitude problem, although the Board of Longitude had to contend with a succession of ludicrous proposals. The first useful consequence of the prize was the invention in 1731 of the reflecting quadrant (Figure 6) by John Hadley (1682-1744), a country gentleman of a scientific turn of mind and a Fellow of the Royal Society.[24]

Hadley published a description of his quadrant in the *Philosophical Transactions* of the Royal Society in 1731.[25] In 1732, the instrument was tested on board the yacht *Chatham,* where it performed remarkably well, and further improvements were effected two years later by the fitting of an artificial horizon. Hadley patented his quadrant in 1734.

A period of continual development and refinement followed, as recorded by Professor Anthony Shepherd (fl. 1763-1772) in his preface to the *Tables for Correcting the Apparent Distance of the Moon and a Star:*

> This Instrument has been several Times tried at Sea, and found to answer those Expectations: the best Observers differing only one Minute in their Observations of the Sun's meridian Altitude. Considerable Improvements have lately been made in this Instrument by the Astronomer Royal and *Mr. Dollond:* the former having obviated the Errors of the Planes of the Mirrors; the latter, for which he has taken out a Patent, having exhibited a new Method of adjusting the Instrument for the back Observation with the same Ease and Certainty as for the fore Observation.[26]

Hadley's quadrant, as it was generally called by seamen, measured up to one quarter of a circle, 90°, on the principle of double reflection. The actual arc of the instrument, therefore, was only one-eighth of a circle, so that it is sometimes called an octant (Figure 7). It made use of two mirrors so that the Sun (or star) and the sea horizon could both be seen at the same instant. This allowed accurate measurements to be taken at sea for the first time, whatever the movement of the vessel, and Hadley was confident that when the theory of lunar prediction was perfected, his octant could be used to measure the necessary lunar distances.

In 1755, Tobias Mayer (1723-1762), a professor of geography at Göttingen, sent a manuscript copy of his lunar tables to England to be considered for the longitude prize. They were examined by James Bradley (1693-1762), the Astronomer Royal at that time, who found that they compared favourably with his own observations at Greenwich.

Between 1757 and 1759, the tables were tried at sea by Captain John Campbell (1720-1790) on the *Royal George*, together with a reflecting circle also designed by Mayer. Mayer believed that Hadley's quadrant was not sufficiently accurate to take the exact measurements between a star and the Moon and designed an instrument on Hadley's principle, but with a scale extended to a full circle. It was also sometimes necessary to measure angles in excess of 90°. By taking a mean of several observations measured around an eighteen-inch circle, Mayer hoped to reduce instrumental and observational errors, but Captain Campbell found it a clumsy instrument to use.

**Figure 7. An octant, inscribed 'Geo: Sterrop Fecit. No. 5259'. England, *ca.* 1750. Overall height: 56.5 cm. (22.2 in.). Courtesy of the Collection of Historical Scientific Instruments, Harvard University. Inv. no. 5250.**

In 1757, Campbell arranged, through Bradley, for the scientific instrument maker John Bird (1709-1776) to construct a Hadley quadrant of similar radius to the diameter

26. Anthony Shepherd, ed., preface to *Tables for Correcting the Apparent Distance of the Moon and a Star from the Effects of Refraction and Parallax....* (Cambridge, 1772), p. ii. The idea for these tables was proposed by George Witchell (1728-1785), headmaster of the Royal Academy, Portsmouth, in 1767, but the tables themselves were computed in Cambridge under Shepherd's supervision, and he wrote the preface.

27. A. N. Stimson, 'Some Board of Longitude Instruments in the Nineteenth Century', in P. R. de Clercq, ed., *Nineteenth-Century Scientific Instruments and their Makers* (Amsterdam: Rodopi/Leiden: Museum Boerhaave, 1985), pp. 95-6. This book is a compilation of papers presented at the Fourth Scientific Instrument Symposium, Amsterdam, 23-26 October 1984.

28. Board of Longitude Confirmed Minutes, 25 June 1774 and 27 May 1775, MS. RGO 14/5, vol. 5(2), Royal Greenwich Observatory Archives, Cambridge University Library.

29. For instructing other workmen, writing illustrated descriptions (published by the Board of Longitude in 1777 and 1779), and allowing the engines to become the property of the Board, Ramsden received £615 in 1774 for the circle dividing engine and £400 in 1778 for the later, straight-line machine. See Derek Howse, *Nevil Maskelyne: The Seaman's Astronomer* (Cambridge: Cambridge University Press, 1989), p. 128. See also Allan Chapman, 'Scientific Instruments and Industrial Innovation: The Achievement of Jesse Ramsden', in R. G. W. Anderson, J. A. Bennett, and W. F. Ryan, eds., *Making Instruments Count, Essays on Historic Scientific Instruments presented to Gerard L'Estrange Turner* (Aldershot, England: Variorum, 1993), pp. 418-23.

Figure 8. Mid-eighteenth-century navigators fixing the ship's noon position. Engraving from John Robertson's *The Elements of Navigation...* (London, 1754), vol. 2, p. 503. Courtesy of the National Maritime Museum, Greenwich.

of Mayer's circle, but with a scale extended to read at least 120°—in other words, a sextant (Figure 9). Bird's sextant answered all the requirements and, with slight modifications and refinement, continues in use today.[27]

Despite the excellence of John Bird's sextant, it had one shortcoming: the length of time it took to graduate the scale by hand. Bird used the same methods he employed to divide the scales of large astronomical instruments, which resulted in large, expensive instruments well beyond the financial means of the average seafarer. However, another London instrument maker, Jesse Ramsden (1735-1800), tackled the problem of rapidly and accurately dividing instrument scales. Ramsden had made his first 'dividing engine' by 1768. A new, improved version, completed in 1774, was examined by the Board of Longitude experts, John Bird among them, one year later.[28] Ramsden's engine was a total success, and the Board of Longitude awarded him £615 providing he divulged its secrets to other instrument makers.[29] Thereby, it became possible to accurately divide a quadrant's scale in twenty minutes and a sextant's in forty, with the consequent reduction in price of the finished product.

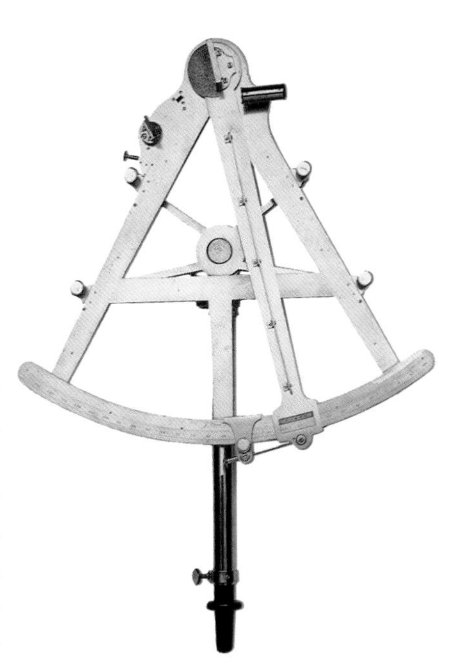

Figure 9. The first sextant, made by John Bird, London, ordered by the Astronomer Royal, James Bradley, in 1757 for the Board of Longitude. The weight of the instrument required that it be supported by a pole that screwed into a frame on the back of the sextant. Overall height: 50.8 cm. (20 in.). Courtesy of the National Maritime Museum, Greenwich. Inv. no. S225/63-22.

Although Ramsden's engine-divided sextants were of undoubted excellence, there was in England a small but steady demand by surveyors and hydrographers for the development of Mayer's reflecting circle. Edward Troughton's (1753-1836) reflecting circle of 1796 sold in small numbers during the early nineteenth century, but in France the reflecting circles (invented in 1774) of Jean-Charles Borda (1733-1799) and other makers were much favoured by naval navigators and hydrographers and continued in use into the twentieth century.

Inexpensive and accurate instruments were now available for the navigator to take full advantage of the new and exciting methods of finding longitude that were made possible by the publication of *The Nautical Almanac* and the increasing availability of chronometers. By 1830, it is known that ten dividing engines were working in London, and the owners of these machines very quickly cornered the market in precision dividing of navigational and surveying instruments.[30]

By the mid-nineteenth century sextants were so numerous that Captain Squire Lecky (1838-1902) could write in his *"Wrinkles" in Practical Navigation:*

> A good Quintant or Sextant costs money—and is worth it. Unless you are a fair judge of one, it is easy to be deceived in purchasing a sextant as in buying a horse. The market is glutted with sextants made for sale. Every pawnbroker's window in a seaport town is half full of them.[31]

Cheap and efficient instruments had certainly arrived, and they were to continue in production until the arrival of satellite navigation in the mid-1960s began the process of gradually ousting astronomical navigation.

30. D. Baxondall, 'The Circular Dividing Engine of Edward Troughton, 1793', *Transactions of the Optical Society* (London), vol. 25.3 (1924), pp. 136-8.

31. Squire T. S. Lecky, *"Wrinkles" in Practical Navigation* (London, 1881), pp. 38-9.

# Longitude and the Satellites of Jupiter

*Albert Van Helden*

**Albert Van Helden was born in The Hague and emigrated to the United States in 1955. He obtained a master's degree in metallurgy at Stevens Institute of Technology and worked as a metallurgical engineer for Ford Motor Company for two years before entering the graduate program in history at the University of Michigan. After transferring to Imperial College, University of London, he received a Ph.D. in the history of science in 1970. In the same year, he joined the History Department of Rice University, where he has served as Chairman and Speaker of the Faculty.**

**His research has focused on the history of astronomy, particularly the invention of the telescope, the replacement of the Galilean telescope by the astronomical telescope, and the development of the compound eyepiece. This work led to his two recent books, *Measuring the Universe* (1985) and a new English translation of Galileo's *Sidereus Nuncius* (1989). He also has published a number of articles on telescopes in the nineteenth and twentieth centuries, the transits of Venus, and seventeenth-century astronomy. He is currently involved in a project with Mario Biagioli on the sunspot dispute between Galileo and Christoph Scheiner.**

# Longitude and the Satellites of Jupiter

*Albert Van Helden*

This conference celebrates the life and work of John Harrison (1693–1776). Harrison made the chronometer that allowed observers on the high seas to fix their longitude by means of observations that were practical on board a ship. For almost two centuries prior to Harrison's chronometers, there were efforts to use observations of the eclipses of Jupiter's satellites for finding longitude at sea. Shortly after the invention of the telescope in 1608,[1] however, it was apparent to practical men that observing conjunctions or eclipses of the satellites from the deck of a ship on the high seas was virtually impossible. We are, therefore, faced with a question: If it was so obvious that this method of determining longitude at sea was not practical, why did intelligent people continue to try to make it work, right up to the time when accurate marine chronometers settled the issue early in the nineteenth century? I think the answer is deceptively simple: it was because this method was so spectacularly successful on land. While sailors were lost on the vast oceans, navigating as best they could by prevailing winds and dead reckoning, astronomers were fixing the longitudes of islands, capes, and ports in an international cooperative effort. After all, the problem of longitude at

1. See Albert Van Helden, *The Invention of the Telescope,* Transactions of the American Philosophical Society, vol. 67, part 4 (Philadelphia, 1977).

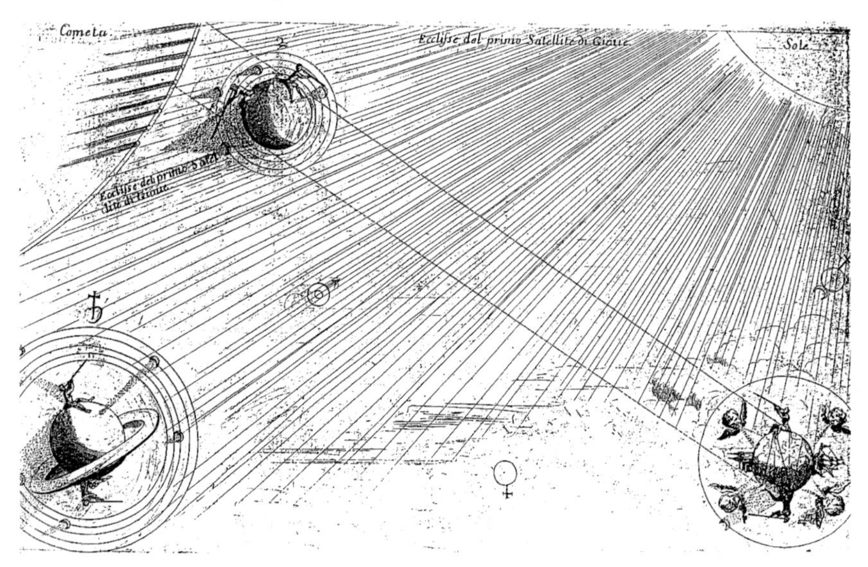


**Figure 1. Engraving illustrating the eclipse of Jupiter's first satellite, published in "Osservationi Delle Comete Che Douranno Seguire, e delle ecclisse del primo satellite di Giove,..." by Cornelius Meyer (Rome, 1696), plate i [9], as the third part of Cornelius Meyer's *L'Arte Di Rendere I Fiumi Navigabili In Varij Modi, con altre nuove inventioni, e varij altri segreti. Divisi In Tre Parti* (Rome, 1696). By permission of the Houghton Library, Harvard University.**

Figure 2. The Gulf Coast at the time of La Salle's expedition. Map prepared by Robert L. Williams. Courtesy of The American Historical Association, Washington, D.C.

sea takes on meaning only when one considers it in the context of longitude on land, for what good would it do a sailor approaching the Carolinas and Virginia if he knew the location of his ship but did not know the location of Cape Hatteras?

There was also the problem of longitude for the explorer on land. An example will illustrate this. On April 9, 1682, René Robert Cavelier, Sieur de La Salle (1643-1687), raised a cross near the mouth of the Mississippi River and claimed the region from Canada (where he had started his exploration) to the Gulf of Mexico in the name of Louis XIV, King of France. Having done this, he returned upriver and eventually made his way back to France through Canada. In 1684, La Salle returned to the Gulf of Mexico with an expedition to establish a military presence and found a settlement at

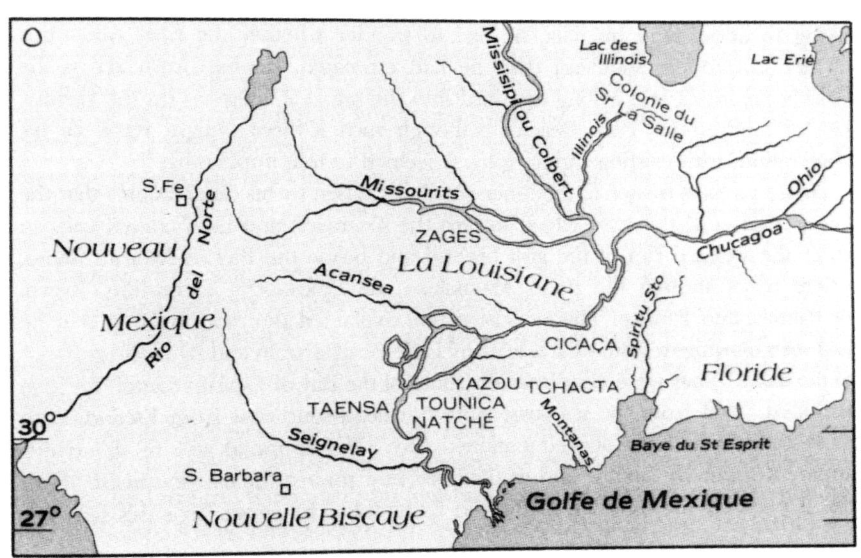


Figure 3. Reconstruction of La Salle's idea about the course of the Mississippi. Map prepared by Robert L. Williams. Courtesy of The American Historical Association.

the mouth of the Mississippi. He arrived at Matagorda Bay on the Texas coast, near modern-day Corpus Christi, thinking he was very close to the mouth of the Mississippi. This error of more than 600 miles caused the failure of the expedition and, eventually, La Salle's own death. Figures 2 and 3 show a modern representation of the region and a reconstruction of the map that must have been in La Salle's head.[2] How could he have been so dreadfully wrong? Explorers such as La Salle traveling in remote regions usually relied on inaccurate maps and had no more idea about their longitude than sailors in the middle of an ocean. As La Salle's end shows (he was murdered by his own men), the dangers caused by an ignorance of longitude on land could be just as great as an ignorance of longitude at sea. Ironically, as La Salle was exploring the New World, his king's astronomers were beginning a revolution in geodesy based on the satellites of Jupiter.

We must, then, take a broader view of the problem of longitude. As Europeans expanded their activities around the globe, their master concept of that globe was one of space and coordinates. To know the globe was to map it, and to map it was to determine the coordinates of specific locations, except at sea where there were no points of reference. How great were the errors in longitude on the maps of the world in 1600 and 1700? I do not want to belabor this point, so let me just note a few examples. Early in the seventeenth century, it was found that the length of the Mediterranean, a sea that had been traveled since time immemorial, was in error by about 10° on the most recent maps and 19° on the maps found in early printed editions of Ptolemy's *Geography*.[3] Later in the century, astronomical observations in the Orient revealed that the locations of China and Japan were in error by as much as 55° on world maps.[4] As late as 1769, a joint French-Spanish expedition to Baja California to observe the transit of Venus found the longitude of the peninsula to be in error by 5° on existing maps.[5] Such errors had disappeared from maps by the beginning of the nineteenth century, when the marine chronometer was just coming into routine use. In the mapping of the world in the seventeenth and eighteenth centuries, longitudes were determined, by and large, by observations of the eclipses of the satellites of Jupiter.

## GALILEO GALILEI (1564-1642) AND NICOLAS-CLAUDE FABRI DE PEIRESC (1580-1637)

Usually historical problems do not have clean and definite beginnings. In the case of the satellites of Jupiter, we can date the start very precisely: it was on January 7, 1610, shortly after sunset. Jupiter had just passed opposition and was the brightest body in the evening sky when Galileo turned toward the planet a telescope that had been adapted for observing bright celestial bodies.[6] What happened then is well known. Before the end of 1610, the satellites of Jupiter were observed by Johannes Kepler (1571-1630) in Prague,[7] by Thomas Harriot (*ca.* 1560-1621) at Syon near London,[8] by Nicolas-Claude Fabri de Peiresc and Joseph Gaultier de la Valette (1564-1647) in Aix en Provence,[9] by the Jesuit mathematicians at the Collegio Romano in Rome,[10] and, it has been argued, by Simon Marius (1573-1624) in Ansbach near Nuremberg.[11]

Important as was the problem of longitude in general—and longitude at sea in particular—the recognition that Jupiter's four moons could act as a celestial timepiece and be used for determining longitude came almost

2. Peter H. Wood, "La Salle: Discovery of a Lost Explorer," *American Historical Review*, vol. 89 (1984), pp. 294-323. The maps are on pp. 306 and 307.

3. This is based on the statement by Pierre Gassendi (1592-1655) in his life of Peiresc. Gassendi was basing his figures on the length of the Mediterranean, 60°, as found in the fifteenth-century editions of Ptolemy's *Geography*. See, *e.g.*, Ptolemy, *Cosmographia* (Rome, 1478). By Gassendi's time that figure was already being improved. In the various editions of Abraham Ortelius's *Theatrum Orbis Terrarum* (Antwerp, 1570, but see also the English edition, London, 1606) and the various editions of the Mercator-Hondius-Janssonius *Atlas* (see, *e.g.*, the English edition, Amsterdam, 1636), the length of the Mediterranean had shrunk to 51°-53°. Its length is given in modern atlases as about 41°.

4. Ortelius (*ibid.*). Note also that, especially in the Orient, longitudes for the same places vary from map to map in the same atlas.

5. Doyce B. Nunis, *The 1769 Transit of Venus: The Baja California Observations of Jean-Baptiste Chappe d'Auteroche, Vicente de Doz, and Joaquín Velázquez Cárdenas de Léon* (Los Angeles: Natural History Museum of Los Angeles County, 1982), pp. 146-50.

6. Galileo Galilei, *Sidereus Nuncius* (1610), in *Le Opere di Galileo Galilei*, ed. Antonio Favaro, Edizione Nazionale, 20 vols. (Florence, 1890-1909; reprinted 1929-39, 1964-6), vol. 3, p. 80. For English translation, see *Sidereus Nuncius or the Sidereal Messenger*, trans. Albert Van Helden (Chicago: University of Chicago Press, 1989), pp. 64-5. See also Galileo's letter to Antonio de' Medici (?), January 7, 1610, *Le Opere*, vol. 10, p. 277.

7. Johannes Kepler, *Narratio de observatis a se quatuor Iovis satellitibus erronibus* (1611), in *Johannes Kepler Gesammelte Werke*, vol. 4, *Kleinere Schriften 1602-1611* and *Dioptrice*, ed. Max Caspar and Frans Hammer (Munich: C. H. Beck, 1941), pp. 315-25.

8. John J. Roche, "Harriot, Galileo, and Jupiter's Satellites," *Archives Internationales d'Histoire des Sciences*, vol. 32 (1982), pp. 9-51.

9. MS. 1803, fols. 189$^r$-223$^r$, Bibliothèque Inguimbertine, Carpentras, France. See also Pierre Humbert, *Un amateur: Peiresc, 1580-1637* (Paris: Desclée de Brouwer et Cie, 1933), pp. 81-9; *idem*, "Joseph Gaultier de la Valette, astronome provençale (1564-1647)," *Revue d'Histoire des Sciences et de leurs Applications*, vol. 1 (1948), pp. 314-22.

10. Christophorus Clavius to Galileo, December 17, 1610, in Galileo, *Le Opere* (*op. cit.*, note 6), vol. 10, pp. 484-5.

11. Simon Marius, *Mundus Iovialis* (Nuremberg, 1614). See also Joachim Schlör, ed., *Die Welt des Jupiter* (Gunzenhausen, Germany: Schrenk-Verlag, 1988), and A. O. Prickard, "The Mundus Jovialis of Simon Marius," *The Observatory*, vol. 39 (1916), pp. 367-81, 403-12, 443-52, 498-503 (incomplete translation). For the arguments for and against Marius's independent discovery, see Joseph Klug, "Simon Marius aus Gunzenhausen und Galileo Galilei," *Abhandelungen der II. Klasse der Königlichen Akademie der Wissenschaften*, vol. 22 (1906), pp. 385-526; J. A. C. Oudemans and J. Bosscha, "Galilée et Marius," *Archives Néerlandaises des Sciences Exactes et Naturelles*, ser. 2, vol. 8 (1903), pp. 115-89; J. Bosscha, "Simon Marius, réhabilitation d'un Astronome Calomnié," *Archives Néerlandaises des Sciences Exactes et*

**Figure 4. Portrait of Nicolas-Claude Fabri de Peiresc, engraved in 1637 by Claude Mellan (1598-1688). 13.7 x 20.8 cm. (5.4 x 8.2 in.). Courtesy of the Musée Boucher de Perthes, Abbeville, France.**

immediately, and studies to determine the elements of their orbits so that tables of their motions could be prepared date from 1610. Peiresc observed Jupiter's moons from November 25, 1610, to June 21, 1612, and the journal of his observations is bound in the same volume of manuscripts that contains the first tables of their motions, prepared by his associate Gaultier. In these tables, which to all appearances were ready for publication, the satellites are assumed to have perfect circular motions about Jupiter and their planes are assumed to lie in the ecliptic.[12] Peiresc used these tables to make ephemerides of the positions of Jupiter's satellites. In 1612 a certain Jean Lombard made a trip to Marseilles, Malta, Cyprus, and Tripoli in Lebanon and observed the formations of Jupiter's satellites whenever he could.[13] But the mere formations of the satellites were found not to be adequate for the purpose of longitude. In his biography of Peiresc, Pierre Gassendi (1592-1655) wrote about these observations:

> ...but they did not sufficiently satisfie him; nor could he conceive, though all the Configurations of these Planets were set down in the Ephemerides, that the invention could prove so generall, as he had hoped. For he knew that Seafaring men could not make any observation, either in the day, or when the skie was cloudy, nor when *Jupiter* was in conjunction with the Sun, or when in the night he should be beneath the Earth; nor for half a year when they should be on one side of the world and *Jupiter* on the other; and such like Cases. Wherefore he laid that care aside, supposing that *Galileus* or *Kepler* at one time or other, would take this Charge upon them, and by their dexterity perfectly finish the same.[14]

In his book *Un amateur: Peiresc*, Pierre Humbert concludes his account of Peiresc's early longitude efforts by speculating that perhaps, during his arduous voyage, Jean Lombard first observed an eclipse of a satellite of Jupiter.[15] The record of Lombard's observations gives no hint that this was the case. But the records of Galileo's observations and calculations show that by 1612 he was observing eclipses.[16] It was at this time that the Florentine, in his calculations, began to take into account "prosthaphaeresis," the angle between the Earth and Sun as seen from Jupiter.[17] By September of that same year, having established the periods of the satellites and having calculated tables of their motions, Galileo proposed to the Spanish Crown to teach Spanish sailors to observe Jupiter's moons and determine longitudes by means of them. This was part of a larger proposal made by the Grand Duke of Tuscany (Cosimo II de' Medici, 1590-1621), one declined by Spain. Efforts to promote the use of Jupiter's satellites as a means of determining longitude at sea occupied Galileo from time to time for the rest of his life.[18]

In 1616, after he had been cautioned by the Inquisition not to defend the Copernican theory,

**Figure 5. Portrait of Galileo Galilei, painted in 1636 by Justus Sustermans (1597–1681). Oil on canvas. 66 x 56 cm. (26 x 22.1 in.). Courtesy of the Uffizi Gallery, Florence. By permission of Scala, Art Resource, ref. no. S0040381.**

*Naturelles*, ser. 2, vol. 12 (1907), pp. 258-307, 490-527; Ernst Zinner, "Zur Ehrenrettung des Simon Marius," *Vierteljahresschrift der Astronomischen Gesellschaft*, vol. 76 (1941), pp. 23-75.

12. MS. 1803, fols. $3^r$-$18^v$, Bibliothèque Inguimbertine, Carpentras, France. See also Suzanne Débarbat and Curtis Wilson, "The Galilean Satellites of Jupiter from Galileo to Cassini, Rømer and Bradley," in René Taton and Curtis Wilson, ed., *Planetary Astronomy from the Renaissance to the Rise of Astrophysics* (1989), pp. 144-57 [146-7], published as volume 2A in the series *The General History of Astronomy*, M. A. Hoskin, ed., 4 vols. (Cambridge: Cambridge University Press, 1984-).

13. MS. 1803, fols. $251^r$-$277^r$, Bibliothèque Inguimbertine, Carpentras, France.

14. Pierre Gassendi, *The mirrour of true nobility and gentility: being the life of the renowned Nicolaus Claudius Fabricius, Lord of Peiresk, Senator of the Parliament at Aix* (London, 1657), p. 147.

15. Humbert, *Un amateur* (*op. cit.*, note 9), p. 90.

16. Observation of December 18, 1612, in Galileo, *Le Opere* (*op. cit.*, note 6), vol. 3, p. 451; observations and calculations of March 18, April 4, and May 4, 1612, vol. 3, pp. 527, 530, 534.

17. *Ibid.*, vol. 3, pp. 521-42.

18. An excellent brief account of Galileo's negotiations with the Spanish Crown, lasting intermittently to the early 1630s, and with the Dutch States-General, from 1636 to 1640, can be found in Silvio A. Bedini, *The Pulse of Time: Galileo Galilei, the determination of longitude, and the pendulum clock*, Biblioteca di Nuncius, Studi e Testi, vol. 3 (Florence: Leo S. Olschki, 1991), pp. 7-21.

Galileo turned his full attention to the problem of longitude. In a proposal submitted to the Spanish Crown, he argued that the method of dead reckoning and sailing on one parallel was hopeless when a vast ocean separated the origin from the destination. Lunar eclipses were the better method and were responsible for improvements in cartography, although there were still very large errors on world maps, especially for the East Indies. Yet the problem with lunar eclipses were several. They were too infrequent to help sailors, and accurately timing the moment the Moon entered or exited the shadow of the Earth was impossible. Errors in time of more than a quarter of an hour, amounting to about 4° of longitude, were common. In contrast, the satellites of Jupiter experienced "a thousand" eclipses per year and could be timed with an accuracy of about one minute, which corresponds to an error in longitude of only 15'. Moreover, he argued, the tables of the Sun and Moon were imperfect, and a prediction of a lunar eclipse from them could easily err by a quarter of an hour. Such an error, added to errors in determining the local time of the eclipse, could result in an error in longitude of 8°. He went on: "It is therefore manifest that when our eclipses and other appearances will be subjected to, and regulated by, tables so exquisite that there will not be an error of one minute of an hour, then the entire matter will be, one may say, reduced to a total perfection, as far as our knowledge can go."[19]

Galileo proposed to make a hundred telescopes that magnified "forty or fifty times" and to bring these to Spain, distributing them among those who needed them. He would bring someone skilled in the method who would remain in Spain to teach others. He himself would annually deliver an ephemeris of the aspects of these moons, so that skilled navigators on the high seas would be able to determine their longitudes directly from observations. Finally, Galileo would produce a "copious and clearly explained discourse," so that future astronomers would be able to improve upon it.[20]

In his reply, Count Orso d'Elci (fl. 1608-d. 1636) went directly to the heart of the matter. The method was just not practical for use on board a ship on the high seas:

> From Your discourse, I understand that from the difference in times at which is seen the same aspect of those stars around Jupiter,

19. Galileo, *Le Opere* (*op. cit.*, note 6), vol. 5, pp. 420-1 [citation 421]. This document, not in Galileo's hand, is contained in his papers. Antonio Favaro argued that it was written by Galileo in 1612. See *Le Opere*, vol. 5, pp. 416-17, and Favaro's "Documenti inediti per la storia dei negoziati con la Spagna per la determinazione delle longitudini in mare," *Memorie del Reale Istituto Veneto di Scienze, Lettere ed Arti*, vol. 25 (1891), pp. 101-48 [104-5]. This volume was issued separately (same pagination) as *Nuove Studi Galileiani* (Venice: Tipografia Antonelli, 1891). Stillman Drake has included a complete translation of the document in *Galileo at Work: His Scientific Biography* (Chicago: University of Chicago Press, 1978), pp. 257-9. Drake has dated the document to 1616, and it certainly fits this period better than the earlier period.

20. Galileo to the Count d'Elci, November 13, 1616, in Galileo, *Le Opere* (*op. cit.*, note 6), vol. 12, pp. 292-3.

**Figure 6. Two telescopes attributed to Galileo. The one on top (inv. no. 2427) is 136 cm. (53.4 in.) long and is made of wood covered with paper. The one below (inv. no. 2428) is 92 cm. (36.2 in.) long and made of wood covered with leather. Courtesy of the Istituto e Museo di Storia della Scienza, Florence.**

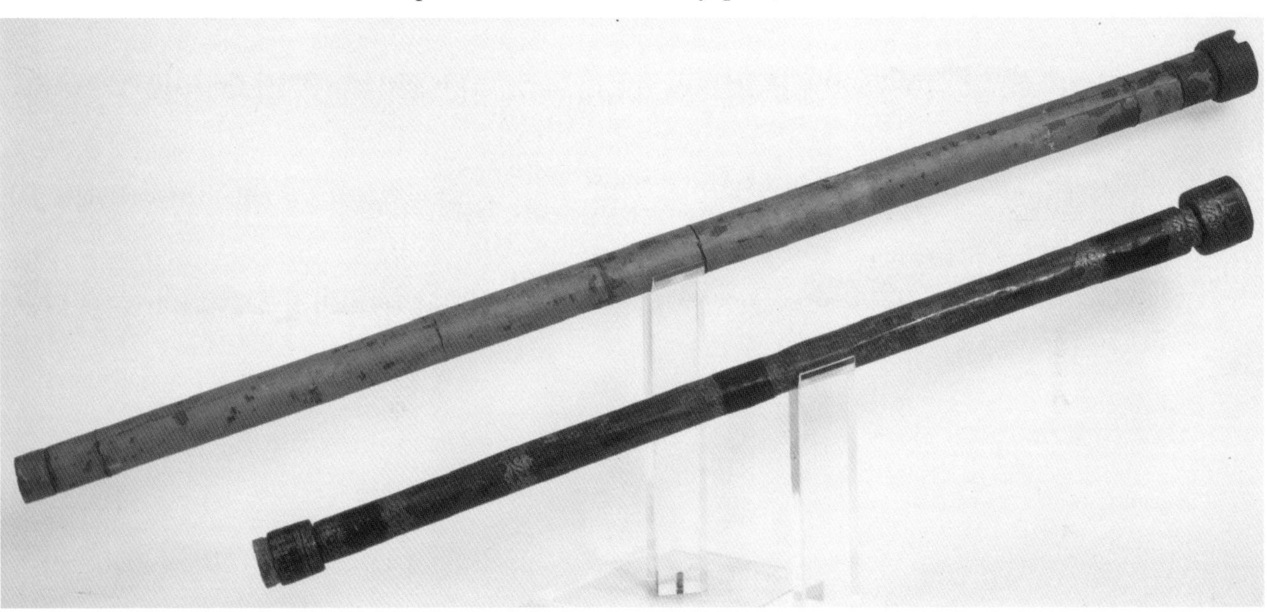

the true longitudes which those cities or places have with respect to each other will quickly be known. But in order to know this, it is compulsory and necessary first to see the said stars and their aspects. I don't know how this can be done at sea, or at least as frequently and quickly as is necessary for the person who navigates. For, leaving aside that the telescope cannot be used in ships because of their motion, even if it could be used it could serve neither during the day nor during overcast weather at night, because the stars are not visible, and the navigator needs to know hour by hour the degree of longitude at which he is.[21]

From these remarks we must not judge that Galileo ignored the practical problems. The opposite is closer to the mark. First, although accurate tables that could predict satellite eclipses a long period in advance did not exist in 1616, Galileo had been making observations and calculations for five years, and he could predict their motions in the near future. Second, he was directly confronting the problem of making observations on a ship. He had designed a fitted helmet with an attached telescope, which he called *"celatone."* One of these devices had been made in the Grand Duke's arsenal, and Galileo tried it on the deck of one of the galleys moored in the harbor at Livorno. A few months later, his friend and student Benedetto Castelli (1578-1643) tried it out at sea. This instrument employed a low-powered telescope, inadequate for observing Jupiter's moons, but the *celatone* made the use of the Galilean telescope possible on board ship.[22] The military commander Giovanni de' Medici (1567-1621), no friend of Galileo's, "judged this invention more important than the discovery of the telescope itself."[23]

It should be realized that the Galilean telescopes (with their concave oculars), in use before about 1650, were not optimum for those observations. Their fields of view were extremely small, so that it was difficult to find the planet and its satellites and almost as difficult to keep it in the field of view. Moreover, as Galileo's own early efforts show, telescope lenses were not exactly interchangeable parts, and instruments that could show the satellites were not off-the-shelf items. It is to be doubted that in 1616 there were more than a few telescopes in Spain that could show the satellites of Jupiter.[24] As late as 1637, during Galileo's negotiations with the Dutch Republic, Martinus Hortensius (1605-1639) admitted that the best telescopes made in the Netherlands showed Jupiter's disk "hairy" and not well defined, so that the moons could not be observed accurately when close to the planet.[25]

In his reply to the Count of Elci, Galileo tried his best to dispose of the practical objections against his proposed method and made a plea that inventors have made through the ages:

> ...this is an entire art and still developing, founded on new principles and methods, but worthy and most noble; and it needs to be embraced, cultivated, and fostered, so that with practice and with time those fruits are drawn from it of which it contains the seeds and roots.[26]

Both Peiresc and Galileo retained their interest in the problem of longitude for the rest of their lives. From 1633 to 1636 Peiresc operated a veritable "Bureau of Longitudes," recruiting observers in remote places to observe

21. The Count d'Elci to Curzio Picchena, November 30, 1617, *ibid.*, vol. 12, p. 353.

22. Annibale Guiducci to Galileo, September 11, 1617, *ibid.*, vol. 12, p. 344; Benedetto Castelli to Galileo, September 18, 1617, *ibid.*, vol. 12, p. 346.

23. Benedetto Castelli to Galileo, February 7, 1618, *ibid.*, vol. 12, p. 372.

24. In March 1610, Galileo wrote to the Grand Duke's secretary that of the 60 telescopes he had made, only a handful were good enough to show his discoveries (*ibid.,* vol. 10, p. 301). The skill of making good telescopes spread only slowly through Europe.

25. Hortensius to Galileo, January 26, 1637, *ibid.,* vol. 17, p. 18.

26. Galileo to the Count d'Elci, December 26, 1617, *ibid.,* vol. 12, p. 361.

eclipses of the Sun and Moon;[27] Galileo did not give up his efforts to convince the Spanish Crown until about 1630 and then, in 1636, entered negotiations with the States-General of the Dutch Republic on the same subject, this time for a rumored prize of 30,000 florins.[28] Neither Peiresc nor Galileo, however, published their proposals for using Jupiter's moons for determining longitude. Simon Marius, the first to publish tables of the satellites, did not mention the longitude connection.[29] Vincenzo Renieri (1606-1647), who took over the observations from his master when Galileo became blind, announced in his planetary tables of 1639 that he would soon publish tables of the motions of Jupiter's moons that would be useful for determining longitude at sea.[30] This was the first mention of this method in print.

A few years later, Pierre Hérigone (d. *ca.* 1643)[31] mentioned the method in his popular *Cursus mathematicus.* Hérigone was aware that Galileo had made proposals about this method, but he apparently knew no details. Protesting that he had had the idea independently two years before he heard about Galileo's efforts, he pointed to conjunctions of the moons with the planet as the events to be timed.[32] Since he was not a practicing astronomer, he was not aware that the timing of these events depended heavily on the quality of the telescope.

In his influential *Selenographia* of 1647, in which he systematized the timing of lunar eclipses, Johannes Hevelius (1611-1687) reviewed the observations and arguments about the satellites of Jupiter but only briefly mentioned the use of their conjunctions with each other and with the planet for determining longitudes.[33] In his *Almagestum Novum* of 1651, Giovanni Battista Riccioli (1598-1671) mentioned (besides Hérigone and Renieri) Galileo's efforts, about which he had heard from Michel Florent van Langren (*ca.* 1600-1675), the Spanish king's cosmographer, and Galileo's pupil, Buonaventura Cavalieri (*ca.* 1598-1647), who, like Riccioli, taught in Bologna.[34] Riccioli was not very impressed with this method, giving it only slight coverage; in his *Geographia et Hydrographia Reformata* of 1661 (second edition 1672), he clearly thought it was much less promising than the method of magnetic deviation.[35]

In the meantime Giovanni Battista Hodierna (1597-1660) in Palermo had taken up the project and published his tables in 1656. Both Galileo and Simon Marius (in his tables of 1614) took prosthaphaeresis (or the annual parallax of the Earth's or Sun's orbit) into account, but both erred on the latitudes of the satellites. They also assumed that with respect to some reference system the motions were uniform and circular. Hodierna's tables were a signal improvement over those of Marius. He referred Jupiter's motion to the Sun, so that synodic periods should be calculated with respect to the midpoint of Jupiter's shadow. Hodierna's latitudes were also much improved, and his tables included ephemerides of the central times of satellite eclipses from 1650 to 1682. It soon became apparent, however, that these tables were still not sufficiently accurate.[36] At about the same time, Laurence Rooke (1622-1662) took up the project in London, but he died before his efforts could bear fruit.[37] By the 1660s, astronomers were still not using the satellites of Jupiter in their efforts to determine longitude differences between the locations of their observatories scattered over Europe; lunar eclipses remained the preferred method.

27. Humbert, *Un amateur* (*op. cit.,* note 9), pp. 217-31.

28. Antonio Favaro, "Documenti inediti per la Storia dei Negoziati con gli Stati Generali d'Olanda per la Determinazione dell Longitudini," *Memorie del Reale Istituto Veneto di Scienze, Lettere ed Arti,* vol. 25 (1891), pp. 289-338; G. Vanpaemel, "Science Disdained: Galileo and the Problem of Longitude," in C. S. Maffeoli and L. C. Palm, eds., *Italian Scientists in the Low Countries in the XVIIth and XVIIIth Centuries* (Amsterdam: Rodopi, 1989), pp. 111-29. See also note 18.

29. Marius, *Mundus Iovialis* (*op. cit.,* note 11). The tables are included in the German but not in the English translation.

30. Giovanni Battista Riccioli, *Almagestum Novum* (Bologna, 1651), part 2, p. 610.

31. Literary name used by Clément Cyriaque de Mangin.

32. Pierre Hérigone, *Cursus mathematicus nova, brevi et clara methodo demonstratus,* 6 vols. (Paris, 1634-42), vol. 5, p. 872.

33. Johannes Hevelius, *Selenographia* (Gdansk, 1647), p. 46.

34. Riccioli, *Almagestum* (*op. cit.,* note 30).

35. Giovanni Battista Riccioli, *Geographia et Hydrographia Reformata,* 2nd ed. (Venice, 1672), pp. 317, 326-8.

36. See Débarbat and Wilson, "Galilean Satellites" (*op. cit.,* note 12), pp. 147-8.

37. Colin A. Ronan, "Laurence Rooke (1622-1662)," *Notes and Records of the Royal Society of London,* vol. 15 (1960), pp. 113-18.

38. Giovanni Domenico Cassini, *Ephemerides Bononienses Mediceorum Syderum ex Hypothesibus, et Tabulis Io: Dominici Cassini* (Bologna, 1668).

39. Jean Picard, *Voyage d'Uranibourg, ou Observations Astronomiques Faites en Dannemarck par Monsieur Picard de l'Académie des Sciences* (Paris, 1680), in *Memoires de l'Académie Royale des Sciences. Depuis 1666. jusqu'à 1699* (Paris, 1729), vol. 7, part 1, p. 227.

40. It is interesting to compare Cassini's prediction for the immersion of the first satellite on October 25, 1671, with Picard's actual observation:
Picard's observation at Uraniborg: 6:57:20 a.m.
Cassini's calculation for Uraniborg: 6:41 a.m.
Cassini's observation in Paris: 6:15 a.m.
Cassini made this prediction before Picard had measured the longitudinal difference between Paris and Uraniborg, but from Picard's results that difference was 42 minutes, and Kepler and Riccioli had estimated it to be 40 and 45½ minutes, respectively. Of the sixteen-minute difference between prediction and observation, no more than a few minutes can, therefore, be ascribed to errors in longitude: in 1671 Cassini's tables erred by as much as fifteen minutes for the first (and most accurately determined) satellite. This is confirmed by the errors that led Ole Rømer (1644–1710) to formulate his thoughts on the speed of light, five years later. Clearly, only simultaneous observations could lead to useful results at this point.

41. Johannes Hevelius, "Occultatio Primi Jovialium ab umbra Jovis," *Philosophical Transactions* of the Royal Society, London *(Phil. Trans.),* vol. 6, no. 78 (1671), p. 3030.

## GIOVANNI DOMENICO CASSINI (1625-1712)

**Figure 7. Portrait of Giovanni Domenico Cassini, *ca.* 1690. Artist unknown. Oil on canvas. 81 x 67 cm. (31.8 x 26.4 in.). Courtesy of Civica Biblioteca Aprosiana, Ventimiglia, Italy.**

Giovanni Cassini was 25 years old in 1650, when he became professor of astronomy at the University of Bologna. His first research project was on the connection between refraction and parallax, and in this area he made his most important contributions to astronomy. In the mid-1660s, however, he came into contact with the Roman telescope maker Giuseppe Campani (1635-1715), in whose hands the telescope was raised to a new level of perfection. With a Campani telescope, Cassini made a series of spectacular discoveries of surface markings and rotations of both Mars and Jupiter and of the transits of Jupiter's satellites across the planet's disk. He observed the satellites for a number of years and published his tables in 1668 (Figure 8).[38] It is from this publication that the method dates. Yet, although Cassini's tables were an improvement over previous efforts, they were by no means perfect. The use of satellite eclipses had as much to do with Cassini's powerful position as it did with the quality of the tables.

Cassini moved to Paris with his tables in 1669. There he quickly took charge of the astronomical endeavors of the Académie Royale des Sciences. Although his tables were by no means accurate enough to allow observers in remote places to determine their longitudes from them, they did allow the observer to know roughly when eclipses would happen, so that observations could be planned. The Académie had the funds to give these efforts the necessary organization. The first celebrated instance of this was on the expedition made in 1671-2 by Jean Picard (1620-1682) to Denmark to redetermine the longitude of the observatory of Tycho Brahe (1546-1601). Predictions of satellite eclipses were circulated as Picard made his way to Denmark, where he observed a number of eclipses, five of which were observed simultaneously by Cassini and his staff in Paris. Picard became a convert to Cassini's method and wrote in his report:

> ...besides the fact that eclipses of the Moon are not frequent, there is nothing more convenient and more precise for the discovery of longitude than the observations of the first satellite of Jupiter, either when that satellite is eclipsed while plunging itself into the shadow of Jupiter, or when it comes out of it and begins to recover its brightness, for this happens with respect to us so quickly that during clear weather, with a telescope of from 14 to 20 feet, one can be sure of the quality of the observation to the nearest few seconds [à peu de secondes prés]; added to the fact that by means of the tables that Mr. Cassini has provided one can easily forecast the observations that are to be made and to confine oneself to them.[39]

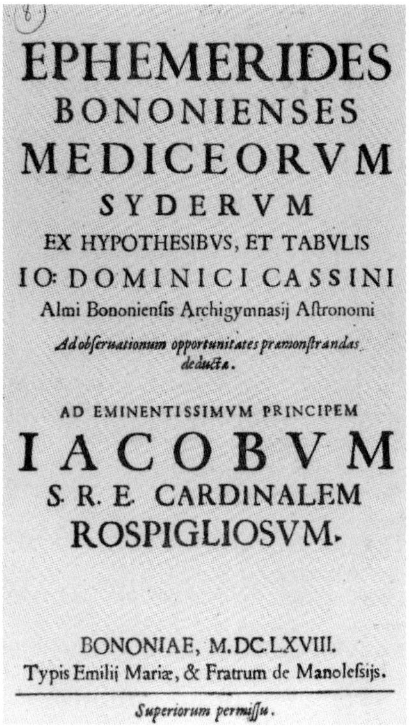

Whereas in the timing of these eclipses Galileo had claimed an accuracy of about a minute, five decades later Picard claimed an accuracy of a few seconds.[40] But where Galileo was speaking about tables, Picard, more realistically, was speaking about simultaneous observations. Others were not so impressed with the method. Johannes Hevelius, one of Europe's foremost observers, who had observed the eclipses at the same time, wrote: "Whether these phenomena are of equal use in determining the difference between meridians than occultations of fixed stars by the Moon, I strongly doubt."[41] Among the French astronomers, the method was confirmed two years later

EPHEMERIDES
BONONIENSES
MEDICEORVM
SYDERVM
EX HYPOTHESIBVS, ET TABVLIS
IO: DOMINICI CASSINI
Almi Bononiensis Archigymnasij Astronomi
*Ad observationum opportunitates praemonstrandas deducta.*
AD EMINENTISSIMVM PRINCIPEM
IACOBVM
S. R. E. CARDINALEM
ROSPIGLIOSVM.

BONONIAE, M.DC.LXVIII.
Typis Emilij Mariae, & Fratrum de Manolessijs.
*Superiorum permissu.*

**Figure 8. Title page of Cassini's *Ephemerides Bononienses Mediceorum Syderum...* (Bologna, 1668). Courtesy of the Observatoire de Paris.**

during the expedition of Jean Richer (1630-1696) to Cayenne, the longitude of which from Paris was found to be 3 hours 29 minutes by simultaneous observations of a lunar eclipse and a conjunction of the first satellite with Jupiter.[42] In the meantime, the research of Cassini and his associates at the Paris Observatory on the theory of the satellites continued. In 1674 Ole Rømer (1644-1710) found the systematic error in the tables of the first satellite that led to his theory about the speed of light. In 1676 Rømer's theory was confirmed by observations,[43] and the following year Cassini was able to verify his own hypothesis concerning the latitudes of the satellites when they appeared exactly in a straight line at the time he had predicted.[44]

Under Cassini's guidance, the method continued to be refined and checked. Cassini wrote:

> It was not until after a large number of experiments made in observing these satellites in concert with other observers, first in the same place and then in two places remote from each other, that we found out which are the most appropriate phases for determining longitudes. These experiments informed us that to be preferred to all other phases are the eclipses that the satellites suffer in passing through the shadow of Jupiter of which one can see the entry and the exit, and sometimes both, without two observers differing between them by a quarter of a minute of an hour (which is an exactitude much greater than which one could obtain previously by means of eclipses of the Moon), and that the eclipses of the first satellite, which is faster than the others and which enters more directly into the shadow, can be determined with an even greater precision; that after these eclipses of the satellites one can use their apparent conjunctions with Jupiter and each other, and particularly when they encounter each other going in opposite directions; and that the shadows that they cast on the disk of Jupiter when they pass between the planet and the Sun, which we discovered to be frequently very visible, are useful in this design, as are also the permanent spots which often appear on the surface of Jupiter and which make a revolution around it which is the most rapid that we have thus far discovered in the heavens, although the instant of the passage of these spots through the middle of Jupiter cannot be determined with the same subtlety as the instant of the eclipses of the satellites.[45]

In their initial mapping of the coastal regions of France, Picard, Cassini, and their colleagues relied exclusively on eclipses of Jupiter's moons for their longitude determinations. In 1693, the Académie published a map designed to show in a single figure the coastlines of France according to the results of Picard and Philippe de La Hire (1640-1718), compared to the best previous results, the map of Sanson that was presented to the Dauphin in 1679 (Figure 9). The legend about this map is that, upon seeing it, King Louis XIV complained that he had lost more territory to his astronomers than to his enemies. The position of Brest was moved to the east by no less than an entire degree, or some 50 miles.[46] It was now clear that the satellites of Jupiter were the key to the reform of astronomy, and they became the royal means of establishing longitudes and reforming geography. As Cassini phrased it,

42. Giovanni Domenico Cassini, *Les Elemens de l'Astronomie Verifiez par Monsieur Cassini par rapport de ses Tables aux Observations de M. Richer faites en l'Isle de Cayenne* (1684), in *Memoires de l'Académie Royale des Sciences. Depuis 1666. jusqu'à 1699* (Paris, 1730), vol. 8, pp. 55-79 [69-70].

43. See Débarbat and Wilson, "Galilean Satellites" (*op. cit.*, note 12), pp. 152-4.

44. Giovanni Domenico Cassini, "An Extract of a Letter written by Signor Cassini to the Author of the Journal des Scavans, containing some Advertisements to Astronomers about the Configurations, by him given of the Satellites of Jupiter, for the years 1676, and 1677, for the verification of their Hypotheses," *Phil. Trans.*, vol. 11, no. 128 (1676), pp. 681-3.

45. Giovanni Domenico Cassini, *Les Hypotheses et les Tables des Satellites de Jupiter Reformées sur de Nouvelles Observations. Par Monsieur Cassini de l'Académie Royale des Sciences* (1693), in *Memoires de l'Académie* (*op. cit.*, note 42), pp. 321-2.

46. Jean Picard and Philippe de la Hire, "Pour la Carte de France corrigée sur les Observations de MM. Picard & de la Hire." This passage was first published in *Recueil d'Observations Faites en Plusieurs Voyages par Ordre de Sa Majesté, pour Perfectionner l'Astronomie et la Geographie...* (Paris, 1693), pp. 91-2 of the third section concerning observations made by Picard and de La Hire. The map is mounted on a stub bound into the volume following this passage. It was later published in *Memoires de l'Académie* (*op. cit.*, note 39), pp. 429-30, with the map bound in as a fold-out plate opposite p. 1.

47. Giovanni Domenico Cassini, *Les Elemens de l'Astronomie Verifiez par Monsieur Cassini par le rapport de ses Tables aux Observations de M. Richer faites en l'Isle de Caïenne. Avec les Observations de MM. Varin, des Hayes, et de Glos faites en Afrique & en Amerique* (Paris, 1684), in *Recueil d'Observations* (*ibid.*), p. 51. This work was later published in *Memoires de l'Académie* (*op. cit.*, note 42). Note that the appendix concerning the observations of Varin *et al.* is not included in this later reprint.

48. *Ibid.*, pp. 52-8.

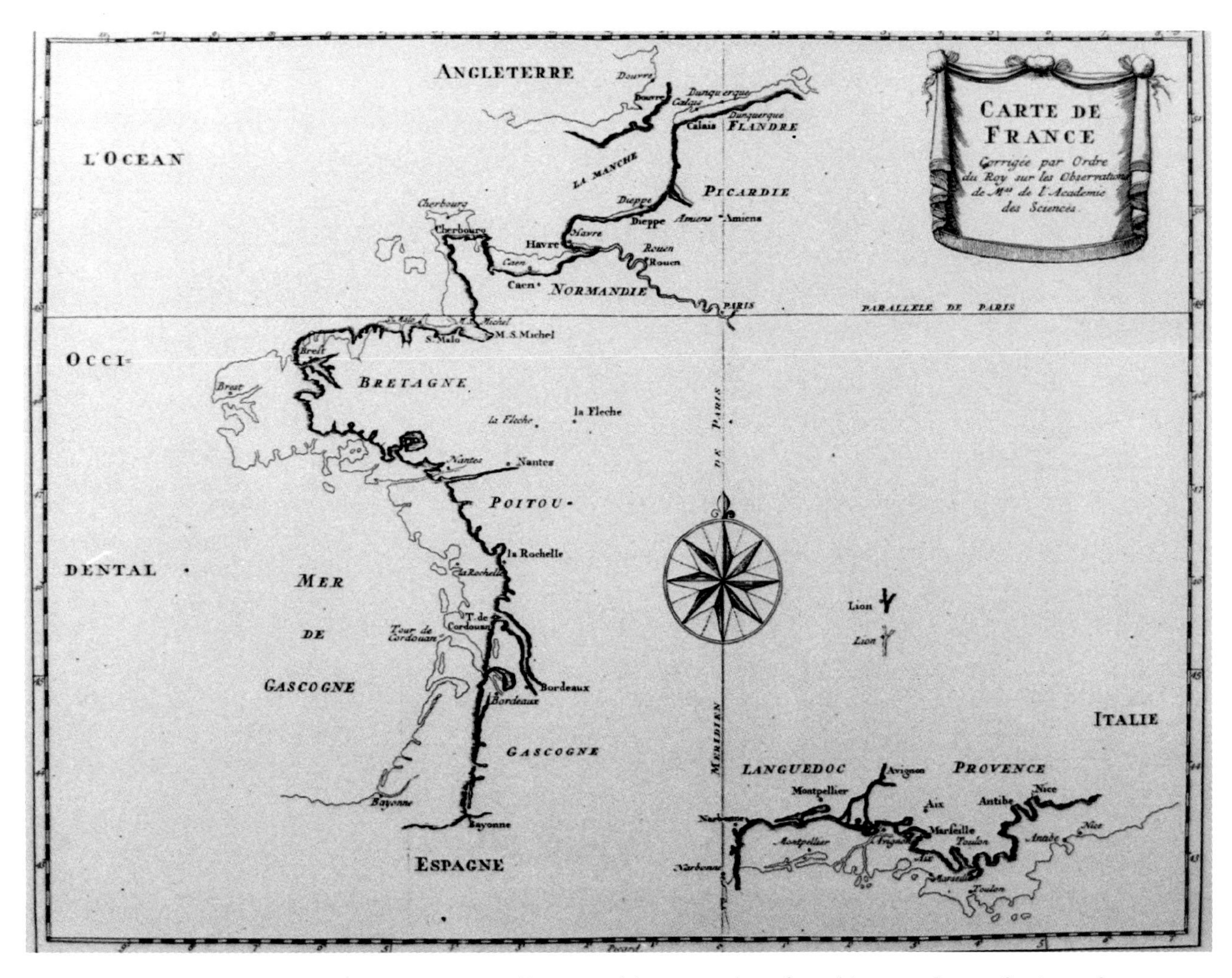


Figure 9. Map of the coastline of France, showing the corrections from the surveys of Jean Picard and Philippe de La Hire (shaded line), superimposed on the map by Sanson that was presented to the Dauphin in 1679 (fine line). From Jean Picard and Philippe de La Hire, "Pour la Carte de France corrigée sur les Observations de MM. Picard & de la Hire," published in *Recueil d'Observations...* (Paris, 1693).

> Because this enterprise of working on the perfection of geography in a way that is new and more perfect than those that had been imagined until this time, conformed to the intent of His Majesty in founding his Academy of Sciences, he ordered that persons be chosen capable of executing it in various places following the instructions that were to be given to them, and that suitable occasions be found for sending them to distant lands.[47]

This statement makes all the more sense when we remember that La Salle had just claimed a large part of the New World, the Louisiana territory, for the French Crown and had only a vague idea of its geographical extent. At the same time that Cassini was writing this, La Salle was entering Matagorda Bay thinking he was near the mouth of the Mississippi.

When, in 1681, the Académie sent three observers to Gorée, a small island in the Cape Verde group, they were first trained in the necessary observations and then given a set of "General Instructions for Geographical and Astronomical Observations to be made during Voyages."[48] Cassini devoted about half of this document to adjusting the clocks so as to be able accurately to measure the time of an immersion or emersion. Then there were instructions for measuring meridian altitudes of the Sun and fixed stars, and finally on timing eclipses of Jupiter's moons:

> The most suitable observations for determining longitude are the immersions and emersions of the first satellite of Jupiter into or from

its shadow. Before total immersion one sees it diminish little by little. If possible, one should count the seconds of time that pass from the time when one begins clearly to see it diminish until it disappears entirely. At the instant it disappears, one must begin to count anew, and if it is found that after having begun counting it still appears, which happens sometimes, one should recommence counting when it ceases to appear. And when one is assured that it does not appear any longer, one continues counting until one sees that the clock marks the seconds. Then one subtracts from it what one has counted since the last time that the satellite will have disappeared, which is to be noted principally. And if one remembers the interval between the sensible diminution and the total occultation, one can note this as well; otherwise one should not give oneself the trouble for they are only for marking how precise the observation is.[49]

There were similar instructions for timing the emersion, the start of which was even more difficult to determine. Cassini also pointed out that for eight days before and after Jupiter's opposition, immersions and emersions of the first satellite could not be timed accurately because they happened too close to the planet. He went on to mention conjunctions of satellites with Jupiter and each other, giving instructions in each case on how to time these events. Almost as an afterthought he mentioned the method of lunar distance: "When the Moon can pass through the telescope of the stationary quadrant with a principal fixed star, either before or after it, it will be useful to observe the times of passage of both limbs of the Moon and the center of the star."[50]

By the 1680s, then, the satellites of Jupiter had become the official method of longitude determination of the French Crown and the Académie, and it was being routinized as much as could be done. When, in 1693, Cassini published revised tables,[51] they were based on almost three decades of observations. The accuracy of the motion of the first satellite was now good enough to allow the explorer who was a conscientious observer to determine his longitude with an error of less than a degree. The motions of the other satellites still left a great deal to be desired.[52] In 1690, the Académie began publishing *Connoissance des Temps,* the annual issues of which contained predictions of the eclipses of all four satellites, daily configurations of the satellites (when they were visible), and instructions for their observations.

The instructions were addressed to non-experts. In the eighteenth century, it would appear that almost anyone could make the observations: "A pendulum [clock], a simple telescope of 15 to 18 feet, and a quadrant that anyone can make out of wood without difficulty or skill, suffice for making very good observations of the satellites of Jupiter."[53] The instructions were limited to some information on where Jupiter's shadow is located before and after opposition and the use of the configuration diagrams. As far as the observations themselves were concerned, the reader was told only:

For finding the longitude of any place on Earth, it suffices to observe any immersion or emersion. One compares the true time of the observation with the hour and minute of the same immersion or emersion calculated for Paris, or observed there the same day. The difference in time, reduced to degrees, minutes, and seconds, will

49. *Ibid.*, pp. 56-7.

50. *Ibid.*, p. 57.

51. Cassini, *Les Hypotheses* (*op. cit.*, note 45), vol. 8, pp. 317-505.

52. Rather than accepting Rømer's explanation of the retardation of the eclipses of the first satellite, the finite velocity of light, Cassini accounted for the correction by incorporating an annual variation depending on the angular distance between Jupiter and the Sun. The maximum this correction attains is 2°. While the correction works fine for the first satellite, it does not correspond to the phenomena of the others. See Débarbat and Wilson, "Galilean Satellites" (*op. cit.*, note 12), pp. 152-4.

53. *Connoissance des Temps, Pour l'Année Commune 1773....* (Paris, 1771), p. 236.

54. *Ibid.*, p. 237.

55. *Ibid.*, p. 237.

56. Ronan, "Laurence Rooke" (*op. cit.*, note 37).

57. Laurence Rooke, "Discourse Concerning the Observations of the Eclipses of Jupiter's Satellites," in Thomas Sprat, *The History of the Royal-Society of London, for the Improving of Natural Knowledge* (London, 1667), pp. 183-9.

58. *Ibid.*, p. 183.

59. John Flamsteed, "An account of the Eclipses or ingresses of Jupiters Satellits into his shadow and such Emersions of them from it as will be visible at the Observatory at Greenwich in the three last Months of this year 1683...," *Phil. Trans.*, vol. 13, no. 151 (1683), pp. 322-3.

60. John Flamsteed, "A Letter from Mr. Flamsteed concerning the Eclipses of Saturns [misprint; should be Jupiter's] Satellit's for the year following 1684,...," *Phil. Trans.*, vol. 13, no. 154 (1683), pp. 404-15.

be the difference between the meridian of this place and the meridian of Paris.[54]

There was one caveat, however: the observer was not to expect adequate precision from the calculated eclipse times, except in the case of the first satellite. The eclipses of the three other satellites could not, even in the 1770s, be predicted with a precision of one minute of time.[55]

## ENGLISH EFFORTS

One of the founding members of the Royal Society, Laurence Rooke, Gresham Professor of Astronomy and then Geometry,[56] made observations of Jupiter's satellites at Gresham College but had not progressed to tables by the time of his death in 1662. A short, general discourse on this subject was published by Thomas Sprat (1635-1713) in his *History of the Royal-Society*.[57] Rooke was not optimistic about the use of satellite eclipses for navigation because they required long telescopes that could not be used on board ship: "A nautical science of longitude is scarcely to be hoped for from the heavens; the geographical science of longitude, however, is especially to be sought by means of eclipses of celestial bodies."[58] Rooke's efforts were not followed up for two decades.

**Figure 10. Portrait of John Flamsteed, painted in 1712 by Thomas Gibson (*ca.* 1680-1751). Oil on canvas. 124.5 x 101.6 cm. (49 x 40 in.). By permission of the President and Council of the Royal Society, London.**

In the early 1680s, Cassini's counterpart in England, John Flamsteed (1646-1719), began promoting the use of satellite eclipses for longitude determinations as well. In the summer of 1683, Flamsteed finished new tables of the motions of Jupiter's satellites, and in September of that year he published his predictions (Figure 11). He was confident that the eclipses of the first satellite would err no more than five minutes, those of the third "but little more," but for those of the second and fourth he made no promises, for he found their motions "evidently intangled with inequalitys," which could only be determined by long observation.[59] Later that year, he published a "catalogue" of eclipses for 1684 in the *Philosophical Transactions* and gave some very helpful instructions for observers.[60] But Flamsteed took sailors to task for ignoring this method. Even if it could not be used on the high sea (and Flamsteed was not convinced of this), sailors could use it to chart the shores. His blast at these unwilling wretches illustrates the Puritan approach to finding longitude:

> And I must confess it is some part of my design, to make our more knowing Seamen ashamed of that refuge of Ignorance, their Idle and Impudent assertion *that the longitude is not to be found*, by offering them an expedient that will assuredly afford it, if their Ignorance,

Sloth, Covetousness, or Ill-nature, forbid them not to make use of what is proposed.[61]

Berating sailors was, of course, no substitute for the social organization necessary to make this method a success. Only astronomers responded to his call, especially after Flamsteed issued the predictions for 1685 in Latin for "the benefit of Foreigners."[62] He went on to publish catalogues for 1686 (English), 1687 (English), and 1688 (Latin).[63] In his ephemeris for 1687 he assessed the accuracy of the predicted times as follows:

> In my last observations, the Eclipses of the 2d and 3d Satellit have anticipated my Calculations something more than I expected in so short a time; of which I thought it convenient to acquaint the Reader, that hee may attend them one quarter of an hour earlyer

*A* Catalogue *of the* Viſible Eclipſes *of* ♄ Satellits, *ſhewing the apparent times of their* Ingreſſes *into* ♃ *ſhadow and* Emerſions, *from it under the* Meridian *of the* Obſervatory *in the year 1684. Calculated from new Tables of their* Motions. *by* John Flamſteed M. *R & R. S. S.*

1684

| *Jan.* | h ' | | | |
|---|---|---|---|---|
| ♂ 1 | 13-03 | 1 | * | i |
| ☿ 2 | 23-30 | 3 | | i |
| ♃ 3 | 7-31 | 1 | | i |
| | 19-09 | 2 | | i |
| ♄ 5 | 1-58 | 1 | | i |
| ☉ 6 | 12-29 | 4 | * | i |
| | 15-39 | 4 | * | e |
| | 20-26 | 1 | | i |
| ☽ 7 | 8 25 | 2 | | i |
| ♂ 8 | 14-54 | 1 | * | i |
| ♃ 10 | 3-36 | 3 | | i |
| | 10-22 | 1 | * | i |
| | 21-42 | 2 | | i |
| ♄ 12 | 4-49 | 1 | | i |
| ☉ 13 | 22-17 | 1 | | i |
| ☾ 14 | 10-58 | 2 | * | i |
| ♂ 15 | 16-40 | 1 | * | i |
| ♃ 17 | 7-22 | 3 | | i |
| | 11-13 | 1 | * | i |
| ♀ 18 | 0-16 | 2 | | i |
| ♄ 19 | 5-41 | 1 | | i |
| ☾ 21 | 0-10 | 1 | | i |
| | 13-32 | 2 | * | i |
| ♂ 22 | 18-38 | 1 | * | i |
| ☿ 23 | 6-27 | 4 | | i |
| | 9 2[illegible] | 4 | * | e |
| ♃ 24 | 11-19 | 3 | * | i |
| | 13-06 | 1 | * | i |
| ♀ 25 | 2-47 | 2 | | i |
| ♄ 26 | 7-34 | 1 | | i |
| ☾ 28 | 2-0[illegible] | 1 | | i |
| | 16-05 | 2 | * | i |
| ♂ 29 | 21-30 | 1 | | i |
| ♃ 31 | 14-5 | 1 | * | i |
| | 15-16 | 3 | * | i |

| *Feb.* | h ' | | | |
|---|---|---|---|---|
| ♀ 1 | 5 22 | 2 | | i |
| ♄ 2 | 9-47 | 1 | * | i |
| ☾ 4 | 3-[illegible]6 | 1 | | i |
| | 19-39 | 2 | | i |
| ♂ 5 | 22-22 | 1 | | i |
| ♃ 7 | 16-53 | 1 | * | i |
| | 19-13 | 3 | | i |
| ♀ 8 | 7-57 | 2 | * | i |
| ♄ 9 | 0-26 | 4 | | i |
| | 3-19 | 4 | | e |
| | 11-21 | 1 | * | i |
| ☾ 11 | 5-50 | 1 | | i |
| | 21-15 | 2 | | i |
| ☿ 13 | 0-19 | 1 | | i |
| ♃ 14 | 18-47 | 1 | | i |
| | 23-13 | 3 | | i |
| ♀ 15 | 10-34 | 2 | * | i |
| ♄ 16 | 13-16 | 1 | * | i |
| ☾ 18 | 7-45 | 1 | * | i |
| | 23-52 | 2 | | i |
| ☿ 20 | 2-14 | 1 | | i |
| ♃ 21 | 20-42 | 1 | | i |
| ♀ 22 | 3-13 | 3 | | i |
| | 13-11 | 2 | * | i |
| ♄ 23 | 15-12 | 1 | * | i |
| ☾ 25 | 9-41 | 1 | * | i |
| ♂ 26 | ☍ ☉ ♃ | | | |
| ☿ 27 | 6-20 | 1 | | e |
| ♀ 29 | 0-55 | 1 | | e |
| | 10-24 | 3 | * | e |
| | 18-35 | 2 | | e |

| *Mar.* | h ' | | | |
|---|---|---|---|---|
| ♄ 1 | 19-24 | 1 | | e |
| ☾ 3 | 13-53 | 1 | * | e |
| ♂ 4 | 7-54 | 2 | * | e |
| ☿ 5 | 8-22 | 1 | * | e |
| ♀ 7 | 2-51 | 1 | | e |
| | 14-24 | 3 | * | e |
| | 21-12 | 2 | | e |
| ♄ 8 | 21-20 | 1 | | e |
| ☾ 10 | 15-50 | 1 | * | e |
| ♂ 11 | 10-30 | 2 | * | e |
| ☿ 12 | 11-19 | 1 | * | e |
| ♃ 13 | 15-13 | 4 | * | e |
| ♀ 14 | 4-48 | 1 | | e |
| | 18-25 | 3 | | e |
| | 23-49 | 2 | | e |
| ♄ 15 | 23-17 | 1 | | e |
| ☾ 17 | 17-40 | 1 | | e |
| ♂ 18 | 13-09 | 2 | * | e |
| ☿ 19 | 12-15 | 1 | * | e |
| ♀ 21 | 6-44 | 1 | * | e |
| | 22-26 | 3 | | e |
| ♄ 22 | 2-28 | 2 | | e |
| ☉ 23 | 1-13 | 1 | | e |
| ☾ 24 | 19-43 | 1 | | e |
| ♂ 25 | 15-46 | 2 | * | e |
| ☿ 26 | 14-12 | 1 | * | e |
| ♀ 28 | 8-41 | 1 | * | e |
| ♄ 29 | 2-26 | 3 | | e |
| | 5-05 | 2 | | e |
| ☉ 30 | 3-10 | 1 | | e |
| | 9-09 | 4 | * | e |
| ☾ 31 | 21-39 | 1 | | e |

| *Apr.* | h ' | | | |
|---|---|---|---|---|
| ♂ 1 | 18-23 | 2 | | e |
| ☿ 2 | 16-09 | 1 | | e |
| ♀ 4 | 10-38 | 1 | * | e |
| ♄ 5 | 6-26 | 3 | | e |
| | 7-42 | 2 | * | e |
| ☉ 6 | 5-07 | 1 | | e |
| ☾ 7 | 23-36 | 1 | | e |
| ♂ 8 | 21-01 | 2 | | e |
| ☿ 9 | 18-05 | 1 | | e |
| ♀ 11 | 12-34 | 1 | * | e |
| ♄ 12 | 10-19 | 2 | * | e |
| | 10-26 | 3 | * | e |
| ☉ 13 | 7-03 | 1 | | e |
| ♂ 15 | 1-32 | 1 | | e |
| | 23-38 | 2 | | e |
| ☿ 16 | 0-54 | 4 | | i |
| | 3-06 | 4 | | e |
| | 20-01 | 1 | | e |
| ♀ 18 | 14-30 | 1 | * | e |
| ♄ 19 | 12-46 | 2 | | e |
| | 14-26 | 3 | * | e |
| ☉ 20 | 18-59 | 1 | | e |
| ♂ 22 | 3-28 | 1 | | e |
| ☿ 23 | 2-15 | 2 | | e |
| | 21-57 | 1 | | e |
| ♀ 25 | 16-26 | 1 | | e |
| ♄ 26 | 15-33 | 2 | | e |
| | 18-26 | 3 | | e |
| ☉ 27 | 10-55 | 1 | * | e |
| ♂ 29 | 5-24 | 1 | | e |
| 30 | 4-51 | 2 | | e |
| | 23-53 | 1 | | e |

*May*

**Figure 11. A page from Flamsteed's predictions of eclipses of Jupiter's satellites for 1684. *Philosophical Transactions* of the Royal Society, London, vol. 13, no. 154 (1683), p. 413. Courtesy of the Fondren Library, Rice University, Houston.**

61. *Ibid.*, p. 405.

62. John Flamsteed, "A Letter from the learned Mr John Flamsteed, Astron. Reg. concerning the Eclipses of Jupiters Satellit's for the Year following 1685...," *Phil. Trans.*, vol. 14, no. 165 (1684), pp. 760-5.

63. John Flamsteed, "An Abstract of a Letter from Mr. J. Flamsteed Math. Reg. & F. of the R.S. giving an account of the Eclipses of ♃s [Jupiter's] Satellits, anno 1686;...," *Phil. Trans.*, vol. 15, no. 177 (1685), pp. 1215-25; *idem*, "An Extract of a Letter from Mr. J. Flamsteed, Astr. Reg. and Reg. Soc. S. giving his calculation of the Eclipses of Jupiters Satellites for the Year 1687...,"*Phil. Trans.*, vol. 16, no. 184 (1686), pp. 196-206; Edmond Halley, "Catalogus Eclipsium omnium Satellitum Jovialium Anno 1688 per universam Terram Visibilium;...Supputante E. H.," *Phil. Trans.*, vol. 16, no. 191 (1687), pp. 435-9.

64. Flamsteed, "An Extract of a Letter" (*ibid.*), p. 197.

65. Norman J. W. Thrower, ed., *The Three Voyages of Edmond Halley in the* Paramore, *1698-1701*, 2 vols., Hakluyt Society Publications, 2nd series, no. 156-7 (London, 1981), vol. 1, p. 102.

66. *Ibid.*, p. 106.

67. Edmond Halley, "A Correction of the Theory of the Motion of the Satelite of Saturn, by that Ingenious Astronomer Mr. Edmund Hally," *Phil. Trans.*, vol. 13, no. 145 (1682/3), pp. 82-8 [82].

68. For example, *Connoissance des Temps...1773* (*op. cit.*, note 53), p. 236. See also Jérôme de Lalande, *Astronomie*, 2nd ed., 4 vols. (Paris, 1771-81), vol. 3, p. 10 (article 2493).

69. Edmond Halley, "Monsieur Cassini his New and Exact Tables for the Eclipses of the First Satellite of Jupiter, reduced to the Julian Stile, and Meridian of London," *Phil. Trans.*, vol. 18, no. 214 (1694), pp. 237-56.

70. James Pound, "New and accurate Tables for the ready Computing of the Eclipses of the first Satellite of Jupiter, by Addition only...," *Phil. Trans.*, vol. 30, no. 361 (1719), pp. 1021-34. For Bradley's remarks about his tables, see *Miscellaneous Works and Correspondence of James Bradley*, ed. Stephen Peter Rigaud (Oxford, England, 1832; reprinted New York: Johnson Reprint Corp., 1972), pp. 81-3.

71. James Bradley, "The Longitude of Lisbon, and the Fort of New York, from Wansted and London, determin'd by Eclipses of the First Satellite of Jupiter...," *Phil. Trans.*, vol. 34, no. 394 (1726), pp. 85-90; reprinted in Bradley (*ibid.*), pp. 58-61.

than the times noted in the Catalogue. In 2 or 3 years more wee may expect opportunitys of observations, which will afford us such a correction for the error, that the Numbers shall fail no more in them, than in the 1st whose Eclipses have not yet differed above 3 minutes from the Calculations.[64]

We must remember, of course, that Flamsteed's accuracy in the case of the first satellite was predicated on his own painstaking work in the Greenwich Observatory. Observers who had to use makeshift observatories in distant lands might not attain such precision. During his voyages in the Atlantic to map the deviation of the magnetic needle (with the hope of using it for longitude determination), Edmond Halley (*ca.* 1656-1742) used a long telescope to observe the satellites of Jupiter. On the coast of Brazil, in March 1699, he missed a satellite eclipse because of "the great hight of the Planet, and want of a convenient support of my long Telescope made it impracticable."[65] A month later, in Barbados, he was in the middle of an observation of an immersion of the first satellite when, "the wind shakeing my Tube, I was willing to gett a more coverd place to observe in, that I might be more Certain, but when I again gott sight of the Planett the Satellite appeared no more."[66] Halley presumably used a simple astronomical telescope, consisting of a convex objective and a convex ocular, and it may very well have been his 24-foot instrument[67] that he took with him on his voyages. Jérôme de Lalande (1732-1807) stated that for these observations a simple astronomical telescope of fifteen to eighteen feet was needed.[68]

**Figure 12. John Hadley's reflector, 1723. *Philosophical Transactions* of the Royal Society, London, vol. 32, no. 376 (1723), pp. 280-1. Courtesy of the Fondren Library, Rice University, Houston.**

In the November/December 1694 issue of the *Philosophical Transactions*, Edmond Halley printed a copy of Cassini's tables with his own comments. He criticized Cassini severely for not accepting Rømer's theory of light because Giovanni Domenico Cassini's stubbornness had led to large errors in the tables of all but the first satellite, the only ones Halley printed, with Julian dates, reduced to the meridian of London. The errors in the motions of the first satellite still ranged up to three minutes of time, which would correspond to 45' of longitude.[69]

English efforts, however, remained haphazard. In 1719, Halley printed tables of Jupiter's first satellite, corrected by James Pound (1669-1724) but based on Cassini's 1693 tables. In the same year, Pound's nephew James Bradley (1693-1762) drew up tables of all four satellites, which were included in Halley's *Tabulae astronomicae* (published in 1749, seven years after Halley's death).[70] In 1726, Bradley mentioned in his calculations of the longitudes of Lisbon and New York from simultaneous observations of the first satellite that there were errors as great as 5 minutes 10 seconds between the time predicted for immersions and emersions and their actual observations in his Wansted observatory. He observed with both a fifteen-foot refractor and Hadley's new reflector (Figure 12) and found that with the latter one could see an emersion about fifteen seconds before it was visible with the former.[71] If such errors could still occur at observatories, where the local times were known with great accuracy, the instruments were optimum, and the observer was an expert in these observations, then what errors would one expect when a traveler in a remote part of the world stopped his expedition for a day or so to observe the first satellite of Jupiter and compare the times of its eclipses with the times predicted in tables? Clearly, in 1726, accurate longitudes using the satellites of Jupiter could still be obtained only by comparing simultaneous observations using similar telescopes.

A typical attempt to interest the Board of Longitude in the use of Jupiter's satellites for longitude at sea was published in 1734 by a certain George Gordon (fl. 1719-1726). The main reason why this method had not been practicable up to then, according to Gordon, was that it required twenty-foot refractors, which could not be managed on board a ship. In his appendix to the 1710 edition of Thomas Streete's *Astronomia Carolina*, Halley had stated that with practice one can make observations from the deck of a moving ship in moderate weather with telescopes of five to six feet in length. But a five- to six-foot reflector is equivalent in magnification to a twenty-foot refractor. Therefore, the use of an appropriate reflector—and it had to be one of Gordon's reflectors, which doubled the magnification of those of the same length made by others—should make the method feasible. In addition, Gordon advocated looking with one eye through an instrument of low power and large field, and with the other through one of high power and small field of view, thus combining high magnification and a large field. Finally, he addressed the problem of the ship's motion. The observer and his instrument should be hung (presumably in a seat) near the center of gravity of the ship, while the ship is allowed to run before the wind in a straight line with a loose helm.[72]

While by the early eighteenth century it was clear that eclipses of Jupiter's moons provided a wonderful method of establishing longitudes on land, it was also becoming clear that they held little promise for doing so at sea. Because of the fairly high magnifications needed to observe the eclipses, the shorter reflecting telescopes could not solve the problem of managing the instrument. Efforts to stabilize the observer by means of "marine chairs" failed. Trials conducted by Nevil Maskelyne (1732-1811) of the observing chair designed by Christopher Irwin (fl. 1758-1763) during the expedition to Barbados in 1773-4 effectively put an end to all hopes. As Maskelyne wrote (in the third person) in his autobiographical notes, "He found Mr. Irwin's marine chair was too much disturbed by the motion of the ship to allow the management of a telescope sitting in it to observe the eclipses of Jupiter's Satellites. Harrison's watch was found to give the Longitude of the Island with great exactness."[73] But if this method was not practical at sea, we must not forget that it was at the center of the revolution in mapping of the seventeenth, eighteenth, and nineteenth centuries.

72. George Gordon, *A Compleat Discovery of a Method of Observing the Longitude at Sea* (London, 1724), pp. 7-28.

73. Derek Howse, *Nevil Maskelyne: The Seaman's Astronomer* (Cambridge: Cambridge University Press, 1989), p. 221.

# The Longitude Timekeepers of Christiaan Huygens

*J. H. Leopold*

**Before his retirement in 1995, John H. Leopold was an Assistant Keeper in the British Museum, in charge of the horological collection. For many years prior to this, he was curator of silver at the Groninger Museum in the Netherlands. His field of interest predominantly concerns early clockwork and how it relates to the astronomy of the period. He is the author of several books, including *The Almanus Manuscript* (1971), *Die grosse astronomische Tischuhr des Johann Reinhold* (1974), *Der kleine Himmelsglobus von Jost Bürgi* (with K. Pechstein, 1977), and *Astronomen, Sterne, Geräte—Landgraf Wilhelm IV. und seine sich selbst bewegenden Globen* (1986). He has published several articles on horology concerning developments in precision timekeeping during the seventeenth century; three of these relate to the work of Christiaan Huygens.**

# The Longitude Timekeepers of Christiaan Huygens


## *J. H. Leopold*


'Vous faites beaucoup d'honneur à la Geometrie, lorsque vous trouvés les plus beaux usages des lignes qu'elle peut fournir'.
('You greatly honour Geometry, when you find the most beautiful applications for the curves she provides'.)

Letter from Leibniz to Huygens, 11 October 1693[1]

Christiaan Huygens (1629-1695) is a truly remarkable figure in horology. A superb scholar, his achievements range over the fields of mathematics, physics, and astronomy. He excelled in the mathematical analysis of physical problems, and his work in horology bears this out: the pendulum clock and the balance spring, the two inventions that revolutionised clocks and watches, were both based on a solid understanding of the fundamental principles involved. Indeed, to Huygens the essence of his inventions was more in the application of a physical or mathematical principle than in the details of the actual device. This attitude set him apart from his contemporaries in horology. On the one hand, it gave him a deep understanding of the principles involved, but it was also at the root of his many disagreements with others in the field—such as Alexander Bruce (d. 1681), Robert Hooke (1635-1702/3), and most of his clockmakers—because he tended to under-estimate the difficulties involved in perfecting a working machine.[2]

Huygens's work in establishing longitude by timekeepers is not nearly so well known as his two great horological inventions.[3] Very little of it was published during his lifetime (in fact, it was a matter of some secrecy), and it did not lead to a major break-through. Yet this work forms a recurrent theme throughout his life, and no other pioneer had so many timekeepers tried at sea. For Huygens was not only a much-respected scholar, he belonged to a very influential family, which undoubtedly made it easier for him than for most to get things done, especially when those things were expensive.

Many of Huygens's longitude experiments involved pendulum clocks of various kinds. To the modern mind this seems unrealistic, but there are several points to be born in mind. Firstly, until 1675 the pendulum was the only

1. *Oeuvres Complètes de Christiaan Huygens* (hereafter referred to as *O.C.)*, 22 vols. (The Hague: Martinus Nijhoff, 1888-1950), vol. 10, no. 2829, p. 539. Gottfried Wilhelm Leibniz (1646-1716).

2. For this attitude of Huygens, see J. H. Leopold, 'Christiaan Huygens and his instrument makers', in H. J. M. Bos *et al.,* eds., *Studies on Christiaan Huygens* (Lisse, the Netherlands: Swets, 1980), pp. 221-33.

3. Two recent studies approach the subject from the mathematical rather than from the horological side: Michael S. Mahoney, 'Christiaan Huygens: The measurement of time and longitude at sea', in H. J. M. Bos *et. al.,* eds., *Studies on Christiaan Huygens* (Lisse, the Netherlands: Swets, 1980), pp. 234-70, and particularly J. G. Yoder, *Unrolling Time: Christiaan Huygens and the mathematization of nature* (Cambridge, Cambridge University Press, 1988), pp. 152ff.

4. Huygens was aware that large sailing ships were better suited for his timekeepers than small ones: *O.C.*, vol. 4, no. 1093, p. 296; no. 1095, p. 301. *O.C.*, vol. 8, no. 2287, p. 406. *O.C.*, vol. 9, no. 2615, pp. 478-9. Kincardine, fol. 162$^{r-v}$ (the Kincardine papers are a privately owned collection of letters; copies in the Royal Society, London).

5. *O.C.*, vol. 2, no. 443, p. 109.

6. *O.C.*, vol. 2, no. 368, p. 5; no. 370, p. 7; no. 382, p. 22; no. 480, p. 166. Christiaan Huygens, *Horologium* (The Hague, 1658), p. 5 (= *O.C.*, vol. 17, pp. 56-7).

**Figure 1. Portrait of Christiaan Huygens, 1671, by Caspar Netscher (1639-1684). Huygens was in The Hague from September 1670 until June 1671, recovering from illness. It was during those months that he would have sat for the fashionable Hague painter. *O.C.*, vol. 22, frontispiece, and pp. 660ff. Oil on panel. 30 x 24 cm. (11.8 x 9.5 in.). Courtesy of The Hague Historical Museum. Inv. no. 12-1926.**

known isochronic oscillator. Secondly, a large sailing ship is less tossed about by the seas than a modern vessel, and under most conditions these seventeenth-century ships appear to have provided a surprisingly steady base for the clocks.[4] And, lastly, no absolute accuracy was expected: any helpful result would be better than nothing.

When Christiaan Huygens invented the pendulum clock (Christmas, 1656),[5] he immediately realised that this improved timekeeper could provide the solution to the problem of establishing longitude at sea. In his very first letter announcing the invention, this possibility is already mentioned.[6] In fact, it could be argued that the longitude problem was at the basis of the invention of the pendulum clock: in his *Horologium* of 1658, Huygens specifically mentions that the invention derived from an attempt to improve on Galilei's tiresome counting of the excursions of a free pendulum,[7] a clear reference to Galilei's project for finding longitude proposed to the Dutch States General in 1636. Huygens's father, the statesman and poet Constantyn Huygens (1596-1687), had been involved in this matter.[8]

In the course of the next few years, the possibility of solving the longitude problem by means of the pendulum clock is occasionally mentioned in Huygens's correspondence, but no practical steps were taken,[9] although in December 1660 he asked his brother Lodewyk (1631-1699) about the per-

7. Huygens, *Horologium* (*ibid.*), pp. 3-4 (= O.C., vol. 17, pp. 54-5). A remark in Huygens's much later 'Anecdota' points in the same direction: 'De Horologio oscillatorio, quomodo primum invenerim ex hodometro' ('About the pendulum-clock, how I first came to invent [it] from the Hodometer') (O.C., vol. 17, p. 36; O.C., vol. 18, p. 666).

8. See Silvio A. Bedini, *The Pulse of Time: Galileo Galilei, the determination of longitude, and the pendulum clock*, Biblioteca di Nuncius, Studi e Testi, vol. 3 (Florence: Leo S. Olschki, 1991). For an edition of the relevant documents, see *Journal tenu par Isaac Beeckman de 1604 à 1634*, ed. C. de Waard, 4 vols. (The Hague: Martinus Nijhoff, 1939-53), vol. 4, pp. 235-77, 285-9. Beeckman became one of the commissioners appointed by the States General for evaluating Galilei's proposal. Constantyn Huygens Sr. was more closely involved with this project than Bedini suggests; for a survey, see W. Ploeg, *Constantyn Huygens en de Natuurwetenschappen* (Rotterdam: Nijgh & van Ditmar, 1934), pp. 49-58.

9. O.C., vol. 2, no. 518, p. 224; no. 543, p. 266; no. 665, p. 480. O.C., vol. 3, no. 789, p. 144.

formance of a pendulum clock during a sea voyage to Spain.[10] The main reason for this delay appears to have been that Huygens thought that sufficient accuracy could only be obtained with a three-foot seconds-beating pendulum,[11] and there were obvious problems about using such a device in a ship.

That was the state of affairs when in April 1661 Huygens visited England and discussed the matter with the Scottish nobleman Alexander Bruce.[12] The two men had known each other for some years: they met in 1658, apparently in The Hague and on purely social grounds (in the next year Bruce married a daughter of close friends of the Huygens family), and in September 1658 Bruce was one of those to whom Huygens sent a copy of his newly published *Horologium*.[13] This probably turned Bruce's attention to the use of such clocks for establishing longitude, and, in 1661 in London, he showed Huygens an experimental clock with a short pendulum, adapted for use at sea. Huygens, who still thought in terms of long pendulums, was not greatly impressed. Bruce, undeterred, tested his clock on a voyage from Scotland to The Hague in early 1662, and the results seem to have convinced Huygens that shorter pendulums could effectively be used for this purpose.[14] There followed some months of joint experiments. In March, one clock was nearly finished, and, in June, they were trying it out; by December, two identical clocks were ready, made by the Hague clockmaker Severyn Oosterwyck (mentioned 1656-1690)[15] and paid for by Bruce.[16]

In December 1662, Bruce took the two clocks to England. During the crossing they ran into a storm and Bruce became violently sick. Worse, one of the clocks (the 'old' one) became dislodged from its suspension and was damaged in the fall; the other (the 'new' one) stopped but could be started again when the weather became quieter. All in all, the results were felt to be encouraging.[17] The damaged timekeeper was replaced by a similar one, made in London by John Hilderson (mentioned 1657-1663),[18] and more experiments followed. In April 1663, the two clocks were taken by Captain Robert Holmes (1622-1692) on a voyage to Lisbon.[19]

The large amount of correspondence that survives about these early experiments reveals some technical details about the timekeepers. They incorporated two adaptations of Bruce: the so-called double crutch, to keep the pendulum moving in the proper plane, and the suspension of the clocks (a sort of ball-and-socket arrangement); otherwise we know that they were spring-driven and had seven-inch pendulums with cycloidal cheeks.[20] An unexpected and interesting point about these early clocks is that they were triangular in shape.[21]

In the summer of 1663, Huygens visited England again and was made a Fellow of the Royal Society. He probably hoped to be at hand when Holmes returned, but this was not the case, as Holmes only returned in September. His report was entirely favourable: the Dutch clock had performed better than the English one, which greatly pleased Oosterwyck.[22] Generally it was felt that the problem of longitude was well on the way to being solved. Patents in England, the Netherlands, and France were discussed, but there were also the first signs of disagreements between Huygens and Bruce over the invention.[23] Meanwhile, Hilderson made some changes in his clock (he replaced the spring and made more room for the pendulum to move),[24] and, in November, Holmes took the two clocks to Gambia on a voyage that was to last about a year.

10. *O.C.*, vol. 3, no. 823, pp. 209-10.

11. Clearly expressed in a letter to Jean Chapelain (1594-1674) of June 1658: *O.C.*, vol. 2, no. 488, p. 181. Measures expressed in feet and inches in the Huygens papers are often difficult to interpret exactly, for they may relate to several different foot-measures. Moreover, in many instances the content makes clear that only approximate values are given.

12. For Alexander Bruce, subsequently Earl of Kincardine, see A. L. Youngson, 'Alexander Bruce, F.R.S., Second Earl of Kincardine (1629-1681)', in H. Hartley, ed., *The Royal Society, its Origins and Founders* (London: Royal Society, 1960), pp. 251-8. For Huygens's cooperation with Bruce and his contact with the Royal Society in general, see J. H. Leopold, 'Christiaan Huygens, the Royal Society and Horology', *Antiquarian Horology*, vol. 21, no. 1 (Autumn 1993), pp. 37-42.

13. *O.C.*, vol. 2, no. 511, p. 209.

14. *O.C.*, vol. 5, no. 1201, pp. 8-9. Kincardine (see note 4), fols. $178^{r}$-$9^{r}$ (printed *O.C.*, vol. 22, pp. 605-7 [606], and Youngson, 'Bruce' [*op. cit.*, note 12], pp. 253-4). See also *O.C.*, vol. 4, no. 984, p. 65.

15. *O.C.*, vol. 4, no. 986, p. 68; no. 988, p. 72; no. 1022, p. 151; no. 1023, p. 153; no. 1067, p. 244; no. 1073, p. 256; no. 1082, p. 278. The clockmaker is simply referred to as Severyn, an unusual Christian name that leaves no doubt about the identity: *O.C.*, vol. 4, no. 1104, p. 324; no. 1175, pp. 452-3.

16. *O.C.*, vol. 4, no. 1080, pp. 274-5.

17. *O.C.*, vol. 4, no. 1082, pp. 278, 280; no. 1083, pp. 280-1; no. 1085, p. 284; no. 1086, pp. 284-5; no. 1088, pp. 287-8; no. 1090, pp. 290-1; no. 1093, p. 296; no. 1095, p. 301; no. 1097, pp. 304, 306. Kincardine (see note 4), fol. $162^{r-v}$.

18. *O.C.*, vol. 4, no. 1102, p. 318; no. 1173, pp. 443-4. Kincardine (see note 4), fol. $170^{v}$.

19. *O.C.*, vol. 4, no. 1174, pp. 446-51.

20. Some additional information derives from slightly later sources:
The double crutch: *O.C.*, vol. 5, no. 1201, pp. 8-9; no. 1252, p. 104; no. 1269, p. 140. Christiaan Huygens, *Horologium Oscillatorium Sive De Motu Pendulorum Ad Horlogia Aptato Demonstrationes Geometricae* (Paris, 1673), p. 17. For a drawing: *O.C.*, vol. 17, p. 166. The English translation of this portion (see bibliography) is rather free and should be used with some caution.
The suspension of the clocks: *O.C.*, vol. 4, no. 986, p. 68; no. 1090, pp. 290-1. *Horologium Oscillatorium*, p. 17. (There is some discussion about these clocks themselves acting as large pendulums: *O.C.*, vol. 4, no. 1093, p. 296; no. 1095, p. 301; no. 1097, p. 304.)
Spring-driven: Clearest in *Horologium Oscillatorium*, p. 16.
Seven-inch pendulum: *O.C.*, vol. 4, no. 1086, pp. 284-5 (*Horologium Oscillatorium* gives the length as six inches, but this is probably an exaggeration).
Cycloidal cheeks: *O.C.*, vol. 4, no. 1188, p. 476. *O.C.*, vol. 5, no. 1269, p. 140.

21. *O.C.*, vol. 17, p. 168. This describes, in 1663, a clock made before those of the newest invention (*i.e.*, the remontoire clocks, see later) as 'het drykantighe slingerwerck' ('the triangular pendulum clock'). There can be no doubt that

In fact, there seems to have been much more interest in these clocks in England than in the Netherlands. There was a great deal of attention from the side of the Royal Society and particularly from Sir Robert Moray (1608?-1673), a mutual friend of Huygens and Bruce, who interested some very important people in the invention, notably the Duke of York (the subsequent James II, 1633-1701) and Prince Rupert of the Rhine (1619-1682). In 1664, several more timekeepers of this type were made, two by John Fromanteel (b. 1638, last mentioned 1682), one by Edward East (1602-1697), and one by an unnamed maker;[25] these all appear to have had their own variations, those of Fromanteel having bushes of bell-metal 'that wear not in 20 yeeres',[26] and the clock by East 'hauing 2 springs to turne one pinion, but with no cheeks'.[27] Some of these clocks were weight-driven.[28] It appears that the clocks by Fromanteel were tried by the Duke of York and Prince Rupert, who both declared themselves highly satisfied.[29] In fact, Moray at this time was thinking about producing these clocks in considerable numbers.[30]

However, while all these things happened in London, Huygens had further developed his timekeepers. Already during the last days of December 1662, just after Bruce's departure, he ordered a clock with a ten-inch pendulum; in April 1663, it was finished but still giving trouble.[31] Little more is heard about this clock; in fact, Huygens spent most of the year and the first half of the next in Paris, where he did no practical work in horology, although, in December 1663, we hear about yet another invention to improve the longitude timekeepers.[32] Huygens kept the details of this improvement a secret for a long time,[33] and it was not until it had been executed in practice, again by Oosterwyck in August 1664,[34] that he finally disclosed it to Moray: Huygens had invented a remontoire clock (Figure 2), both trains being weight-driven, with the secondary weight attached to the 'scape-wheel and with a period of a half-minute.[35] Moray ordered a clock to be made for him, but it took several years to arrive in London and, by that time, it made no impact.[36]

Early in December 1664, Holmes returned from his voyage to Gambia, with a very favourable report: once again the Dutch clock had per-

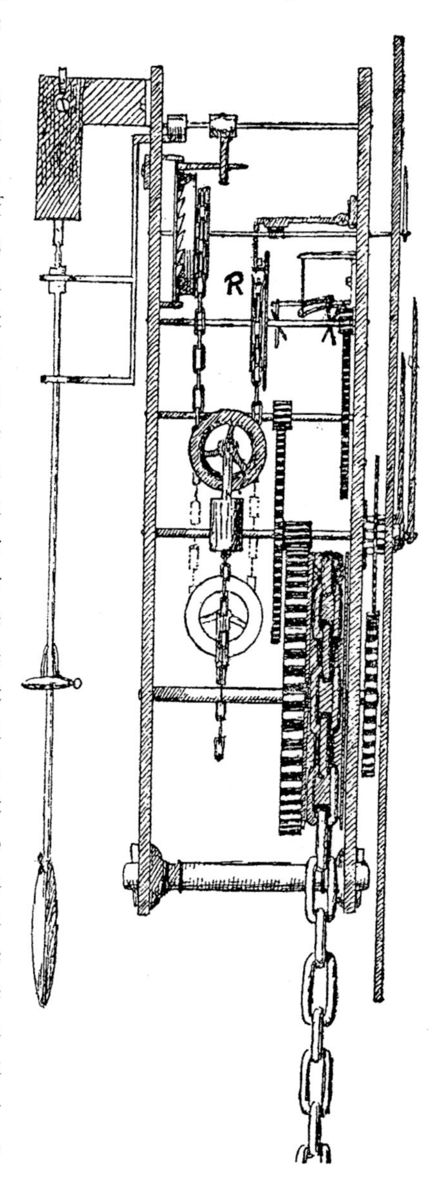


**Figure 2. Movement with double crutch and remontoire, pendulum with cycloidal cheeks and cursor weight. Drawing dating from the end of 1664 or early 1665. O. C., vol. 17, p. 178.**

this refers to the outer shape of the clock and not to a triangular pendulum, a construction not yet invented. The passage allows the identification of a few drawings (*O.C.*, vol. 17, p. 165) and helps to identify the clock in the frontispiece of T. Sprat, *The History of the Royal-Society of London, for the Improving of Natural Knowledge* (London, 1667).

22. *O.C.*, vol. 4, no. 1163, p. 426; no. 1165, pp. 427-8; no. 1167, pp. 431-2; no. 1173, pp. 443-4; no. 1174, pp. 446-51; no. 1175, pp. 452-3; no. 1177, p. 456.

23. *O.C.*, vol. 4, no. 1095, p. 301; no. 1167, pp. 431-2; no. 1175, pp. 452-3; no. 1178, pp. 458-60; no. 1187, p. 474. Many letters in *O.C.*, vol. 5, the more important ones being: no. 1200, p. 7; no. 1239, p. 79; no. 1247, pp. 93-4; no. 1261, p. 126; no. 1269, p. 140; no. 1274, pp. 148-9; no. 1280, pp. 157-8; no. 1301, p. 186. Also *O.C.*, vol. 22, no. XLIV, pp. 80-1; no. XII, p. 174. Kincardine (see note 4), fols. $162^{r-v}$, $172^{r-v}$, $174^{r-v}$, $176^{r}$-$7^{r}$, $178^{r}$-$9^{r}$ (= *O.C.*, vol. 22, p. 606; also Youngson, 'Bruce' [*op. cit.*, note 12], pp. 253-4), $180^{r}$-$1^{v}$, $184^{r}$, $192^{v}$. For the actual patents, see note 43.

24. Kincardine (see note 4), fols. $170^{v}$, $176^{r}$.

25. For the unnamed maker, see *O.C.*, vol. 5, no. 1252, pp. 103-4.

26. *O.C.*, vol. 5, no. 1243, pp. 85-8; no. 1247, p. 93. Kincardine (see note 4), fol. $181^{v}$ (quotation). John Fromanteel took four months to produce these two clocks and estimated that any further clocks would take a month each.

27. *O.C.*, vol. 5, no. 1252, pp. 103-4. Kincardine (see note 4), fol. $184^{r}$.

28. *O.C.*, vol. 5, no. 1247, p. 93; no. 1250, pp. 98-9; no. 1252, pp. 103-4; no. 1255, p. 114.

29. *O.C.*, vol. 5, no. 1252, pp. 103-4; no. 1255, p. 114; no. 1287, p. 168; no. 1301, p. 186; no. 1421, p. 378.

30. *O.C.*, vol. 5, no. 1243, pp. 85-8; no. 1250, pp. 98-9; no. 1252, pp. 103-4. The main problem was felt to be the long time it took to produce these clocks (see note 26).

31. *O.C.*, vol. 4, no. 1086, pp. 284-5; no. 1104, p. 324; no. 1097, pp. 304, 306. *O.C.*, vol. 5, no. 1212, p. 27; no. 1221, p. 47. Kincardine (see note 4), fol. $162^{r-v}$.

32. *O.C.*, vol. 4, no. 1186, p. 473.

33. *O.C.*, vol. 4, no. 1188, p. 476; no. 1189, p. 478; no. 1190, p. 479. *O.C.*, vol. 5, no. 1208, pp. 19-20; no. 1212, p. 27; no. 1240, p. 81.

34. *O.C.*, vol. 5, no. 1242, p. 84; no. 1250, pp. 98-9.

35. *O.C.*, vol. 5, no. 1253, p. 108; no. 1255, pp. 113-14; no. 1258, pp. 119-20; no. 1268, pp. 136-7; no. 1274, pp. 148-9; no. 1280, pp. 157-8. *The Correspondence of Henry Oldenburg*, ed. A. R. and M. B. Hall, 13 vols., 1965-86 (Madison: University of Wisconsin Press [vols. 1-9], 1965-73; other volumes by other publishers), vol. 2, no. 325, p. 224. Kincardine (see note 4), fols. $188^{v}$, $195^{r}$ (Moray misunderstood the period to be three to four minutes).

36. *O.C.*, vol. 5, no. 1255, pp. 113-14; no. 1258, pp. 119-20; no. 1280, pp. 157-8; no. 1301, p. 186; no. 1336, p. 245; no. 1345, p. 256; no. 1362, p. 282; no. 1386, pp. 321-2; no. 1400, p. 344; no. 1411, pp. 360-2; no. 1421, p. 377;

formed better than the English one.[37] Altogether things looked good, and Huygens set about writing his directions for the use of marine timekeepers, which were published in February 1665;[38] these contain his table for the equation of time (see Figure 6) and announce that the new clocks were being made by Oosterwyck.[39] In the same month, experimenting with two remontoire clocks, Huygens discovered the sympathetic motion of two pendulums.[40] Oosterwyck at this time was planning a clock with a period of a quarter-minute.[41] In March, Henry Oldenburg (*ca.* 1618-1677)[42] published an account of Holmes's voyage in the first issue of *Philosophical Transactions.* A few days before, the Royal Society had been granted a patent for the marine timekeepers; this had been negotiated by Moray, who had persuaded Huygens and Bruce to donate half of any proceeds to the Society and share the remainder equally.[43]

By September 1665, however, it had become clear that there was no great enthusiasm for the new clocks in England or the Netherlands,[44] and the activity shifts to France, where Huygens moved in April 1666 in order to join the recently established Académie Royale des Sciences. His new position meant that, for the next ten years, we have far less information. He realised that establishing longitude at sea was a matter of strategic value,[45] and the Académie was an advisory body to the French Crown. In fact, Huygens's work fitted well into the activities of the Académie, which was interested in establishing longitudes (mainly by Jupiter's satellites) for the purpose of mapmaking.[46]

Huygens brought two remontoire clocks with him to Paris, having already sent three from The Hague, but they were a disappointment: they were too delicate and all gave trouble,[47] which appears to have soured his relationship with Oosterwyck.[48] So the clocks subsequently made in Paris had conventional going trains.[49] Their maker is not specifically mentioned, but there need be little doubt that it was Isaac Thuret (d. 1706), with whom Huygens had already negotiated the rights to make the new clocks, and who, in fact, had designed a remontoire clock of his own.[50]

We have little information about the French experiments. Indeed, there is some confusion about the number of trials that were made. There seem to have been three sea voyages. The first one was a trip to Lisbon made by François de Vendôme, Duc de Beaufort (1616-1669) in April-May 1668; the observer on this voyage was a man called de la Voye (d. 1687). It was reported that during this trip the clocks never stopped, even during a storm, but otherwise no observations appear to have been made.[51] After their return Huygens made some changes in the clocks. He gave them larger cases (four feet tall), bigger weights, and longer cheeks, and improved the ball-and-socket suspension; these clocks had nine-inch pendulums.[52] Next came the most successful voyage, also under de Beaufort, in the Mediterranean in May-September 1669; again de la Voye went as the observer, and this time he did a good job and the results were highly satisfactory.[53] Huygens arranged that de la Voye was given money for more experiments, but this he appears to have frittered away.[54] Huygens subsequently made changes to the timekeepers; the main change was that he abolished Bruce's double crutch, suspended the pendulum further away from the movement (about two inches), and made the pendulum-bobs heavier (one-half to three-quarter pound).[55] A third trial was made in 1670, during a voyage to Cayenne; at first the clocks

no. 1436, pp. 426-8; no. 1466, p. 486; no. 1508, p. 551. *O.C.*, vol. 6, no. 1512, p. 1; no. 1518, p. 11; no. 1530, pp. 23-4; no. 1540, pp. 35-6; no. 1546, p. 47; no. 1563, p. 86; no. 1576, p. 108; no. 1594, p. 135; no. 1666, p. 268; no. 1708, p. 370; no. 1719, p. 392; no. 1721, p. 396; no. 1726, p. 417; no. 1729, pp. 421-2; no. 1730, p. 424. Although the clock was finished by March 1665, it apparently did not arrive in London until 1668; the delay was partly caused by the second Dutch War (1665-7).

37. *O.C.*, vol. 5, no. 1280, pp. 157-8; no. 1287, p. 168; no. 1315, pp. 204-6; no. 1318, pp. 212-13; no. 1329, pp. 234-5; no. 1336, p. 245; no. 1353, pp. 269-71; no. 1363, pp. 284-6. Robert Holmes, 'A Narrative concerning the success of Pendulum-Watches at Sea for the Longitudes', *Philosophical Transactions* of the Royal Society, London *(Phil. Trans.),* vol. 1, no. 1 (1664/5), pp. 13-15 [14]. T. Birch, *The History of the Royal Society of London...*, 4 vols. (London, 1756-7),vol. 2, pp. 21-2.

38. Christiaan Huygens, *Kort Onderwys Aengaende het gebruyck Der Horlogien Tot het vinden der Lenghten van Oost en West* ([The Hague], [1665]); copies of this extremely rare booklet are in the collections of the Pulkovo Observatory, near St. Petersburg, Russia, and The Time Museum, Rockford, Illinois. *O.C.*, vol. 5, no. 1301, p. 187; no. 1331, p. 240; no. 1338, p. 247; no. 1344, p. 254.

39. Last paragraph of Huygens, *Kort Onderwys* (*ibid.*). *O.C.*, vol. 5, no. 1325, pp. 224-5.

40. *O.C.*, vol. 5, no. 1335, pp. 243-4; no. 1338, pp. 147-8; no. 1345, p. 256; no. 1370, pp. 301-2. *O.C.*, vol. 17, pp. 183-7. Huygens, *Horologium Oscillatorium* (*op. cit.*, note 20), pp. 18-19.

41. *O.C.*, vol. 5, no. 1331, p. 240.

42. Henry Oldenburg, the first secretary of the Royal Society, began publishing the *Philosophical Transactions* of the Society in 1665.

43. For the publication, see note 37. For the English patent, see *O.C.*, vol. 17, p. 176. For the division of the proceeds, see Kincardine (see note 4), fol. $172^{r-v}$ (*cf. O.C.*, vol. 4, no. 1167, p. 432; no. 1187, p. 474; *O.C.*, vol. 5, no. 1201, pp. 8-9; no. 1261, p. 126). Huygens obtained patents in his own name in the Netherlands (December 1664, *O.C.*, vol. 17, p. 175; *O.C.*, vol. 5, no. 1286, pp. 166-7) and in France (March 1665, *O.C.*, vol. 5, no. 1346, p. 257; *O.C.*, vol. 18, p. 20).

44. *O.C.*, vol. 5, no. 1385, pp. 319-20; no. 1386, pp. 321-2. *O.C.*, vol. 22, no. XLVII, pp. 86-7 (= Oldenburg, *Correspondence* [*op. cit.*, note 35], vol. 2, no. 427, p. 549). Earlier in the year, Huygens had found leading Dutch navigation experts sceptical: *O.C.*, vol. 5, no. 1356, p. 277; no. 1358, p. 278; no. 1386, pp. 321-2; see also note 77. A positive reaction came from the Nymegen clockmaker Jan van Call: *O.C.*, vol. 5, no. 1372, p. 303 (April 1665). In *Horologium Oscillatorium* (*op. cit.*, note 20), p. 17, Huygens claims that the Dutch had experimented with his clocks; he may refer to the proposal by Lt.-Adm. van Gent to try Huygens's clocks in 1667, but this apparently was not followed up: *O.C.*, vol. 6, no. 1614, p. 167; no. 1618, pp. 171-2; also C. A. Davids, *Zeewezen en Wetenschap, de Wetenschap en de ontwikkeling van de Navigatietechniek in Nederland tussen 1585 en 1815* (Amsterdam/Dieren: De Bataafsche Leeuw,

seem to have performed well (even during a storm), but then the observer, Jean Richer (1630-1696), became seasick and apparently neglected the clocks entirely, to the utter disgust of Huygens.[56] Early in 1672, a fourth trial, again on a voyage with Richer but this time with a clock of a new construction, was abandoned because Huygens did not prepare the machine in time.[57] A letter in May of that year makes clear that this new construction was in fact the triangular pendulum:[58] a pendulum suspended from two points and thus forced to move in one plane only.

News of the French experiments also came to England, which is probably why the Royal Society decided to publish, in 1669, an English translation of Huygens's *Kort Onderwys,* entitled 'Instructions Concerning the Use of Pendulum-Watches, for finding the Longitude at Sea'.[59]

Huygens devoted much of the year 1672 to writing the final version of his *Horologium Oscillatorium*, which was published in April 1673.[60] Its main portion is a masterly mathematical analysis of evolute curves (such as the cycloid), the cycloidal pendulum, and the compound pendulum. It also briefly describes the longitude work up to that point. This portion includes a description with illustrations of his new (and as yet untried) marine timekeeper with triangular pendulum, and we see that he had finally abolished the ball-and-socket suspension in favour of gimbals. Detailed drawings of this clock survive in manuscript: it was spring-driven with a going barrel, it had what would now be called a regulator-dial, and from the train-count it is clear that it ticked two-fifths seconds, which makes for a pendulum of about 6¼ inches.[61]

With the publication of the *Horologium Oscillatorium,* Huygens did not end his research into the properties of the cycloid. In the second half of 1673, this research led him to a remarkable discovery. He found that in a cycloidal pendulum, the returning force is proportional to the distance from the neutral position, and he realised that if this condition obtains in any other mechanical system, then that system will be isochronous too.[62] In other words, from here on Huygens was no longer dependent on the cycloidal pendulum as the only source of isochronism for his timekeepers, but was free to search for other constructions.

The first construction based on Huygens's new understanding of the condition of isochronism was the balance spring of 1675. No contemporary proof of its isochronism survives among Huygens's papers, but in his description in the *Journal des Sçavans*, he states clearly that in the new construction 'the motion is regulated by a principle of egality, just as is that of a pendulum corrected by a cycloid', which shows that he knew why a balance with a spiral balance spring is theoretically isochronous: in this construction, the distortion (and therefore the tension) of the spring is proportional to the displacement of the balance.[63]

There is no need to go deeply into the introduction of the balance spring in 1675, for at that time the work had little to do with marine timekeepers.[64] Suffice it to say that Huygens from the first realised that here he had a construction that could serve to find longitude, and in the course of the next years he occasionally refers to this.[65] Particularly interesting are his remarks when, in 1680-1, he planned his clockwork-driven planetarium: it was at first designed with a pendulum, but he later decided to have a balance with a spiral spring (and suspended by a silk string), adding the remark that

[1985]), pp. 135ff. A report, which reached Isaac Thuret (d. 1706) in 1668, held that Adm. de Ruyter had a pendulum clock on board his ship during a voyage in 1665, but without special suspension: *O.C.*, vol. 6, no. 1618, p. 171. No confirmation of this has been found (*cf.* Davids, pp. 135ff.); the clock may well have been an ordinary one, not meant for longitude observations.

45. *O.C.*, vol. 6, no. 1614, p. 167; no. 1711, pp. 378-9; no. 1770, pp. 514-15.

46. For a survey of this work, see Lloyd A. Brown, *The Story of Maps* (Boston: Little, Brown, 1949), pp. 212ff.

47. *O.C.*, vol. 5, no. 1350, p. 265; no. 1358, p. 278; no. 1370, pp. 301-2; no. 1385, pp. 319-20; no. 1408, pp. 357-8; no. 1413, p. 363; no. 1430, pp. 398-9; no. 1444, p. 438; no. 1445, pp. 439-40; no. 1454, pp. 475-6; no. 1455, p. 477; no. 1461, p. 483; no. 1485, pp. 510-11. *O.C.*, vol. 6, no. 1601, p. 149; no. 1614, p. 167; no. 1721, p. 396. Another disadvantage of these remontoire clocks was that they were very expensive: *O.C.*, vol. 5, no. 1325, pp. 224-5; no. 1329, pp. 234-5. *O.C.*, vol. 6, no. 1721, p. 396. See also *O.C.*, vol. 22, no. XLVII, pp. 86-7 (see note 44).

48. *O.C.*, vol. 6, no. 1601, p. 149; no. 1606, p. 155. Another reason for their disagreement was that Huygens owed Oosterwyck rather a lot of money: *O.C.*, vol. 6, no. 1546, p. 47; no. 1563, p. 86; no. 1576, p. 108; no. 1594, p. 135; no. 1636, p. 211; no. 1665, pp. 266-7.

49. *O.C.*, vol. 6, no. 1614, p. 167.

50. Huygens first came into contact with Isaac Thuret in 1664 (*O.C.*, vol. 5, no. 1227, p. 58). *O.C.*, vol. 5, no. 1260, pp. 124-5; no. 1265, p. 129; no. 1331, p. 240; no. 1352, pp. 267-8; no. 1361, p. 281; no. 1370, pp. 301-2; no. 1398, p. 341; no. 1408, pp. 357-8; no. 1409, pp. 358-9; no. 1411, pp. 360-2; no. 1417, pp. 370-1; no. 1429, pp. 396-7; no. 1430, pp. 398-9; no. 1435, p. 425; no. 1445, pp. 439-40; no. 1454, pp. 475-6; no. 1485, pp. 510-11; no. 1491, p. 525. *O.C.*, vol. 6, no. 1618, pp. 171-2.

51. Oldenburg, *Correspondence* (*op. cit.*, note 35), vol. 4, no. 778, p. 174; no. 807, p. 226. *O.C.*, vol. 6, no. 1630, p. 200; no. 1639, p. 218. Little is known about de la Voye.

52. *O.C.*, vol. 6, no. 1711, pp. 378-9. *O.C.*, vol. 18, pp. 21-2 (see also 23-4). Huygens, *Horologium Oscillatorium* (*op. cit.,* note 20), pp. 17-18.

53. *O.C.*, vol. 6, no. 1645, p. 226; no. 1721, p. 396; no. 1757, p. 486; no. 1765, p. 500; no. 1766, pp. 501-3. *O.C.*, vol. 18, pp. 633-4, p. 635, note 4.

54. *O.C.*, vol. 7, no. 1806, pp. 26-7. *O.C.*, vol. 18, p. 633, note 2. In 1675, de la Voye tried to attract Huygens's attention again: *O.C.*, vol. 7, no. 2002, p. 398; no. 2033, pp. 465-6.

55. *O.C.*, vol. 7, no. 1806, pp. 26-7. *O.C.*, vol. 18, pp. 21-2 (see also 23-4).

56. *O.C.*, vol. 6, no. 1738, p. 440; no. 1757, p. 486. *O.C.*, vol. 7, no. 1793, p. 4; no. 1824, pp. 54-5.

57. *O.C.*, vol. 7, no. 1853, p. 117; no. 1866, p. 142. On this trip, Jean Richer established the difference in length of a seconds-pendulum in Paris and near the equator.

by making this balance large he would have a test of the accuracy of such timekeepers.[66]

It was, however, not until after Huygens had returned from Paris to The Hague in 1682 that practical experiments with longitude timekeepers were resumed. In October of that year, we first hear about this, and on the last day of the year Huygens obtained funding from the Dutch East India Company, his collaborator being the Hague clockmaker Johannes van Ceulen (first mentioned 1675, d. 1715).[67] The work took most of the next year. Two timekeepers were made, and, in July 1683, they were said to perform very well; however, there were also signs of friction between Huygens and van Ceulen. The invention, in November, of epicyclic winding-gear (which provides maintaining power during winding) was clearly part of this work.[68] Then they came to a stumbling block: in early December, Huygens became aware that the spring was greatly influenced by changes in temperature, a problem for which he had no solution.[69]

Thus, Huygens was forced to constructions in which the regulating force was once again gravity, with all the problems that entailed. The first construction was the tri-cordal pendulum, an ingenious device consisting of a ring suspended at three points by threads and made to oscillate around its centre. Huygens invented this in December 1683 and gave the proof of its near-isochronism by showing that it approaches the condition for isochronous vibration.[70] On 17 December, Huygens took his model of the tri-cordal pendulum to van Ceulen. He took his brother Constantyn (1628-1697) along to be a witness, because the clockmaker had claimed to have invented something similar. Their relations were apparently becoming more and more strained. More work followed: both balance-spring timekeepers were converted to the new tri-cordals. A sketch shows that Huygens considered using cheeks to correct the motion of the threads, but this was given up, and, by June 1684, we hear that long threads are being used: the analogy with short and long pendulums is clear.[71]

In spite of all the work, the timekeepers were still not satisfactory, and, by September, Huygens decided to return to his old, but as yet untried, concept of the triangular pendulum published almost ten years before. At first, Huygens decided to convert one clock only, turning its movement upside-down. But it soon became clear that the new construction was far better, and so the other clock was also converted.[72]

Little is known about the work that followed, but, by August 1685, Huygens was ready to test the clocks in an actual ship, which took place on the Zuiderzee in September, with the assistance of Johannes de Graaf. The results were satisfactory: they ran into a storm that stopped one of the clocks, but Huygens reckoned he could improve its suspension.[73] He next set about training the man who was to look after the clocks, Thomas Helder (d. 1687), and, in December, he composed written instructions.[74]

These instructions contain a lot of technical information that would otherwise be lost; there is even a drawing (see Figure 3). The clocks were practically identical. They were suspended in gimbals and had a main dial for hours and minutes and a smaller dial under it for the seconds. They each had a remontoire, a small spring of hammered brass in a small barrel being wound every minute on the minute, with a warned construction, the warning taking place at the 30th second. The clocks were to be wound twice per day,

58. *O.C.*, vol. 7, no. 1901, p. 210 (with drawing); this letter specifies a six-inch pendulum.

59. Oldenburg, *Correspondence* (*op. cit.*, note 35), vol. 4, no. 778, p. 174; no. 807, p. 226. The text was published in *Phil. Trans.*, vol. 4, no. 47 (1669), pp. 937-53; see *O.C.*, vol. 6, no. 1743, pp. 446-59. See also *O.C.*, vol. 6, no. 1732, p. 427; no. 1738, p. 440; no. 1742, p. 444.

60. *O.C.*, vol. 7, no. 1912, p. 229; no. 1927, pp. 259-61; no. 1933, p. 269. The longitude work is described in Huygens, *Horologium Oscillatorium* (*op. cit.*, note 20), pp. 16-20. A list, apparently drawn from memory, of those presented with a copy exists (*O.C.*, vol. 7, p. 321, note 2). It does not mention a copy in the Houghton Library at Harvard (f*NC6 H9846 673h[B]). This is inscribed on the title page 'Ex Libris Lomenianis Dono Authoris Amici colendiss: an. M D C L XXVI Brienne'. Louis-Henri de Loménie, Comte de Brienne (1635-1698) was the son of Henri-Auguste de Loménie, Comte de Brienne (1594-1666), who was Secretaire d'État des affaires étrangères. Huygens met both in February 1662 during his visit to Paris and was then not impressed by the younger Loménie's intelligence. In December 1662, the son was in Stockholm, where he was considered an elegant man, and saw a great deal of Constantyn Huygens Sr., but there does not appear to have been any further contact with Christiaan, who may have presented the book at the instigation of his father (*O.C.*, vol. 4, no. 986, p. 69; no. 1081, p. 277; no. 1094, p. 300; *O.C.*, vol. 22, pp. 545, 552, 557). The copy has the errata corrected in manuscript, and there are a few additions (notably to the lower figure of p. 78).

61. *O.C.*, vol. 18, pp. 15-16.

62. *O.C.*, vol. 18, pp. 489-95.

63. *Journal des Sçavans*, 20 February 1675 (= *O.C.*, vol. 7, no. 2014, p. 424). The present argument assumes that Huygens was aware of Hooke's law: there is some evidence that the Académie Royale had by this time studied the properties of springs and the force they exert (*O.C.*, vol. 19, pp. 24, 25, 27); two passages in the much later correspondence between Huygens and Leibniz also suggest this (*O.C.*, vol. 10, no. 2664, p. 52; no. 2667, p. 58). It should be noted that, some ten years later, Huygens mentions this proof for the isochronism of the balance-spring construction as something obvious (*O.C.*, vol. 18, p. 537).

64. For an account of Huygens's work on the balance spring, see J. H. Leopold, 'L'Invention par Christiaan Huygens du Ressort Spiral réglant pour les Montres', in *Huygens et la France* (Paris: Vrin, 1981), pp. 153-7. A shorter version is in *ANCAHA*, vol. 25 (1979), pp. 9-11.

65. *O.C.*, vol. 7, no. 2008, p. 409; no. 2011, p. 419; no. 2013, p. 422; no. 2014, p. 424; no. 2023, p. 437. *O.C.*, vol. 8, no. 2091, p. 11; no. 2185, p. 197.

66. *O.C.*, vol. 18, p. 525. *O.C.*, vol. 21, p. 161. The balance in the actual planetarium (Museum Boerhaave, Leiden) beats one-half seconds.

67. *O.C.*, vol. 8, no. 2279, p. 394; General Archives, The Hague, VOC 241, 31 December 1682; Davids, *Zeewezen* (*op. cit.*, note 44), p. 136. The original document mentions van Ceulen as the maker, Huygens merely as adviser: there appears to be a parallel with the original patent

for the pendulum clock, taken out in the name of the clockmaker Salomon Coster (d. 1659).

68. *O.C.*, vol. 8, no. 2307, p. 429; no. 2314, p. 439; no. 2319, pp. 453-4. *O.C.*, vol. 18, pp. 621-2 (invention of epicyclic winding).

69. *O.C.*, vol. 18, p. 527. It is interesting to note that the winter of 1682-3 was very cold, with frost starting in November; see C. Easton, 'Klimatologische Studies', *Tijdschrift van het Kon. Ned. Aardrijkskundig Genootschap*, 2ᵉ ser., vol. 45 (1928), pp. 248ff.

70. *O.C.*, vol. 18, pp. 527-8. Huygens proves that the system will be isochronous if any one point of the ring moves along a parabola curved around the cylinder defined by the ring. He does this by analogy with the conical pendulum (*Horologium Oscillatorium* [*op. cit.*, note 20], pp. 157-9), which probably accounts for the fact that he calls this construction a pendulum, although it actually is a gravity-controlled balance.

71. *O.C.*, vol. 18, pp. 528-34. *O.C.*, vol. 8, no. 2327, p. 475.

72. *O.C.*, vol. 18, p. 533: the editors suggest a date in 1685 for this piece, but, in view of the following letter, early September 1684 is more likely. *O.C.*, vol. 8, no. 2368, p. 541: this clearly refers to the triangular pendulum.

73. *O.C.*, vol. 9, no. 2392, p. 17; no. 2394, p. 20; no. 2396, p. 24; no. 2397, p. 25; no. 2398, pp. 26-7; no. 2401, pp. 30-1. *O.C.*, vol. 18, pp. 534-5, 539. Johannes de Graaf was a son of Abraham de Graaf, the East India Company's examiner of mariners from 1679 until his death in 1714 (*O.C.*, vol. 9, p. 27, notes 3 and 4; p. 266, note 1; Davids, *Zeewezen* [*op. cit.*, note 44], p. 399).

74. *O.C.*, vol. 9, no. 2406, p. 37; no. 2407, p. 37; no. 2423, pp. 55-76 (French translation of the technical parts: *O.C.*, vol. 18, pp. 539-43). See also *O.C.*, vol. 9, no. 2520, pp. 292-3.

75. Several specimen calculations in the instructions are taken from de la Voye's observations during the second French voyage. The warned remontoire (a construction similar to a striking train with partial unlocking near the hour) had been mentioned in September(?) 1684 (*O.C.*, vol. 18, p. 533). A few details derive from Huygens's subsequent report (see note 78). The remontoire spring is described as being of beaten *koper* or copper, which is assumed to be short for *geel koper* or brass. The cursor weights are not shown in the drawing.

76. *O.C.*, vol. 9, no. 2424, p. 77; no. 2445, p. 110. See also Huygens's report, note 78. The clockmaker was probably Willem van der Dussen of Dordrecht, who died in 1689 (*O.C.*, vol. 9, no. 2445, p. 110; no. 2488, pp. 222-3; *O.C.*, vol. 18, p. 515, note 6).

77. *O.C.*, vol. 9, no. 2481, p. 208; no. 2488, pp. 222-3; no. 2492, p. 230; no. 2510, p. 255. *O.C.*, vol. 18, pp. 544, 637-8. It appears that the observers were not taken seriously on board: *O.C.*, vol. 9, no. 2519, p. 289; no. 2615, p. 479; see also note 44.

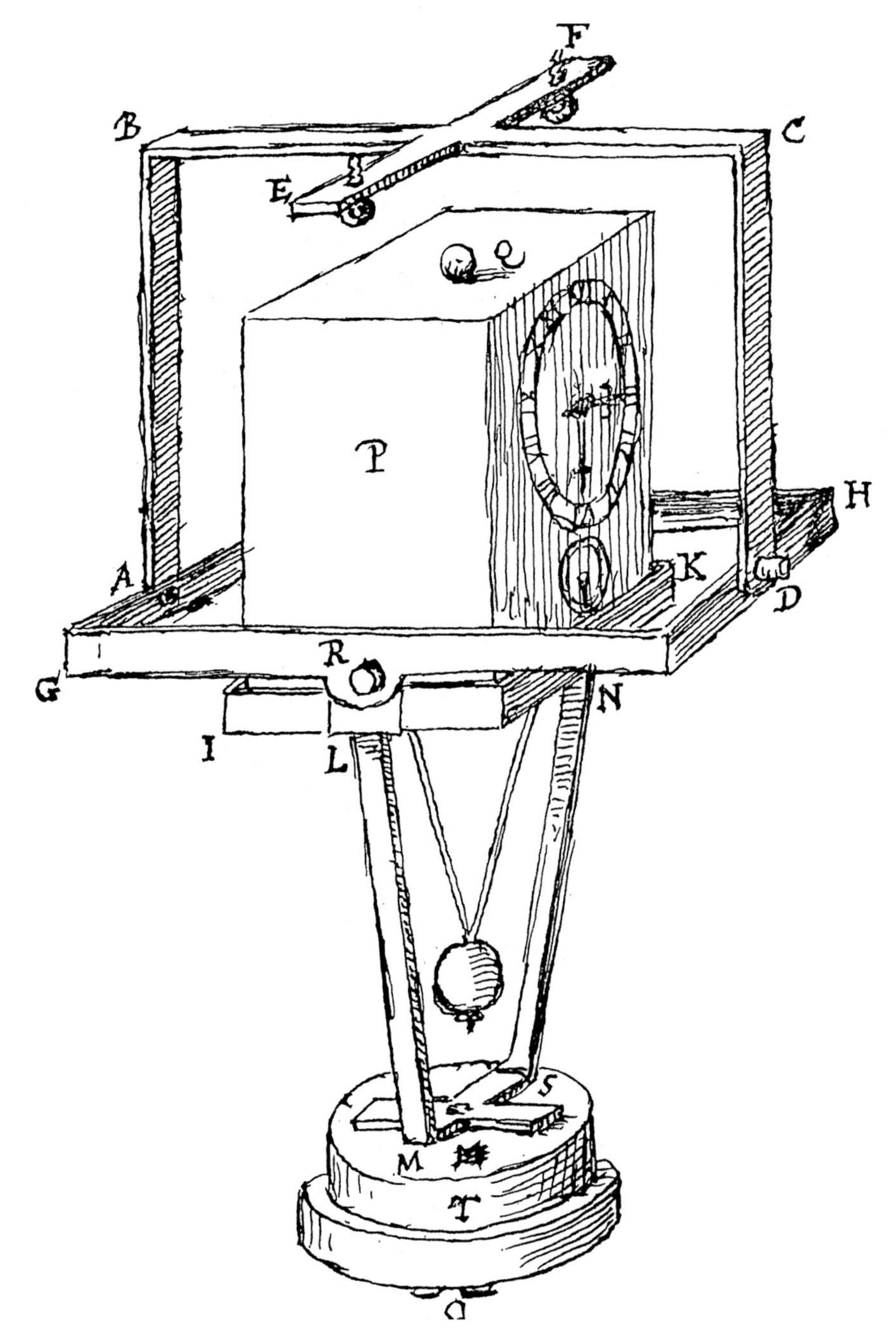


**Figure 3. One of the clocks with triangular pendulum, tried in 1686-7 and 1690-2. *O. C.*, vol. 9, no. 2423, p. 56; vol. 18, p. 540.**

and care was to be taken not to wind during warning or winding of the remontoire. The main train was also spring-driven and had fusee and chain. The triangular pendulum could be unhooked and was to be removed during transportation, and there were cursor weights.[75]

In April 1686, the clocks were sent to Amsterdam, and, in June, the ship *Alcmaer* left with Helder, Johannes de Graaf as second observer, the clockmaker van der Dussen, and the clocks.[76] They were only to go as far as Cape of Good Hope and return from there, but even so it took until August 1687 for the clocks to return—without Helder, who died on the return journey. The clocks had kept going but the results had not been as good as hoped for, and Huygens set about examining the reports.[77]

Huygens received all the papers that returned with the clock, and, in

April 1688, he sent the East India Company his comments. No proper observations to establish local time had been made in Texel, and so he concentrated on the return journey. He corrected the results of the observers by including the recently discovered difference in the length of the pendulum near the equator. He enclosed a chart on which he plotted the course, established by dead reckoning, as calculated by de Graaf, and the corrected course, calculated by himself. This corrected course gave fairly accurate results, and Huygens concluded that after a voyage of four months an accuracy of about 1° could be reached with his clocks.[78]

All this may have confused the Company's directors somewhat, and they submitted the report to Professor Burchard de Volder (1643-1709) in Leiden for his consideration. De Volder suggested that the irregular going of the clocks might in part have been caused by temperature changes (see below) but felt on the whole that the results were encouraging and recommended more trials.[79] The directors decided accordingly. The clocks were returned to Huygens for minor changes. In his report, Huygens had suggested, for example, that the suspension be strengthened and the remontoire springs replaced by steel ones. By May 1690, the clocks were fully prepared. Surprisingly, the work had been done not by van Ceulen (apparently Huygens had broken off relation with this superb clockmaker), but by Pieter Visbagh (d. 1722).[80]

In the summer of 1690, Huygens instructed Johannes de Graaf. There were some more experiments, particularly with the suspension, and, by November, everything was ready for a new trial.[81] In the end of December, the clocks left for Cape of Good Hope on the ship *Brandenburgh* with de Graaf, the second observer van Laer, and the clockmaker Gilles Meybos.[82]

Again there was a long time to wait: the clocks arrived at the Cape in

78. *O.C.*, vol. 9, no. 2516, p. 266; no. 2517, pp. 267-8; for the actual report, see *O.C.*, vol. 9, no. 2519, pp. 272-91 and the map with the three courses at the end of this volume; *O.C.*, vol. 18, pp. 544, 639-42. The variation in length of the pendulum was discovered by Richer (see note 57). Huygens explained this effect out of the changing centrifugal force and included a table for this variation in the report; he calculated that a pendulum clock going accurately at the poles will lose 2.5 minutes per day at the equator. The difference in longitude between Texel and Cape of Good Hope had recently been established (by Jupiter's moons) as 14° 25'.

79. *O.C.*, vol. 9, no. 2538, p. 317; no. 2539, pp. 319-20; no. 2547, pp. 339-43. *O.C.*, vol. 18, pp. 544-5. Burchard de Volder was professor of mathematics at Leiden University.

80. *O.C.*, vol. 9, no. 2545, p. 337; no. 2546, p. 338; no. 2588, pp. 418-19; no. 2615, pp. 477-9 (see correction in *O.C.*, vol. 18, p. 516, note 3); no. 2602, pp. 452-3. *O.C.*, vol. 10, no. 2670, p. 72. Huygens had also suggested that the duration be increased to 24 hours and the period of the remontoire slowed down to two minutes, but it is doubtful that such major changes were executed. Pieter Visbagh took over the shop of Salomon Coster in 1660.

81. *O.C.*, vol. 9, no. 2602, pp. 452-3; no. 2609, pp. 467-8; no. 2615, pp. 477-9; no. 2620, p. 491; no. 2621, pp. 492-3; no. 2622, pp. 494-5; no. 2637, p. 553.

82. *O.C.*, vol. 9, no. 2638, p. 554; no. 2642, pp. 567-8; no. 2645, p. 577; no. 2646, p. 578; no. 2647, p. 579; no. 2648, p. 580; no. 2649, p. 581; no. 2650, p. 582; no. 2651, p. 583; no. 2652, p. 583; no. 2653, p. 584. *O.C.*, vol. 10, no. 2656, p. 2. *O.C.*, vol. 18, p. 545. Gilles Meybos was presumably related to the Amsterdam clockmaker Carel Meybos (b. 1657, mentioned 1690).

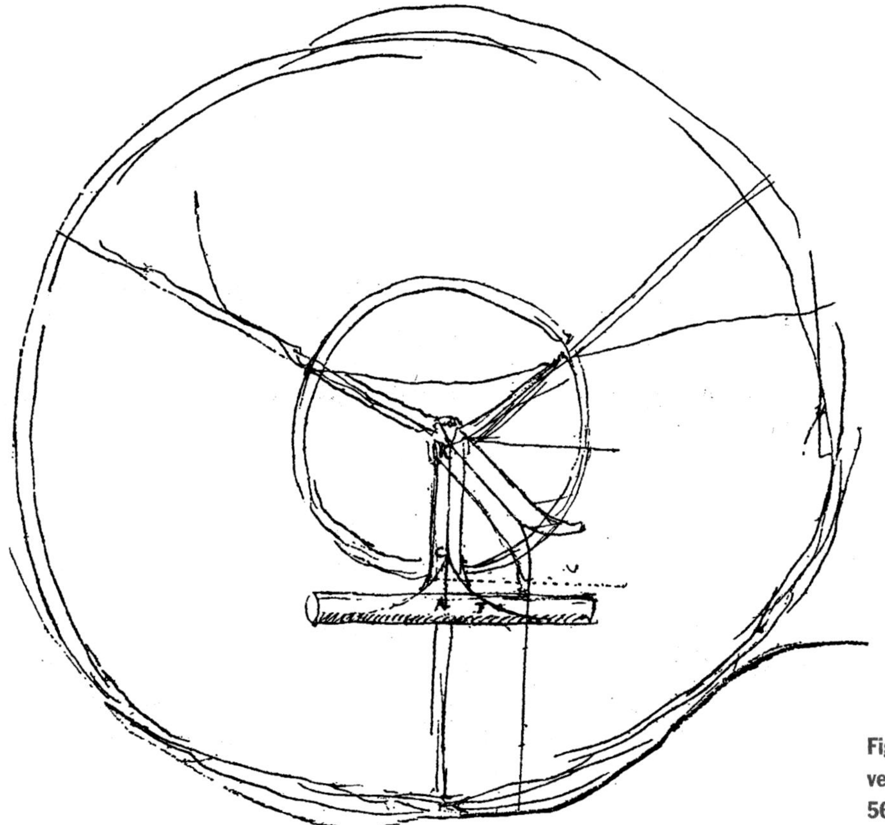

**Figure 4. The Perfect Marine Balance, version of March 1693. *O. C.*, vol. 18, p. 562.**

June, but illness of de Graaf caused delays, and the clocks did not return until October 1692.[83] De Graaf's first report was unfavourable. They had had trouble with the spring of one of the clocks, and the results were poor. All the papers were again sent to Huygens for his comments.[84] He found that much of the supposed error was due to a mistake in de Graaf's calculations: south of the equator, he had added the correction for the latitude where it should have been subtracted. But even so, and after consultation with de Volder, Huygens had to admit that the results were mostly inconclusive.[85]

That was the end of the involvement of the East India Company. And Huygens, too, by this time had lost interest in these timekeepers with triangular pendulums, for he had picked up again an earlier invention: a construction known as the Perfect Marine Balance *(Balancier Marin Parfait)*. This is essentially a balance moving in a vertical plane and controlled by various devices based on gravity. The earliest notes on this appear to date from 1684, shortly after he abandoned the tri-cordal pendulum; these sketches involved a weighted chain, arranged in such a way that the returning force was proportional to the angular distance from the neutral position, thus rendering the construction isochronous (as Huygens clearly stated in his notes). A second set of notes at this time replaced the weighted chain by a float, partly submerged in oil or mercury with the same effect.[86]

Huygens returned to this construction in early 1693. First, he wondered about the best way to suspend the balance, whether on knife-edges or on rollers. Next, in early February, he began practical experiments, both with various types of chains and with a float in water (which he found to involve too much friction).[87] Then, in early March, he hit upon the best solution: he made a large balance vibrate under the controlling influence of a small weight suspended between curved cheeks mounted on the staff of the balance (compare Figure 4). Huygens set about calculating the required shape of the cheeks[88] and notified the East India Company of a

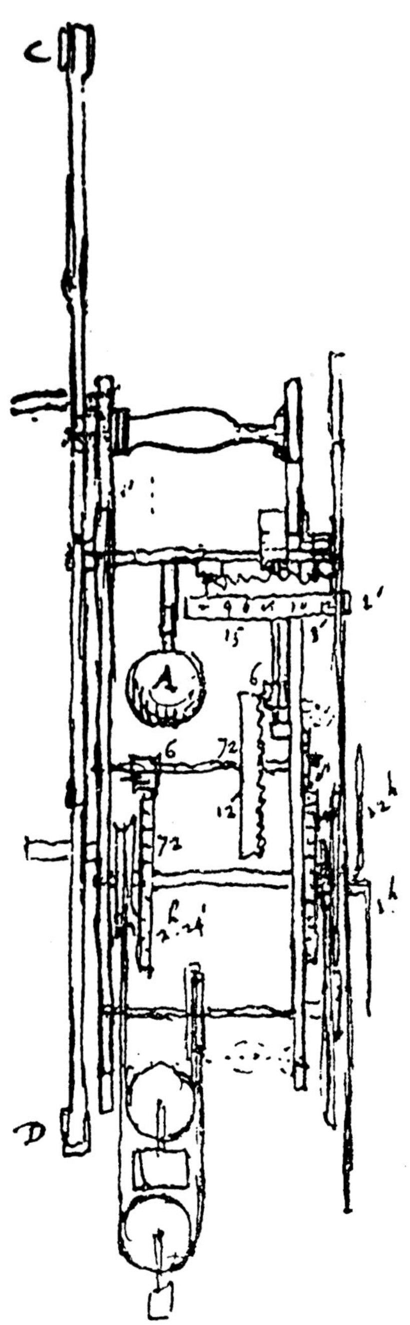


**Figure 5. Movement of a clock with Perfect Marine Balance, March 1693. The train-count shows that the balance was to beat double-seconds. *O.C.*, vol. 18, p. 569.**

83. *O.C.*, vol. 10, no. 2674, pp. 79-80; no. 2715, pp. 205-6; no. 2720, pp. 207-8; no. 2744, p. 269; no. 2767, p. 323; no. 2703, p. 166 (initially misdated to 1691, but see no. 2767).

84. *O.C.*, vol. 10, no. 2772, p. 339; no. 2773, p. 340; no. 2774, p. 341; no. 2789, pp. 396-8. *O.C.*, vol. 18, pp. 642-51.

85. *O.C.*, vol. 10, no. 2786, pp. 389-90; no. 2789, pp. 396-8; no. 2795, pp. 422-3; no. 2796, pp. 423-4; no. 2798, pp. 433-4; no. 2799, p. 435; no. 2800, pp. 435-6; no. 2802, p. 442; no. 2803, pp. 443-4.

86. *O.C.*, vol. 18, pp. 536-8.

87. *O.C.*, vol. 18, pp. 546, 548-61.

88. *O.C.*, vol. 18, pp. 562-70.

new invention, but they do not appear to have been particularly enthusiastic after the previous experiments.[89]

In September 1693, Huygens informed the scholarly world that he had discovered a new curve useful for horology. He kept its nature secret (it was the involute of a circle) but sent an anagram to the *Acta Eruditorum;* Leibniz, familiar with Huygens's work on the cycloid, reacted with the remark quoted at the beginning of this paper.[90] In March 1694, Huygens continued this work and increased his understanding of the theoretical problems involved (it is only at this point that he comes to a true distinction between mass and weight). In the practical work, too, he made significant advances. The balance is now a two-armed, seconds-ticking assembly; for the suspension of the balance Huygens considered knife-edges, rollers, and ribbons.[91] By mid-March, the new clock was under construction; a month later, it was finished, and Huygens began to experiment with it. It had been made by yet another clockmaker, Barent van der Cloese, but it appears that Huygens did most of the final adjusting himself. The clock was weight-driven and had a conventional dial for hours and minutes with the seconds, appearing in an aperture over the dial, apparently engraved on the 'scape-wheel. The balance had knife-edge suspension and, at each end, a weight suspended between cheeks.[92] One of those to whom Huygens showed the new clock appears to have been Crown Prince Friedrich von Hessen-Kassel (1676-1751). The Prince's father, Landgraf Karl, expressed the wish to have one like it, but Huygens politely declined.[93]

It appears that in March 1695, Huygens had an existing clock (with equation of time) converted to the new balance,[94] but about this we have no further details. Indeed, it is the last we hear from Huygens about horology, for he died on 8 July 1695. A publication on the new clock, which he had announced shortly before his death, was never written.[95]

★ ★ ★ ★ ★

Having surveyed the entire range of Huygens's experiments with marine timekeepers, we are faced with two questions: What did he achieve and where did he go wrong?

What he achieved has often been under-estimated, because there were no lasting practical results. There is no doubt, however, that Huygens cleared a lot of the initial problems. The *Kort Onderwys* is not only the first set of directions for the use of marine timepieces: it explains the problems of establishing local time,[96] including the correction for the equation of time. His timekeepers demonstrated that the solution for the longitude problem was not entirely out of reach, even though his machines failed to yield consistently reliable results. And most importantly, his theoretical work was the basis for all subsequent work and resulted, for example, in the construction that was ultimately to provide the solution: the balance with balance spring.

Where did he go wrong? There were two main points. First, there was the temperature influence. At the end of the seventeenth century, the effect of heat on solid bodies was little understood, and Huygens never fully realised that temperature changes affected his pendulum clocks. As late as 1690, he dismissed temperature influence as of no consequence for moderate temperatures.[97] This also frustrated his attempts with the balance spring. When he realised that the spring was very sensitive to temperature changes, he had

89. *O.C.*, vol. 10, no. 2796, pp. 423-4; no. 2878, pp. 684-5.

90. *O.C.*, vol. 10, no. 2823, pp. 514-15 (to Leibniz for the *Acta* [Leipzig, 1682-]); the anagram solved in the margin of Huygens's copy: 'Flexilis ambitum cum linea deserit orbem', 'the ambit of the flexilis when the line unwraps a circle' ('linea flexilis' is the name Huygens uses for the unwrapping line that generates the involute; see *Horologium Oscillatorium* [*op. cit.*, note 20], p. 59, and Yoder, *Unrolling Time* [*op. cit.*, note 3], p. 6). *Acta*, October 1693. The correspondence with Leibniz and with de l'Hospital: *O.C.*, vol. 10, no. 2820, p. 499; no. 2828, p. 538; no. 2829, pp. 539-40; no. 2830, p. 544; no. 2876, p. 682.

91. *O.C.*, vol. 18, pp. 571-91.

92. *O.C.*, vol. 10, no. 2846, p. 584; no. 2850, p. 598; no. 2854, pp. 609-10; no. 2859, p. 626; no. 2863, p. 639. *O.C.*, vol. 18, pp. 592-6. Barent van der Cloese is mentioned in The Hague 1688-1719.

93. *O.C.*, vol. 10, no. 2878, pp. 684-5 (October 1694). Crown Prince Friedrich, subsequently Landgraf Friedrich I and also King of Sweden, lived in The Hague from June 1692 until May 1694, so it is likely that it was he who told Landgraf Karl (1654-1730) about the clock; Alexander Rolas de Rosey, to whom Huygens's letter is addressed, was Chamberlain to the Prince during his stay in The Hague (for this voyage, see Hessisches Staatsarchiv Marburg, 4a 57.24, 25; 4a 72.2). Another who saw the clock was Christiaan's brother Constantyn; see *Journaal van C. Huygens, den zoon 1670-78, 1688-96,* Werken uitgegeven door het historisch gezelschap te Utrecht, Nieuwe Serie nos. 23, 25, 32, 46 (Utrecht, 1876-88), 22 May, 8 October 1694 (vol. 4, pp. 347, 422).

94. *O.C.*, vol. 10, no. 2891, pp. 709-10 (last surviving letter of Huygens).

95. *O.C.*, vol. 10, no. 2891, pp. 709-10; no. 2892, p. 711.

96. Realising the difficulty of establishing local noon directly, Huygens recommended middling the time of sunrise and sunset (a method already used at the end of the sixteenth century: Ploeg, *Constantyn Huygens* [*op. cit.*, note 8], p. 55), or taking observations of equal altitudes in the early morning and late afternoon.

no way of compensating for the effect, and so he had to abandon this promising construction.

The other point where Huygens went consistently wrong is one in which all experimenters of the next generation followed him: he was convinced that heavy, slow-moving constructions were better than light, briskly moving ones. There are many indications for this. His early insistence on long pendulums and his subsequent attempts to increase the length of the pendulum in his short-pendulum clocks all point in this direction. In fact, it is not until 1672 that he finally became reconciled to short pendulums, realising that his first priority was to make the clocks reliable rather than ideally accurate.[98] But that did not change his underlying conviction, and, with the balance-spring construction, he mentioned on several occasions that the balances for marine timekeepers were to be large for greater precision.[99] The same applies to the last experiments, where again he insisted on a large, seconds-beating balance.[100] It was a prejudice that was only overcome by H.4 in 1759.

Huygens's invention of the pendulum clock and of the balance spring revolutionised horology. His *Horologium Oscillatorium* remained the standard text-book (together with the *Horologium* and Latin translations of the 'Instructions', a letter on the sympathetic motion of pendulums, and the description of the balance spring, it was reprinted in 1724 and again in 1751).[101] But it is clear that other contributions of Huygens also became part of the clock-making tradition. Epicyclic winding, an invention claimed by several French makers in the eighteenth century[102] and subsequently used by Arnold and others, would appear to be one of them. The influence of Huygens's work on Harrison has long been recognised.[103] And the longitude machine of Henry Sully (1680-1728) shows too many points of similarity with Huygens's *Balancier Marin Parfait* to deny a direct connection.[104]

How did this transfer of practical information take place? Some of it had been printed and thus was freely available. Otherwise, there are several possibilities to be considered. Huygens left his scholarly papers, including his notes on the last experiments in horology, to the library of Leiden University, where they could be consulted. Henry Sully, who lived in Leiden for some years, may well have seen these notes; in addition, he may have talked to the maker of the last clock, van der Cloese, who was still active in The Hague. Furthermore, a marine timekeeper (most likely that of 1694) remained in the Huygens family until 1754, when it was auctioned and disappeared.[105] Until then, anybody with a good introduction could have seen it. And finally, there was direct transmission by word of mouth. Huygens talked freely to his brother Constantyn, who was a secretary to King William III and lived in London in the years 1689-96. We know that Constantyn discussed his brother's work: in January 1692, he asked Christiaan about the trial of the sea watches, because several of his acquaintances had asked about this.[106] Constantyn was on friendly terms with Thomas Tompion (1639-1713), and we know that Christiaan expected his brother to talk about work in the Netherlands. In his very last surviving letter, written in March 1695, Huygens assumes as a matter of course that Constantyn has told Tompion about the Perfect Marine Balance.[107]

97. Correspondence with Denis Papin: *O.C.*, vol. 9, no. 2595 p. 432; no. 2617, p. 485. See also de Volder's suggestion in 1689 that temperature changes might have made the going of these clocks irregular: *O.C.*, vol. 9, no. 2547, p. 343. This inability to provide compensation for temperature changes (and other suspected influences of the air) resulted in a number of longitude proposals in which a clock was enclosed in an air-tight vessel; the earliest of these appears to have been by Campani of Rome in 1669 (*O.C.*, vol. 6, no. 1761, p. 495; no. 1779, p. 533).

98. *O.C.*, vol. 7, no. 1901, p. 210; *cf.* Huygens, *Horologium Oscillatorium* (*op. cit.*, note 20), p. 19.

99. *O.C.*, vol. 7, no. 2023, p. 437. *O.C.*, vol. 8, no. 2091, p. 11; no. 2185, p. 197 ('construits en grand volume parce que la justesse croit a mesure'). *O.C.*, vol. 21, p. 161 (the balance in the planetarium beats one-half seconds, *O.C.*, vol. 18, p. 525).

100. *O.C.*, vol. 18, p. 569. *O.C.*, vol. 10, no. 2859, p. 626.

101. Christiaan Huygens, *Opera Varia* (Leiden, 1724), vol. 1; Christiaan Huygens, *Opera Mechanica* (Leiden, 1751), vol. 1.

102. Described in A. Thiout, *Traité de l'Horlogerie*, 2 vols. (Paris, 1741; facsimile reprint with extra material, Paris, 1972), vol. 2, p. 383 and plate 38, fig. 14; subsequently claimed by Massoteau de St. Vincent, 'Lettre', in *Memoires pour l'Histoire des Sciences et des Beaux Arts* (September 1742), pp. 1667-70 (reprinted at the end of the facsimile edition of Thiout).

103. See W. S. Laycock, *The Lost Science of John "Longitude" Harrison* (Ashford, England: Brant Wright Associates, 1976), pp. 84-94.

104. First noted in *O.C.*, vol. 18, pp. 520, 546-7; also p. 701.

105. *O.C.*, vol. 15, pp. 19, 21; *cf.* p. 22, note 1.

106. *O.C.*, vol. 10, no. 2725, p. 220 ('Quelques curieux icy m'en ont demandé des nouvelles').

107. *O.C.*, vol. 10, no. 2891, pp. 709-10. Christiaan had met Tompion during his London visit of 1689: *O.C.*, vol. 22, p. 746.

(6)

Tafel van vereffening des Tijdts.

| Dagen. | Januar. Min. | Januar. Sec. | Febr. Min. | Febr. Sec. | Mart. Min. | Mart. Sec. | Apr. Min. | Apr. Sec. | May. Min. | May. Sec. | Jun. Min. | Jun. Sec. |
|---|---|---|---|---|---|---|---|---|---|---|---|---|
| 1 | 10 | 40 | 0 | 32 | 2 | 15 | 11 | 18 | 18 | 32 | 18 | 10 |
| 2 | 10 | 10 | 0 | 24 | 2 | 28 | 11 | 37 | 18 | 39 | 18 | 1 |
| 3 | 9 | 41 | 0 | 18 | 2 | 42 | 11 | 56 | 18 | 46 | 17 | 51 |
| 4 | 9 | 13 | 0 | 13 | 2 | 56 | 12 | 15 | 18 | 53 | 17 | 41 |
| 5 | 8 | 45 | 0 | 9 | 3 | 11 | 12 | 34 | 18 | 59 | 17 | 30 |
| 6 | 8 | 17 | 0 | 6 | 3 | 26 | 12 | 53 | 19 | 4 | 17 | 19 |
| 7 | 7 | 50 | 0 | 3 | 3 | 41 | 13 | 12 | 19 | 9 | 17 | 8 |
| 8 | 7 | 23 | 0 | 1 | 3 | 56 | 13 | 31 | 19 | 14 | 16 | 57 |
| 9 | 6 | 58 | 0 | 0 | 4 | 12 | 13 | 49 | 19 | 18 | 16 | 46 |
| 10 | 6 | 34 | 0 | 0 | 4 | 29 | 14 | 6 | 19 | 22 | 16 | 35 |
| 11 | 6 | 10 | 0 | 0 | 4 | 46 | 14 | 23 | 19 | 25 | 16 | 24 |
| 12 | 5 | 47 | 0 | 2 | 5 | 4 | 14 | 39 | 19 | 28 | 16 | 13 |
| 13 | 5 | 24 | 0 | 4 | 5 | 22 | 14 | 55 | 19 | 29 | 16 | 1 |
| 14 | 5 | 2 | 0 | 8 | 5 | 40 | 15 | 10 | 19 | 29 | 15 | 49 |
| 15 | 4 | 41 | 0 | 12 | 5 | 58 | 15 | 25 | 19 | 29 | 15 | 37 |
| 16 | 4 | 21 | 0 | 16 | 6 | 16 | 15 | 39 | 19 | 28 | 15 | 24 |
| 17 | 4 | 2 | 0 | 21 | 6 | 33 | 15 | 53 | 19 | 26 | 15 | 11 |
| 18 | 3 | 44 | 0 | 26 | 6 | 51 | 16 | 7 | 19 | 24 | 14 | 58 |
| 19 | 3 | 27 | 0 | 32 | 7 | 9 | 16 | 21 | 19 | 21 | 14 | 45 |
| 20 | 3 | 11 | 0 | 40 | 7 | 27 | 16 | 34 | 19 | 18 | 14 | 32 |
| 21 | 2 | 55 | 0 | 48 | 7 | 45 | 16 | 47 | 19 | 15 | 14 | 19 |
| 22 | 2 | 39 | 0 | 57 | 8 | 3 | 16 | 59 | 19 | 11 | 14 | 6 |
| 23 | 2 | 23 | 1 | 6 | 8 | 22 | 17 | 11 | 19 | 7 | 13 | 53 |
| 24 | 2 | 7 | 1 | 16 | 8 | 41 | 17 | 22 | 19 | 2 | 13 | 40 |
| 25 | 1 | 52 | 1 | 26 | 9 | 1 | 17 | 33 | 18 | 57 | 13 | 27 |
| 26 | 1 | 38 | 1 | 37 | 9 | 21 | 17 | 43 | 18 | 51 | 13 | 15 |
| 27 | 1 | 25 | 1 | 49 | 9 | 41 | 17 | 53 | 18 | 45 | 13 | 3 |
| 28 | 1 | 13 | 2 | 2 | 10 | 1 | 18 | 3 | 18 | 39 | 12 | 52 |
| 29 | 1 | 2 | | | 10 | 21 | 18 | 13 | 18 | 33 | 12 | 41 |
| 30 | 0 | 51 | | | 10 | 40 | 18 | 23 | 18 | 26 | 12 | 30 |
| 31 | 0 | 41 | | | 10 | 59 | | | 18 | 18 | | |

(7)

Tafel van vereffening des Tijdts.

| Dagen. | Jul. Min. | Jul. Sec. | Aug. Min. | Aug. Sec. | Sept. Min. | Sept. Sec. | Octob. Min. | Octob. Sec. | Nov. Min. | Nov. Sec. | Dec. Min. | Dec. Sec. |
|---|---|---|---|---|---|---|---|---|---|---|---|---|
| 1 | 12 | 19 | 10 | 4 | 16 | 23 | 26 | 30 | 31 | 55 | 25 | 34 |
| 2 | 12 | 8 | 10 | 8 | 16 | 42 | 26 | 49 | 31 | 55 | 25 | 10 |
| 3 | 11 | 58 | 10 | 13 | 17 | 1 | 27 | 8 | 31 | 54 | 24 | 45 |
| 4 | 11 | 48 | 10 | 18 | 17 | 21 | 27 | 26 | 31 | 52 | 24 | 20 |
| 5 | 11 | 38 | 10 | 23 | 17 | 41 | 27 | 43 | 31 | 50 | 23 | 55 |
| 6 | 11 | 28 | 10 | 28 | 18 | 1 | 28 | 0 | 31 | 47 | 23 | 30 |
| 7 | 11 | 18 | 10 | 34 | 18 | 21 | 28 | 16 | 31 | 43 | 23 | 4 |
| 8 | 11 | 9 | 10 | 41 | 18 | 41 | 28 | 32 | 31 | 37 | 22 | 38 |
| 9 | 11 | 0 | 10 | 49 | 19 | 1 | 28 | 47 | 31 | 30 | 22 | 11 |
| 10 | 10 | 52 | 10 | 58 | 19 | 21 | 29 | 2 | 31 | 22 | 21 | 43 |
| 11 | 10 | 47 | 11 | 7 | 19 | 41 | 29 | 16 | 31 | 13 | 21 | 14 |
| 12 | 10 | 38 | 11 | 16 | 20 | 1 | 29 | 30 | 31 | 3 | 20 | 44 |
| 13 | 10 | 31 | 11 | 25 | 20 | 22 | 29 | 43 | 30 | 53 | 20 | 14 |
| 14 | 10 | 25 | 11 | 36 | 20 | 43 | 29 | 56 | 30 | 43 | 19 | 44 |
| 15 | 10 | 19 | 11 | 48 | 21 | 4 | 30 | 9 | 30 | 32 | 19 | 14 |
| 16 | 10 | 13 | 12 | 1 | 21 | 25 | 30 | 22 | 30 | 20 | 18 | 44 |
| 17 | 10 | 7 | 12 | 14 | 21 | 47 | 30 | 34 | 30 | 8 | 18 | 14 |
| 18 | 10 | 2 | 12 | 28 | 22 | 9 | 30 | 45 | 29 | 55 | 17 | 44 |
| 19 | 9 | 58 | 12 | 42 | 22 | 31 | 30 | 55 | 29 | 40 | 17 | 14 |
| 20 | 9 | 54 | 12 | 57 | 22 | 52 | 31 | 4 | 29 | 23 | 16 | 44 |
| 21 | 9 | 51 | 13 | 12 | 23 | 13 | 31 | 12 | 29 | 6 | 16 | 14 |
| 22 | 9 | 49 | 13 | 27 | 23 | 33 | 31 | 19 | 28 | 48 | 15 | 44 |
| 23 | 9 | 47 | 13 | 43 | 23 | 53 | 31 | 26 | 28 | 30 | 15 | 14 |
| 24 | 9 | 46 | 13 | 59 | 24 | 13 | 31 | 32 | 28 | 11 | 14 | 43 |
| 25 | 9 | 46 | 14 | 16 | 24 | 33 | 31 | 38 | 27 | 51 | 14 | 12 |
| 26 | 9 | 46 | 14 | 33 | 24 | 53 | 31 | 43 | 27 | 30 | 13 | 41 |
| 27 | 9 | 47 | 14 | 50 | 25 | 13 | 31 | 47 | 27 | 8 | 13 | 10 |
| 28 | 9 | 49 | 15 | 8 | 25 | 33 | 31 | 50 | 26 | 45 | 12 | 40 |
| 29 | 9 | 52 | 15 | 26 | 25 | 52 | 31 | 53 | 26 | 22 | 12 | 10 |
| 30 | 9 | 56 | 15 | 45 | 26 | 11 | 31 | 55 | 25 | 58 | 11 | 40 |
| 31 | 10 | 0 | 16 | 4 | | | 31 | 55 | | | 11 | 10 |

**Figure 6. Table for the equation of time, published by Christiaan Huygens in 1665 in *Kort Onderwys...* ([The Hague]), pp. 6 and 7. This table, which shows, for each day of the year, the number of minutes and seconds that must be added to Sun time in order to set a clock correctly to mean time, was the first to be calculated specifically for use with a timekeeper. A full discussion of Huygens's work on the equation of time will be the subject of a forthcoming article. See also Appendix D in this volume. Courtesy of The Time Museum, Rockford, Illinois.**

# In the Wake of the Act, but Mainly Before

*A. J. Turner*

**After studying history at Oxford, Anthony Turner concentrated his research in the fields of early scientific instruments, horology, and the social history of science. He has worked in several museums in Great Britain and is the author of numerous books and articles, including *The Clockwork of the Heavens* (1973), *Science and Music in 18th Century Bath* (1977), *Early Scientific Instruments: Europe 1400-1800* (1987), *From Pleasure and Profit to Science and Security: Etienne Lenoir and the transformation of Precision Instrument-making in France, 1760-1830* (1989), *Time* (1990), *Pierre Gassendi, explorateur des sciences* (1992), *Of Time and Measurement: Studies in the History of Horology and Fine Technology* (1993), and *Mathematical Instruments in antiquity and the Middle Ages: an Introduction* (1994). When not engaged in historical research, Turner organizes exhibitions and is a partner in an antiquarian book business based in Greenwich, England. He has lived in France since 1979.**

# In the Wake of the Act, but Mainly Before

*A. J. Turner*

The 1714 Longitude Act must surely rank as one of the most uncontroversial pieces of major legislation ever to pass through Parliament. Between the date when the petition that provoked the Act was first presented and the date when the bill was accepted by the House of Lords, a mere 70 days elapsed. Beginning as an attempt to obtain funding for the proposal by Humphry Ditton (1675-1715) and William Whiston (1667-1752) to moor ships along the trade routes that would fire projectiles when it was midnight at the Pic de Tenerife, thus marking a base time, the question was quickly generalised in committee. As Dr. Samuel Clarke (1675-1729) remarked, 'there could no Discredit arise to the Government, in promising a Reward in general, without respect to any particular Project, to such Person or Persons who should discover the Longitude at Sea'.[1] Adopting from the original petition[2] the suggestion that the size of the reward should be proportional to the degree of accuracy achieved, the expert committee composed of Isaac Newton (1642-1727), Samuel Clarke, Edmond Halley (*ca.* 1656-1742), and Roger Cotes (1682-1716) recommended to the House 'that a Reward be settled by Parliament upon such Person or Persons as shall discover a more certain and practicable Method of ascertaining Longitude than any yet in Practice; and the said Reward to be proportional to the Degree of Exactitude to which the said Method shall reach'.[3] The report was accepted on 11 June 1714. The bill was drafted by 17 June, when it received its first reading, and committed on 22 June, when it passed its second reading. On 1 July, the committee's report was accepted with only one amendment. On 3 July, the bill was sent to the Lords, who accepted it on 8 July, Royal Assent being given the next day.[4]

The speed and ease with which the bill passed are a reflection not only of its uncontroversial nature, but also of the fact that it could be seen to be urgent. No one in the House needed to be convinced of the necessity of encouraging research into ways of finding longitude following a series of disastrous losses at sea.[5] The subject was already present in the consciousness of the nation and was already the subject of active research. But, although the ferment of activity in England in the last decade of the seventeenth century

1. *Journals of the House of Commons,* vol. 17 (1711-14), 11 June 1714, p. 677.

2. A printed copy of what seems to be Whiston and Ditton's petition is preserved in the Houghton Library, Harvard University. See Appendix 1 at the end of this paper. For the method, see the authors' own description in William Whiston and Humphry Ditton, *A New Method For Discovering the Longitude Both At Sea and Land,...* (London, 1714). *Cf.* the comments in Derek Howse, *Greenwich time and the discovery of the longitude* (Oxford: Oxford University Press, 1980), pp. 47-50; the account in J. E. Force, *William Whiston, Honest Newtonian* (Cambridge: Cambridge University Press, 1985), p. 22; and the recent discussion by Larry Stewart, *The Rise of Public Science: Rhetoric, Technology, and Natural Philosophy in Newtonian Britain, 1660-1750* (Cambridge: Cambridge University Press, 1993), chap. 6, which came to my attention only after the present paper had been completed.

3. *Journals of the House of Commons,* vol. 17 (1711-14), 11 June 1714, p. 677.

4. *Ibid.*, p. 721. The stages of the progression of the Act through Parliament can also be followed with slightly greater detail in the reports printed daily in the Commons' proceedings, *Votes of the House of Commons,* 1714.

5. Howse, *Greenwich time* (*op. cit.*, note 2), pp. 45-7.

6. *Ibid.*, pp. 6-10.

7. W. G. L. Randles, 'Portuguese and Spanish attempts to measure Longitude in the 16th Century', *Vistas in Astronomy,* vol. 28, parts 1/2 (1985), pp. 235-41; reprinted with an additional appendix in *Boletim da Biblioteca da Universidade de Coimbra,* vol. 39 (Centro de Estudos de História e Cartografia Antiga).

8. Rupert T. Gould, *The Marine Chronometer: Its History and Development* (London: J. D. Potter, 1923), pp. 11-12; Silvio A. Bedini, *The Pulse of Time: Galileo Galilei, the determination of longitude, and the pendulum clock,* Biblioteca di Nuncius, Studi e Testi, vol. 3 (Florence: Leo S. Olschki, 1991), pp. 9-14.

and the first decades of the eighteenth that gave rise to the 1714 Act can be delineated, it should always be remembered that longitude was not just an English problem, but one that affected all the nations of Europe that engaged in trans-oceanic trade. Unfortunately, because success was ultimately achieved by English clockmakers and because the subject has been most often handled by English historians, accounts of the 'longitude story' are heavily Anglocentric. They are also whiggish, concentrating on the successful method by chronometers and discussing far less the equally successful, but less popular, method of lunar distances, and relegating to footnotes and patronising quips the numerous methods tried that contributed only the fact of their failure to the ultimate success. Yet these unsuccessful efforts are just as much part of the development of understanding of the longitude problem as their more successful rivals. And they are just as worthy of study.

Attempts to find ways of determining longitude accurately were coterminous with oceanic travel. Already in the sixteenth century the basic methods had been adumbrated—comparing the time of the occurrence of a single celestial phenomenon in two different places, or comparing local time with the time of a reference place carried by a timekeeper.[6] Attempts to fix the longitude of several places were made by Spanish scholars in the sixteenth and early seventeenth centuries.[7] For the finding of longitude at sea, prizes were offered either formally by the state, as in Spain,[8] or informally by allowing it to become common knowledge that there would be reward for a viable method. That generous reward would be given to he who found a practical solution was evident, and it was this desire of gain that underlies not only the well-known efforts of Galileo and Huygens,[9] but also the now less well-known, though notorious in the seventeenth century, claims by Jean-Baptiste Morin (1583-1656),[10] as well as those of a host of now virtually forgotten men. But these men, it should be noted, were, in the first part of the seventeenth century, rather seldom English.

For in the early seventeenth century, finding longitude at sea was not yet a matter of vital concern to English statesmen. Not until the third quarter of the century would trans-oceanic trade preponderate in English sea-borne commerce.[11] In consequence, finding a solution to the longitude problem was less urgent than it would become, and English mathematical writers of the period tend rather to rehearse the theoretical possibilities with appropriate exclamations of despair that no practical solution has been achieved,[12] than actually to set about tackling the problem. There is no parallel in England during the first half of the century either for the thorough-going programme, theoretical and practical, proposed by Morin for using lunar-distance methods, nor for the official commission that was set up by Richelieu in 1634 to examine it, nor for the quarrel between Morin and the government that dragged on for thirteen years as a result of the commission's negative verdict, nor for the pension that was eventually awarded to Morin as recompense for his work.[13] When eventually in 1664 an English scholar, Thomas Streete (1622-1689), did develop a longitude method in some detail (also using lunar distances), rather than just referring in passing to possible ways of doing it,[14] he published his claim in print but refused to divulge details of it in the absence of a suitable reward.[15] Despite his access to influential men, Streete, unlike Morin, clearly did not have expectations that he could provoke a reward from the government. It is partly this difference of

9. See Albert Van Helden's and John Leopold's papers in this volume.

10. Jean Parès, 'Jean-Baptiste Morin (1583-1656) et la querelle des longitudes, 1634-1647' (thèse de 3[e] cycle, l'Université de Paris, n.d.).

11. A. J. Turner, 'France, Britain and the Resolution of the Longitude Problem in the 18th Century', *Vistas in Astronomy*, vol. 28, parts 1/2 (1985), pp. 315-19; A. J. Turner, 'L'Angleterre, la France et la navigation: le contexte historique de l'oeuvre chronométrique de Ferdinand Berthoud', in Catherine Cardinal, ed., *Ferdinand Berthoud, 1727-1807: Horloger mécanicien du Roi et de la Marine* (La Chaux-de-Fonds, Switzerland: Musée International d'Horlogerie, 1984), pp. 143-64, reprinted in A. J. Turner, *Of Time and Measurement: Studies in the History of Horology and Fine Technology* (Aldershot, England: Variorum, 1993), chap. 14.

12. J. A. Bennett, 'The Longitude and the New Science', *Vistas in Astronomy*, vol. 28, parts 1/2 (1985), pp. 219-25 [220].

13. For a detailed discussion of Morin and his ideas, see Parès, 'Jean-Baptiste Morin' (*op. cit.*, note 10), *passim*.

14. For example, the suggestions of William Bedwell (State Papers Domestic 12/153/27, Public Record Office, London) and William Oughtred that water-clocks should be used. I am grateful to Stephen Johnson of the Science Museum, London, for a full transcription and discussion of the Bedwell reference.

15. Thomas Streete, *An Appendix to Astronomia Carolina containing a Proposition touching the Discovery of true Longitude, and the observation of three Lunar Eclipses* (London, 1664).

climate that explains why, before the third quarter of the century, relatively few English scholars, although not without interest in the longitude problem, were prepared to invest time and research in working out the practical details of their ideas.

In contrast to this situation in England, Morin in France was not an isolated figure. For example, in 1603 Guillaume de Nautonnier (1560-1620) had published his vast and magnificent volume on the use of magnetic variation for finding longitude.[16] This work, which did elicit some response in England in that it was perhaps partially plagiarised in 1609 by Anthony Linton (fl. 1609)[17] and refuted in 1610 by Edward Wright (1561-1615),[18] was a major contribution to study of the use of geomagnetism for finding longitude and provoked several further treatises.

If a report on a timepiece method (sand-glass and other good clock) presented by the 'sieur Boulenger mathematicien' to the Comte de Soissons, which was recorded among his mathematical papers by Nicolas-Claude Fabri de Peiresc (1580-1637),[19] is neither dated (pre-1636/7) nor provides any detail, it is of interest as perhaps being by the same Boulenger mathematician who was a member of the commission appointed to examine Morin's ideas.

More immediately based in practical experience was the proposal of the Recollect friar in Guienne, Leonard Duliris (fl. 1647-1655), of using the Moon's position in the zodiac and its height above the horizon as a function of date for longitude determination. Duliris spent eight years developing his method, and he records that he made:

> 88 observations at La Rochelle in 1644;
> 73 observations during a voyage to Canada in 1645;
> 12 observations on the St. Lawrence River, also in 1645, where he obtained a result of 304° 15', from which he derived a time difference of $5^{h}$ $15/20^{m}$ with Paris;
> and 30 observations on his return to Paris in 1646.

Duliris's book is realistic, practical, and convincing, although he is more concerned to find a usable longitude method than with high precision.[20] In 1655, he published his *Ephemeride*, specially conceived for use with his longitude theory.[21] In this, he provides a glimpse of the degree of attention that was being paid to the longitude problem in the upper levels of Parisian society. In 1648, Duliris's longitude method was discussed at a meeting of the mathematical club, which met in the house of the Duchesse d'Aiguillon (1604-1675), Petit Luxembourg. The club was presided over by Étienne Pascal (1588-1651, father of Blaise), and it was as a result of Étienne's good opinion of the method that the Duchess decided to underwrite the publication of Duliris's emphemeris. Like Boulenger, Pascal had been a member of the committee that had examined Morin's work in 1634. Duliris was fully aware of his predecessor,

> ...qui a tout travaillé, pour donner les moyens d'auoir la precision des mouvemens celestes, & fourny de si belle regles pour observer la longitude,... ie luy accorde fort librement, ce qu'il a touiours dit & moy avec luy, qu'il falloit esperer que le temps & l'estude des scavans Astronomes donneroit enfin, sinon vne totale, pour le moins

16. Guillaume de Nautonnier, *Meycometrie de leymant c'est a dire la maniere de mesvrer les longitudes par le moyen de l'eymant...* (Venès, 1603).

17. David W. Waters, *The Art of Navigation in England in Elizabethan and Early Stuart Times* (London: Hollis & Carter, 1958), pp. 274-5.

18. *Ibid.*, p. 275.

19. MS. fr. 9531, fol. 116$^{r}$, Bibliothèque Nationale, Paris.

20. Leonard Duliris, *La Theorie des longitudes redvite en pratique sur le globe celeste, extraordinairment appareillé pour coignoistre facilement en Mer, combien l'on est eslogné de toutes les terres du monde...* (Paris, 1647).

21. Leonard Duliris, *Ephemeride maritime dressé pour observer en mer la longitude et latitude selon l'invention du Pere Leonard Duliris Recollet. Avec un nouveau moyen de perpeuer l'ephemeride du soleil, pour avoir tousiours exactement sa declinaison...* (Paris, 1655).

vne svffisante precision, des mouvemens celestes, pour observer exactement la longitude. Et mon esperance n'a esté fondée que sur l'experience que i'ay faite en observant en mer la Longitude en telle sorte, que ie trouvois la distance itineraire beaucoup plus precisement, & asseurement que les gens de mer en leurs estimation.[22]

Thus, by the mid-century in France, advances had been made both in the provision of better lunar tables and in finding ways of making lunar methods applicable at sea. What should be underlined here is that at this stage the aim was to bring the new methods into use and to show that they were an improvement over the prevailing empirical dead-reckoning method. The aim was not to attain a prescribed degree of precision in longitude determination such as would be laid down in the 1714 Act. Such a prescription was a novelty, although as we shall see, not original to the Act. Seventeenth-century scholars were aiming for improvement in general, and their achievement was to show that the new methods could be made to work. They knew, however, the limitations of the tools at their disposal and constantly investigated other methods, whether by developing variations on the main methods of timekeepers, magnetic variation, and lunar distances, or by more original suggestions, such as that of César d'Arcons (fl. mid-17th century), advocate at the Parlement of Bordeaux, of using the movement of the tides for this purpose.[23]

D'Arcons's proposal, unlike Duliris's, seems not to have been followed up, but perhaps the examples cited are sufficient to give some impression of the considerable activity taking place in France concerning the longitude. It is valuable to bear in mind the devastatingly sceptical resumé of Georges Fournier (1595-1652) concerning the numerous methods proposed.[24] But that Fournier was moved to write such a denunciation is in itself further evidence of the interest provoked by the subject.

From the mid-century onwards, with the re-establishment of relatively stable social conditions and a rapid expansion in long-distance overseas trade, the rhythm of interest in the longitude problem in England also quickened. In 1654, Nicolas Mercator (1620-1687) adumbrated a way of finding longitude that Samuel Hartlib (*ca.* 1600-1662) preserved among his papers.[25] Henry Bond (*ca.* 1600-1678), during the middle decades of the century, was working on his magnetic theory for resolving the problem, while in the early 1660s, Thomas Streete was publishing his Moon's zodiacal position method. In 1662, Alexander Bruce (d. 1681) carried out his sea trials of clocks designed by Christiaan Huygens (1629-1695). In 1665, Robert Theaker (fl. 1665-d. 1687) published an instrument that he claimed would facilitate the affair.[26] In the early 1670s, two Royal Commissions were set up to examine two different methods: the geomagnetic solution of Henry Bond and a new proposal from the still unidentified Sieur de St. Pierre.[27]

In the light of what has been said earlier of the lively French interest in longitude determination during the first half of the century, it is not astonishing that a French proposal should have been received at Charles II's court, especially if, as John Flamsteed (1646-1719) suggested, St. Pierre was supported by 'a French Lady then in favour at Court' who is presumed to be Charles's mistress, Louise de Kéroualle (1649-1734), Duchess of Portsmouth.[28] For St. Pierre's proposal, we have only Flamsteed's largely negative

22. '...who has left no stone unturned to supply ways of having precise knowledge of the celestial movements, & furnished such fine rules for longitude observations,...I freely grant him, what he has always said and I with him, that we have to hope that time and the studies of learned astronomers will eventually give, if not total, at least sufficient accuracy in the celestial movements, that the longitude can be observed exactly. And my hope is based only on the experience that I had observing the longitude at sea in such a way, that I found the way made much more accurately and certainly than the seamen by their dead reckoning'. *Ibid.*, pp. 121-2.

23. César d'Arcons, *Le secret du flux et reflux de la mer et des longitudes, delivré a la sapience eternelle* (Rouen, 1655).

24. Georges Fournier, *Hydrographie Contenant la Theorie et la Practique de Toutes les Parties de la Navigation* (Paris, 1643) and several later editions.

25. Wilbur Applebaum, 'A Descriptive Catalogue of the Manuscripts of Nicolas Mercator, F.R.S. (1620-1687) in Sheffield University Library', *Notes and Records of the Royal Society of London*, vol. 41, no. 1 (October 1986), pp. 27-37 [31].

26. Robert Theaker, *A Light to the Longitude or the Use of an Instrument call'd the Seaman's Director speadily resolving all Astronomical Cases and Questions concerning the Sun, Moon, Stars with several propositions whereby sea-men may find at what meridian and Longitude they are at, in parts of the world* (London, 1665). Theaker's instrument was basically a geographical astrolabe.

27. For a survey of English activity from 1660 to 1680, see Bennett, 'Longitude and the New Science' (*op. cit.*, note 12). For Bond, see D. J. Bryden, 'Magnetic Inclinatory Needles: approved by the Royal Society', *Notes and Records of the Royal Society of London*, vol. 47, no. 1 (January 1993), pp. 17-31.

28. Howse, *Greenwich time* (*op. cit.*, note 2), pp. 24-6; Eric G. Forbes, *Greenwich Observatory, The Royal Observatory at Greenwich and Herstmonceux, 1675-1975*, vol. 1, *Origins and Early History (1675-1835)* (London: Taylor and Francis, 1975).

evidence, and, though the proposal was certainly defective, it was perhaps not as jejune as Flamsteed made it out to be. However, what does emerge from the discussions is the growing recognition that what is sought is no longer a method to find the longitude, but the means to make one of the known methods usable. According to Flamsteed, 'it was agreed that observations of the moon's distances from fixed stars were the most proper expedient for the discovery of it [the longitude]',[29] but, as Flamsteed was at pains to point out, this implied a true theory of the Moon and precise star catalogues. The crucial realisation that occurred in the second half of the seventeenth century was that longitude could be found, but that finding it required an enormous investment of time and energy in patient research. The best expression of this is to be found slightly later in the writings of Henry Sully (1680-1728), but Edward Harrison (fl. 1686-1700) expressed the same realisation more pithily in 1696: 'Some Blockheads are apt to say, the Longitude cannot be found; no, no it cannot Accidentally...but by Care, Diligence and Industry; it may be found, without which it cannot be understood'.[30]

The realisation that finding the longitude was less a matter of inventing a new method and more a matter of long-term research to make one of the known methods viable coincided, as we have seen, at the end of the seventeenth century with a period of renewed activity in England born of a sense of urgency. It was this latter sense that perhaps led Parliament to pitch the prize so high, the former understanding that gave the Act its originality in the clause that empowered the Commissioners to provide interim support and progress payments for projects that looked as if they had a real chance of success. Even before the government acted, however, the situation had led at least one private individual, Thomas Axe (d. 1691), to offer his own private prize for the promotion of research.

Thomas Axe was a Somerset estate steward charged with the administration of the lands of Sir William Portman (1641?-1690)[31] at Orchard. In view of the importance of Portman's estates, the description of Axe as 'Gentleman' in his will is probably not inaccurate, and Axe was himself a relatively wealthy man. He was born in Ottery St. Mary, Devon, the son of either Thomas[32] or George Axe[33] of the same town. Whichever of these was the father, according to John Aubrey (1626-1697), who derived his information from Axe himself, taught Thomas Jr. 'the Table of Multiplication, when he was seven years old'.[34] This early training in arithmetic seems to have given Axe a lifelong interest in commercial arithmetic, rents, and statistical analysis. He collected information on variations in the price of corn during the previous hundred years,[35] recorded the chronological incidence of outbreaks of smallpox at Sherborne and Taunton,[36] and was consulted by Aubrey about the relative interest of different kinds of rents and leases.[37] Aubrey's friend Andrew Paschall (*ca.* 1631/2-1696)[38] consulted Axe on the advisability of purchasing the manor of Weston, Somerset; in his reply, Axe clearly shows how he applied the results of his statistical analyses to everyday affairs.[39] It is also clear from these letters that Axe acted as an adviser in the investment of money in land. In 1672, he so advised Sir John Cutler (*ca.* 1608-1693), and through his statistical interests, Axe was in contact with William Petty (1623-1687) and probably with John Graunt (1620-1674),[40] both pioneers in such methods of economic analysis. Cutler, Petty, Graunt, and Aubrey were all members of the Royal Society, as was Robert Hooke

29. Flamsteed to Edward Sherburne, 12 July 1682, in Francis Baily, *An Account of the Revd. John Flamsteed, the first Astronomer Royal...* (London, 1835; reprinted 1960), p. 126.

30. Edward Harrison, *Idea Longitudinis: Being, a brief Definition Of the best known Axioms For finding the Longitude...* (London, 1696), Dedication.

31. The most influential Tory in the West of England after Sir Edward Seymour, Portman was the captor of the Duke of Monmouth after his defeat at Sedgmoor. Elected to the Royal Society on 29 December 1664, he was an inactive member who was expelled for non-payment of dues in 1685. See Michael Hunter, *The Royal Society and its Fellows, 1660-1700: The Morphology of an Early Scientific Institution,* British Society for the History of Science, monograph 4 (Chalfont St. Giles, England, 1982), pp. 192-3.

32. A son of Thomas Axe (the elder) of Ottery, Robert, matriculated at Wadham College, Oxford, on 22 June 1638. See Robert Barlow Gardiner, *The Registers of Wadham College, Oxford. (Part 1.) From 1613 to 1719* (London: George Bell and Sons, 1889), p. 142.

33. John Buchanan-Brown, editor of *John Aubrey: Three Prose Works. Miscellanies, Remaines of Gentilisme and Judaisme, Observations* (Fontwell, England, 1972), p. 514, states without evidence that Axe was an attorney admitted to the Middle Temple in 1664. The Thomas Axe who interests us, however, had a brother George living in Ottery, so the Middle Temple Thomas may be his son and nephew to the Thomas Axe of Orchard. Ambrose Axe, who matriculated at Trinity College, Oxford, in 1655/6, aged 17, son of George Axe of Ottery, is also clearly a relative.

34. Aubrey MS. 10, fol. 29$^{r}$, Bodleian Library, Oxford University.

35. Michael Hunter, *John Aubrey and the World of Learning* (London: Duckworth, 1975), p. 194, note 2.

36. John Aubrey, *Miscellanies* (London, 1696), pp. 32-3; Aubrey, *Three Prose Works* (*op. cit.*, note 33), p. 24; *Dorset & Somerset Notes & Queries,* vol. 20 (1930-2), p. 124.

37. Aubrey MS. 10, fol. 38 (*op. cit.*, note 34).

38. For Paschall, see A. J. Turner, 'Learning & Language in the Somerset Levels, Andrew Paschall at Chedsey', in W. D. Hackmann and A. J. Turner, eds., *Learning, Language and Invention* (Paris/Aldershot, England: Variorum, 1994), pp. 297-308.

(1635-1702/3) with whom Axe was also in contact. It is clear, therefore, that despite his being based in the provinces, Axe was well informed about the new scientific ideas and activities in London and the Royal Society. Indeed, he himself was not infrequently in London,[41] where his kinsman, another Thomas Axe, bookseller and binder at the Blew Bull in Duck Lane, bound well, held auctions, and sold books and globes.[42] He also helped his namesake to recover outstanding debts and doubtless supplied him with books and cultural information. That Axe had wider interests than just those that served his professional needs is suggested by the fact that he kept a daily weather diary for at least 25 years, beginning in 1661.[43]

On 13 September 1691, Richard Lapthorne (fl. 1687-1697) noted in a letter to Richard Coffin (d. 1700) in Portledge, Devon, that 'Mr Axe Sir William Portman's servant, whom I knew well, lately dyed and by his will hath given 1000 li to incourage some Mathematical students to finde out the true longitude, for the benefit of Navigators'.[44] Axe's legacy has, to my knowledge, been mentioned only three times in the modern literature of horology.[45] It is, however, worthy of some attention.[46] Axe provided that in the event of his wife Dorothy and his son Henry dying without issue, if any person should 'make such a perfect discovery how men of mean capacity may find out the Longitude at Sea, soe as thatt they can truely pronounce upon observations if within halfe a degree of the true Longitude...', then his executors or their successors were empowered to sell certain houses and land in Southwark in order to supply such a person with a £1,000 prize. Axe's reason, as stated in his will, for setting up this prize was that he had been convinced by several competent people that they could teach mariners how to find longitude but did not 'communicate this usefull secret to the world seeing it hath and will cost them much money and paines by making Instruments and therefore doe expect a reward from the public or others for the discovery thereof'.[47]

'Money and paines'—the realisation that making the known ways of finding longitude usable would be a long and arduous exercise—focused attention on the necessity for stimulus and support. Axe was clear-headed in his provisions. A contender for his prize should demonstrate it to the professors of astronomy and geometry at Oxford and Cambridge and provide certificates from 'at least Twenty able masters of Shippes that shall have made several experiments thereof in long voyages', which had been sworn before all or most of the twelve judges of England. Such certificates being produced, the executors could proceed to the sale of the property and make the award. But Axe had also recognised another need. He empowered his trustees to make payments of up to £40 per year on the rents of the Southwark properties to those who occupied themselves in making observations of the positions of places 'in order to the making of a correct Map of the world as to Latitude and Longitude'.

In the distribution of the £40, as of the full prize of £1,000, Axe demanded that 'my good friends Mr Edmond Halley and Mr Robert Hooke be consulted'. This condition makes it clear that at least one member, Halley, of the commission set up to consider the Whiston-Ditton proposal from which emerged the 1714 Act, was aware of Axe's prize. Between the two prizes, there are a few direct similarities and more things in common. Of these the most important is the measure of accuracy required. The 1714 Act

39. Axe to Paschall, 16 August 1684 and 4 September 1684, Aubrey MS. 12, fols. 11$^r$-14$^v$ (*op. cit.*, note 34). Axe to Aubrey, 7 September 1684 and 22 September 1684, Aubrey MS. 12, fols. 15-16 (*op. cit.*, note 34).

40. It was Graunt who proposed Axe's employer Portman as a Fellow of the Royal Society. Hunter, *The Royal Society* (*op. cit.*, note 31), p. 192.

41. There are several mentions of him, for example in Hooke's diary.

42. F. T. Wood, 'Notes on London Booksellers and Publishers', *Notes & Queries*, vol. 161 (1931), pp. 39, 93. One of Axe's auction sale catalogues is preserved in the British Library, pressmark 824 e 58.

43. John Aubrey, *The Natural History of Wiltshire, edited...by John Britton* (London, 1847; reprinted 1969), p. 14.

44. *Appendix 5th Report*, p. 382a, Historical Manuscripts Commission (HMC), London. However, the date of the letter is given as 19 September in Russell J. Kerr and Ida Coffin Duncan, eds., *The Portledge Papers being extracts from the Letters of Richard Lapthorne...to Richard Coffin...from December 10th 1687-August 7th 1697* (London, 1928), p. 121.

45. By Gould, *The Marine Chronometer* (*op. cit.*, note 8), pp. 12-13, in a dismissive manner; by J. J. Hall, 'An Unrecorded Episode in the History of the Marine Chronometer', *Horological Journal*, vol. 73, no. 868 (December 1930), pp. 65-9; and by G. H. Baillie, *Clocks and Watches: An Historical Bibliography* (London: N.A.G. Press, 1951), p. 117, following Hall.

46. All of the following details derive from the will of Thomas Axe. I am grateful to Mr. S. C. Hornsby, Clerk of the Trustees of Thomas Axe's Charity, for lending me a solicitor's copy of the will retained by the Charity at Ottery St. Mary. For the text of that part which concerns the longitude, see Appendix 2 at the end of this paper.

47. Given that Axe in his will uses the phrase 'true longitude' that was also used by Thomas Streete, it is not impossible that it was Streete's ideas that Axe had in mind, or even that Streete was one of the 'several competent people' who had convinced Axe that they did really know how to find longitude. Streete had stated that he would not divulge his method unless a reward were forthcoming, because of the pains and expense involved in making the method usable (see note 15).

offered the full prize of £20,000 only if the longitude could be determined to an accuracy of half a degree. This is the same value as was required by Thomas Axe before his prize of £1,000 could be awarded. Similar also is the requirement that the method should have been tried at sea and shown to work, and that the theory of it should have been revealed to competent judges. The university professors specified by Axe would also be *ex officio* members of the Board of Longitude. That Axe employed judicial testimony to validate claims, which the 1714 Act did not, is perhaps no more than a reflection of the difference between Axe's environment and that of the academics who reported to the House of Commons. In any case, affirmation before justices was a relatively common way of validating documents or actions in seventeenth-century England.

Although no direct proof can be offered that Axe's prize influenced the framing of the 1714 Act, both witness to the understanding of their creators that any proposal must be very carefully tested, and both are products of the increasingly generalised sense in the decades around the turn of the century that serious action needed to be taken about longitude determination. From what has been said above, it seems clear that Axe's prize was provoked by the concern felt permanently, though not constantly expressed, by the scientific community in London, and the 1714 Act can be seen as the eventual result of this concern. Axe's prize was not unknown, although its terms were criticised. In 1696, Edward Harrison gave a negative, but accurate, account of the conditions of the prize ('never to be paid I think'),[48] but was equally scathing about Venetian and Dutch prizes of which he had heard. More sanguine was a claimant in 1706 whose activities were reported by Elias Smith (fl. *ca.* 1700) to Thomas Hearne (1678-1735):

> ...I'll tell you of a great discovery. There's one pretends yt he has found the ye long-studied secret of taking the Longitude by sea. He has communicated with Sir Is. Newton & Mr Flamstead, and is preparing a large instrument to make ye experimt, & will go down to York to try it, it is not impossible yt he may succeed at sea if what he says is true yt he can do it wth ease & certainty as he walks on foot by looking at some stars thro his instrument. When he has sufficiently experienced it he designs to go to my Ld Pembrook to gett a Proclamacion for a publickk reward, & to appoint him wt Judges ye Governt pleases to try it & publish it to. Besides his expectations from ye Pub. he is sure of a praemium of 1000 l. left in will by a Gent. for that purpose, this account I had from one of ye Trustees of that will, who offer'd to shew it me.[49]

That this claimant was 'sure' of the £1,000 premium in 1706 implies that the trustees were interpreting the terms of Axe's will with some flexibility, ten years, for example, not from Axe's death, but perhaps from that of his son. But Smith's letter also witnesses to the continuing ferment of activity in longitude research during the period 1690-1714. Activity seems indeed not to have slackened throughout the later seventeenth century. In 1689, several members of the Royal Society made what seem to have been a series of quite official trials of a longitude method proposed by another French mathematical practitioner, the Poitevin clockmaker René Grillet (fl. 1670-1686).[50] In 1699, Samuel Fyler proposed a scheme to establish lists of stars on the

48. Harrison, *Idea longitudinis* (*op. cit.*, note 30).

49. *Letters addressed to Thomas Hearne M.A....* (London, 1874), pp. 8-9.

50. For these, see several entries in Robert Hooke's diary for December 1689, printed in R. T. Gunther, *Early Science in Oxford*, vol. 10, *The Life and Work of Robert Hooke,* part 4 (Oxford, England, 1935), pp. 172-3, 176-7. The method had been discussed at the Académie Royale des Sciences *ca.* 1673. For Grillet, a Huguenot who took refuge in the Low Countries at Groningen and who also developed a form of calculating machine among other inventions, see 'Tardy' [Henri Lengellé], *Dictionnaire des Horlogers Français* (Paris, 1972), p. 277, and Enrico Morpurgo, *Nederlandse klokken- en horlogemakers vanaf 1300* (Amsterdam: Scheltema en Holkema, N. V., 1970), p. 49.

prime meridian passing through the Azores every fourteenth midnight in the year. Then, by noting on a specially made map which stars were on the meridian for a given date, the user could work out his longitude following an observation of what stars occurred on his own local meridian.[51] Similarly, in 1710 Edmond Halley published, for what they were worth, a number of newly made and more accurate lunar observations that could be useful in finding longitude by observation of the Moon's 'Appulses or Occultations of the *Fix'd Stars* by the *Moon'*, a method he claimed to have used with success during his magnetic voyages.[52]

That the Whiston-Ditton longitude proposal in 1713/14 should have provoked the formation of a parliamentary committee and issued in the 1714 Act may be seen then as, if not inevitable, at least highly probable, even predictable. It was so because of the general understanding of the problem, the presence of influential astronomers on the committee, and the obvious disasters that too frequently occurred at sea. What was perhaps not predictable was the decision to offer such a staggeringly large prize. Inevitably, such an unparalleled sum provoked a marked response from projectors, mathematical practitioners, savants, and cranks. In 1714, at least fourteen pamphlets were published, and, by 1720, the number had reached 28.[53] Of the wilder solutions offered, something is said elsewhere in this volume,[54] but several of the pamphlets show a good understanding of the problem even if they fail to make proposals sufficiently precise to be usefully implemented.[55] Among the tracts, however, there are a few that go further, and it is to the examination of one of these that the rest of this paper will be devoted.

**Figure 1. Engraved portrait of John Ward, used as frontispiece to the sixth edition (1734) of his *The Young Mathematician's Guide....***

The author of the tract that concerns us, John Ward (1648-*ca.*1727), of Chester, was a mathematical practitioner in the fullest sense of the term. Teacher of mathematics, designer and inventor of mathematical instruments, chief surveyor and gauger to the Excise, writer of effective mathematical text-books, he was no 'nutty opportunist', but a pragmatic, highly competent, professional mathematical practitioner. Born in Chester, he spent his professional life in London before returning in July 1710 to his native county for reasons of health.[56] In London, Ward had carried on the parallel careers of gauger to the Excise and mathematics teacher. He wrote several text-books for his students[57] and also invented a number of new instruments that he claims to have had made by John Rowley (fl. 1698-1728), one of the best instrument makers then working in London.[58] Since I have recently been able to identify a surviving instrument signed by Rowley with one invented

51. S[amuel] F[yler], *Longitudinis Inventæ Explicatio Non Longa, Or, Fixing the Volatilis'd, And Taking Time on Tiptoe, Briefly Explain'd;...* (London, 1699).

52. Edmond Halley, 'Observationes Syderum & imprimis Lunae, In Suburbio Londinensi apud Islington. Annis 1682, 1683 & 1684...', appendix to Thomas Streete, *Astronomia Carolina: A New Theory of the Coelestial Motions,* 2nd ed. corrected (London, 1710), p. 46.

53. See Appendix 3 at the end of this paper.

54. See Owen Gingerich's paper in this volume.

55. See, for example, Stephen Plank, *An Introduction To The Only Method For Discovering Longitude....* (London, 1714), which provides a very clear description of the way in which watches should be used in combination on board ship. The only problem with his suggestions is that he believed that watchmakers were already capable of producing watches 'that can keep time'. So they could, but Plank neglected to examine within what limits they could keep it.

56. John Ward, *A Practical Method To Discover the Longitude at Sea, By a New Contrived Automaton* (London, 1714), p. 32.

57. See Appendix 4 at the end of this paper.

58. John Ward, *The Young Mathematician's Guide being a plain and easy Introduction to the Mathematicks in five Parts...* (London, 1707), 'Advertisement'. Also in Ward, *A Practical Method* (*op. cit.,* note 56), p. 6.

and described by Ward,[59] it is clear that he did exactly what he claimed.

Shortly before the onset of the illness that caused him to leave London, Ward had begun serious work on the development of an automaton for use at sea in finding longitude. Not only is the fact that development started in London vouchsafed for in an affidavit sworn by a friend of Ward's, William Bird, citizen of Chester,[60] but the fact that Ward explained the design of his machine to a London clockmaker also indicates that Ward was still in London when he conceived his longitude clock. The date of the first stages of Ward's work may therefore be set before 1710. This dating is important, since it shows that Ward's activity, like the 1714 Act itself, was a product of the intense concentration on the longitude problem, which characterises the years around the turn of the century briefly described above, and was not provoked by the Act itself. Ward's publication date is clearly related to the promulgation of the Act, but his work on the problem predates it. What he published on 10 January 1714/15[61] was *A Practical Method to Discover the Longitude at Sea by a new contrived Automaton*.

Ward's book has to my knowledge never been discussed anywhere in the ancient or modern literature of horology, and it has only occasionally been mentioned in bibliographical works.[62] This is perhaps not astonishing as only four copies of the work are so far known to have survived.[63] This being the case, it is worth describing the contents in some detail. The tract is dedicated to the Commissioners of the Longitude, and, after a preface explaining the genesis of the work that has been partly used above, Ward settles down to the description of his method. This, he emphasises, is practical, in contrast to the many methods already published, for example in the *Philosophical Transactions*, 'all of them true in *Theory*, but either so Abstruse, or so attended with such Difficulties, as have hitherto render'd them Impracticable at Sea'.[64] After a basic description of the elementary cosmography needed by mariners, Ward describes a new instrument that he has invented and had made by John Rowley, before describing the main prevailing methods of finding longitude: lunar eclipses, the occultation of a star by the Moon, Jupiter's rotation about its axis, the eclipses or the immersions and emersions of Jupiter's satellites, and the method proposed by Ditton and Whiston. All these, however, are very difficult and really all that is needed is an accurate timekeeper, 'an *Artificial Movement*, made either by Wheel-work, or some other contrivance equivalent to it'.[65] If such a device usable at sea could be made, then the difference between two meridians could be found directly and more easily than by any other method.

Like Henry Sully and other horologists seriously interested in longitude timekeepers, Ward's starting point, as he indicates, was the work of Huygens. But Huygens had not foreseen the full consequences of two major problems: one was climatic variation in different geographical regions or at different times of the year and the effect that '*Saline Effluvia* or *Particles of* Sea Air have upon *Metals* in causing them to *Rust*...';[66] the other was the irregular motion of a ship at sea. It was on these two problems that Ward's attention concentrated. He describes the machine that he developed to overcome them:

> The *Frame* of the *Movement* should be at Least Six Inches Broad, and its height the same, that so the *Divisions* on the *Dial-Plate* may be seen distinct and plain: The First, or *Inner Circle* thereon, next the

59. A. J. Turner, 'One of "Two Brass Semicircles": an unidentified instrument in the Orrery collection identified' (in preparation).

60. Ward, *A Practical Method* (*op. cit.*, note 56), Preface.

61. This is an old style date, so Ward's work appeared in January 1715 by our reckoning. If it was also a Julian calendar reckoning, then the date of publication in the modern (Gregorian) calendar is 21 January. It was advertised in the *Monthly Catalogues*, i, 76a (April 1715).

62. Notably by E. G. R. Taylor, *The Mathematical Practitioners of Tudor and Stuart England* [1485-1714] (Cambridge: Cambridge University Press, 1954), p. 295 (No. 508). A full listing of Ward's works will eventually be included in R. V. and P. J. Wallis, *Bibliography of British Mathematics and its Applications* (Newcastle-upon-Tyne, England: Phibb), but the appropriate volume has yet to be published. I should like to thank Mrs. Wallis for her kindness in making available to me information in the files concerning Ward.

63. The copies are located in the following institutions: the Bodleian Library, Oxford University; the National Library of Scotland (*ex* library of the Faculty of Advocates, which acquired it in 1721); University Library, Göttingen (*ex* Academia Georgia Augustae); the John Carter Brown Library, Providence, Rhode Island (*ex* Sion College Library, to which it was presented in 1721 by Joseph Hodgson, bookseller in Chester, who may have received it from Ward himself [the inscription is illegible]). The collation of the work, which is an 8° in 4s, is $A^4$, $B$-$E^4$, $F^2$. All the copies conform. None are annotated. The copies in the Bodleian and the John Carter Brown libraries have been personally examined by the writer. For reporting on that in Edinburgh, I thank David Bryden of the Royal Museum of Scotland; for that in Göttingen, I thank Dr. Wesley Steven.

64. Ward, *A Practical Method* (*op. cit.*, note 56), p. 1.

65. *Ibid.*, p. 7.

66. *Ibid.*, p. 8.

67. *Ibid.*, pp. 8-9.

68. *Ibid.*, pp. 10-11; see also Robert Boyle, *New Experiments Physico-Mechanicall Touching the Spring of the Air and its Effects*... (London, 1660) and several subsequent, enlarged editions detailed in John F. Fulton, *A Bibliography of the Honourable Robert Boyle*..., 2nd ed. (Oxford: Oxford University Press, 1961), pp. 13-20. It must remain an open question as to whether Ward

*Center,* is to be Divided into *Twelve* equal Parts, *(or Hours,)* and every one of those into *Four* equal Parts (*or Quarters,)* to which the *Index* must Point, as in common *Clocks.* The next or outward *Circle,* is to be Divided into *Fifteen* equal Parts, and its *Index* to move Round that Circle every *Quarter* of an *Hour;* consequently that *Index* will shew the *Minutes* of Time; every one of those *Minutes,* I would have *Subdivided* into other *Fifteen* equal Parts, then will the same *Index* as it moves over the Minutes, shew every *Four Seconds* of *Time,* which is equal to one Minute of a *Degree* in the *Equator.* That is, this last *Index* will by that Means pass over 21600 of those smaller Divisions in a *Natural Day,* or 24 Hours, and just so many Minutes or Geographical Miles are contained in the 360 *Degrees* of the *Equator.* (viz, 360 x 60 = 21600 *the Minutes or Miles in a great Circle.)*

The *Inner* parts of the *Movement,* as the *Fusy,* the *Main-Spring,* the *Chain,* and *Wheels,* I would have them all made in full Proportion to the *Size* of the *Frame;* And their Motion to be Govern'd by a pretty Weighty *Round Ballance,* and that to be *Regulated* by Two *Spiral Springs,* which must be so fitted to the *Verge* of the Ballance, as to *Bend,* and Unbend Alternately, *viz.* Whilst one of those *Springs* is Folding up by the Motion and Force of the Ballance, the other Spring must be at *Liberty* to unfold it Self; and by that Means those Two *Springs* being well *Adjusted* to perform their Office, will keep the *Ballance* to a very Regular *Motion,* not much, *(if any,)* Inferior to that of a *Long Pendulum,* and in some Respects to be Prefer'd before it.[67]

Such a movement, Ward claimed, if correctly made by a good craftsman, would go in any position and would not 'be Subject to any Disorder, by the Irregular Motion of a *Coach*, or that of a *Ship* at *Sea'*. Thus one of the two basic problems would be overcome. To deal with the second, Ward proposed putting and keeping the timekeeper in a vacuum, grounding his reasons on the results of the experiments of Robert Boyle (1627-1691) on pendula vibrating in a vacuum.[68] After replying to some possible objections to this idea, Ward then describes: the preparation of the machine for a voyage, explaining how to rate it against a meridian line (of which he describes the construction);[69] the equation of time, for which he gives tables based on those of Flamsteed, recalculated by himself; how to find local time at sea; and finally, how actually to use the timekeeper at sea for finding longitude, with the corrections and calculations necessary, which he explains by way of a worked example.[70]

It is not possible here to go into greater detail about Ward's methods and ideas, but it is necessary to underline a number of points. In the first place, Ward's timekeeper does offer definite advantages. His fifteen-minute dial, and the double division into fifteen, allows degrees as well as time to be read directly, thus simplifying conversion between the two. His double balance spring, which could have been influenced by some knowledge of Hooke's use of double springs for controlling balances,[71] will within limits do what he claims, and it anticipates Harrison's use of two springs acting reciprocally in H.1. Secondly, if Ward was indeed developing his ideas as early as 1710, then his suggestion of enclosing the timekeeper in a vacuum predates Jeremy

was aware of the 'Experiments about the Motion of Pendulums in Vacuo' by William Derham (1657-1735), published in the *Philosophical Transactions* of the Royal Society, London, vol. 24, no. 294 (1704), pp. 1785-9, and summarised in Charles K. Aked, 'William Derham and the "Artificial Clockmaker" ', part 1, *Antiquarian Horology*, vol. 6, no. 6 (March 1970), pp. 362-72. Since Ward specifies that the timepiece has to be 'truly Adjusted, to keep *equal Time'* in the vacuum-jar exhausted by the air-pump, he may have done. Derham's trials, however, were not directed to the same ends as Ward's, and there was no strict need for Ward to mention them. In characteristic style, however, Ward recommends (p. 10) that the air-pump to be used should be one made by the leading maker thereof in the London of his generation, Francis Hawksbee Sr. (d. 1713). That Ward notes that Hawksbee is deceased is evidence that he was still at work on the composition of his tract in late 1713 and later. For Hawksbee's pump, see Anne C. van Helden, 'The Age of the Air-pump', *Tractrix. Yearbook for the History of Science, Medicine, Technology and Mathematics,* vol. 3 (1991), pp. 149-72 [166-9] and references there given.

69. Ward, *A Practical Method* (*op. cit.*, note 56), pp. 14-17, but Ward's meridian method is unusual in being based on the reflection method of dialling. He claims an accuracy of four to five seconds for his readings with such a meridian.

70. *Ibid.*, pp. 25-31. It is worth noting here Ward's early appeal for the establishment of a prime meridian: 'And when once all *Geographers*, shall agree upon and *Pitch* upon a *proper* Place, (as suppose Pico Teneriff, &c.) for the *Prime* or First *Meridian*, from whence the *Degrees* of *Longitude* must be Accounted; Then may the *Longitude* of any Place on either side of that *Meridian*, be as certainly known, as now the Degrees of Latitude are on each side of the Equinoctial, which is all that's wanting to Compleat both Navigation and Geography' (p. 31).

71. Although Hooke does not seem to have used spiral balance springs. See Michael Wright, 'Robert Hooke's Longitude Timekeeper', in Michael Hunter and Simon Schaffer, eds., *Robert Hooke: New Studies* (Woodbridge, England: D. S. Brewer, 1989), pp. 63-118 [88-92].

Thacker (fl. 1714), who published the idea shortly before Ward in a pamphlet that appeared in November 1714, but who, Ward claims, had simply plagiarised the idea.[72] Thirdly, we should note how Ward's ideas developed out of the contemporary, practical, and philosophical science developed by members of the Royal Society, with which Ward became familiar during his years in London, and how Ward tested his ideas by having the instruments he imagined made. He tells us that he had instruments of his devising made by John Rowley, and at least one such instrument has survived. There seems no reason to doubt that Ward had indeed described his timekeeper to a London clockmaker as he says, and perhaps had one built.

And indeed, we can go a little further. Figure 2 illustrates one of a small group of surviving watches by Richard Street (fl. 1687-*ca.* 1715). Although round, not rectangular, and far smaller than the dial of Ward's automaton, the layout of these dials corresponds with that of Ward's machine with the single exception that the minutes are divided into six divisions of ten seconds, instead of into fifteen divisions of four seconds each—this perhaps for reasons of clarity in reading the smaller dial. That there is an influence seems possible, even probable, given the smallness of the technical and learned world of early eighteenth-century London. But Richard Street, the maker of these two watches, was the maker to whom Ward tells us *'the making of this [i.e.,* his, Ward's] *Movement is already well known'*,[73] so the probability becomes overwhelming and is further increased by the fact that two of the watches have an equation scale engraved on the outer case. Ward had been at pains to explain the importance of the equation of time.[74]

To claim that these watches were longitude timekeepers would be to press the evidence too far, but their existence suggests that Ward did indeed explain his automaton to Street, who made one. In so doing, Street seems to have realised the advantages of the fifteen-minute dial, which could be read with greater precision at the same time as it allowed the motion work

72. 'When I found that Author proposed a Machine to Move in a Glass Receiver, as a means to Discover the Longitude; I had Reason to Suspect, as well from private Circumstances, from the crude indigested Notion he seems to have...of so Useful a thing as a Recipient rightly Applied, will undoubtedly prove; That the knowledge he pretends to have in this affair, came by *Trap*, and not the way he so Boastingly tells the world of them'. Thacker's treatise is *The Longitudes Examin'd...* (London, 1714). Ward received a copy of it on 29 November 1714.

73. Ward, *A Practical Method* (*op. cit.*, note 56), p. 9.

74. The watches are No. 214, a large silver pair-cased verge watch sold at Sotheby's on 15 November 1971 (lot 125), now in The Time Museum, Rockford, Illinois, and No. 408, a silver-cased verge watch with reversed fusee, now in the collection of the Clockmakers' Company at the Guildhall Library, London. See Cecil Clutton and George Daniels, *Clocks and Watches: The collection of the Worshipful Company of Clock-*

**Figure 2. Silver pair-cased pocket watch, signed on the back plate 'Rich'd Street, London 214'. On the dial, the minute ring is divided into fifteen parts, marked with arabic numerals. Each minute is subdivided into six parts, representing ten-second intervals. This arrangement allows the minute hand, which makes four revolutions every hour, to indicate the time to within a few seconds. The back of the watch's pair case, on the left, is engraved with a calendar indicating the equation of time. Diameter of outer case: 7.3 cm. (2.9 in.). Courtesy of The Time Museum, Rockford, Illinois. Inv. no. S214.**

to be simplified, whereupon he used it in some of his high-quality precision pieces.[75] That this is the case is supported by the existence of a longcase clock by Street[76] in which a similar system is used, the minute hand making one revolution every ten minutes. Street's use of the equation table on his watch cases may also be a result of influence from Ward.

The pamphlet in which Ward explained his ideas is characterised by a reserved, serious style, clarity of concept and explanation, and great practicality. This practicality of Ward's conceptions is shown by the fact that his designs could be realised, and his whole tract stands in sharp contrast to the fanciful assertions and bombastic language of many of the ephemeral publications evoked by the 1714 Act. But Ward's small treatise, as we have seen, was not stimulated by the Act. Rather it arose, like the Act itself, from the ferment of activity in the first decades of the eighteenth century around the longitude problem in the circles that combined scholars and craftsmen in the pursuit of technical innovation and the experimental testing of theories and ideas. In London, Ward moved in such circles and was stimulated by them; and he pursued his ideas to realisation. He may claim a place among the pioneers of English attempts to produce a usable marine timekeeper.

I am grateful to Jeremy Evans of The British Museum, London, for information concerning Richard Street; to Patricia Atwood for photographs of the Street watch in The Time Museum; and to William Andrewes for a variety of things.

*makers* (London: Sotheby Parke Bernet, 1975), Cat. No. 102, pp. 22-3. A further silver-cased, fifteen-minute dial watch by Street, No. 234, is in the Noel Terry Collection, Fairfax House, York. See Peter Brown, *The Noel Terry Collection of Furniture and Clocks* (York, England, 1987), Cat. No. W1, p. 26, and illustration, p. 25. It is also briefly discussed by Paul Tuck, 'Fine Examples of Antiquarian Horology seen during the Scottish Tour, 1988', *Antiquarian Horology,* vol. 18, no. 2 (Summer 1989), pp. 181-9 [184-5].

75. In the absence of any knowledge of the early history of the regulator by Fromanteel with fifteen-minute dial (now in the Fogg Art Museum, Harvard University), in which the minute hand revolves four times an hour over a chapter ring divided to five seconds, it is impossible to say whether this, or another clock of similar arrangement, could have been known to Ward, nor if the design was sufficiently widespread in seventeenth-century London for Ward to have adopted it from a common stock of knowledge. For the Fromanteel clock, see Ronald A. Lee, *The First Twelve Years of the English Pendulum Clock, or the Fromanteel Family and their Contemporaries, 1658-1670* (London, 1969), 'General Notes' and plates 92-8. The clock is also illustrated in Percy G. Dawson, C. B. Drover, and D. W. Parkes, *Early English Clocks* (Woodbridge, England: Antique Collectors' Club, 1982), p. 79.

76. The clock is now in the National Maritime Museum, Greenwich.

## APPENDIX 1

### 'A PETITION ABOUT THE LONGITUDE'

Printed sheet with title on back, not signed, but probably the petition of Humphry Ditton and William Whiston. Dated 29 April 1714.

Whereas Her Majesty has been pleased this very session of Parliament, particularly to recommend the Improvement of the *Trade* and *Naval-Force* of *Great-Britain* from the Throne:[77] and whereas it is known, that nothing can be either at Home or Abroad more for the common Benefit of Trade and Navigation than the Discovery of the LONGITUDE at Sea; which has been so long desir'd in vain, and for want of which so many Ships and Men have been lost: Whereas also a Proposal for that Purpose has now been offered to the World for some time, and has met with Approbation among some of the best Juudges, to whom it has been Privately discover'd, but for want of suitable Encouragement could not hitherto be communicated to the Publick: it is humbly desir'd that a Bill, or Clause of a Bill, may be brought in this Parlament, to Appoint a suitable Reward for such as shall first lay before the Publick any sure Method for the Discovery of that LONGITUDE; to be then due when the most proper Judges, who may be appointed in the Bill, shall declare that such Method is both True in it self, and is also Practicable at Sea. That the lowest Reward may be allotted to the discovering the same within one whole Degree of a great Circle, or 70 Measur'd Miles; a greater to the discovering it within one half; and a still greater to the discovering it within one Quarter of that Measure: and that withall if it be thought fit, proper Rewards may also be allotted to such as shall afterwards make any further considerable Improvements for the perfecting so important a Discovery. This is the humble Desire of the Authors of this Invention, as well as of many others, who are very unwilling that this their Native Country of *Great-Britain* should thus lose the Honour and Advantage of its first Discovery, Practice, and Encouragement.

77. The speech from the throne at the opening of the session of Parliament on 2 March 1713/14 (12 Anne) included the remark, 'Our situation points out to us our true Interest; for this Country can flourish only by Trade; and will be most formidable by the right application of our Naval Force'. *Journals of the House of Commons,* vol. 17 (1711-14), 2 March 1713/14, p. 474.

APPENDIX 2

## EXTRACT FROM THE WILL OF THOMAS AXE ESTABLISHING A PRIZE FOR THE DISCOVERY OF A PRACTICAL METHOD OF FINDING LONGITUDE

Whereas I am convinced by severall of great knowledge in astronomy and cosmography that they know how to make the masters of ships and others of the meanest capacity to find out the Longitude as readily and almost as fast as they can find the Latitude but doe not communicate this usefull secret to the world seeing it hath and will cost them much money and paines by making Instruments and therefore doe expect a reward from the public or others for the discovery thereof. Wherefore if any person or persons shall within Ten years next after my decease as aforesaid make such a perfect discovery how men of mean capacity may find out the Longitude att sea soe as att they can truely pronounce upon observations if within halfe a degree of the true Longitude and shall demonstrate how it may be done to the two professors of astronomy and to the two professors of Geometry in Oxford and Cambridge for the time being and shall make ample proofe thereof by the Affadavit in writing of att least Twenty able masters of Shippes that shall have made severall experiments thereof in long voyages which affadavits are to be made before the Twelve Judges of England or before the major part of them of w^ch^ the Lords Chief Justices to be two and shall deliver a certificate therof signed and sealed by the said major parte of the said Justices unto the said Samuel Keeble, William Pratt and Henry Hooper[78] or to the survivors and survivor of them or to the heirs of such Survivor. That in such case upon the receipt of such a certificate the said Samuel Keeble, William Pratt and Henry Hooper and the survivors and survivor of them and the heires of such survivor shall with al convenient speed next after the death of me my wife and sonne childless sell the fee of all my said houses in the said parish of St Olaves and St Saviour's for as much money as can be gotten for the same and that they shall pay the sume of One thousand pounds of such purchase moneyes unto such person or persons as shall have soe discovered such a ready and easy way of finding the true Longitude att sea and shall have made such proofe as aforesaid for the benefitt of mankind. Which said sume of One thousand pounds on the conditions and in the case aforesaid I Give and bequeath to such person or persons accordingly. And moreover my will is That from and immediately after the death of my said wife and of myselfe and said sonne leaving noe issue of either of our bodyes behind us as aforesaid That the said Samuel Keeble William Pratt and Henry Hooper and the survivors and survivor of them and the heires of such survivor shall once every year untill perfect proofe of the discovery of the true longitude shall be made as aforesaid and the six houses remain unsold pay forty pounds parte of the rents of the said houses unto and amongst such person and persons for their encouragement as shall have employed their time to make observation at severall capes islands and noted places in Europe Asia Africa and America according to the rules and instructions of such said person or persons as shal have made such discovery how all men may find out the Longitude in order to the making of a correct Map of the world as to the Latitude and Longitude And my will is that in all things concerning the said Longitude the disposal of the said sume of One thousand pounds and of the said Forty pounds by the yeare my good friends Mr Edmond Halley and Mr Robert Hooke be consulted And likewise I give and bequeath Thirty pounds of the remainder of the said purchase moneyes unto the said Samuel Keeble, William Pratt and Henry Hooper and their respective executors to be equally distributed between them that shall be concerned in the execution of this part of my will.

78. Axe's executors and trustees.

APPENDIX 3

## A CHECKLIST OF LONGITUDE PAMPHLETS PUBLISHED BETWEEN 1714 AND 1726

Alimari, Dorotheo. 1714. *The New Method Propos'd by...to discover the Longitude. To which are added, Proper Figures of some Instruments which he hath Invented to that Purpose: With a plain Description of Them....* 2 eds. London.

———. 1715. *Longitudinis uut terra aut mari Investigandae Methodus. Adjectis insuper Demonstrationibus aut Instrumentoru Iconismis.* London.

Anon. 1726. *A Sailor's Proposal for Finding his Longitude by the Moon....* 2nd ed. London. (The first edition, also published in London, has no date of publication.)

B., R. 1714. *Longitude To be found out with A new Invented Instrument, Both By Sea and Land.... With A better Method for discovering Longitude, than that lately propos'd by Mr. Whiston and Mr. Ditton.* London.

Billingsley, Case. 1714. *A Letter to the Commissioners appointed by the Parliament of Great Britain, for discovering the Longitude at Sea..., Further to Explain his late Proposal for that Discovery, by the Sun, Moon and Stars; and a true Time-keeper.* London.

———. 1714. *The Longitude At Sea, Not to be found by Firing Guns, nor by the Most Curious Spring-Clocks or Watches....* London.

Browne, Robert. 1714. *Methods, Propositions, and Problems, for finding the Latitude; with the Degree and Minute of the equator upon the Meridian. And the Longitude at Sea, by Cœlestial Observations only. And also by Watches, Clocks, &c. and to correct them and to know their Alterations.* London.

Clarke, James. 1714. *An Essay wherein a Method is Humbly propos'd for Measuring Equal Time with the utmost Exactness; Without the Necessity of being confin'd to Clocks, Watches or any other Horological Movements; in order to Discover the Longitude at Sea.* London.

———. 1715. *The Mercurial Chronometer improv'd: or a Supplement to a Book entitled An Essay...In which all objections that are in the least bit rational are remov'd, and the Method confirm'd....* London.

French, John. 1715. *A Perfect Discovery Of The Longitude at Sea; In Compliance with what's propos'd In a late Act of Parliament. Being the Product of Nine Years Study, and frequent Amendments of a Mathematician, who has us'd the Sea upwards of Twenty five Years, and has had the Experience thereof at Sea. And his Projection hath been View'd and Approv'd on by most of the Mathematicians about London.* London.

Gordon, George. 1724. *A Compleat Discovery of a Method of Observing the Longitude at Sea.* London.

Haldanby, Francis. 1714. *An Attempt To Discover the Longitude At Sea, Pursuant To what is Proposed in a Late Act of Parliament.* London.

Hall, William. 1714. *A New and True Method to Find the Longitude, much more Exacter than that of Latitude by Quadrant also a new Method for the Latitude....* London.

Hanna, John. 1725. *An Astronomical Creed, In Ten Articles, With Natural and Moral Reflections from them; To which is subjoyned, the new Method of discovering the Longitude, by the appulse of the Moon to the fixed Stars:....* Dublin.

Hawkins, Isaac. 1714. *An Essay For The Discovery Of The Longitude at Sea, By several New Methods fully and Particularly laid before the Publick.* London.

Hobbs, William. 1714. *A New Discovery For Finding the Longitude. Humbly Submitted to the Approbation of the Right Honourable the Lords Spiritual and Temporal,...* London.

Mel, C. 1719. *'Pantometrum nauticum seu Machina pro invenienda Longitudinae et latitudine locorum in Mari', in Antiquarius Sacer, quamplurima dubia atque obscuriaria sacroe idicta, ex statu ecclesiastico, politico, militari atque conomoco, Hebraoeorum, Romanorum, Graoecorum Illustris & explicans. Cum Mantisia dissertationum....* Frankfurt-am-Main.

Palmer, William. 1715. *A great Improvement in Watchwork; which may be of great use at Sea, for discovering the Longitude. Humbly offer'd to the Consideration of the Learned. With some Remarks on another Way of Discovering the Longitude....* York, England.

Pitot, Allain. 1716. *L'Automate De Longitude. Nouveau Systéme d'Hydrométrie; par les Périodes d'un Mouvement Nautique, qui marque à un Cadran, les Lieuës qu'un Navire fait dans sa Route. Présenté à nos Seigneurs les Commissaires de la Grande Bretagne pour l'Examen des Découvertes sur la Longitude.* London.

Plank, Stephen. 1714. *An Introduction To The Only Method For Discovering Longitude. Humbly Presented, for the Good of the Publick.* London.

———. 1720. *An Introduction To a true Method For the Discovery of Longitude at Sea....* London.

Thacker, Jeremy. 1714. *The Longitudes Examin'd. Beginning with a short Epistle to the Longitudinarians, and Ending with the Description of a smart, pretty Machine of my own which I am (almost) sure will do for the Longitude, and procure me The Twenty Thousand Pounds.* London.

Ward, John. 1714. *A Practical Method to Discover the Longitude at Sea, By a New Contrived Automaton....* London.

Whiston, William. 1721. *The Longitude And Latitude Found by the Inclinatory Or Dipping Needle; Wherein the Laws of Magnetism are also discover'd. To which is prefix'd, An Historical Preface; and to which is subjoin'd, Mr. Robert Norman's New Attractive, or Account of the first Invention of the Dipping Needle.* London.

Whiston, William, and Ditton, Humphry. 1714. *A New Method For Discovering the Longitude Both At Sea and Land, Humbly Proposed to the Consideration of the Publick.* London.

———. 1715. *A New Method For Discovering the Longitude Both At Sea and Land, Humbly Proposed to the Consideration of the Publick.* 2nd ed. London.

APPENDIX 4

## THE WORKS OF JOHN WARD

**1695**

*A Compendium of Algebra. Consisting of plain, easie and Concise Rules for the Speedy attaining to that Art*.... London. 2nd ed., enlarged. London, 1724.

**1707**

*The Young Mathematician's Guide being a plain and easy Introduction to the Mathematicks in five Parts*.... London. Reissue with cancel title page, 1709; 2nd ed. corrected, 1713; 3rd ed. corrected, 1719; 4th ed., 1722; 5th ed. corrected, 1728; 6th ed., 1734; 7th ed., 1740 (the first edition with the additional *History of Logarithms*...); 8th ed., 1747; 9th ed., 1752; 10th ed., 1758; 11th ed., 1762; 12th ed., 1771 (by Samuel Clarke).

**1714**

*Clavis usurae or, a Key to Interest both simple and compound*.... London. 2nd ed., London, 1740.

*A Practical Method to Discover the Longitude at Sea, By a New Contrived Automaton*.... London.

**1730**

*The Posthumous Works of Mr. John Ward...published by a particular Friend of the author's from the original Manuscript and revisd by Mr George Gordon, Mathematician in London*. London. 2nd ed., 1765.

# Cranks and Opportunists: "Nutty" Solutions to the Longitude Problem

## *Owen Gingerich*

Owen Gingerich is a Senior Astronomer at the Harvard-Smithsonian Astrophysical Observatory. He is also Professor of Astronomy and History of Science at Harvard University. In 1984, he won the Harvard-Radcliffe Phi Beta Kappa prize for excellence in teaching. He has been Vice President of the American Philosophical Society and has served as Chairman of the United States National Committee of the International Astronomical Union.

Gingerich's research interests have ranged from the recomputation of an ancient Babylonian mathematical table to the interpretation of stellar spectra. He is co-author of two successive standard models for the solar atmosphere. In the past two decades, he has become a leading authority on Nicholas Copernicus and Johannes Kepler. He was awarded the Polish government's Order of Merit in 1981, and more recently an asteroid has been named in his honor.

Besides numerous technical articles and reviews, Professor Gingerich has written more popularly on astronomy in several encyclopedias and journals. Two anthologies of his essays have recently been published: *The Great Copernicus Chase and Other Adventures in Astronomical History* (1992) and *The Eye of Heaven: Ptolemy, Copernicus, Kepler* (1993).

# Cranks and Opportunists: "Nutty" Solutions to the Longitude Problem

*Owen Gingerich*

A year ago I prepared for *Scientific American* an article on astronomy in the age of Columbus.[1] Shortly thereafter I received a letter from a reader who assured me that Columbus kept two logs, one for himself with his personal thoughts, and a second public one for the world to see, and he cautioned me that Columbus's 2½-hour error in determining the longitude of the Indies from a lunar eclipse could have been a deliberate political act related to the division of lands between Spain and Portugal. I was fascinated by the idea but at a loss to decide whether or not the claim had any merit. However, I was even more intrigued by the reader's urgent emphasis, elsewhere in his letter, on the north polar star clock and the use of sidereal time. In response, I assured him that simply knowing sidereal time does not solve the longitude problem.

Within a few days I got a reply, and I quote, "Note that local sidereal and solar times can easily be obtained from reading either Ursa Major or Cassiopeia's dials anywhere in the northern hemisphere. Given such a set of averaged REAL TIME times, how could Crisobol Colon have made an error of 2½ hours in his public ship's log?" In yet a third letter, the reader was still more explicit: "The fact [is] that star clocks, after say 200 BC, could have been read to one minute of accuracy. These combinations of instruments would have been efficient chronometers on land or sea....Your evasion may be unintentional," he continued, "but the point is that sidereal time could have technically allowed maps to be drawn much as we know today. Why could you not find the energy to aid this obvious hypothesis?"

With this, I was prepared to end the discussion, so I wrote, "Next fall we will be celebrating the anniversary of John Harrison, and I have been scheduled to give a paper on nutty theories of longitude determination. Thank you very much for your assistance." I suppose you have already guessed that he didn't even catch on!

It is, I think, rather unusual to encounter a nutty longitude theory today. But the early eighteenth century was a particularly verdant time for such fantasies, and I shall report on a group of thirteen books or pamphlets spanning the period from 1699 to 1743. The majority date from 1714, the year in which the longitude prize was constituted by an act of Parliament. They are all found in Harvard's Houghton Library, which has for many years taken a special interest in the longitude problem, and more recently the collection has been enriched by Mrs. Harrison D. Horblit. Several of the pamphlets are exceedingly rare. Not all of them are nutty, but at least they are impractical. Some are deliciously naïve; others are outlandishly funny. All claim to be putting forward their ideas for the enlightenment of mankind, but today's dream of winning the lottery certainly had its parallel with these often impoverished authors. I shall describe them here more or less in chronological order.

But first I must remark on what is undoubtedly the most bizarre longitude solution of them all, now rather well known because of its mention in Gould's *The Marine Chronometer* and Landes's *Revolution in Time*.[2] Harvard does not have the rare anonymous pamphlet in question, called *Curious Enquiries* and published in London in 1688, but there is a copy at Brown University. The method is based, entirely tongue-in-cheek, on the "powder of sympathy" of Sir Kenelm Digby (1603-1665). Despite Gould's claim that the inventor of this method appears to have been serious, the satirical quality of the proposal is self-evident from the full text:

> Sir Kenelm Digby in his Discourse of the Sympathetick Powder, tells us how he made Mr. Howel start, upon his putting a Bloody Garter, with which Mr. Howel's wounded Hand had been bound up withal, into a Bason of Water mixed with that Powder: If such a starting could be made to any Inferiour Creature at a great distance, and by often doing it, it would not in two or three months lose its power, we might at Sea with great Ease and Pleasure know when the Sun was upon the Meridian at London, or any other appointed time; and consequently by the difference of Time, the difference of Longitude. Fye! says one, or other, you would not sure put a Dog to the misery of having always a Wound about him to serve you, would you? Why not as well as to keep a Dog two or three days together starving, that he may give his Master an hour or two's Pleasure the better after it. Or for a Seaman to put out his right Eye to serve the Merchant: As before the Back-Quadrants were Invented, when the Forestaff was most in use, there was not one Old Master of a Ship amongst Twenty, but what a Blind in one Eye by daily staring in the Sun to find his Way.[3]

1. Owen Gingerich, "Astronomy in the Age of Copernicus," *Scientific American*, vol. 267, no. 5 (November 1992), pp. 100-5.

2. Rupert T. Gould, *The Marine Chronometer: Its History and Development* (London: J. D. Potter, 1923), pp. 10-11. David S. Landes, *Revolution in Time* (Cambridge: Harvard University Press, 1983), pp. 145-6.

3. *Curious Enquiries. Being Six Brief Discourses, viz. I. Of the Longitude. II. The Tricks of Astrological Quacks. III. Of the Depth of the Sea. IV. Of Tobacco. V. Of Europes being too full of People. VI. The various Opinions concerning the Time of Keeping the Sabbath* (London, 1688). I wish to thank John H. Stanley, Head of Special Collections of the Brown University Library, for providing a copy of this rare text.

### SAMUEL FYLER

*Longitudinis Inventae Explicatio Non Longa, Or, Fixing the Volatilis'd, And Taking Time On Tiptoe, Briefly Explain'd;...* (London, 1699).

One of the most charming nutty theories of longitude, put forth in 1699, is the method of "time on tiptoe" expounded by one Samuel Fyler, Rector of Stockton in Wiltshire, M.A. from Baillol in Oxford in 1654 (Figure 1). For many years a schoolmaster at Heylesburg Grammar School, the 70-year-old Fyler urges that his method be not less esteemed just because it isn't taken out of the writings of Ptolemy, Copernicus, Tycho Brahe, or Galileo. Yet he allows that he has not handled globes since he was at the university some 45

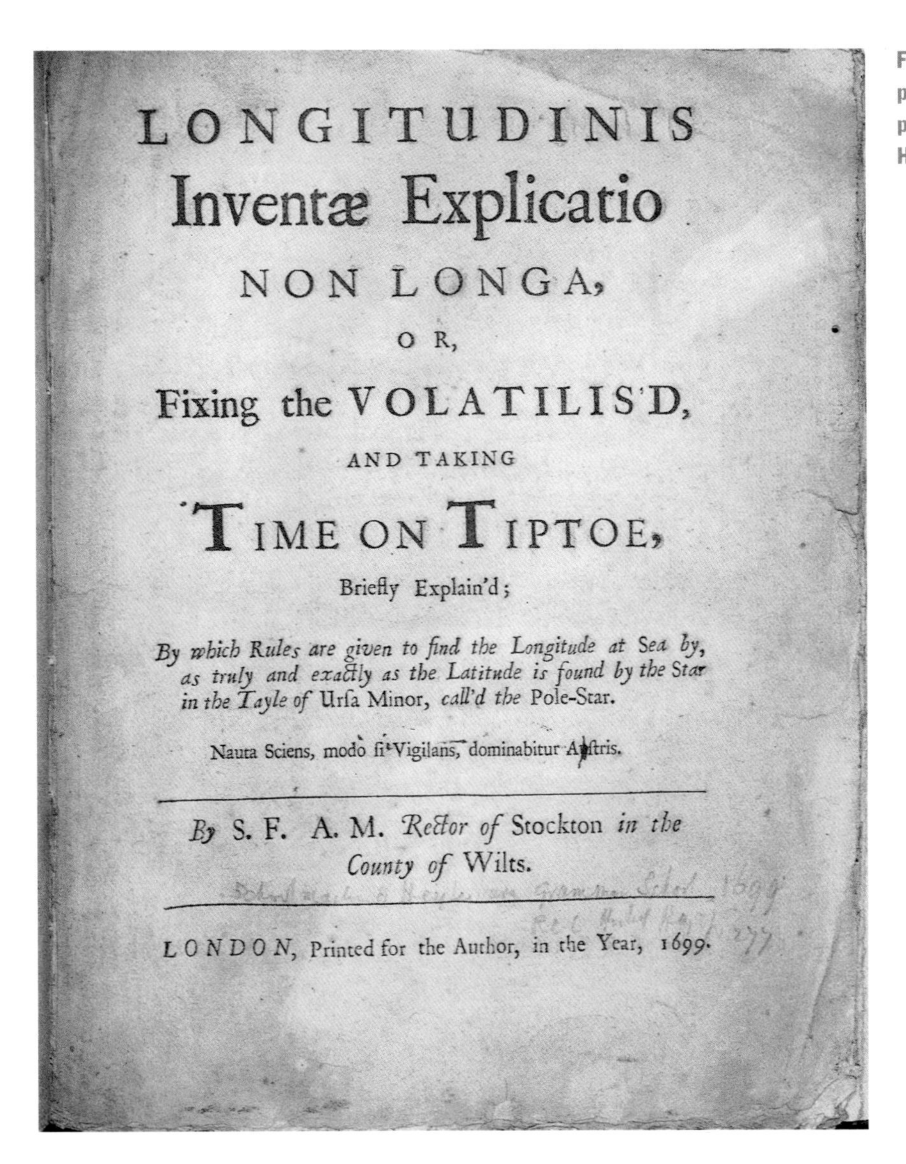

LONGITUDINIS
Inventæ Explicatio
NON LONGA,
OR,
Fixing the VOLATILIS'D,
AND TAKING
TIME ON TIPTOE,
Briefly Explain'd;

*By which Rules are given to find the Longitude at Sea by, as truly and exactly as the Latitude is found by the Star in the Tayle of* Urſa Minor, *call'd the* Pole-Star.

Nauta Sciens, modo ſit Vigilans, dominabitur Aſtris.

*By* S. F. A. M. *Rector of* Stockton *in the County of* Wilts.

LONDON, Printed for the Author, in the Year, 1699.

Figure 1. Title page of Samuel Fyler's pamphlet, published in 1699. By permission of the Houghton Library, Harvard University.

years earlier, so that some skillful artist might have to amend or improve on his method.[4]

The Rev. Mr. Fyler expounds his scheme somewhat circuitously, in a delightfully wrong-headed dialogue with a noble seaman whom he hopes to lure away from the lusty punch bowl and to become a water-drinker like himself.[5] Briefly, the well-intentioned rector proposes to map a row of stars along series of equidistant celestial meridians, that is, as if he is laying out 24 lines of right ascension in the sky, though by no means does he describe it that clearly. Then, at an appropriate moment, he would observe which row is closest to his apparent meridian, and compare this with the one that would be on the meridian at midnight at the reference longitude, and from the difference calculate the longitude.

If you are confused, perhaps you need his example: "If you are Five Rows off from the First Meridian, you will see, of necessity, but one Row beyond it, which will seem to you to be at, or near to the Eastern Horizon, and consequently the Fifth is the First Meridian Row, which is vertical to them at the Azores; if you are Four Rows off from the First Meridian, you can see but Two Rows beyond it, the sight of the rest being lost," and so on.[6]

But why on tiptoe? The stars will stay at a point but a minute or two, being enough for the purpose. "Then by taking Time on Tiptoe, I mean a speedy and quick Observation in finding agen those Stars, which I have thus fix'd to a Point, when there shall be occasion." The questioner then asks, "By what you have said yet, Sir, I do not well understand what your meaning is; will you pleas'd to make your Hypothesis yet more plain?"[7] Now this comes at the beginning of his dialogue, but from our point of view it could just as well have come at the end. Clearly he implicitly assumes some knowledge of the time back at the Azores when he is imagining his observation in the western Atlantic, and if he could have been any more lucid in his description, he himself would have seen the circular argument and the futility of his scheme. We can all agree that "Time being as a Bird taking Wing every Moment, he that will take hold of it, had need be speedy."[8] As a rector, he had perhaps really written his booklet just to quote Psalm 107: "They that go down to the Sea in Ships, that do business in great waters, these see the works of the Lord, and his wonders in the deep."[9]

Fyler's scheme, alas, is not one of the wonders of the deep.

4. Fyler, *Longitudinis Inventae,* "To the Reader."

5. *Ibid.*, p. 7.

6. *Ibid.*, p. 5.

7. *Ibid.*, p. 1.

8. *Ibid.*, p. 3.

9. *Ibid.*, p. 11.

10. Howard, *Copernicans,* pp. 28-9.

11. *Ibid.*, p. 31.

### EDWARD HOWARD

*Copernicans Of all Sorts, Convicted:...* (London, 1705).

The idea that the difference between the true north and the magnetic north might solve the longitude problem remained a tantalizing will-o'-the-wisp throughout this entire period. The practical application remained elusive even though in principle it could have provided the essential information. Consequently, books espousing this method should not be immediately dismissed as cranky. But when the book comes with the title *Copernicans Of all Sorts, Convicted: By Proving, that the Earth hath no Diurnal or Annual Motion, as is suppos'd by Copernicans*, our suspicions are perhaps justifiably aroused. The honorable Edward Howard of Windsor praises Copernicus as "a Person of Excellent Science; who finding that *Ptolemy*, as he conceiv'd, had made a supernumerary use of Circles and Epicircles, as he methodiz'd his Computations, *Copernicus* thought fit to Reject, and invent a System more facily Practical; and this he judg'd might be more plausibly perform'd, if he reviv'd the Opinion of *Pythagoras*,"[10] but for the Earth to be in motion is absurdly incoherent "since neither from *Copernicus*, or any Adherents to his Hypothesis, there is Extant, at this time, any avow'd Demonstration, where it might be, with any certainty, imply'd, that the Earth had, of necessity, a diurnal and yearly Revolution in the room of the *Sun*."[11]

To be candid, Mr. Howard was quite right that, as of 1705, no observational proof had as yet been adduced for the motion of the Earth, although Isaac Newton's powerful *Principia Mathematica,* published not quite two decades earlier, made little sense without this assumption. Nevertheless, it would be anachronistic to categorize Howard as nutty or even as a crank, curious as his position strikes us today. Therefore, when we find appended to his book "A useful Memorandum to the precedent Book and Appendix, in reference to the practical finding of the Longitude at Land or Sea," we must approach it open-mindedly. In fact, he gives a very level-headed discussion of the longitude differences between Windsor, London, and Greenwich as related to the variation of magnetic north from true north, and he presents at great length the problem of the annual variation of magnetic north. Certainly the honorable Edward Howard has outlined a credible approach in a scientific fashion, antediluvian as his Copernican views appear from a twentieth-century vantage point.

### FRANCIS HALDANBY
*An Attempt To Discover the Longitude At Sea,...* (London, 1714).

Francis Haldanby's attempt to discover longitude at sea (Figure 2), printed in a tiny book of ten pages priced at two pence (and which has been marked up to three pence in Harvard's copy), is wonderful by its sheer naïveté. It now is so scarce that E. G. R. Taylor (1879-1966) was unable to trace a copy of the work. Published in 1714, the tract announced on its title page that it was a response to the Act of Parliament. In dedicating his little work to James Stanhope (*ca.* 1673-1721), one of the secretaries of state, "without the Ceremony of an introducer," Haldanby writes, "Sir, Your penetrating Judgment, and Skill in the Mathematicks will soon determine the fate of this Attempt. The Method is obvious, and soars not to that height, which the more learned have flown, but yet is Mathematical, and may easily be apprehended by the skill in Navigation." Haldanby had noticed that if you are near the equator and are sailing northwest by the compass, that is, at 45° to the lines of latitude, you will gain one degree of longitude for every degree of latitude. At other compass points of sail, the ratio can easily be found from the grid he has provided, and presumably he intended eventually to provide grids for other latitudes. In the absence of ocean currents his scheme would have much to commend it, but then again, marvelous navigation would be possible if there were no currents. "Let it's Doom be what it will,...Sir, Your most Obedient, and most Humble Servant, Fr. Haldanby." Need I say more?

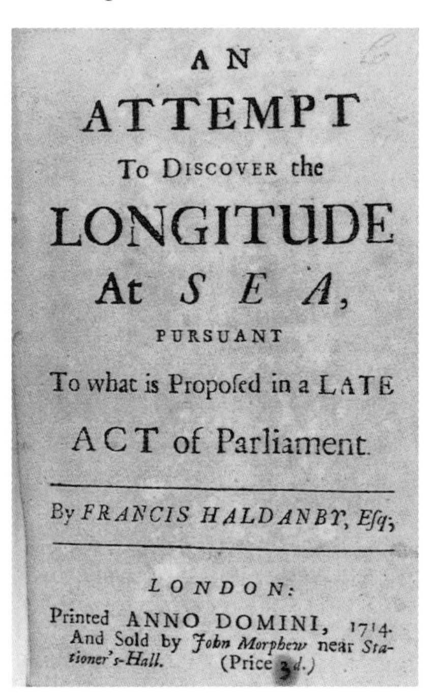
AN
ATTEMPT
To DISCOVER the
LONGITUDE
At SEA,
PURSUANT
To what is Propoſed in a LATE
ACT of Parliament.

By *FRANCIS HALDANBY, Eſq;*

*LONDON:*
Printed ANNO DOMINI, 1714.
And Sold by *John Morphew* near *Stationer's-Hall.* (Price 3*d.*)

**Figure 2. Title page of Francis Haldanby's pamphlet of 1714. By permission of the Houghton Library, Harvard University.**

### JOHN FRENCH
*A Perfect Discovery Of The Longitude at Sea;...* (London, 1715).

Another of these attempts is by one John French, a former seaman who became a teacher of mathematics in Grays-Inn-Lane, Holborn, London. In her book, *The Mathematical Practitioners of Tudor and Stuart England*, E. G. R. Taylor reports that she had been unable to locate a copy of this pamphlet by French,[12] but in the sequel, *The Mathematical Practitioners of Hanoverian England,* she inserts an extra entry to report that the work is extremely rare,[13] perhaps referring to the copy that has since come to the Houghton Library at Harvard. French's method is based on an instrument that seems as preposterous as a perpetual motion machine. The device is shown in Figure 3.

12. E. G. R. Taylor, *The Mathematical Practitioners of Tudor and Stuart England* [1485-1714] (Cambridge: Cambridge University Press, 1954), p. 431.

13. E. G. R. Taylor, *The Mathematical Practitioners of Hanoverian England, 1714-1840* (Cambridge: Cambridge University Press, 1966), p. 119; besides Harvard, copies are found at the New York Public Library and Cambridge University Library.

14. French, *A Perfect Discovery,* p. 6.

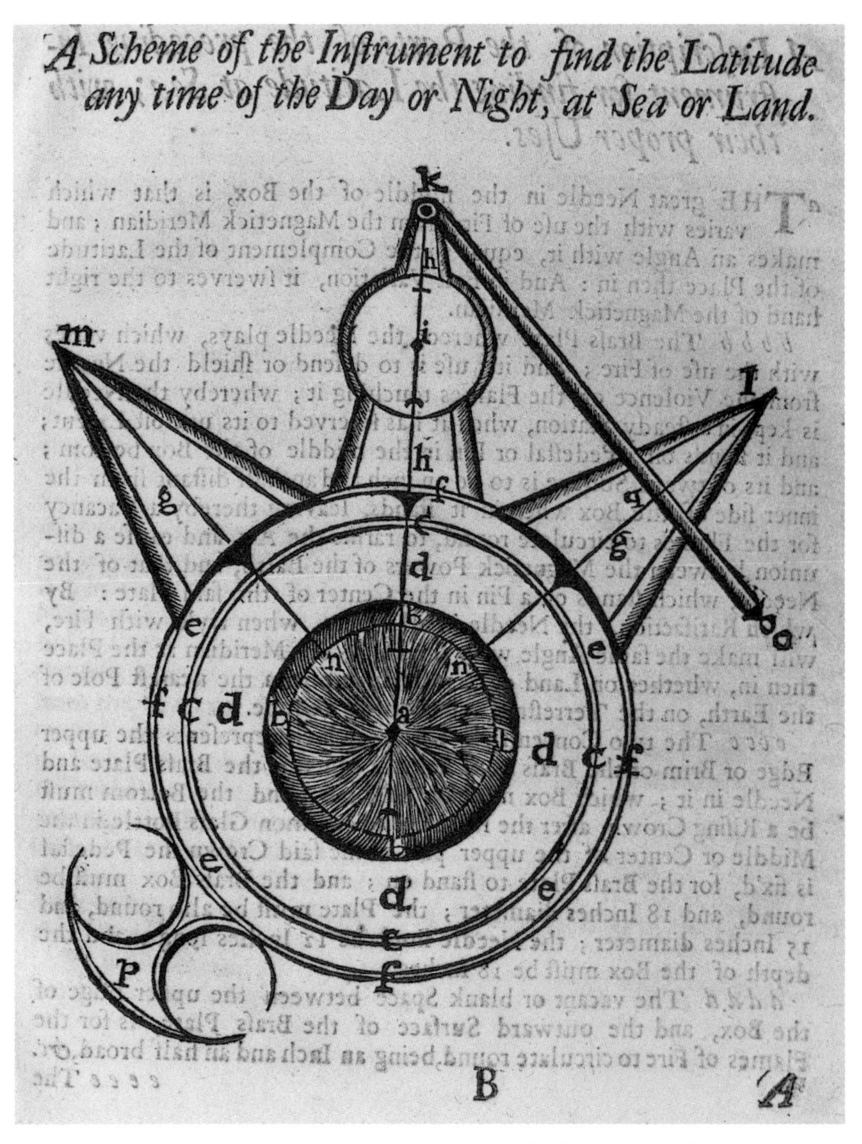


Figure 3. Illustration from John French's pamphlet of 1714. By permission of the Houghton Library, Harvard University.

The dark central portion is a brass plate upon which a needle plays. The needle is heated, apparently to drive any magnetization out of it, and this allows that "the Needle in all Places, when try'd with Fire, will make the same Angle with the Magnetick Meridian as the Place then in, whether on Land or Sea, is distant from the nearest Pole of the Earth, on the Terrestrial and Aqueous Globe"—in other words, a proposed scheme to find latitude, not longitude.[14] *dddd* is an open space for the flames (one hopes that the ship is not set on fire!). *p* is a handle to turn the brass ring, *ffff,* around so that *n* or *n* matches the point of the needle for comparison with the magnetic needle, *hh.* A 50-inch-long brass rule, *ko,* is used for making the necessary calculations graphically. Of course, since the demagnetized needle will respond randomly, the calculations will be random, and it would be a waste of our time to refine them further.

Nevertheless, it is interesting to see the ingenuity of French's next step. French claims that his instrument yields the complement of the latitude. If he now observes the altitude of the Sun at noon, knowing his latitude, he can determine the Sun's declination, and from this he can find the Greenwich time from a table of the Sun's declinations as a function of date. At the time of the equinox, the Sun's declination is changing by 1' per hour, so if he could measure the noon solar altitude to 1' and the latitude by his bogus device to the same accuracy, in principle he would have his longitude to within 15°, hardly very helpful, though in principle the procedure does work. Hence French's scheme is a remarkable mixture of perception, impracticality, and phoniness!

French also proposes an even more convoluted scheme to find the longitude from a series of lunar observations that would establish the place of the lunar nodes, but if nothing else, the scheme falters for the lack of adequate lunar tables:

> With God's Permission, I do also propose to make an ample and perfect Theorax of the Moon;...if I meet with an agreeable Reception and proper Encouragement, but otherwise, being a poor Man, cannot proceed any farther; for by my long and tedious Endeavours, together with my lost Time and extraordinary Charges of Alterations and Amendments, have reduced my self and my Family, to so mean and low a Degree, that I cannot endeavour farther the Promotion of the General Good of this my Country; and run my self and Family into inevitable Ruin, which I am upon the brinks of already;...but must desist, till rewarded for these Endeavours hereby publish'd.[15]

Unfortunately, the Commissioners did not see fit to award French a prize, and it took another 30 to 40 years for good lunar tables to be produced.

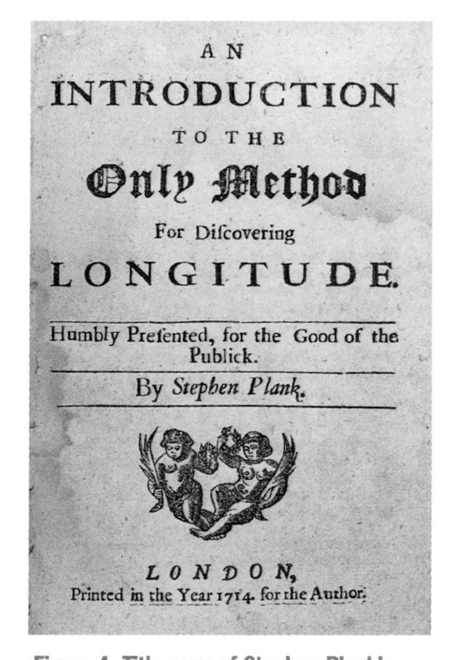
AN
INTRODUCTION
TO THE
Only Method
For Discovering
LONGITUDE.
Humbly Presented, for the Good of the Publick.
By Stephen Plank.
LONDON,
Printed in the Year 1714. for the Author.

Figure 4. Title page of Stephen Plank's first pamphlet, dated 1714. By permission of the Houghton Library, Harvard University.

### STEPHEN PLANK

*An Introduction To The* **Only Method** *For Discovering Longitude...* (London, 1714) and *An Introduction To a true Method For the Discovery of Longitude at Sea...* (London, 1720).

Some authors proposed more than one scheme to interest the Commissioners of Longitude. In 1714 one Stephen Plank from Spittlefield, about whom nothing more seems known, offered an eight-page pamphlet entitled *An Introduction To The* **Only Method** *For Discovering Longitude* (Figure 4), in which he earnestly explained that all that was required was a dial (to find the local time) and some good watches set at the port of departure. These watches should be kept in cotton in a brass box placed on a small stove to ameliorate the effects of temperature changes.

Six years later Stephen Plank tried again, this time with the "true Method" for discovering longitude at sea. Now he describes the well-known lunar method, and like French and others before him, he recognizes the lack of adequate tables and urges "a continued Series of Observation."[16] In addition, he realizes that an accurate device is required to measure the lunar positions, but now ensues a tale of woe: "The Benefits and Advantages I imagin'd might accrue from these my Humble Labours, to all Mankind in general and my Country in Particular, induc'd me to the Thoughts, of Contriving an Instrument fit to take Observation at Sea; which I having caused to be made, to my Great Trouble and Expense."[17] But not having heard from the Commissioners and unwilling to let slip any opportunity of doing his country service,

> I sacrific'd all my other Engagements on Shore, to accomplish it, left my Habitation and Family, and took a Voyage to the Island of

> Jamaica, with my said Instrument, intending by Reason of the Serenity of that Air, to make a Series of Observation, on this preceding Method...but in my Passage thither, had the Misfortune to have my said Instrument broken on board the ship I was in, in a Storm; thereupon was oblig'd to return back to *England*, without the Satisfaction of making one Observation; by reason there was no Person in that Island, that could fit up my Instrument again.[18]

His sad story told, he indicated his continued readiness with utmost alacrity to communicate to whatever worthy persons were ready to listen. With that, Stephen Plank and his undescribed instrument seem to have vanished from the scene.

15. *Ibid.*, pp. 19-20.

16. Plank, *An Introduction To a true Method,* p. 5.

17. *Ibid.*, p. 7.

18. *Ibid.*, pp. 7-8.

19. Hawkins, *An Essay For The Discovery,* "The Epistole Dedicatory."

20. Taylor, *Tudor and Stuart Practitioners* (*op. cit.*, note 12), pp. 306, 428-9.

ISAAC HAWKINS
*An Essay For The Discovery Of The Longitude at Sea,...* (London, 1714).

An instrument of an entirely different sort is described by a Londoner, Isaac Hawkins, who is persuaded "that the Foundation at least of this great Discovery is herein laid; and therefore cannot, but reckon it a Duty, to run all Ventures, rather than conceal one distant Hint from a better Projector, who may improve it to the desired Effect."[19] Mr. Hawkins proceeds to describe a barometer, but why a barometer? He has very ingeniously noted that high tides circulate around the Earth and reach a particular longitude at a particular local time determined by the motion of the Moon. In principle, the time of high tide determines where the Moon is, so that finding the local time of high tide would be similar to observing where the Moon is. But how can you know the time of high tide at sea? By the rise in altitude, which should be measurable with a barometer! Hawkins makes no effort to discuss the potential accuracy of this determination. As a serious proposal, it goes straight into the crazy bin. His Yorkshire contemporary, Jeremy Thacker (fl. 1714), who was the first to use in print the word "chronometer," took space in his pamphlet *The Longitudes Examin'd* to scoff at Hawkins's impracticable suggestion.[20] But Hawkins's detailed and wordy proposal runs for 97 pages—clearly a much more ambitious publishing project than some of the humble little tracts of a dozen or two small pages. In fact, after commending his tidal method to those who would have the patience to improve it, he devotes the remaining 73 pages to the requirements of a marine timekeeper. He knows the problems of temperature, and of "the temper of oyl," and he worries about changes of weather and air pressure. His solutions, alas, are obscure or impractical. Thus, although his prescriptions are not off the wall, his contribution was scarcely worth the printing.

ALLAIN PITOT
*L'Automate De Longitude...* (London, 1716).

As the parliamentary contest was not limited to British citizens, others got into the act. Writing in French, one Allain Pitot proposed *l'automate* for measuring longitude. In his "new system of hydrometry," he carefully described how terrestrial distances could be measured with a rolling wheel or waywiser. The ocean is, of course, not a solid, but a fluid, as is the air. Pitot thus describes how the turns of a windmill on a boat pulled through a canal would in effect measure the distance traveled. Finally, by analogy, a windmill-like device—his hydraulic automat—could measure the movement in water and in various ways could be calibrated. The absence of details here should not be held against him, he implies, because "time and experience can perfect this new invention that is only now undergoing its birth."[21] Like all beautiful mechanical inventions, as in all the other sciences, one should not let some small imperfection deter the development of something so perfect in its nature.

But surely, has not something important been left out? Pitot has now reached page 19 of his small 24-page tract, when he allows "la seule objection d'un peu importance," that is to say, the ocean currents. Ah yes, he admits, but these are for the most part well known. Lack of information about them can in time be remedied, so he can see nothing that could retard the development of his automat, which must be much more practical than eclipses, observations of Jupiter's satellites, or clocks, all of which are unknown in actual maritime usage. And with the ingenuity of the clockmakers in London, they can find out the optimum size and shape of the water-mill for measuring the distances traversed at sea. "Thus, the automat of longitude can some day be brought to perfection by the industry of good horologic artists and by the judicious observations of pilots."[22]

Meanwhile, Pitot reports, he is not going to abandon his other speculations concerning another method, which involves making a longitude compass similar to an ordinary compass, but since he has not yet had an occasion to test his ideas on this matter, the second part of his tract will have to await later publication. Alas, there seems to be no trace of it.

### WILLIAM WHISTON AND HUMPHRY DITTON

*A New Method For Discovering the Longitude Both At Sea and Land, Humbly Proposed to the Consideration of the Publick* (London, 1714).

By 1714, the career of Dr. William Whiston (1667-1752) had already gone over the top. A young protégé of Isaac Newton's, he had succeeded his master as Lucasian Professor at Cambridge, where he had served for a decade before being expelled in 1710 for religious heterodoxy. Now leading an impoverished and somewhat Bohemian life, Whiston saw in possible government intervention a way of improving his diminished lot. What began as a casual afternoon and evening of conversation with his friend Humphry Ditton (1675-1715), master of the new mathematics school at Christ's Hospital in London, became an increasingly serious, and increasingly harebrained, suggestion for the determination of longitude.

Mr. Ditton took an occasion, among other common discourse (so they inform us), to observe that "[t]he nature of Sounds would afford a method, true at least in Theory, for the discovery of the Longitude."[23] The idea was that if you know the standard time of a loud sound—for instance, a cannon fired at midnight—and your own local time, you can calculate the difference in longitude of the two stations after correcting for the time of propagation of sound. "Mr. Whiston immediately own'd the truth of the Proposition and added, 'That as to the Propagation of Sounds, he remembred to have himself plainly heard the Explosion of great Guns about 90 or 100 Miles, viz. when the French Fleet was engaged with Ours, off Beachy-head in Sussex; and he himself was at Cambridge; and that he had been inform'd, that in one of the Dutch Wars, the sound of the like Explosions had been heard into the very middle of England, at a much greater distance.' "[24] Realizing that they might be onto something, they obtained pledges of secrecy from the rest of the company, and within two days Whiston had prepared a short paper on the subject.

However, discreet inquiries with seafaring men failed to show evidence of long-range propagation of sound at sea, and the matter was temporarily dropped until Whiston dreamed up what he believed to be a great improvement upon this method, namely, that the guns might also carry shells to a great height, something over a mile, and there explode, thus joining the eye and ear together for the same purpose. But would the flash be seen at a sufficient distance? This was answered by a day of extraordinary fireworks, the Thanksgiving day for the Peace, celebrated July 7, 1713.[25] Whiston thereafter convinced himself that an explosion at 6,440 feet could be visible nearly a hundred miles away.

But how could these sound theoretical considerations be forged into a practical plan? Here is where the scheme slides down the slippery slope into a genuinely crank proposal. Whiston and Ditton, in their *New Method For Discovering the Longitude Both At Sea and Land*, proposed that stationary gunships, called hulls, be spaced about 600 miles apart in the oceans. At local midnight each night, each ship would fire a projectile 6,440 feet into the air, which would explode at the apex of its trajectory. Getting a shell to that height is very easy, the authors note, because the same charge of gunpowder that will carry a shell 12,880 feet distant from a cannon set at 45° will cast the shell perpendicularly upward 6,440 feet.[26] The sight and sound, or at least the sound, would enable ships within a hundred-mile radius to determine their positions. Precisely how this would be done depended, of course, on the amount of information obtained, but the authors systematically outlined the various options. Even in cloudy weather, the sound could give an approximate position of a vessel. As for the vaster spaces between the hulls, ships would find their way by dead reckoning, but could get a fix on their locations on two consecutive nights each week.

A great variety of objections can be brought against the Whiston-Ditton proposal, and these they attempted to address in advance. For example, how could the hulls be securely fixed in place? If not by anchors, then, in the extraordinary cases of very deep ocean, by weights let down from the hulls through the upper currents into the still waters below. "This Matter belongs to Tryal and Experiments, and is not to be here particularly demonstrated,"[27] the authors state. And how could the positions of the hulls be established? By the standard methods of solar or lunar eclipses, eclipses of Jupiter's satellites, the Moon's close approaches to fixed stars, or even by actual triangulation of the lights shot up from the hulls and surveying ships.[28]

Finally, what about terrorism upon the high seas? The hulls, minus rigging and sails, would be sitting ducks for piracy. "As this Method ought to be put in Practice by the Consent of all Trading Nations; so ought every one of the Hulls employ'd therein to have a legal Protection from them all; And it ought to be a great Crime with every one of them, if any other Ships either injure them, or endeavour to imitate their Explosions, for the Amusement and Deception of any."[29] The practical problems of recruiting reliable crews, of keeping the hulls safe and in place, and even of financing

21. "Le temps & l'experience pourront perfection une nouvelle invention qui ne fait encore que de naitre." Pitot, *L'Automate*, p. 19.

22. *Ibid.*, p. 23.

23. Whiston and Ditton, *A New Method For Discovering the Longitude*, p. 18.

24. *Ibid.*, pp. 18-19.

25. *Ibid.*, p. 23.

26. *Ibid.*, p. 72.

27. *Ibid.*, p. 52.

28. *Ibid.*, p. 54.

29. *Ibid.*, p. 75.

such an ambitious scheme would surely have given any parliamentary committee considerable pause. Nevertheless, Whiston had Newton's ear, and Newton, while maintaining his scepticism about the practicality of any longitude method, allowed before the investigating committee that the Whiston-Ditton scheme might be useful near the shoreline.[30] The House of Commons committee reported favorably on June 7, 1714, and the bill to offer the reward was given Royal Assent early in July. The Whiston-Ditton preface is dated July 7, and the Houghton Library copy of the book contains a handwritten date, July 15. In principle their scheme could have given useful information on positions, but somehow it seems just wild enough as a proposal to qualify for the category of a nutty method.

CASE BILLINGSLEY

*The Longitude At Sea, Not to be found by Firing Guns, nor by the Most Curious Spring-Clocks or Watches. But The only True Method for Discovering that Valuable Secret by the Sun, Moon or Stars, and an Exact Time-keeper,...* (London, 1714).

A casual or superficial reading of Case Billingsley's tract of the same year, 1714, might lead to tossing it also into the stack of nutty proposals. After all, he mentions the needle of a new and wonderful magnet that could point out the degrees of longitude, but he goes on to say, "And tho' even this is not only possible, but very easy to Him that can do all Things; yet we may with as much reason hope e're long to see the Heavens drawn with curious Black Meridians through every Tenth Degree of the Equator, which shall meet in the Poles, as to find a Stone that shall help us to the Longitude."[31] And so we recognize that Billingsley (fl. 1714-1724) is here writing tongue-in-cheek; indeed, his opinions throughout are expressed with wit and clarity.

Billingsley takes time out from his own proposal to castigate "[t]hose Gentlemen that propose to do it by Hulls of Ships and Guns," pointing out that "it would be a difficult Matter to collect every one's Quota (especially in a Time of War) or to chastise a Potent Enemy, if by the help of these Hulls of Ships, and by firing Guns contrary to the Standing Rules agreed on, he should deceive us to the Ruin of our whole Fleet."[32] There must be more hulls than they imagine, he declares, and it would be impossible to fix them at sea because of the actions of swells upon the cables, not to mention the hard life of the sailors who could face starvation or pirates and the problem of keeping the sailors sober.[33]

Billingsley recognizes that the problem of longitude at sea is the problem of finding an appropriate timekeeper, and it is here that his proposal loses ground. He wants a pendulum rod that won't change its length with temperature and that will last for more than 30 years.[34] The long pendulum must be free from damps and all extraneous air. "The Contrivance of my Movement, differs as much from a common Clock, as a *Ship* differs from a *House*," he writes, and it is "freed from every thing that may Clog it, or interrupt its *regular Motion*, for near a Year and a Quarter, without Winding up; the whole Engine about Four Foot and a Half long: I doubt not, I have hit upon such Ways to secure its exact Motion as that I shall be able to shew the **Longitude** thereby at Sea, **to Half a Degree, or Thirty Geographical Miles**."[35]

Billingsley, writing from his house at St. John's, Wapping, volunteered

30. See the committee report from the *Journals of the House of Commons* as reprinted in *The Correspondence of Isaac Newton*, ed. A. Rupert Hall and Laura Tilling (Cambridge: Cambridge University Press, 1976), vol. 6, pp. 161-3.

31. Billingsley, *The Longitude At Sea*, pp. 12-13.

32. *Ibid.*, pp. 20-1.

33. *Ibid.*, pp. 22-3.

34. *Ibid.*, p. 25.

35. *Ibid.*, p. 27.

to demonstrate that a regular and steady motion could be preserved in the midst of a very uncertain and uneven one. We might guess that he had a stabilization scheme of sorts, but no actual clock. Not quite in the category of nutty, these visionary, but unrealized, notions share company with a host of crank theories.

R. B.
*Longitude To be found out with A new Invented Instrument, Both By Sea and Land.... With A better Method for discovering Longitude, than that lately propos'd by Mr. Whiston and Mr. Ditton* (London, 1714).

Yet another author who attacked the Ditton-Whiston proposal was one R. B., possibly Robert Burleigh, secretary to the Admiral and General Sir

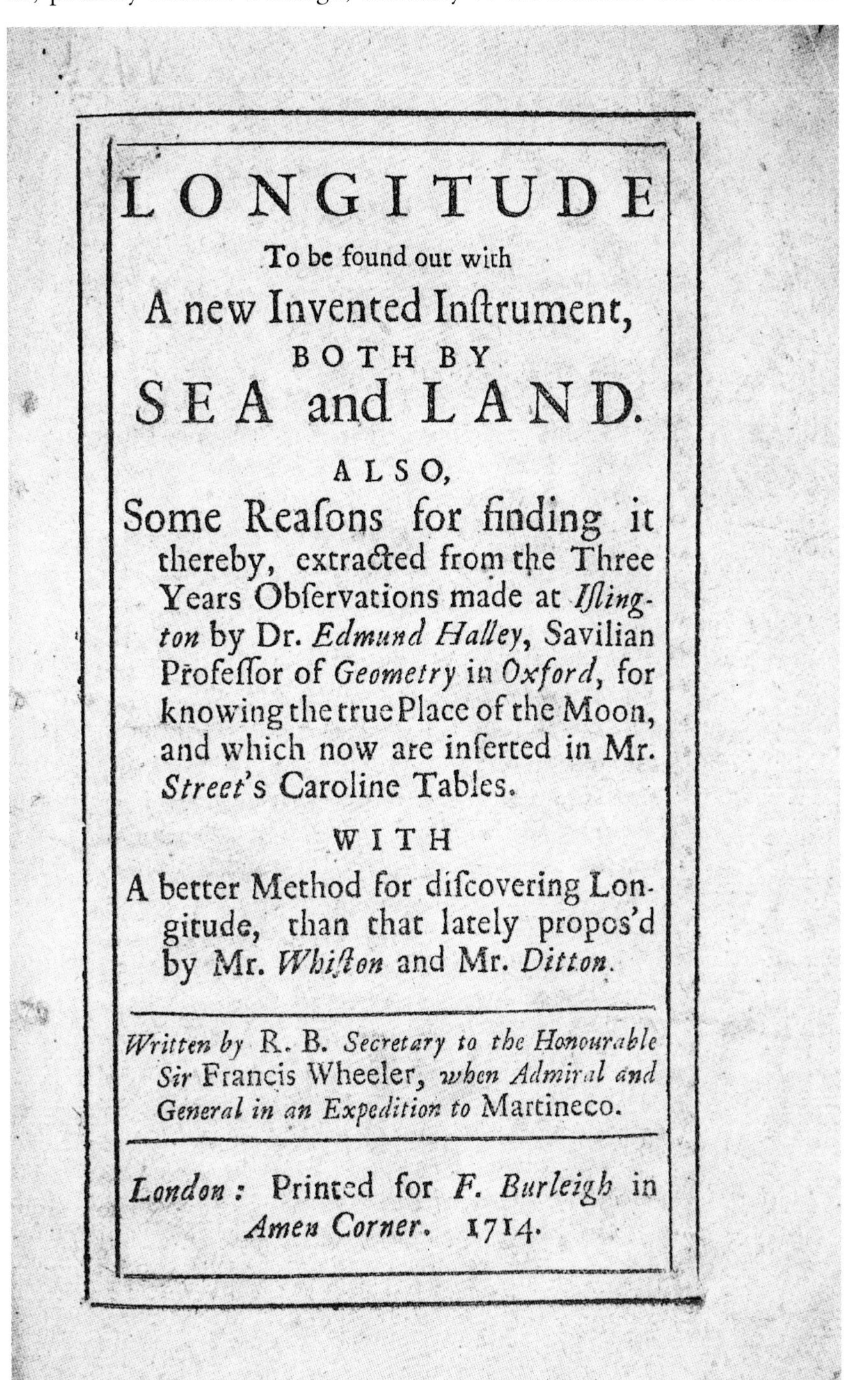

LONGITUDE
To be found out with
A new Invented Inftrument,
BOTH BY
SEA and LAND.
ALSO,
Some Reafons for finding it thereby, extracted from the Three Years Obfervations made at *Ifling-ton* by Dr. *Edmund Halley*, Savilian Profeffor of *Geometry* in *Oxford*, for knowing the true Place of the Moon, and which now are inferted in Mr. *Street*'s Caroline Tables.
WITH
A better Method for difcovering Longitude, than that lately propos'd by Mr. *Whifton* and Mr. *Ditton*.

*Written by* R. B. *Secretary to the Honourable Sir* Francis Wheeler, *when Admiral and General in an Expedition to* Martineco.

*London:* Printed for *F. Burleigh* in *Amen Corner*. 1714.

**Figure 5. Title page of R. B.'s pamphlet of 1714. By permission of the Houghton Library, Harvard University.**

Francis Wheeler (fl. 1694). "Since an Act of Parliament is past for the great Encouragement of that Person who shall first find out the Longitude of Places by Sea or Land, several Eminent Men excellently Skill'd in Mathematicks are putting their Wits to work for a Discovery of that Secret, which will be highly Beneficial to the Whole World," he writes.[36] But he reserves a great deal of scorn for the scheme to position hulls and guns across the high seas. "Truly this is a very whimsical Notion, looking very Ridiculous in Mr. Ditton and Mr. Whiston; the first of which Gentlemen I do not know, but as for the other, People says he is a little beside himself, or rather, if he has any such Thing as Brains, they are really crackt."[37]

On the title page of this very rare pamphlet (Figure 5), R. B. advertises a "new Invented Instrument" whereby longitude is to be found out both by sea and land, but the only device he refers to is a traditional cross-staff for determining the Moon's position. His is, in principle, a reasonable procedure for establishing longitude, but as we know, both accurate instrumentation and far more accurate lunar tables would be required before the method could have practical application. R. B. suggests that the Carolinian Tables of Thomas Streete (1622-1689) would be good enough, and these were in fact the best of that day.[38] My guess is, however, that R. B. would not have felt the urge to produce a 32-page booklet on longitude except for his indignation with Whiston and Ditton, so his tract really belongs in the category of anti-crank works.

### JOHN HANNA

*An Astronomical Creed, In Ten Articles, With Natural and Moral Reflections from them; To which is subjoyned, the new Method of discovering the Longitude, by the appulse of the Moon to the fixed Stars:...* (Dublin, 1725).

Scattered among these pamphlets are the occasional perceptive and astute tracts that do not belong in the slightest to the class of crank theories. Such is the 1725 book by John Hanna of Dublin.[39] Hanna scolds Whiston and Ditton for their impracticable, expensive, and dangerous scheme of stationing gun-ships at fixed intervals in the ocean. He points out that the variation of the compass will be too small to make the method useful, and he objects that the observations of Jupiter's satellites, while excellent on land, would require a six-foot telescope that would be unmanageable at sea. In contrast, he argues, the method of lunar positions offers great promise. Using Flamsteed's catalogue of stars and the full complexity of Newton's lunar theory, he calculates enough sample cases to persuade us that he knows what he is writing about. Hanna even discusses whether Rømer's observations of Jupiter's satellites, which suggest a finite speed for light, have any relevance to the positions of the stars and concludes that it has no adverse effect on his method. It is splendid to see at least a few clear-headed reports among the many proposals of the early eighteenth century.

### JANE SQUIRE

*A Proposal To Determine our Longitude,* 2nd ed. (London, 1743).

And finally, to complete our roster of nutty longitude theorists, we have Jane Squire (fl. 1731-1743). Her book was published later than the others, in 1743, but, from the many letters prefaced to the work, it is clear that she had

36. R. B., *Longitude To be found out,* p. iii.

37. *Ibid.,* p. 26.

38. Thomas Streete, *Astronomia Carolina. A New Theorie of the Coelestial Motions...* (London, 1661 and later editions); see Owen Gingerich and Barbara L. Welther, "The Accuracy of Historical Ephemerides," in their *Planetary, Lunar and Solar Positions, New and Full Moons, A.D. 1650-1805,* American Philosophical Society *Memoirs* series, vol. 59S (Philadelphia, 1983).

39. This may be the same John Hanna, M.A., "Teacher of the Mathematicks," who translated into English Voltaire's account, *The Elements of Sir Isaac Newton's Philosophy* (London, 1738).

40. Jane Squire, *A Proposal To Determine our Longitude,* pp. 30-2.

41. *Ibid.,* p. 33.

42. Taylor, *Hanoverian Practitioners* (*op. cit.,* note 13), p. 19.

been advocating her system for over a decade. In a letter to Sir Thomas Hanmer, one of the Commissioners of Longitude, she writes,

> I cannot help taking this Opportunity of answering two Objections I hear are made; not against the Proposal I have made, but against my making such a Proposal.
>
> The first is, that Mathematicks are not the proper Study of Women.
>
> I do not remember any Play-thing, that does not appear to me a mathematical Instrument; nor any mathematical Instrument, that does not appear to me a Play-thing: I see not, therefore, why I should confine myself to Needles, Cards, and Dice; much less to such Sorts of them only, as are at present in Use.
>
> The second Objection is, against my invading the Property of the Mathematicians, by pretending in case I should discover the Longitude, to be myself intitled to the Reward offered by Parliament.
>
> To this, I must say,...[it] still appears to me; as fair a Prize as any Plate given to be run for at *Newmarket* or elsewhere; the Pursuit more entertaining, the Victory more glorious, and the Attempt free to all; without the least Apprehension of encroaching upon the Property of any one.[40]

In reply, Sir Thomas urged her to publish her thoughts, "and if they carry that Demonstration which you think they do, Mankind will greedily receive them, the Advantages of such a Discovery will be too great to be stifled, nor will it be difficult afterwards to determine to whom the Merit and the Prize belong."[41]

So publish she did, in an elaborate book of 160 pages, selling for four shillings bound. The copy in Houghton Library comes in the original binding, specially stamped with symbols from her elaborate and, I must confess, sadly goofy proposal.

Squire recognizes that every point on Earth corresponds to a point on the celestial sphere that is zenithal to it. She divides the globe and sky into a million lozenges, or "cloves" as she calls them. Since only 3,000 stars are known, she states, there will be only one chance in 300 to find a star in one of the fields, but more stars can be discovered, and eventually there will also be glasses so that stars can be seen in the daytime. She next proposes that the zero of terrestrial longitude be drawn through Bethlehem, and that "astral time" be reckoned from there, and she requires accurate tables of sunrise there in order to compare the astral time with local solar time at sea. As E. G. R. Taylor says, "Her fantastic scheme required the rejection of many of the basic principles as well as the terminology, of current astronomy, and the mastering of a new international system of numeration of Miss Jane's invention....Halley was the expert at the moment, and as he could only consider the 'cloves' and the new language as beneath contempt, Miss Squire had the mortification of receiving no reply."[42]

Jane Squire's book is a long one, and I shall not summarize every step of finding the zenithal field and recognizing its star, or the calculations relating astral and solar time. However, one further small detail must be noted: an accurate clock will be needed! When Miss Squire learned that the Board had voted financial assistance to John Harrison (1693-1776) for a clock at Halley's recommendation, she at once jumped to the conclusion that her idea had been stolen. Not only did mankind fail to receive her plan greedily, but it did not even determine that the merit and prize belonged to her.

## SUMMARY

What, then, can we learn from this eclectic baker's dozen of proposals? As a group, they don't even come close to solving the longitude problem. "Of no value" and "cannot be taken seriously" are Taylor's judgements on two of them.[43] Are they for the most part comparable to today's proposals for perpetual motion machines, or to the output of circle squarers, and simply precious because they are rare ephemera of yesteryear?

I think it is difficult to identify another scientific or technological problem so clearly formulated, so eagerly desired by society as a whole, and potentially within the hopes of solution by Everyman, whether in the Renaissance, the Enlightenment, or today. The offer of a substantial prize—and it was truly princely, millions of dollars by today's standards—of course flushed opportunists from the bushes. I don't doubt that a number of these hopeful souls impoverished themselves devising their instruments or printing their pamphlets, and I feel a little sorry for those who scarcely dreamed that quarry was so far beyond their reach. I suspect that if we had a comparably focused project and similar prize, for example, to save the Tower of Pisa, we would have an outpouring of nutty proposals. That some of the solutions were proposed by cranks is clear enough—at least to Americans, for according to the *Oxford English Dictionary,* "crank" is an American colloquialism referring to someone enthusiastic about his or her hobby or willing to take up impractical projects. Visionaries such as Charles Babbage would fit the definition. And in a sense even the ultimate winner, John Harrison, was stamped more from the mold of the crank-enthusiast than of the scientific establishment that would arbitrate the prize. The world would be much the poorer without such enthusiasts.

We can be happy that this odd assortment of longitude proposals does in fact survive, for they give us a delightful window on a time long ago and yet not all that different from our own. Last spring one of my colleagues in the History of Science Department asked me to save some samples of the crank literature that comes to any observatory. I have been surprised at how many pieces have arrived in the past six months. Would we be smart enough to recognize a kernel in the chaff that continually flows in upon us? Perhaps not, though some of it is very interesting and provocative—but at least we can take heart that now, as then, most of it is chaff.

43. Taylor, *Tudor and Stuart Practitioners* (*op. cit.*, note 12), p. 412 (No. 518, Fyler); p. 420 (No. 570, Howard).

# The Lunar-Distance Method of Measuring Longitude

## *Derek Howse*

Derek Howse was a regular officer in the Royal Navy, serving at sea throughout the Second World War, latterly as a specialist navigating officer. Retiring from the Navy in 1958, he became a curator in the Department of Navigation and Astronomy of the National Maritime Museum at Greenwich in 1963. He was in charge of the astronomical and horological collections, which include the original timekeepers of John Harrison and Larcum Kendall. Retiring from the Museum in 1982, he was appointed Clark Library Professor at the University of California, Los Angeles, during the academic year 1983-4.

His publications include *The Clocks and Watches of Captain James Cook, 1769–1969* (with Beresford Hutchinson, 1969), *The Tompion Clocks at Greenwich and the Dead-Beat Escapement* (1970–1), *The Sea Chart* (with Michael Sanderson, 1973), *Greenwich Observatory,* vol. 3, *The Buildings and Instruments* (1975), *Francis Place and the Early History of Greenwich Observatory* (1975), *Greenwich time and the discovery of the longitude* (1980), *Nevil Maskelyne: The Seaman's Astronomer* (1989), *Background to Discovery: Pacific Exploration from Dampier to Cook* (ed., 1990), *A Buccaneer's Atlas* (ed. with Norman Thrower, 1992), and *Radar at Sea* (1993).

# The Lunar-Distance Method of Measuring Longitude

*Derek Howse*

To explain how lunar distances come into the longitude story, I must briefly go back to first principles. The theory of the astronomical solution to the problem of finding longitude lies in the measurement of time, because longitude and time are inexorably bound up with each other and with the rotation of the Earth on its axis. Therefore, to find the difference of longitude between one's own meridian and that of some reference meridian—shall we call that Greenwich?—all one has to do is to find the difference between the local time at the ship and the local time on the reference meridian (that is, Greenwich time) at that very same moment.

Simple in principle, but not easy in practice. Finding local time can be done fairly easily by measuring the height of some heavenly body above the horizon and then doing a little trigonometry. But how do you know what time it is at Greenwich at that same instant? The obvious answer is for the ship to carry Greenwich time with her in the form of some clock or watch. But, as you all know, the technology to do this just did not exist until the second half of the eighteenth century. The other possibility is for the navigator to observe some astronomical phenomenon, the Greenwich time of which can be accurately predicted. The solution eventually adopted was to make use of the fact that the Moon appears to move comparatively fast against the background of the stars, approximately her own diameter in an hour. Actually, while the stars move 15° westward each hour, the Moon only moves about 14½°.

This motion can be used as a clock, the Moon herself being the hand of the clock, the Sun, stars, and planets being the time markers on the dial. This is the basis of the lunar-distance method of longitude measurement. Figure 1 shows how the angular distance between the Moon and the Sun increases with time

1. Johann Werner, *In hoc opere haec cōtinentur Noua translatio primi libri geographiae Cl'. Ptolomaei:...* (Nuremberg, 1514). The passage concerning Werner's description of the lunar-distance method of finding longitude is in *cap.* 4, note 8, sig. dv$^{v}$. (See Appendix C.)

2. Petrus Apianus, *Cosmographicus Liber Petri Apiani Mathematici studiose collectus* (Landshut, 1524), fols. 30-1.

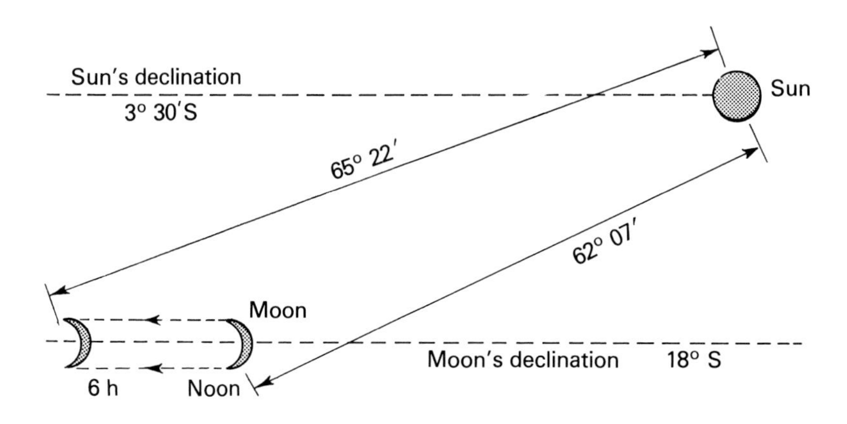


**Figure 1. How the lunar distance—in this case, the distance between the Moon and Sun—changes with time. The figures are for noon and 6 p.m., Greenwich Apparent Time, on 1 October 1772, as tabulated in *The Nautical Almanac* for 1772, the relevant page of which is reproduced in Figure 8.**

**Figure 2. The cross-staff in use on land, both for observing a lunar distance from a star and for measuring the heights of buildings. From the title page of Peter Apian's *Introductio Geographica Petri Apiani in Doctissimas Verneri Annotationes,...* (Ingolstadt, 1533).**

because the Sun is to the west of the Moon; it decreases if it is to the east. The changing lunar distance can therefore be a measure of time.

While the principle is simple, one of the big disadvantages of this method is that the arithmetic needed to compute the longitude is formidable, largely because one has to make allowance for atmospheric refraction—the bending of the light rays from outer space as they enter the dense terrestrial atmosphere—and for lunar parallax, because the observations are made from the Earth's surface, not its centre. The net result of these two is that the Sun and stars always appear to be higher than they really are, whereas the Moon always appears lower.

The lunar-distance method was first suggested by Johann Werner (1468-1522) of Nuremberg, who wrote the following in 1514:

> Therefore the geographer goes to one of the given places and from there observes, by means of this observational rod [the cross-staff] at any known moment, the distance between the Moon and one of the fixed stars which diverges little or nothing from the ecliptic.[1]

Figure 2 shows this being done on land. Werner went on to say that, with the aid of astronomical tables for the star's position and an almanac for the predicted Moon's position, he could find his difference of longitude from whatever place the almanac was based on. But it was not Werner's description that made the lunar-distance method known to seamen and scholars, but that of Peter Apian (1495-1552), whose *Cosmographicus* of 1524 gave a clearer description as well as a picture.[2] However, these descriptions were only theory, the most fundamental practical difficulties being:

(a) that the Moon's motion, 'the going of the clock', is highly irregular and very difficult to predict;

(b) that the star's positions, 'the time markers', were not known with sufficient precision; and

(c) that an angle-measuring instrument accurate enough to be useful at sea had still to be invented.

Two and a half centuries elapsed before these difficulties were overcome.

The next development was due to Jean-Baptiste Morin (1583-1656), doctor of medicine and professor of mathematics at the Collège Royal in Paris. Somewhat eccentric, an astrologer, and a believer in the Ptolemaic Earth-centred universe, he distrusted clocks and is reported to have said that, though the Devil might succeed in making a longitude timekeeper, it would be folly for a man to try.[3] However, in 1634 he told Cardinal Richelieu (1585-1642) that he had discovered the secret of longitude, but not by using a timekeeper. Basically, he demonstrated the observations and mathematical calculations needed for the lunar-distance method I have outlined above, taking account of lunar parallax and refraction, which Werner and Apian had not done. Morin's method was geometrically sound, but the commission Richelieu set up to examine the claim considered he had not found the longitude because the imperfections in tables of the positions of the Moon and stars meant that it could not be used in practice. If that is so, said Morin, why not set up an observatory to provide the data?

**Figure 3. Louise de Kéroualle, Duchess of Portsmouth, one of King Charles II's mistresses, who was involved in the foundation of Greenwich Observatory. Detail from a portrait by Sir Peter Lely (1618-1680), *ca.* 1671-4. Oil on canvas. 122 x 101.5 cm. (48 x 40 in.). Courtesy of the Collection of the J. Paul Getty Museum, Malibu, California. Inv. no. 78.PA.223.**

Paris Observatory was founded 33 years later, but the first really positive step taken to find an astronomical solution to the longitude problem took place in 1675, when Morin's advice was followed in England—when Greenwich Observatory was set up specifically to solve the longitude problem.

It came about in this way. In 1674, another Frenchman, the Sieur de St. Pierre, also claimed to have discovered the secret of longitude by measurements of the Moon and stars. Failing to sell his ideas in France, he came to England and called upon his compatriot, Louise de Kéroualle (1649-1734)—King Charles II's current mistress who had recently been made the Duchess of Portsmouth. He asked her to present his method to the King, presumably in the hope of receiving a reward.

A Royal Commission, already investigating a claim that longitude could be found by measuring magnetic variation and dip, was asked by the King to look into St. Pierre's claim as well. The Commission, whose members included Christopher Wren (1632-1723) and Robert Hooke (1635-1702/3), co-opted the 27-year-old astronomer John Flamsteed (1646-1719), who said that, while St. Pierre's method was not without some merit—just a little—it was not practicable at that moment for precisely the same reasons that Morin had been turned down 40 years earlier: lack of data to predict the motions of the Sun and Moon, and inaccurate star positions. Flamsteed then repeated Morin's advice: set up an observatory with large instruments and

3. F. Marguet, *Histoire de la longitude à la mer au XVIII^e siècle en France* (Paris, 1917), p. 7.

4. State Papers Domestic 29/368, fol. 299, and State Papers Domestic 44, p. 10, Public Record Office, London.

5. Act 12 Anne *cap.* 15 (1714): *An Act for Providing a Publick Reward for such Person or Persons as shall Discover the Longitude at Sea.*

**Figure 4. The Royal Observatory from Croom's Hill, *ca.* 1680, soon after its completion, by an unknown artist. Shown within the walls of the Observatory grounds is an 80-foot mast that was used for supporting and adjusting the height of a 60-foot refracting telescope. Due to instability, this apparatus was removed shortly after 1690. Oil on canvas. 102 x 168 cm. (40 x 66 in.). Courtesy of the National Maritime Museum, Greenwich. Inv. no. BHC1812.**

telescopic sights so that the necessary data could be obtained, though it might take many years to do this. The search for the data demanded by Flamsteed was to occupy astronomers the world over for the next 150 years.

So in 1675, the Royal Observatory at Greenwich (Figure 4) was founded specifically to make the lunar-distance method practicable. The King appointed Flamsteed himself to be Astronomer Royal, enjoining him forthwith to employ himself 'with the most exact Care and Diligence to the rectifying the Tables of the Motions of the Heavens, and the Places of the fixed Stars, in order to find out the so much desired Longitude at Sea, for the perfecting the art of Navigation'.[4]

The next important development in the longitude story also took place in England: the passing of the Longitude Act of 1714, offering rewards of up to £20,000 'for a due and sufficient Encouragement to any such Person or Persons as shall Discover a proper Method of Finding the said Longitude'. It was to be payable regardless of nationality 'as soon as such method for the Discovery of the said Longitude shall have been Tried and found Practicable and useful at Sea'.[5] Commissioners were appointed, familiarly known as the Board of Longitude, to administer the provisions of the Act and to recommend what sums should be awarded and to whom. The Astronomer Royal of the day was an *ex officio* member of the Board.

The results of this Act have been, and will be, referred to extensively in papers in this symposium, so I will not discuss them in detail. Suffice it to say, first, that the Act polarised the subject worldwide, the huge rewards offered stimulating the thoughts of natural philosophers, astronomers, mathematicians, and clockmakers the world over, just as Parliament had intended. And secondly that, despite the 'nutty' solutions described in Owen

Gingerich's paper,[6] by the early eighteenth century most thinking people realised that the only two basic approaches to the problem of longitude at sea that had any chance of success were the method using a timekeeper (chronometer) and the method involving lunar distances.

You will remember that there were three ingredients still needed to make the lunar-distance method viable: an instrument capable of making the angular measurements with the required precision, an accurate star catalogue, and the ability to predict the motion of the Sun and Moon accurately several years in advance.

The first of these was provided in the 1730s by John Hadley (1682-1744) with his invention of the reflecting quadrant. This instrument and its development into the sextant still in use today are described in the paper by Alan Stimson.[7] The second ingredient was partially satisfied when the star catalogue of the first Astronomer Royal, John Flamsteed, was published posthumously in 1725.[8] Though better than anything that had gone before, his positions were later refined by others, particularly by the Abbé Nicolas-Louis de Lacaille (1713-1762), who in 1757 published a catalogue of 398 stars, based on his own observations in Paris and at the Cape of Good Hope.[9]

But the most fundamental problem was predicting the motion of the Sun and Moon. The solution was largely the result of work by continental mathematicians and astronomers, particularly Alexis-Claude Clairaut (1713-1765), Leonhard Euler (1707-1783), and Lacaille himself. But it was a practical astronomer who finally came up with solar and lunar tables of the required accuracy, tables from which the Moon's position relative to the Sun and stars in the zodiac could be predicted for any moment with the required precision. This was Tobias Mayer (1723-1762) of Göttingen, who, in 1755, sent his tables to Admiral Lord Anson, First Lord of the Admiralty, who laid them before the Board of Longitude in March 1756. Because of the constraints of the Seven Years' War (1756-1763), the first sea trials of these tables were not conclusive, but the Astronomer Royal, James Bradley (1693-1762), after comparing predictions from Mayer's tables with actual observations made at Greenwich, was most enthusiastic.[10]

**Figure 5. Nevil Maskelyne, fifth Astronomer Royal, aged about 44, originator and first editor of the annual *Nautical Almanac.* and its companion, *Tables Requisite.* From 1767, these publications made finding longitude at sea by the lunar-distance method practicable for the ordinary navigator. From a drawing in black and red chalks on blue paper attributed to John Russell (1745-1806), drawn *ca.* 1776. 33 x 29 cm. (13 x 11.5 in.). In the possession of Mrs. H. C. Arnold-Forster. Photograph by permission of the National Maritime Museum, Greenwich.**

It was not until 1761 that Mayer's tables were properly tested by Nevil Maskelyne (1732-1811), the future Astronomer Royal, on his voyage to and from the island of St. Helena, off the west coast of Africa, in 1761-2 to observe the Transit of Venus. Using a Hadley quadrant (not a sextant) and Mayer's first tables, he made some very successful lunar-distance observations, generally achieving an accuracy in longitude of better than 1°. Immediately on his return, he published his *British Mariner's Guide,*[11] explaining in simple terms lunar observations at sea.

In fact, Maskelyne was not the first person to use lunar distances successfully at sea. In 1753-4, Lacaille had made such observations on his way home to France from the Cape of Good Hope, though the lunar tables he used were less precise than those of Mayer. Lacaille explained his methods in the French almanac *Connoissance des Temps* for 1761 (published in 1759), giving also diagrams for graphical solutions and tables of pre-computed lunar distances every four hours for the month of July 1761. Maskelyne had a copy of this on his voyage to and from St. Helena, as did the French astronomer Alexandre-Guy Pingré (1711-1796), who also made successful lunar-distance observations on passage to observe the Transit of Venus on Rodriguez Island.

It was Lacaille's method of working (though not the diagrams) that Maskelyne recommended in his own book, while the pre-computed lunar-distance tables he eventually published in successive editions of *The Nautical Almanac* were based on Lacaille's model.

Shortly before Mayer died in 1762 at the early age of 39, he had prepared a more accurate set of tables that were tested at sea by the Danish scholar Karsten Niebuhr (1733-1815) in 1760. Mayer's widow sent them to the Board of Longitude, and they were tried out with success by Maskelyne on voyages that he made to and from Barbados in 1763-4 in connection with the trial of Harrison's fourth, and ultimately prize-winning, timekeeper.

On 8 February 1765, Nevil Maskelyne became fifth Astronomer Royal and, as such, an *ex officio* member of the Board of Longitude. The very next day, the Board met at the Admiralty in London for what was probably the most important meeting in its 114-year history. After considering the results of the Barbados trials of Harrison's timekeeper, the Board went on to consider a long memorial from Maskelyne praising the lunar method using Mayer's last tables, which he himself had proved at sea. Four officers from the East India Company were present to testify to the Board that they had each independently followed the instructions in Maskelyne's *British Mariner's Guide* and had consistently found their longitude to within 1°. The only difficulty, they said, was the complexity of the calculations, which had taken Maskelyne himself up to four hours to complete—and he was an expert. The solution, said Maskelyne, was the publication of a nautical ephemeris or almanac based on Mayer's last tables, containing pre-computed predictions of lunar distances.

The Board's recommendation resulted in Parliament passing a new Longitude Act[12] in which the first £10,000 of the £20,000 prize was promised to Harrison as soon as he had explained the principles of his watch. £3,000 was awarded to Mayer's heirs and £300 to the Swiss mathematician Leonhard Euler, who had developed the theoretical equations on which Mayer's tables were based.

The Act also directed the Board of Longitude to publish a nautical almanac, as Maskelyne had suggested at the March meeting. Planned and executed by him with characteristic energy—he was still only 33—*The Nautical Almanac and Astronomical Ephemeris for the year 1767* and its companion, *Tables Requisite to be used with the Astronomical and Nautical Ephemeris,* were published in January 1767, the almanac being an annual publication, the *Tables Requisite* containing those tables that do not change from year to year, such as dip, parallax, and refraction, or that change only slowly, like star positions. With these books and a set of simple trigonometrical tables, the navigator had all that was needed to reduce any observations, whether for latitude, longitude by lunars, or longitude by chronometer. After 1781, he did not even need the trigonometrical tables because they were included in the second and subsequent editions of the *Tables Requisite.*

The almanac also caused another fundamental change in the navigator's practice. Up to that time, seamen had usually expressed their longitude as a certain number of degrees and minutes (or leagues) east or west of the departure point or of their destination—3° 47' west of the Lizard, for example. But now, any navigator using Maskelyne's *Nautical Almanac* to find longitude astronomically—and a very high proportion of the world's deep-sea navigators began to do so from 1767—must end up with an answer based on the

6. See Owen Gingerich's paper in this volume.

7. See Alan Stimson's paper and Appendix D in this volume.

8. John Flamsteed, *Historia Coelestis Britannica*, 3 vols. (London, 1725).

9. Nicolas-Louis de Lacaille, *Astronomiae Fundamenta* (Paris, 1757).

10. For an account of the development of solar and lunar tables, see Eric G. Forbes, *Greenwich Observatory, The Royal Observatory at Greenwich and Herstmonceux, 1675-1975,* vol. 1, *Origins and Early History (1675-1835)* (London: Taylor and Francis, 1975), pp. 109ff.

11. Nevil Maskelyne, *The British Mariner's Guide containing Complete and Easy Instructions for the Discovery of the Longitude at Sea and Land,...* (London, 1763).

12. Act 5 George III *cap.* 20 (1765): *An Act for explaining and rendering more effectual Two Acts, One made in the Twelfth Year of the Reign of Queen Anne,...and the other in the Twenty sixth Year of the Reign of King George the* Second,....

Greenwich meridian. He then needed to plot his position on a chart, so map and chart publishers the world over began to provide longitude graduations based on Greenwich, so much so that, when the need eventually arose for a prime meridian for longitude and time to be agreed upon internationally—at the International Meridian Conference in Washington, D.C., in 1884—it was Greenwich that was chosen, largely because at that time no less than 72 per cent of the world's shipping tonnage was using charts based on Greenwich. And it was the publication of the first annual *Nautical Almanac* in December 1766 that started this chain of events.

Before 1767, probably not more than a score of navigators of any nationality had succeeded in measuring their longitude when out of sight of land. Now, any competent mariner could do so quickly and comparatively easily, provided he could afford a Hadley quadrant at about three guineas[13] (a new-fangled brass sextant would be much better though more expensive at about twelve guineas),[14] a watch accurate to a minute or so in six hours, a set of simple trigonometrical tables at about ten shillings, and the almanac and *Tables Requisite* at six shillings the pair.[15] Of course, if he could afford one or more reliable chronometers (at least 40 guineas each),[16] he would find it easier still—but that was to be at least 40 years ahead for all but the favoured few.

Though the means for measuring longitude by lunar distance were now available, it naturally took many years for the practice to become commonplace, for mariners are a conservative breed. In Britain, the officers of the East India Company and of the Royal Navy showed the way. Even before 1767, some of the former had begun to find their longitude the hard way, using the methods described in *The British Mariner's Guide;* now they thankfully turned to the new almanac, cutting the time for computation from four hours to about 30 minutes.

They say that imitation is the sincerest form of flattery. In 1772, Joseph-Jérôme Lefrançais de Lalande (1732-1807) published, in the French almanac *Connoissance des Temps pour l'Année 1774*, lunar-distance tables copied directly from the British *Nautical Almanac*, but with instructions and table headings translated into French. These tables were based on the Greenwich meridian, though all the other tables in the almanac were based on the Paris meridian. The English tables continued to be used until the *Connoissance des Temps* for 1790, when French-computed tables began to be published.[17] The cooperation between Maskelyne and his French colleagues continued despite

13. For example, see 'A Catalogue of mathematical philosophical and optical instruments, made and sold by John Troughton, successor to Benjamin Cole...', in the back of Joseph Harris, *The Description and Use of the Globes and the Orrery...*, 12th ed. (London, 1783).

14. 'Best 10-inch sextant by Ramsden, £12 12s 0d' in 1787 (MS. RGO 14/18, fol. 257, Royal Greenwich Observatory Archives, Cambridge University Library).

15. Priced on title pages: *Nautical Almanac* for 1772, 3s 6d; *Tables Requisite* for 1767, 2s 6d.

16. Arnold's chronometers sold from 25 to 120 guineas *(Report from the Select Committee of the House of Commons,...to whom the Petition of Thomas Mudge, Watchmaker, was referred;...* [June 1793], para. 82).

17. F. Marguet, *Histoire générale de la Navigation du XV$^{e}$ au XX$^{e}$ Siècle* (Paris, 1931), p. 131.

18. A running fix is the position obtained by combining two observations not taken simultaneously, allowance being made for the ship's estimated run between the two.

19. MS. RGO 14/67, fol. 46$^{v}$, Royal Greenwich Observatory Archives, Cambridge University Library.

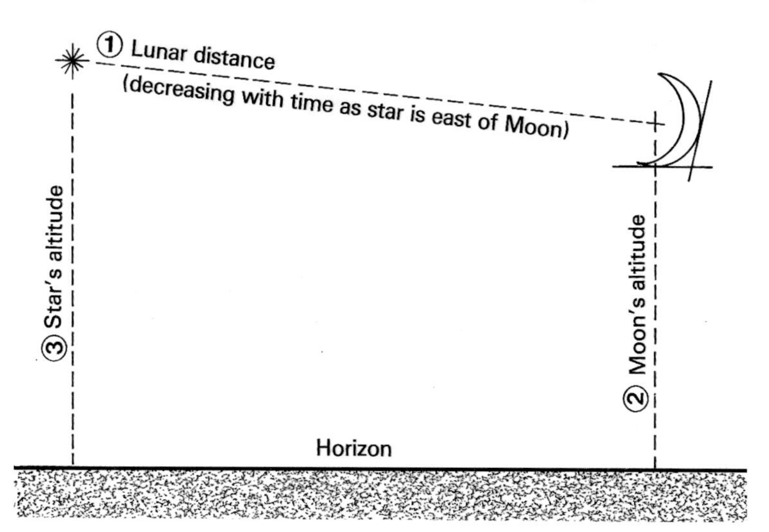


**Figure 6. The three near-simultaneous observations needed to find longitude by lunar-distance: (1) the angular distance between the Moon and a selected star in the zodiac (or the Sun); (2) the altitude of the Moon above the horizon; and (3) the altitude of the star (or the Sun). Note that what are actually measured with the sextant are the angles from the limb (edge) of the Moon (and of the Sun if used, but not a star), whereas what are needed are the measurements from the centres. The observed measurements have therefore to be corrected for semi-diameter, to get the readings for the centres of the Sun and Moon. The observed lunar distance has then to be 'cleared' from the effects of refraction and parallax.**

Computations for finding the Longitude by Observations taken 4th Oct.r 1772

| Time by Watch | Dist. ☉ or ✶ & ☽ | Altitude ☉ or ✶ | Altitude ☽ |
|---|---|---|---|
| 4: 20. 15 | 102. 24. 45 | 19. 14 | 31. 40 |
| 4 . 26 . 0 | 102. 27 | 10. 00 | 32. 42 |
| 4 . 32. 45 | 102. 29 | 16. 40 | 33. 44 |
| 13 . 19 | 307. 20. 45 | 54. 02 | 98. 14 |
| 4. 26. 20 Mean | 102. 26. 55 Mean | 18. 01 Mean | 32. 45 Mean |

For rectifying the Watch Observ'd at ........ 4. 26. 20
Altitude ☉ lower limb 10. 01 Watch by
Suppos'd Apparent time
Latitude comp.d 27. 22 | Longitude comp.d 8
-From 90. Co Lat. 62. 38 | + or - in time = p.6. 7. 40
Time Observ'd by Watch ........ 4. 26. 20
Suppos'd Apparent time at Greenwich ........ 6. 14. 20

For Computing the Apparent time.

☉ Declination 4. 4. 39. 42 | 4. 39. 42
D.o 5 6. 2. 50
Difference in 24 hours 23. 08 | 6
☉ Declination at the time of Observation ........ 4. 45. 42
+ or - from 90. is Polar Distance ........ 94. 46
Altitude ☉ lower limb Observ'd ........ 10. 01
+ ☉ Semidiameter less by dip & Refraction ........ 7
Altitude ☉ center corr.d ........ 10. 08
-From 90. is Zenith Distance ........ 71. 52
Zenith Distance 71. 52
Polar Distance 94. 46 Ar. Comp.t Sine 001150
Co Latitude 62. 38 Ar. Comp.t Sine 05154
Sum 229. 16
½ Sum 114. 38 = 65. 22 Sine 95856
½ Sum-Zenith Distance = 42. 46 Sine 83100
Sum ........ 1845140
½ Sum is Co Sine 38. 22 ........ 92174
Doubled 2
Horary ∠ 66. 44 p.6. = 4. 26. 16
Time by Watch when the Altitude ☉ was taken ........ 4. 26. 20
Difference is Watch by ........ 36

To Compute from the Observations above

Time by the Watch when Dist ☽ was taken ........ 4. 26. 20
Watch being by ........ 36
Apparent time at taking the Distance ☽ ........ 4. 26. 56
+ or - for Longitude from Greenwich comp.d
Apparent time at Greenwich ........
Mean of the Observ'd Altitude ☉ or ✶ ........ 10. 01
+ Semidiameter less by Dip ........ 10
Altitude of the ☉ or ✶ corr.d ........ 10. 11
Mean of Observ'd Altitude ☽ ........ 32. 45
+ or - Semidiam.r according which limb is Observ'd ........ 21
Altitude of ☽ corr.d ........ 33. 24

By the Ephemeris.

Semidiameter ☽ at 15. 26 | 15. 26
D.o at 15. 20
12 hours Difference 6 | 3
| 15. 23
+ For Increase of Altitude ☽ ........ p.153. = 0
Apparent Semidiameter ☽ ........ 15. 31
+ Semidiameter ☉ ........ 16. 4
Sum of Apparent Semidiameter ☉ & ☽ ........ 31. 35

Computations continued

Hor.l Par. ☽ at 56. 37 Prop.l Logarithm 5023
At 56. 14 D.o 5053
Prop.l Logarithm ........ 5023 | 12 Difference 30
Prop.l part + or - ........ 15 | D.o is Prop.l part 15
Prop.l Logarithm 5038 = Hor.l Parallax
Distance ☉ or ✶ & ☽ Observ'd ........ 102. 26. 55
+ or - Semidiameters ☉ & ☽ ........ 31. 35
Distance of the ☉ or ✶ and ☽ centers ........ 102. 58. 30

Computation of Refraction by M.r Lyons Table.

Alt. ☉ corr.d 10. 1 32. 10 T.N. 1136 D.o T.N. 1136
Alt. ☽ D.o 32. 45 - 33. 10 1146 1031
T.N. 1136 1.st Diff.ce 60 2.d Diff.ce 105
+ 1.st Prop.l part 45 1.st Pro.l part 45 2.d Pro.l part 2
Sum ........ 1181
- 2.d Prop.l part 2
to this ........ 1179 Number + for an Index 2 . 1179
+ Logarithm Co Sec.t Distance ........ 0112
Logarithm ........ 195 ........ 2. 1291
By Tab. 2.d with Dist. & less Alt. 26 Distance less 90 - more +
Sum or Difference is the ........ 161 = 2. 41 Effect of Refrac.n
Distance of ☉ or ✶ & ☽ centers ........ 102. 58. 30
Effect of Refraction ........ + 2. 41
Distance clear'd of Refraction ........ 103. 01. 11

For Parallax

Altitude ☉ or ✶ corr.d 10. 11
- Refraction p.2 ........ 3
Alt. ☉ or ✶ corr.d ........ 10. 8 Co. Sec.t 10. 5069
Dist ☉ or ✶ & ☽ clear'd Ref.n 103. 1 Sine 9. 9887
Prop.l Log Hor.l Parallax ........ 5038
Prop.l Log Arch 1.st ........ 10. 2 ........ 9994
Altitude ☽ corr.d ........ 33. 24
- Refraction p.2 ........ 1
Alt. ☽ corr.d ........ 33. 23 Co. Sec.t 10. 2712
Dist. ☉ or ✶ & ☽ ........ 103. 1 Tang.t 10. 6360
Prop.l Log Hor.l Parallax ........ 5038
Prop.l Log Arch 2.d ........ 6. 59 ........ 1. 4110
Arch 1.st ........ 10. 2
Prin.l Effect of Parallax ........ 25. 01 or Parallax in Distance
Distance clear'd of Refraction ........ 103. 1. 11
Prin.l Effect of Parallax ........ 25. 1
Distance clear'd of Principal Effect of Parallax ........ 102. 36. 10
By Table 4.th for second corr.n of Parallax ........ 2
Reduced Dist. clear'd of Refraction & Parallax ........ 102. 36. 8

By the Ephemeris

Dist. ☉ or ✶ & ☽ 4. at 6 102. 23. 14 D.o 102. 23. 14
D.o at 103. 51. 20 Red.d Dist. 102. 36. 0
in 3 hours 1.st Diff.ce 1. 38. 00 2.d Diff.ce 12. 54
Proportional Logarithm of 1.st Difference ........ 2634
D.o 2.d Difference ........ 1. 1447
Proportional Log. 23. 09 ........ Diff.ce 0818
+ hour of the 1.st Dist. 6
Gives Apparent time 6. 23. 09 at Greenwich
Apparent time 4. 26. 56 at taking the Dist. ☉ or ✶ & ☽
Difference ........ 1. 56. 13 In time = p.6. 29. 11
is Longitude between the Place of Observation and Greenwich

N.B. Distance clear'd of Refraction { less 90. take the Diff.ce of the two Arches / More the Sum of the two Arches } is Principal Effect of Parallax { Arch first greatest - contra + from the Dist. clear'd of Refr.n }
By the Requisite Tables find the Parallax in Alt.de & by Table 4.th with Distance & Parallax in { Alt.de / Dist. } Difference sec.d corr.n Parallax
which is to be + if distance is less 90. but more -
By the requisite Tables page 3 4 & 5 with App.t Altitude ☽ and Hor.l Parallax gives the Parallax in Altitude.
By Robert Bishop.

Figure 7. The computation of an actual lunar-distance observation made on 4 October 1772 in an unknown ship, probably an East-Indiaman, some 500 miles west of the Canary Isles. The navigation teacher Robert Bishop published these forms in 1768 for use with the newly published *Nautical Almanac* and *Tables Requisite*. From the Board of Longitude papers, MS. RGO 14/67, fol. 46v, Royal Greenwich Observatory Archives, Cambridge University Library.

France's position on opposite sides to England in the American and subsequent wars from 1778. Maskelyne always managed to get his tables to Paris in time to be translated for the *Connoissance des Temps*.

Let me now show briefly what the navigator had to do, step by step, using as an example an actual lunar-distance observation taken in the North Atlantic in October 1772 by an unknown navigator:

(1) *Take three more or less simultaneous OBSERVATIONS* (as shown in Figure 6), preferably with at least two observers, timed with a pocket watch (not necessarily a chronometer); the three observations needed were: (a) *the lunar distance,* between the Moon and the Sun or a star tabulated in the almanac; (b) *the altitude of the Moon* above the horizon; (c) *the altitude of the other body* (Sun or star).

To improve accuracy, it is usual to make each observation three or perhaps five times and to take the mean. In this case, the angles were observed three times at approximately five-minute intervals. A brass sextant or circle should be used to measure the lunar distances as the highest precision is needed, an error of only one arc minute in the observation resulting in an error in the longitude of 30 arc minutes, or ½°. For the altitudes, however, a wooden Hadley octant is adequate. With the Sun, observations are taken in daylight when both bodies are at least 10° above the horizon and between 35° and 120° apart. This can only be done on about fifteen days each lunar month, so a star must be used on other days, preferably during morning or evening twilight when both the horizon and the star are visible at the same time. If this is not possible, a separate altitude observation to find local time must be made at some time when the horizon is visible, and a running fix[18] used.

Figure 7 shows the printed form that our unknown navigator used for his computation in 1772.[19] The computation involved the following steps:

(2) *Calculate the MEAN OF MEASURED ANGLES AND TIMES of the observations made.*

(3) *Find the LOCAL APPARENT TIME from the measured altitude of the Sun or star,* which must be at least two hours from the meridian. Having

obtained the Sun's place from the almanac, or the star's place from the *Tables Requisite*, this is done by solving a spherical triangle by trigonometry.

(4) *Obtain TRUE LUNAR DISTANCE by clearing the mean of the measured lunar distances in (1a) of the effects of refraction and parallax*. Many different ways of doing this were published over the years. This printed form was designed for use with the method of Israel Lyons (1739-1775) from Maskelyne's *Tables Requisite* of 1766.

Another factor that has to be taken into account at this stage is that what is needed is the lunar distance between the centres of the Sun and Moon, whereas what is actually observed with the sextant is the distance between their limbs, or outer edges (see Figure 6).

This step occupied nearly 60 per cent of the whole computation. Looking at the computations, the mean lunar distance as observed with the sextant was 102° 26' 55". When cleared of semi-diameter, refraction, and parallax, the true lunar distance comes to 102° 36' 08". Remember, every arc minute error in the lunar distance makes an error of half a degree in the longitude found.

(5) *Find GREENWICH APPARENT TIME from the True Lunar Distance in (4)* by interpolation in the lunar-distance tables in the almanac (Figure 8). Until 1834, the almanac tabulated apparent time (or time as shown by the sundial) rather than mean time (or time as shown by the clock). With a True Lunar Distance of 102° 36' 08", our navigator's interpolation gives the Greenwich Apparent Time as $6^h\ 23^m\ 39^s$.

(6) *The LONGITUDE is the difference between Local Apparent Time in (3) and Greenwich Apparent Time in (5)*, generally expressed in degrees, minutes, and seconds of arc. From the computations, this difference was $1^h\ 56^m\ 43^s$, which is the equivalent of 29° 11', west because 'Greenwich Time [is] best'.[20]

[118] OCTOBER 1772.

Diſtances of ☽'s Center from ☉, and from Stars weſt of her.

| Days. | Stars Names. | Noon. D. M. S. | 3 Hours. D. M. S. | 6 Hours. D. M. S. | 9 Hours. D. M. S. |
|---|---|---|---|---|---|
| 1 | The Sun. | 62. 6. 55 | 63. 44. 49 | 65. 22. 18 | 66. 59. 22 |
| 2 | | 74. 58. 25 | 76. 32. 59 | 78. 7. 10 | 79. 40. 56 |
| 3 | | 87. 24. 0 | 88. 55. 28 | 90. 26. 35 | 91. 57. 21 |
| 4 | | 99. 26. 2 | 100. 54. 47 | 102. 23. 14 | 103. 51. 23 |
| 5 | | 111. 7. 52 | 112. 34. 22 | 114. 0. 37 | 115. 26. 38 |
| 3 | Antares. | 33. 0. 51 | 34. 36. 17 | 36. 11. 37 | 37. 46. 49 |
| 4 | | 45. 40. 30 | 47. 14. 40 | 48. 48. 38 | 50. 22. 24 |
| 5 | | 58. 8. 6 | 59. 40. 36 | 61. 12. 55 | 62. 45. 2 |
| 6 | | 70. 22. 45 | | | |
| 6 | β Capricorni. | 15. 30. 17 | 17. 2. 20 | 18. 34. 12 | 20. 5. 52 |
| 7 | | 27. 41. 48 | 29. 12. 32 | 30. 43. 8 | 32. 13. 37 |
| 8 | | 39. 44. 19 | 41. 14. 8 | 42. 43. 52 | 44. 13. 31 |
| 9 | | 51. 40. 38 | | | |
| 9 | α Aquilæ. | 57. 50. 30 | 59. 8. 51 | 60. 27. 28 | 61. 46. 19 |
| 10 | | 68. 23. 45 | 69. 43. 44 | 71. 3. 50 | 72. 24. 5 |
| 11 | | 79. 6. 58 | 80. 27. 49 | 81. 48. 43 | 83. 9. 41 |
| 12 | Fomalhaut. | 59. 29. 53 | 60. 47. 28 | 62. 5. 21 | 63. 23. 34 |
| 13 | | 69. 58. 45 | 71. 18. 29 | 72. 38. 25 | 73. 58. 33 |
| 14 | | 80. 41. 47 | | | |
| 14 | α Pegaſi. | 64. 16. 48 | 65. 41. 57 | 67. 7. 19 | 68. 32. 50 |
| 15 | | 75. 43. 16 | 77. 9. 54 | 78. 36. 43 | 80. 3. 42 |
| 16 | | 87. 21. 14 | | | |
| 16 | α Arietis. | 43. 43. 39 | 45. 11. 59 | 46. 40. 39 | 48. 9. 40 |
| 17 | | 55. 39. 44 | 57. 10. 44 | 58. 42. 4 | 60. 13. 44 |
| 18 | | 67. 56. 57 | 69. 30. 35 | 71. 4. 33 | 72. 38. 50 |
| 19 | Aldebaran. | 47. 6. 49 | 48. 45. 7 | 50. 23. 46 | 52. 2. 47 |
| 20 | | 60. 23. 26 | 62. 4. 44 | 63. 46. 26 | 65. 28. 32 |
| 21 | | 74. 5. 12 | 75. 49. 46 | 77. 34. 46 | 79. 20. 10 |
| 22 | Pollux. | 46. 12. 29 | 47. 56. 0 | 49. 40. 7 | 51. 24. 49 |
| 23 | | 60. 16. 20 | 62. 4. 6 | 63. 52. 16 | 65. 40. 51 |
| 24 | | 74. 49. 8 | 76. 39. 41 | 78. 30. 30 | 80. 21. 31 |
| 29 | The Sun. | 42. 58. 46 | 44. 38. 3 | 46. 16. 55 | 47. 55. 19 |
| 30 | | 56. 0. 32 | 57. 36. 15 | 59. 11. 31 | 60. 46. 20 |
| 31 | | 68. 33. 43 | 70. 5. 53 | 71. 37. 39 | 73. [illegible]. 59 |
| N. 1 | | 80. 39. 43 | | | |

**Figure 8. The lunar-distance tables for October 1772 from *The Nautical Almanac*, used in the computations in Figure 7. The reduced Moon-Sun distance cleared of refraction and parallax was 102° 36' 08". Interpolating in the table between the sixth and ninth hours on 4 October gives the Greenwich Apparent Time of the observation as $6^h\ 23^m\ 39^s$.**

Figure 9 shows a somewhat fanciful picture of a Victorian navigator taking a lunar observation. Figure 10 presents a very much more authentic picture, a sketch from life of a navigator in an East-Indiaman doing the same thing sitting on a chair, which is a sensible thing to do to get the greatest posssible accuracy measuring angles of up to 120°, high in the sky. This sketch comes from second officer J. L. Kirby's log of the Blackwall frigate *Owen Glendower*, on passage from England to Bombay in 1846-7.[21] On a later page, when the ship was in the tropics and the Sun and Moon were

20. 'Longitude east, Greenwich time least; longitude west, Greenwich time best'. This was a mnemonic taught to many generations of budding navigators.

21. MS. log M1, 27 August 1846, National Maritime Museum, Greenwich.

22. *Ibid.*, 8 February 1847.

23. W. E. May, 'How the Chronometer went to Sea', *Antiquarian Horology*, vol. 9, no. 6 (March 1976), p. 649.

Figure 9. Taking a lunar-distance observation. A somewhat fanciful Victorian representation from E. Dunkin, *The Midnight Sky*, 2nd ed. (London, [1879]), p. 256.

even higher in the sky, Kirby sketched the navigator taking an observation lying on his back on deck.[22]

The heyday of 'lunars' (the lunar-distance method) was probably from about 1780 to 1840. However, fairly early in that period, chronometers began to become cheap enough to be carried in the better-found ships, particularly in East-Indiamen and warships on foreign stations. Though longitude by chronometer was intrinsically more accurate by a factor of two and was easier for the navigator to observe and calculate, chronometers were not always reliable, though this could be overcome by ships carrying three or more (some surveying ships had as many as twenty).[23] But they still had to be set to accurate time in the first place. However much the navigator might prefer the chronometer method, in these early days lunars were still necessary to check the going of the chronometers.

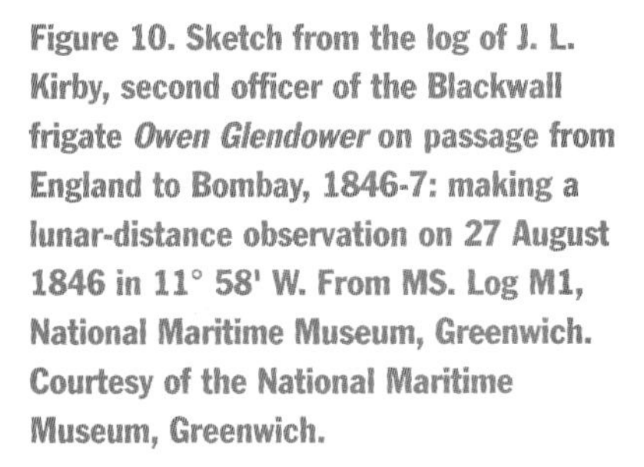

Figure 10. Sketch from the log of J. L. Kirby, second officer of the Blackwall frigate *Owen Glendower* on passage from England to Bombay, 1846-7: making a lunar-distance observation on 27 August 1846 in 11° 58' W. From MS. Log M1, National Maritime Museum, Greenwich. Courtesy of the National Maritime Museum, Greenwich.

Figure 11. Another sketch from the log of J. L. Kirby, 1846-7:

The Sun: *I say Old Boy, do you smoke?*

The Moon: *No you brute, so keep your distance.*

27 September 1846 in 24° 56' W. From MS. Log M1, National Maritime Museum, Greenwich. Courtesy of the National Maritime Museum, Greenwich.

However, time balls and time guns became more common worldwide in the 1850s, so that ships could easily set their chronometers to time before sailing from any of the larger ports. As a result, the lunar-distance method began to be used less and less, and then only as a check on the chronometers after some weeks at sea. Just as happened in the 1950s, when there were many calls not to abandon astronomical navigation because of the advent of radio navigational aids, there was in the 1850s a reaction in the nautical press urging the continued teaching and practising of lunars. Let me quote the splendid Victorian prose of Captain H. Toynbee, commander of the East-Indiaman *Glorian*, writing in 1859:

> In sending another paper on lunars to the Nautical Magazine, my object is not to bolster up a tottering subject, which the advancement of science causes to be no more requisite, but to maintain the fact that lunars are the only method of finding the longitude, available at sea, independent of delicate machinery so liable to suffer a stroke of the sea, changes in climate, or the effects of iron, as is the chronometer.[24]

The final nail in the lunar-distance coffin came when the first regular radio time signals began to be broadcast from Washington, D.C., in 1905, and from Norddeich Radio near Emden in 1910. At last it was possible for a ship to check her chronometers out of sight of land without taking lunar-distance observations.

Lunar-distance tables continued to be published in the French *Connoissance des Temps* until 1904 and in the British *Nautical Almanac* until 1906, and instructions on how to compute and reduce them were included in both until the 1920s. But for a practical epitaph let us listen to a practical seaman, Squire Thornton Stratford Lecky (1838-1902), Master Mariner, writing in the second edition of his *"Wrinkles" in Practical Navigation* in 1884:

> Once upon a time, Lunars used to be the crucial test of a good navigator, but that was in the "good old days" when ships were made snug for the night, and the East India "Tea-waggons" took a couple of years to make the round voyage.
>
> The writer of these pages, during a long experience at sea in all manner of vessels, from a collier to a first-class Royal Mail steamer, has not fallen in with a dozen men who had themselves taken Lunars or had even seen others do so. Whether Lunars are worth cultivating or not may, in the minds of some people, still be open to question, but certain it is that they have fallen into disuse, and, without in the least being endued with the mantle of prophesy, the writer ventures to say they will never be resurrectionised, for the best of all reasons—they are no longer required.[25]

24. H. Toynbee, 'A few more words on lunars', *Nautical Magazine*, vol. 28 (October 1859), p. 505.

25. Squire T. S. Lecky, *"Wrinkles" in Practical Navigation*, 2nd ed. (London, 1884), p. 280.

APPENDIX 1

## FURTHER READING

Cotter, Charles H. 1968. *A History of Nautical Astronomy*. London: Hollis & Carter.

Howse, Derek. 1980. *Greenwich time and the discovery of the longitude.* Oxford: Oxford University Press.

———. 1989. *Nevil Maskelyne: The Seaman's Astronomer.* Cambridge: Cambridge University Press.

Sadler, D. H. 1968. *Man is not Lost: A record of two hundred years of astronomical navigation with the Nautical Almanac, 1767-1967*. London: Her Majesty's Stationery Office.

# John Harrison

See Andrewes Figure 28 for caption for this page.

# John Harrison's Family Tree

Showing the ancestors and descendants of John and his brother James. Based on the work of Colonel Humphrey Quill (1897-1987), with amendments and corrections by Andrew King.

**KEY**
Bp.=Date of baptism
Bu.=Date of burial
M.=Date of marriage
D.=Date of death
31.3.1693 = 31 March 1693

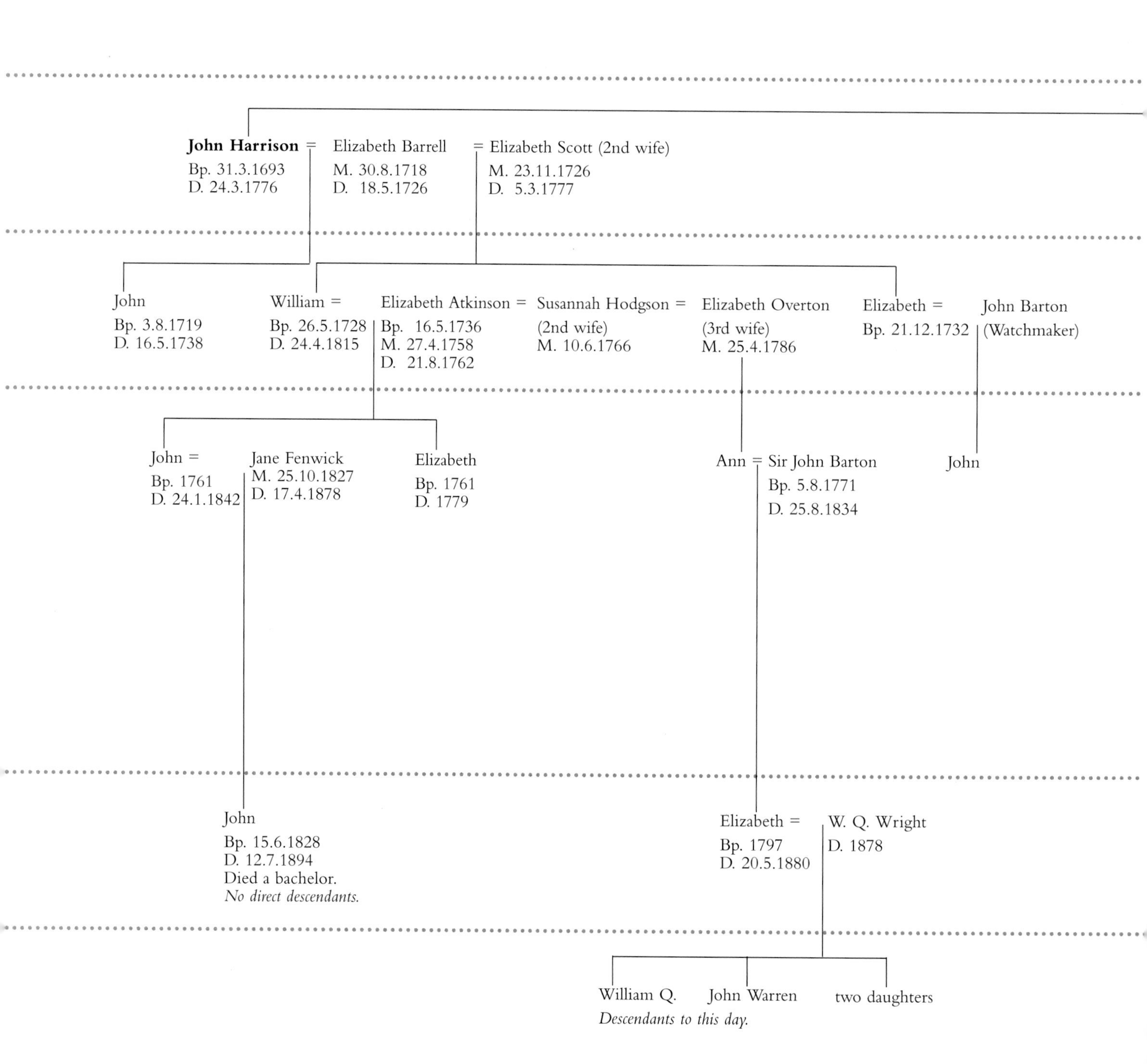

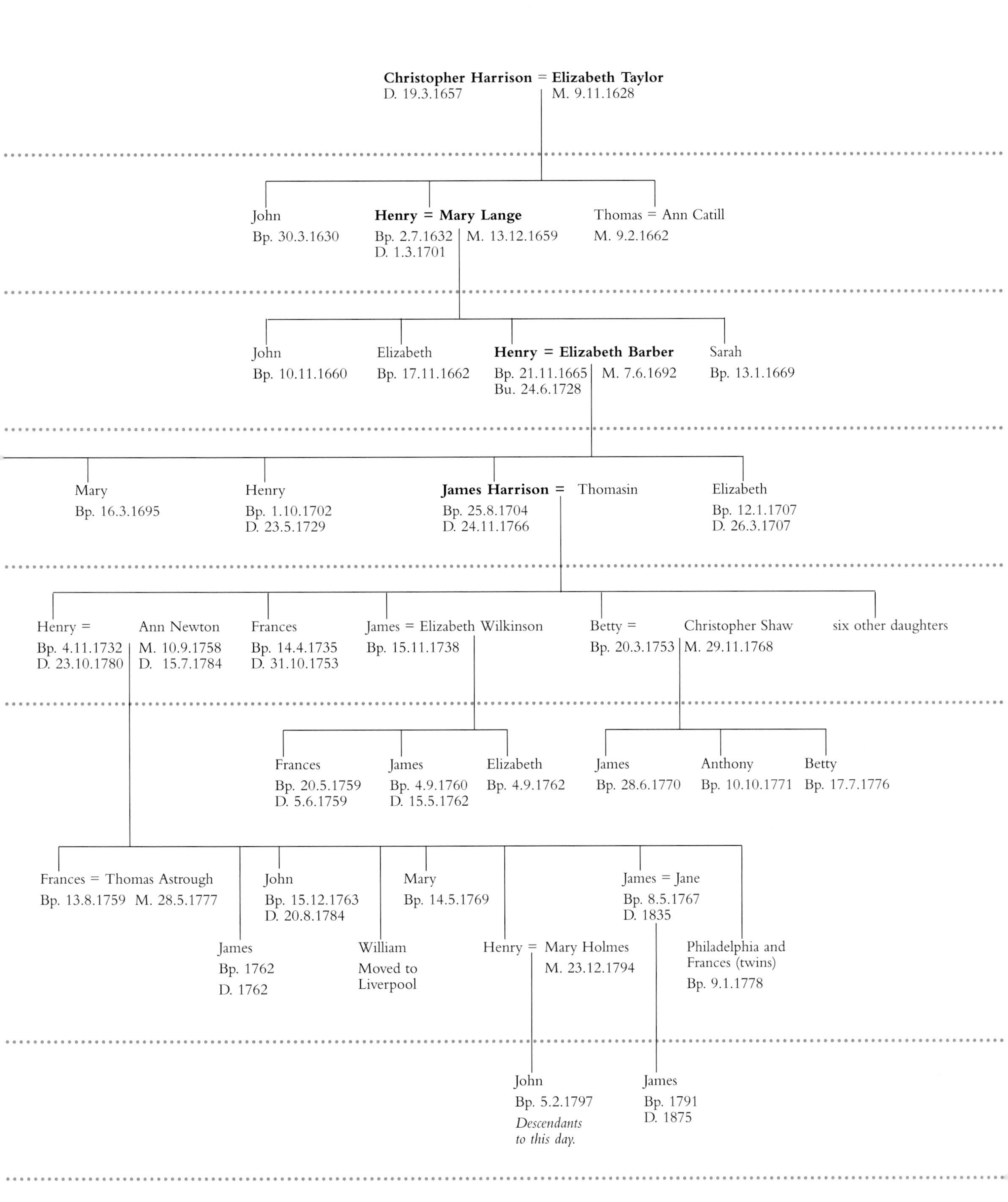
Christopher Harrison = Elizabeth Taylor
D. 19.3.1657
M. 9.11.1628
John
Bp. 30.3.1630
Henry = Mary Lange
Bp. 2.7.1632
D. 1.3.1701
M. 13.12.1659
Thomas = Ann Catill
M. 9.2.1662
John
Bp. 10.11.1660
Elizabeth
Bp. 17.11.1662
Henry = Elizabeth Barber
Bp. 21.11.1665
Bu. 24.6.1728
M. 7.6.1692
Sarah
Bp. 13.1.1669
Mary
Bp. 16.3.1695
Henry
Bp. 1.10.1702
D. 23.5.1729
James Harrison = Thomasin
Bp. 25.8.1704
D. 24.11.1766
Elizabeth
Bp. 12.1.1707
D. 26.3.1707
Henry = Ann Newton
Bp. 4.11.1732
D. 23.10.1780
M. 10.9.1758
D. 15.7.1784
Frances
Bp. 14.4.1735
D. 31.10.1753
James = Elizabeth Wilkinson
Bp. 15.11.1738
Betty = Christopher Shaw
Bp. 20.3.1753
M. 29.11.1768
six other daughters
Frances
Bp. 20.5.1759
D. 5.6.1759
James
Bp. 4.9.1760
D. 15.5.1762
Elizabeth
Bp. 4.9.1762
James
Bp. 28.6.1770
Anthony
Bp. 10.10.1771
Betty
Bp. 17.7.1776
Frances = Thomas Astrough
Bp. 13.8.1759 M. 28.5.1777
John
Bp. 15.12.1763
D. 20.8.1784
Mary
Bp. 14.5.1769
James = Jane
Bp. 8.5.1767
D. 1835
James
Bp. 1762
D. 1762
William
Moved to Liverpool
Henry = Mary Holmes
M. 23.12.1794
Philadelphia and Frances (twins)
Bp. 9.1.1778
John
Bp. 5.2.1797
Descendants to this day.
James
Bp. 1791
D. 1875

Mezzotint of John Harrison by Philippe Joseph Tassaert (1732-1803), published in 1768, after the portrait by Thomas King (d. *ca.* 1769). In this print, the hand-held watch shown in King's portrait is replaced with the famous prize-winning watch, H.4, lying on the table beside its celebrated maker. A later engraving made by William Holl II (1807-1871) after this mezzotint (published in 1835 in vol. 5 of *The Gallery of Portraits: with Memoirs,* 7 vols. [London: Charles Knight], opposite p. 153) simplifies the composition to just Harrison and H.4, the piece for which he is chiefly remembered. Dimensions of image: 39.4 x 28.6 cm. (15.5 x 11.25 in.). Courtesy of The Time Museum, Rockford, Illinois. Inv. 2494.

# 'John Harrison, Clockmaker at Barrow; Near Barton upon Humber; Lincolnshire': The Wooden Clocks, 1713-1730

*Andrew L. King*

Andrew King is a clockmaker who has spent much of his time during the last twenty years studying the life and work of John Harrison. In addition to solving some of the mysteries surrounding the design, construction, and operation of Harrison's wooden clocks, he has established an entirely new approach to understanding Harrison's methods.

Educated at Eastbourne College and in Oxford, King entered a horological apprenticeship in Chelmsford, Essex, before establishing his present workshop in Beckenham, Kent. He works for museums and has a private clientele but is principally involved with special projects and commissions as well as writing on technical subjects covering the period from the eighteenth to the twentieth century. As a member of the Harrison Research Group, he contributed biographical background as well as results from his research into the early phase of Harrison's work in precision timekeeping. He has lectured on the subject throughout the United Kingdom and on visits to the United States. King's fascination with Harrison's work has influenced the design of the mechanical clocks he currently makes. He fervently believes that the future of mechanical horology lies in its potential as a work of art.

# 'John Harrison, Clockmaker at Barrow; Near Barton upon Humber; Lincolnshire': The Wooden Clocks, 1713-1730

*Andrew L. King*

The 1714 Longitude Act was a determined attempt by a government made aware that navigational problems were the main contributing factor to the loss of shipping in a world heavily dependent on its navy and its merchant fleet. The immense awards offered under the Act are testimony to the urgency of the problem. Inevitably they led to a plethora of mostly ill-conceived proposals from applicants hopeful of an award that would make them rich beyond their wildest dreams. Yet some of these proposals were based on theoretically possible concepts, one of which was the invention of an accurate marine timekeeper. What, then, did this mean to horologists, and how did they meet the challenge?

Under the terms of the Act, in order to obtain the full £20,000 it was required that the method, whatever it might be, must determine longitude to within a distance of 30 miles during a voyage from England to the West Indies. To achieve this, a mechanical timekeeper would have to be accurate to within a total of just two minutes during the proposed six-week trial. Every clockmaker knew that this was impossible with the technology then available. In 1714, even the best pendulum clocks were capable of maintaining a rate of no more than several seconds a week. No one had yet solved the problem of the influence of temperature variation and barometric pressure on pendulums; moreover, changes in the viscosity of available lubricating oils caused by fluctuations in temperature affected the performance of clocks. Apart from these, there were, as everyone knew, many problems in taking a pendulum clock to sea. As for the established watchmakers, the challenge was apparently so difficult that very few of them seriously looked at the problem. Even by the middle of the eighteenth century the best rate obtainable from the finest quality watches anywhere in the world was no better than about a minute per day.

There are, of course, exceptions to everything in life. This time, the inventive genius came from a very remote corner of England. If John Harrison (1693-1776) had been born in any of the established clock-making areas of the country, it is probable that his thorough, scientific approach to precision timekeeping might well have been lost forever. It was essential that this unique and individual character was able to think and work in isolation, free from the shackles of contemporary traditional thought. Until the age of 44, John Harrison lived and worked in Barrow upon Humber, a village in the north of Lincolnshire on the south bank of the river Humber. It was here that he developed all of the fundamental ideas that became the basis of his life's work. It was also here that he made all of his early wooden clocks and designed and made the world's first successful marine timekeeper.

His father, Henry Harrison (1665-1728), was a joiner. Soon after 1697, when he arrived in Barrow with his family, Henry Harrison was appointed parish clerk, a responsible position that required a reasonable education. This position made him a notable character in local affairs. As was the accepted practice of the times, he would have been responsible for the instruction of his own children. Therefore, by the standards of the early eighteenth century, John Harrison would have received a basic education, but his enquiring mind seems to have led him to further his knowledge considerably, as revealed by a close study of his writing, most of which remains only in manuscript. At this point in history, education in natural philosophy, or science, would have been virtually non-existent in remote areas of England, such as Barrow, so he was fortunate in being lent, in his youth, a copy of the notable lectures on Newtonian philosophy delivered by Nicholas Saunderson (1682-1739), the Lucasian Professor of Mathematics at Cambridge University. Harrison made his own copy of these lectures, which was of immeasurable assistance to him throughout his life.[1]

**Figure 1. Engraved portrait of Nicholas Saunderson holding an armillary sphere, from 'The Life of Dr. Nicholas Saunderson,...' in *The Universal Magazine of Knowledge and Pleasure* (London), vol. 10, no. 69 (May 1752), p. 192. Private collection.**

John Harrison was brought up to be a joiner like his father, which is, quite simply, the reason that all of his early clocks were made almost entirely of wood. Also like his father, he took an active interest in the village life of Barrow. He was a juror in the Court of the Lord of the Manor, the local jurisdiction,[2] as well as a village overseer or constable. In addition, his interest in music resulted in his taking charge of the training of the church choir and, more importantly, in becoming a bell-ringer.[3] Indeed, it is possible to argue that the Church of the Holy Trinity, Barrow upon Humber, was the very cradle of the world's first successful precision timekeeper.[4]

If there is a particular point in Harrison's life when he became aware of the properties of bell-ringing, and thus was introduced to oscillator theory, it was not later than the year 1713. In this year, Harrison was twenty years old and had completed his first clock. It was also in 1713 that an additional

1. Although Saunderson's lectures were never published, copies were made by students. The lectures followed a set pattern, but the surviving copies do not necessarily contain all the subjects covered by Saunderson. The full lectures included 'Astronomy', 'the Barometer', 'Hydrostatics', 'the Tides', 'Optics', 'Mechanics', and a discussion of the effects of 'Heat and Cold'—an excellent insight into the science of the day. Harrison's own copy was listed as item 8960 in *Bibliotheca Chemico-Mathematica: Catalogue of Works in Many Tongues on Exact and Applied Science,* vol. 2 (London: Henry Sotheran and Co., 1921), p. 453. From the catalogue description it is clear that Harrison included some manuscript additions later in his life. This and other important Harrison manuscripts listed in the Sotheran catalogue have since disappeared. See Humphrey Quill, *John Harrison, the Man who found Longitude* (London: John Baker, 1966), pp. 13-15, concerning the visiting clergyman who may have been responsible for lending Harrison one of these copies.

2. Barrow Court Rolls, 1720-36, LR3/36/2,3, 4,5,6,7, Public Record Office, London.

3. Nevil Maskelyne, 'Notes taken at the Discovery of Mr. Harrison's Time-keeper', MS. RGO 257, Royal Greenwich Observatory Archives, Cambridge University Library. This manuscript was used as the basis for the publication of *The Principles of Mr. Harrison's Time-keeper, with Plates of the Same* (London, 1767) and was published after the preface in that volume; the reference to bell-ringing may be found on p. xiv. A facsimile reprint of *Principles* has been published by the British Horological Institute in *Principles and Explanations of Timekeepers by Harrison, Arnold and Earnshaw* ([Upton, Notts., England], 1984).

4. Much has been written about Harrison's life and work, in particular by Lt.-Comdr. Rupert T. Gould (1890-1948) and Col. Humphrey Quill (1897-1987). A brief summary of Harrison's activities, with some new information concerning his family, is contained in an exhibition catalogue entitled *From a Peal of Bells: John Harrison, 1693-1776* ([Lincoln, England]: Lincolnshire County Council, 1993), which was researched and written by the author of this paper. This exhibition was organised in 1993 by the Usher Gallery in Lincoln to commemorate the tercentenary of John Harrison's birth.

bell was installed in the church at Barrow.[5] Until then, there had been a peal of just five bells. The new addition, the tenor bell, was bigger than any of the others. Henry Harrison's duty as parish clerk included the care of the bells, and so he would have taken a close interest to ensure that the new bell was hung correctly. The inauguration of a new bell was a notable occasion, and it is inconceivable that John Harrison would not have been aware of it. But it is the practice of bell-ringing itself that is important. From Harrison's short statement, 'I from being a Ringer (or taking a hint therefrom)...',[6] it is possible to understand what Harrison might have experienced. A bell has an enormous arc of swing, 250° or more. As a ringer, Harrison would have been able to feel the speed of the bell slowing as the arc increased. This he may have likened to the action of a pendulum.[7] He must also have appreciated the effects of the energy that he put into the swinging bell. Perhaps the first experiment he applied to his clock was to see what happens with an increase in the driving weight; in this case, he probably would have expected the clock to go slower as the arc increased. But no, he would have found that it went faster. And if he did carry out this experiment, he would have discovered the effect of circular error[8] as well as the hastening effect of the escapement.

Between the years 1713 and 1730, Harrison produced at least eight clocks.[9] This inventory does not include any early test pieces that Harrison may have made before he was twenty years old, in 1713. He was first and foremost a joiner, not a clockmaker, and the first clock that he made, in 1713, is really quite conventional in its layout and construction, offering evidence of Harrison's original trade. All of his longcase clocks share certain basic fea-

**Figure 2. Dial and movement of the early wooden clock dated 1715 by John Harrison. The clock is wound by removing the lower spandrels and inserting, in the hole behind them, a geared winding key. Courtesy of the Science Museum, London. Inv. no. 1834-80.**

5. The inscription on this bell at Barrow upon Humber, checked by the author, confirms this date.

6. John Harrison, 'An Explanation of my Watch or Timekeeper for the Longitude...', 7 April 1763, MS. 3972/1, p. 4, Library of the Worshipful Company of Clockmakers, Guildhall Library, London.

7. Maskelyne, 'Notes taken' (*op. cit.*, note 3).

8. Circular error is the difference in the amount of time that it takes a pendulum to describe a large arc as opposed to a small arc. The arc of swing of a pendulum cannot be maintained precisely due to fluctuations in the driving force, barometric pressure, etc. There were two ways that were used to overcome circular error: one was to employ an escapement that required a very small arc of swing—2° or less—at which point circular error becomes negligible, and the other was to introduce a device that caused the pendulum to swing in the arc of a cycloid, in which theoretically, regardless of its amplitude, it performs its oscillations in a constant period of time. Cycloidal cheeks were devised about 1657 by Christiaan Huygens (1629-1695), who subsequently perfected them mathematically. Harrison claimed that he came upon the idea of using cycloidal cheeks without any prior knowledge of Huygens's work: 'I from being a Ringer, or taking a hint therefrom had made use of an Artificial Cycloid (but had no name for it) some years before I had so much as heard of Mr. Huygen's name'. Harrison, 'An Explanation of my Watch' (*op. cit.*, note 6), p. 4.

Editor's note: The escapement is the device that provides the connection between the wheels of the timekeeper and the pendulum or balance. The train of wheels, through which the time is indicated, transmits the force provided by the driving weight, coiled spring, or other power source to maintain the oscillation of the pendulum or balance, the frequency of which

tures: wooden dial plates, which in the early clocks were probably gilded; construction of the frame with three-way mortice-and-tenon joints; wooden wheel construction; wheel trains that are supported by plates let into the main frame; and construction mostly using carefully selected oak with a brass escape wheel. In addition, the movements of all of these clocks are rather large and, where their cases have survived, are a very close fit.[10]

The first three clocks have an unusual winding arrangement. The lower spandrels can be removed to allow a geared winding key to be inserted, as shown in Figure 2, thereby turning the barrels through a reduction gear. This arrangement is an early example of Harrison's attention to detail design, in this instance to achieve a perfectly symmetrical and uncluttered dial. With the dial removed, the motion work is shown to be driven by the second wheel of the going train, which is to the right; the striking train is on the left (Figure 3). The signature and date, as with all of Harrison's wooden long-case clocks, appear on the calendar disc. The wheels of the first three clocks have both the arbors and the solid leaf pinions of boxwood. The wheel pivots are steel, and they run in brass bushes that are let into the wooden plates.

The first two of these clocks have the type of anchor escapement that was common in the early eighteenth century, but the third clock, dated 1717, has a rather different escapement (Figure 4). It is badly worn and the pallet frame has a repaired fracture, but fortunately everything has been preserved. There is an indication—and I stress that it is only an indication—that Harrison was carrying out escapement experiments with this clock. To begin with, the shape of the teeth of the escape wheel, which appears to be orig-

regulates the speed at which the mechanism operates. The action of the escapement in releasing and locking the wheels of the train and delivering the impulse required to maintain the oscillation causes the familiar ticking sound heard in every mechanical clock or watch.

9. (1) Longcase clock, dated 1713 (movement only survives; collection of the Worshipful Company of Clockmakers, London).
(2) Longcase clock, dated 1715 (movement only survives; Science Museum, London).
(3) Longcase clock, dated 1717 (later oak case; Nostell Priory, Wakefield, Yorkshire).
(4) Turret clock, *ca.* 1722 (Brocklesby Park, Lincolnshire).
(5) Longcase clock, *ca.* 1725-6 (present grass-hopper escapement and gridiron pendulum not original; The Time Museum, Rockford, Illinois).
(6) Longcase clock, dated 1727 (converted to anchor escapement and missing original pendulum; private collection, England).
(7) Longcase clock, dated 1728 (collection of the Worshipful Company of Clockmakers, London).
(8) Longcase clock (movement only survived in an incomplete state; restored in the style of the 1725-30 period; collection of the Worshipful Company of Clockmakers, London).
In addition, a longcase movement rumoured to be by Harrison but not positively identified, was reported in Sheffield, England, *ca.* 1970-1. Formerly in the collection of William Joshua Shaw.

10. Descriptions of Harrison's wooden clocks have been published on two previous occasions: Quill, *John Harrison* (*op. cit.*, note 1) and William Andrewes, 'John Harrison: A Study of His Early Work', *Horological Dialogues,* vol. 1 (1979), pp. 11-38. The dial of the 1715 and 1717 clocks have remains of original gilding. The 1713 dial has been extensively cleaned; it is quite likely that it too was gilded.

**Figure 3. Movement of the 1713 clock with dial removed to show the motion work. John Harrison's signature and the date can be seen on the date wheel. The holes in which the winding key is inserted to engage the winding pinion can be seen at the bottom corners of the front plate. Courtesy of the National Trust, England.**

**Figure 4. Detail of the escapement of the 1717 clock. Courtesy of the National Trust, England.**

inal, differs from those in the earlier clocks. In addition, the reshaped impulse plane (shown on the left in Figure 5), with the escape wheel turning clockwise, is possible evidence that Harrison was trying to reduce the fierce friction in recoil to equalise the much lower friction on the exit pallet on the right. A feature of all Harrison's clocks, and indeed his watches, is the clear evidence of his layout lines, as visible here. If the geometry of the escapement of this clock still adheres to Harrison's original layout, it would be the first instance of his altering the shape of the pallets to suit the long and short arcs of swing with varying amounts of energy. One thing is certain: it never had a dead-beat escapement since it is clearly evident that the existing impulse planes of this clock are nowhere near correct for a dead-beat action.[11]

Apart from these three clocks, it is unlikely that any other examples from this early period exist, although it is just conceivable that there could have been one more dating from around 1719.[12] I do reiterate that, at this stage in his life, John Harrison was first and foremost a joiner, not a clockmaker. The clocks he had made must have gained him something of a reputation, however, because he was soon to be given a most important commission by Sir Charles Pelham (1679-1763) of Brocklesby Park. Brocklesby Park, about nine miles from Barrow, is still the largest estate in North Lincolnshire, extending to more than 30,000 acres. The original house of 1603 was completely rebuilt in the very early years of the eighteenth century, and alongside the house, a large stable block was built around 1720. We do not know exactly when Harrison was approached to make the clock for the stable turret, nor do we know for certain when the construction took place, because there was a serious fire at Brocklesby Park in 1898 when a considerable amount of archival material was lost.[13] I think, however, that the clock, which can still be seen today, was made around 1722.[14]

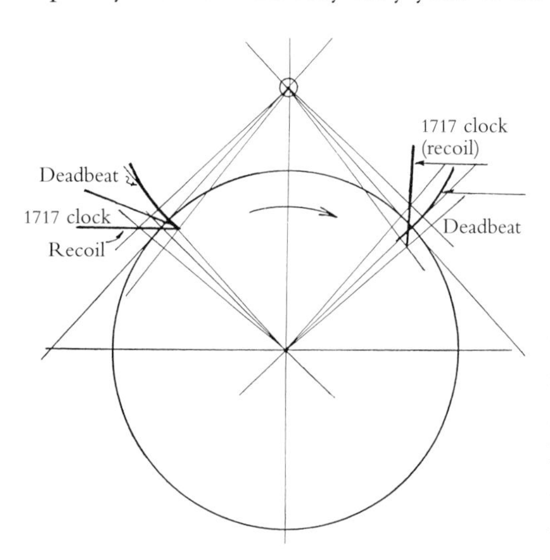


**Figure 5. Drawing to illustrate the geometry of the escapement of the 1717 clock, comparing the layout of a conventional anchor (recoil) escapement with that of a standard dead-beat escapement. The existing impulse planes are indicated together with the geometry for a conventional anchor escapement and the positions required for the locking planes of a dead-beat escapement. Drawing by the author.**

This commission was a big step forward for Harrison. He had never made a turret clock before, but he was certainly aware of the problems, not the least of which was the need for a good surplus of energy to combat the varying weather conditions. Among the

11. On the entry side, to the left, the present position could be associated with an original experiment by Harrison, but, of course, it could also be the result of continual filing and stoning to eliminate wear. However, the existing position of the exit impulse plane coincides exactly with that of a conventional anchor escapement, an indication that perhaps the escapement has not suffered much wear after all.

12. The surviving early clocks of 1713, 1715, and 1717 indicate that Harrison was making a clock every two years, and so it is possible that there could have been one made in 1719. The next surviving example of Harrison's work, the Brocklesby Park turret clock, was probably constructed around 1722 (see also note 14), and after 1725 he began working on his precision regulators. The 1713 clock came into the possession of Mr. A. Riley of Leeds (*Horological Journal* [August 1890], p. 194), from whom the collector Evan Roberts acquired it. Roberts's description of this clock appeared in *Horological Journal* in 1894, and it was from Roberts's daughter, Mrs. Williamson, that Col. Quill acquired the clock in 1960 for the Worshipful Company of Clockmakers, London. The 1715 clock was presented to the Patent Museum (later to be incorporated into the Science Museum, London) in 1864 by George Empringham (1812-1877) of Barton-upon-Humber, who in turn had acquired it from Robert Minto, a watchmaker, also from Barton-upon-Humber. The 1717 clock was discovered and purchased around 1870 by Canon Edward Cross (1821-1897), who made several attempts to find other Harrison clocks. It was from his widow that Sir Rowland Winn of Nostell Priory acquired the clock in 1910. Correspondence and references relating to the history of this clock are given by Dennis Jones in 'E. T. Cottingham, F.R.A.S.', *Antiquarian Horology*, vol. 19, no. 6 (Winter 1991), pp. 594-6.

problems were high winds buffeting the clock hands and, of course, pigeons roosting and, worse still, nesting around the area of the motion work. Harrison met the challenge with a bold design (Figure 6). The movement sits on a very substantial oak beam, which tends to dominate and certainly contrasts with the very finely made cabinet. The cabinet is fully panelled, with pegged joints, and the hung edges of the doors are mitered with matching edges to the case. Everything fits very closely. The frame is heavily constructed but broadly follows Harrison's previous practice, with copious mortice-and-tenon joints and ladder-style plates let into the frame. All the wheels are of an oak construction.

Within this bold design, Harrison was to incorporate five important innovations. First, he employed lignum vitae[15] (rather than brass, as had been used in the earlier clocks) for the bushes (bearings) of this clock, thereby overcoming the need for lubricating the pivots. By doing this, Harrison was able to make the pivots of brass (rather than steel), making this his first clock to contain no ferrous metals at all. Thus he eliminated, as far as possible, any problems resulting from corrosion, ever present in the unheated and spartan environment in which turret clocks have to function—an environment that would also have a deleterious effect on the unstable lubricants available to him. This combination of brass pivots in lignum vitae bushes is an almost

13. The significance of the Brocklesby Park turret clock was not recognised until, about 1952, Col. Quill found a mention of a clock owned by the Earl of Yarborough at Brocklesby Park in Lincolnshire. This information came from an undated manuscript scrapbook owned by a direct descendant of James Harrison (1704-1766). The Pelham family was granted the Earldom of Yarborough in 1837.

14. The date of the turret clock at Brocklesby Park was given as 1727 by Col. Quill, who published a description of the clock in 'A James Harrison Turret Clock at Brocklesby Park, Lincolnshire', *Horological Journal,* vol. 96, no. 1146 (March 1954), pp. 156-9, and vol. 96, no. 1147 (April 1954), pp. 234-6. This date was based on a statement by James Harrison (1767-1835), the grandson of John Harrison's brother James, that the clock was built 'about 1727' (letter dated 6 February 1829 to *Mechanics' Magazine,* vol. 11, no. 304 [1829], p. 264). However, I think that this date is untenable and that the clock was probably installed around 1722, soon after the completion of the stable building.

15. Lignum vitae *(Guaiacum officinale/G. sanctum)*, an extremely dense hardwood about 70 per cent heavier than oak and heavier than water, is found in the Caribbean and South America. The dark heart-wood, easily distinguished from the pale sap-wood, contains natural resins that never dry out, making it an excellent bearing material. It is immensely strong and can be worked well when turned on a lathe. Traditionally, it was these properties that were exploited: from use as the bearing material for the stern gland on the propeller shaft on ships, to sheaves in pulley blocks aboard ships where it was used principally for its strength, to the fine turnings and carvings beloved of the English Victorians.

**Figure 6. Movement of the turret clock at Brocklesby Park, Lincolnshire, England. Courtesy of the Earl of Yarborough.**

perfect bearing combination: resistant to corrosion and subject to minimal wear. Indeed, in more than 270 years of almost continuous operation, the brass pivots have never become scored and there is record of the clock being extensively rebushed on just one occasion, in 1884.[16]

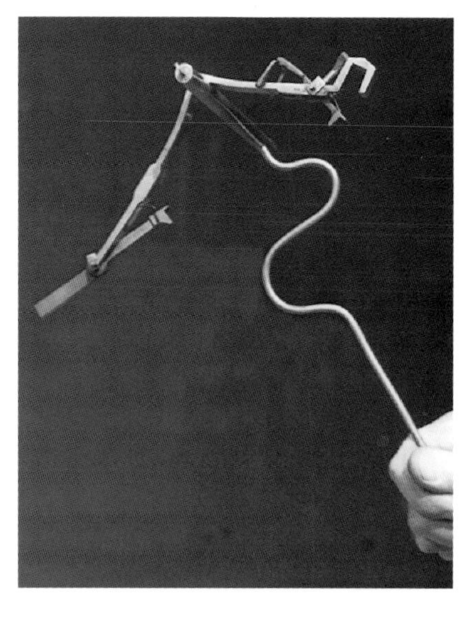

**Figure 7. Pallets from the Brocklesby Park turret clock. The mortice on the right arm of the frame is evidence of the original escapement. Courtesy of the Earl of Yarborough.**

The next improvement involved the escapement, which is the heart of any clock movement. Harrison knew that he needed an escapement that could absorb the surplus of energy required by all turret clocks,[17] but although he obviously gave the matter deep thought, it took him two attempts to achieve a satisfactory long-term result. It is almost certain that his first design employed an anchor (recoil) escapement. Harrison knew, however, that one theoretically incorrect feature of the conventional anchor escapement was that it gave unequal lift and thus unequal impulse between the acting faces of the pallets. On his first attempt, therefore, he redrew the anchor escapement to give equal impulse; the evidence for this is the curved face on the exit pallet (Figure 7). Although now theoretically correct, the result considerably increased the friction. For once in his life, however, Harrison actually wanted more friction on the pallets, because it enabled him to provide the clock with more driving weight for the surplus of energy needed for the motion work, etc. The escape wheel (Figure 9), which turns anti-clockwise, has the curved teeth typical of Harrison's work, and the wheel is thick to provide wide contact. If the surviving pallet frame is imagined without the later modifications, it is quite possible to see how the original escapement was envisaged. The mortice to the right is in the correct position for an exit pallet for an anchor recoil layout spanning 8½ spaces. The mortice is also cross-drilled in exactly the position required to secure a pallet pad. The pallet frame to the left has been cut and extended, removing any evidence of a mortice to accept an entry pallet, but it is not difficult to see where this should be. To avoid friction on the escapement pivots, the whole escapement oscillates on knife-edges resting on glass plates, which complete the escapement unit. The pendulum itself is a plain brass rod with a lead disc-shaped bob and a brass suspension spring. This is how the Brocklesby clock was almost certainly first equipped. However, within a few months of the installation it was clear that there were problems: apparently Harrison had to make several journeys, without rec-

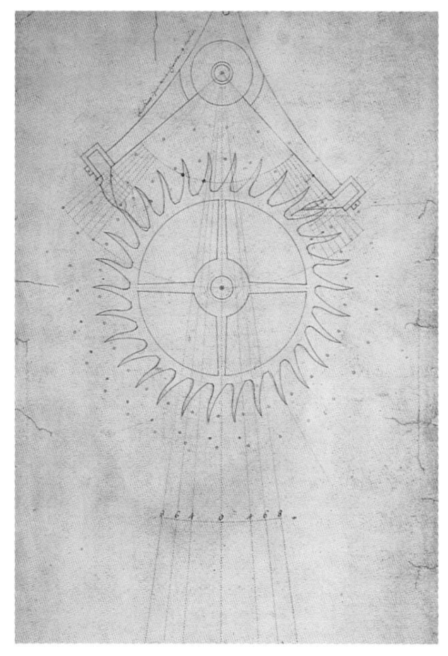

**Figure 8. Drawing by John Harrison of the equal-impulse anchor escapement that he designed for the turret clock at Trinity College, Cambridge, *ca.* 1755. Both the pallet frame and the escape wheel teeth closely resemble those used on the Brocklesby Park turret clock. The layout of the brass pallet frame of this clock, with inset detachable pallets, is a Harrison design feature adopted by other turret clockmakers. Pen and ink. 24.5 x 37 cm. (9.6 x 14.6 in.). Courtesy of the National Maritime Museum, Greenwich. Ref. no. Ch.354-4.**

16. While the brass pivots never show any appreciable sign of wear, the lignum vitae does wear and requires replacement, but only at intervals of perhaps as long as 100 years or even longer. On the inside of the sliding door of the pendulum box, below the clock, is a list of repair dates from 1882 to 1936, when the clock was maintained by Alfred Fryer (d. 1948) of Ulceby, a village adjacent to Brocklesby Park. In this list of dates, for 1884, is the word 'BOSSED', and from this it is believed that Fryer rebushed most of the movement at this time. This could well be the first time that rebushing was carried out. It has certainly not been rebushed since then. In 1954 the clock was dismantled by Col. Quill and Tom Hyde. It was brush-cleaned, and soft string was used to clean the lignum vitae bushes. Nothing further was required, and no maintenance has been needed since.

17. A drawing exists of just such an escapement, which Harrison designed for the turret clock at Trinity College, Cambridge, *ca.* 1755 (see Figure 8). Although the Trinity College clock no longer survives, Jonathan Betts has discovered three other turret clocks fitted with this escape-

**Figure 9. Detail of the escape wheel from the Brocklesby Park turret clock, showing Harrison's first roller pinion and the curved teeth of the escape wheel. Courtesy of the Earl of Yarborough.**

ompense for his trouble, to re-oil the parts of the clock that required lubrication.[18]

Perhaps the clock had been installed during the warmer months of the middle of the year and the poor oils, perhaps goose fat, had congealed during the on-coming winter months. Whatever the cause, it was an infuriating problem. The inherent high-frictional properties of the escapement, although intentional, had resulted in too much resistance and, even worse, a resistance that was both variable and unpredictable. It was reputed that, while trudging home on these occasions, he pondered how to make a clock go without oil. After several experiments he contrived a new escapement, now commonly known as the 'grasshopper' escapement, as well as the present roller pinion and antifriction rollers.[19] Through these innovations, coupled with the self-lubricating properties of lignum vitae, friction was reduced to a minimum, and the need for oiling was eliminated.

The roller pinion (also shown in Figure 9), which is on the escape wheel arbor, consists of lignum vitae rollers rotating on fixed brass pins; thus it reduces to a minimum the friction in the transmission of power to the escape wheel. The antifriction rollers, made of brass with a lignum vitae bush, support the escape wheel arbor and are intended to reduce the friction on the escape wheel pivots. These rollers, although correct in general principle, are not entirely satisfactory in the Brocklesby clock: because Harrison used only a single roller at each end, the escape wheel pivots tend to fall to one side of the centre of the tangential point of the roller and thus cause unnecessary friction by making contact with the stationary lignum vitae bush. On the clocks he made after this, Harrison made far better use of antifriction rollers.

With the grasshopper escapement, we really see the brilliant, ingenious, and resourceful mind of John Harrison at work. The original pallet frame was retained; the solid pallet outline was divided and each half separately pivoted, each pallet describing a small sector of a large circle. The recoiling action allows each pallet in turn to be released, while its partner remains in contact with the escape wheel, thus providing impulse to the pendulum. The escapement has no 'drop', the action of an escape wheel tooth being released

ment. Two of these are signed by William Smith, the maker of the Trinity College clock, and the third, although signed by Colley and Priest, was almost certainly made by Smith as well.

18. John Robison, *A System of Mechanical Philosophy,* 4 vols. (Edinburgh, 1822), vol. 4, p. 574, note †. This passage, which mentions problems that Harrison experienced with a clock he had made 'for a turret in a gentleman's house', almost certainly refers to the Brocklesby Park clock.

19. *Ibid.* Robison's account suggests that Harrison did not incorporate the roller pinion and the antifriction rollers—although the latter are not specifically mentioned—until this time. It is also possible that the lignum vitae bushes and the knife-edge escapement pivots were not introduced either until he encountered the problems with lubrication.

by one pallet and falling onto the surface of the other. There is also no sliding friction at the pallet faces—thus no need for the grease pot—and minimal friction at the pallet pivot points. Once fitted to the clock, however, Harrison immediately discovered, as he had probably suspected, that his new low-friction escapement with its minimal energy losses resulted in a huge increase in the arc of swing of the pendulum. To reduce the arc without altering the driving weight, Harrison applied vanes to the bob of the pendulum (Figure 10). These vanes have a marked effect.[20]

To maintain the non-ferrous content of the clock, the pendulum is suspended from a brass suspension spring, and on either side of this, we see for the first time Harrison's suspension cheeks (Figure 11), the last innovation that he applied to this clock. These are now commonly referred to as 'cycloidal cheeks' because, in theory, they cause the pendulum to swing in a cycloidal arc rather than a circular arc, thereby allowing its period of oscillation to remain constant despite fluctuations in its amplitude of swing. Although the cheeks of this clock are adjustable and thereby appear to be an early attempt to overcome circular error,[21] the acting curve is rather arbitrary; Harrison was to improve on the cycloidal cheeks and use them in a very effective manner in the precision regulators that he subsequently made. I think it is fair to suggest that the cheeks also served to protect the brass suspension spring by easing the necking action as the spring is continually deflected, a property that Harrison himself was strongly aware of.[22]

**Figure 10. Detail showing one of the vanes on the pendulum-bob of the Brocklesby Park turret clock. Vanes were fitted to each side of the bob in order to reduce the pendulum's arc of swing when Harrison incorporated his new grasshopper escapement. Courtesy of the Earl of Yarborough.**

The Brocklesby turret clock was both Harrison's first major landmark and an immensely important research-and-development vehicle. It displays so many improvements and refinements over his earlier clocks that it is fair to claim that it is this clock that was the spring-board, the real break-through that pushed his work forward in the development of precision timekeeping. With it, Harrison had introduced five significant innovations,[23] although three of these—the antifriction rollers, the grasshopper escapement, and the cycloidal cheeks—show strong evidence of being very much in embryo. As previously noted, it has been suggested that this clock was made as late as 1727.[24] However, all the evidence, including the fact that these three innovations appear in very much further developed states in the precision wooden clocks, which were started about 1725, suggests an earlier date of about 1722, closer to the completion of the Brocklesby stable building.

At this point, mention should be made of John Harrison's second brother, James (1704-1766). Eleven years younger than John, who was his eldest brother, James Harrison is a character we know little about, but he has

20. Few horologists have experimented with the vanes on the Brocklesby clock, but their effectiveness is further evidence of Harrison's great ingenuity.

21. See note 8.

22. John Harrison, *A Description Concerning Such Mechanism as will Afford a Nice, or True Mensuration of Time;...* (London, 1775), p. 45.

23. By way of summary, these innovations were (a) brass pivots with lignum vitae bearings, (b) the roller pinion, (c) antifriction rollers, (d) the grasshopper escapement, and (e) cycloidal cheeks.

24. See note 14.

always been considered important because his name appears on many of the innovative Harrison clocks. There is very clear evidence that he was a skilled joiner and that during the 1720s the two brothers worked together. However, it is also clear that James did not have the innovative mind, the commitment, or the determination and drive of his brother John. As we have seen, from around 1720 John Harrison was very active in the village life at Barrow. His name continually appears in the court rolls, recording his activity. In contrast, James Harrison's name appears but once between the years 1720 and 1736. On the very day that John Harrison embarked on the first voyage with his first 'Sea Clock', 14 May 1736, James was summoned before the Court of the Manor of Barrow and fined the sum of one shilling for allowing his swine to trespass into a cornfield.[25] This appears to be the sum total of his recorded activities in the village.

In 1722, the period of the work on the clock for Brocklesby Park, James would have been eighteen years old. It is very likely that he had a hand in its construction, and, indeed, this could be the reason why it is signed with just the surname, Harrison; the Brocklesby clock is, in fact, the only Harrison clock signed in this manner. The longcase clocks that were to follow are signed by James Harrison alone, and this has led to the belief that he was responsible for their design as well. This supposition, however, is just not acceptable, as there is not the remotest evidence to suggest that he was anything more than a skilled workman. When John Harrison moved to London in 1737, there is evidence that James was there as well, but this can only have been for a short period during the year 1738, because there are records showing that James was busy making bell frames in Lincolnshire and Yorkshire throughout 1737.[26] The period that he remained in London, in any case, was certainly not long enough for him to have made any worthwhile contribution, as accepted by tradition, to the second sea clock, which was made between 1737 and 1739. By 1739, James had returned to Barrow, where he continued for the rest of his life as a general carpenter and joiner, specialising particularly in bell hanging, bell frame making, and even the manufacture of church furniture.[27] A steady record of similar work in Lincolnshire and Yorkshire continues to within two years of his death. Toward the end of his life, he started a bell foundry with his son in the village of Barrow. It was perhaps inevitable that when John Harrison achieved notability and eventual fortune in London, jealousy crept in back home, in the remoteness of Barrow upon Humber, leading to the belief handed down through later generations that James had been *'the DUPE of his more cunning brother'*.[28] There is, however, no evidence to suggest that this was the case.

**Figure 11. Pendulum suspension cheeks of the Brocklesby Park turret clock. Courtesy of the Earl of Yarborough.**

With the Brocklesby turret clock at last working satisfactorily with its new escapement, John Harrison must have watched its performance very

25. Barrow Court Rolls, 1720-36, LR3/36/6,7, Public Record Office, London.

26. In 1737, he made one bell frame, which still exists, for the church at Waithe near Grimsby in Lincolnshire, and he made another, somewhat larger, for the church at Barnsley in West Yorkshire. In the year 1739, he travelled to Wakefield in Yorkshire to make a frame for the chime barrel in All Saints Church, and he made other bell frames at Pocklington and Richmond in Yorkshire and at Great Coates in Lincolnshire. For further information, see [King], *A Peal of Bells* (*op. cit.*, note 4).

27. Evidence of this work survives at the Church of St. Laurence at Aylesby, near Grimsby, where James Harrison made all the pews and the pulpit in 1759.

28. *The Barrow Monthly Monitor*, no. 140 (March 1890).

**Figure 12. Modified single-pivot grasshopper escapement of *ca.* 1728. This illustration also shows details of the construction of the wooden wheels and the roller pinions, the lignum vitae antifriction rollers used for supporting the escape wheel arbor, and the lignum vitae bushes that serve as bearings for the pivots of the other wheels. Illustration by David Penney.**

closely. He was obviously very interested to find out what performance could be achieved if he could make one on a scale of his earlier longcase clocks where the energy input could be balanced exactly to the requirements of the escapement. So, with a clean sheet of paper in front of him and with the experience of the turret clock to guide him, he designed a new longcase movement. This was John Harrison's first 'fully integrated' mechanism, in which he was able to incorporate all of his ideas into an overall concept that required no further major revisions as at Brocklesby.

The precision wooden clocks display all the virtues of integration derived from earlier research, and they are therefore his first really successful precision timekeepers.[29] With these clocks, Harrison was able to design the movement around his new escapement and suspension cheek arrangement, together with a much improved and much lower-friction wheel train. This time, he would use roller pinions throughout the movement, with brass pivots and bushes of lignum vitae. The antifriction rollers to support the escape wheel were now to be mounted in pairs to support the pivots totally, in both load and thrust conditions. The only part of the movement itself that he would later need to modify was the design of the escapement, which at first was the same as the one used on the Brocklesby clock. The design was ready in 1725, and it is believed that he could have been planning to make a small batch of perhaps as many as six.[30]

Within a year, two of these clocks may have been completed, but only one has survived.[31] The second surviving clock from this period, dated 1727, was sold by Harrison soon after.[32] The third surviving clock, shown in Figure 20, was finished by 1728,[33] and this time the layout of the grasshopper escapement was modified to a single pivot point for both pallets (Figure 12). Harrison kept the first and the third clocks for experimental and development work, and he soon fitted the first with the new escapement layout. After tests and adjustments, he must have realised that the clocks he had pro-

29. Precision longcase clocks are often referred to as 'regulators', because they serve as the standard by which other clocks can be regulated.

30. It is not known exactly how many of these clocks Harrison completed or planned to complete. Apart from the three extant complete clocks described here, there is a restored movement in the collection of the Worshipful Company of Clockmakers in London (see note 9). This movement, which formerly belonged to the Duke of Sussex and was sold at Christie's for £4. 7. 0. upon his death in 1843, was presented to the Worshipful Company of Clockmakers by W. Thoms in 1875. It appears to have been an unfinished movement that was completed in a rather crude manner after Harrison's death, and it contains features representative of both the earlier Harrison clocks and the period of the regulators. This movement, which was restored by William Andrewes between 1969 and 1972, is illustrated in its condition both before and after the reconstruction and described in his article 'John Harrison' (*op. cit.*, note 10). There are also very strong rumours of what could have been another movement in Sheffield as late as 1970-1 (see note 9).

31. In a paper read to the Royal Society on 9 November 1752 by Harrison's friend James Short (1710-1768), it was stated that 'He [Harrison] also made a drawing of a clock, in which the wheels are disposed in a different manner from those then in use; which drawing I have seen, signed by himself in the year 1725. Two of these clocks with pendulums, as described above [the gridiron pendulums], were finished in the year 1726'. James Short, 'A Letter of Mr. James Short, F.R.S. to the Royal Society, concerning the Inventor of the Contrivance in the Pendulum of a Clock...', *Philosophical Transactions* of the Royal Society *(Phil. Trans.)*, vol. 47, no. 88 (1752), p. 519. The surviving clock, although missing its date wheel, and therefore both its date and signature, is believed to be one of the clocks that Short refers to. It was never sold, and after being used for experimental purposes, it came into the possession of James Harrison, who kept it throughout his life and passed it on to his son Henry (1732-1780). Henry's wife Ann (d. 1784) later sold it to a surgeon from Barton-upon-Humber, Mr. B. M. W. Benton (d. *ca.* 1800). John Harrison's grandson, John Harrison (1761-1842), acquired the clock around 1826. Around 1860 it was cleaned by a Hull clockmaker, Andreas Koenig (d. 1925), who found it in the cellar of a public house in Hull; Koenig became a British citizen and Anglicised his name to Andrew King (no relation to the author). The clock passed down through the King family until Col. Quill bought it in Hull in 1954 for £50 in a much-altered state. Long years of meticulous restoration followed,

duced were rather special. It appears that Harrison first heard of the Longitude Act in 1726,[34] and thereafter he concentrated on perfecting precision timekeeping on land with the intention of then developing his inventions for use in a marine timekeeper. With this, any thought of making further clocks for sale must have been abandoned.

Perhaps the most remarkable new feature seen in these clocks is their pendulum assembly, which automatically compensated for variations in temperature by using metals with different coefficients of expansion. Harrison designed the assembly in such a manner that the expansion of brass rods compensated for the expansion of steel rods. Known as the gridiron pendulum (shown in Figure 19), this was perhaps Harrison's greatest invention. To have discovered or noticed the expansion of metals was very observant, but to have appreciated that metals have varying thermal rates and then proceed to accurately measure them is an even greater achievement. When the first of these clocks was made in 1725-6, it was almost certainly designed to have not only the same grasshopper escapement as at Brocklesby, but in addition, despite what James Short stated,[35] a Brocklesby-type pendulum—a plain brass rod. This type of pendulum was also fitted to the clock dated 1727. It was not until 1727-8 that the gridiron pendulum was fully developed after a period of observation and trials. Of the three surviving clocks that are known to have been made during this period, only the one dated 1728, I believe, was originally fitted with a gridiron pendulum.

These clocks retain the unique and idiosyncratic stamp of their maker. The wooden frames, like those of their predecessors, are fairly complex affairs, with 22 mortice-and-tenon joints (Figure 13). The three-way joints are to an extent self-locking, but nevertheless they are glued to add to their rigidity. The frames are large, but they completely contain the wheel trains as well as the motion work. This permits the dial to be fitted directly onto the front of the frame without the aid of traditional dial pillars. Tenons extending from the rear of the frame locate the movement in a fore-and-aft plane in the case. The front plate has to be lowered evenly into position and is secured by four brass latches. The motion work is driven by the second wheel of the going train, and the large calendar wheel, supported on the bell-shaped rollers, is pin driven. In addition to the signature on their dials, these clocks were signed, dated, and numbered on their calendar rings. Only the second and third clocks retain their original calendar rings.

**Figure 13. Movement of the 1727 clock with dial removed to show the motion work. The date wheel is signed '2nd James Harrison Barrow 1727'. Private collection.**

The wheel construction is yet another example of the great care and thought that John Harrison put into his wooden clocks. The wheels are light but strong, and the oak has been selected according to its precise role in the construction of the wheel.[36] The main body of the wheel is deeply slotted around the periphery, and individual segments are let in so that the grain is radial all around the wheel. This ensures the

and eventually Seth Atwood acquired it in 1980 for The Time Museum in Rockford, Illinois.

32. This clock is signed on the date wheel '2nd James Harrison Barrow 1727'. The fact that it is signed '2nd', and the clock made in the following year was signed '3rd' suggests that James Short was mistaken when he referred to two clocks instead of one being made in 1726 (see note 31). Little is known of the history of the 1727 clock, but it is almost certainly the clock Harrison mentions in his manuscript of 1730 as the one without a temperature-compensated pendulum that he had sold '3 Years ago' (untitled manuscript signed 'John Harrison, Clockmaker at Barrow; Near Barton upon Humber; Lincolnshire. June 10. 1730', MS. 6026/1, p. 12, sect. 21, Library of the Worshipful Company of Clockmakers, Guildhall Library, London). When found, the remnants of the escapement suggested that it was originally of a similar layout to the escapement of the Brocklesby Park clock. Nothing is known of its subsequent history, although it would appear that it always remained in Yorkshire until its discovery in 1976 in an original but neglected state.

33. The third clock, signed 'James Harrison 3rd 1728 Barrow', was retained by John Harrison throughout his life. It then passed to his son William (1728-1815), whose grand-daughter, Elizabeth Wright (1797-1880), sold it with other Harrison relics, including H.5, to Sir Robert Napier (1810-1890) in 1869, having failed to gain the interest of the Patent Museum (later to become the Science Museum, London). The clock was acquired in 1877 by the Worshipful Company of Clockmakers, London, after the auction of the Napier collection.

34. Harrison states this fact in the 1770 reissue of a broadsheet entitled *The Case of Mr. John Harrison;* the original version, published *ca.* 1766, does not mention when Harrison first heard of the Act.

35. See note 31.

36. The construction of the wheels was first discovered by William Andrewes and George Daniels during the reconstruction of the restored Harrison movement, now in the collection of the Worshipful Company of Clockmakers (see note 30). Col. Quill subsequently confirmed this finding by having an original wheel X-rayed at the British Museum, London. Although the initial question was answered, not everything was explained. When I started my investigation into the methods of construction utilised by Harrison, I had all the wheels of the 1727 clock X-rayed, as well as the front plate, one of the arbors, and the cranked winding key. From this I was able to deduce much further information.

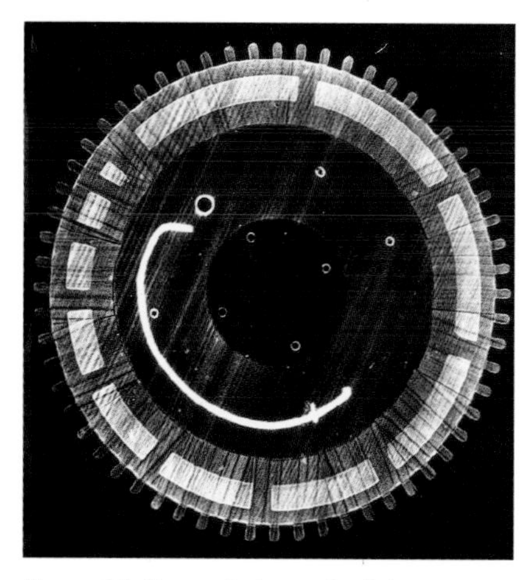

Figure 14. X-ray photograph of the great wheel of one of the precision wooden long-case clocks. Photograph by the author.

strongest possible wheel teeth. The oak selected for the wheel tooth segments has been taken from a fast-growing tree, where the annual growth rings are furthest apart, thus producing the maximum area of new wood, where the oak is strongest.[37] The main body of the wheel, on the other hand, is a large, low-stressed mass, so the oak for this has been taken from a slow-growing tree, where the annual growth rings are much closer together. When highly magnified, these growth rings are seen to be of a honeycomb structure. This means that, while the body of the wheel is not so strong, because of a high ratio of growth rings to new wood, it has, theoretically, less mass and is therefore lighter in weight. The oak for the segments set into the perimeter of the wheel, however, is the strongest possible at the expense of a small increase in weight. An X-ray examination of all the wheels (Figure 14) from one of these clocks shows a consistency in the selection of the oak that confirms that Harrison, as a joiner, was extremely knowledgeable about the material of his trade. The arbors are made of oak as well, and they are drilled to accept short lengths of brass rod inserted from each end to form the wheel pivots (Figure 15).

The unusual nature of these clocks does not stop at this point. The dials are also unique (Figure 16). Harrison was not only a fine craftsman and a scientist, he was also an artist. Made of oak, the dials are painted black and decorated with a delightful floral design. The dial of the 1727 clock is not without humour: the oval below XII on the chapter ring, when viewed closely, shows a pair of grotesque masks; when viewed from further away, there appears to be the silhouette of a man with a full-bottomed wig (Figure 17). The overall design is both original and striking. The very soft gold of the decoration is complemented by a lightly polished brass chapter ring, which shows no trace of ever having been silvered. Indeed, unlike most clocks of the period, the chapter rings of these precision regulators were almost certainly never intended to be silvered. Although those of the other surviving precision clocks are now silvered, this is believed to have been carried out by subsequent clock repairers.

37. When examined anatomically, it can be seen that the annual growth rings in an oak tree are of a honeycomb structure, while the new wood between these growth rings is very much denser and heavier but also stronger. Under good growing conditions the tree will grow faster, the annual growth rings will be further apart, and thus there will be more new wood. It is clearly evident that Harrison knew this because in all the highly stressed wheels of the 1727 clock, it can be seen that the oak chosen for the segments contains a high degree of new wood, while the oak chosen for the body of the wheels displays very close growth rings for the lightest possible construction. Throughout the movement there is evidence that Harrison paid the closest possible attention to the materials he used.

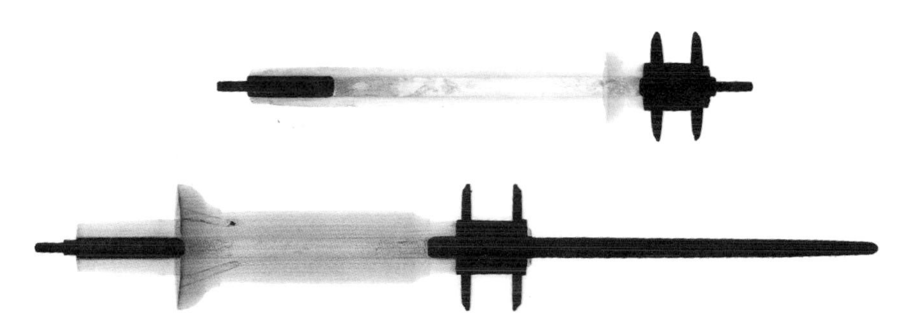

Figure 15. X-ray photograph of the fly arbor (upper) and escape wheel arbor (lower) of one of the precision wooden longcase clocks, showing that, although the wooden arbors were drilled right through, the brass pivot rod was inserted at each end and does not run through the entire length. Photograph by the author.

Figure 16. The 1727 clock dial, which has survived in an entirely original state. Private collection.

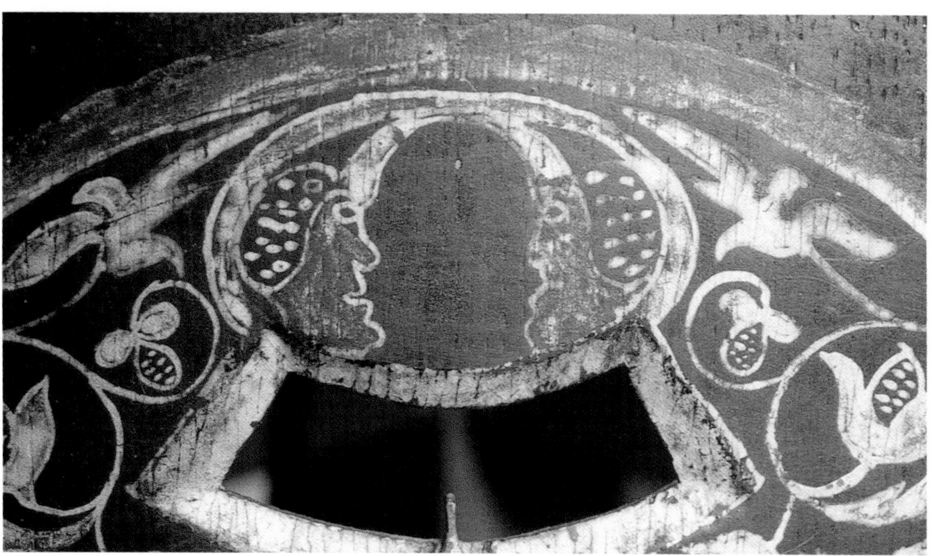

Figure 17. Detail of the oval design above the seconds ring aperture on the dial of the 1727 clock. Private collection.

John Harrison is one of the few English horologists to have written extensively about his own work. Very little was ever published, but a considerable amount fortunately survives in manuscript. The earliest of these manuscripts is the only example of his written work from the time when he was living in Lincolnshire.[38] This very important document extends to 22 pages and is neatly set out in 29 sections. The first 24 sections include an explanation of the precision wooden clocks, how they were made and tested; the last five sections outline the proposals for adapting the technology to a portable 'Sea Clock'. However, although there is much detail, much is also left out. Ever fearful, almost paranoid, of plagiarism, Harrison provided just sufficient information to gain priority but little else.

It is almost certain that this paper with its accompanying drawings was produced when Harrison was planning a journey to London to gain support for his proposal to make the first sea clock. This visit could have been as early as 1728, but it is more likely to have been during the summer of 1730, soon after he had written his paper.[39] Harrison, 37 years old at that time, signed the paper with a great flourish, making authorship, and thus priority, unequivocally clear, and by this date he defined himself as a clockmaker. This manuscript may appear to be rather quaint, looking at it from a distance of a quarter of a millennium. But let there be no doubt about it: it represents the most advanced thinking in precision timekeeping during the eighteenth century. Surely this is the paper referred to by Harrison when he met the renowned George Graham (*ca.* 1674-1751) for the first time in 1730 and discussed—or, as Harrison says, 'debated'—horological topics for ten hours non-stop:

> ...we reasoned the cases, or upon the Principles, more than once; nay once, and that in a very extraordinary Manner, was at the very first Time I saw him, and our reasoning, or as it were sometimes debating, (but still, as in the main, understanding one another very well) then held from about Ten o'Clock in the Forenoon, 'till about Eight at Night, the Time which Dinner took up included, for he invited me to stay to dine, &c. Now it is to be understood, that I had along with me (as affording the Principles upon which we reasoned) the Descriptions, with some Drawings of the principle Parts

38. Harrison, 1730 manuscript (*op. cit.*, note 32).

39. Since it is not known exactly when Harrison first visited London, it is necessary to consider all possibilities. 'An Account of the Life of the late Mr. Harrison' (under 'Characters') in *The Annual Register* (London, 1777), p. 25, states 1728, but this date is almost certainly too early. By 1728 Harrison had only just finished making the third of the precision wooden clocks. It is also tolerably clear that Harrison did not fully develop his gridiron temperature-compensated pendulum until the summer of 1728 (ref. unpublished manuscript, 'Harrison's Pendulum Horology', by W. S. Laycock [1973], an analysis of all Harrison's work with fixed pendulum clocks, drawn from manuscript sources and discussions with Col. Quill and Martin Burgess). It is far more likely that Harrison visited London in the latter half of 1730 after completing the manuscript dated 10 June 1730 (*op. cit.*, note 32).

40. Harrison, *A Description Concerning Such Mechanism* (*op. cit.*, note 22), p. 19.

41. *Ibid.*, p. 20.

42. *Ibid.*, p. 20.

43. *Ibid.*, p. 21.

44. George Graham, 'A Contrivance to avoid the Irregularities in a Clock's Motion, occasion'd by the Action of Heat and Cold upon the Rod of the Pendulum', *Phil. Trans.*, vol. 34, no. 392 (1726), pp. 40-4.

45. Another notable clockmaker, John Ellicott (1706-1772), carried out similar experiments in 1732, and he too failed. Harrison's fear of plagiarism was confirmed in 1738, when Ellicott suddenly produced a description and drawing of a compensated pendulum after seeing Harrison's work. When Ellicott submitted his idea to the Royal Society in 1752, Harrison immediately objected to it. See John Ellicott, 'A Description of Two Methods, by which the Irregularity of the Motion of a Clock, arising from the Influence of Heat and Cold upon the Rod of the Pendulum, may be prevented', *Phil. Trans.*, vol. 47, no. 81 (1752), pp. 479-94.

46. Harrison, *A Description Concerning Such Mechanism* (*op. cit.*, note 22), p. 21.

47. *Ibid.*, p. 23.



**Figure 18 (left). John Harrison's signature on the final page of his earliest surviving manuscript. Courtesy of the Worshipful Company of Clockmakers, London.**

**Figure 19 (right). Drawing of the gridiron pendulum from Harrison's 1730 manuscript, alongside the gridiron pendulum from the 1728 regulator. Courtesy of the Worshipful Company of Clockmakers, London.**

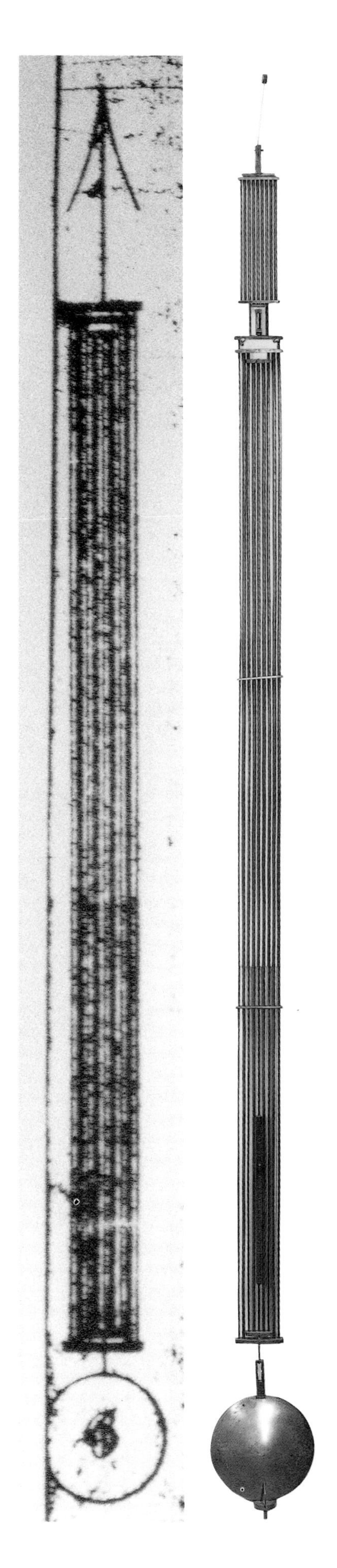

of the Pendulum-Clock which I had made, and as also of them of my then intended Time-Keeper for the Longitude at Sea.[40]

Not wanting to reveal his proposals to another horologist, Harrison had been extremely reluctant to call on George Graham, who was Harrison's senior by almost twenty years and was undoubtedly the greatest horologist and instrument maker in England at that time. On his arrival in London, however, Harrison had wasted no time in going to see the Astronomer Royal, Edmond Halley (*ca.* 1656-1742), and it was he who persuaded Harrison to visit Graham. Harrison remarks

> ...that [Halley's] Advice went hard with me, for I thought it as a Step very improper to be taken; but he told me, that in the Way in which I was in, viz. as by Machinery (for the Longitude) more than Astronomy, that I should certainly be sent to Mr. Graham, and therefore advised me to go without any farther to do; certifying me also, that Mr. Graham was a very honest Man, and would do me no harm, viz. as by pirating any Thing from me, but that on the contrary, would certainly do me Good if it was in his Power;....[41]

The meeting with George Graham was the most important step that Harrison could have made. Graham provided unswerving support; there is no better example to illustrate the epitaph 'Honest George Graham'. He appears to have been the only horologist to have appreciated Harrison's work, although there was evidently a misunderstanding in the beginning. Harrison was probably an irascible character and certainly verbose. Halley had warned him to get to the point with Graham as quickly as possible, but Harrison admits that, although he tried his best, he still had a problem. 'Mr. Graham. began, as I thought it, very roughly with me, and the which had like to have occasioned me to become rough too'.[42] I am sure that this contretemps was due entirely to Harrison's apparently abrasive approach, but his later friendship with Graham seems to have helped him to become a shrewd and skilful negotiator. Harrison then continues, 'however, we got the Ice broke'.[43] What broke the ice? I suggest that it was when Graham saw the drawing of the gridiron pendulum (Figure 19).

Graham, perhaps motivated by the 1714 Longitude Act, had tried as early as 1715 to devise a pendulum utilising what Harrison had discovered for himself alone in the remoteness of Lincolnshire: the varying expansion rate of metals.[44] He did not succeed in his attempts to use metals such as brass and steel, but in 1721 he devised his famous mercury pendulum.[45] So the sight of Harrison's gridiron pendulum was a true revelation to Graham, and it is small wonder that, as Harrison said,

> ...and indeed he became as at last vastly surprised at the Thoughts or Methods I had taken, or had found Occasion to take, and as thence found Reason enough to believe that my Clock might go to a Second a Month,....[46]

Indeed, some years later, when Graham had actually seen the 1728 clock, which Harrison set up in his home in London, Harrison reported that 'Mr. Graham said to several Gentlemen, that for my Improvement in Clock-Work, I deserved 20,000 l. [£20,000]—was no Longitude to be concerned'.[47]

At that first meeting, Graham must have become ever more interested in and impressed with this strange man from Barrow upon Humber. We can assume that those ten long hours of discussion and debate circled around the 29 sections of Harrison's manuscript. Right at the beginning of that work, Harrison had acknowledged the difficulty of using a clock at sea due to the fact that the motion of a ship

> ...has render'd the Motion of all Machines that have been try'd so irregular, as to be of no service to the Sea-Men in the Matter of Longitude.

Even on land, however, Harrison judged the existing technology to be insufficiently precise:

> But Query; wou'd any of these Machines that have been try'd, go so true as is there requir'd, if fix'd on Land: If not, the Motion of the Ship need not much be blam'd, tho it might make it worse.[48]

A similar observation had been made by Isaac Newton (1642-1727) in 1721, in a letter to Josiah Burchett (*ca.* 1666-1746), secretary to the Admiralty, when discussing a proposal for finding the longitude at sea by a Mr. Laurans, who advocated watch-work. Newton had stated that longitude 'is not to be found at sea by any method by wch it cannot be found at land. And it is not yet found at land by Watch-work'.[49] It is most important to appreciate that, elsewhere in this letter, Newton distinguished between finding the longitude, which can be done only by astronomical means, and keeping the longitude: a timekeeper cannot find longitude, but it can keep it. Newton knew, however, that no watch or timekeeper had yet been made that was anywhere near accurate enough to do so. Therefore, he was correct when he insisted that timekeepers could not find longitude, but this does not mean that he was necessarily opposed to a horological solution to the problem of determining longitude at sea. However, it is probable that he did confuse the issue for those of lesser intellect, who might not have appreciated the subtle difference between finding the longitude and keeping the longitude, and it is certainly true that he advocated an astronomical solution.

By 1730, there was still no timekeeper anywhere in the world, not even a pendulum clock, that was accurate enough in the long term—over very extended periods—to meet the requirements of the 1714 Longitude Act. That is, with the exception of John Harrison's precision regulators. Describing the development of these clocks, as well as his hopes for further improvements for a sea clock, Harrison stated:

> Some Years ago I made several alterations in order to render the Motion of Clocks more exact than heretofore, but when I came to try them by strict observation as below, I judg'd the best performance of the best Pendulum Clock I ever saw, made, or heard of, to be incapable of this Matter, wou'd it go as well in a Ship at Sea in any part of the World, as in any one fix'd place on the Land. Yet from several observations, I still endeavour'd to make farther Corrections in this Motion; and in these 3 last Years have brought a Clock to go nearer the truth, than can be well imagin'd, considering the vast Number of seconds of Time there is in a Month, in

48. Harrison, 1730 manuscript (*op. cit.*, note 32), p. 1, sect. 1-2.

49. Newton to Burchett, October 1721, *The Correspondence of Isaac Newton*, vol. 7, ed. A. Rupert Hall and Laura Tilling (Cambridge: Cambridge University Press, 1977), no. 1377, p. 172.

> which space of time, it does not vary above one second, & that mostly the way I expect: so that I am sure I can bring it to the Nicety of 2 or 3 Seconds in a Year.[50]

George Graham must have been enthralled, if not stunned, by this claim of accuracy. If what Harrison said really was true, he was way ahead of anyone else. Graham knew perfectly well that his own clocks, the finest generally available, were not as good. For one thing, his required oil, and this, if nothing else, would eliminate them from contention as long-term precision timekeepers due to the ever-changing viscosity of the oil. Harrison continued:

> But a Pendulum moving in a Cycloid to Regulate a Clock true, wch is not in Vacuo, must not always keep it's same Length, unless the Air gave always the same resistance; which agrees both wth reason & Experiment. Viz the Pendulum must be rather Shorter in Warm than in Cold Weather, wch is contrary to the Operation of the Wire.[51]

50. Harrison, 1730 manuscript (*op. cit.*, note 32), pp. 1-2, sect. 4.

51. *Ibid.*, p. 4, sect. 8.

52. *Ibid.*, p. 11, sect. 19.

Here, Harrison gives a concise description of the concept of barometric error, and he reasons that the clock will lose time with an increase in the resistance of the air. As with his determination of the differing rates of the expansion of metal, Harrison again demonstrates his great powers of perception and his ability to understand the problem. With the precision wooden clocks, the cycloidal cheeks are put under similar scrutiny. Although he says that he had designed an engine that would cut perfect cycloids, it appears that, even by 1730, he had realised that such a cycloid was not what was required, because he had discovered that the curve of the cheeks had to be matched to the demands of the grasshopper escapement.

Throughout the 1730 manuscript, Harrison never reveals too much about the construction of his clocks, but, when it comes to how he actually tested them, it is another matter. He had two basic methods. In section nineteen, he describes how he checks the clocks against one of the fixed stars by taking daily sidereal sightings. With an assistant calling out the seconds as the fixed star approaches the line of his meridian, Harrison explains how he uses

> ...a very large sort of an Instrument, of about 25 Yards Radius, Composed of the East side of my Neighbours Chimney (wch is situated from my House towards the South) & the West side of an exact place of some one of the upright parts of my Own Window Frames; by which the Rays of a Star are taken from my sight almost in an Instant:....[52]

With constant practice, Harrison seems to have been able to use this system with great accuracy. It is, however, the second method of testing that is really interesting, because it illustrates Harrison's true genius in identifying all aspects of a problem:

> But, my other way is the better part of the Completion: and that is the two Clocks plac'd one in one Room & the other in another, yet so, that I can stand in the Doorstead, & hear the beats of both the Pendulums, when the Clock Case heads are of; & before or after the hearing can see the seconds of one Clock, whilst another Person

count the seconds of the other: by which Means I can have the difference of the Clocks to a small part of a second. And in very Cold and Frosty Weather I sometimes make one Room very warm, wth a great Fire, whilst the other is very Cold. And again the Contrary. And sometimes the like in Summer by the Sun's Rays in at the Windows of one Room, & also a Fire, whilst the other is close shut up and Cool. Thus I prove the Operation of the Pendulum Wires....And to prove or adjust the Cycloid to Vibrations perform'd in different Arches as requir'd,...I cause the Pendulum to describe such by increasing & decreasing the draught of the Wheels, & that by adding to & taking from the Weight; by wch I can make 8 or 10 Times more difference, than Nature ever will, & yet the effect be nearly the same...as if Nature it self had alter'd the Weight of the Air so much.[53]

Figure 20. Precision longcase clock, signed 'James Harrison 3rd 1728 Barrow'. This clock was retained by John Harrison throughout his life. Courtesy of the Worshipful Company of Clockmakers, London/Bridgeman Art Library, London.

53. *Ibid.*, pp. 11-12, sect. 20.

It is this particular testing method, which Harrison explains in such detail, that was the real key to his success. Indeed, I would suggest that without these particular testing methods, which Harrison evolved before 1730 when he still lived in Lincolnshire, he would never, ever, have achieved his claimed results with the precision wooden clocks, and the marine timekeepers could never have been made. This testing method is based on two principles that are described in the passage above. First, he placed two clocks in adjoining rooms and adjusted them to keep accurate time under the same environmental conditions. Then, by closing the door between the rooms, he changed the temperature of one room artificially and observed the effect by comparing the time shown by the clock in that room with that of the control clock in the other room. By adjusting the amount of compensation given by the gridiron pendulum, he was able to alter the rate of the clock being tested so that, under changing conditions, it was able to maintain the same time as the control clock in a more stable environment. The second principle, although it is so vitally important, Harrison dismissed with a mere four words: 'And again the Contrary...'. Perhaps better known to us as reciprocal correction, it involved reversing the situation above, making the test clock the control and the control clock becoming the clock under observation. Through this method, Harrison was able to refine the performance of both clocks and thereby achieve the remarkable accuracy of one second in a month. Here I would like to quote from Bill Laycock (1924-1976):

> His [Harrison's] realisation of these two principles and their possibilities in combination was a stroke of sheer genius, and I mean really original thought....For one of the two principles involved has only come into prominence in ANY field of research during the second half of the twentieth century, whilst although the other has been known and practised in slightly different forms for a longer time, I know of no other instance in which the two have been combined so soundly, elegantly, and <u>EFFECTIVELY</u> as they were by this remarkable character almost 250 years ago.[54]

54. Laycock, 'Harrison's Pendulum Horology' (*op. cit.*, note 39), p. 34.

By the time Harrison bid his farewell to Graham at eight o'clock in the evening at the end of that first meeting, poor George Graham must have been exhausted. Never before could he have been on the receiving end of such a technological barrage, but he was certainly mightily impressed. He was to become Harrison's most important supporter. In fact, along with Halley, Graham was absolutely crucial and pivotal to Harrison's later success, and Harrison readily acknowledged this.

By 1730, John Harrison had discovered all of the fundamental knowledge that he was to need for the rest of his working life. He had moved the goal-posts of precision timekeeping by a Herculean margin. His achievement is at least equal to that of Christiaan Huygens (1629-1695), who, through the application of the pendulum to clocks in 1656, had reduced timekeeping rates from minutes to seconds in a week. By 1730, Harrison had reduced this to within one second in a month with his precision wooden clocks. And yet, in 1730, he was only at the threshold of his working life.

**Enamel paste portrait of John Harrison, modelled by James Tassie (1735-1799) about 1775. This portrait is not included in either of the two catalogues of Tassie's works (1775 and 1791), and only two examples for certain appear to have survived. Neither is dated, but, because Tassie is known to have worked from life, this portrait must have been made before Harrison died in 1776. Dimensions within frame: 9.5 x 7.6 cm. (3.7 x 3.0 in.). Courtesy of the National Portrait Gallery, London. Inv. no. 4599.**

# Even Newton Could Be Wrong: The Story of Harrison's First Three Sea Clocks

*William J. H. Andrewes*

William Andrewes is the David P. Wheatland Curator of the Collection of Historical Scientific Instruments and Preceptor in the History of Science at Harvard University. He also directs the Collection's catalogue series, the first of which was published in August 1992, and coordinates the development of some interactive multimedia applications for teaching and research.

Andrewes's interest in horology began in 1966. Under the guidance of the watchmaker George Daniels, he restored one of John Harrison's early clocks between 1970 and 1972. After receiving his degree in three-dimensional design in 1972, he won a competition to design and model the tercentenary medals for the Old Royal Observatory at Greenwich. From 1974 to 1977, he was responsible for the conservation and restoration of the clocks at the Old Royal Observatory and the National Maritime Museum and was commissioned by the former to make a reconstruction of a Shelton journeyman clock. In 1977, soon after being elected a Freeman of the Worshipful Company of Clockmakers, he came to the United States to be Curator of The Time Museum in Rockford, Illinois. His appointment at Harvard began in 1987.

Andrewes has appeared on the BBC, NBC, CBS, and PBS in various programs relating to horology and scientific instruments and has published articles on the history of timekeeping devices and scientific instruments in both popular and scholarly publications.

# Even Newton Could Be Wrong: The Story of Harrison's First Three Sea Clocks

*William J. H. Andrewes*

> The Longitude will scarce be found at sea without pursuing those methods by which it may be found at land. And those methods are hitherto only two: one by the motion of the Moon, the other by that of the innermost Satellit of Jupiter.[1]

So begins a letter dated August 26, 1725, from Isaac Newton (1642-1727) to the Admiralty, responding to yet another proposal concerning the finding of longitude at sea. Although a special committee, the Board of Longitude, had been established by the 1714 Act of Queen Anne[2] to review such proposals, the enormous sums offered by the Act attracted so many far-fetched solutions that, doubtless, until an idea had the support of some notable individual in the scientific community, it was not worth the time and expense of holding a meeting.[3] Surviving records seem to indicate that, in fact, almost a quarter of a century passed before the Board's first meeting occurred. Meanwhile, Newton, who was in his seventies when the Act was passed, became the unfortunate academic to whom the Secretary of the Admiralty referred many of these harebrained schemes.[4]

Newton's influence in this matter was due in part to the fact that he had been responsible for establishing the three levels of accuracy specified by the Act, a fact that is noted in the draft of a letter he wrote less than nine months after the Act was passed:

> Upon my representing that the Longitude might be found by the motion of the Moon without the error of above two or three degrees, & that if it could be found to a degree it would be useful[,] if to ⅔ of a degree it would be more useful, if to ½ a degree it would be as much as could be desired[,] the Committee of Parliament set Premiums upon finding it to these degrees of exactness, & thereby the Act of Parliament points at the finding it by the Moons motion.[5]

By the early eighteenth century, there were three methods that were considered promising. Two, as Newton noted in the first passage quoted, were astronomical solutions—the lunar-distance method and that which used

1. Letter dated August 26, 1725, from Isaac Newton to Josiah Burchett (*ca.* 1666-1746), the Secretary of the Admiralty, in reply to a request for his opinion concerning the ideas of Jacob Rowe (fl. 1725-1734) for improving navigation, in particular solving the longitude problem with improved hourglasses employing sands of tin. *Correspondence of Isaac Newton*, ed. H. W. Turnbull *et al.*, 7 vols. (Cambridge: Cambridge University Press, 1959-77), vol. 7, no. 1476, pp. 330-2.

2. Act 12 Anne *cap.* 15 (1714): *An Act for Providing a Publick Reward for such Person or Persons as shall Discover the Longitude at Sea.*

3. The 1714 Act stipulated that at least five of the 24 commissioners empowered to serve on the Board of Longitude had to be in attendance for a meeting. The commissioners, all distinguished, high-ranking officials, were not paid for this appointment, so the Admiralty had an obligation not to occupy their time unnecessarily. Later, when meetings were held on a more regular basis, there is evidence that the *ex officio* commissioners from Oxford and Cambridge were granted an allowance of £15 per meeting. Harrison, *Remarks on a Pamphlet Lately published by the Rev. Mr. Maskelyne Under the Authority of the Board of Longitude* (London, 1767), p. 27.

4. Newton was not known for his humor, but the continuing stream of ludicrous proposals evidently made an impression: "[Mr. French's] Project for the Longitude is as impracticable as to make a perpetual motion like that of the heart but much more uniform or to observe the Sun's meridional altitude to a second or to deduce the Longitude from the complement of the Latitude, or to find that complement by burning brandy." Draft of letter dated March 22, 1715, in Newton, *Correspondence* (*op. cit.*, note 1), vol. 6, no. 1137, p. 211. John French's proposal and other "nutty" solutions are described by Owen Gingerich in his paper in this volume.

5. *Ibid.*, vol. 6, no. 1137, pp. 211-12. On June 11, 1714, Newton had stated that the lunar-distance method was "exact enough to determine her Longitude within Two or Three Degrees, but not within a Degree." *Journals of the House of Commons,* vol. 17 (1711-14), p. 677. For the

the eclipses of the satellites of Jupiter. The third was the horological, or mechanical, solution. In the passage above, Newton revealed that he believed that only the lunar-distance method would provide a practicable solution. Continuing, he juxtaposed the astronomical solution in general with the horological, forcefully voicing his support for the former, and, to emphasize this point, he distinguished between finding the longitude and keeping the longitude:

> And I have told you oftener then once that it [the longitude] is not to be found by Clock-work alone. Clockwork may be subservient to Astronomy but without Astronomy the longitude is not to be found. Exact instruments for keeping of time can be usefull only for keeping the Longitude while you have it. If it be on[c]e lost it cannot be found again by such Instruments. Nothing but Astronomy is sufficient for this purpose. But if you are unwilling to meddle with Astronomy (the only right method & the method pointed at by the Act of Parliament) I am unwilling to meddle with any other methods then the right one.[6]

This strong opinion made a lasting impression upon the academic community. The solution to the problem of finding longitude at sea would evidently be made "not by Watchmakers or teachers of Navigation...but by the ablest Astronomers."[7]

Newton's negative attitude toward proposals involving timekeepers was not unjustified: a clock, as he correctly noted, only keeps time, and, by the first quarter of the eighteenth century, no portable timekeeper had been made that was able to keep accurate time on land, let alone at sea. The best timekeepers available were those that were fixed in one location. The accuracy even of these, however, could be relied upon only over a short period; their performance during a long period was unstable owing to the deterioration of the lubricants and the resulting wear of the mechanism, as well as to continual changes in temperature and barometric pressure.[8] In addition to overcoming these problems, a timekeeper on board a ship would have to remain unaffected by the corrosive saline atmosphere, the pitching and rolling in a rough sea, and the variation of gravity in different latitudes, which, as Newton commented, was "not yet sufficiently known."[9]

Added to all this was the fact that the standards set by the 1714 Act were very exacting. To qualify for the rewards offered, all proposed methods had to undergo a trial on board a ship bound for the West Indies, a voyage that could last about 60 days. To win the largest prize of £20,000, a timekeeper could gain or lose no more than a total of two minutes—an average of only two seconds per day—during this entire period.[10] To contend for even the lowest prize of £10,000, its total error could not exceed four minutes, an average of four seconds per day. With 86,400 seconds in a day, a variation of only nine seconds per day represents 99.99 per cent accuracy—and this is more than twice the variation permitted for the lowest reward.

There had, of course, been several notable attempts to devise a marine clock, but all of these had failed.[11] The best known had been made by the celebrated Dutch mathematician and astronomer Christiaan Huygens (1629-1695). Although Huygens had managed, with the application of the pendulum and the invention of the spiral balance spring, to make significant

three levels of accuracy recommended by Newton, rewards were set at £10,000 (1°), £15,000 (⅔°), and £20,000 (½°).

6. Newton, *Correspondence* (*op. cit.,* note 1), vol. 6, no. 1137, p. 212.

7. *Ibid.,* vol. 7, no. 1377, p. 172. This letter was written in October 1721 to Josiah Burchett, in response to a proposal submitted by a Mr. Laurans for "finding the Longitude at sea by Watch work."

8. By 1722, the clockmaker George Graham (*ca.* 1674-1751) had produced a regulator (fitted presumably with his dead-beat escapement and his mercury temperature-compensated pendulum) that maintained a remarkably accurate rate over a short period. His notes dated June 7, 1722, record the timing of one of his clocks compared with a fixed star: between May 2 and June 2, the clock lost an average of 7.5 seconds per day, but from this steady losing rate it did not vary one second during the entire trial. MS. by George Graham entitled "M^r^ Graham's acc^t^ of Col^l^ Moleworth's Hourglass," June 7, 1722, ref. no. Cl.P.III(2)15, Royal Society Records, London.

9. Newton, *Correspondence* (*op. cit.*, note 1), vol. 7, no. 1476, p. 331.

10. The Earth rotates on its axis one degree every four minutes, so to determine the longitude to half a degree, a timekeeper could not vary more than two minutes in 60 days. The period of 60 days for a voyage to the West Indies is based on the average duration of the voyages to and from Jamaica in the first trial of Harrison's ultimately successful timekeeper: the outward journey took 62 days, the return 58 days.

11. Detailed and amusing descriptions of several early efforts are contained in Lt.-Comdr. Rupert T. Gould's classic work, *The Marine Chronometer: Its History and Development* (London: J. D. Potter, 1923), pp. 27-39A. Readers should be aware that the pagination of the most recent edition of this work (Woodbridge, England: Antique Collectors' Club, 1990) is different from that of the original and its 1960 reprint.

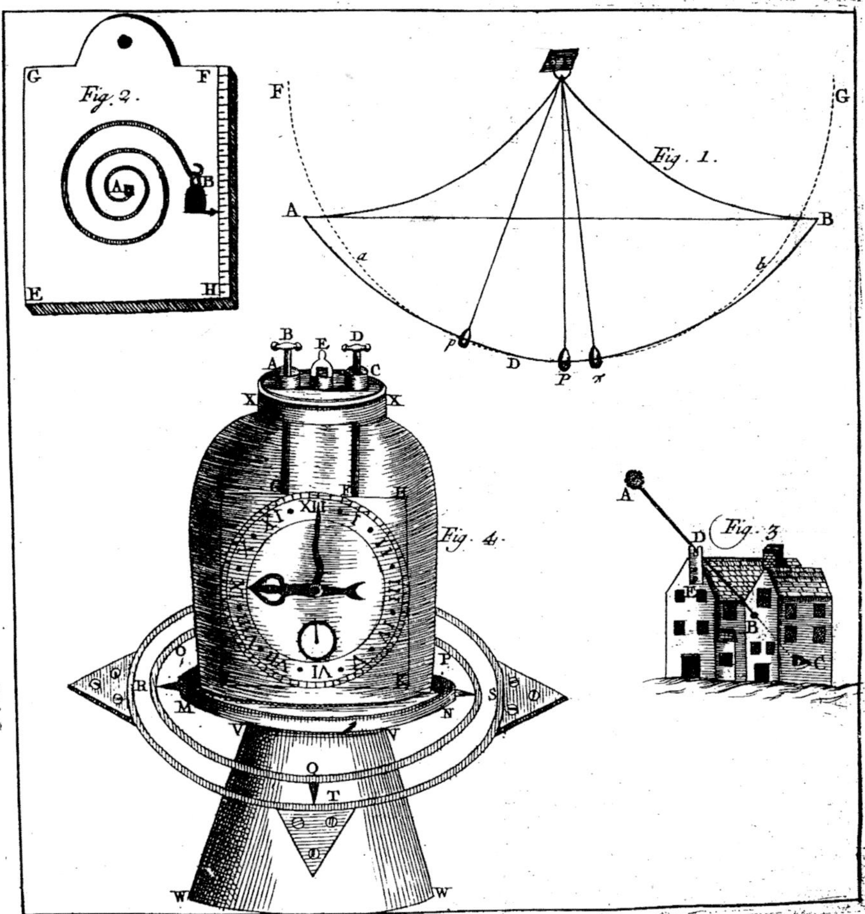


**Figure 1. Illustrations from the plate in Jeremy Thacker's *The Longitudes Examin'd:...* (London, 1714), showing: in *Fig. 1*, a diagram to explain the difference between a circular and a cycloidal arc of a pendulum, the latter being performed in the same period of time regardless of the amplitude of swing; in *Fig. 2*, his ingenious weighted spiral-spring thermometer; in *Fig. 3*, the transit instrument he devised to rate his timekeepers, a method similar to that later employed by Harrison (by sighting through a half-inch hole in the wall of his house to the notch in the iron bracket attached to the gable, he could observe the transit of a star from one night to the next); and in *Fig. 4*, his chronometer in its evacuated jar suspended in gimbals.**

advances in precision timekeeping, he was unable to master the problems of keeping time at sea.[12] The next effort worthy of note was made in 1714 by the English horologist Jeremy Thacker, from Beverley in Yorkshire (interestingly, less than fifteen miles from Barrow upon Humber, where John Harrison had just completed his first clock). Nothing is known about this inventor apart from what is contained in the pamphlet describing his "chronometer."[13] From Thacker's description, this balance-spring timekeeper (shown in the last illustration in Figure 1), which was suspended in gimbals, incorporated some unusual and interesting features. First, it was contained in an evacuated jar, but his idea of using this was not, as might be expected, to overcome barometric pressure, but "to settle the Matter of the Moisture and Dryness" of the air.[14] Second, it employed a form of spring-driven maintaining power, a secondary spring to keep the clock running while it was being wound.[15] To solve the problem of temperature variation, Thacker produced a table showing the clock's rate at different temperatures, so that its variation due to temperature change could be predicted. The thermometer that he devised to make the required observations (shown in the second illustration in Figure 1) was ingenious in its simplicity and, in certain respects, foreshadowed the later development of the bimetallic strip. Thacker claimed that, during a trial on land, his clock never exceeded a variation greater than six seconds per day. Its trial at sea, however, was doubtless disappointing, because nothing more was heard of either the machine or its maker.

About ten years later, another notable attempt was made by Henry Sully (1680-1728), an English horologist who worked most of his life on the continent. In order to gain the interest of the Royal Society and thereby the

12. For details of Huygens's marine timekeepers, see John Leopold's paper in this volume. Although the spiral balance spring, invented by Huygens in January 1675, represented a major advance in the accuracy of portable timekeepers, it was not the first attempt to use a spring to control the oscillation of a balance in the same way that gravity controls the oscillation of a pendulum. Huygens mentioned such an application being made in Paris in 1660 (*Oeuvres Complètes de Christiaan Huygens*, 22 vols. [The Hague: Martinus Nijhoff, 1888-1950], vol. 5, p. 486); a few straight balance-spring watches that almost certainly predate 1675 survive.

13. "In a Word, I am satisfy'd that my Reader begins to think that the *Phonometers*, *Pyrometers*, *Selenometers*, *Heliometers*, *Barometers*, and all the *Meters* are not worthy to be compar'd with my *Chronometer*." Jeremy Thacker, *The Longitudes Examin'd...* (London, 1714), p. 23. Although Thacker appears to have been the first to use the word "chronometer" in print, the English horological writer, the Rev. William Derham (1657-1735), also used it in his paper, dated August 4, 1714 (but not read to the Royal Society until November 4, and not published), entitled "Observations concerning the Motions of Chronometers," ref. no. Cl.P.III(2)10, Royal Society Records, London. Thacker's book might have appeared after Derham's paper, because the former included appraisals of several proposals that would not have appeared before the 1714 Act was granted Royal Assent in July. The word was also used by John Desaguliers (1683-1744) in *A Course of Experimental Philosophy*, 2 vols. (London, 1734), vol. 1, pp. 375-6, to describe a small short-duration timer made for him about 1730 by George Graham. Graham's "chronometer," which had a verge escapement and a short-bob pendulum (2.45 in. long), was fitted with a dial divided into quarter seconds and with a quadrant to arrest the pendulum to within the nearest quarter of its vibration, thereby providing a measurement of time to one-sixteenth of a second. Desaguliers used this chronometer for timing astronomical observations, the descent of falling bodies, and velocities, and for other experimental purposes. The same type of *"chronometre"* is referred to in Diderot and d'Alembert's

Board of Longitude, Sully sent his marine clock (Figure 2) in 1724 to George Graham (*ca.* 1674-1751), an eminent English maker who had been elected a Fellow of the Society four years earlier, and, in 1726, he published a description of the clock's mechanism and an account of its trials.[16] The movement of this clock (Figure 3) incorporates antifriction wheels and Sully's frictional-rest escapement. It shows a strong influence of Huygens's work, particularly in the design of the balance, which closely resembles Huygens's Perfect Marine Balance (see John Leopold's paper, Figure 4). This ingenious device was an attempt to combine the accuracy of a pendulum with the portability of a balance. To avoid the pronounced effect that temperature variation has on the elasticity of a balance spring, the motion of the balance

**Figure 2. Illustration showing Henry Sully's marine clock (suspended in gimbals from above), his meridian-finding instrument (lower right), and his marine watch (lower left). While the marine clock reveals the influence of Christiaan Huygens (see John Leopold's paper, Figure 3), the marine watch in its gimballed frame, which Sully first described and illustrated in 1716, is highly original in its design, being very similar in general appearance to a modern marine chronometer. From Henry Sully, *Description Abrégée d'une Horloge d'une nouvelle invention, Pour la juste mesure du Temps sur Mer*, 2nd ed. (Paris, 1726), plate 3. Courtesy of Giuseppe Brusa.**

*Encyclopédie, ou Dictionnaire Raisonné des Sciences, des Arts, et des Métiers,* vol. 3 (Paris, 1753), p. 402.

14. Thacker (*ibid.*), p. 16. The idea of placing a clock in a vacuum, although probably unknown to Thacker, had been mentioned by the Italian clockmaker Matteo Campani (1620-1687) nearly 50 years earlier. In 1704, William Derham had also conducted experiments with a clock in an evacuated chamber. See also note 66.

15. Thacker's maintaining power used an auxiliary spring, which had to be set each time the clock's mainspring was wound. In the illustration of Thacker's timekeeper in Figure 1, the key inserted at point G winds the clock, while that inserted at point F activates the maintaining power. In 1683, Huygens had illustrated a portable maintaining power mechanism that operated automatically, but no example has survived. See also note 43.

16. Henry Sully, *Description Abrégée d'une Horloge d'une nouvelle invention, Pour la juste mesure du Temps sur Mer* (Paris, 1726). There are two parts to this work, but, because they were produced about a year apart, several copies contain only the first part. This has a preface dated January 31, 1726, and contains 48 pages describing his marine clock. The second part, which has a preface dated December 31, 1726, and an additional 242 pages, comprises most of the information about the history of Sully's timekeepers. It contains further details of the clock and a description of his marine watch, which appears to have had a more promising design. His watch is also described in an article (no. 177 for the year 1716) entitled "Montre pour la mer, inventée par M. Sully," published in M. Gallon, *Machines et Inventions approuvées par l'Académie Royale des Sciences,...* (Paris, 1735), vol. 3, pp. 93-4. No examples of this timekeeper, which from its size and shape in the engraving (see Figure 2) closely resembled the design of the modern marine chronometer, are known to have survived.

is controlled by the force of gravity with a weight suspended by a thread from cycloidal cheeks, which, oscillating with the balance, were intended to make the balance's oscillations, whether large or small, isochronous—in other words, to be performed in the same period of time.[17] Sully controlled the motion of the balance with an unusual type of pendulum, pivoted horizontally, so that it oscillated up and down instead of from side to side. Graham pointed out, however, that the ship's motion would influence the inertia of the pendulum. His concerns were justified: although during tests in the sheltered waters at the mouth of the river Garonne the clock's performance was promising, a trial in the open sea revealed that the timekeeper was no match for the longitude problem. Sully might have continued working to overcome the deficiencies of his timekeeper, but he died of a sudden illness in 1728.

**Figure 3. Movement of Henry Sully's timekeeper, illustrating the going barrel, his frictional-rest escapement, the balance staff mounted horizontally on antifriction wheels, and the pendulum pivoted in a horizontal plane and attached by a thread to cycloidal cheeks on the balance. Height to top of balance: 27.9 cm. (11 in.). Courtesy of the Worshipful Company of Clockmakers, London/Bridgeman Art Library, London. Inv. no. 597.**

These failures, combined with Newton's continual rejection of any proposal connected with timekeepers, seem to have convinced most theoreticians that only astronomy would provide a viable method for solving the longitude problem. But all these odds did not deter a resolute young Yorkshireman named John Harrison (1693-1776). Harrison was a practical man, not a theoretician, and Newton's opinion on the subject, if it ever reached the remote village of Barrow upon Humber in north Lincolnshire, where Harrison worked as a joiner and clockmaker, certainly did not discourage him from pursuing a mechanical solution. Having begun making wooden clocks around 1713, he embarked on a quest during the 1720s to perfect accuracy and reliability in clockwork.[18] By 1730—three years after Newton died—Harrison had developed not only a clear understanding of the problems associated with precision timekeeping but was also satisfied that he had overcome the difficulties of keeping time on land:

> ...in these 3 last Years [I] have brought a Clock to go nearer the truth, than can be well imagin'd, considering the vast Number of Seconds of Time there is in a Month, in which Space of time, it does not vary above one Second, & that mostly the way I expect: So that I am Sure I can bring it to the Nicety of 2 or 3 Seconds in a Year. And 'twill also continue this exactness for 40 or 50 Years or more; however so as not to vary above 2 or 3 Seconds from what it

17. The power that maintains the motion of a balance or a pendulum is subject to variation because of fluctuations in the power source, problems with lubrication, or inaccuracies in the meshing of the wheels and pinions. When the power increases, the motion of the balance will likewise increase, and the period that it takes to perform its oscillation will therefore take longer. This would cause a clock to lose time. A decrease in power would have the opposite effect. To overcome this problem, Huygens devised cycloidal cheeks to automatically decrease the effective length of a pendulum as its amplitude of swing increased. Because a short pendulum oscillates faster than a long pendulum, cycloidal cheeks, at least in theory, ensure that all arcs, large or small, are performed in the same period of time. Thinking that the cheeks would thus nullify the effects of any fluctuations in power, Sully employed a going barrel, a power source that houses the mainspring and contains the first wheel of the train. The going barrel, which was commonly used in French clocks, provides a continuous source of power, because when the inner coils of the spring are being wound, the outer coils continue to drive the clock. Thus there is no need for maintaining power to keep the clock running while it is being wound. Obviously, when fully wound, the spring will deliver more power than when it needs to be wound, but this, thought Sully like many clockmakers in France at that time, would be overcome by the cheeks.

18. Details of Harrison's early clocks are given in Andrew King's paper in this volume.

19. Untitled manuscript signed "John Harrison, Clockmaker at Barrow; Near Barton upon Humber; Lincolnshire. June 10. 1730," MS. 6026/1, p. 2, sect. 4, Library of the Worshipful Company of Clockmakers, Guildhall Library, London. Because this document is untitled, it is usually referred to as Harrison's "1730 manuscript."

20. Martin Folkes, speech to the Royal Society dated November 30, 1749, ref. no. JBC.XX, Minutes of the Royal Society, London, pp. 187-8. This claim of accuracy was also mentioned in a document entitled "Some Account of Mr. Harrison's Invention for determining the Longitude at Sea, and for correcting the Charts of the Coasts," dated January 16, 1741/2, and signed by twelve distinguished members of the Royal Society, including the president, the Astronomer Royal, and professors from Oxford and Cambridge. Here it was stated that Harrison had "made a Pendulum-Clock that keeps Time so exactly with the Heavens, as to not err above

did the Year next before; for 'twill not want Cleaning, & the little it wears can but alter it insensibly little.[19]

In order to achieve this degree of accuracy and long-term stability, Harrison had to understand exactly what caused variations in timekeeping and discover ways to overcome them. By 1730 he had already devised ways to compensate for changes in temperature and barometric pressure and to overcome the problems of lubrication. No one else, not even the celebrated London clockmaker George Graham, had attempted to resolve the last two problems. Instead of inhibiting Harrison's progress, the limitations of materials and information about the traditional practices of clockmaking that existed in Barrow upon Humber appear to have stimulated his ingenuity and caused him to adopt his original and scientific approach to solving problems—an approach unlike that of other clockmakers, whose professional training limited their creativity. Harrison's achievements, although later forgotten, were to be recognized by the scientific community over the course of the next two decades. Upon presenting Harrison with the Copley Gold Medal on November 30, 1749, the president of the Royal Society, Martin Folkes (1690-1754), stated, in reference to his early work, that

> ...two clocks, in different parts of his house, kept time together, without the variation of more than one single second in a month: and that one of them, which he kept for his own use, and still has by him, and which he constantly compared with a fixed Star, had not varied from the Heavens so much as one whole minute in Ten Years,....[20]

It would appear that Harrison first heard about the longitude prize in 1726 and that soon after he turned his attention to the problems of keeping time at sea.[21] A document signed by members of the Royal Society in 1741/2[22] and prepared no doubt with Harrison's assistance states that it was in 1727 that he began to consider adapting the technology being developed for his precision regulators for use at sea.[23] Harrison was 34 at the time. His first wife, Elizabeth (née Barrell), had died in May of the previous year, leaving him with their seven-year-old son, John (1719-1738). He remarried the following November. His second wife, also named Elizabeth (née Scott) (d. 1777), gave birth in the following year to his second son, William (1728-1815), about a month after which his father, Henry Harrison (1665-1728), died; four years later, his last child and only daughter, Elizabeth, was born.[24]

Finding longitude at sea was a problem that had eluded the greatest scientific minds for more than two centuries. Harrison thought of the challenge in mechanical terms: What was needed to overcome the seemingly impossible problem of producing a machine capable of keeping accurate time at sea? He recognized that, first, the solutions to the problems of keeping time on land, which he was able to resolve in his precision regulators, had to be adapted for use at sea; specifically, these involved a constant power source to drive the mechanism, the reduction of friction and the elimination of the need for oil, and the ability to compensate for temperature variation and for changes in barometric pressure. Second, he had to overcome the problems specific to the ocean environment—in particular the sudden shocks and continual rocking motion to which the timekeeper would be subjected.

one Second in a Month, for ten Years together." This document was published in Appendix 5 of an unsigned pamphlet entitled *An Account of the Proceedings, in order to the Discovery of Longitude:...* (London, 1763), p. 20. See also note 28 concerning the authors of this work.

21. "In the year 1726, Mr. Harrison...was first informed of the Contents of the above-mentioned Act of Parliament [the 1714 Longitude Act], and, upon the Faith thereof, applied his whole Attention to adapt the Principles of his Clock to the Motion of a Ship at Sea." Broadsheet entitled *The Case of Mr. John Harrison*, 2nd ed. (1770), p. 1. This passage is not included in the first edition (*ca.* 1766) of this work.

22. This document ("Some Account of Mr. Harrison's Invention" [*op. cit.*, note 20]) is dated January 16, 1741/2. The two years referred to relate to the two calendars in use at that time. England and its colonies continued until September 1752 to use the Old Style Julian calendar and celebrated the New Year on March 25. In contrast, most of Europe by this time had converted to the Gregorian calendar, introduced in 1582 by Pope Gregory XIII, and celebrated the New Year on January 1. Thus, to avoid misunderstanding as the use of the Gregorian calendar spread, dates between January 1 and March 25 in the Julian calendar were often written in the form above, with 1741 denoting the Julian calendar year and 1742 the Gregorian calendar year.

23. *Ibid.*, p. 20.

24. For further information on Harrison's family, see the Harrison family tree and Andrew King's paper at the beginning of this section. Other related details can be found in Humphrey Quill, *John Harrison: The Man who found Longitude* (London: John Baker, 1966), and in [Andrew King], *From a Peal of Bells: John Harrison, 1693-1776*, Usher Gallery exhibition catalogue ([Lincoln, England]: Lincolnshire County Council, 1993).

In addition to being an ingenious mechanic, Harrison was one of the first material scientists. He conducted numerous experiments to determine the most suitable material for each part of his timekeepers. For example, in order to establish which metals to use for his gridiron pendulum, he "prepar'd a Convenience on the outside of the Wall of my House, where the Sun at 1 or 2 a Clock makes it very warm." There, "in the Cool of the Mornings & Evenings, & in the heat of the Days," he compared the variation in the expansion and contraction of particular metals: steel to iron, steel to brass from Sheffield, from London, and from Holland—the last both hard and annealed—and steel to silver and copper.[25] For his marine timekeepers, he developed alloys of tin and copper, adjusting the proportions to suit particular purposes.[26] In short, his mechanisms were based on thorough experimentation, and, like a scientific instrument, the design of every part was dictated by its function. He made no unnecessary embellishments. The unusual appearance of his first three marine timekeepers, which may seem extraordinary at first sight, can be seen upon investigation to have been developed from the function of their integral parts.

There is evidence that Harrison began the construction of his first marine timekeeper, commonly known as H.1,[27] about 1729 and that the first drawings were made in 1727.[28] The earliest surviving document describing this masterpiece, however, is a manuscript signed by John Harrison and dated June 10, 1730.[29] In this work, Harrison both provided a detailed account of the technical solutions and methods that he had employed in perfecting his precision longcase clocks and described how this technology could be incorporated into a marine timekeeper. His expectations for the success of his proposed "Sea Clock" were summarized in the conclusion, which announced a precision unheard of at the time: "in the Ships they shou'd vary 4 or 5 seconds in a month."[30]

There can be little doubt that Harrison's main reason for writing this document was his need for financial support to make the proposed marine timekeeper. As far as we know, the design and construction of his earlier clocks had been done without outside financial assistance. As a joiner, Harrison had access to a plentiful supply of the well-seasoned oak and other woods needed for these clocks, and once he had finished experimenting, they could be sold.[31] Producing his marine timekeeper, however, would require other materials: because of the motion and shocks to which a ship at sea is subjected, he realized that wood was an unsuitable material for most parts of a marine timekeeper. Thus, the substantial cost of purchasing brass and steel, as well as specialized tools for working in metal, must have made Harrison recognize that outside financial support would be needed.[32] And London being the obvious place where such support would most likely be forthcoming, to London he planned to go. He would not go unprepared. As he later wrote,

> ...I had along with me...the Descriptions, with some Drawings of the principal Parts of the Pendulum-Clock which I had made, and as also of them of my then intended Timekeeper for the Longitude at Sea.[33]

It is highly probable that the "Descriptions" were those contained in the manuscript dated June 10, 1730, and it is evident from the manner in which

25. Harrison, 1730 manuscript (*op. cit.*, note 19), pp. 4-5, sect. 9.

26. For instance, Harrison developed alloys of high- and low-tin bronze to minimize the use of ferrous metals in his sea clocks. Analysis of parts of his third timekeeper (H.3) made under the supervision of Jonathan Betts in the early 1980s reveals that Harrison used a low-tin bronze, a reddish alloy with high tensile strength, for arbors, clicks, and levers, and a high-tin bronze, a whiter, bell metal-like alloy with high compressive strength, for the faces of his balance bearers and the rollers and race of H.3's two caged roller bearings. H.1 and H.2 employ similar alloys. As will be described later (see note 46 and related text), Harrison employed the wood lignum vitae to eliminate the need for oil.

27. The abbreviations "H.1," "H.2," "H.3," "H.4," and "H.5" have been commonly adopted as the manner of referring to Harrison's five marine timekeepers. In his book *The Marine Chronometer* (*op. cit.*, note 11), Gould referred to the Harrison timekeepers as "No. 1," "No. 2," "No. 3," "No. 4," and "No. 5," although he designated Kendall's copies as "K1," "K2," and "K3." This mode of referring to the timekeepers appears to have been continued for many years; Jonathan Betts informs me that the abbreviation "H.4" was used in the correspondence between Gould and David Evans (1919-1984), of the Ministry of Defence's chronometer workshops, during the restoration of that timekeeper in 1946, and D. W. Fletcher (1891–1965) adopted the abbreviation "H.1" in his article "Restoration of John Harrison's First Marine Timekeeper," *Horological Journal*, vol. 94, no. 1125 (June 1952), pp. 366-9.

28. "Some Account of Mr. Harrison's Invention" (*op. cit.*, note 20), p. 20. Although Harrison did not write this document, there can be no doubt that he was consulted in its preparation: the French astronomer, Jérôme de Lalande (1732-1807) states in his *Bibliographie Astronomique* (Paris, 1803), p. 483, that *An Account of the Proceedings* (*op. cit.*, note 20), in which this essay was published, was written for Harrison as a favor by two of his friends, the optician James Short (1710-1768) and a well-known barrister, Taylor White (1701-1771), who was one of the early treasurers of the Foundling Hospital, with which John Harrison's son William was later involved. Lalande states in the journal he titled "Voyage d'Angleterre" that White composed the text and Short was responsible for the calculations. This is mentioned in the entry for Tuesday, April 26, 1763, the day he was given a copy of the pamphlet by Short. *Jérôme Lalande: Journal d'un Voyage en Angleterre*, transcribed by Hélène Monod-Cassidy (Oxford, England: Voltaire Foundation at the Taylor Institution, 1980), p. 52; the original is preserved in the Bibliothèque Mazarine, Paris (MS. 4345 [3322]).

29. Harrison, 1730 manuscript (*op. cit.*, note 19). This document is of great importance to our understanding of the technology that Harrison developed. It was left by William Harrison, along with H.5 and other relics that had belonged to his father, to his granddaughter, Elizabeth Wright (1797-1880), who, in 1869, offered them unsuccessfully to the Patent Museum (later the Science Museum) in London for £100. They were subsequently purchased for £75 by Robert Napier (1810-1890) of Gairloch, Scotland, and finally came into the possession of the Worshipful Company of Clockmakers shortly after his collection was sold at Christie's in 1877. The manuscript remained neglected until 1950, when G. H. Baillie (1873-1951) published a complete transcript entitled "The 1730 Harrison M.S.," in the *Horological Journal*, vol. 92, no. 1102 (July 1950), pp. 448-50, and vol. 92, no. 1103 (August 1950), pp. 504-6.

30. Harrison (*ibid.*), p. 22, sect. 29. In this document, Harrison refers to his marine timekeeper as a "Sea Clock" (p. 12, sect. 21 and p. 15, sect. 24). The use of the plural "they" in this passage implies that Harrison was anticipating making more than one.

31. Harrison mentions selling one of his clocks in 1727 (*ibid.*, p. 12, sect. 21). His surviving clocks from this period all conform to a standard design, suggesting that they were part of some limited production. They are housed in cases that reflect the fashionable style of the day and strike the hours, a function that, although of domestic use, would disturb their performance and serve no purpose for a precision timekeeper. With these additions, however, his clocks could serve not only as a test bed for his ideas and experiments, but also as part of a practical and necessary commercial venture through which he could earn his living. For further information, see Andrew King's paper in this volume and William Andrewes, "John Harrison: A Study of his Early Work," *Horological Dialogues*, vol. 1 (1979), p. 22.

32. The substantial cost was mentioned by members of the Royal Society in 1735, when the machine was ready for its trial at sea, in an untitled document beginning *"John Harrison,* having with great Labour and Expence, contrived and executed a Machine for measuring Time at Sea,...." This document is listed as "NUMB. 1" in the appendix of *An Account of the Proceedings* (*op. cit.*, note 20), p. 17.

33. John Harrison, *A Description Concerning Such Mechanism as will Afford a Nice, or True Mensuration of Time;...* (London, 1775), p. 19.

this document is signed—"John Harrison, Clockmaker at Barrow, Near Barton upon Humber; Lincolnshire."—that Harrison intended to establish himself as sole originator and inventor of the precision timekeepers described.

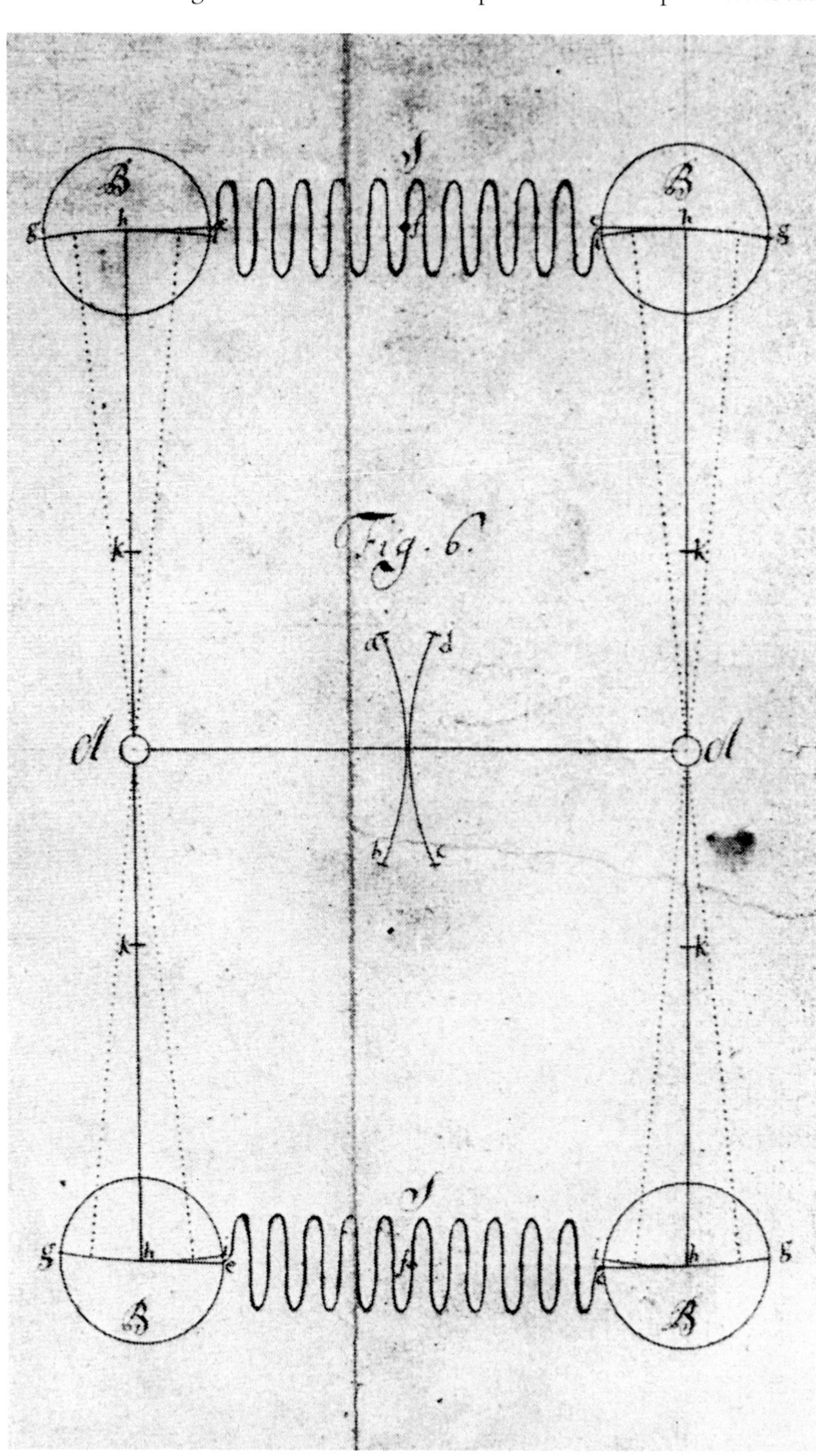


**Figure 4. John Harrison's drawing of the balances, or artificial pendula, for H.1. Letter *B* represents the brass spheres. *S*, the helical springs, or worm springs as Harrison called them, are connected to the arcs *g*, *h*, *i* on each balance; these arcs, which prevent distortion of the springs when the balances oscillate, are altered when isochronal adjustments are needed. *A* represents the axis of each balance, and *a*, *b*, *c*, *d* the arc along which the flat springs connecting the two balance bars are stretched, one from *a* to *c*, the other from *b* to *d*. From John Harrison's 1730 manuscript, fig. 6, p. 24. Courtesy of the Library of the Worshipful Company of Clockmakers, London.**

It is also clear from the way he writes that he had carefully considered how to make his proposal, as the Act of Queen Anne had stipulated it should be both "Practicable and Useful at Sea."[34] Although it has been stated that Harrison's visit to London was made in 1728,[35] it is far more likely, particularly from the above reference, that he made the journey after June 10, 1730.

The first person whom Harrison contacted was Edmond Halley (*ca.* 1656-1742). By virtue of his appointment as Astronomer Royal, Halley was an *ex officio* member of the Board of Longitude and thus an influential figure in this matter. Upon his advice, Harrison went to see George Graham, who, Halley told him, was a very honest man and would certainly do him "Good if was in his Power."[36] After a somewhat rough start, the meeting did indeed do him good, for Graham became, and remained, his close ally. Recommending as an initial step that Harrison should not apply to the Board of Longitude until his ideas had been put into practice,[37] Graham appears to have assisted Harrison by lending him money without security or interest.[38] As a member of the Royal Society, Graham also used his influence to encourage both the trial of the first sea clock and the financial support of the Board of Longitude. Without Graham, the construction and testing of Harrison's marine timekeepers would not have progressed as they did. Indeed, in regard to Graham and others who had helped him, Harrison acknowledged near the end of his life that "from the Encouragement of the Public alone, I could never have gone through what I did go, nor consequently have made a Completion of the Matter."[39]

34. Act 12 Anne (*op. cit.*, note 2). In summarizing the 1730 manuscript ([*op. cit.*, note 19], pp. 22-3, sect. 29), Harrison tried to emphasize how his regulators and his sea clocks would be both practicable and useful for finding longitude at sea. While his regulators could be "fix'd at Sundry Ports in the World" to serve as "good Standards to Set the Sea Clocks by," sea clocks of the design he proposed could be relied upon to carry the time between each Port: "such little variation cannot deceive the Sea Men much in the Time they sail to a far Port, or to where there is another fix'd Clock." Thirty-five years later, however, when Harrison's fourth timekeeper had actually qualified for the £20,000 prize, the problems of reproducing it in quantity stood in the way of it being officially recognized as being both "Practicable and Useful."

35. This is the date given in "Memoirs of the late Mr. Harrison," published in *The Universal Magazine of Knowledge and Pleasure* (London), vol. 63 (November 1778), pp. 266-7. This article provides a fairly detailed and accurate account of Harrison's life and appears to have been written by or with the assistance of someone who knew him well.

36. Harrison, *A Description Concerning Such Mechanism* (*op. cit.*, note 33), p. 20. A very readable account of this meeting is given in chapter 9 ("The Man Who Stayed to Dinner") of David S. Landes's *Revolution in Time* (Cambridge: Harvard University Press, 1983).

37. "An Account of the Life of the late Mr. Harrison" (under "Characters"), *The Annual Register* (London, 1777), p. 25. No doubt

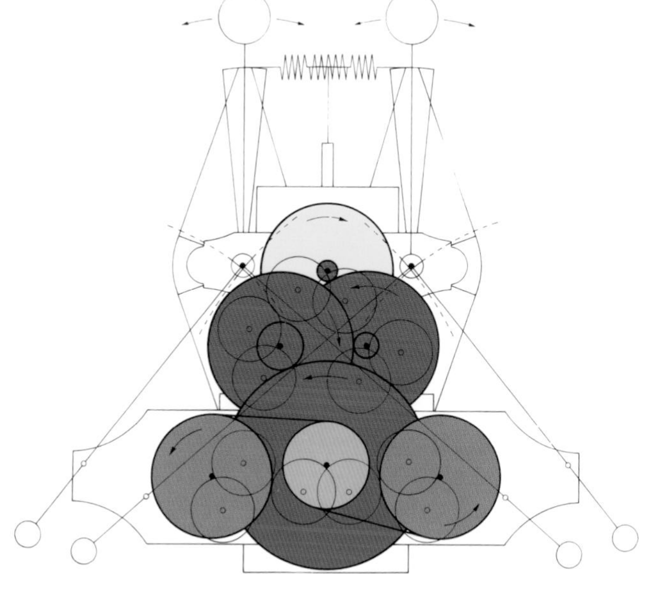

**Figure 5. General view of Harrison's first "sea clock," H.1. Overall height: 63 cm. (24.8 in.). Courtesy of the National Maritime Museum, Greenwich. Inv. no. Ch.35.**

**Figure 6. Layout of H.1; the gray-colored parts represent the two barrels, which are connected to the fusee (marked in dark yellow) by the fusee chains. The wooden wheels are shaded in brown and the brass escape wheel in light yellow. Note that the pivots are supported on antifriction wheels. Illustration by David Penney, based on the drawing published in D. W. Fletcher's article, "Restoration of John Harrison's First Marine Timekeeper,"** ***Horological Journal*****, vol. 94, no. 1125 (June 1952), p. 367.**

Figure 7. Harrison's maintaining power. Inside the fusee, a secondary spring is held in tension by the force of the mainspring. When the mainspring is rewound, a click acting on the ratchet wheel to which the secondary spring is attached prevents the secondary spring from recoiling and losing its tension. In this way, the secondary spring continues to provide the power to drive the clock while the mainspring is being wound. Illustration by David Penney, based on the drawing by Rupert Gould in *The Marine Chronometer*, fig. 13, pp. 43-4.

Graham's generosity toward Harrison is all the more remarkable in light of the fact that the technology for precision timekeeping that Graham had developed was in complete opposition to the principles that Harrison recommended, yet he was able to rise above the critical remarks that Harrison leveled at its very roots. No doubt his trait of encouraging others accounts, at least in part, for Graham's enduring influence: after both men had died, the technology for precision timekeeping continued upon the principles that Graham had established, not upon Harrison's which actually produced better results.[40] In addition, although Harrison was one of the few clockmakers to write about his work, he had difficulty communicating his ideas effectively in words—sometimes because words for what he was trying to say did not exist. Therefore, when Harrison died, his technology died with him.

Harrison thus returned to Barrow with the support he needed to continue the construction of his sea clock, a task in which he was aided by his brother James (1704-1766).[41] Completed in 1735, H.1 represents a natural progression from his precision regulators. In place of the weight that had served as the power source in those clocks, he introduced two coil springs housed inside barrels positioned on either side of the fusee[42] and connected to the fusee by two chains (Figure 6). The clock, which runs for about 38 hours, is wound by pulling a cord that winds onto the fusee arbor as the fusee unwinds. This unusual arrangement permitted winding to take place without removing the glazed case that originally protected the mechanism from moisture and the other elements experienced at sea. To prevent the mechanism from being overwound, Harrison designed a pivoted link in the drum to serve as "stopwork."

When a clock is wound, there is a loss of power unless some supplementary driving force is provided. In the regulators, this supplementary driving force, known as "maintaining power," was provided by a weighted lever, acting upon the third wheel of the train. Harrison recognized, however, that this traditional method, known as "bolt and shutter" maintaining power, was unsuitable for a portable timekeeper because the bolt that provided the supplementary power while the clock was being wound could easily become disengaged from the wheel if the mechanism received a sudden jolt. Another disadvantage of this mechanism was that it had to be manually set; hence the introduction of the winding hole shutter as a reminder. Using this device would also have required the outer casing to be removed—an almost impossible task in a rocking ship.

Graham made this suggestion because, at that time, it is doubtful if the Board of Longitude would have considered a timekeeper to be a viable solution to the longitude problem, and it is even more unlikely that the members of the Board would have understood his plans; after all the impractical ideas that had been proposed to solve the longitude problem, a working timekeeper would, Graham recognized, be the only way of enlisting their support for Harrison's remarkable work.

38. It is not entirely clear whether Harrison received funds from Graham for the construction of H.1 or for one of his later timekeepers. The information about the financial support that Harrison received is given in a brief passage in *A Description Concerning Such Mechanism* (*op. cit.*, note 33), p. 21. Here he listed some of the support he received over an unspecified period for the construction of his marine timekeepers. In addition to the loan from George Graham, there were "indeed...many others [by whom] I was encouraged": from Charles Stanhope, Esq.—possibly a younger brother of Philip Stanhope (1717-1786)—he received £80 in four installments of £20 each; from the East India Company, £100; from George Graham, Martin Folkes, a Dr. Heberden—probably Dr. William Heberden (1710-1801)—and James Short, 10 guineas each; and from a Lord Barrington—probably William Wildman Barrington (1717-1793), who served on the Admiralty board—5 guineas.

39. *Ibid.*, pp. 21-2.

40. Graham, upholding the reputation established by his master, Thomas Tompion (1639-1713), maintained the highest standards of workmanship, and he expanded the business from clocks and watches to scientific instruments, engaging apprentices and experts in various specialized branches of his work. Several of the individuals whom he employed, such as Thomas Mudge (1715?-1794), John Bird (1709-1776), and John Shelton (fl. 1737-1769), were to become well-known makers in their own right. The technology that Graham developed thereby survived him through those who continued the traditions he had established. In contrast, although Harrison is known to have employed workmen, such as John Jefferys (fl. 1726-1753), for specific purposes, he did not train any apprentices in the complicated and scientifically based technology he developed.

41. The amount that James Harrison contributed toward the design and construction of the first sea clock is not known. The fact that all of the complete surviving wooden regulators bear his, not John's, name on the dial suggests that James was responsible at least for their construction, if not their design. It was John, however, who described their development in the 1730 manuscript (with no mention of any contribution by his younger brother) and who appears to be alone responsible for the production of the marine timekeepers. For further information on this, see Andrew King's paper in this volume; Quill, *John Harrison* (*op. cit.*, note 24), p. 28; and Andrewes, "John Harrison" (*op. cit.*, note 31), p. 14.

42. The fusee was used to equalize the varying force of a spring-driven timekeeper's mainspring. The force of this spring, from maximum strength when fully wound to minimal strength when nearly unwound, is equalized by the changing diameter of the fusee's spiral track, which guides the chain connecting the fusee to the mainspring barrel. The principle of the fusee appears to have originated in Italy in the early fifteenth century and was commonly employed in spring-driven timekeepers in Harrison's day.

In order to overcome these problems and provide the requisite constant source of power that his delicate form of escapement required, Harrison devised an ingenious form of maintaining power that operated entirely automatically. While not the first form of portable maintaining power to be invented,[43] Harrison's maintaining power, shown in Figure 7, was the most successful and was later to become a standard feature of every marine chronometer and high-grade portable timekeeper. Because it is not mentioned in the 1730 manuscript, this device must have been invented between 1730 and 1735, during the construction of the timekeeper. No original drawing of it is known to survive.

While Harrison made the barrels, fusee, and escape wheel of brass, he constructed the other wheels, including the motion work, of wood, with teeth cut in groups of three or four into the segments inset into the rim of the wheel.[44] This is the same method that he had employed in his precision regulators, with the exception that, in H.1, he secured each tooth segment with two brass pins.[45] Also in the same manner as the regulators, Harrison incorporated several features to reduce friction to a minimum and allow the mechanism to run without oil (Figure 8): brass lantern pinions are fitted with rollers of lignum vitae[46] pivoted on brass wire; all the wheels of the train, including the mainspring barrels, are supported at each end by a pair of antifriction wheels, carefully designed and positioned to share the thrust of the wheels and pinions;[47] and the antifriction wheels themselves are fitted with lignum vitae bushes that pivot on fixed brass pins. Combined with these features that eliminate the need for oil, the high-numbered wheel train permits a smooth transmission of power from the mainsprings to the escapement.

An enlarged arrangement of antifriction wheels is also used to support the two bar balances (discussed below) that Harrison used in place of a pen-

**Figure 8. The roller pinion and antifriction wheels, two devices that Harrison used to reduce friction in the gearing. Illustration by David Penney.**

43. Christiaan Huygens had described and illustrated a portable maintaining power mechanism in 1683 (*Oeuvres Complètes* [*op. cit.*, note 12], vol. 18, pp. 621-2). His design, in essence, was the system that became known as "sun and planet" maintaining power, later employed by John Arnold (1736-1799) in some of his chronometers. The small gear wheels used inside the fusee were subject to wear and sometimes caused the mechanism to jam as winding took place. Harrison's maintaining power, elegant in its simplicity, was extremely efficient and reliable in its operation.

44. I have relied upon D. W. Fletcher's article, "Restoration of John Harrison's First Marine Timekeeper" (*op. cit.*, note 27), as a source for some of the technical details of H.1.

45. The use of brass pins to secure the teeth reveals that Harrison was concerned that the humidity at sea might affect the glue and cause the wheel segments to work themselves loose. For the design and construction of the wooden wheels, see Andrew King's paper in this volume and Humphrey Quill, *John Harrison: Copley Medallist and the £20,000 Longitude Prize*, Monograph No. 11 (Ticehurst, England: Antiquarian Horological Society, 1976), pp. 3-5; also Andrewes, "John Harrison" (*op. cit.*, note 31), pp. 28-30, concerning the discovery of this construction.

46. Lignum vitae *(Guaiacum officinale/G. sanctum)* named the "wood of life" on account of its reputation as a medicinal agent—principally in the treatment of syphilis—was introduced into Europe by the Spaniards in 1508. Lignum vitae's properties of hardness, self-lubrication, and resistance to the destructive effects of seawater caused it to be used for bearings and pulley blocks on ships. It is unclear if it was employed in this manner in Harrison's day, but Harrison appears to have discovered these properties on his own: "I find by experience that these Rolls of Wood move so freely as to never need any Oyl." Harrison, 1730 manuscript (*op cit.*, note 19), pp. 2-3, sect. 5. Earlier in this same section, either concealing information or feeling that it would not be of interest, he states, "I need not mention what wood or how I order it." For further information on this wood, see the article on "Guaiacum" in the *Encyclopaedia Britannica*, 11th ed. (Cambridge: Cambridge University Press, 1910-11), vol. 12, pp. 646-7, as well as "Lignum vitae," Hardwood: Timber leaflet 67 (London: Timber Development Association, n.d.).

dulum. These balance bearers, designed to minimize friction in the oscillation of the balances, require the use of only about an inch of their circumferences as a bearing surface.[48] The radial arms have steel pivots riveted into the end of their arbors and are counterbalanced by small weights. Stretching diagonally across the frame, these arms, or "Roll[s] of great radii," as Harrison described them,[49] give the timekeeper its distinctive appearance. Harrison eliminated end-shake (forward or backward movement) in the balance arbors by using axial torsion wires:

> Let them [the balance arbors] have at each end in the Centers of the Pevets a Brass Wire, of a competent thickness & length, fix'd fast, & then stretch'd stark in the same direction with the Axis, by w^ch^ means the Axes cannot shove end-way, and the Wires will twist with an elastick force to the Vibration.[50]

It was through the successful testing and the long-term potential of these various techniques for reducing friction to a minimum and eliminating the need for oil that Harrison made his claim—scarcely believable at the time—that his clocks would operate continuously for over 40 or 50 years without cleaning.[51] This extraordinarily efficient operation continues today: H.1 was last cleaned in 1951, 44 years ago. Jonathan Betts, who currently looks after these timekeepers, shared with me that this is somewhat frustrating, because the clock has performed so well that there has been no excuse to take it apart to study it in detail. Most clocks need to be cleaned and reoiled every five or ten years.

H.1's escapement, also virtually frictionless, is a variant of the "grasshopper" escapement that he had developed in the precision regulators.[52] In order to adapt the escapement to work with the balances, Harrison divided the pallets, mounting one on each balance arbor on opposite sides of the escape wheel; this new design (Figures 9 and 10) also served to prevent derangement in its action that might occur from the violent shocks experienced at sea. Comparing H.1's grasshopper escapement to the design that he used in his regulators, Harrison claimed that "these Pallats will have less Friction than the other."[53]

**Figure 9. View of the entry pallet of H.1's grasshopper escapement. From the model made by Len Salzer between 1981 and 1984. Courtesy of The Time Museum, Rockford, Illinois. Inv. no. 2368.**

The design and successful application of an oscillator suitable for a marine timekeeper evidently caused Harrison much concern. The oscillator would have to operate without being disturbed either by the motion of the ship or by changes in temperature and barometric pressure, and it was clear to him that a pendulum, controlled by gravity, would not be appropriate for this

47. It is not known how Harrison acquired his sophisticated knowledge of mechanics, but about 1713, he made a copy of a lecture delivered by Nicholas Saunderson (1682-1739), the famous blind Lucasian Professor of Mathematics at Cambridge. This copy, which Harrison entitled "Mr. Sanderson's Mechanicks," represented one of the few fragments of knowledge that had survived concerning Harrison's education. It was last heard of in 1921, listed with other Harrison documents in vol. 2 of Henry Sotheran and Co.'s *Bibliotheca Chemico-Mathematica: Catalogue of Works in Many Tongues on Exact and Applied Science* (London: Henry Sotheran and Co., 1921), items 8955-70, pp. 453-4. The buyer of these lots has not been traced, and the present whereabouts of the items is unknown; both Quill (*John Harrison* [*op. cit.*, note 24], p. 233), and Gould (*The Marine Chronometer* [*op. cit.*, note 11], p. 70) searched for them in vain. A valuable contribution to our knowledge of Harrison's education will be made if these missing documents are found.

48. In H.1, there are no stops to prevent the balances from rolling off the balance bearers, so presumably Harrison did not find this to be a problem: the balance bearers have centering springs on their arbors to maintain their position. However, Fletcher, in his article on the restoration of H.1 ([*op. cit.*, note 27], pp. 367-8), relates that, after Gould's reconstruction in the early 1930s, there was a tendency for this problem to occur. He states, however, that the problem disappeared when he replaced the excessively rusty steel balance pivots. Harrison fitted stops to the balance bearers of his second timekeeper, H.2.

49. Harrison, 1730 manuscript (*op. cit.*, note 19), p. 16, sect. 25.

50. *Ibid.*, pp. 15-16, sect. 25.

51. *Ibid.*, p. 2, sect. 4. On page 22 of *Remarks on a Pamphlet* (*op. cit.*, note 3), Harrison states that before H.1's removal from his house to the Greenwich Observatory in 1766, it had been "going more than thirty years."

52. The name "grasshopper" was given to this escapement because the action of the pallets resembles the kicking of the hind legs of that insect. The earliest known use of this descriptive name is in an editorial comment in the *Horological Journal*, vol. 40 (July 1898), p. 152, and it became the standard term for this escapement after Gould adopted it in his book, *The Marine Chronometer* (*op. cit.*, note 11). Harrison had referred to the escapement simply as "the pallats" (1730 manuscript [*op. cit.*, note 19], p. 13, sect. 23) or "my pallats" (*A Description Concerning Such Mechanism* [*op. cit.*, note 33], p. 8). Before the term "grasshopper" was coined, authors referred to it in a variety of ways: Thomas Hatton (fl. *ca.* 1740-1774), in *An Introduction to the Mechanical Part of Clock and Watch Work* (London, 1773), p. 25, called it "the curious pallets." In Abraham Rees's *The Cyclopedia; or Universal Dictionary of Arts, Sciences and Literature* (London, 1819-20), reprinted in *Rees's Clocks Watches and Chronometers* (Rutland, Vermont: Charles E. Tuttle Co., 1970), p. 215, the escapement is called "Harrison's" (and, incidentally, illustrated it incorrectly).

53. Harrison, 1730 manuscript (*op. cit.*, note 19), p. 17, sect. 25.

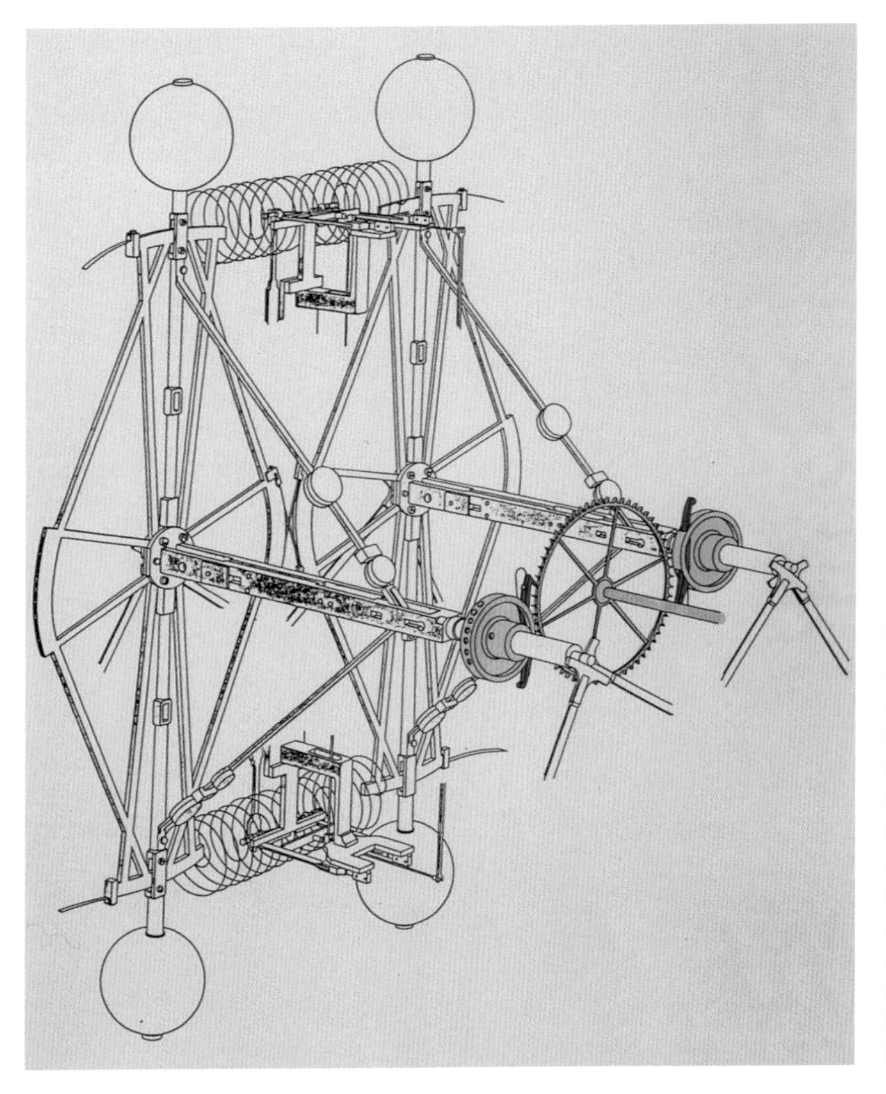

**Figure 10. Illustration of H.1's balances and its grasshopper escapement, the latter identified in color. The pallets, mounted on opposite sides of the escape wheel, engage the wheel alternately; by controlling its rotation, they transmit the power required to maintain the oscillation of the balances. As one pallet engages the wheel, it causes a recoil in the wheel that releases the opposite pallet. The sudden springing action of the tail of the pallet as it is released resembles the kicking of the hind legs of a grasshopper—hence the name. Illustration by David Hirst.**

application. Because this was crucial to the successful operation of the sea clock, he carefully described his solution in the 1730 manuscript.[54] Many of the details, however, were to be worked out by trial and error.

To counteract the rocking of the ship, Harrison devised an oscillator composed of two bar balances of equal length, linked together, with polished brass balls of equal weight at each end (Figure 10). The balances, acting in opposition to each other, are linked together in a frictionless manner using two flat brass wires, similar to a pendulum suspension spring or a watch mainspring. Although Harrison was not the first to use the concept of a double balance,[55] his application was highly original and effective. He describes the action thus:

> ...So consequently if the whole Body of the Clock be turned one way (whether it be slowly, or faster than ever the Ship can turn it,) any portion of the Circle, (whether coinciding with the plain, in which the Ballances vibrate, or inclining thereto) it cannot alter the relative position of the Ballances, but they will still remain as if the Clock has not been moved.[56]

To control the oscillating motion, Harrison employed four helical balance springs,[57] two at the top and two at the bottom of each balance. Each of these was secured in the center after being bent in a reverse curve. The

54. *Ibid.*, pp. 18-21, sect. 27.

55. The idea of employing two balances had been used by the celebrated Swiss clockmaker Jost Bürgi (1552-1632) in his cross-beat escapement and by Christiaan Huygens, whose sketch of 1675 embodies the same arrangement (Huygens, *Oeuvres Complètes* [*op. cit.*, note 43], vol. 18, p. 525). Both Bürgi and Huygens, however, geared the balances together, while Harrison eliminated this unnecessary friction by linking the balances together with steel ribbons. Jonathan Betts drew my attention to the fact that, in theory, because H.1's two balances are pivoted at different points, the ship's motion could influence the motion of each balance differently. In practice, however, the effect would be minimal because the timekeeper would have been placed in the center of a ship and would have been small compared to the size of the hull.

56. Harrison, 1730 manuscript (*op. cit.*, note 19), p. 20, sect. 27.

57. Harrison called these "Worm Springs." *Ibid.*, p. 19, sect. 27.

58. "And here whether greater Vibrations take more or less Time than less ones, they may be reduced to exact equality by part of the Arches g h i,...." *Ibid.*, p. 21, sect. 27.

59. *Ibid.*, p. 20, sect. 27. The phrasing of this whole passage in the future tense suggests that

opposite ends of each spring were attached to a flat piece of spring that acted on the radial arc of the brass frame. In this way, with minimal friction, the springs would be free from distortion, and the expansion and compression of the springs at the top of the clock would be counteracted by the compression and expansion of the springs at the bottom. In addition, he shaped the inner portion of each radial arc to serve as a cycloidal cheek, automatically adjusting the tension of the balance springs in the same way that in his regulators the cheeks altered the effective length of the pendulum (see Figure 4). Thus, regardless of whether the balances performed a large arc or a small arc, the time taken to perform them would be the same.[58] The rate of the timekeeper could be altered by adjusting the length of the wires connecting the balance to the balance springs and, for fine adjustments, by moving the sliding weights on the diagonal struts of the balances. Although the balances were originally intended to perform "2 Vibrations in one Second,"[59] Harrison apparently decided to change this, because H.1's balances oscillate only once every second.

To overcome variations in temperature, Harrison made several attempts to modify the successful temperature-compensated gridiron pendulum that he had developed for his regulators.[60] The gridiron is a framework of brass and steel rods designed so that the expansion or contraction of the brass would be counteracted by the expansion or contraction of the steel. Using this principle, a pendulum could be constructed to automatically maintain a constant length as the temperature changed; alternatively, by adjusting the proportion of brass and steel (as Harrison later did), the gridiron could be made to overcompensate or undercompensate for temperature in order to take into account other variables, such as barometric pressure. The 1730 manuscript reveals that Harrison's original idea for his sea clock appears to have been to use a form of gridiron to support the balance balls:

> Let them [the balls] be communicated to their Axes with Wires, such as in Fig the 5 [a drawing of the gridiron pendulum]. So that they be nearer the Center of Motion when Warmer; but rather somewhat nearer than in the Natural Pendulum; because some part of their support, (or rather a bigger part of it than in a Natural Pendulum) will not by this means be brought nearer. But of their Support in particulars I shall not here enlarge:....[61]

This would appear to be the first proposal to incorporate the temperature compensation into the balance of a timekeeper, and it is a detail that became of great significance in the subsequent evolution of the marine chronometer.[62] Adapting this idea to the large balances used in H.1, however, evidently presented considerable difficulties, because, as Harrison must have realized, the violent shocks experienced on board ship would cause the brass and steel rods to flex. Existing holes and slots suggest that he first attempted to overcome this problem by mounting a gridiron on each balance arbor in such a way that it would move weighted levers toward or away from the balance arbor to counteract the expansion or contraction of the balances and the changing elasticity of their springs. Recognizing, however, that this method was also impracticable, he abandoned the idea of trying to incorporate the compensation in the balance and instead designed the gridiron to act upon the balance springs.[63]

the balances were still in the planning stages in 1730, when the manuscript was written.

60. In his speech to the Royal Society on November 30, 1749, when he presented Harrison with the Copley Medal, Martin Folkes described Harrison's temperature-compensated pendulum as "somewhat resembling a gridiron" (Minutes of the Royal Society [*op. cit.*, note 20], p. 187), and H.1's compensation as being "Gridiron like frames of brass and steel" (*ibid.*, p. 190). In a letter to James Short dated December 12, 1752 (included in MS. 3972/2, pp. 9-10, the Library of the Worshipful Company of Clockmakers, Guildhall Library, London), Harrison states that this term had been commonly adopted. The development of the gridiron pendulum is discussed in Andrew King's and Martin Burgess's papers in this volume.

61. Harrison, 1730 manuscript (*op. cit.*, note 19), pp. 18-19, sect. 27. This passage suggests that, at the time of writing, Harrison had neither considered the problems of using gridirons to support the balance balls nor tried them in practice.

62. One of the most important elements of the marine chronometer in the standard design that was developed by 1790 was that the temperature compensation was contained in the balance itself. Although Harrison never employed this feature successfully—applying instead the temperature-compensation mechanism to the balance springs—he was the first to suggest this arrangement and later admitted that it were best if the temperature compensation "could properly be in the Balance itself." Harrison, *A Description Concerning Such Mechanism* (*op. cit.*, note 33), p. 103. There is some, although by no means firm, evidence that Harrison was conducting experiments along these lines with another marine timekeeper (possibly an H.6), which was under construction at the time of his death. Scant details are given by Herman Bush, a jeweller working in Hull, in a letter to the editor entitled "Re Harrison's Invention of the Marine Chronometer" in the *Horological Journal*, vol. 28, no. 327 (November 1885), p. 45. Bush relates that "an uncompleted chronometer was found amongst his effects, which was composed of brass, bell-metal, tutenage, and specially prepared hard wood, intended to be used for the arms of the balance, to avoid friction, rust, and magnetism, and to impart a peculiar mode of compensation against variation of temperature."

63. These developments were, in fact, later described by the celebrated watchmaker Thomas Mudge (1715?-1794), who stated that Harrison, against the advice of George Graham, carried his original design, incorporating the temperature compensation in the balance, into execution "but found by experience, that Mr. Graham was right, and was forced to throw it all away, and to contrive his method of applying it to the balance springs." Thomas Mudge Jr., *A Description, with plates, of The Time-keeper invented by the late Thomas Mudge...*, (London, 1799), p. 150. Based on the surviving evidence mentioned, D. W. Fletcher (*op. cit.*, note 27), p. 367, illustrated the arrangement that Harrison might have originally tried.

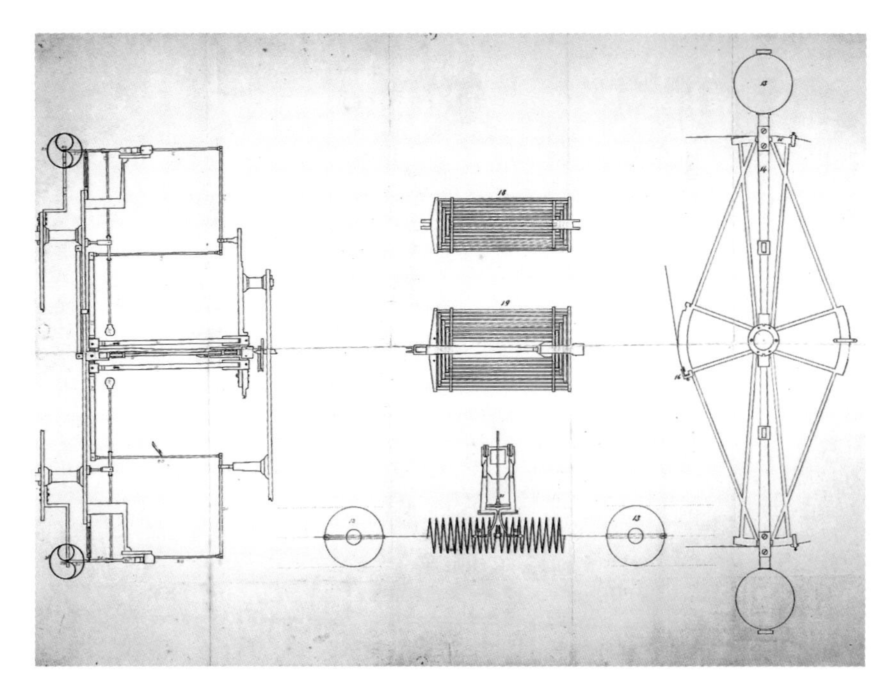

**Figure 11. Pen-and-ink drawing, probably made by Thomas Bradley *ca.* 1840, depicting individual parts of H.1's temperature compensation and one of the balances. On the left is a side cross-section illustrating the positions of the gridirons in the center and the levers above and below that magnify their movement and connect them via the wedges to the balance springs at top and bottom left. In the center are plan views: above, of two of the gridirons and, beneath, of the balances, springs, and one of the wedges through which the levers adjust the stops that alter the effective length of the balance springs. Drawing dimensions: approx. 91 x 61 cm. (36 x 24 in.). By permission of the Syndics of Cambridge University Library and of the Director of the Royal Greenwich Observatory. Ref. no. MS. RGO 6/586 f214ʳ.**

The present temperature-compensation arrangement of three gridirons employs a system of levers, shown in Figures 11 and 12, that advance or retract a wedge on both the upper and lower balance springs. The advance or retraction of the wedges against the tapered surface of the check arms adjusts the position of the check-arm points, which control the effective length of the balance springs. In this way, a rise in temperature, which causes the gridiron lever system to retract the wedges, increases the distance between the check arms so that the effective length of the balance springs is shortened. This counteracts the natural weakening of the springs and expansion of the balances, which would otherwise cause the timekeeper to lose time. Conversely, with a fall in temperature, the wedges are advanced, allowing the check arms to move together and cause the effective length of the spring to be increased. This counteracts the stiffening of the springs and the contraction of the balances, which would otherwise cause the timekeeper to gain time.[64] The lever system magnifies the movement of the gridirons so effectively that even the heat of a person's hand can cause a visible motion of the wedges.[65] Nevertheless, he must have experienced difficulties in fine-tuning this sensitive mechanism, because in his next timekeeper he altered the design to make the gridiron arrangement easier to adjust.

By the time he composed the 1730 manuscript, it is clear that Harrison had a good understanding of the effects of barometric pressure. He had noticed with his regulators how changes in the air's resistance influenced the period of oscillation of a pendulum: the lower the air pressure, the less the resistance. While others had sought to overcome this variation by placing the timekeeper in a vacuum,[66] Harrison countered the effect by changing the amount of compensation, or, as he put it, "ordering and adjusting the wires" of the gridiron.[67] Also recognizing that, due to the influence of barometric pressure and other variables, the amplitude of the balances would not always remain constant, he adjusted the radius of the inner part of the curve on which the springs are attached to the balance arms so that regardless of the amplitude, the oscillation of the balances would be isochronous—in other words, would be performed in the same period of time.

64. A detailed explanation of H.1's temperature compensation is contained in D. W. Fletcher's article, "Temperature Compensation of Harrison's First 'Marine Timekeeper,' " *Horological Journal*, vol. 93, no. 1113 (June 1951), pp. 377-9.

65. This observation was made by Fletcher (*ibid.*, p. 379), who restored this part of the mechanism in 1951. Gould, who had undertaken H.1's reconstruction two decades earlier, had not attempted to put the gridirons into working order because they were in such a rusty state. Gould's and Fletcher's restoration and reconstruction work are discussed in greater detail in the postscript at the end of this paper.

66. Probably the earliest mention of this was in 1668 by Antimo Tempera in his book *L'Oriuolo Giusto D'Antimo Tempera Utilissimo a'Naviganti* [The Accurate Clock of Antimo Tempera, Most Useful to Navigators] (Rome, 1668). Silvio Bedini informed me (personal correspondence dated November 4, 1994) that Antimo Tempera was a pseudonym used by Matteo Campani degli Alimeni, the eldest of the three Campani brothers, who were makers of clocks and scientific instruments. Further information on early Italian attempts to produce a longitude timekeeper will be available in Bedini's forthcoming book, *Uncommon Genius. The Campani Brothers of Rome*. In 1704, experiments that involved placing a clock in an evacuated chamber were conducted by William Derham, whose paper, "Experiments about the Motion of Pendulums in Vacuo," was delivered to the Royal Society in December of that year and published in *Philosophical Transactions* of the Royal Society, London *(Phil. Trans.)*, vol. 24, no. 294 (1704), pp. 1785-9. As discussed earlier, in 1714, Jeremy Thacker used an evacuated jar to protect his chronometer from the effects of humidity (*The Longitudes Examin'd* [*op. cit.*, note 13]).

67. Harrison, 1730 manuscript (*op. cit.*, note 19), pp. 20-1, sect. 27.

In summary, Harrison developed a technology through which he was able to control and adjust—and thereby predict—the influence that variations in temperature and barometric pressure would have upon the going of his timekeeper. Then, by making fine adjustments to the circular error of the balances—the difference between their period of oscillation in a large arc or a small arc—he was able to negate these environmental effects. Thus, Harrison's timekeepers were able to maintain a remarkably constant rate under changing atmospheric conditions, and, because they were designed to run without oil, their accuracy could be relied upon for decades. By any standards, this was an extraordinary achievement.

To bring H.1 to time, adjustments are made to the length of the wires connecting the balances to the balance springs: if the clock is slow, the wires are slackened, allowing the inner turns of the spring to be stationary for a longer period of time. Small adjustments to the length of the balance spring cause large variations in the timekeeping: an alteration of half a millimeter makes a difference of several minutes a day. No doubt this is the reason that Harrison also chose the balance springs as the part to which his temperature compensation should be applied. Finer adjustments for timekeeping are made by moving the sliding weights on the diagonal struts supporting the balances.[68]

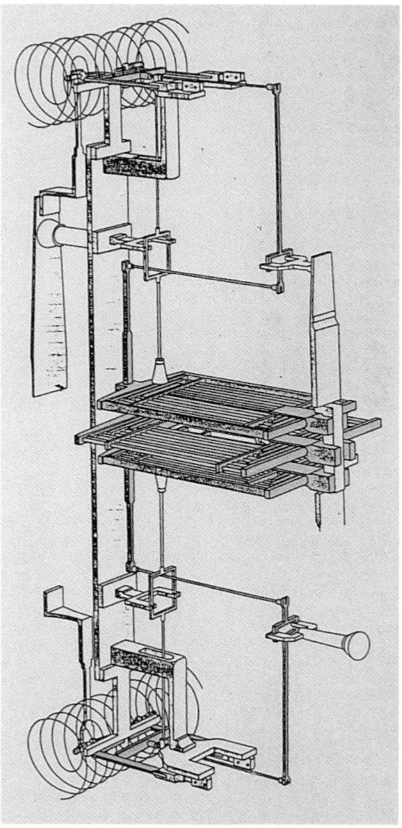

Figure 12. Illustration showing the assembly of H.1's temperature compensation: the three gridirons, the lever system that magnifies their expansion and contraction, and the wedge assemblies that transmit this movement to alter the effective length of the balance springs. Illustration by David Hirst.

The dial that Harrison designed for his first sea clock gave indications of seconds, minutes, hours, and days, the latter being manually adjusted for short months. The indication of the day evidently proved unnecessary in practice, probably because this information was recorded in the ship's log, and this feature was therefore omitted on his later marine timekeepers. Although the dial is the part of a clock that normally carries the maker's signature, this is not the case with H.1, which in fact has no signature. Since it is unlikely that a piece of such importance to its maker would not have been signed, it is possible that the signature appeared instead on the outer casing, which has since been lost. His two subsequent timekeepers, H.2 and H.3, were both signed on their dials.

H.1 was originally mounted in a glazed case to keep out moisture.[69] The case was slung in gimbals similar to those used in a mariners' compass, "so that it alter it's *[sic]* position but very little, tho' the Ship Toss much, nor receive any great Shocks from the Waves."[70] The machine itself weighs 75

68. These details are taken from Fletcher, "Restoration of John Harrison's First Marine Timekeeper" (*op. cit.*, note 27), p. 369. In June 1952, after completing the restoration, Fletcher obtained a rate of eight seconds per day with a variation during the 24 hours of eight seconds. He believed that Harrison could have achieved a daily rate that would not exceed three seconds per day; if this estimate is correct, H.1 would have qualified for the £15,000 prize.

69. Although the case no longer survives, it appears that it was still in existence at the beginning of this century. F. J. Britten's *Old Clocks and Watches & their Makers,* 3rd ed. (London, 1911), p. 327, refers to H.1 as "a cumbersome affair in a wooden frame," suggesting that, at the time of writing, the case, and perhaps also the gimbals, still existed. Gould described his first view of the timekeeper in the spring of 1920 thus: "Covered with a glazed penthouse resembling a cucumber-frame, it was standing loose upon a rough wooden baseboard." Rupert T. Gould, "The Reconstruction of Harrison's First Timekeeper," *The Observatory, A Monthly Review of Astronomy*, vol. 56, no. 709 (June 1933), p. 193.

70. Harrison, 1730 manuscript (*op. cit.*, note 19), p. 21, sect. 28. Another description of the gimbals was provided by Martin Folkes, who spoke of H.1 as "being placed in a sort of moveable frame, somewhat resembling what the Sailors call a Compass Jimber, but much more artificially and curiously made and disposed." Minutes of the Royal Society (*op. cit.*, note 20), p. 191.

**Figure 13. View of H.1 from above. Courtesy of the National Maritime Museum, Greenwich.**

lb. (34 kg.) and stands 24.8 in. (63 cm.) tall, on an area 27.5 in. (70 cm.) wide by 17.7 in. (45 cm.) deep. With its case and gimbals, it must have occupied a considerable amount of space, particularly in the confines of a small cabin.

After testing H.1 on a barge on the river Humber,[71] Harrison was satisfied with its performance and brought it to London to request an official trial.[72] Impressed with its potential, George Graham, Edmond Halley, and three other distinguished members of the Royal Society issued a certificate to the Admiralty in 1735 strongly recommending a trial at sea.[73] Harrison had begun to earn himself a good reputation: his clock was recognized as being the best that had ever been made for measuring time at sea, and he was considered to be "a very sober, a very industrious, and withal, a very modest Man" who was "capable of finding out something more than he has already, if he can find Encouragement."[74]

Why H.1 was not tried on a voyage to the West Indies to compete for the £20,000 prize is not known, but perhaps a shorter voyage to a well-known destination was deemed preferable as a preliminary test. The destination chosen was Lisbon, a normal port of call for Royal Navy ships en route to and from the Mediterranean, probably because its longitude had been established only a decade earlier, in 1726, by observations of the eclipses of Jupiter's satellites by James Bradley (1693-1762) and others.[75] The chief assignment given to Harrison was to measure the difference in time between

71. This test was mentioned by Martin Folkes in his speech to the Royal Society on November 30, 1749 (*ibid.*, p. 191). The river Humber was only a mile or two from Harrison's home, and on its opposite bank was Kingston upon Hull, one of the principal seaports in England at the time.

72. Whether H.1 was transported by land or by sea is not known. In his journal written between 1761 and 1766, Harrison stated that he generally moved H.1 in pieces, suggesting transportation by land. Harrison's "Journal," pp. 113-14, in memorandum by William Frodsham and Walter Williams.

73. This document (*op. cit.*, note 32), although not dated, is believed to have been drafted in 1735. In addition to George Graham and Edmond Halley, it was signed by James Bradley (1693-1762), Savilian Professor of Astronomy at Oxford, who became Astronomer Royal on Halley's death in 1742; John Machin (1680-1751), Gresham Professor of Astronomy and Secretary of the Royal Society; and Robert Smith (1689-1768), Plumian Professor of Astronomy at Trinity College, Cambridge, whom Harrison was later to accuse of pirating his scale of music (see note 122).

74. Harrison was so described in two letters dated May 14 and May 17, 1736, between Sir Charles Wager (fl. 1708-1737), the first Lord of the Admiralty, and George Proctor (d. 1736), the captain of the *Centurion*. These are published as the second and third appendices of *An Account of the Proceedings* (*op. cit.*, note 20), p. 18.

75. In his paper to the Royal Society, "The Longitude of Lisbon, and the Fort of New York, from Wansted and London, determin'd by Eclipses of the First Satellite of Jupiter" (*Phil. Trans.* vol. 34, no. 394 [1726], pp. 85-90), James Bradley calculated that the longitude of Lisbon was 9° 7' 30" west of London. His calculations were based on a comparison of the observations made by his uncle, James Pound (1669-1724), at his private observatory at Wanstead (about ten miles north of London) with those made by Father Jean-Baptiste Carboni (1694-1750) at the Jesuit's College in Lisbon. The latitude of Lisbon, also calculated by Father Carboni, was stated as 38° 42' 30" ("Observationes Astronomicae habitae Ulyssipone, Anno 1725, & sub init. 1726," *Phil. Trans.*, vol. 34, no. 394 [1726], p. 99).

London and Lisbon, no doubt so that the results could be compared to the established calculations.[76] After some delay, Harrison was instructed to take his timekeeper aboard the *Centurion*, a warship of 60 guns that lay anchored at Spithead,[77] in preparation for its voyage to join the main fleet at Lisbon. The clock was duly placed in the captain's cabin, and the *Centurion* set sail on May 19, 1736.

Although the weather was bad, the voyage was quick and they reached Lisbon in a week. The rough weather made Harrison seasick but did not affect the performance of the clock. No record of a comparison of the time at Lisbon has survived, but we know from a later report that the clock was set to Lisbon time before departure for the return voyage.[78] By June 3, Harrison and his timekeeper were aboard the *Orford* on their way back to England. Amid alternating periods of gales and calm weather, this voyage took one month. During the third week, as land was sighted at the entrance to the English Channel, not too far from where Sir Clowdisley Shovell (*ca.* 1650-1707) had been shipwrecked 29 years earlier,[79] there was a disagreement as to the ship's position: the Master of the *Orford* and other members of his crew claimed it to be Start Point, while Harrison, basing his reckoning on his timekeeper's performance, said it was the Lizard, a point about 68 miles, or 1.5°, to the west. Harrison turned out to be right. His timekeeper was proved to be a practical invention.

Harrison soon became well known for his remarkable machine. Many curious visitors, including nobility, came to see H.1—and later his subsequent timekeepers—and he showed them how the rocking of the machine on its gimbals would not affect the motion of the balances. After attending one of these demonstrations, John Gainsborough (1711-1785), the elder brother of the artist Thomas Gainsborough (1727-1788), remarked, "Harrison made no account of me in my shabby coat, for he had Lords and Dukes with him, but after he had shewn the Lords that a great motion to the machine, would no ways effect its regularity, I whispered him to give it a gentle motion." Surprised by this comment, Harrison apparently "whispered him to stay, as he wanted to speak to him after the rest of the company were gone."[80] The antiquarian Reverend William Stukeley (1687-1765) commented in 1740 during his travels in Lincolnshire that

> I passed by Mr. Harrison's house at Barrow, that extraordinary genius at clock-making, who bids fair for the golden prize due to the discovery of the longitude. I saw his clock last winter at Mr. Geo. Graham's. The sweetness of its motion, the contrivances to take off friction, to defeat the lengthening and shortning of the pendulum through heat and cold, and to prevent the disturbance of motion by that of the ship, cannot be sufficiently admired.[81]

The English artist William Hogarth (1697-1764) described Harrison's third marine timekeeper, H.3, as "one of the most exquisite movements ever made," but, finding the appearance of its parts to be displeasingly shaped and recognizing that this was of no consequence to its maker, he qualified this remark by saying that it did not begin to compare with nature's machines, where "beauty and use go hand in hand."[82] Another notable visitor to Harrison's house was Benjamin Franklin (1706-1790), who visited on December 1, 1757. Franklin's account book indicates that he paid Harrison

76. This was related by Capt. Proctor and his chief officers in a document (MS. Adm.1/379, Public Record Office, London) that was discovered and written about by Daniel Baugh in his article "The Sea-trial of John Harrison's Chronometer, 1736," *The Mariner's Mirror,* vol. 64, no. 3 (August 1978), pp. 235-40.

77. Spithead is the strait between Portsmouth and the Isle of Wight.

78. In his speech to the Royal Society in 1749 (Minutes of the Royal Society [*op. cit.*, note 20], p. 191), Martin Folkes noted that "at the entrance of the channel, it [H.1] gave to good exactness the distance of the present meridian from that of Lisbon [underlined in the manuscript]." To provide this measurement, the clock would have had to be set to Lisbon time. That this was done is not surprising, because, due to its size, H.1 would have been extremely difficult to carry off one vessel and onto another without stopping the balances, thereby requiring it to be reset to local time. As mentioned above (see note 72), Harrison also stated that H.1 was usually moved in pieces.

79. This disaster took the lives of 800 men, including that of the admiral. Although possibly not directly related to an ignorance of longitude, this event emphasized the importance of solving the longitude problem and served to stimulate the British government's decision to offer a large reward for a solution.

80. This visit by Thomas Gainsborough's rather eccentric older brother John, a watchmaker aspiring to solve the longitude problem, is described by Philip Thicknesse in *A Sketch of the Life and Paintings of Thomas Gainsborough, Esq.* (London, 1788), pp. 59-60; further information is given by Leonard F. Miller in his article "The inventive Gainsborough brothers," *Horological Journal,* vol. 119, no. 11 (May 1977), pp. 5-8. Because this reference relates to a later period in John Gainsborough's life, probably about 1760 but certainly after 1745 when he was married, the machine in question could have been any of the first three machines.

81. Diary entry for the year 1740, *The Family Memoirs of the Rev. William Stukeley, M.D. and the Antiquarian and other Correspondence of William Stukeley, Roger & Samuel Gale, Etc.,* Surtees Society, publication no. 76, vol. 2 (Durham, England, 1883), p. 298. It is not known if the clock he saw was H.1 or H.2, although the latter—having just been completed—would seem most likely.

82. William Hogarth, *The Analysis of Beauty* (London, 1753), pp. 70-1.

10s. 6d. "to see his Longitude Clock."[83] It is not known whether this was a charge or a donation.

Although most of Harrison's visitors came out of idle curiosity, there were some who sought to learn the secrets of his technology. However, it is one thing to be shown something, but quite another to comprehend it sufficiently to duplicate it. Upon his return to Paris after seeing Harrison's clocks, Jean-Baptiste Le Roy (1719-1800), the second son of the famous French clock-maker Julien Le Roy (1686-1759), described some of their ingenious parts to his older brother, Pierre. Pierre Le Roy (1717-1785), who was to become one of France's most celebrated horologists for his contributions toward the development of the marine chronometer, admitted in a somewhat guarded fashion that he had been influenced by the information he had received:

> I agree that the suspension of the balance, called foliot, by two threads which twist themselves can have, just as was pointed out to me, some similarity to mine. I acknowledge even that when I conceived the idea of suspending my regulator thus, I had knowledge of Mr. Harrison's [balance, which was] controlled by metal threads that twist with each vibration—My brother, who had seen it in London, described it to me. But I hope that in comparing this same part of my work with that which could have been executed in this manner, one will find it new in many respects....[84]

On May 9, 1763, Pierre Le Roy's great rival, Ferdinand Berthoud (1727-1807), accompanied two well-known French *savants*, the astronomer Jérôme de Lalande (1732-1807) and the mathematician Charles Camus (1699-1768), on a visit to Harrison's house in the hope of learning the secrets of H.4, Harrison's prize-winning watch, the accuracy of which was difficult for anyone to believe at that time. Lalande recorded in his diary that Berthoud became even more anxious to see it after examining the three large machines, H.1, H.2, and H.3, which he found "très belles très ingenieuses, très bien éxécutées," but there is no mention that he actually saw H.4. Alluding no doubt to financial considerations, Harrison informed the three Frenchmen that the Spanish ambassador was prepared to pay him "2000 pieces" for the regulator he made in 1726.[85]

On June 30, 1737, a year after H.1's sea trial, the Board of Longitude met, as far as we can tell, for the first time since being established by the Act of 1714.[86] The subject of the meeting, however, was not about any further trial for H.1 to qualify for the £20,000 prize, but a request by its maker for funding to construct a second machine, smaller and improved, which he stated could be completed within two years. The requested sum of £500 was granted, half to be paid immediately and half to be paid when the clock had been delivered on board a ship bound for the West Indies.[87] The minutes of the meeting also stated that, when the ship returned from the West Indies, both H.1 and H.2 would remain in the Board of Longitude's possession "for the Use of the Public," a stipulation that would later cause Harrison much concern. Harrison appears to have viewed the Board's financial encouragement as a form of royal commission, because, upon completion, his "Second Machine," as he called it, was engraved with the following inscription: "*Made For His Majesty* GEORGE THE II$^{ND}$ *By order of a Committee Held the 30$^{th}$ of June 1737.*"

83. [Benjamin Franklin], "Account Book of Benjamin Franklin kept by him during his First Mission to England as Provincial Agent 1757-1762," *The Pennsylvania Magazine of History and Biography*, vol. 55, no. 2 (1931), p. 106. As a member of the Royal Society, Franklin later showed his support for John Harrison by nominating his son William, on February 7, 1765, for election as a Fellow of the Royal Society.

84. Translated from a manuscript in Pierre Le Roy's hand dated December 7, 1763, entitled "Memoire Sur une nouvelle Montre Marine." In the original French, the passage reads exactly as follows: "Je conviens que la suspension du balancier appellé foliot par deux fils qui se tordent peut avoir ainsi qu'on me la fait remarquer quelque rapport avec la mienne. j'avouerai même que quand je conçus l'idée de suspendre ainsi mon regulateur j'avoit connaissance de celui de Mr. Harisson contenu par des fils de metal qui se tordaient dans chaque vibration—Mon frere qui l'avoit vu a Londres m'en aiant fait la description. mais j'espere qu'en comparant cette partie même de mon ouvrage avec ce qui peut avoir été executé dans ce genre on la trouvera nouvelle a beaucoup d'égards...." MS. 104, p. 72, Archives de l'Académie des Sciences de l'Institut de France, Paris. I am most grateful to Giuseppe Brusa for bringing the above reference to my attention. For information on Le Roy's work, see Catherine Cardinal's paper in this volume. Rupert Gould, in the *The Marine Chronometer* (*op. cit.*, note 11), p. 45, footnote §, and also on p. 91, states that Pierre Le Roy saw H.1 in London in 1738 and remarked that it was "d'un construction fort ingenieuse" [spelling as quoted]. Gould, however, does not give a reference, and I have been unable to find any evidence that Pierre Le Roy ever went to London.

85. *Jérôme Lalande: Journal* (*op. cit.*, note 28), p. 62. I have transcribed both quotes, with all their imperfect punctuation, exactly as written on page 81 of Lalande's original manuscript. The number "84" at the top of the following page of the manuscript is probably a mistake in pagination, although it suggests that pages 82 and 83 contained more information about their visit. While there is no record of Berthoud seeing H.4 on that occasion, it would seem from the outward appearance of his *Montre Marine* No. 3—made upon his return to Paris (see Anthony Randall's paper, Figure 14)—either that he saw the watch in its case, or that Lalande, who had seen it three weeks earlier on April 22, gave his colleague a careful description.

86. As previously noted, the fact that this was their first meeting indicates that Harrison was the first person to be seriously considered as a contender for the prize.

87. Board of Longitude Confirmed Minutes, June 30, 1737, MS. RGO 14/5, vol. 5(2), Royal Greenwich Observatory Archives, Cambridge University Library. Three of the eight commissioners who, by virtue of their academic appointments, attended this meeting—Edmond Halley, James Bradley, and Robert Smith—were among the members of the Royal Society who had signed the document (*op. cit.*, note 32) recommending the trial of H.1. The minutes of this meeting are reproduced in Quill, *John Harrison* (*op. cit.*, note 24), Appendix 2, pp. 228-9. H.2, it should be noted, was not tried at sea, so Harrison never received the £250 balance. I

The trial to Lisbon must have revealed defects in H.1 that Harrison believed could be remedied only by the construction of a second machine. The most obvious problem was its size. This had not been a matter of consideration when he built H.1: the wooden wheels, the balances, and the balance bearers, all of which were relatively large, had dictated the dimensions. In its case and gimbals, H.1 must have occupied an area of approximately four feet by four feet and therefore cannot have been a welcome addition to the limited confines of the captain's cabin.[88] Thus, H.2, his second machine, was made narrower and taller. Without its case, it occupies less than half the space, measuring 18.5 in. (47 cm.) wide, 12.6 in. (32 cm.) deep, and 26 in. (66 cm.) tall. Its weight of 86 lb. (39 kg.),[89] compared to H.1's 75 lb. (34 kg.), is due largely to the fact that H.2's plates are made of a heavier gauge brass and that the wheels are made of brass instead of wood. Its case and gimbal suspension, both of which have been lost, reportedly weighed 62 lb. (28 kg.).

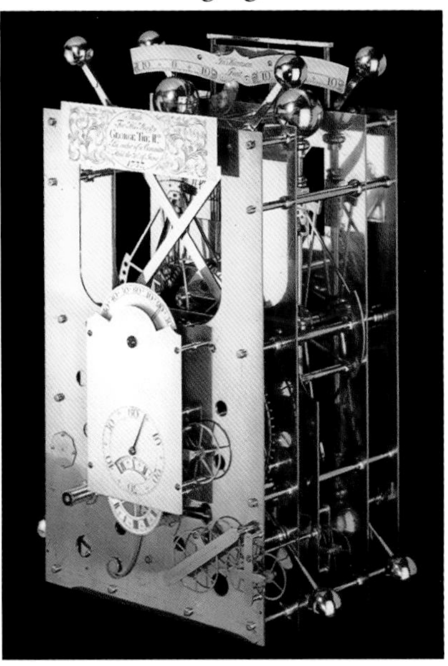

**Figure 14. Harrison's second marine timekeeper, H.2. Overall height: 66 cm. (26 in.). Courtesy of the National Maritime Museum, Greenwich. Inv. no. Ch.36.**

Mechanically, H.2 is a refined version of its predecessor. Like H.1, it has two mainsprings, but instead of being positioned on opposite sides of the fusee, they are mounted in tandem on the same arbor, each with its own set-up, ratchet, and click—presumably to provide finer adjustment of the driving torque.[90] H.2's fusee is wound with a key, rather than a cord as in H.1, and is fitted with a sophisticated geared form of stopwork to prevent the mechanism from being overwound.[91] In addition, the maintaining power is positioned not on the fusee, as in H.1, but on the second arbor of the train, where it is more accessible and allows the use of a finer spring to drive the clock during winding.

The use of this secondary spring to keep the clock running during rewinding may have influenced Harrison to develop a mechanism for H.2 that would provide a more direct drive to the escapement and thereby overcome fluctuations in the power transmitted through the wheel train. This device, called a "remontoire,"[92] is automatically rewound by the mainspring at regular intervals (Figure 15). Harrison was not the first to use such a mechanism,[93] but his design was original and ingenious: he attached the remontoire springs to "cheeks" of changing radius, which equalized the force of the remontoire springs as they relaxed. H.2's remontoire was mounted on the third arbor and rewound every $3\frac{3}{4}$ minutes, sixteen times every hour. Eliminating any fluctuation in the power delivered to the escapement, and thereby to the balances, greatly assisted the final adjustments to compensate for the errors caused by changes in temperature and barometric pressure.

The escapement and the balance bars are similar to those in H.1, except that the escape wheel is larger in diameter and the pallets are longer. Stops have been added to the bearing surfaces for the balance arbors to prevent any

am most grateful to Adam Perkins of the Royal Greenwich Observatory Archives for his help in giving me access to the Board of Longitude papers.

88. In his speech to the Royal Society in 1749, Martin Folkes related that the Board of Longitude was pleased to encourage Harrison "to make for the publick a second clock, somewhat less if possible than the first, as that had been thought, through the room that it took up somewhat cumbersome on board." He then proceeded to say that H.2 hardly took up "so much as half the space of the former." Minutes of the Royal Society (*op. cit.*, note 20), p. 192.

89. Gould, *The Marine Chronometer* (*op. cit.*, note 11), p. 47 states that H.2 weighed 103 lb., a discrepancy that might be accounted for if he had weighed the timekeeper while it was seated on its base, to which part of the gimbals were attached. The timekeeper is illustrated in this manner in plate 8 (opposite p. 45). During the intervening years, both the case and gimbal suspension have disappeared and are presumed lost. Some details of the particulars of this and the other timekeepers may be found in the Harrison tercentenary booklet by Jonathan Betts, *John Harrison* (Greenwich, England: National Maritime Museum, 1993).

90. My description of the mechanisms of H.2 and H.3 have been aided by Anthony Randall's excellent two-part article, "The Technology of John Harrison's Portable Timekeepers," *Antiquarian Horology,* vol. 18, no. 2 (Summer 1989), pp. 145-60, and vol. 18, no. 3 (Autumn 1989), pp. 261-77.

91. The design of H.2's stopwork is highly efficient. It has been used by Peter Hastings of the Royal Observatory in Edinburgh for a mechanism inside an infrared spectrograph, presently under construction. In case of a malfunction in the switches or the software that turn off the drive motor, the stopwork will automatically lock the mechanism after 55 turns and thereby protect the instrument's delicate optical components. In his letter dated June 14, 1995, in which he described this to me, Hastings mentioned that no other form of stopwork has this kind of range while remaining compact.

92. *"Remontoir d'égalité,"* from which the English word "remontoire" is derived, is the traditional French name for this device—*"remonter"* means to rewind, and hence "remontoir" is used as a generic term in French for winding mechanisms. The use of the term in England does not appear to have been adopted until the second half of the eighteenth century. Harrison's remontoire was described by Martin Folkes in his address to the Royal Society in 1749 without a specific name (Minutes of the Royal Society [*op. cit.*, note 20], pp. 192-3), and, in a holograph manuscript dated April 7, 1763, Harrison referred to it as "a larger controller or corrector of the force" (John Harrison, "An Explanation of my Watch or Timekeeper for the Longitude...," MS. 3972/1, pp. 1-102, Library of the Worshipful Company of Clockmakers, Guildhall Library, London). But by 1773, Thomas Hatton related that "they call it a train by remantwaur, a far fetched French word, which, from its etymology, means a rewinding or doing the same thing over again" (*Introduction to... Clock and Watch Work* [*op. cit.*, note 52], p. 150).

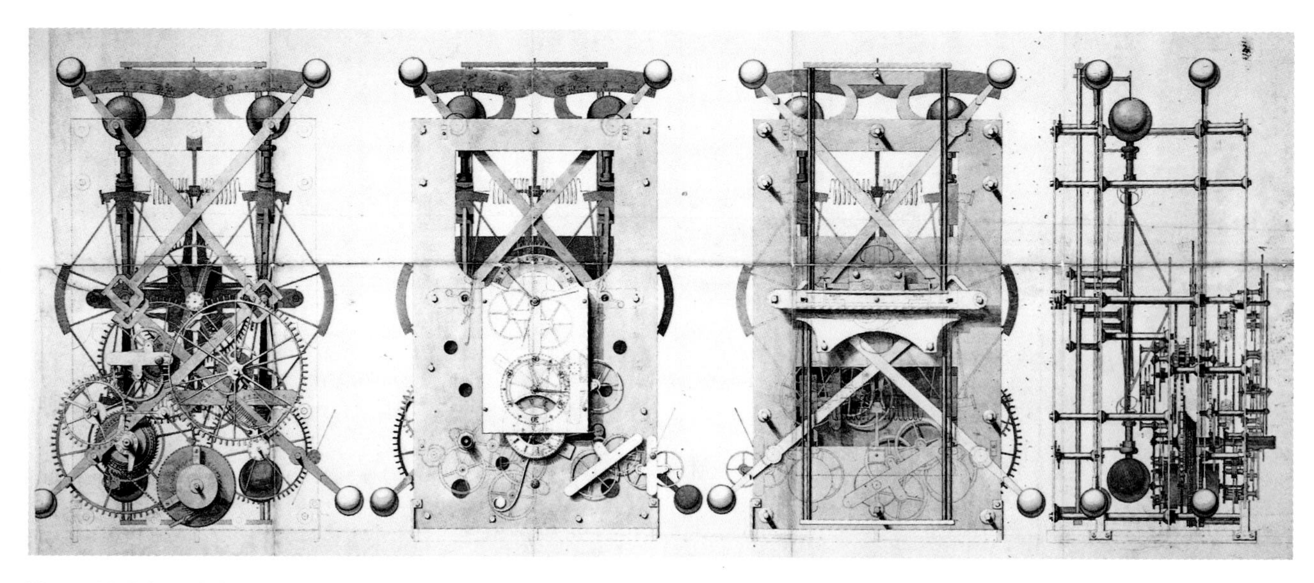

**Figure 15. Ink-wash drawing showing four views of the mechanism of H.2, signed by Thomas Bradley and dated 1840. The ingenious constant-force remontoire that Harrison devised for this timekeeper is shown in the cross-section of the mechanism on the left: mounted on the third arbor, it provides a direct drive to the escape wheel pinion. Drawn on paper approx. 91 x 61 cm. (36 x 24 in.). By permission of the Syndics of Cambridge University Library and of the Director of the Royal Greenwich Observatory. Ref. no. MS. RGO 6/586 f215$^r$.**

possibility of the balance pivots rolling off. Harrison also changed the design of the temperature compensation so that it would be more rigid and easier to adjust without stopping the clock. He accomplished this by mounting it on the back plate, employing two thick steel rods running vertically down the length of the movement (Figure 16). Four thinner brass rods were mounted on either side of the steel rods, two on the inside at the top and two on the outside at the bottom. These smaller rods operate a lever system that adjusts the tension and free motion of the balance springs to compensate for the effects of temperature change on the length of the balances and the elasticity of the balance springs.

It is tempting to imagine that Harrison would have designed the temperature compensation so that the brass and steel rods would react in the same period of time. Indeed, he understood that brass has a higher conductivity and a lower specific heat than steel, and he explained that, in order to make the two metals react at the same time, the brass rods would have to be thicker than the steel.[94] So what was he trying to achieve by making the brass rods thinner than the steel, an arrangement that would result in gross overcompensation until the steel had completed its reaction to the temperature change? This large transient temperature error appears to be an ingenious method of compensating for variations in the elasticity of the steel remontoire springs, which, because of their fineness, respond almost immediately to changes in temperature. A rise in temperature, for example, would weaken the springs and thereby cause a reduction in the power delivered through the escapement to the balances; this would cause the balances to move more slowly, and thus the clock would lose time. Therefore, by introducing transient temperature error in the design of the gridiron, Harrison was able to overcome the error introduced by changes in the elasticity of the remontoire springs.[95]

H.2 was completed about 1739,[96] and during the next two years its rate was compared to that of H.1. The machine underwent tests of heating and cooling in addition to being, as stated in the Royal Society report of 1741/2, "agitated for many Hours together, with greater Violence than what it could

93. The invention of the train remontoire—as opposed to the escapement remontoire, a later device in which the remontoire is incorporated in the escapement itself—is ascribed to Jost Bürgi (see Klaus Maurice, "Jost Bürgi, or on Innovation," in Klaus Maurice and Otto Mayr, eds., *The Clockwork Universe* [New York: Neale Watson Academic Publications, 1980], pp. 93-4). Christiaan Huygens described his independent invention of this device in a letter dated August 29, 1664, to Robert Moray (1608?-1673), the first president of the Royal Society: "le contrepoids qui fait aller la roue de rencontre est pendu sur la roue mesme et est remontè *[sic]* chaque demie minute par la force du grand contrepoids [the counterweight (the weight-driven form of maintaining power, known as "endless rope," invented by Huygens about 1658) which drives the escape wheel is mounted on the same wheel and is rewound every half minute by the force of the large counterweight (the clock's main driving weight)]" (Huygens, *Oeuvres Complètes* [*op. cit.*, note 43], vol. 5, no. 1253, p. 108). Being weight-driven, Huygens's remontoire was never used successfully in his marine timekeepers, but the idea was adopted by several French makers, including Alexandre Le Bon (fl. 1714-1750) and Pierre Gaudron (fl. 1690-1730). It is not known if Harrison had any prior knowledge of these developments, but the design of his constant-force spring-driven remontoire was, in any case, entirely original. Its significance, however, was not understood by many of his contemporaries: while Hatton came to appreciate it, Alexander Cumming (1733-1814) "even says, it is useless"! (Hatton, *Introduction to...Clock and Watch Work* [*op. cit.*, note 52], p. 150).

94. Harrison, "An Explanation of my Watch" (*op. cit.*, note 92). This manuscript has been transcribed by Andrew King, who kindly gave me a copy. I have been unable to check the location of this passage in the original manuscript. As a guide for those interested, this reference occurs about two-thirds of the way through the document.

**Figure 16. Back view of H.2 with the balances in motion, showing the brass and steel temperature-compensation rods just inside the back plate. These rods and the mechanism through which they adjust the motion of the balances can be seen in the second view from the right in the drawing by Thomas Bradley (Figure 15). Courtesy of the National Maritime Museum, Greenwich.**

receive from the Motion of a Ship in a Storm. And the Result of these Experiments, is this; that *(as far as can be determined without making a Voyage to Sea)* the motion is sufficiently regular and exact, for finding the Longitude of a Ship within the *nearest Limits proposed by Parliament* and *probably* much *nearer*."[97] Yet despite this recommendation, H.2 was never tried at sea. The reason given in 1740 by George Graham was the "Hazard of Capture...(a War being then open with Spain)." Halley suggested at the same meeting that, to avoid the perils of a voyage, the timekeeper should be tried on the Sussex Downs, where conditions experienced at sea could be simulated and its rate compared by signal to that of one of Graham's best clocks, which could be kept regulated in Deal Castle.[98] However, Harrison's notes about his second machine, which were never published, show that he had misgivings about its performance during a long voyage.

Even before the machine was finished and set in motion, he had discovered by an accidental experiment that the oscillation of the balances could be seriously affected by centrifugal force; this he explains with the drawing shown in Figure 17. If the motion of the ship causes the timekeeper to be swung on its side gimbal hinges so that the dial inclines and reclines in a ver-

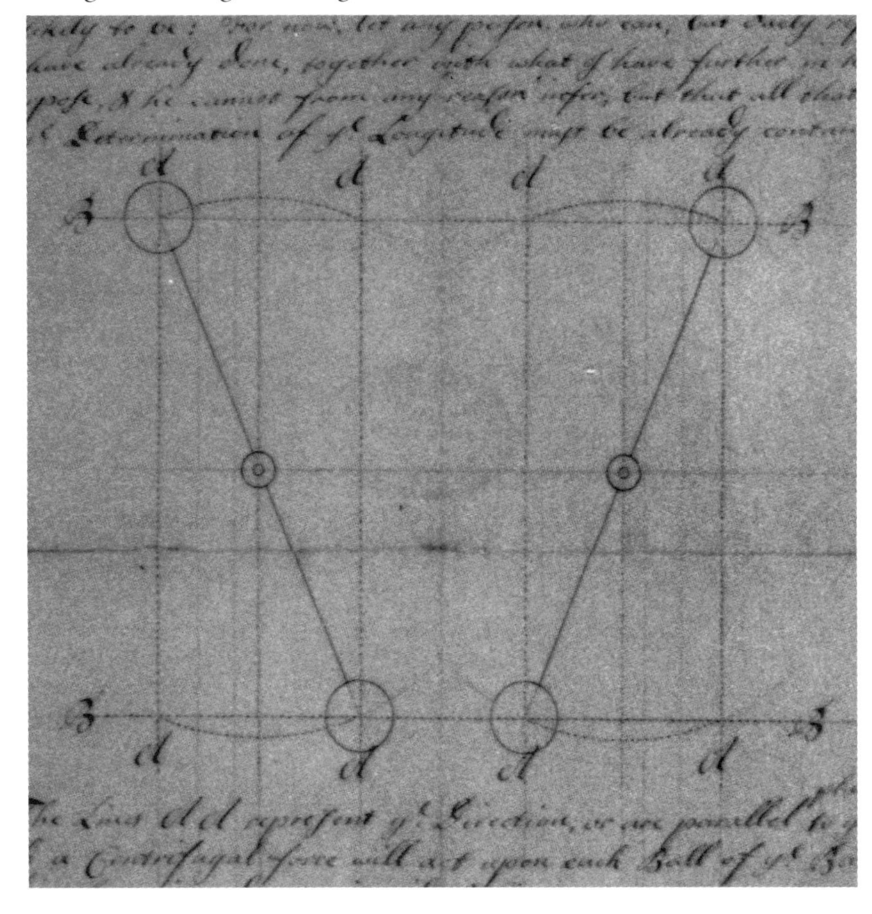

**Figure 17. John Harrison's drawing, *ca.* 1740-1, accompanying his description of two serious problems with the design of the balances of H.1 and H.2 and his explanation as to why he had to set about making a third timekeeper. From "That the Ballances of my Second Machine are, from their Figure or Construction unfit for their intended purpose...," MS. 3972/2, p. 3. Courtesy of the Library of the Worshipful Company of Clockmakers, London.**

95. I am grateful to Martin Burgess, who helped to identify and explain the reason behind the unusual arrangement of H.2's compensation. In his letter to me dated April 13, 1993, Burgess also noted the fact that Harrison would have been able to detect such a change immediately by observing a variation in the amplitude of H.2's balances against their degree scales and to pinpoint the source of the error if the arc of swing of his pendulum clock remained constant during this observation.

96. "The second Machine was finished about two Years ago,...." This reference is given in "Some Account of Mr. Harrison's Invention" (*op. cit.*, note 20), p. 20, a document dated January 16, 1741/2.

97. *Ibid.*, p. 20.

98. "Proposal for examining Mr. Harrison's Time-Keeper at Sea," Appendix 20 of *An Account of the Proceedings* (*op. cit.*, note 20), p. 45. It was during this war, the Spanish war that broke out in October 1739 in consequence of conflicting claims in America, that the cost of failure in determining longitude was once again made apparent. While attempting to round Cape Horn to plunder Spanish interests in the Pacific, Commodore George Anson (1697-1762) aboard his flagship *Centurion*—ironically the same ship in which H.1 had been sent to Lisbon—and five other vessels became hopelessly lost; by the time the *Centurion* reached the island of Juan Fernandez, the squadron of six ships had been reduced to three and the strength of Anson's crews from 961 to 335. In a prolonged voyage, supplies would run short and serious outbreaks of scurvy were not uncommon: time meant lives.

tical plane, *AA*, then the resulting centrifugal force would cause the motion of the balance balls to be accelerated as they are pushed into a vertical position. Conversely, if the timekeeper is, by chance, pivoted about the axes of the balances in a horizontal plane, *BB*, then the resulting centrifugal force would retard the motion of the balances. Even though the disturbance that the machine would have experienced when suspended in gimbals on board ship would never have been as severe as the artificial agitation to which he subjected it in his workshop, Harrison realized that the chance of these errors ever balancing each other was too remote. He then discovered another error after H.2 had been running for a while. This was caused by the diamond surfaces that acted as the stops for the points of the balance springs: in hot and dry weather, he found that the points tended to slip on these surfaces, while at other times they tended to stick, causing a disturbing tremor in the balance springs. In admitting these inherent problems in the design of his first two machines, he stated two significant and easily overlooked facts about how the errors were detected: the first was that he had a regulator—one of his precision wooden clocks—against which he could constantly check H.2's performance; and the second was that he would not have been able to identify the source of these problems without the experience of having made H.1.[99]

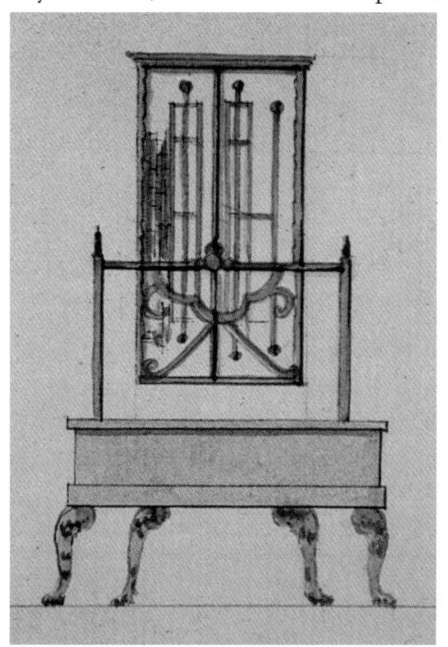

**Figure 18. Small color-wash drawing of H.2 made about 1785 by John Charnock (1756-1807), showing the case, the gimbals, and the stand, none of which have survived. Approx. 5.1 x 5.9 cm. (2 x 2.2 in.). Courtesy of the National Maritime Museum, Greenwich. Ref. no. PAF 2941, "Charnock's Views," vol. 4, fol. 13.**

For the construction of his second machine, John Harrison had moved to London, where he had been joined by his younger brother James. Despite claims made in the nineteenth century that James had a great share in the invention and construction of H.1 and H.2,[100] recent evidence found by Andrew King shows that James was increasingly involved in his bell-founding business in Lincolnshire during this period.[101] His name is never mentioned in any of his brother's writings, and John's name alone appears as the maker of H.2. Certainly by the time of the completion of H.2, and probably the year before, the partnership between John and James Harrison had ended. The reasons are not known, but it would not be surprising if James felt excluded from the developments with the marine timekeepers.

John Harrison was 46 years old when H.2 was finished. His eldest son, John, the only child by his first wife, had died the previous year. Of his two remaining children, William, who was now eleven, was to play an important role in his father's struggle to obtain the longitude prize. In this year, 1739, John Harrison and his family moved from Leather Lane, where he was recorded as residing in 1737,[102] to a house on Orange Street, off the northwest corner of Red Lion Square. They remained there until 1752, when they

99. Harrison described these problems in two separate, but similar, documents, neither dated but both believed to have been composed about 1740-1: "That the Ballances of my Second Machine are, from their Figure or Construction unfit for their intended purpose,..." and a slightly longer version, "Some account of two of the most particular Circumstances whereby I was induced to Set about to make a third Machine." These are both included in MS. 3972/2, pp. 1-3 and 5-7, Library of the Worshipful Company of Clockmakers, Guildhall Library, London. Harrison's statements about the importance of his regulator and H.1 in making these observations is included on the last page of the latter document. My reading of these and three other Harrison manuscripts has been greatly aided by a typewritten transcript prepared by Andrew King, a copy of which he kindly gave me. King published "Some account of two of the most particular Circumstances..." in his article "A Manuscript of John Harrison, Circa 1740" in *Horological Journal*, vol. 119, no. 2 (August 1976), pp. 4-6.

100. "My grandfather, James Harrison, was the workman who made the two regulators for the time-keeper for ascertaining the longitude at sea; and the two first machines or time-keepers were likewise almost entirely made by him—the first being made at Barrow, and the second in London." James Harrison (1767-1835) mentioned this in his article, "Observations on Winn's Improvements in Church and Turret Clocks," *Mechanics' Magazine*, vol. 10, no. 260 (1828), p. 3. See also note 41.

101. [King], *From a Peal of Bells* (*op. cit.*, note 24), p. 22. King relates that, during the period of the construction of H.1 and H.2, James Harrison made a standard type of horizontal sundial in 1732 for the Church of the Holy Trinity in Barrow upon Humber, a bell frame for York Minster in 1733, bell frames for the churches at Waithe and Barnsley in 1737, and a chime barrel and four other church bell frames in 1739. On the day that John sailed with H.1 for Lisbon aboard the *Centurion*, James was brought before the Barrow court and fined one shilling for allowing his swine to trespass into a cornfield.

102. This address is given in the "Historical Chronicle" section of *The Gentleman's Magazine* (July 1737), p. 448, in a short entry for Thursday, June 30, announcing the award to Harrison of £250 for the construction of H.2.

Figure 19. Map showing the location of the places where Harrison and some of his well-known contemporaries worked. The side of each square on the map represents a distance of 1,000 feet. After 1739, Harrison lived near the outskirts of the London suburbs in an area that today is considered part of central London. From *London Surveyed or a new Map of...* printed for John Bowles (London, 1742). Courtesy of the Harvard Map Collection. Ref. no. 200.1742

| | | | |
|---|---|---|---|
| 1. | John Harrison | *ca.* 1737-9 | Leather Lane |
| 2. | John Harrison | 1739-52 | Orange Street |
| 3. | John Harrison | 1752-76? | corner of Lee Street and Red Lion Square |
| 4. | John Jefferys | *ca.* 1726-53 | Holborn |
| 5. | Larcum Kendall | | Furnival's Inn Court |
| 6. | James Short | 1738-68 | Surrey Street |
| 7. | George Graham | 1720-51 | Dial and One Crown, Fleet Street (no. 135) |
| 8. | Thomas Mudge | 1751-70? | Fleet Street (no. 151) |
| 9. | Royal Society | 1710-80 | Crane Court |

moved to a house on the northwest corner of Lee Street (later Leigh, now Dane Street) and Red Lion Square.[103] At that time, Red Lion Square was near the outskirts of London, but it was close to the area in which many people in the watch, clock, and scientific instrument trade worked (Figure 19).

In the explanation that Harrison wrote about 1740 concerning the problems with H.2, he admitted that he thought it "improper to go to the Sea to try it, because a much better Machine, not only in this, but in some other respects...is already in great part made."[104] Putting behind him the thousands of hours he had devoted to the construction of H.2, he began on his own initiative, before receiving funding from the Board of Longitude, to construct a third timekeeper. To avoid the problems he had experienced with H.2, he replaced the bar balances with balance wheels.

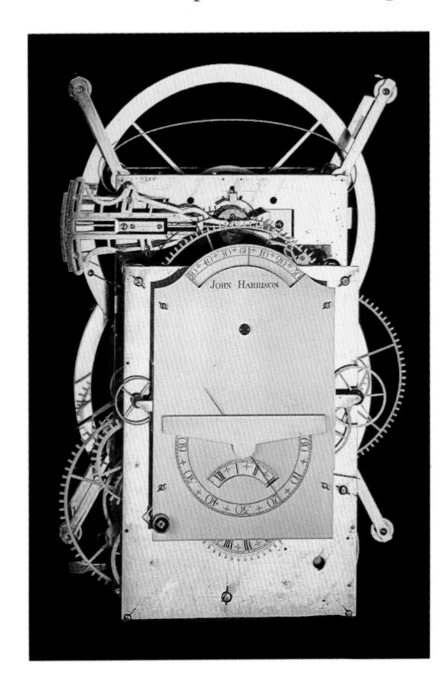


**Figure 20. Harrison's third marine timekeeper, H.3. Overall height: 59 cm. (23.2 in.). Courtesy of the National Maritime Museum, Greenwich. Inv. no. Ch.37.**

Nevertheless, it was not long before he needed additional funds to continue building this machine. On January 16, 1741/2, therefore, the Board of Longitude met for a second time to consider the matter. With the reputation he had earned, Harrison had many influential friends, who were prepared to do what they could to ensure that he obtained the support he needed. On the same day as the meeting, twelve members of the Royal Society, five of whom served as *ex officio* Commissioners of the Longitude, signed a report recommending that the Board of Longitude continue funding Harrison's work. In this document, it was stated that the construction of this timekeeper was well under way and that, when completed, it would have the following advantages over the earlier machines:

> That being of a *smaller Size*, it will be more commodious; that being of a plainer and simpler Construction, it will be less subject to any Disorder, will be easier adjusted, and will serve as a *better Model for Workmen* to make others by; *so that Ships may be furnished with the like Machines at a cheaper Rate.*[105]

The Board of Longitude minutes indicate that Harrison expected to have this timekeeper ready for trial by August 1, 1743.[106] But although he was granted the £500 that he requested, this prediction and his hopes for a plainer, simpler, and more easily adjustable machine, unfortunately, did not materialize: H.3 also contained design flaws, and these embodied problems that were to occupy him for another fifteen years. However, during this long and agonizing period of analyzing and overcoming the problems he encountered, Harrison made two of his most important inventions and acquired the knowledge that allowed him to produce the radically different design of his

103. Red Lion Square was less than half a mile from the Foundling Hospital, at which his son William was to play such an active role later on. Quill (*John Harrison* [*op. cit.*, note 24], p. 231), who tracked down Harrison's other moves from Rate Books in the Holborn Public Library, states that Harrison's address on Lee Street was changed to Red Lion Square in 1767.

104. [Harrison], "That the Ballances of my Second Machine" (*op. cit.*, note 99), p. 1.

105. "Some Account of Mr. Harrison's Invention" (*op. cit.*, note 20), p. 21.

106. Board of Longitude Confirmed Minutes (*op. cit.*, note 87), January 16, 1741/2.

107. Martin Folkes noted that "with all its apparatus, [i.e. its case and gimbals, it stood] upon an Area of only four square feet." Minutes of the Royal Society (*op. cit.*, note 20), p. 192. More descriptively, another observer said that its case was "in figure like the Body of a Coach and would hold a Child of 4 years old within it." Anonymous manuscript entitled "Description of Mr. Harrison's Clock," *ca.* 1740-8; I am grateful to Andrew King for sending me a copy of this document, which is in a private collection.

108. In its case, H.3 weighs 95 lb. (43 kg.). Of the original cases that Harrison built for his machines, only H.3's survives. Unfortunately, the gimbal suspensions for all three timekeepers have been lost.

109. According to the anonymous manuscript describing H.3 ("Description of Mr. Harrison's Clock," [*op. cit.*, note 107]), it appears that the first remontoire fitted to this clock rewound every twenty seconds. Martin Folkes's account of the clock, however, indicates that in 1749 the clock had its present half-minute remontoire (Minutes of the Royal Society [*op. cit.*, note 20], p. 193). Because there is no sign that a conversion ever took place, it is thought that the anonymous writer was mistaken. A detailed analysis of H.3's remontoire appears in Peter Hastings's article, "John Harrison—Spacecraft Engineer?," *Horological Journal*, vol. 136, no. 6 (December 1993), pp. 193-7. See Burgess Figure 17.

110. Harrison discussed the size and mass of these balances in an undated manuscript (*ca.* 1760) entitled "Of the Nature or Phaenomenon of Ballances, or the Solution of a seemingly Paradox therein," MS. 3972/2, Library of the Worshipful Company of Clockmakers, Guildhall Library, London, pp. 11-16. A description of this document was published in Charles Aked's article, "John Harrison's Paradox," *Horological Journal*, vol. 119, no. 1 (July 1976), pp. 3-5. Another longer manuscript on the same subject exists, entitled "A Description of the Nature or Phenomenon of Ballances, as I found from Experience in my 3rd Machine: Or the Solution of a seemingly Paradox therein"; this is the second of three parts of MS. 3972/1 (*op. cit.*, note 92), pp. 103-19.

prize-winning timekeeper, H.4. Ultimately, sheer determination combined with great ingenuity would allow him to succeed where others had failed.

H.3 is smaller than H.2, standing 23.2 in. (59 cm.) high on an area 15.3 in. (39 cm.) wide by 8.7 in. (22 cm.) deep.[107] The timekeeper itself therefore occupies less than one-third of the space of H.1. At 59.5 lb. (27 kg.), it also weighs considerably less than either of its predecessors.[108] Harrison positioned the maintaining power again on the center arbor, but he redesigned the remontoire (shown, with other components, in Figure 21) to fit on the escape wheel arbor, where it provides a more constant and uninterrupted power source to the grasshopper escapement, being rewound every 30 seconds.[109] H.3's remontoire has a five-step cam for adjusting the tension of the springs and providing five fixed positions for conducting tests. By setting the cam in the middle to establish the normal running arc of the balances, there remain two positions for increasing the arc and two positions for decreasing the arc. The ability to provide a series of repeatable adjustments to the remontoire's constant force was essential during the tests to perfect the isochronism of the balance spring. As with H.1 and H.2, no lubricant is required in this timekeeper, except on the mainspring and the fusee chain.

The most obvious difference in the appearance of the clock is due to the fact that, for the reasons stated earlier, Harrison replaced the bar balances with two large and heavy circular seconds-beating balances, mounted one above the other and linked together by thin metal ribbons.[110] While this new arrangement successfully addressed the problems that he had encountered with H.2, the balances introduced serious problems in the spring that controlled their oscillating motion.

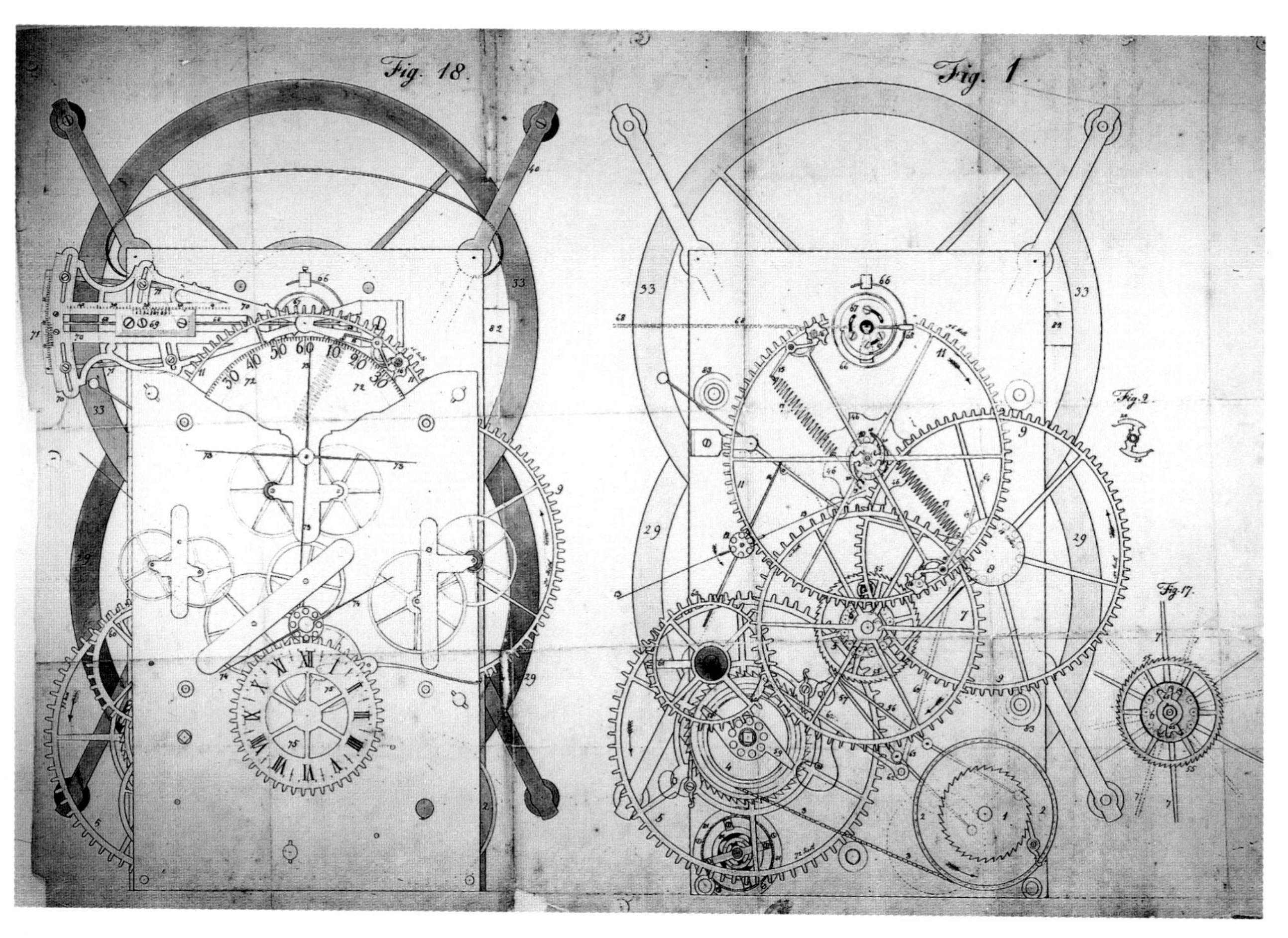


**Figure 21. Two cross-section views of H.3, probably drawn by Thomas Bradley *ca.* 1840, to show the component parts of the timekeeper. The illustration on the left includes the compensation curb and anti-friction wheels behind the dial, while on the right, the wheelwork, the remontoire (with its cam for adjusting the tension of the springs), and the balance spring are shown. Drawing dimensions: approx. 91 x 61 cm. (36 x 24 in.). By permission of the Syndics of Cambridge University Library and of the Director of the Royal Greenwich Observatory. Ref. no. MS. RGO 6/586 f217$^{r}$.**

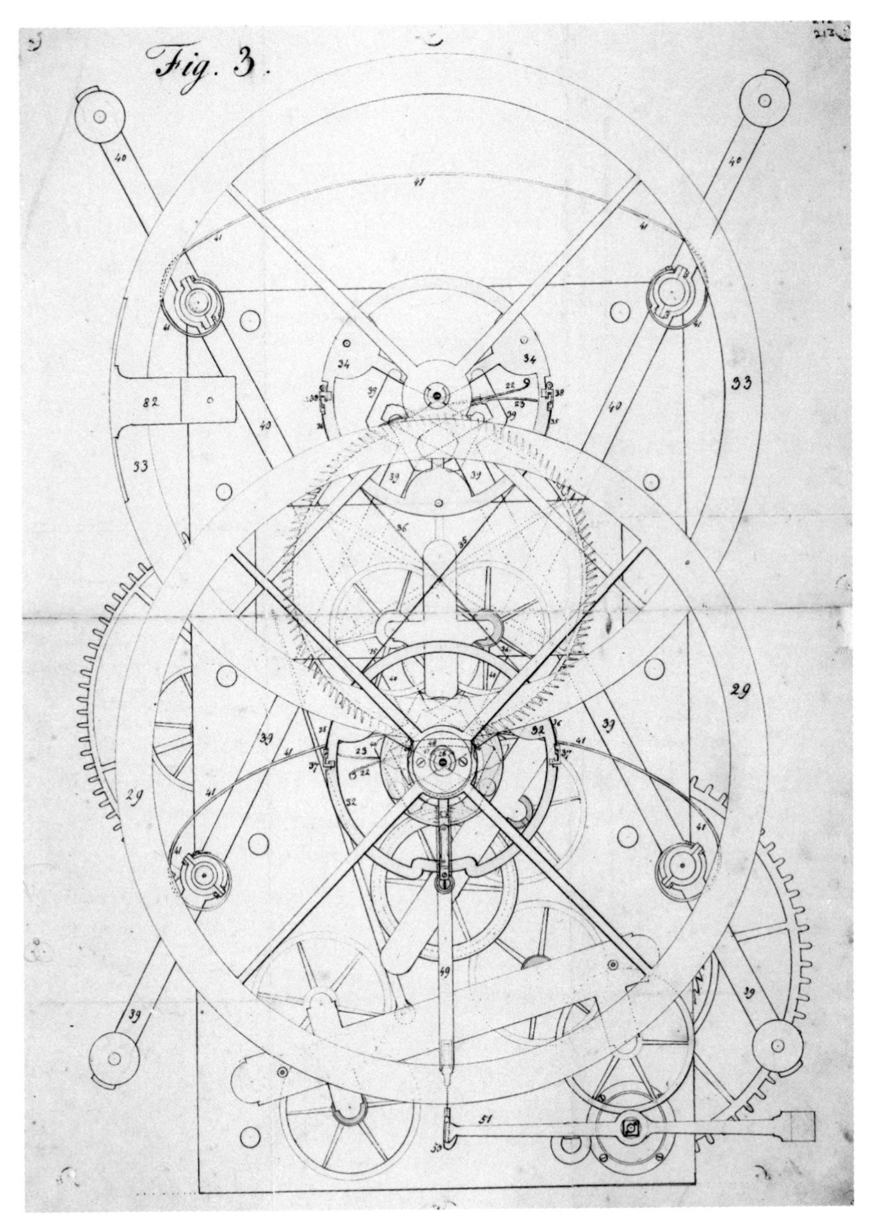


**Figure 22. Illustration of H.3, probably drawn by Thomas Bradley *ca.* 1840, showing the component parts of the isochronal corrector, which extends down from the center of the lower balance wheel and across to the pivot (the roller bearing, not shown) in the lower right corner. The anti-friction arms stretching across the plates to support the balance pivots are also visible. Drawing dimensions: approx. 91 x 61 cm. (36 x 24 in.). By permission of the Syndics of Cambridge University Library and of the Director of the Royal Greenwich Observatory. Ref. no. MS. RGO 6/586 f213ʳ.**

The use of a single spring—a short, stiff, spiral spring attached to the upper balance—overcame the problems induced by the rocking of the ship as well as the difficulties of making four balance springs of equal strength. It posed, however, a greater obstacle in making the spring isochronous, since the spring had to provide a restoring force proportional to the amplitude of the balance's oscillation. In other words, the spring had to control the oscillation of the balances, so that, regardless of the amplitude of their motion, their oscillations would always be performed in precisely the same period of time. It appears that Harrison was still engaged in this problem when the Board of Longitude met on June 4, 1746, to consider his request for a further £500 to allow him to continue his research; in the minutes for that meeting it was stated:

> That, after much time & thought spent in making many new experiments upon springs and in various methods of tempering them, in order to remedy the imperfection in the part aforesaid he has, at last, gained so much knowledge in the Properties of Springs, as he doubts not will enable him to remove the said Imperfection & render the

motion of the Machine entirely answerable to the degree of exactness which he first expected to find in it.[111]

In searching for a solution, he devised the first of the two significant inventions that H.3 was to yield. To counteract the nonlinear character of the spring's force, Harrison introduced an isochronal corrector (Figures 22 and 24), a spring-loaded mechanism designed to slow the oscillation of the balances as their amplitude increased. The problems in minimizing the friction of this device must have been considerable. Harrison's solution was to devise a special pivot for the corrector, composed of four concentrically mounted antifriction rollers made of high-tin bronze. This brilliant innovation, his caged roller bearing, is the forerunner of the caged ball bearing (Figures 23 and 24), which is used in almost every machine today.

The other exceptional invention that Harrison made in H.3 was the bimetallic strip. An anonymous document describing H.3, probably composed between 1740 and 1748, describes the first form of temperature compensation that Harrison used:

> A finger upon each ballance, at every other stroke, comes against a piece fix't to the end of a long slender piece of Watch spring, which very easily gives way to it. These springs for the two Ballances open wider by means of the stretching of iron rods in warm weather, and so by meeting the ballances sooner on, that account make the springs narrower to compensate for the affections of the heat, &, in cold weather the rod by contracting brings the springs nearer each other, which gives the ballances room to swing so much the wider. In short it is a little world of itself independent of the difference of gravity, heat, or cold of this our Globe.[112]

111. Board of Longitude Confirmed Minutes (*op. cit.*, note 87), June 4, 1746. This document also mentions that this work has "so entirely ingrossed his time & thoughts for many years past as to render him quite incapable of following any gainfull employment for the support of himself & family." The Board granted him the funds he requested.

112. "Description of Mr. Harrison's Clock" (*op. cit.*, note 107).

**Figure 23. Detail of one of H.3's two caged roller bearings, the forerunner of the ball-bearing, which Harrison developed to support the pivot of the spring-loaded isochronal corrector. Note the high-tin bronze alloy that Harrison used for the rollers, because of its good compressive strength (see note 26). Courtesy of the National Maritime Museum, Greenwich.**

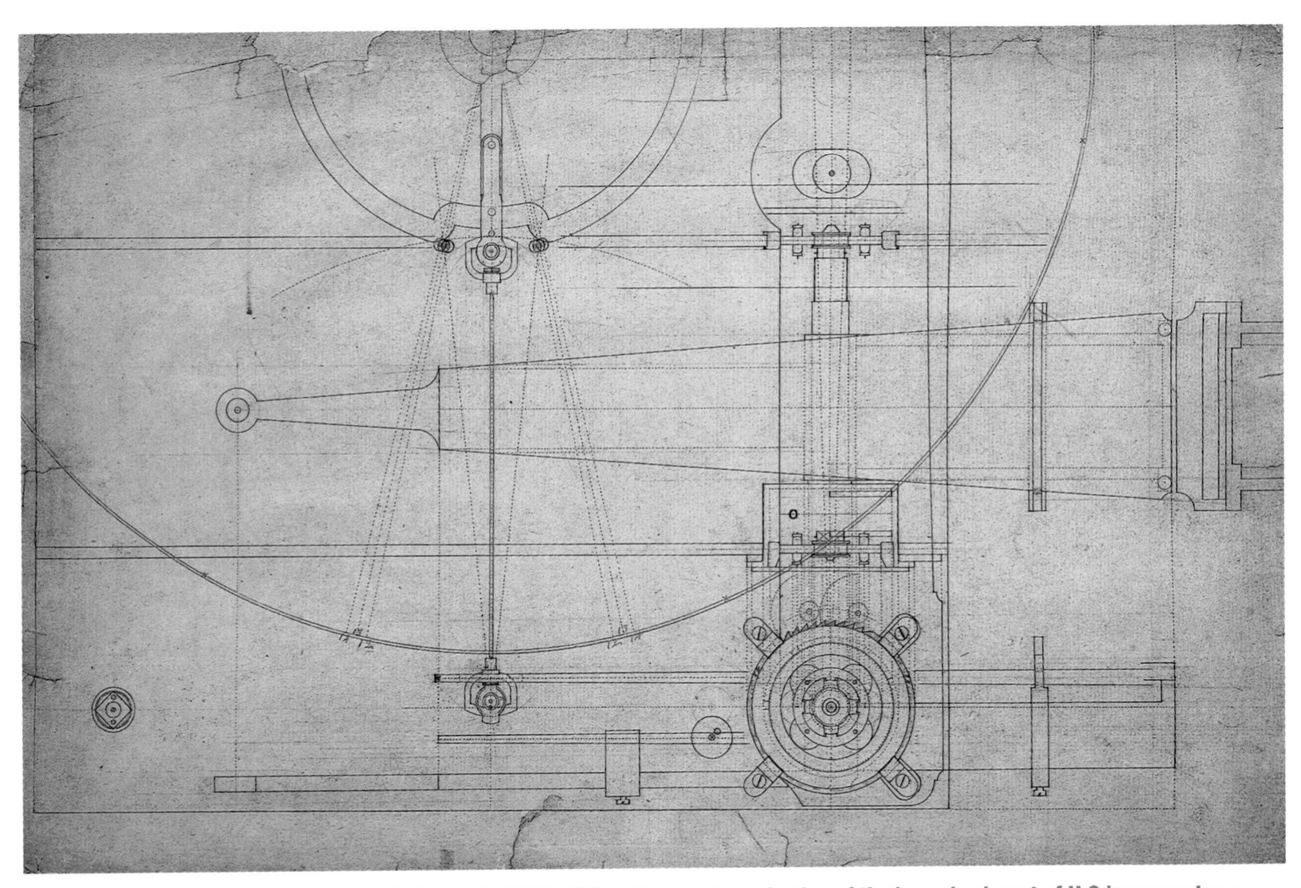

**Fig. 24. Drawing made by John Harrison in the early 1740s. This orthographic projection of the lower back part of H.3 incorporates a combination of front and plan views of the component parts of the isochronal corrector that Harrison fitted to the lower balance arbor. The circle crossing the image represents the outer rim of the lower balance, and the caged roller bearing is shown in elevation in the lower right corner with its plan view projected immediately above. 38 x 24.5 cm. (15 x 9.6 in.). Courtesy of the National Maritime Museum, Greenwich.**

From this description, it appears that H.3's first bimetallic temperature compensation was adapted from the type of gridiron used in H.2. However, this arrangement evidently did not prove satisfactory, because Harrison set about devising a method that would not only respond more immediately to changes in temperature but would also be easier to adjust. He riveted together two thin sheets of steel and brass, fixing one end in an adjustable fiddle-shaped frame, which pivoted about the arbor of the upper balance; on the other end of this brass and steel strip, he positioned two pins to embrace the balance spring. The different coefficients of expansion of brass and steel cause this bimetallic strip to bend one way or the other with changes in temperature; thereby, through the pins that embrace the balance spring, this device automatically adjusts the effective length of the spring to compensate for variations in temperature (Figure 25). The small mass of the bimetallic strip makes this compensation more sensitive to changes in temperature and therefore more effective than gridiron rods. Furthermore, the amount of compensation can be easily altered by moving the slide on its graduated scale, a very useful feature for making the minute adjustments required to compensate for errors introduced by changes in the elasticity of the remontoire springs. The timekeeper can be made to go faster or slower by rotating the fiddle-shaped frame against its scale to change the effective length of the balance spring. The bimetallic strip is one of the most widely used of Harrison's inventions, being generally employed today in thermostats and other tem-

113. Minutes of the Royal Society (*op. cit.*, note 20), p. 194.

114. Board of Longitude Confirmed Minutes (*op. cit.*, note 87), November 28, 1757, citing Harrison's letter to George Anson (now Lord Anson) regarding the completion of H.3. When Gould restored this clock in the 1920s, he noted that H.3's mainspring was marked "2 Ap 1757 Nightingale," indicating that Harrison was still working on the clock earlier in the year. Nightingale was the spring maker. This information is given on p. 9 of a British Horological Institute monograph of Gould's lecture "The Restoration of John Harrison's Third Timekeeper." This lecture was also published in five parts in the *Horological Journal*, vol. 74 (September 1931-August 1932): March, pp. 105-8; April, pp. 120-7; May, pp. 151-3, 148; June, pp. 166-7; July, pp. 178-81.

115. Harrison, *A Description Concerning Such Mechanism* (*op. cit.*, note 33), p. 102.

116. According to the Board of Longitude Confirmed Minutes (*op. cit.*, note 87), the following payments totaling £2,250 were made during the construction of H.2 and H.3: £250 on June 30, 1737 (£500 was authorized but only £250 was paid since H.2 was never tested at sea); £500 on January 16, 1741/2; £500 on June 4, 1746; £500 on July 17, 1753; and £500 on June 19, 1755. In order to complete his work on his ultimately successful timekeeper, the prize-winning H.4, Harrison was granted a further

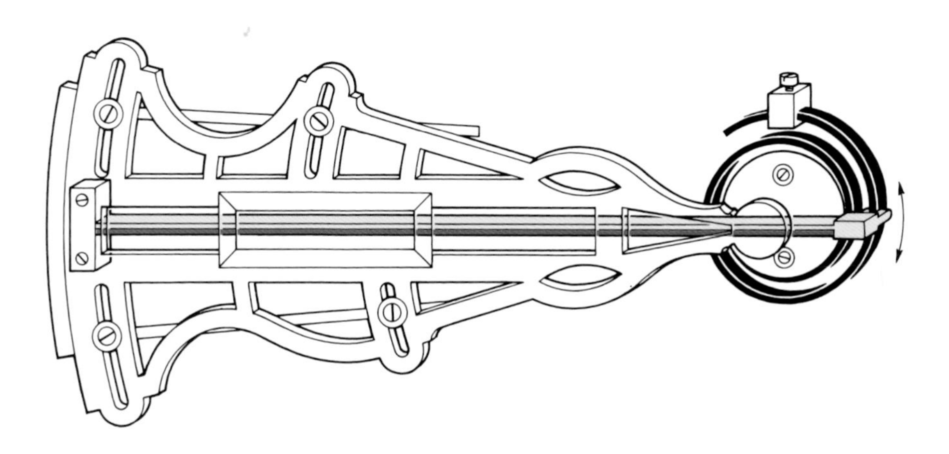

**Figure 25. Detail of H.3's bimetallic strip. Courtesy of the National Maritime Museum, Greenwich.**

four payments totaling £1,500: £500 on November 28, 1757; £500 on July 18, 1760; £200 on October 13, 1761 (£500 was voted, but £200 was all that remained at their disposal at the time); and £300 (the remainder of the previous £500) on June 3, 1762. A further £250 was paid on March 12, 1761, to outfit his son William for the voyage to Jamaica, £300 was granted on August 9, 1763, to cover his expenses on the second trial of H.4 to Barbados, and £15 was granted on February 9, 1765, to cover expenses incurred for computations following that trial. Therefore, excluding the sums he was awarded specifically for the longitude prize, the support that Harrison received for the development of H.2, H.3, and H.4 and the trials of the latter totaled £4,315. Nevertheless, the first three payments totaling £1,250 were eventually deducted from the final payment of the prize, as had been stipulated at the Board's meeting on January 16, 1741/2. (See also note 130 and Appendix 2 in this paper).

117. In 1763, Jérôme de Lalande recorded that there were 50 tax collectors in England, employed at an annual salary of £250 each, 50 supervisors at a salary of £100, and 100 excisemen at a salary of £50 (*Jérôme Lalande: Journal* [*op. cit.*, note 28], p. 94). He further stated that his colleague, Charles Camus, paid 18 shillings per week (about £47 per annum) for his housekeeper (p. 93), while the French diplomat Duc de Nivernais (1716-1798) happily dispensed with 40,000 livres each month—the equivalent of about £870 sterling per month or £21,000 per annum—half for servants and half for the rent of his house (p. 100). The exchange rate between France and England of 22.97 livres to £1 sterling is given by the mention that the £360 cost of an eight-foot quadrant by John Bird was equivalent to 8270 livres (p. 92).

perature-control devices. From the fact that Martin Folkes referred to Harrison's "new metalline Thermometer"[113] at the presentation of the Copley Gold Medal, it is clear that this important invention had been made by November 1749.

Although H.3 had seemed almost complete in 1741, Harrison was not satisfied with its operation until 1757.[114] Later in life, he referred to it as "my curious third machine"[115]—curious, perhaps, because it did not always operate in the manner he anticipated. The 1740s and 1750s must have been a difficult period as his expectations for this machine's performance remained unfulfilled and the need for financial support continued to be an essential factor in his struggle for perfection. To be sure, moral support from his friends in the Royal Society provided much-needed encouragement: in 1749, Harrison became the seventeenth recipient of the Society's highest distinction, the Copley Medal. But while this honor illustrates the high regard in which he was recognized by the scientific community at that time, it carried no financial reward. The Board of Longitude, having several *ex officio* members with strong ties to the Royal Society, continued to support his work financially throughout this period. Yet by the time that Harrison announced the completion of H.3 at the Board's meeting on November 28, 1757, the development of H.2 and H.3 had cost the British government only £2,250.[116] Over the period from 1737 to 1757, this represents an average of £112-10-0 per year, less than half the salary of a collector of excise.[117] For this small sum, England had sponsored some of the most valuable scientific research conducted during the eighteenth century—research that, with a small amount of additional support, would lead to the marine timekeeper that solved the longitude problem. And, as a by-product of this research, the world was enriched with several inventions that would be of great technological significance in the future. These two machines had cost Harrison twenty years of his life; if the vital experience gained from the wooden regulators and H.1 is included, over 30 years had been devoted to this work.

**Figure 26. H.3 seen from behind inside its case. This is the only case belonging to the three timekeepers to have survived. The gimbals of all the machines have been lost. Courtesy of the National Maritime Museum, Greenwich.**

There is evidence that about this time, John Harrison became engaged in other projects, possibly to supplement his income. In October 1756, he was paid three guineas for the design of a new escapement for the tower clock at Trinity College, Cambridge (see Andrew King, Figure 8).[118] In the summer of the following year, the Society for the Encouragement of Arts, Manufactures, and Commerce offered a premium of £50 to anyone who could devise a mill by which poor people could grind their own corn. Among the contestants were John Harrison and the astronomer James Ferguson (1710-1776), both of whose mills were favored during the trials in December 1757. Neither, however, was successful, Harrison's mill breaking because it was worked too hard.[119] A few years after this, in 1765, Harrison appears to have produced a bracket clock, although, because it is so unlike any other surviving clocks made by him, there is some doubt about its

**Figure 27. Bracket clock in a mahogany case, signed on the dial "John Harrison, Red Lion Square, London, 1765." The present whereabouts of this clock is unknown. Height: 54.6 cm. (21.5 in.). Photograph from the *Catalogue of the Vernay Collection...* (New York, *ca.* 1940).**

authenticity. Nothing is known of its details, apart from the fact that it had gridiron temperature compensation, as shown in Figure 27. Following its sale by the New York antique dealer Arthur S. Vernay about 1940, the clock vanished into a private collection and has not been heard of since.[120]

In 1757, after so many years of dedicated work in bringing H.3 to completion, one would like to think that the longitude prize would have been within easy reach for Harrison. But, alas, this was not to be the case. Time was taking its toll on his most influential supporters: Edmond Halley had died in 1742,[121] George Graham in 1751, and Martin Folkes, the former president of the Royal Society, in 1754;[122] James Bradley, the Astronomer Royal after Halley, would die in 1762. Without the support of these individuals and the appearance of a timekeeper ready for trial, the reputation he had established with the success of H.1 was in jeopardy. Now that many of those who had believed in him were no longer there to support him, how long would it be before the financial support he needed to complete his task would be withheld?

In the place of the old régime of friends in the scientific community came a new generation of astronomers and natural philosophers whose eagerness to make their own mark on the world led them to think little of the mechanical achievements of an aging clockmaker. And for this new generation, the Moon was rising, while the tide was turning against Harrison: by the middle of the century, the eminent Swiss mathematician Leonhard Euler (1707-1783) had devised a lunar theory based on which, during the 1750s, the German astronomer Tobias Mayer (1723-1762) calculated new tables of the Moon's motion.[123] These developments gave astronomers and academicians renewed hope of solving the longitude problem with the only method they trusted, and thus they began to view the mechanical solution with more suspicion than they might have done had no breakthrough been forthcoming. After the lunar-distance method had been tried at sea and a special version of Hadley's quadrant—the sextant—had been developed for making the necessary observations, astronomers seemed set to bring to fruition the solution to the longitude problem that Newton had so firmly recommended. The rising young star of these developments was Nevil Maskelyne (1732-1811), who made a reputation for himself during the 1761 Transit of Venus expedition by employing lunar-distance observations to determine longitude. Subsequent expeditions brought him into favor with the Earl of Morton (1702-1768), a powerful ally who, after becoming president of the Royal Society in 1764, helped to secure Maskelyne's appointment as Astronomer Royal in the following year. This, combined with his involvement over the trial of H.4, made Maskelyne the *bête noire* during the remainder of Harrison's long struggle for the longitude prize.

Many of the difficulties that Harrison experienced might have been avoided had the position of Astronomer Royal been granted to someone who recognized the advantages of the timekeeper method. Maskelyne, it should be noted, was not the only candidate for this eminent and influential post: following the death of Nathaniel Bliss (1700-1764), who died only two years after succeeding James Bradley, Harrison's long-time friend and supporter James Short (1710-1768) became one of the prime choices. Short, however, fell out of favor with the Earl of Morton precisely because of his staunch support for Harrison:

118. College records, Trinity College, Cambridge, quoted in Quill, *John Harrison* (*op. cit.*, note 24), p. 81. Harrison probably received this commission from Robert Smith, who had known Harrison for twenty years and was by that time the Master of Trinity College. The fact that Harrison agreed to undertake this work suggests that, despite accusing Smith of plagiarizing his scale of music, he was still on speaking terms with the professor (see note 122).

119. This story is related in a letter dated January 17, 1758, from James Ferguson to the Rev. Alexander Irvine of Elgin, published in E. Henderson, *The Life of James Ferguson, F.R.S.* (Edinburgh, 1867), pp. 224-8. Ferguson noted that poor people were often abused by millers who, instead of grinding their corn for them, forced them to exchange it for adulterated meal—hence the need for the competition.

120. This clock was offered for sale on Arthur S. Vernay's retirement from business in 1940 along with more than 800 other pieces, which are described in the *Catalogue of the Vernay Collection of Early English Furniture, Porcelains, Old English Silver, Glassware and other Art Objects* (New York, n.d.). The clock, item 392, is described and illustrated on p. 107.

In the Royal Institution, London, there is a large chemical balance (5 ft. 4 in. high including its stand, 2 ft. 7 in. wide) that was reputedly made by John Harrison for Henry Cavendish (1731-1810), the celebrated English chemist and natural philosopher. There seems to be no substantiating evidence for this attribution to Harrison, which was made in a note written by J. G. Children, who presented the instrument to the Royal Institution in 1830. Jonathan Betts kindly went to inspect the instrument on my behalf and remarked that the workmanship, being of somewhat inferior quality and of a later style, appeared to have nothing to do with Harrison.

121. In *A Description Concerning Such Mechanism* (*op. cit.*, note 33), p. 59, Harrison recollected what a good support Halley had been, promising Graham that he would attend the Board of Longitude meeting come "rain, snow, or blow." The only meeting Halley was able to attend, however, was the first, held on June 30, 1737.

122. Another of his early influential supporters, one of the first from the Royal Society, was Robert Smith. But this friendship was put in jeopardy in 1749, when Smith published in his *Harmonics, or the Philosophy of Musical Sounds* (Cambridge, England, 1749) a scale of music supposedly of his own invention. Harrison had invented a similar scale of music some years earlier and was deeply offended that his friend should have pirated his ideas. Although, in the introduction, Smith admitted discussing the scale with Harrison, he gave him no credit. Harrison describes his scale of music and his association with Dr. Smith in *A Description Concerning Such Mechanism* (*op. cit.*, note 33), pp. 68-108.

123. For further details on the development of the lunar-distance method during this period, see Bruce Chandler's and Derek Howse's papers in this volume.

Mr. Short is a Candidate for Greenwich but having opposed Lord Morton in the £5000 afair [concerning Harrison's reward], Lord Morton now opposes him and gives it as a reason, that Mr. Short is a Scotch Man, though he acknowledges that he is the fittest for it of any Man. These two [the other being John Michell (1724-1793)] who would have done honour to the place, being thus laid aside, I beleive the Tory Interest, at preşent all powerful, will get it for an Oxonian, who never made an Observation. What Candide Patrons we are of the Sciences![124]

It was during the bleak years of the early 1750s, in the midst of the seemingly endless struggle to perfect H.3's performance, that Harrison designed a pocket watch—made by one of his workmen, John Jefferys (fl. 1726-*ca.* 1753)—and it was this piece that pointed the way toward his ultimately successful timekeeper.[125] Unable to resolve the problems he encountered with H.3 to his complete satisfaction, Harrison ultimately placed all his hopes for the longitude prize on a totally different type of timekeeper—a large watch that contained only a few of the innovations that he had developed during the previous 35 years. This prize-winning timekeeper, H.4, could not, however, have been made without the years of experience that he had gained from the three earlier machines.

Great achievements can look easy and obvious in retrospect. When it comes to icing the cake, there are always those who wish to take or share the glory, although they have played no part in its making and baking. One of the first occasions on which Harrison suffered this injustice was in June 1752, when a well-known London clockmaker, John Ellicott (1706-1772), delivered a paper to the Royal Society claiming the invention of a temperature-compensated pendulum using the different rates of expansion and contraction of brass and steel.[126] Although Ellicott's design was different from Harrison's gridiron pendulum, it worked on exactly the same principle. That Ellicott totally ignored Harrison's earlier work was strange because only three years earlier the president of the Royal Society, Martin Folkes, had mentioned it in his speech when presenting Harrison with the Copley Medal. However, no description concerning Harrison's discovery had been published, either in the *Philosophical Transactions* or elsewhere, so perhaps Ellicott felt, now that Harrison's long-time supporter George Graham had died, that he could claim priority in this discovery. Fortunately, James Short, in one of his first gestures of friendship toward Harrison, set the record straight in a paper he delivered five months later.[127]

Twelve years later, as soon as the successful trials of H.4 became well known, another respected maker, Alexander Cumming (1733-1814), attempted to belittle Harrison's achievement in a bid to gain recognition for his own abilities:

Rival geniuses are apt to be highly Jealous of one another. This is the Case between Cumming the Watch Maker; (a very ingenious Man, and excellent Mechanick) and Mr. Harrison. They were formerly very intimate, and Mr. Harrison, who is frank and open, freely answered many questions that Cumming asked him. Whether from this Source, or from his own Noddle, I shall not determine, but Cumming thinks that he can make a Watch which shall answer

124. Letter dated December 1, 1764, from Alexander Small (1710-1794) to Benjamin Franklin, published in *The Papers of Benjamin Franklin*, ed. Leonard W. Labaree *et al.*, 30 vols. (New Haven: Yale University Press, 1959-93), vol. 11, pp. 481-2. The post did, of course, go to an Oxonian, but, in all fairness, Nevil Maskelyne was an experienced observer and played a very important role in the development of the Greenwich Observatory during his long and distinguished career.

The "£5000 afair" mentioned in this quote relates to the sum offered to Harrison by an Act of Parliament (Act 3 George III *cap.* 14, passed on March 31, 1763) for disclosing the mechanism of H.4 before a subcommittee. However, the conditions laid down by the subcommittee—which included the requirement of two or more copies of the watch to be made—proved to be of such concern to Harrison that, after four months of negotiation, he decided to abandon the £5,000 in the hope that a second trial of the timekeeper would result in a more just reward for his work.

125. For details of the Jefferys watch and the subsequent production of H.4, see Anthony Randall's paper in this volume.

126. John Ellicott, "A Description of Two Methods, by which the Irregularity of the Motion of a Clock, arising from the Influence of Heat and Cold upon the Rod of a Pendulum, may be prevented," *Phil. Trans.*, vol. 47, no. 81 (1752), pp. 479-94. Harrison never forgave Ellicott for this injustice. In 1763, Ellicott told Lalande that Harrison would rather renounce his reward than tell him the secrets of his longitude watch. *Jérôme Lalande: Journal* (*op. cit.*, note 28), p. 88.

127. James Short, "A Letter of Mr. James Short F.R.S. to the Royal Society, concerning the Inventor of the Contrivance in the Pendulum of a Clock, to prevent the Irregularities of its Motion by Heat and Cold," *Phil. Trans.*, vol. 47, no. 88 (1752), pp. 517-24.

128. Letter dated December 1, 1764, from Small to Franklin in *The Papers of Benjamin Franklin* (*op. cit.*, note 124), vol. 11, p. 481. Alexander Cumming was one of the experts appointed in 1763 by the Board of Longitude to attend Harrison's proposed disclosure of the mechanism of H.4; this disclosure, in fact, did not occur until August 1765, and by this time Cumming was no longer involved. Cumming was a talented maker, chiefly remembered for the barograph clock he made for King George III in 1765 and for his book, *The Elements of Clock and Watch-work* (London, 1766), but he was not in Harrison's league.

**Figure 28. Portrait of John Harrison, painted by Thomas King (d. *ca.* 1769) probably early in 1766, when Harrison was about 73 years old. Behind him are H.3 and a regulator with a sophisticated design of gridiron pendulum. In his right hand is a pocket watch, probably the one made for him by John Jefferys in 1753, that served as a model for his prize-winning timekeeper, H.4. That H.4 itself is not included can be explained by the fact that Harrison had delivered it to the Board of Longitude on October 28, 1765. Because H.3 was removed from Harrison's house on May 23, 1766, it is almost certain that Harrison sat for the portrait between these dates. The mezzotint published in 1768 (shown on p. 166), which Philippe Joseph Tassaert (1732-1803) made from this portrait, replaces the hand-held watch with the famous H.4, lying on the table beside its celebrated maker. Oil on canvas. 127 x 102 cm. (50 x 40.2 in.). Courtesy of the Science Museum/Science and Society Picture Library, London. Inv. no. 1884-217.**

> the purpose of the Longitude, and is become an Enemy to Harrison, or at least wants much to come in for an equal share of the glory, and therefore throws every rub in the way that he can. I do beleive that it was he who put the Commissioners of [off] the £5000 on demanding Conditions of Mr. Harrison, which the Act did not warrant them, in the Opinion of most people.[128]

Less than a year later, the clockmaker John Whitehurst (1713-1785) had similar ideas:

> Since Mr. Ludlams report of Mr. Harrisons Time-keeper, has been made public, I have been considering what further improvements are wanting to render such machines of more general Use and flatter myself with Some hopes of contributing towards that end.
>
> First, by reducing the price. Secondly, by reducing the Machinery to a more easie construction, so, as to bring the Executive parts within the Compass [of] any common workman. Thirdley [the?] machine may be taken to peices [and put] together without altering the time regulation, of it. Fourthly that it shall not be liable to stop, as Mr. Ludlam observes is the Case with Mr. Harrisons. Such are the ends I propose to myself, as improvements

> but the event must be left to tryal to determine the reallity of them.[129]

Needless to say, the reality proved too much. Those who attempted to follow in Harrison's footsteps would recognize through experience the magnitude of his achievement.

Despite the fact that he ultimately won the £20,000 longitude prize,[130] only a few of Harrison's inventions were adopted by other makers. The significance of his work is perhaps as little understood today as it was during his lifetime. Yet although the marine chronometer developed by 1790 superficially bore little resemblance to his prize-winning timekeeper, it was based upon many elements of H.4. When Harrison began his work, there was no path to follow: many of the problems that he was to encounter were still unresolved, while others had not even been recognized. His knowledge was based on practical experience combined with an unrelenting desire to discover the truth; his long and arduous struggle to perfect each timekeeper eventually provided him with the experience to design and construct H.4. Historians of horology have frequently viewed Harrison's contribution as a dead end, giving most of the praise to Pierre Le Roy and others who followed in Harrison's wake. Yet the very size of H.4 and the whole concept of its high-frequency balance, not to mention maintaining power and the principle of the bimetallic strip, were adopted as basic elements in the design of the marine chronometer. Perhaps of greatest significance is the fact that, despite what Newton had stated, John Harrison showed the path to follow by proving beyond any shadow of doubt that a timekeeper was a reliable solution to the problem of determining longitude at sea. As his son William stated in a letter to the Board of Longitude in February 1764, "it is well known to be much harder to beat out a new road, than it is to follow that road when made."[131] For the road he constructed—against all odds of scientific thought of the time—and for the impact that this road was to have upon the subsequent growth of Western civilization, John Harrison deserves more credit than history has given him. In the words of one of his contemporaries, "It may be said of this gentleman,...as it was said of Shakespear in poetry, that he is nature's mechanic."[132]

## POSTSCRIPT

The subsequent history of Harrison's first three timekeepers is worth mentioning, because it provides a valuable lesson in the importance of preserving the past for the inspiration of future generations and demonstrates how easily artifacts that have helped to shape our civilization can be neglected and lost through ignorance of their significance.

After the second trial of H.4 in 1764, when Harrison's timekeeper was officially recognized as being able to perform within the limits of the 1714 Act, the question arose as to whether it was, as the Act had stipulated, "Practicable and Useful at Sea." How easily could this miracle be duplicated? The lunar-distance method was, by then, being used with success and was thought by many to be a more practicable and useful solution, not least because of the cost and difficulty of producing reliable and accurate timekeepers in quantity. Furthermore, as Newton had stated, a timekeeper was only good for keeping the longitude; to find the longitude, only an astro-

129. Letter dated October 20, 1765, from Whitehurst to Franklin in *The Papers of Benjamin Franklin* (*op. cit.*, note 124), vol. 12, p. 327.

130. Following the dispute over the performance of H.4 on its first voyage (November 18, 1761, to March 27, 1762), on August 17, 1762, the Board of Longitude authorized payment to Harrison of £1,500 immediately and a further £1,000 after a second trial to the West Indies had taken place (this was conducted from March 18 to July 18, 1764). This sum, totaling £2,500, was subsequently deducted from the £10,000 that Harrison was awarded on October 28, 1765. Following the intervention of George III, Harrison was granted a final reward of £8,750 on June 19, 1773. This final part of the £20,000 prize that H.4 had qualified for would have been £10,000, except that the first three payments, totaling £1,250 (made in 1737, 1741/2, and 1746), had been granted with the proviso, first mentioned at the Board's meeting in 1741/2, that they should be deducted from any subsequent reward. In 1773, the government saw fit to uphold this stipulation and thereby reduced his final reward to £8,750. Between 1753 and 1763, however, Harrison had also received other grants amounting to £3,050 to enable him to complete H.3 and H.4. In all, therefore, he was paid £23,065. (See also note 116 and Appendix 2 in this paper.)

131. "Copy of Mr. Harrison's Declaration of the Rate of the Going of the Timekeeper, to the Board of Longitude. Dated March 26, 1764." Published within the appendix "The Original Observations of the Going of the Watch from Day to Day" in Nevil Maskelyne's *An Account of the Going of Mr. John Harrison's Watch,...* (London, 1767), p. liii.

132. Hatton, *An Introduction to...Clock and Watch Work* (*op. cit.*, note 52), p. 22.

nomical method would suffice. What Newton and many others chose not to mention, however, were the inherent problems of the lunar-distance method: it could not be used for two or three days at the time of each new Moon, would be rendered useless when clouds obscured the sky, and involved tables and calculations that were subject to human error. Moreover, it required an accurate portable timekeeper to time the necessary observations. Nevertheless, the Board chose to overlook the difficulties. While they could not deny Harrison any reward, neither could they bring themselves to pay him the full amount. Thus, in October 1765, through manipulations and changes in the wording of the Longitude Act, Harrison received, much to his anger and frustration, a balance-due payment that brought his reward to only £10,000, half the amount for which his timekeeper had qualified under a just interpretation of the requirements.

To add insult to injury, the Board upheld the stipulation, maintained ever since their first support of his work in 1737, that, in order to receive this reward, he would be required to surrender not only the prize-winning watch, but the first three marine timekeepers as well.[133] Thus it was that on Friday, May 23, 1766, Nevil Maskelyne and his workmen arrived at Harrison's home to transport the three large timekeepers to the Royal Observatory at Greenwich. The encounter of the two champions of the rival methods was terse and cold, Maskelyne attending to an unwanted obligation of his office, Harrison fulfilling a condition for a reward that he felt was incomplete. It did not take long for the tension to be ignited: Were the machines in perfect order, as stipulated by the Board? Harrison said he would sign an affidavit to this effect and requested the Astronomer Royal to do the same. Playing it absolutely safe with the exactitude of a cautious administrator, Maskelyne agreed only to certify that they appeared to be in good condition. About 3:00 p.m., the papers were exchanged.

Now the Astronomer Royal had to decide how to move the three large and sensitive timekeepers from Harrison's house to the observatory, ten miles away. Harrison was in no mood to help. Adopting Maskelyne's cautionary attitude, he stated that if the Astronomer Royal had been empowered to move them, then it would be improper for him to help; if he did, he might be held liable for any accident that should befall the machines during the journey. Maskelyne and his colleague Professor Roger Long (1680-1770)—the Master of Pembroke College, Cambridge, who had arrived in the interim to assist in the negotiations—were infuriated by this obstinacy. Reluctantly, Harrison informed them that he usually moved his first machine in pieces and the two later machines as they were. Then, unable to suffer the situation any longer, he withdrew from the room, foolishly leaving the task of taking to pieces one of the most unusual and delicate timekeepers ever made entirely in the hands of someone who had no knowledge, nor cared to have any knowledge, of how to take apart a clock. In due course, a smith arrived with a pair of pincers. About 4:00 p.m., while Maskelyne was supervising the workmen he had brought with him to carry the machine and assisting with the disassembly, H.1 was dropped and badly damaged.[134] Following this, the other two timekeepers were then "so far abused in the Carriage by Land to Greenwich, as to be rendered quite incorrect" and, as far as Harrison could determine, "incapable of being repaired without having some essential Parts made anew."[135] Thus ended the first chapter in the history of the first three sea clocks.

133. Some writers have suggested that the Board of Longitude demanded the surrender of Harrison's earlier machines without just reason, but, because of the stipulations laid down by the Board when funding Harrison's work, this could not have come as a total surprise. It was unfair, however, that it took so long for the government to pay him the amount that was, by all reasonable interpretation of the 1714 Act, clearly his due and that the removal of the timekeepers from his house was conducted in a thoughtless and confrontational manner without respect either for the machines or for their maker. Furthermore, it should be remarked that, rather than making the timekeepers available "for the Use of the Public," as the Board had stipulated in 1737, Harrison's machines were neglected and allowed to fall into a state of decay.

134. This event was recorded by William Frodsham and Walter Williams in Harrison's "Journal" (*op. cit.*, note 72), pp. 113-14.

135. Harrison, *Remarks on a Pamphlet* (*op. cit.*, note 3), pp. 22-3.

Between October of that year and the following June, the timekeepers were tested at the Royal Observatory. H.1 had suffered badly from its fall, and, even after being repaired by Larcum Kendall (1719-1790),[136] its performance over the trial of 30 weeks was highly erratic. H.2's rate was better, but it was still not impressive. H.3, however, before stopping during its first trial, performed almost as well as H.4.[137] During a visit to Greenwich, probably between April and June 1769, the astronomer Jean Bernoulli III (1744-1807), the nephew of the mathematician Daniel Bernoulli (1700-1782), noted the location of two of the timekeepers: one was in the "chambre des muraux" (probably the mural quadrant room, now located at the west end of the Meridian Building) along with a regulator by John Shelton (fl. 1737-1769); another was in the library.[138] It cannot have been long after this that the three timekeepers were stored away. Compared to the new technology, they were large and awkward. Maskelyne undoubtedly had no desire to see them again. Forgotten and neglected during the remainder of his long administration, the machines slowly deteriorated.

**Figure 29. H.1 in 1920 before restoration, showing its deterioration from decades of neglect. Courtesy of the British Horological Institute, Upton, Notts., England.**

When Maskelyne died in 1811, he was succeeded as Astronomer Royal by John Pond (1767-1836), who recognized that some action should be taken to repair the machines. In June 1824, he ordered that two of the clocks should be repaired. Apparently these were not returned for several years because, on October 29, 1835, the newly appointed successor, George Airy (1801-1892), wrote to the chronometer makers Arnold and Dent concerning the whereabouts of the timekeepers and requesting their return. Two days later, Arnold and Dent replied that they had all three of Harrison's early marine timekeepers in their possession, "the first of which, from neglect, is in a complete state of decay. The second will not perform, but in other respects is perfect. The third is injured by rust in consequence of it being left with the glass broken, which has admitted moisture, and the chronometer has partly been taken to pieces and many parts are wanting."[139] Arnold and Dent requested permission to retain the timekeepers and offered to restore them at no charge, "feeling it to be of great importance that so valuable a specimen of national production should be preserved."[140] Another five years were to pass, however, before their work was completed. On May 26, 1840, Airy wrote again to Arnold and Dent to ask if they could say when the timekeepers would be returned. As a result of this prompting, Dent set to work to complete the restoration, and on July 30 he requested a two-week extension, stating that the "cases are now being painted."[141] The problems of restoration were compounded by the fact that no drawings or explanation of the timekeepers could be found; therefore, to facilitate the work, Dent had a set of five drawings made by Thomas Bradley[142] (four of which are shown in Figures 11, 15, 21, and 22). The timekeepers were made to look respectable but were not restored to going

136. Larcum Kendall had been apprenticed to John Jefferys, the maker of Harrison's pocket watch, whose performance encouraged Harrison to construct H.4. In 1767, Kendall was commissioned by the Board of Longitude to make an exact copy of H.4. Further information on these timekeepers is given in Anthony Randall's paper in this volume.

137. The records of these trials are printed in Appendix I of Gould's *The Marine Chronometer* (*op. cit.*, note 11), pp. 259-60.

138. Jean Bernoulli, *Lettres Astronomiques où l'on donne une idée de l'état actuel de l'astronomie pratique...* (Berlin, 1771), pp. 95-6. Based on his information that the regulator by John Shelton was due to go to the North Cape of Norway, Bernoulli's visit to Greenwich can be dated to shortly before the Transit of Venus expedition in June 1769. Further on in the description of his visit (p. 110), while regretting that he had never had the chance to visit Harrison's house, Bernoulli mentions that Harrison had a beautiful collection of astronomical instruments, some of which were purchased at the sale of James Short's estate, which took place on April 5 and 6, 1769.

139. Vaudrey Mercer, *The Life and Letters of Edward John Dent, Chronometer Maker, and some account of his Successors* (London: Antiquarian Horological Society, 1977), p. 99. Mercer provides a valuable account of this period in the history of Harrison's first three marine timekeepers, quoting some of the original correspondence, but unfortunately without indication as to the whereabouts of the documents.

140. *Ibid.*, p. 99.

141. *Ibid.*, p. 99.

order,[143] and they were returned to Greenwich on August 25, 1840.

Eighty years passed before any further attention was paid to the timekeepers. Of the three, H.1 was in by far the worst condition (Figure 29). Rupert Gould (1890-1948) saw it for the first time in the spring of 1920:

> Covered with a glazed penthouse resembling a cucumber-frame, it was standing loose upon a rough wooden baseboard. Removal of the cover disclosed a mass of mechanism which might once have been polished brass, but which now offered to the eye a superb bluish-green patina, rivalling any Etruscan bronze in the British Museum. Had it sunk with the 'Royal George' and been recovered from Spithead after the War, it could scarcely have achieved a more perfect degree of corrosion.[144]

The Astronomer Royal, Sir Frank Dyson (1868-1939), gave permission for Gould to undertake some initial restoration, but he was not allowed to do any reconstruction at this time in order to put the machine in going order. Gould undertook this work without payment, although for subsequent work on the timekeepers he was reimbursed for out-of-pocket expenses. It took him about a year "to get what remained of the machine moderately clean and covered in a gold-tint lacquer." Although the decision not to allow him to reconstruct the machine must have been rather frustrating at the time, he later admitted that it was a "GOOD THING" because "my knowledge of Harrison's mechanisms was then quite inadequate."[145] In 1921, following this work, H.1 was elevated from its former neglected surroundings to an airtight showcase in the Octagon Room at the Royal Observatory, where it remained for ten years, until Gould undertook the reconstruction necessary to make it operable.

Gould was then granted permission to restore Harrison's other timekeepers to working order. He attended first to H.4, which was in fairly good condition, having been cleaned in 1890 for display in the Naval Exhibition. Then, about 1923, he tackled H.2, which had remained in the best condition of the three large machines. Within a year, it was operating once again and was exhibited, first at Wembley in 1925 and then at the Science Museum, where it remained until 1935. After this, Gould undertook the restoration of H.3 (Figures 30 and 31): the machine that had taken Harrison about seventeen years to perfect took Gould seven years to restore. The work was begun in 1924 and, with various interruptions, was not completed until 1931, being returned to Greenwich on March 23 of that year.[146]

The only timekeeper then remaining to be placed in working order was H.1. Having had the experience of restoring H.2, H.3, and H.4 in the intervening years, Gould was granted permission in the summer of 1931 to reconstruct H.1—restore would be too light a word because so many parts were missing. He admitted that the job looked

> pretty formidable....There were no mainsprings, no mainspring-barrels, no chains, no escapements, no balance-springs, no banking-springs, and no winding-gear. Many parts of the complicated (gridiron) compensation were missing, and most of the others defective. One of the eight long balance-bearers was gone, and another had shed its counterpoise. The seconds-hand was missing, and the

142. Bradley's superbly detailed drawings of the Harrison timekeepers were originally kept in a volume, dated 1844-8, in the papers of George Airy entitled "Correspondence with clockmakers," MS. RGO 6/586, under sect. 24, "Five drawings of Harrison's three chronometers (furnished to the Board of the Admiralty by Messr^s^ Arnold & Dent)." They are now preserved in the Royal Greenwich Observatory Archives at the Cambridge University Library (MS. RGO 6/586 f213^r^-f217^r^). Very little is known about Thomas Bradley: it appears that Dent's correspondence with Airy mentions him on only one other occasion, in 1842, this time in connection with a drawing of a clock escapement.

143. Gould, "The Restoration" (*op. cit.*, note 114), p. 2. Gould mentions that broken parts were filed off to make them look neat, leaving no indication as to the size and shape of the original.

144. Gould, "The Reconstruction" (*op. cit.*, note 69), p. 193.

145. *Ibid.*, pp. 193-4.

146. The above information was taken from Gould's well-known lecture, "John Harrison and his Timekeepers," which was delivered to the Society of Nautical Research on February 21, 1935, at Drapers' Hall in London, where all five of Harrison's marine timekeepers were exhibited together for the first time. The lecture was published in *The Mariner's Mirror,* vol. 21, no. 2 (April 1935), pp. 115-39, and subsequently by the National Maritime Museum in a pamphlet that has been reprinted many times since 1958.

Figure 30. H.3 before it was restored by Rupert Gould. Courtesy of the National Maritime Museum, Greenwich.

> hour-hand broken. As for small parts—pins, screws, etc.—scarcely one in ten remained.[147]

Because the temperature compensation had rusted so badly, Gould did not attempt to restore it to a functional condition. From what remained of the timekeeper, however, some of the symmetrical parts could be duplicated from their surviving counterpart, while the shape and size of others could be determined on general principles. Although he attempted to stay as close to the original as possible, he did introduce several features that did not exist in the original, for example the four small supporting posts that are shown in many photographs of the timekeeper and a new key winding mechanism, which was later converted back to its original pull-wind form. Gould stated that, in general, the assembly was not as difficult as H.2 or H.3 except for the last process, that of adjusting the steel check-pieces on the balance springs, which, he said, could "only be likened to threading a needle stuck into the tailboard of a motor-lorry which you are chasing on a bicycle."[148] About 4:00 p.m. on February 1, 1933, he finally succeeded in "threading the needle," and soon after H.1 came to life, perhaps for the first time since its trial at the Royal Observatory ended on June 1, 1767. It is largely due to Gould that these wonderful machines survived, for without his years of dedicated restoration, research, and publication, their importance might not have been recognized, and, like so many others treasures of technology that have been allowed to deteriorate, they might have been discarded.

During the Second World War, all the timekeepers except H.3 were hurriedly evacuated for safekeeping from Greenwich to the Cambridge Observatory; H.3 was sent about 1940 to Gould, who had moved away from London at the beginning of the war to Upper Hurdcott, near Salisbury. H.1 unfortunately suffered again from being stored in a damp location. In 1945, the timekeepers that had been stored in Cambridge were returned to the National Maritime Museum and then, beginning with H.4 in December 1945, were sent to the Ministry of Defence's Chronometer Department in Bradford-on-Avon to be restored by David Evans (1919-1984). Gould, who by June 1946 had had a heart attack and was suffering from Parkinson's disease, realized that he could no longer work on H.3 and therefore sent it to Evans as well. Evans completed the restoration on H.2, H.3, and H.4, but he returned H.1 to Greenwich because the amount of work would occupy more of his time than the Ministry of Defence would allow.

147. Gould, "The Reconstruction" (*op. cit.*, note 69), p. 194.

148. *Ibid.*, p. 195.

149. Fletcher described his work in two articles: "Temperature Compensation of Harrison's First 'Marine Timekeeper' " (*op. cit.*, note 64) and "Restoration of John Harrison's First Marine Timekeeper" (*op. cit.*, note 27).

In 1951, the Astronomer Royal, Sir Harold Spencer Jones (1890-1960), entrusted Douglas W. Fletcher (1891–1965), a civil engineer, to restore H.1 to going order in time for the British Clockmaker's Heritage Exhibition that was due to open in May 1952. Fletcher did this work in his spare time, initially making frequent visits to Herstmonceux Castle in Sussex, where the Royal Greenwich Observatory was by this time located, to collect and return various parts of the timekeeper, and eventually taking all the parts in order to complete the restoration. Because Gould had made certain improvements that Harrison had not incorporated in H.1, Fletcher undertook to restore the timekeeper as close as possible to its original condition and to rebuild the gridiron temperature compensation.[149]

Since the early 1950s, the Harrison timekeepers have all been on exhibition at the National Maritime Museum at Greenwich, but until 1980, their care remained the responsibility of the Chronometer Department of the

**Figure 31. Lt.-Comdr. Rupert T. Gould with H.3 in 1920. Gould devoted about fifteen years of his life to the restoration of Harrison's timekeepers. Courtesy of the National Maritime Museum, Greenwich.**

Ministry of Defence's Hydrographic Office. In that year, when the Chronometer Department closed, the preservation of the timekeepers became the responsibility of the National Maritime Museum and soon after were moved up the hill to the Old Royal Observatory, where they are now on display. Thus it is that Harrison's timekeepers have finally been recognized as national treasures and are now enjoyed by thousands of visitors every day.[150] Long may they continue to serve as a memorial to their maker.

In researching and writing a work of this nature, one inevitably benefits from others whose studies serve as foundations to build upon. I owe a debt of gratitude, in particular, to Colonel Humphrey Quill (1897-1987), who was kind enough to give me guidance and encouragement when I first became intrigued by Harrison's work about 26 years ago, and to Martin Burgess and George Daniels, who cultivated my interest. I am also grateful to Jonathan Betts and Andrew King for assisting in a variety of ways, supplying copies of original documents when needed and reading the text before publication, and to David Penney for providing several of the illustrations and copies of reference articles. In addition, my colleague David Landes made useful comments on the text. Assisting with the search for references and original documents have been my assistant Martha Richardson, Mariana Oller of the Houghton Library, and my work-study students, Hans Stöhrer and Amita Shukla. For some references, I have relied upon Adam Perkins, archivist at the Royal Greenwich Observatory, and upon The Time Museum in Rockford, Illinois, where former colleagues—Karon Anderson, Dorothy Mastricola Verstynen, and John Shallcross—have been a never-failing support. I also wish to thank my long-time friend Bruce Chandler and our ever-diligent copy editor, Peggy Liversidge, for their helpful comments. To these, and indeed many others who have enriched this paper, I am indebted.

150. Since it went on public display in the early 1950s, several people have been inspired to make copies of H.1: not only is it of great historical significance, being the first marine timekeeper to be tried successfully on a voyage of some duration, but it is also fascinating to watch in action. None of the four reconstructions made, however, have been produced with official approval or assistance. The first, a half-scale model, was made by Herbert Davies (1916-1993), a watch repairer in Portsmouth, England. Davies received detailed notes and drawings from D. W. Fletcher, who had restored H.1 in the early 1950s and was planning to make a one-third scale model himself. He also received some friendly assistance from the Astronomer Royal, Sir Harold Spencer Jones (1890-1960), who gave him access to the drawings that Gould had made in the early 1930s. Davies's reconstruction took about twelve years to make and was completed in the early 1960s. It was purchased from his widow in 1994 by the Worshipful Company of Clockmakers and is now on display in their museum at the Guildhall Library in London. For details, see Derek Pratt's article, "The first replica of John Harrison Sea Clock H1," in the *Horological Journal*, vol. 136, no. 3 (September 1993), pp. 88-91. The second, a full-sized model now owned by the National Maritime Museum, was made by an ingenious toolmaker and designer named Len Salzer (1922-1990), who lived in Biggin Hill, Kent, England. Salzer's first replica was begun in 1969 and was completed in 1977. It was made without any assistance—dimensions and details being gleaned by careful examination of the original machine through its glazed display case during his journeys home from work. Col. Quill, who served in intelligence during the war, told me that he considered Salzer's model to be equivalent to the greatest achievements in industrial espionage. Between 1981 and 1984, Salzer was commissioned by Seth Atwood to make another copy for The Time Museum in Rockford, Illinois, this time with the author's guidance and the invaluable assistance of Charles Allix and Jonathan Betts. (Well do I remember watching Len and Charles on that Monday afternoon—May 7, 1984—carrying "S.2" off the Boeing 747 at Chicago's O'Hare Airport.) For details of both reconstructions, see Len Salzer's article, "The Making of a Full Scale Model of No. 1," in the *Horological Journal*, vol. 125, no. 9 (March 1983), pp. 16-18, and Derek Pratt's article, "Len Salzer—the only man to make two replicas of John Harrison's Sea Clock, H1," in the *Horological Journal*, vol. 136, no. 8 (February 1994), pp. 262-5. A fourth reproduction, commissioned by Jürgen Abeler and made by Karl Wilhelm Bangert, Werner Bause, and Wilhelm Ehm (apparently in a period of only nine months!), was completed in October 1983 to celebrate the 25th anniversary of the Wuppertaler Uhrenmuseum in Wuppertal, Germany. Details may be found in Jürgen Abeler's booklet, *Die Longitudo zur See* (Wuppertal, Germany: Wuppertaler Uhrenmuseum, 1983), pp. 29-45, and in an article by the same author entitled "Die Nachbildung von Harrisons Seechronometer »H1«" in *Alte Uhren*, vol. 1/1984 (January 1984), pp. 11-22. A copy of H.2 is presently being made in England by Malcolm Leach.

APPENDIX 1

## CHRONOLOGY OF JOHN HARRISON'S PRINCIPAL INNOVATIONS AND INVENTIONS

| *Date* | *Innovation/Invention* | *Associated work/evidence* |
|---|---|---|
| *ca.* 1713 | Introduces the following innovations:<br>The wooden wheel construction (also used later in his regulators and in his first sea clock, H.1).<br>Reduction gearing for winding. The clock is wound with a wooden pinion-shaped key through holes behind the lower spandrels of the dial. | Longcase clock |
| *ca.* 1717 | Experiments with a modified anchor escapement. | Longcase clock |
| *ca.* 1720 | Introduces the following innovations:<br>The use of lignum vitae, a naturally oily wood. (The use of this material at points of friction overcame the need for oil. As a result Harrison's clocks could run for at least 30 years without need for cleaning.)<br>Bushes (bearings) made of lignum vitae and wheel pivots made of brass.<br>Roller pinion, using lignum vitae rollers.<br>Antifriction wheels made of lignum vitae.<br>Adjustable cycloidal cheeks. (Harrison recognized the phenomenon of circular error from his experience as a bell-ringer. He claimed he designed his own solution before he heard about Huygens's work.)<br>Invents the grasshopper escapement, a virtually frictionless escapement. | Brocklesby Park turret clock |
| 1720-30? | Invents a new scale of music, in which the relationship of an octave to a major third is in the same proportion as the circumference of a circle is to its diameter. (The date of this invention is not known for certain, but it was probably during this period.) | Playing the viol, training the choir, and bell-ringing; Harrison describes his scale in *A Description Concerning Such Mechanism* (*op. cit.,* note 33), pp. 68-108 |
| 1725-6 | Develops the following:<br>Improved use of antifriction wheels.<br>Improved grasshopper escapement.<br>Improved design for making the cycloidal cheeks adjustable. | Longcase regulators |
| 1727-8 | Invents the gridiron pendulum, a pendulum that could maintain a constant length despite changes in temperature (an effect that would otherwise influence the performance of a clock). Harrison designed this pendulum so that the amount of temperature compensation was adjustable.<br>Devises a method of overcoming variations in timekeeping caused by changes in barometric pressure, by adjusting the curvature of the cycloidal cheeks and altering the amount of compensation given by the gridiron pendulum.<br>Introduces the method of reciprocal correction for perfecting the performance of his regulators, testing one clock against another in rooms of different temperature and then reversing the procedure. | Longcase regulators |

| *Date* | *Innovation/Invention* | *Associated work/evidence* |
|---|---|---|
| 1729-35 | Develops the following improvements:<br>Modified grasshopper escapement.<br>Linked arrangement for the balances, so that their oscillation would not be affected by the motion of a ship at sea.<br>High- and low-tin bronze alloys for use in place of steel.<br>Antifriction arms to support the balances.<br>Pull-winding mechanism to allow the timekeeper to be wound without removing its case.<br>Gridiron temperature compensation to act upon the balance springs, resulting in the first temperature-compensated balance.<br>Invents the following:<br>A form of maintaining power that operates automatically.<br>Stopwork, to prevent the clock from being overwound. | H.1 |
| 1737-9 | Devises the following innovations:<br>Constant-force, spring-driven remontoire.<br>Improved design of adjustable gridiron temperature compensation.<br>Improved geared form of stopwork. | H.2 |
| 1739-49 | Devises the following innovations:<br>Constant-force remontoire with adjustable tension.<br>Isochronal corrector.<br>Invents the following:<br>Caged roller bearing.<br>Bimetallic strip. | H.3 |
| *ca.* 1753 | Develops the following improvements:<br>Maintaining power (the first for a watch).<br>Internally cut teeth (on third wheel—to reduce friction when the escapement causes the train to recoil).<br>Invents the following:<br>Modified verge escapement with diamond pallets (with isochronal adjustments to suit the balance spring).<br>Method of shaping diamond pallets.<br>Bimetallic curb temperature compensation (earliest surviving—and probably first successful—example of temperature compensation in a pocket watch). | Jefferys watch |
| *ca.* 1755 | Invents an equal-impulse anchor escapement. | Tower clock at Trinity College, Cambridge |

| Date | Innovation/Invention | Associated work/evidence |
|---|---|---|
| 1755-9 | Introduces the following:<br><br>A reduction in the size of his marine timekeeper. (Before details of H.4 became widely known, makers of the period envisaged that, to be successful, a marine timekeeper had to be large [see, for example, Le Roy's and Berthoud's early work]. The compact size of H.4—about 5 in. in overall diameter—changed this vision and thereby influenced the future size of the marine chronometer.)<br><br>Use of pierced jewels as bearings for most of the pivots, in order to reduce friction. H.4 is the first example of a timekeeper with an extensive use of jewels.<br><br>"Quick train." (H.4 was also the first marine timekeeper to use a fast-moving balance, oscillating 18,000 times per hour [five times every second]. His former timekeepers and the early ones made by Le Roy and Berthoud had employed a balance beating 3,600 times per hour [once every second]. Harrison recommended a 21,600 train [the balance beating six times every second] for timekeepers smaller than H.4. Eventually, a 14,400 train [the balance beating four times every second] became the standard for marine chronometers.) | H.4 |
| Before 1766 | Gridiron pendulum with further auxiliary adjustable temperature compensation. | Shown in the portrait of Harrison by T. King (Figure 28) |

APPENDIX 2

## SUMMARY OF PAYMENTS MADE TO JOHN HARRISON BY THE BOARD OF LONGITUDE

| *Amount* | *Date authorized* | *Purpose (Remarks)* |
|---|---|---|
| £250 | June 30, 1737 | Construction of H.2 (£500 was granted but £250 was contingent upon the successful trial of H.2 which was never carried out) |
| £500 | January 16, 1741/2 | Construction of H.3 |
| £500 | June 4, 1746 | Construction of H.3 |
| £500 | July 17, 1753 | Construction of H.3 |
| £500 | June 19, 1755 | Completion of H.3 and two watches, "one of such a size as may be worn in the Pocket & the other bigger"; the latter was to become H.4 |
| £500 | November 28, 1757 | Final adjusting of H.3 and completion of the above-mentioned watches (H.4 and the pocket-watch) |
| £500 | July 18, 1760 | Final adjusting of H.4 |
| £250 | March 12, 1761 | Costs related to first sea trial of H.4 |
| £200 | October 13, 1761 | Construction of H.4 (£500 was authorized but only £200 was paid at this time due to lack of funds) |
| £300 | June 3, 1762 | Construction of H.4 (balance of previous payment) |
| £1,500 | August 17, 1762 | Award following the first sea trial of H.4, with stipulation of a further £1,000 to be paid after a second sea trial |
| £300 | August 9, 1763 | Expenses related to second sea trial of H.4 |
| £1,000 | September 18, 1764 | Award following the second sea trial of H.4 (as stipulated on August 17, 1762) |
| £15 | February 9, 1765 | Expenses incurred for computations following the trial of H.4 to Barbardos. |
| £7,500 | October 28, 1765 | First half of longitude prize (£10,000), minus £2,500 paid in 1762 and 1764 (as stipulated on August 17, 1762) |
| £8,750 | June 19, 1773 | Second half of longitude prize (£10,000), minus £1,250 paid in 1737, 1741/2, and 1746 (as stipulated on January 16, 1741/2; this part of the prize was granted through the intercession of George III) |
| £23,065 | | TOTAL RECEIVED BY JOHN HARRISON FROM THE BOARD OF LONGITUDE |

# The Timekeeper that Won the Longitude Prize

## *Anthony G. Randall*

Anthony G. Randall is a watchmaker who specializes in the design and construction of precision timekeepers that employ new ideas. After receiving a B.Sc. in applied physics from Manchester University, England, in 1960, he studied watchmaking for eighteen months at the Technicum Neuchâtelois in La Chaux-de-Fonds, Switzerland (1963-4). On return to England, he worked for a few months with George Daniels in London and then moved to Birmingham, where he taught horology at the School of Jewellery and Silversmithing from 1965 to 1971. Randall was made a Fellow of the British Horological Institute (1964) and has been awarded the following honors: the Victor Kullberg Medal of the Stockholm Watchmakers' Guild (1983), the Worshipful Company of Clockmakers Certificate of Excellence (1985), and the Barrett Medal of the British Horological Institute (1991). His publications include *Catalogue of Watches in the British Museum* (with Richard Good, 1990) and *The Time Museum Catalogue of Chronometers* (1992).

# The Timekeeper that Won the Longitude Prize

*Anthony G. Randall*

As so often happens in life, we learn far more from our failures than from our successes. Even what appear to be failures may turn out—in the long run—to have been the very stepping stones to success. This was surely the situation for John Harrison (1693-1776) with H.3—though whether or not he was in a position to appreciate the fact, toward the end of those twenty-odd years of purgatory while he struggled with its vagaries, seems a little unlikely! What H.3 taught him, above all, was the strange properties and complex behaviour of balance and spring combinations, and how one can obtain that almost mystical property known to horologists as isochronism.[1] As Will Andrewes has explained, Harrison was unable to obtain isochronism for H.3's double balance system with its single short, stiff balance spring.[2] Even if Harrison had succeeded, it is extremely doubtful whether H.3, with all its complexity, could ever have been reproduced in quantity, at a price that navigators could have afforded.

Harrison clearly needed some sort of 'break'. This came in the early 1750s, in the middle of his Herculean struggle with H.3, when he designed an extraordinary pocket watch for his own use. This design was to provide a fresh approach to the whole problem he was facing. Until then, he seems to have assumed that a timekeeper for the longitude, in order to be not only accurate but stable, would, of necessity, have to be big. At the time, of course, nobody knew how big or small a successful timekeeper might need to be—indeed, this was all part of the problem to be solved.

## THE JEFFERYS POCKET WATCH

Having designed the new watch, Harrison employed a watchmaker, called John Jefferys (active 1726-1753), to make it for him. We do not know under what terms Jefferys worked on this project, or how much he may have contributed to it. Nor do we know much else about Jefferys, except that he had an apprentice, Larcum Kendall (1719-1790), who, it is assumed, also worked for Harrison. Curiously, Harrison seems to have allowed other workmen to sign timekeepers of his design when they had been responsible for the actual construction.[3]

1. Isochronism simply means that whether the arc of swing of a pendulum or balance be large, small, or even varying in between, the time taken for each swing remains constant and unvarying.

2. See William Andrewes's paper in this volume.

3. Another similar instance is James Harrison's signature on the early regulators, at a time when the two brothers were working together. For details, see Andrew King's paper in this volume.

Figure 1. John Harrison's own pocket watch, known as the Jefferys watch. It was severely damaged during the Second World War, when it was roasted in a burning building. Diameter of case: 6.2 cm. (2.4 in.). Courtesy of the Worshipful Company of Clockmakers, London, and Trinity House, Hull. WCC cat. no. 187.

Figure 2. View inside the inner silver case of the Jefferys watch, showing the back plate and the dust cap that bears the signature and the date. The original gilding has almost disappeared due to the heat of the fire. Diameter of back plate: 4.1 cm. (1.62 in.). Courtesy of the Worshipful Company of Clockmakers, London/Bridgeman Art Library, London, and Trinity House, Hull.

In external appearance the Jefferys watch resembles many an English silver pair-cased pocket watch of the mid-eighteenth century (Figures 1 and 2). The movement also has features that are similar to other watches of the period. On John Harrison's death, the watch passed to his son William (1728-1815), thence via a grandson to a great grandson who died in 1894. In 1905 another relative sold the watch to a Mr. Rust, a jeweller, who had a shop in Hull. During the Second World War the shop was bombed and the safe containing the watch was baked for several days. Miraculously, the watch was not completely destroyed, and it was discovered by Colonel Humphrey Quill (1897-1987) in 1954. It was subsequently partially restored, preserving most of the original parts, by David Evans (1919-1984), when he was working at the Admiralty chronometry workshops at Herstmonceux Castle. Unfortunately, the enamel dial has darkened, almost obscuring the figures, and only faint traces remain of the gilding on the movement. The steel components have all been annealed to the strength and consistency of soft iron, parts like the hands losing all trace of blueing in the process. The brass parts have also become very soft. As a result the watch, though complete, cannot be run. David Evans once declared to the writer that he had replaced one of the pallets, which was missing. If so, he did a superb job, as they now appear to be identical.

When compared with a conventional verge pocket watch of the mid-eighteenth century, the Jefferys watch contains no less than four remarkable—and revolutionary—features, none of which exist in any previous watch. It has maintaining power, to prevent stopping or even reversal of the train during winding (Figure 3). The third wheel has internally cut teeth, to allow the train to recoil with the escapement, with far less friction than the conventional type of gearing (Figure 4). The escapement (Figure 5) is a derivative of the verge, but it allows at least double the amplitude and, perhaps more significantly, enables isochronal adjustments to be made to suit the balance spring. Finally, the watch has thermal compensation in the form of a straight bimetallic curb, similar to that in H.3 (Figures 6 and 7). It is the earliest surviving watch with compensation and probably the first successful one. John Ellicott (1706-1772) made a watch with compensation at a slightly earlier date, but it seems likely that the compensation did not work properly.[4] Neither the watch nor any details about it appear to have survived.

In addition to these four features, the Jefferys watch incorporates jewelling at all the pivots of the escapement and train as far as the third arbor, all with endstones. The balance is made of gold and is remarkably massive.

No records of the actual going, or performance, of this watch have survived, but we learn that it must have done well from remarks made by John Harrison's grandson, also named John (1761-1842). Writing under the pseudonym Johan Horrins, an anagram of his name, he said:

> The smaller Watch intended for the pocket, and the original of the chronometers used for finding the Longitude at this day, is now in the Author's possession. It was made under the Inventor's inspection by a clever workman, whom he allowed to put his name on it, *viz.* John Jefferys, which is repeated on the cap, with the addition of the date, and this being 1753, shows it to have been constructed two years prior to its being brought forward. It was always John Harrison's pocket watch, except when Admiral Campbell[5] borrowed it, to find his Longitude by, for which it answered nearly as well as the larger but more expensive Timekeepers.[6]

It has been assumed from these remarks of Horrins that it was the success of the Jefferys watch that encouraged Harrison to persevere by designing an improved version. That is how he came to design H.4. Initially, H.4 was intended as a deck watch for use on board ship to carry the time from H.3 up to the deck, where the actual observations of heavenly bodies were to be made.[7] However, as Harrison's hopes for the ultimate success of H.3 gradually ebbed away, H.4 was found to fill the gap.

## THE CONSTRUCTION OF H.4

The earliest known mention of H.4 (Figures 8-11 and 18-19) occurred at a meeting of the Board of Longitude on the 19th of June 1755. At that same meeting, Harrison presumably showed them his own pocket watch for the first time. This is probably what Horrins meant by 'its being brought forward'. Harrison then asked for an advance of money to enable him to do further work on H.3 and to undertake the construction of two more watches, one of pocket size, the other larger. He was awarded £500, 'to adjust his Machine [H.3] & to complete & bring the said two Watches to as great Perfection as possible'.[8] Of the two watches, only the larger one, H.4, was ever completed, as far as we know.

By 1759, H.4 was probably completed but not adjusted. That year, the movement was engraved and the outer case hallmarked. The inner case, marked 'IH', was probably made by Harrison himself. It bears the date letter for 1758. The outer case has the maker's mark 'HT', attributed to Henry Thompson of Silver Street, London.

The Board of Longitude first saw H.4 at their meeting on the 18th of July 1760, but at that time the temperature compensation was still not fully

4. '...in the year 1748, I made a model of a contrivance to be added to a pocket-watch, founded upon the same principles, and intended to answer the like purpose, as the pendulum above described [his temperature-compensated pendulum]. And, at a meeting of a council of this Society, on February 15 last, I produced a watch (which I had made for a gentleman) with this contrivance added to it, and likewise the model, by which was shewn to the gentlemen then present what effect a small degree of heat would have upon it. But, as I have not yet had sufficient trial of this watch, I shall defer giving a particular description of this contrivance, till I am fully satisfied to what degree of exactness it can be made to answer the end proposed'. John Ellicott, 'A Description of Two Methods, by which the Irregularity of the Motion of a Clock, arising from the Influence of Heat and Cold upon the Rod of the Pendulum, may be prevented', *Philosophical Transactions* of the Royal Society, London, vol. 47, no. 81 (1752), pp. 493-4. This paper was read on 4 June 1752.

5. Adm. John Campbell (1720-1790), a celebrated sailor and navigator, was also one of the mathematicians engaged by the Board of Longitude to calculate the official rate of H.4 after its second trial voyage to Barbados, in 1764-5 (see note 28).

6. Johan Horrins, *Memoirs of a Trait in the Character of George III...* (London, 1835), p. 166.

7. Humphrey Quill, *John Harrison, the Man who found Longitude* (London: John Baker, 1966), p. 77.

8. Board of Longitude Confirmed Minutes, 19 June 1755, MS. RGO 14/5, Royal Greenwich Observatory Archives, Cambridge University Library.

3

4

5

Figure 3 (above left). Under-dial work of the Jefferys watch, showing the drive to the centre seconds and the maintaining power ratchet and clicks. Courtesy of the Worshipful Company of Clockmakers, London, and Trinity House, Hull.

Figure 4 (top right). Part of the gear train and escape wheel of the Jefferys watch. The internally cut third wheel greatly reduces the friction when the escapement causes the train to turn backward against the force of the mainspring. Courtesy of the Worshipful Company of Clockmakers, London, and Trinity House, Hull.

Figure 5 (above right). Detail of the lower pallet assembly on the balance staff of the Jefferys watch, showing the polished steel pivot and the brass mount with its diamond pallet. The escapement is very similar to that in H.4. Courtesy of the Worshipful Company of Clockmakers, London, and Trinity House, Hull.

Figure 6 (below left). Detail of the back plate of the Jefferys watch, with the balance cock removed to show the heavy gold balance, the balance spring (replaced), and the compensation curb. Courtesy of the Worshipful Company of Clockmakers, London, and Trinity House, Hull.

Figure 7 (below right). Detail of the remains of the compensation curb of the Jefferys watch. The rack and pinion are for mean time adjustment. Courtesy of the Worshipful Company of Clockmakers, London, and Trinity House, Hull.

6

7

**Figure 8. Front view of H.4, Harrison's prize-winning watch. The decorations around the edge of the enamel dial were designed by John Harrison. Diameter of case: 13.3 cm. (5.25 in.). Courtesy of the National Maritime Museum, Greenwich. Inv. no. Ch.38.**

adjusted. By the following year, the watch was ready for testing, and, in February, Harrison asked that both H.3 and H.4 should be given a trial at sea. Unfortunately, the Board of Longitude had not made proper preparation for this, though they knew that such a request would be forthcoming. They sent William Harrison, John Harrison's son, to Portsmouth with H.3 to await further instructions; John was to follow with H.4. But the instructions never came, and, after five wasted months, William returned to London. The protracted and pointless stay in Portsmouth must have been particularly hard to bear for William, whose wife was expecting a child. In a letter to him she mentions the row of beans he planted, to while away the time, though hardly expecting to see them grow. Not only did they grow, he had time to harvest them as well![9]

## THE FIRST TRIAL OF H.4

Presumably the Board of Longitude must have been somewhat chastened by this demonstration of incompetence. With the help of suggestions from the Royal Society, they set about organising the trial afresh. Meanwhile, the enforced lull had given John Harrison the time to complete all the adjustments of H.4 and to confirm that it went better than H.3. He therefore withdrew H.3 from contention for the longitude prize. While waiting to embark in the *Deptford* at Portsmouth, William Harrison was to allow John Robertson (1712-1776), master of the Portsmouth Royal Academy, to check

Figure 9. Under the dial of H.4, showing the motion work gearing and part of the maintaining power. Diameter of pillar plate: 10 cm. (3.94 in.). Courtesy of the National Maritime Museum, Greenwich.

the rate of H.4 against local time obtained by astronomical observation, to set the watch to that time, and to send the results under seal to the Admiralty. Robertson obtained successful observations on at least four days prior to sailing and set the watch accordingly. William Harrison and H.4 finally sailed aboard the *Deptford*, bound for Jamaica, on the 18th of November 1761. After encountering contrary winds, they left the English coast on the 28th of November. After ten days at sea, they were in the latitude of Madeira, but were unsure whether the island lay to the east or west. The captain, Dudley Digges (active 1760s), estimated that they were well to the east. William Harrison stated that they were 100 miles nearer to the island and that if they held their course, they would see Madeira the following morning. Captain Digges agreed to abide by the indications of the watch, and all were rewarded with sight of the island, as predicted by it. Not only was the usefulness of the watch clearly demonstrated, but the ship's company was much relieved. They had been forced to consign to the deep 1,057 gallons of beer and 48 pounds of cheese, which had gone bad, and were having to drink stagnant water from the ship's water supply.[10]

William Harrison recounted what happened on the next part of the voyage in a letter to his father-in-law, Robert Atkinson (1704-1789):

> We sailed from Madeira on the 18th December and see no land until 11th. January, when about half an hour after six in the morning we see the island of Deseada [Desirade]. In our voyage from Madeira to this place my reckoning by the watch was kept a secret, for we found that, was it known, all the reckoning in the ship would be corrected by it, if I had made it public; therefore Capn Digges, Mr Robison (which was the Gentleman that was sent along with me) and myself agreed that nobody should know the account the watch gave but ourselves, till we should be near making land; on the 10th. January in the afternoon I had a very good observation, and according to the watch and supposing the wind to continue to blow as it then did, we should see the island of Deseada the next morning about 10 o'clock. The Captain and I agreed that this afternoon (that is the 10th) I should tell all the Officers where we was by the Watch, and

9. Elizabeth Harrison (née Atkinson, 1736-1762) to William Harrison, 10 July 1761, private collection. William Harrison was probably staying with John Howard (1726-1790), of the Howard League for prison reform, who lived near Portsmouth. He and William had a mutual interest in the Foundling Hospital.

10. Captain's log of the *Deptford*, 7 December 1761, MS. Adm.51/241, Public Record Office, London, and Master's journal of the *Deptford*, 9 December 1761, MS. Adm.52/828, Public Record Office, London.

**Figure 10. Part of the gear train and remontoire mechanism of H.4, including the internally cut third wheel. Courtesy of the National Maritime Museum, Greenwich.**

> that I expected to see the land the next day; accordingly I told them which very much surprised them for, according to the nearest of their reckonings, they were 150 miles further from the Island than I was, but did not at all dispute my being right, I having already given them an instance at Madeira. Accordingly, about half an hour after six in the morning of the 11th (as I have before observed) we see the land which was only about three hours sooner than expected; the correctness of the watch amazed them very much. About 12 o'clock the same day we see Guardeloupe and about 2 in the afternoon see the island of Antego [Antigua]....[11]

And so the voyage continued, until they reached Port Royal, Jamaica, on the 19th of January 1762.

Soon after their arrival, they were informed by Commodore Forrest, the acting governor of the island, that no ship would be sailing for the return journey until June. William Harrison and John Robison (1739-1805), the latter having been sent by the Board to make the astronomical observations required to establish a meridian and determine the local time in Jamaica, naturally assumed that there was no rush to complete this task. But after making only a few observations, they were informed suddenly, on the 25th of January, that a ship would leave on the 28th as a matter of urgency. Word that war with Spain had just been declared was the ultimate cause of their precipitate departure.

In contrast to the outward journey, the voyage home in the little sloop *Merlin* was a nightmare. The weather was so rough that, at one stage, the ship lost its rudder. Luckily, conditions eased long enough for a replacement to be fitted. The sloop also sprang leaks on seven or eight occasions, with water frequently six inches deep in the captain's cabin, where the watch was kept. William Harrison decided that he had to keep H.4 going, for fear that it might be considered 'a thing too delicate and tender to go to sea'.[12] His efforts to keep the sea-water out of the watch box are described thus:

11. William Harrison to Robert Atkinson, 10 April 1762, MS. 6206, Library of the Worshipful Company of Clockmakers, Guildhall Library, London. This letter was written after the return to England.

12. *Ibid.*

13. Private 'Journal', in manuscript, maintained for John and William Harrison between 1761 and 1766 by Walter Williams, friend and legal advisor of William Harrison. The original is in private ownership; a contemporary copy exists in the Public Library, Melbourne, Australia.

14. *An Account of the Proceedings, in order to the Discovery of the Longitude:...*, 2nd ed. (London, 1763), p. 80.

15. The rate is the small daily amount that a timekeeper is found to gain or lose when tested before the start of a trial; it is multiplied by the number of days since the timekeeper was last set accurately to time. If the rate is losing, then the figure obtained must be added to the time shown by the hands to find the exact time—and vice versa. Provided that the rate remains constant, the exact time can always be found from the timekeeper. If not, an increasing, unpredictable error accumulates.

Figure 11. Detail of H.4's remontoire releasing mechanism and spring, and the fourth wheel, which drives the remontoire. Courtesy of the National Maritime Museum, Greenwich.

> Mr Harrison had no other method than to keep a blanket about it, and when it had imbibed so much water that he could with his hand squeeze it out...he replaced the Box in another, and had at times no other method to get these blankets dry again but by covering himself up in them when he went to sleep, and this with the sea-sickness threw him into a severe fit of illness.[13]

William Harrison's relief at getting home safely must have been turned to elation when further observations had been made. His father soon announced that the timekeeper had lost only 1 minute 54.5 seconds during the entire voyage to and from the West Indies, lasting 147 days.[14] The Longitude Act of 1714 stipulated that the trial would involve only the outward journey, and that the longitude should be determined to within ½° (the equivalent of a total error of two minutes of time), to win the full £20,000 of the prize. H.4 had not only kept to within the ½° limit, but for twice the distance and duration required! These were, however, unofficial results, based on the assumption that a rate[15] would be applied to the going of the watch. H.4's performance had therefore, as far as the Harrisons were concerned, qualified to win the full £20,000 prize.

The Board of Longitude took a rather different view. Both Robertson's and Robison's figures were sent to three independent mathematicians for calculation, to evaluate the going of the watch. Their results were never made known and have, mysteriously, been lost. One can only presume that they must have been so good that the Board simply could not believe them. There was also the problem of the watch's rate. This should have emerged from Robertson's figures at Portsmouth, but it seems that the whole principle of applying a rate was not acceptable to the Board. Furthermore, the Board began to find fault with the way in which the trial had been carried out, seemingly blaming everyone else involved in general and William Harrison in particular. Even the precise longitude of Jamaica was in doubt. In the end, the only way of resolving the problem appeared to be to order a retrial. Bearing in mind the risks and hardships endured by William Harrison—and by H.4—on the first trial, and the injustice of the Board's behaviour, it is not at all surprising that this proposal was met with dismay and stiff resistance from the Harrisons.

## THE FRENCH ARE INVITED TO SEND 'SPIES' TO SEE H.4

While the Board of Longitude refused to accept that H.4 had won any part of the longitude prize, they did acknowledge that it was 'an invention of considerable utility to the public'.[16] Accordingly, they awarded Harrison £2,500, of which £1,500 was to be paid immediately, the remainder at the conclusion of the projected new trial. Harrison was heartily dissatisfied at this outcome, and so began a period of argument and wrangling in which others, including Parliament, became involved on both sides. At one stage, it was decided that Harrison, to obtain any part of the prize and as a precaution in case the timekeeper should be lost at sea, should be required to disclose full details of the mechanism of H.4. Whilst the Board was attempting to impose this stipulation, an invitation was made to the French, by person or persons connected with the governments on both sides—presumably for political reasons—that observers be sent to hear Harrison's explanation.

In view of the fact that Harrison had still not received any part of the actual prize, had been struggling some 35 years with the technological problems involved, and had demonstrated quite convincingly the accuracy of H.4, we can well imagine that he was not too keen on either of these developments. The French delegation, in the persons of the mathematician and astronomer Charles-Étienne-Louis Camus (1699-1768) and the clockmaker Ferdinand Berthoud (1727-1807), arrived on the 1st of May 1763. Jérôme de Lalande (1732-1807), the well-known French astronomer, was staying in London at the time, and it was through him that a visit was arranged on the 9th of May to Harrison's house, where they were shown H.1, H.2, H.3, and one of the wooden regulators. Harrison refused, however, to show them H.4, and, in order to avoid any further discussion (after playing a cat-and-mouse game with them for some time), he took himself off to Lincolnshire, until they had returned to France. As an aside, it is perhaps worth mentioning a

16. Board of Longitude Confirmed Minutes (*op. cit.*, note 8), 17 August 1762.

17. Lalande's diary of his trip to England in 1763, Bibliothèque Mazarine, Paris; this conversation is described by Seymour L. Chapin in 'Lalande and the Longitude: A Little Known London Voyage of 1763', *Notes and Records of the Royal Society of London*, vol. 32, no. 2 (March 1978), p. 172.

18. Now in the Musée National des Techniques, Conservatoire National des Arts et Métiers, Paris (inv. no. 1387).

19. Also in the Musée National des Techniques, Conservatoire National des Arts et Métiers, Paris (inv. no. 1388).

20. This watch is probably the one he numbered '417', which is now in the British Museum collection (ref. no. CAI-276). For a detailed description of this piece, see A. G. Randall, 'Ferdinand Berthoud: The influence of his contemporaries. Part 1, and his "Première Montre Astronomique" ', *Antiquarian Horology*, vol. 16, no. 2 (June 1986), pp. 149-65.

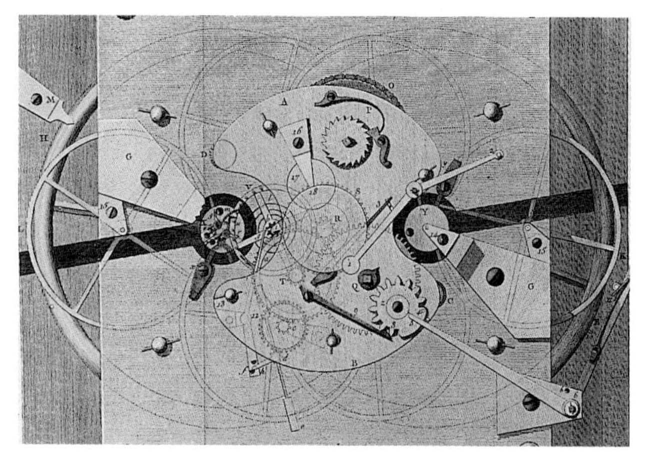

**Figure 12. Plan of Ferdinand Berthoud's *Horloge Marine* No. 2 of 1763, in which the grasshopper escapement and train remontoire are shown. The timekeeper itself is preserved in the Musée National des Techniques, Conservatoire National des Arts et Métiers (inv. no. 1387). From plate 6 in Ferdinand Berthoud, *Traité des Horloges Marines,...* (Paris, 1773). By permission of the Houghton Library, Harvard University.**

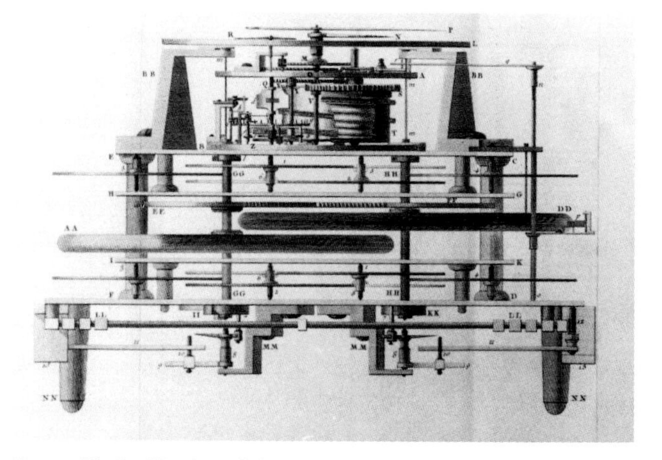

**Figure 13. Profile view of Ferdinand Berthoud's *Horloge Marine* No. 2, clearly inspired by Harrison's H.2 and H.3. It has a train remontoire, a grasshopper escapement, a gridiron compensation, and twin balances geared together with strip suspensions and antifriction wheels. From plate 5 in Ferdinand Berthoud, *Traité des Horloges Marines,...* (Paris, 1773). By permission of the Houghton Library, Harvard University.**

Figure 14. Ferdinand Berthoud's *Montre Marine* No. 3 of 1763, probably inspired by his visit to John Harrison in May of that year. This timekeeper belongs to Berthoud's first series of *montres marines* and should not be confused with his *Montre Marine* No. 3 of 1776. Diameter of case: 16.5 cm. (6.5 in.). Courtesy of the Musée National des Techniques, Conservatoire National des Arts et Métiers, Paris. Inv. no. 1388.

Figure 15. Back plate of Ferdinand Berthoud's *Montre Marine* No. 3 of 1763. It now has a compensation curb (running vertically across the centre) in place of the original gridiron compensation. Courtesy of the Musée National des Techniques, Conservatoire National des Arts et Métiers, Paris.

remark that Lalande recorded in his diary: at a dinner on the 14th of May 1763, Harrison's friend Taylor White (1701-1771) suggested that, at the next Parliament, the Board of Longitude should be terminated and its members reprimanded, for having 'vexed' Harrison![17]

As soon as Berthoud got back to France, he abandoned a timekeeper he had started before leaving and began his *Horloge Marine* No. 2 (Figures 12 and 13),[18] incorporating no less than four of Harrison's inventions. These were gridiron compensation, two balances interconnected, a type of remontoire, and a grasshopper escapement. Berthoud also made a large watch, his *Montre Marine* No. 3 (Figures 14 and 15),[19] whose mechanism was probably based on what he imagined H.4 to contain. He would have been able to gain some idea of the size and shape of H.4, either through Lalande who had seen it on the 22nd of April, or through some other contact. This knowledge, the actual size that a successful timekeeper should be, would have been invaluable to him.

Not content with that, he also made a pocket watch that looks suspiciously as if it was based on ideas inspired by the Jefferys watch. Berthoud called this watch his *Première Montre Astronomique* (Figures 16 and 17).[20] In actual fact, not one of these three timekeepers was very successful, demonstrating that Berthoud had only a superficial grasp of Harrison's concepts. The *Montre Marine* No. 3 was considerably modified on at least two occasions. Berthoud brought the pocket watch to London when he returned in 1766, where it was sold through a Mr. (Christopher?) Pinchbeck to King George III.

**Figure 16. Ferdinand Berthoud's *Première Montre Astronomique* of 1763, possibly inspired by the Jefferys watch. Diameter of case: 4.9 cm. (1.93 in.). Courtesy of the Trustees of the British Museum, London. Ref. no. CAI-276.**

**Figure 17. Detail of the back plate of Ferdinand Berthoud's *Première Montre Astronomique* showing the mean time adjustment and gridiron compensation, the compensation that may have been fitted originally to his *Montre Marine* No. 3 (see Figure 15). The escapement is an ordinary verge. Courtesy of the Trustees of the British Museum, London.**

## THE SECOND TRIAL OF H.4

John Harrison, as we have noted, at first resisted the Board's request for a second trial of H.4. At about the time of the first trial, however, he had begun to realise that the lunar-distance method, the leading alternative method for finding longitude at sea, was becoming a serious rival for the prize. This method involved determining the position of the Moon in relation to the fixed background of stars—in effect using the Moon as the hands of a giant clock.[21] It relied, after all, on astronomy and mathematics, precise subjects that were understood—and trusted—by the scholarly and academic members of the Board of Longitude, who would also have been well aware of the unreliability of ordinary watches. Was the performance of H.4 on its recent trial simply a fluke? It was only natural for Harrison to have assumed that the Board members would be biased in favour of 'lunars'. His fears can only have been heightened by the growing influence of Nevil Maskelyne (1732-1811), one of the principal advocates of the method who, in 1763, published *The British Mariner's Guide,* a work that contained full details of the theory and practice of the lunar-distance technique.[22] Along with this, John Harrison must have weighed the risks of damage to H.4—even its possible loss at sea—quite apart from danger to his son William. In the end, he decided that even though as far as he was concerned a new trial would merely be a rerun of the previous, he had to allow H.4 to go to sea again. As he said at the time, 'if it will end all disputes'.[23] If only! Such was the background to the second trial.

The various negotiations between Harrison, the Board of Longitude, and the government took from March 1762 until March 1764, with little to show for them except the agreement that a second trial should take place. Once the decision had been made, the Board began to organise the trial. The destination of the timekeeper was changed from Jamaica to Barbados, which was considered a healthier island at the time. Two astronomers, Nevil Maskelyne and Charles Green (1735-1771), were sent in advance to make a fresh determination of the longitude of Barbados. Maskelyne had been making no secret of the fact that he preferred the lunar-distance method for determining longitude at sea. The Harrisons, although on their guard, presumably gave Maskelyne the benefit of the doubt that this other interest would not preclude him from carrying out his duties. On the voyage out to Barbados, in advance of H.4, Maskelyne had been instructed by the Board

21. For details of this method, see Derek Howse's paper in this volume.

22. Nevil Maskelyne, *The British Mariner's Guide containing Complete and Easy Instructions for the Discovery of the Longitude at Sea and Land,...* (London, 1763). Maskelyne was to become Astronomer Royal in 1765 and hence an *ex officio* member of the Board of Longitude.

23. Harrison, 'Journal' (*op. cit.*, note 13).

24. *Ibid.*, pp. 112-14.

25. The astronomers would have been required to take 'equal altitudes' of the Sun to determine the precise local time and to observe the eclipses of Jupiter's first satellite and 'as many Occultations of the first Stars and Planets as may be visible at that Island...'. [John Harrison], *A Narrative of the Proceedings relative to The Discovery of the Longitude at Sea;...* [London], [1765], p. 6. By comparing the results of these observations with corresponding observations made at Portsmouth, the longitude of Barbados could be accurately determined.

26. John Bradley was a nephew of James Bradley (1693-1762), the Astronomer Royal from 1742 to 1762. John Bradley was described as having the rank of Purser at Portsmouth Naval Academy while he was assisting with computations relating to the second trial of H.4.

to try out the lunar-distance method, which he proceeded to do. He claimed to have predicted the position of the ship, shortly before reaching the island, to within ½°.[24] It would be only human to have been elated by this success and to have made remarks about it to his hosts on the island.

William Harrison and H.4 sailed from Spithead on the 28th of March 1764, in the *Tartar*. The outward voyage was almost a carbon copy of the previous, with the islands of Madeira and Barbados being accurately predicted as before. However, on arrival at Bridgetown, Barbados, on the 13th of May, William Harrison got to hear what Maskelyne had been saying about the merits of the lunar-distance method. Inevitably a dispute arose, in which Harrison accused Maskelyne of partiality toward his own preferred method and thus of not being unbiased in judging another method. In the end, this dispute resulted in an agreement that Maskelyne should be permitted to make observations by which the accuracy of the timekeeper would be determined, but only on alternate days with the other astronomer, Charles Green.[25] This incident caused Maskelyne considerable resentment and discomposure.

William Harrison sailed for home on the 4th of June, in the merchantman *New Elizabeth*, leaving Maskelyne and Green to continue observations of the eclipses of Jupiter's first satellite, the established method of determining the longitude of a place on land, in order to determine afresh the longitude of Barbados. The result of nine observations made by them at Barbados, compared with five similar observations that had been made at Portsmouth by John Bradley (active 1742-1772),[26] gave the longitude difference to within a distance of 8.77 miles of what it is now known to be. The corresponding difference in time between Portsmouth and Barbados averaged out at 3 hours, 54 minutes, 18.15 seconds.[27] By comparison, the official time difference for H.4, derived from the average of the calculations made by four specially selected individuals,[28] was 3 hours, 54 minutes, 57.27 seconds—a truly amazing result! This calculation was made after applying a rate for the timekeeper, which, in contrast to the previous trial, had been officially declared before leaving Portsmouth.[29] The official error of the timekeeper, after a voyage of 46 days, was, therefore, 39.1 seconds of time or, in the latitude of Barbados, less than ten nautical miles.[30] According to the Act of Queen Anne,[31] in order to win the full prize, the longitude had to be kept to within ½°, or 30 nautical miles. John Harrison could hardly have hoped, even in his wildest dreams, to have succeeded so convincingly. Surely, fame and fortune could not be denied him any longer.

27. Board of Longitude Confirmed Minutes (*op. cit.*, note 8), 19 January 1765.

28. Capt. (later Adm.) John Campbell (see note 5), Dr. John Bevis (1693-1771), George Witchell (1728-1785), and James Short (1710-1768). The first three, respectively a navigator, an astronomer and physician, and a mathematician, were selected by the Board of Longitude. Harrison's only choice was his friend James Short, a well-known optical instrument maker and a Fellow of the Royal Society.

29. The rates that William Harrison gave to the Admiralty on 26 March 1764 included the following: 'When the thermometer is at 42, it will gain 3 seconds in every 24 hours. When the thermometer is at 52, it will gain 2 seconds in every 24 hours. When the thermometer is at 62, it will gain one second in every 24 hours. When the thermometer is at 72, it will neither gain nor lose. When the thermometer is at 82, it will lose one second in every 24 hours'. Presumably, the thermometer was marked in degrees Fahrenheit. Quoted in Quill, *John Harrison* (*op. cit.*, note 7), Appendix 3.

30. There is a certain irony in that this is almost the same as the error now known to have occurred in Maskelyne's astronomical observations.

31. Act 12 Anne *cap.* 15 (1714): *An Act for Providing a Publick Reward for such Person or Persons as shall Discover the Longitude at Sea.*

## THE AFTERMATH OF THE SECOND TRIAL

No sooner had these results been declared and freely acknowledged by the Board of Longitude than fresh difficulties arose. The Board still knew nothing about the mechanism of H.4, or how difficult or expensive it might be for copies to be made. Would copies go as well as the original? Would they cost so much that no ordinary navigator could afford to buy one? Such was the Board's questionable interpretation of the clause 'Practicable and Useful at Sea', as set out in the original Act of Queen Anne. In order to resolve these uncertainties, the Board decided that the rules for the awarding of the prize would have to be changed—or, as they saw it, 'clarified'. The only way to do this legally was to have new rules sanctioned by Parliament. All resis-

**Figure 18. Back plate of H.4. The magnificent piercing and engraving were probably intended as a 'camouflage', to divert attention from the mechanism concealed beneath it. Diameter: 9.7 cm. (3.82 in.). Courtesy of the National Maritime Museum, Greenwich. Inv. no. Ch.38.**

tance from the Harrisons, in the form of petitions and pamphlets, was swept aside, and the Board got its way. In a new Act of Parliament, published in 1765,[32] it was stipulated that the prize should be divided into two parts of £10,000 each. Quite separate new conditions had to be met for the award of either part.

Before being willing to give Harrison even the first £10,000 of the prize, the Board now insisted that he make a full and complete disclosure of the mechanism of H.4 to a small subcommittee of scholars and horologists. As could be expected, Harrison continued to resist. As far as he was concerned he had won the prize, and by a substantial margin. The watch had proved itself in every way 'Practicable and Useful at Sea'. Payment should be made first; then he would disclose all the hard-won secrets of H.4's mechanism. Appeals for redress either to the public or to Parliament were all to no avail. The Board of Longitude would not budge. However, before being forced to capitulate, Harrison did manage to obtain some protection for his inventions from the Board. They decided that the individuals nominated to attend the explanations should be 'enjoined not to make any discovery [disclosure] of the Principles of the Watch to any but the Board without the leave of the Commissioners'.[33] Even so, as we shall see, the Board promptly forgot this undertaking and failed to inform all the members of their subcommittee.

32. Act 5 George III *cap.* 20 (1765): *An Act for explaining and rendering more effectual Two Acts, One made in the Twelfth Year of the Reign of Queen Anne,...and the other in the Twenty sixth Year of the Reign of King George the Second,....*

33. Board of Longitude Confirmed Minutes (*op. cit.*, note 8), 30 May 1765.

34. *Ibid.*, 18 July 1765.

## JOHN HARRISON REVEALS THE SECRETS OF H.4'S MECHANISM

Finding that there was no alternative if he wished to obtain any part of the actual prize, and having been reassured as regards confidentiality by the Board, Harrison finally agreed to take the oath to make a full disclosure and to fully explain the mechanism of H.4.[34] Without more ado, the Board sent a team headed by Maskelyne, including the natural philosopher John Michell (1724-1793), the mathematician William Ludlam (1717-1788), the instrument maker John Bird (1709-1776), and three watchmakers, Thomas Mudge

**Figure 19. Detail showing the bimetallic compensation curb, the balance and spring, and the adjustable cycloidal spring concealed beneath the decorative elements on the back plate of H.4. Diameter of balance: 5.5 cm. (2.2 in.). Courtesy of the National Maritime Museum, Greenwich.**

(1715?-1794), William Mathews (*ca.* 1723-*ca.* 1776), and Larcum Kendall. They all met at Harrison's house in Red Lion Square on the 14th of August 1765. Eight days later, Harrison was given a certificate, signed by them all, stating that his explanations, drawings, and answers to all questions had been given to their entire satisfaction.[35]

Although Michell and Ludlam subsequently expressed to the Board various misgivings about Harrison's explanations,[36] the Board was satisfied and at last, on the 12th of September 1765, agreed that Harrison should apply for the first £10,000, less £2,500 already received—but on condition that he hand over H.4 reassembled and in 'perfect condition' together with his other three timekeepers, H.1, H.2, and H.3. So, on the 28th of October, Harrison was finally forced to give up his most precious timekeeper, H.4 in return for a payment of £7,500. The Board gave instructions that it should be sealed up in its box and deposited at the Admiralty. It was to remain there, utterly useless, until the following year.

The other three machines remained with Harrison, but not for long. On the 23rd of May 1766, Harrison was surprised when Maskelyne and several workmen arrived unexpectedly at his house with an order from the Board for the removal of the other machines. After the initial shock, Harrison was understandably not very cooperative, saying that if he gave advice or assistance and anything went wrong, he would surely be held responsible, and he withdrew to his room on the floor above.[37] Shortly afterwards, while Maskelyne was organising the removal of the timekeepers, H.1 was dropped, while 'Mr. Maskelyne's hand was...on the machine'.[38] To make matters worse, all three machines were then taken to Greenwich Observatory in an unsprung cart, instead of by sedan chair to the river and thence by water, as would have been expected to avoid damage. After a brief trial at the Royal Observatory, all three timekeepers were simply stored away rather than being made available for study or put to any other sensible use. Thus, although it had been clearly stated on several occasions in the minutes of the Board of Longitude that the three timekeepers should come into their possession when the longitude prize had been won, it is hard to find any real justification for their removal from Harrison's workshop, where they had been admired by many.[39]

35. Certificate dated 22 August 1765. Described in Harrison's 'Journal' (*op. cit.*, note 13), p. 96.

36. Michell's letter dated 9 September was sent from Newark, Nottinghamshire, and arrived in time for the meeting of the Board of Longitude on the 12th. Ludlam presented his reservations in person at the meeting. Board of Longitude Confirmed Minutes (*op. cit.*, note 8), 12 September 1765.

37. Quill, *John Harrison* (*op. cit.*, note 7), p. 164.

38. Memorandum by William Frodsham and Walter Williams concerning the removal of the three timekeepers from Harrison's house, included in Harrison's 'Journal' (*op. cit.*, note 13), pp. 113-14.

39. While H.2 and H.3 had been subsidised by the Board during their construction, H.1 was made as a private venture. However, its future possession by the Board of Longitude 'for the use of the public' was clearly stated as a condition for the funding granted during the construction of the other two timekeepers. The subsequent history of these three machines is given in William Andrewes's paper in this volume.

## BERTHOUD OBTAINS THE SECRETS OF H.4

News of the success of H.4 soon reached France, causing a considerable stir. Ferdinand Berthoud was again dispatched to discover its secrets and arrived in London early in 1766. It appears that he first tried to negotiate with Harrison, who had asked the sum of £4,000 to explain the details of the mechanism. Berthoud, however, was authorised by the *Ministre de la Marine* to offer only £500, a proposal that was promptly rejected by Harrison as being too trifling a sum.[40] Berthoud then got in touch with Thomas Mudge, one of the members of the group to whom Harrison had made his disclosure, and Mudge proceeded to reveal everything he knew. The cause of this breach of confidence appears to have been the incompetence of the Board of Longitude, which had neglected to inform Mudge of the agreement made with Harrison on the 30th of May of the previous year. The information Berthoud received, however, was undoubtedly insufficient to allow him to construct an exact copy of H.4, as the French government expected. In fact, he never did so, wriggling out of the undertaking by delay and various excuses. He did, nonetheless, adopt the use of Harrison's straight bimetallic compensation curb, as used in H.4.

## THE SECOND £10,000 PROVES EVEN HARDER TO WIN

Having won the first half of the longitude prize, Harrison was now required to make two more timekeepers in order to fulfil the conditions of the second half of the prize. These timekeepers would then have to undergo an unspecified trial, at the Board's discretion, to prove that they performed as well as H.4. By 1766, however, John Harrison was in his 74th year, and it may have seemed to the ageing inventor that he had indeed been set an impossible task. To make matters worse, when he asked for the return of H.4 to help him make the duplicates, the Board refused. Instead, they sent it off to Greenwich, to undergo still further testing by—of course—Nevil Maskelyne, now the Astronomer Royal. Such an apparently pointless exercise, after the two previous successful sea trials, only lends weight to the assertion that the Board members still did not trust H.4 or believe that its performance was not some sort of fluke, or trick, or lucky chance. Harrison had expressed his frustration with their attitude three years earlier:

> But still, they say a watch is but or can but be a watch and that Mr Ellicott has tried what a watch will do and that the performance of mine (though nearly to truth itself) must be altogether a deception.[41]

When he had reassembled H.4 after the disclosure of its mechanism, Harrison had not prepared it for further tests. As a result, and because no attempt was made to establish its rate and it was subjected to tests for which it was not designed, its performance was predictably bad. Maskelyne wasted no time in publishing them and in stating in his conclusion:

> ...Mr Harrison's watch cannot be depended upon to keep the Longitude within a degree in a West India voyage of six weeks; nor to keep the longitude within half a degree for more than a few days.[42]

40. Berthoud uses the words 'une telle bagatelle' to describe Harrison's remark. This passage is quoted in Jean-Claude Sabrier's article, 'La contribution de Ferdinand Berthoud aux progrès de l'horlogerie de marine', in Catherine Cardinal, ed., *Ferdinand Berthoud, 1727-1807: Horloger mécanicien du Roi et de la Marine* (La Chaux-de-Fonds, Switzerland: Musée International d'Horlogerie, 1984), p. 168. The original reference comes from Ferdinand Berthoud's *Supplément au traité des montres à longitudes...* (Paris, 1807), Appendix, p. 63, article XLVI.

41. John Harrison, 'An Explanation of my Watch or Timekeeper for the Longitude...', 7 April 1763, MS. 3972/1, Library of the Worshipful Company of Clockmakers, Guildhall Library, London.

42. John Nevil Maskelyne, *An Account of the Going of Mr. John Harrison's Watch,...* (London, 1767).

Figure 20. Harrison's last marine timekeeper, H.5, in its original box. Similar in size to H.4, H.5 was tested at King George III's private observatory at Richmond in 1772. Ultimately, the success of this trial led to Harrison receiving a further award of £8,750 in 1773. Diameter of watch case: 13.3 cm. (5.25 in.); diameter of watch dial: 6.4 cm. (2.5 in.). Courtesy of the Worshipful Company of Clockmakers, London/Bridgeman Art Library, London. WCC cat. no. 598.

Figure 21. Detail showing the inside of the inner case and the back plate of H.5. In contrast to H.4, the back plate of this watch is very plain in design. The signature, 'No. 2 John Harrison and Son London 1770', indicates that it was his second longitude watch. Diameter: 9.83 cm. (3.87 in.). Courtesy of the Worshipful Company of Clockmakers, London/Bridgeman Art Library, London.

He then went on to damn it with faint praise, with a final suggestion that it might be a helpful aid in taking lunars! As an example of adding insult to injury, this conclusion surely has few equals. We can hardly be surprised if in his published reply, Harrison went slightly overboard in his criticism of Maskelyne.[43]

Another of the Board of Longitude's directives instructed Maskelyne to prepare, based on the information that Harrison had given under oath, a publication providing the details necessary to construct copies of H.4. Yet, from the information given in this work, *The Principles of Mr. Harrison's Timekeeper,*[44] it is hard to believe that anyone involved can have seriously imagined that this could be done. Not only are the drawings and text totally inadequate, but certain vital information, such as the making of the diamond pallets, is conspicuously absent. Whether this is merely a further example of the incompetence of the Board, aided and abetted by Maskelyne, or whether they were trying quite deliberately to keep certain information back for strategic reasons, or, as seems likely, they were simply out of their depth, we can only speculate.

In spite of all the difficulties and handicaps placed in his way, John Harrison, aided by William, set about making one more timekeeper. Started in 1766 or 1767, the new watch was engraved in 1770 but not fully adjusted until 1772. By then, however, John Harrison was in his 80th year, and it proved to be an impossibility for him to make yet another timekeeper as required by the Board. In desperation, approaches were made to King George III, who readily agreed that the new watch, now known as H.5 (Figures 20 and 21), should undergo a period of testing in his private obser-

43. John Harrison, *Remarks on a Pamphlet Lately published by the Rev. Mr. Maskelyne Under the Authority of the Board of Longitude* (London, 1767).

44. [John Harrison], *The Principles of Mr. Harrison's Time-keeper, with Plates of the Same* (London, 1767); facsimile reprint by the British Horological Institute in *Principles and Explanations of Timekeepers by Harrison, Arnold and Earnshaw* ([Upton, Notts., England], 1984).

vatory at Richmond Park. The result of this test has been given by Johan Horrins as an error of 4½ seconds after ten weeks.[45] The King was most impressed, and, as in a fairy story, when he heard the full extent of the way in which the Harrisons had been treated, he decided to use his influence to help them. A special bill was introduced to Parliament with the King's blessing, resulting in the award to John Harrison of an additional sum of £8,750, as 'a further reward and encouragement'.[46] If this is added to all the previous sums he received from the public purse, at one time or another during his career, the total comes to £23,065,[47] somewhat more than the total amount initially offered by the Act of 1714. Thus, despite his years of difficulties with the Board of Longitude, and despite the fact that he was denied the satisfaction of having won the second half of the prize outright, Harrison did receive his due financially for the 45 years he had devoted to solving the longitude problem.

## KENDALL'S COPY (K.1) OF H.4

Following Maskelyne's trial of H.4 at the Royal Observatory which ended in March 1767, the timekeeper had been handed over to Larcum Kendall, who had agreed to make a copy, although not to be responsible for its rate. He was to be paid £450. This copy, known as K.1 (Figures 22 and 23), took Kendall two and a half years to complete. It was engraved in 1769 and handed over on the 13th of January of the following year. It is in every way a worthy and very close copy of H.4, and Kendall was very deservedly awarded an additional £50 for having completed it so swiftly. This prompted certain members of the Board to make comparisons with the Harrisons, who at the time had still failed to produce any more timekeepers. As a result, the Board approached Parliament in order to initiate legislation that would require Harrison to forfeit any rights he might have had to the remaining £10,000 unless the two additional watches that he was required to make were completed within five years. Attempts to apply legal requirements of this sort to an octogenarian seem nothing short of ludicrous. Harrison, as we have seen, was very fortunate in finding in King George III a champion to help him to finally outmanoeuvre the Board.

On receipt of K.1, the Board sent it to Greenwich, and Maskelyne, for testing. The result was a disappointing performance—similar to H.4's when similarly tested. However, a more practical and dramatic test was about to get under way. Captain James Cook (1728-1779) was preparing to set out on his second voyage of discovery to the Pacific (1772-5). K.1 went with him, and he used it with great success to draw the first charts of Australia and New Zealand. The consistent rate of the watch meant that he could rely on it, which he came to do more and more. It saved the long delays that would have been inevitable if he had depended only on lunars and shore-based observations. Having mentioned K.1 in his log as 'our trusty friend the Watch' and 'our never-failing guide the Watch',[48] Cook paid his own tribute:

> It would not be doing justice to M$^{r}$ Harrison and M$^{r}$ Kendall if I did not own that we have received very great assistance from this usefull and valuable time piece....[49]

How very different were Cook's words from Maskelyne's harsh and big-

45. Horrins, *Memoirs* (*op. cit.*, note 6), p. 18.

46. Recommendations of a special Parliamentary Finance Committee, accepted by the Commons on 19 June 1773, *Journals of the House of Commons*, vol. 34, p. 383.

47. For a complete summary of the payments made to John Harrison by the Board of Longitude, see Appendix 2 in William Andrewes's paper in this volume.

48. J. C. Beaglehole, ed., *The Journals of Captain James Cook on his Voyages of Discovery*, vol. 2, *The Voyage of the* Resolution *and* Adventure, *1772-1775,* Hakluyt Society Publications, extra series no. 35 (Cambridge: Cambridge University Press, 1961), p. cxii.

49. *Ibid.*, p. 654, note 1. This passage, which is followed by Cook's signature, is included in the entry for Tuesday, 21 March 1775, the last entry in MS. Adm.55/108, Public Record Office, London (referred to by Beaglehole as the 'Admiralty MS. A').

**Figure 22. Larcum Kendall's first marine timekeeper, K.1 (the timekeeper copied from Harrison's prize-winning timekeeper, H.4), for which Kendall received £500 from the Board of Longitude. K.1 was taken on Captain Cook's second voyage (1772-5) to the South Pacific, where it performed with such reliability that Cook called it 'our trusty friend the Watch'. Diameter of case: 12.5 cm. (4.92 in.); weight in case: 1.3 kg. (2.87 lb.). Courtesy of the National Maritime Museum, Greenwich. Inv. no. Ch.39.**

**Figure 23. Back plate of K.1, illustrating the same exquisite design and workmanship that was lavished on H.4. Diameter: 9.81 cm. (3.86 in.). Courtesy of the National Maritime Museum, Greenwich.**

oted statements when he sat in judgement on both H.4 and K.1. Yet, not surprisingly, it was the Astronomer Royal, Maskelyne, who prevailed. In 1774, largely on account of the £8,750 award made to John Harrison, the legislation for awarding prizes for improvements in finding longitude was completely revised.[50] Maskelyne, who played a part in drafting the new document, is said to have remarked privately that he 'had given the mechanics a bone to pick that would crack their teeth'.[51] Needless to say, the £10,000 prize offered by the new legislation was never won.

Eight months after Cook's return, John Harrison died at his house at Red Lion Square. Whether he learned of Cook's vindication of his timekeeper, history does not record. We can only hope so. Leaving John Harrison the last word, by way of epitaph:

> ...I heartily thank Almighty God that I have liv'd so long as in some measure to compleat it....I think I may make bold to say that there is neither any other Mechanical nor Mathematical Thing in the World that is more beautiful, or Curious in Texture than this my Watch or Time-keeper for the Longitude....[52]

## CONCLUSION

When John Harrison began his long struggle, no scientist or horologist of the day could offer any practical solution to the longitude problem. Even Newton was baffled; maybe his inactivity in this area discouraged many a

50. Act 14 George III *cap.* 66 (1774): *An Act for the repeal of all former Acts concerning the Longitude at Sea.*

51. Thomas Mudge Jr., *A Narrative of Facts relating to some Time-Keepers, constructed by Mr. Thomas Mudge,...* (London, 1792), p. 5.

52. Harrison, 'An Explanation of my Watch' (*op. cit.*, note 41), pp. 100-1. Quoted in Rupert T. Gould, *The Marine Chronometer: Its History and Development* (London: J. D. Potter, 1923), p. 63.

lesser light and helped to create a psychological barrier. In breaching that barrier, Harrison solved perhaps the most important—and difficult—technological problem that was solved in the whole of the eighteenth century. At the time, even the background science did not exist, so Harrison had to start from scratch and do much of the basic research. For years he was alone in the field, with no other horologist seemingly able to enter the fray. Presumably the problems were just too daunting. Having battled with the technical difficulties, he was then faced with the long struggle with the Board of Longitude. Surely not many individuals would have survived both ordeals. In the end, his success in all these areas opened the way for others to follow. Progress became rapid and a new industry was created, as English chronometers became known and appreciated at home and abroad. Not only were the world's charts and maps drawn or redrawn with accuracy and precision, but the lives of countless sailors were saved. The stage was set for the rapid expansion of the British Empire, and with it the Industrial Revolution. Accident of history or not, few men who have made such an extraordinary contribution are so little known or appreciated as John "Longitude" Harrison.

This paper has been concerned with the historical aspects of Harrison's later work, and for this I have relied greatly on the pioneering work of Colonel Humphrey Quill.[53] The technical details, which I covered in my lecture at the Longitude Symposium, have been described in an article in *Antiquarian Horology* in 1989.[54]

53. Quill, *John Harrison* (*op. cit.*, note 7).

54. A. G. Randall, 'The Technology of John Harrison's Portable Timekeepers', *Antiquarian Horology,* vol. 18, no. 2 (Summer 1989), pp. 145-60, and vol. 18, no. 3 (Autumn 1989), pp. 261-77. The second part of this article is concerned with the development of the Jefferys watch, H.4, and H.5.

# The Scandalous Neglect of Harrison's Regulator Science

*Martin Burgess*

Martin Burgess is a maker of precision timekeepers designed as sculptures to reveal the ingenuity of the mechanism. Strongly influenced by John Harrison's work, he spent many years constructing and perfecting two clocks that have revealed the accuracy achieved by employing the principles Harrison discovered 250 years ago. One of these timekeepers was installed in Norwich, England, in 1987. In 1979, Burgess became one of the founding members of the Harrison Research Group.

Burgess was educated at Greshams School in Norfolk, England, where he constructed his first clock. After studying silversmithing and design at the Central School of Arts and Crafts in London, he restored ancient Egyptian antiques for University College, London. In 1963, he established a workshop at his home in Essex. His best-known piece is the huge world-time wall clock at the merchant banking firm Schroder Wagg in London, about which an internationally acclaimed documentary film was made. Another of his clocks is on permanent display at The Time Museum in Rockford, Illinois.

Burgess is a Fellow of the International Institute for the Conservation of Historic and Artistic Works (1955), The Society of Antiquaries of London (1959), and the British Horological Institute (1976). He received the Barrett Medal of the British Horological Institute (1988) and has published many articles on horology and the manufacture of chain-armour.

# The Scandalous Neglect of Harrison's Regulator Science

*Martin Burgess*

I am often asked what makes the Harrison regulator system better than others of his day. Is there some great secret that gives the high stability? This lecture is an attempt to answer that question. My research into the technology that John Harrison (1693-1776) developed and perfected had its beginning in the publication in 1966 of Harrison's biography by Colonel Humphrey Quill (1897-1987).[1] In reading this book, I became intrigued by the claim that Harrison made in his 1730 manuscript that his precision long-case clocks, or regulators, would keep time to one second in a month and maintain that accuracy for 40 or 50 years without need of cleaning.[2] Quill was doubtful about the reliability of this statement, and other horologists thought that it was utter nonsense—not surprising perhaps, because that kind of accuracy was unheard of in 1730 and, to crown it all, the mechanisms of Harrison's regulators were made largely of wood!

After corresponding with Colonel Quill, my interest in Harrison's early wooden regulators was further stimulated in an unexpected manner. In 1970, I was contacted by William Andrewes, then a young art school student, about my work on sculptural clocks—clocks in which all the working parts are visible and serve as significant elements in the design as well as in the operation of the mechanism. Andrewes had also developed an interest in Harrison and had been in touch with Colonel Quill concerning a thesis he was planning to write on Harrison's early work. Through a series of fortuitous circumstances, Quill placed Andrewes under the guidance of the well-known watchmaker George Daniels to complete one of Harrison's early unfinished wooden regulators.[3] While Andrewes worked with Daniels one day during the week on Harrison's clock, he came on weekends to work under my guidance on a sculptural clock he was making. Our mutual interests drew me closer to Harrison's early wooden regulator technology.

In 1972, William S. Laycock (1924-1976), a brilliant mechanical engineer, approached me about an article I had written on Harrison's clock escapement.[4] Laycock became interested in experiments I had begun concerning Harrison's measurement of the coefficient of expansion of different metals. Laycock and I shared the belief that technical truth requires practical

1. Humphrey Quill, *John Harrison, the Man who found Longitude* (London: John Baker, 1966).

2. John Harrison, untitled manuscript signed 'John Harrison, Clockmaker at Barrow; Near Barton upon Humber; Lincolnshire. June 10. 1730', MS. 6026/1, Library of the Worshipful Company of Clockmakers, Guildhall Library, London, p. 2, sect. 4. A transcription of this document was made by G. H. Baillie and published in 'The 1730 Harrison M.S.', *Horological Journal,* vol. 92, no. 1102 (July 1950), pp. 448-50, and vol. 92, no. 1103 (August 1950), pp. 504-6.

3. An account of the work on this regulator was published by William Andrewes in 'John Harrison: A Study of his Early Work', *Horological Dialogues,* vol. 1 (1979), pp. 11-38.

4. Martin Burgess, 'The Grasshopper Escapement: Its Geometry and its Properties', *Antiquarian Horology,* vol. 7, no. 5 (December 1971), pp. 416-22. This followed Col. Quill's article, 'The Grasshopper Escapement', published in the previous issue, vol. 7, no. 4 (September 1971), pp. 288-96.

5. Harrison, 1730 manucript (*op. cit.,* note 2).

6. John Harrison, 'An Explanation of my Watch or Timekeeper for the Longitude...', 7 April 1763, MS. 3972/1, Library of the Worshipful Company of Clockmakers, Guildhall Library, London. This manuscript has been transcribed by Andrew King but has never been published.

7. John Harrison, *A Description Concerning Such Mechanism as will Afford a Nice, or True Mensuration of Time;...* (London, 1775).

experiment—that the truth cannot be obtained or fully understood by theory alone.

Of the documents written by Harrison that are known to have survived, there are three that serve as the main sources of information about the technology he developed: an untitled manuscript written in 1730,[5] a manuscript dated 1763 entitled 'An Explanation of my Watch',[6] and his only published work of a technical nature, *A Description Concerning Such Mechanism,* issued in 1775, just a year before his death.[7] Through a careful analysis of these works, Laycock began to uncover some fundamental reasons for the accuracy of the regulators. Shortly before he published his findings in 1976,[8] he delivered a lecture about his research to the British Horological Institute.[9] As a result of this lecture, we met two other 'Harrisonophiles', Andrew King and Mervyn Hobden. Following Laycock's death, on 9 December of that year, it became clear that further study and testing of Harrison's work was not a one-man job. It required a team of highly qualified specialists. Bonds of common interest eventually resulted in the formation of such a team—the Harrison Research Group—in 1981.[10] Each member of this group was interested in a specific area of Harrison's work, and, through our combined efforts, an enormous amount of information about Harrison's technology has been uncovered.

As to my own specific interest, I had been approached in 1974 by Gurney's Bank of Norwich—which had been founded, interestingly, in 1775, the year of the publication of *Concerning Such Mechanism*—to build a clock for the city of Norwich commemorating the bank's 200-year association with the city. This commission provided the opportunity to create a modern Harrison regulator, combining Harrison's technology and modern materials. I began by building two identical regulators, but only the one for Gurney's Bank has been completed so far; it is this regulator that is referred to throughout this paper.[11] As a result of this project, I have been able to prove beyond any doubt that, because of the highly sophisticated scientific approach that Harrison adopted, the remarkable claims he made for the accuracy of his regulators were not exaggerated.

★ ★ ★ ★ ★

A pendulum is a good timekeeper. It does not require good craftsmanship for it to keep time well enough to run a household or get people to work on time. A branch out of the hedge with two house bricks taped to one end, pushed and pulled by a simple crank, will do. However, for scientific work, especially astronomy, very much more has to be done. We need an oscillator with a high degree of short- and long-term stability, both to see an error in another instrument quickly and to give us confidence that we are looking at the correct time, months or years later.

John Harrison, in addition to being the inventor of the prize-winning marine timekeeper known as H.4, made significant contributions to the field of precision timekeeping on land. For a variety of reasons, however, the science and technology that he developed and perfected was not understood by those engaged in the mainstream of horological science, and it is only in the last three decades that intensive investigation has been undertaken to elucidate his research. It is with the recent investigations into Harrison's pioneering work in this field that this paper is concerned.

8. W. S. Laycock, *The Lost Science of John "Longitude" Harrison* (Ashford, England: Brant Wright Associates, 1976).

9. Laycock's lecture was delivered at the Royal Society of Arts in London on 13 January 1976 and was subsequently published under the title 'John Harrison, the man who mastered the pendulum' in *Horological Journal*, vol. 118, no. 8 (February 1976), pp. 5-13.

10. The original group consisted of Jonathan Betts (then horological conservator at the Old Royal Observatory at Greenwich), Peter Hastings (a mechanical engineer, now at the Royal Observatory in Edinburgh), Mervyn Hobden (chief engineer at a large electronics firm, who suggested the formation of the group), Beresford Hutchinson (then in charge of horology at the Old Royal Observatory), Andrew King (a specialist in the history of Harrison's life and work who is involved in making two exact copies of Harrison's early wooden regulators), Anthony Randall (a well-known maker of precision portable timekeepers), and myself. Starting in the spring of 1981, this loosely affiliated group met annually at the Old Royal Observatory at Greenwich. David Harrison, a mathematician whose Ph.D. thesis at the University of Leeds on oscillator physics was influenced by Harrison's work, later joined the group. Will Andrewes was never involved in this group, having moved by this time to the United States to work at The Time Museum in Rockford, Illinois.

11. The Gurney clock was installed in a public park in Norwich in 1987 but had to be removed in 1993 due to vandalism to the structure in which it was housed. It is now on display at the British Horological Institute's headquarters at Upton Hall, near Newark, Nottinghamshire.

Harrison began his work with precision clocks, or regulators, in the mid-1720s, when he was a young clockmaker in Barrow upon Humber in Lincolnshire.[12] From experiments made with these early wooden regulators, he gained considerable knowledge and expertise, which, beginning about 1729, he then put to use in his marine timekeepers.[13]

The only known surviving Harrison regulator made during the period that he was working on his marine timekeepers[14] is the R.A.S. regulator (Figure 1), so named because it now belongs to the Royal Astronomical Society.[15] After Harrison died, this regulator remained in the family until it was presented to the Society in 1836.[16] While the clock embodies many features typical of Harrison's work during this period,[17] at the time it was donated it was fitted with a wood-rod pendulum—unlike anything that Harrison ever made—suggesting that the clock was unfinished at the time of his death.[18] The present gridiron pendulum, as well as the case and the driving weight, were made when the clock was restored by Lt.-Commander Rupert T. Gould (1890-1948) in 1928. The remontoire, a rewinding mechanism designed to overcome irregularities in the driving force, was not changed during the restoration, but this too appears to be a part of the clock that was completed after Harrison's death: striving as he always did for perfection, Harrison would certainly have employed the same type of constant-force remontoire that he used in H.3.[19] Evidence that this clock was never finished by Harrison is further supported by a statement he made in 1775: 'my next or second Clock will be somewhat better than if it had been finished sooner'.[20] This suggests that he was in the process of constructing a clock, which, considering that he died in March of the following year, was probably never completed. This unfinished clock may be the R.A.S. regulator.

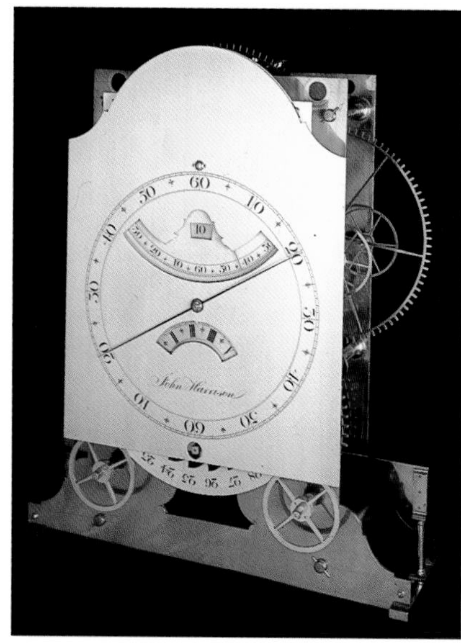

**Figure 1. View of the dial and movement of the Royal Astronomical Society (R.A.S.) regulator. This is probably the last timekeeper made by John Harrison, and it appears to have been unfinished at the time of his death in 1776. The clock, with its recent case, weight, and gridiron pendulum, is currently on display at the Old Royal Observatory at Greenwich. Width of dial: 28 cm. (11 in.). Courtesy of the Royal Astronomical Society, London. Photograph by permission of the National Maritime Museum, Greenwich. Inv. no. R16.**

The above statement also implies that he had already completed one regulator of this type. Evidence that such a piece existed may be found in the well-known portrait of Harrison, painted in 1767. In the background on the right, one can see the lower half of a regulator in a plain wooden (possibly walnut) case with its gridiron pendulum and weight visible because the door has been opened; this detail is shown in Figure 7. That this clock is the R.A.S. regulator in a finished state seems improbable, because it is highly unlikely that the three parts of the clock visible in the painting—the pendulum, the weight, and the case—would all have been altered either before or after the

12. See Andrew King's paper in this volume.

13. See William Andrewes's paper in this volume.

14. There is one other regulator that, although not made by Harrison, bears the inscription in Latin on the dial that it was made under his authority by John Jefferys (fl. 1726-*ca.* 1753) in 1748 and restored by Joseph Barber in 1768. This inscription would appear to have been engraved after the restoration, but accompanying documents imply that John Harrison's son William (1728-1815) testified to its accuracy when he saw the clock in 1800. The clock, as described by Humphrey Quill in *John Harrison* (*op. cit.*, note 1), pp. 56-7 and fig. 19, is housed in a fine-quality walnut case, runs for one month, is fitted with bolt-and-shutter maintaining power, and has a gridiron pendulum. Quill thought, from examining the gridiron pendulum, that the present jewelled dead-beat escapement was not original. The accompanying documents indicate that the clock was probably originally fitted with a grasshopper escapement. There are signs that it also had antifriction wheels.

15. This clock is also sometimes referred to as the 'late' regulator to distinguish it from Harrison's earlier wooden regulators.

16. This information was found in E. T. Cottingham's article, 'A Description of the Royal Astronomical Society's Harrison Clock, with a brief account of the maker', *Horological Journal*, vol. 52 (May 1910), pp. 137-41. The regulator was given by John Barton (1771-1834) of the Royal Mint to his son, W. H. Barton (d. 1868), who presented it in 1836 to the Royal Astronomical Society. It was, however, not delivered to the Society until 1838.

17. These include roller bearings to support the great wheel arbor, antifriction wheels for the other arbors, a grasshopper escapement, and the use of brass alloys that are particularly suitable for certain parts of the timekeeper.

18. Herman Bush, a jeweller working in Hull, mentioned in the *Horological Journal* of November 1885 that after Harrison's death, 'an uncompleted chronometer was found amongst his effects, which was composed of brass, bell-metal, tutenage, and specially prepared hard wood, intended to be used for the arms of the balance'. If Harrison was indeed experimenting with such materials at the end of his life, it is just conceivable that the wood-rod pendulum was fitted by him to the R.A.S. regulator in order to find out the coefficient of linear expansion of that particular wood. If this is true, then the watch mainspring fitted to the regulator as a remontoire spring may well have been a temporary arrangement made by Harrison to conduct this test. Laycock thought that the remontoire spring attachment looked original. The passage above from Bush's article is quoted by Col. Quill in *John Harrison* (*op. cit.*, note 1), p. 218.

clock was donated to the Royal Astronomical Society.[21] The clock shown in Harrison portrait, therefore, must be another gridiron-pendulum regulator that Harrison started to make probably during the period in which he was perfecting his third marine timekeeper (H.3) and had completed by 1767.[22]

It is clear that the regulator technology that Harrison developed was not understood by his contemporaries, and it remained neglected until the 1970s, when the thorough investigation into his work began. The regulator technology that became the mainstream of horological science was based instead on the work of George Graham (*ca.* 1674-1751), who, although being a great admirer and supporter of John Harrison, had very different views on how a precision timekeeper should be constructed. Of the regulator that he was constructing near the end of his life, Harrison claimed that, under optimal conditions, there was more reason to expect this clock to keep time to a second in 100 days than to expect Graham's clocks to keep time to a second in one day,[23] and he stated that Graham once 'said to several Gentlemen, that for my Improvement in Clock-Work, I deserved 20,000 l. [£20,000]—was no Longitude to be concerned'.[24] Although the accuracy of two of Harrison's wooden regulators constructed in 1726 had been attested to by the President of the Royal Society in 1749 as being 'without the variation of more than one single second in a month',[25] even most modern clockmakers have considered these claims to be a lie.

**Figure 2. View of the back plate of the R.A.S. regulator, showing the roller race, the antifriction wheels, and part of the grasshopper escapement. Height of back plate: 46.5 cm. (18.3 in.). Courtesy of the Royal Astronomical Society, London. Photograph by permission of the National Maritime Museum, Greenwich.**

Indeed, a second in 100 days is about the limit for a pendulum in air, as recent research appears to show[26]— it might be considered the Holy Grail of pendulum horology. With a clock as accurate as this, the seconds hand only shows you which second you are in, but Harrison says he could read his clock to one-twentieth of a second or better. To do that, he would have had to look at the instant position of the pendulum.[27] With my own regulator, I have done much the same thing.

Harrison was an old man when he started to build this regulator. Had his energies not been so dissipated by the problems he encountered with the Board of Longitude over the reward for his work on marine timekeepers, he might have completed it before he died. Although the Royal Observatory would have benefited greatly from a much better standard timekeeper, the Astronomer Royal, Nevil Maskelyne (1732-1811), showed no interest in this side of Harrison's work—hardly surprising, considering the bitterness that had arisen over Maskelyne's supervision of the trials of Harrison fourth marine timekeeper, H.4.[28]

From its dial, there is no doubt that the R.A.S. regulator was designed for astronomical use. Mechanically, it is like no other clock before or since

19. The purpose of a remontoire, a device rewound periodically by the clock's main source of power (weight or coiled spring), was to overcome irregularities in the transmission of power through the wheels of the timekeeper by providing a direct source of power especially for the escapement. The present remontoire fitted to this regulator, which is rewound every 30 seconds, has a spiral spring without any provision for equalising the irregularities of its driving force. The ingenious remontoire that Harrison used in H.3—also activated every 30 seconds—incorporates two helical springs fitted with cams that ensure that the power provided remains constant. The remontoire wheel of the R.A.S. regulator has the attachment points for the outside ends of the two helical springs that would have been fitted to it.

20. Harrison, *A Description Concerning Such Mechanism* (*op. cit.,* note 7), p. 52.

21. If any alteration to the clock had occurred, it is far more likely that it would have been made to the grasshopper escapement, which, if mishandled, could have been a source of trouble. The R.A.S. regulator retains its original grasshopper escapement.

22. The design of the case and the sturdy construction of the gridiron pendulum shown in the painting reveal that the clock portrayed is not one of Harrison's early wooden regulators.

23. '...there must be then more Reason...that it shall perform to a Second in 100 Days, yea, I say, more reason, than that Mr. Graham's should perform to a Second in 1'. Harrison, *A Description Concerning Such Mechanism* (*op. cit.,* note 7), pp. 35-6.

24. *Ibid.,* p. 23.

25. Address by Martin Folkes (1690-1754), President of the Royal Society, on the presentation to John Harrison of the Copley Gold Medal on 30 November 1749, Minutes of the Royal Society, London, p. 187. Folkes mentions on the following page that one of these clocks, which Harrison had kept with him, had not altered 'so much as one whole Minute in Ten Years'.

26. Henry Wallman, 'Do Variations in Gravity mean that Harrison approached the Limit of Pendulum Accuracy?', *Horological Journal,* vol. 135, no. 1 (July 1992), pp. 24-6, and J. E. Bigelow, 'Barometric Pressure Changes and Pendulum Clock Error', *Horological Journal,* vol. 135, no. 2 (August 1992), pp. 62-4.

27. In order to adjust, test, and regulate a timekeeper like the prize-winning watch, H.4, observation of the time from the position of the seconds hand would not have been sufficient, because very small differences between the rates of the two timekeepers would have been impossible to detect in this manner. Such an observation could only be made by determining the exact position of the pendulum at a particular instant, and, to help with this, Harrison might well have marked the scale beneath the pendulum in time rather than in degrees. The pendulum's wide arc of swing would greatly assist this precise observation.

28. For details, see Anthony Randall's paper in this volume.

# Harrison's own regulator c. 1760

## History

John Harrison died in 1776 and the clock was passed via family connections to Sir John Barton, Deputy Comptroller of the Royal Mint. Whilst there, the original gridiron pendulum was replaced by a wooden rod type, possibly because the gridiron had become damaged. In 1836 the Barton family presented the clock to its present owners, the Royal Astronomical Society.

Nearly 100 years later, in 1928, the clock was restored by Cdr R T Gould RN assisted by J H Agar-Baugh of Hammersmith. Cdr Gould fitted new pallets, crutch assembly, pulley and weight and a stop mechanism to prevent the remontoire from running down. The wooden rod pendulum was replaced by a gridiron pendulum.

John Harrison intended this clock to be his masterpiece, with an accuracy of one second in 100 days. He incorporated all the devices he had previously used to attain accurate timekeeping: the grass-hopper escapement, a gridiron pendulum with a wide arc of swing, cycloid checks, anti-friction wheels and a "*remontoire*" (a mechanism designed to eliminate uneven power transmission to the escapement) – the first two being his own invention.

*signed on dial* : John Harrison

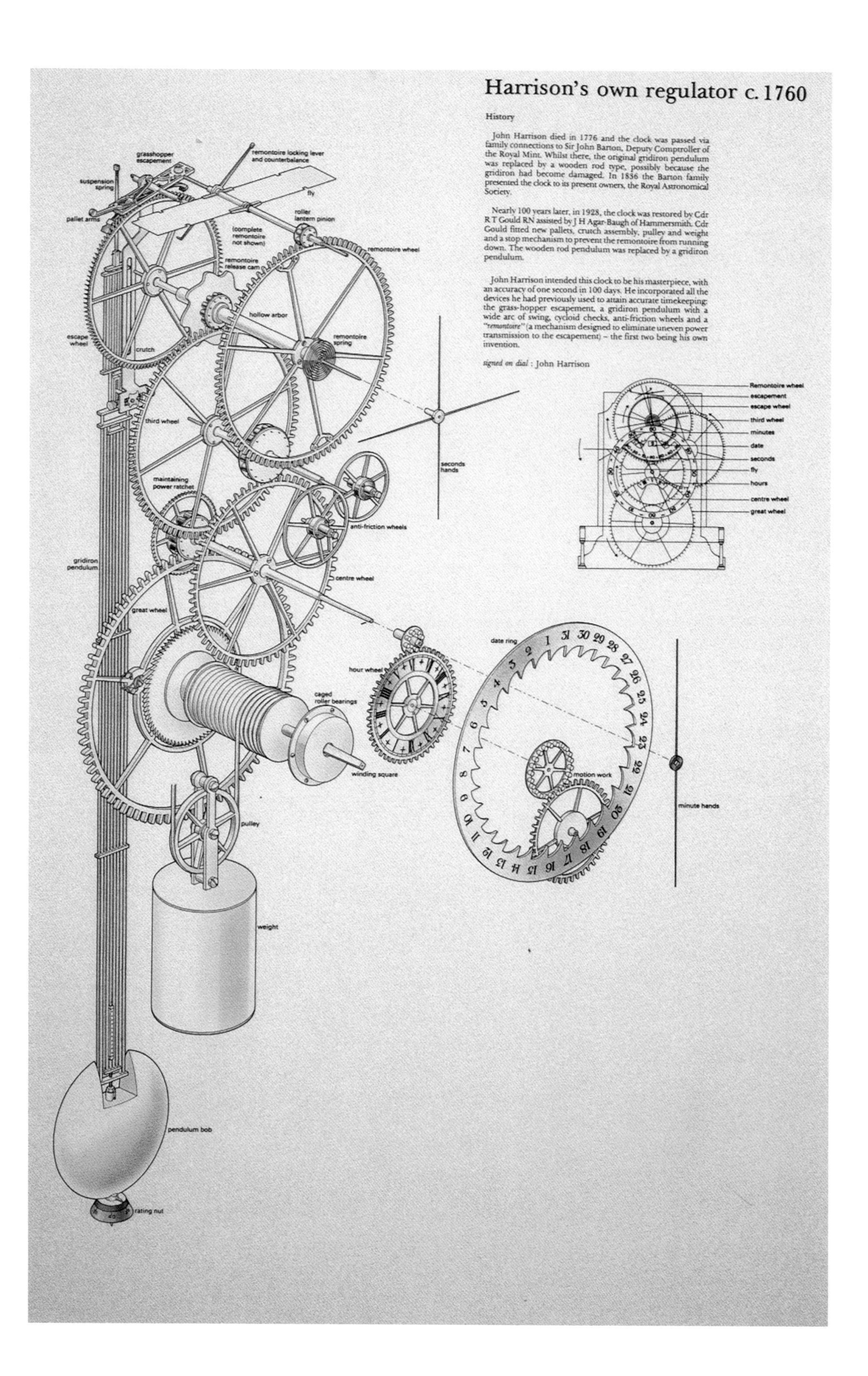

**Figure 3 (left). Illustration showing the construction of the R.A.S. regulator in its present state, with its weight and gridiron pendulum. Drawing by David Penney, commissioned by the National Maritime Museum, Greenwich.**

(Figures 2 and 3). The wheel train is very like that of H.3: antifriction wheels, lignum vitae bushes and roller pinions, and caged roller bearings to support the great wheel arbor. The clock is fitted with a remontoire that is rewound every 30 seconds, but, as stated above, this appears to be have been completed after Harrison's death. Almost all the parts are made of copper alloys of different compositions to prevent microwelding. No oil is used anywhere, so, if all dust could be excluded, the clock would hardly ever need cleaning. Because there is very little friction or wear, a run without stopping of at least 100 years is not out of the question. This is highly advantageous for a precision timekeeper, because regulation and other adjustments take years to get within such fine limits. It is impossible to do this with a clock that requires oil, because such a clock changes its rate whenever it is cleaned. In the eighteenth century, when lubricants were very poor and soon decayed, oil represented a great problem for precision timekeeping. Yet no other clockmaker succeeded in finding a satisfactory solution.

The movement plates spread out into a bridge at the base, and the attachment points for mounting the clock are right at the outside. Harrison stated that he had no good place to mount it in his home.[29] Peter Hastings[30] believes that Harrison intended to mount it in a brick alcove at Greenwich. The pendulum suspension would have been bolted to the back wall, and the movement would have sat on stone ledges in the side walls. This is a very far cry from even modern clockmakers, who think they can get accuracy in a clock case made of wood.[31]

**Figure 4. Regulator made by the author for Gurney's Bank in Norwich between 1974 and 1987. This clock was designed upon the principles of Harrison's regulator science, but I chose to take advantage of modern materials and methods for its construction. Wall plate on which clock is mounted: 61 x 137.2 cm. (24 x 54 in.). Diameter of great wheel: 61 cm. (24 in.). Courtesy of the author.**

My own clock (Figure 4) looks nothing like the R.A.S. regulator, but all the same functions are there. As with Harrison's clock, there is no oil. Stainless steel, dry-lubricated ball bearings, however, replace the antifriction wheels and the lignum vitae rollers, and the wheels are made of aluminium alloy for lightness, rather than the copper alloy that Harrison used. The pendulum suspension spring and the remontoire spring are made of Durinval C, a material that does not change its elasticity with temperature change. Just as Harrison always used the materials he thought best, even if they were new to horology, I have tried to do the same. The aim was to find out what his method was, not to make a copy, from which we might learn little.

A pendulum in a perfect vacuum, hung on a frictionless suspension from a totally rigid support in an environment with constant temperature and

29. 'I have for some time had such a Clock to the purpose in great part made; but as not designing to fix it up in the not rightly convenient Place or House in which I live, I did not hasten its finishing'. Harrison, *A Description Concerning Such Mechanism* (*op. cit.*, note 7), p. 32, note 2.

30. Peter Hastings, who works at the Royal Observatory, Edinburgh, is the mechanical engineer of the Harrison Research Group.

31. Harrison described the problems as follows: 'I had, after some disagreeable Experiments, discovered that if Wood was concerned in the Suspension of the Pendulum,...the Clock would as thence go faster in moist Weather than in dry, the Strength of such a Suspension becoming as thereby, viz. by Moisture increased....Now, at some Years after I had communicated this to Mr. Graham, he upon some Occasion removed a Clock from one Side of his Room to the other, and when fixed there, he found it to go about 6 Seconds a Day different to what it went before'. Harrison, *A Description Concerning Such Mechanism* (*op. cit.*, note 7), pp. 30-1.

gravity and without light, would keep perfect time once it was set moving. Yet it would be of no use. To attempt to get near some of these ideal conditions might improve the timekeeping but would only be a palliative. It would not abolish the errors at their root.

Leaving aside vibrations and draughts, there are five things that change the rate of a pendulum:

(1) *Gravity:* The gravity at every place on the surface of our planet changes with the relative positions of the Sun and Moon. These changes, however, are cyclic and are never enough to produce an error of a second in any period of 100 days.

(2) *Energy:* Any oscillator must have an energy input to overcome energy absorption. The input magnitude can change: if you push something harder, it moves faster.

(3) *Period of oscillation:* A pendulum is not isochronous if it is swinging in a circular arc. The period increases with increased arc. The function is non-linear: for a unit change of arc, the rate changes more at large arcs than at small ones.

(4) *Temperature:* Temperature variation changes the length of the pendulum rod. It also changes the resistance of the air through which the pendulum swings and affects the viscosity of the oil, if oil is used. The most serious error is caused by the changing length of the rod, but if we compensate for this, we must also compensate for all the other temperature-caused changes. A rod that does not expand or contract at all would be quite useless. But if we want an error of not more than one second in 100 days, we have to be very careful: in all conditions, the pendulum's theoretical length must not vary more than nine-millionths of an inch—an amount that fits about 220 times into the thickness of a sheet of carbon paper. This fact by itself demonstrates the enormous difficulties of achieving the rate that Harrison set out to accomplish.

(5) *Barometric pressure:* Changes in barometric pressure affect the air's density. Harrison called it the 'weight of the air', which is exactly what it is. If the barometric pressure rises, the air gets denser and the pendulum moves more slowly. As a result, the arc of swing gets smaller, because more energy is being consumed at the original arc and additional energy is not available. The same thing happens at lower temperatures. But in higher pressure, the pendulum also gets lighter, and this also causes the clock to lose.

All of these factors, with the exception of gravity, change the arc of swing. The variation of time in its period of oscillation caused by a change in its arc of swing is called 'circular deviation'. In Figure 5, you will see that the sign of circular deviation for each of the environmental changes is opposite to the sign for the change of the rate directly caused by that change in the environment. From the point of view of the pendulum, a change of energy input is also a change in its environment. From this, it is but a small step to see that the circular deviation can be used to counter all the other errors and wipe them out.

| ENVIRONMENTAL CHANGE | | CLOCK RATE | CIRCULAR DEVIATION |
|---|---|---|---|
| Gravity | Up | + | 0 |
| | Down | - | 0 |
| Temperature up | Air thinner; | small + | - |
| | rod expands | - | 0 |
| Temperature down | Air thicker; | small - | + |
| | rod contracts | + | 0 |
| Energy input | Up | + | - |
| | Down | - | + |
| Barometer up | Air thicker; | - | + |
| | pendulum lighter | - | 0 |
| Barometer down | Air thinner; | + | - |
| | pendulum heavier | + | 0 |
| (With pendulum over-compensated) | Temperature up | + | - |
| | Temperature down | - | + |

**Figure 5. Table showing the fundamental effects of environmental changes and circular deviation upon the rate of a pendulum. Because of the opposing results of gain and loss, the circular deviation can be used to negate the effects of environmental changes on the rate of the clock.**

Harrison evolved a design that permitted him to adjust the magnitude of each of these changes, either by direct adjustment or by rebuilding certain parts, until all the functions could be brought together in agreement at or close to the running arc of the pendulum. It is this that is the fundamental secret of the Harrison system, though I must stress that Harrison made no secret of it. He published it![32]

Only a great technical genius could have devised such a method at this early date. Even then, it was an evolution that took 50 years of rigorous testing and observation—thousands upon thousands of hours of unpaid work in order to get a .01 per cent better result. By comparison, all other precision pendulum horologists were unable to advance because they lacked this rigorous, experimental, scientific approach. Please remember that Harrison was never paid anything for his pendulum research.

The first Harrison clock we know of was made when he was twenty. In theory, it is just a normal longcase clock with an oiled anchor recoil escapement. After completing two other similar clocks, he began to perform tests on pendulums. These tests might have been inspired by his observation and knowledge of church bells: because of circular deviation, the rate of strike slows as the arc increases as the bell is got up.[33] Harrison discovered, however, that the clock did not behave in the same way. If he increased the arc of the pendulum by increasing the energy input, the clock went faster. Too much escapement deviation! Not enough circular deviation! A clockmaker wanting to improve the stability would say that the pendulum does not have enough dominion over the escapement. There are four things he could do:

(1) The recoil escapement is pushing the pendulum backward and forward, so the temptation would be to considerably increase the weight of the pendulum-bob.

32. In his book, *A Description Concerning Such Mechanism* (*op. cit.*, note 7).

33. Laycock, *The Lost Science* (*op. cit.*, note 8), p. 43.

(2) The recoil might be made less fierce by a modification of the shape of the pallets.

(3) The arc of swing could be greatly increased to create a large circular deviation.

(4) The length of moment arm of torque about the pallet arbor might be reduced in relation to the length of the pendulum; this would be the same as making the pendulum longer.[34]

George Graham did the first two things. He increased the weight of the pendulum and abolished recoil.[35] The next most pressing improvement was to compensate for temperature error. After some unsuccessful attempts to make use of the different coefficients of expansion of different metals, he used a large volume of mercury expanding and contracting in a glass jar.[36] It did not keep time to much better than a second per week. Graham regulators still don't!

Harrison's path was quite different. The pallets on the 1717 clock at Nostell Priory look as if he was beginning to modify the pallets to reduce recoil. But then came the turret clock that he made for Brocklesby Park.[37] It appears that the original escapement, which is thought to have been an anchor (recoil) type, caused the clock to stop, so Harrison modified it in order to eliminate the need for oil. The modified escapement, commonly known as the grasshopper escapement, consumed so little energy that he had to nail copper vanes onto the pendulum-bob to consume the surplus. A turret clock must have surplus energy to continue to work in adverse conditions. The clock was in an unheated room and the pendulum rod and suspension spring were of brass. Temperature error, between winter and summer, was now obvious. This huge error could not be blamed on changed energy input, because the arc had not changed. Harrison had started down the road to precision.

Harrison's gridiron pendulum, which uses the different coefficients of expansion of brass and steel rods to counteract temperature variation, is well known.[38] Other clockmakers seized upon it as a cheap way of counteracting the change of the pendulum rod's length in heat and cold, and it is a pity they did not understand it. The rods are very thin and therefore respond quickly to changes in temperature.[39] The conductivity and specific heat of the brass and steel are different, so their cross-sections are different: the brass rods are thicker. The rods are riveted, not pinned, into the bridges. Where they pass through a bridge, they are loose to prevent stick slip. The bridges are narrow, so the pumping action of more than a 12° arc causes the centrifugal force to keep the whole grid fully live. The air trapped between the rods is also pumped downward, so there is a lot of air washing over the rods. The vital factor, however, is the compensation adjustment.

Bill Laycock called Harrison's device for compensation adjustment the 'tin whistle adjuster' (Figure 6). It breaks the central rod, which is steel, with a sandwich in which the meat is steel and the bread is brass. A pin passed through one of the holes unites them. Laycock calculated that the shift of one hole by the pin would change the proportion of brass and steel so that half a pin shift (the nearest you can get to correct) would compensate for a temperature shift of 13° C (23.4° F)—comparable, perhaps, to the difference

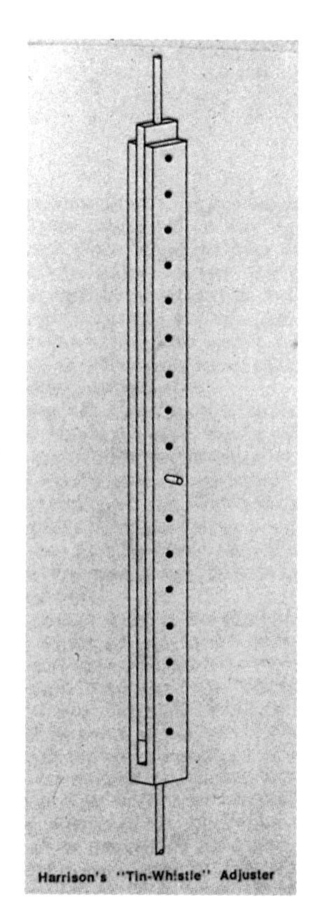


**Figure 6. Drawing of the 'tin whistle' adjuster that Harrison devised to adjust the amount of temperature compensation in the gridiron pendulum. Drawing by William Laycock.**

34. A small force exerted over a large distance fluctuates in proportion to itself more than a large force acting over a small distance. This is true especially for mechanical devices with intermittent movement, such as clocks, in which the inertia of the moving parts and the friction play a great part in the energy delivered. Harrison describes this in his 1730 manuscript (*op. cit.*, note 2), p. 3, sect. 6, and pp. 7-10, sect. 17, and in his book *A Description Concerning Such Mechanism* (*op. cit.*, note 7), pp. 3-12.

35. Graham accomplished this by using the dead-beat escapement, which he began to use around 1720. Although, because of his influence and involvement in its development, this type of escapement is usually associated with him, he was not its inventor. Correspondence dated 11 December 1675 between the first Astronomer Royal, John Flamsteed (1646-1719), and Richard Towneley (1629-1707) illustrates the dead-beat escapement that Graham's master, Thomas Tompion (1639-1713), had made for a clock 45 years earlier. For further information, see A. J. Turner, 'The Introduction of the Dead-beat Escapement: A New Document', *Antiquarian Horology*, vol. 8, no. 1 (December 1972), p. 71, and Derek Howse, 'The Tompion Clocks at Greenwich and the Dead-beat Escapement', *Antiquarian Horology*, vol. 7, no. 1 (December 1970), pp. 22-3.

36. Graham began experimenting with different metals around 1715 and devised his mercury-compensated pendulum in December 1721. These developments are described by Graham in 'A Contrivance to avoid the Irregularities in a Clock's Motion, occasion'd by the Action of Heat and Cold upon the Rod of the Pendulum', *Philosophical Transactions* of the Royal Society, London, vol. 34, no. 392 (1726), pp. 40-4.

37. For details of this clock and its escapement, see Andrew King's paper in this volume.

38. The invention of this pendulum is described in Andrew King's paper in this volume; the development of gridiron temperature compensation in the sea clocks is included in William Andrewes's paper.

39. This was an important feature of the gridiron pendulum. In Graham's temperature-compensated pendulum, the mercury in the the jar takes longer to adjust to a temperature change than its supporting steel pendulum rod.

40. Rupert T. Gould, *The Marine Chronometer: Its History and Development* (London: J. D. Potter, 1923), p. 41, footnote †.

41. Laycock, *The Lost Science* (*op. cit.*, note 8), pp. 68-70.

between average winter and summer temperatures in England—and still allow the clock to stay within one second in 100 days.

Now look at the detail of the clock in the portrait of Harrison, shown in Figure 7. The two outer brass rods on the pendulum are short and are unlike anything on any other gridiron. Gould thought they were figments of the artist's imagination.[40] But with The Master in the room—a tough Yorkshireman, remember—no one would have been permitted to paint figments of the imagination. Laycock thought that these two brass rods were intended to provide a fine temperature-compensation adjustment in heat by lifting the bridge that unites them.[41] The picture had not been cleaned when he made this observation. Now that it has been cleaned, we can see that there are a few 'tin whistle' holes near the bridge, not only in the brass rods but also in the outer steel rods next to them. Thus, the heavy bridge that carries the pendulum-bob has the two short brass rods fixed to its outer ends. The bridge between them unites them by shiftable pins to the outer steel rods, thus providing two 'tin whistles' to give a fine adjustment to the temperature compensation. This pendulum is clearly much stronger and of better design than Harrison's earlier version of gridiron pendulum, in which the bob is only fixed to the single central steel rod. It is the weight of the bob that causes the dangerous stresses when the pendulum is taken off the clock and moved. With the design in the portrait it would even be possible to remove the bob unit altogether while the gridiron was still hanging on the clock. Of course, moving the bridge would require the clock to be re-regulated, but such a clock would require re-regulating after any adjustment. The fact that all these little holes have been shown in the picture demonstrates how much care the artist was instructed to take in order to give an accurate representation.

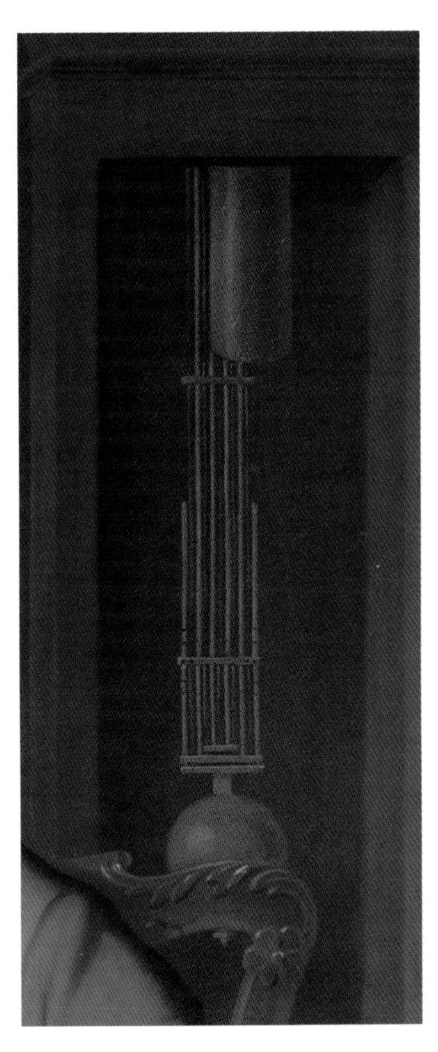

**Figure 7. Detail of the regulator in Harrison's portrait, painted by Thomas King (d. *ca.* 1769) in 1767. (The full portrait is shown in Figure 28 of William Andrewes's paper in this volume.) The gridiron pendulum is fitted with a rising rod compensator: the small holes in the two short additional brass rods on the outside would allow the cross-piece to be moved up or down in order to alter the amount of temperature compensation that the gridiron provides. Oil on canvas. 127 x 102 cm. (50 x 40.2 in.). Courtesy of the Science Museum/Science and Society Picture Library, London. Inv. no. 1884-217.**

At what distance from the centre of suspension a pendulum is impulsed is very important. The length of a pendulum is measured approximately from the centre of its suspension to the centre of the bob. A given force applied to the bob will cause a certain deflection. To achieve the same deflection at one-tenth of the distance from the centre of suspension, ten times the force will be required. Any variation in the impulse at the bottom of the pendulum will have a much greater effect than any variation at the top. This fundamental principle of mechanics had made itself felt at Greenwich before Harrison was born. The two clocks by Thomas Tompion (1639-1713) in

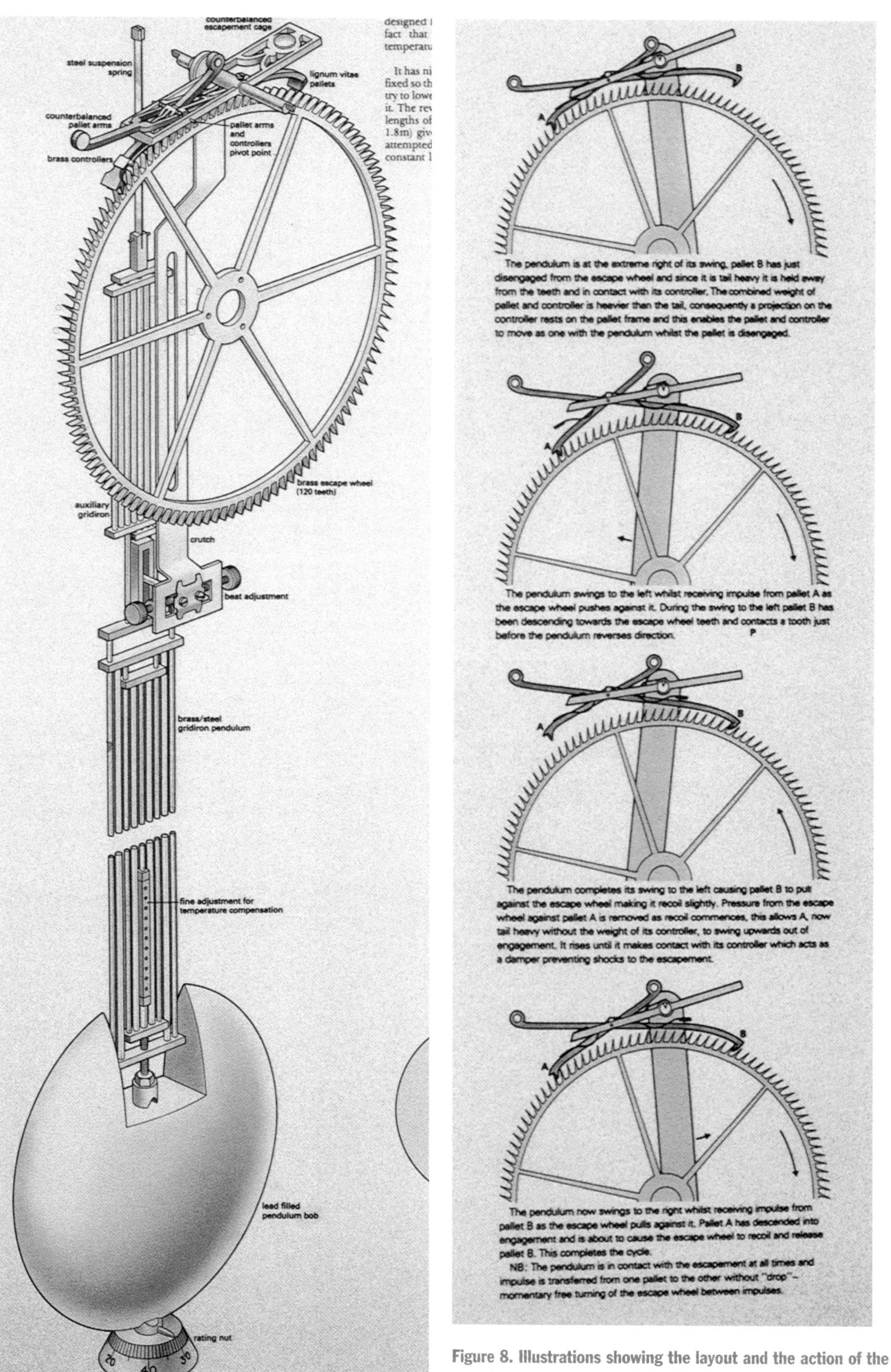


**Figure 8. Illustrations showing the layout and the action of the grasshopper escapement in the R.A.S. regulator. Drawings by David Penney, commissioned by the National Maritime Museum, Greenwich.**

the Octagon Room at Greenwich had their two-second pendulums impulsed at the bottom. Because the escapement deviation with a recoil escapement would have been huge, especially as the arc was so small, a form of dead-beat escapement had to be made.[42]

I still have great difficulty convincing clockmakers that the moment arm of torque, which Harrison called the 'length of the pallets', matters as far as the dominion of the pendulum over the escapement is concerned. I have come to the conclusion that most clockmakers are neither mechanics nor scientists, but are labourers who confuse finish with function. Harrison evidently had the same problem, which is perhaps why he spent so long attempting to make it clear in the early part of *Concerning Such Mechanism*.[43] Many well-known clockmakers have used a mechanical escapement to impulse a pendulum at the bottom: Edward John Dent (1790-1853), William George Schoof (1830-1901), and Abraham-Louis Breguet (1747-1823), to name but three. Although this arrangement might be visually tempting, it is mechanically unsound. If Gould had studied *Concerning Such Mechanism* more carefully and with greater humility, he might not have been so pleased with the work of Pierre Le Roy (1717-1785).[44] What modern watchmaker would impulse a balance at the rim?

Much has been written about Harrison's 'grasshopper' escapement. Yet, until recently, due to lack of research, those who have written about it have not understood it. I didn't in 1971.[45] The escapement works with little friction and wear, and with no oil. It is often said that it likes to work at a large arc. Nonsense! It can work at any arc for which it is designed. The whole point is that it can work at a large arc with a wide fluctuation of arc, so necessary for the energy input tests that Harrison conducted. The best paper to date on this subject was published in 1993 by Peter Hastings.[46] All I need to say here is that the arc simply has to be large or there will not be enough circular deviation to counter the environmental changes that cause the variation of arc.

Most grasshopper escapements made since Harrison's day were for entertainment—the action of the escapement is fascinating to watch. They confer no accuracy on an ordinary clock. They may even make it worse. Benjamin Vulliamy (1747-1811) no doubt expected improvements from it, but I don't think he could have tested it at different arcs. From George Daniels's very clear drawing of Vulliamy's grasshopper escapement,[47] it is obvious that Vulliamy's design, in fact, negates everything Harrison set out to do. The drawing of Harrison's grasshopper escapement that Gould published in his book *The Marine Chronometer* [48] was based on the illustration published in Rees's *Cyclopedia* in 1820.[49] Unfortunately, it is unlike anything Harrison ever made. This gross error has led many people astray, myself included. The correct construction of this escapement, drawn by the great horological illustrator David Penney, is shown in Figure 8.[50] The two pallets and the two composers all turn about the same pivot, so there is no rubbing anywhere.

Figure 9 shows a detail of Harrison's drawing for the construction of the grasshopper escapement that he used in the R.A.S. regulator.[51] I copied the dimensions exactly but made a large drawing to make sure that I got the growing impulse right. Harrison has drawn the four torque moment circles around the centre of the crutch arbor pivot. These show the torque at the

42. The history of these clocks and the introduction of the dead-beat escapement has been superbly researched by Derek Howse in 'The Tompion Clocks' (*op. cit.*, note 35), pp. 18-34.

43. Harrison, *A Description Concerning Such Mechanism* (*op. cit.*, note 7).

44. In comparing the work of Pierre Le Roy and John Harrison in his famous book, *The Marine Chronometer*, Gould states: 'The difference in their machines is fundamental—Harrison built a wonderful house on the sand; but Le Roy dug down to the rock'. Gould, *The Marine Chronometer* (*op. cit.*, note 40), p. 91. Had he understood *Concerning Such Mechanism*, he would never have made such a statement. Nevertheless, we are enormously indebted to Gould, for it was he who was responsible for bringing Harrison's work to the attention of the general public through his restoration of the marine timekeepers and through the publication of several related articles that have inspired many to take an interest in Harrison's work.

45. Burgess, 'The Grasshopper Escapement' (*op. cit.*, note 4).

46. Peter Hastings, 'A Look at the Grasshopper Escapement', *Horological Journal*, vol. 136, no. 2 (August 1993), pp. 48-53.

47. Daniels's drawing is shown on p. 294 of Quill, 'The Grasshopper Escapement' (*op. cit.*, note 4). The escapement illustrated comes from the Vulliamy clock in the Science Museum in London. Three other Vulliamy clocks with this type of escapement are known.

48. Gould, *The Marine Chronometer* (*op. cit.*, note 40), fig. 12 (between pp. 43 and 44).

49. Article on Horology (Escapements), plate 33, p. 215, in Abraham Rees, *The Cyclopedia; or Universal Dictionary of Arts, Sciences and Literature* (London, 1819-20); reprinted in *Rees's Clocks Watches and Chronometers (1819-20)* (Rutland, Vermont: Charles E. Tuttle Co., 1970).

50. First published in Quill, 'The Grasshopper Escapement' (*op. cit.*, note 4), p. 289.

51. This drawing is included on p. 5 of MS. 3972/3 in the Library of the Worshipful Company of Clockmakers, Guildhall Library, London. There is no proof that it was a working drawing for the R.A.S. regulator, but the dimensions are exactly the same in the drawing as in the clock.

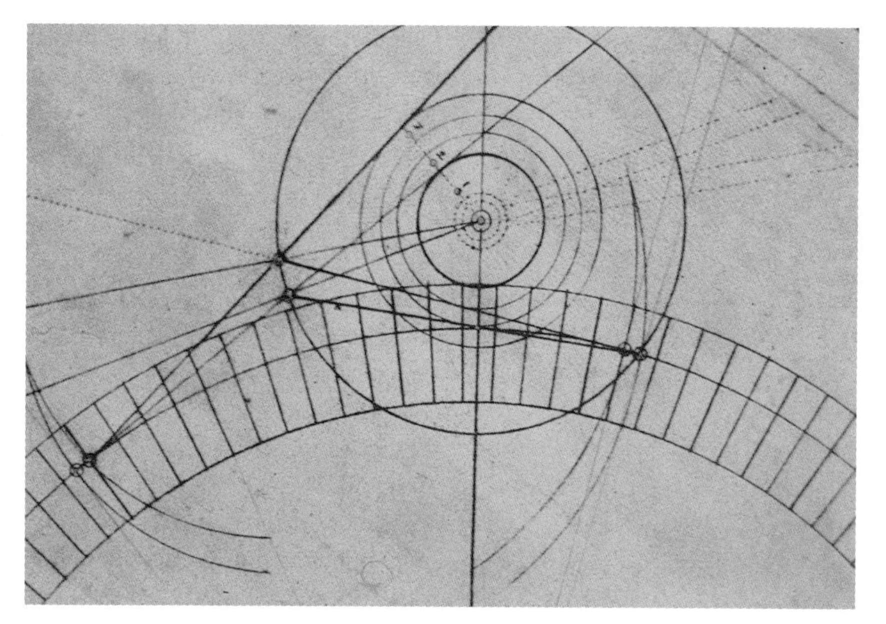

**Figure 9. Working drawing, made by John Harrison, for the type of grasshopper escapement used in the R.A.S. regulator. MS. 3972/3, p. 5, Library of the Worshipful Company of Clockmakers, London.**

beginning and end of the escaping arc for each pallet. He tells us that the two minor moments summed must be about two-thirds of the two major moments summed.[52] The mean moment is one-hundredth of the length of the pendulum, just as he tells us it should be. The energy input grows from two units to three units as the escaping arc progresses. At the start of recoil, the force drops to two units and declines further throughout recoil. This abolishes escapement deviation, and we are left only with the hastening effect of pushing the pendulum. The weight of the composers[53] can be adjusted to fine tune the pattern of the energy input to the pendulum from the grasshopper escapement.

A pendulum is subjected to many random outside forces: draughts, vibrations from traffic or wind on the building, distant stationary engines, earthquakes (almost never in London). Figure 10 shows what happens if a fan blows air along the wall toward my pendulum. Measurements were made with a Beckman averaging counter expressing to one-hundredth of a microsecond per swing. The fluctuations are shown here as error per day. Before the fan is turned on, the clock is keeping good time. The short-term gains and losses are due to the noise from the air. The noise increases considerably with the increased air resistance when the fan is turned on: this

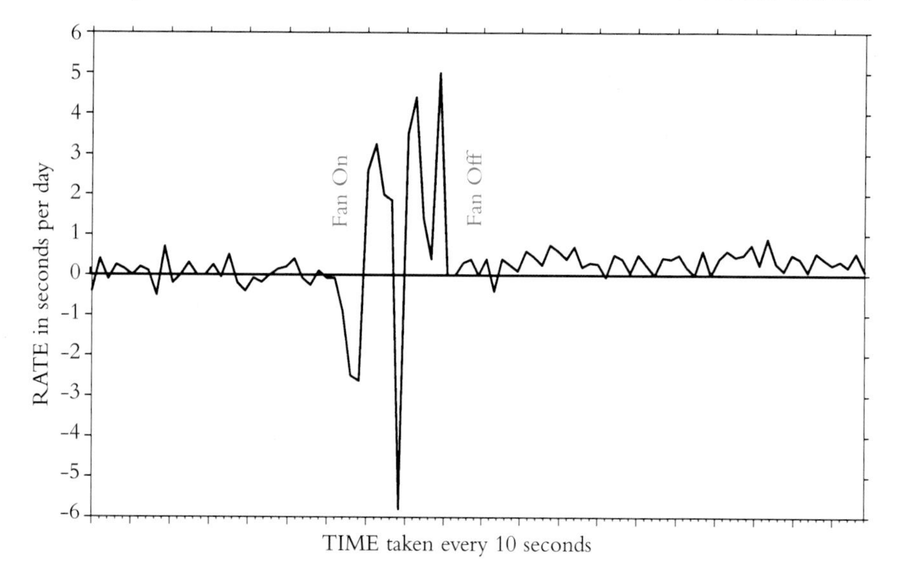


**Figure 10. Graph showing the effect on the Gurney clock's pendulum of an increase and decrease of air resistance. The variation was created by an electric fan. This test was made on Monday, 1 December 1986.**

**Figure 11. Graph showing the variation of the Gurney clock during a short run of 40 hours in conditions of quiet temperature and pressure. The graph, which represents eighteen-hundredths of a second on the vertical scale, shows how the clock varied 65 milliseconds during this period. The dotted line demonstrates how lazy I was: when you do an hourly test like this, you have no business going to bed.**

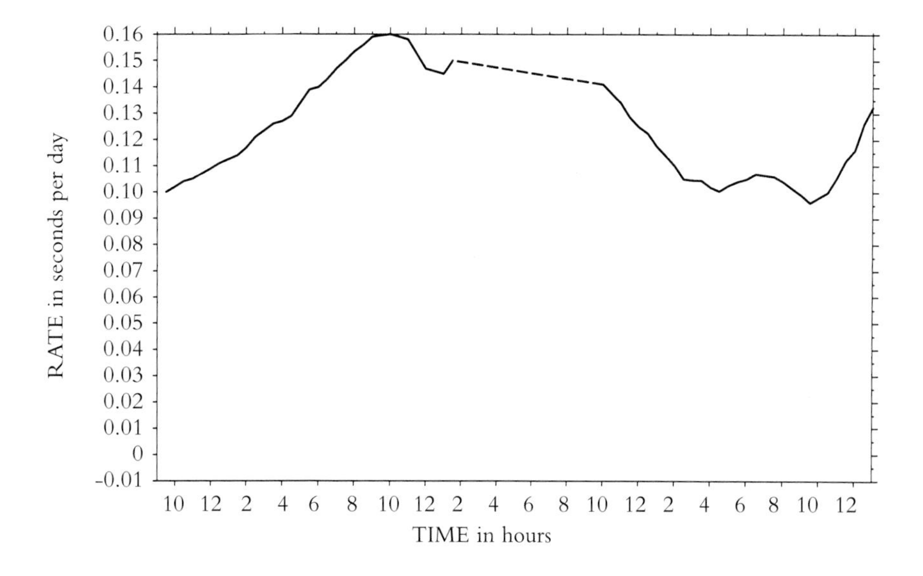


causes the arc to drop from 12½° to about 11° Then the fan is turned off, and now there is a surplus of energy over energy consumption: the arc builds back, but the surplus has to go somewhere, so there is a small gain. If you push something harder it moves faster. At the end of the test, the difference of time from the beginning was seven milliseconds. Figure 11 shows a 40-hour run in conditions of quiet temperature and pressure.

A light pendulum with a big energy throughput remembers a disturbance for a much shorter time than a heavy one. But a lot of stored energy is needed to resist disturbance. The energy stored in a swinging pendulum equals half the mass multiplied by the square of the velocity.[54] That is why a little rifle bullet can punch through a steel sheet. Harrison wrote in 1763 and 1775 that a three-pound pendulum swinging in a 12° arc has the same stored energy as a 48-pound pendulum swinging in a 3° arc.[55] The large arc gives us a massive stored energy. It also gives us a lot of air resistance and, with a gridiron, air friction—that, also, is just what we want.[56]

Harrison stated, 'neither, as by any Means, does a suitable Matter of this, viz. of the Air's Resistance, want to be avoided, as many have foolishly imagined, but is of real or great Use'.[57] It is natural to think that a machine that consumes less energy is better. That does not make it more accurate. The only things that limit the arc of a pendulum are air resistance and air friction. If you reduce them, the clock becomes more unstable. As the air resistance of the pendulum's design is reduced, changes of energy input change the arc more, but changes of air density change it less. Here is the method of bringing change of air and change of energy input together. By 'a suitable matter of this', The Master is telling us to adjust the air resistance. If he had got it almost right, how would the final adjustment be done? Look again at the two outer brass rods on the pendulum in the portrait (Figure 7). The artist has shown only a few tin whistle holes. I suggest that we should start with rods that are too long, so that the arc changes too much with pressure change, and then cut the rods down until the arc is correct. In my first attempt at a Harrison regulator, I failed to adjust the air resistance.

We now have a small range of control over the changes. We have plenty of circular deviation to wipe these changes out, so we must be able to adjust that as well. Harrison used circular arc brass cheeks to constrain the suspension spring. They are very long, so that the active part has the temperature

52. Harrison, *A Description Concerning Such Mechanism* (*op. cit.*, note 7), pp. 25-6. The same statement is made in Harrison's 'An Explanation of my Watch' (*op. cit.*, note 6), but, until recently, no clockmaker ever took any notice of it.

53. The composers are the parts of the grasshopper escapement that both absorb the upward force of the heads of the pallets and settle ('compose') them in the correct position.

54. $E = \frac{1}{2} MV^2$.

55. Harrison, 'An Explanation of my Watch' (*op. cit.*, note 6), and Harrison, *A Description Concerning Such Mechanism* (*op. cit.*, note 7), p. 22.

56. A gridiron pendulum is affected both by air resistance and air friction. Air resistance can be likened to a flat sheet of metal being pushed against the wind broadside on. Air friction can be likened to a metal grill being pushed against the wind edgeways on, where the resistance is caused by still air rubbing against moving air.

57. Harrison, *A Description Concerning Such Mechanism* (*op. cit.*, note 7), p. 27.

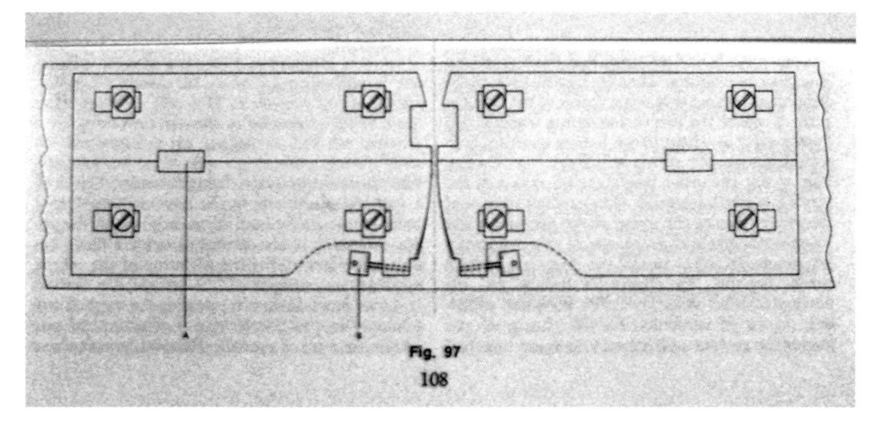


Figure 12. Illustration of the pendulum suspension bracket of the R.A.S. regulator. The cheeks are designed to be easily removed, so that they can be remachined for making adjustments to the pendulum's circular deviation. Drawing by William Laycock.

of the room and not that of the wall behind. It matters! The spring must be a very close fit between the cheeks, and the fit must remain exactly the same. I am talking in terms of one-millionth of an inch. It follows that the coefficient of expansion of the spring must be the same as the cheeks and their support block. Harrison tells us to use a work-hardened, high-copper gold spring.[58] We did not understand why, so I used a spring of Durinval C. It is most helpful that this alloy does not change its elasticity, but its almost zero coefficient of expansion is a problem not yet properly overcome.

Note the unused central slot and the micrometer screws on the pendulum suspension cheeks shown in Figure 12. Laycock devised a shaping fixture that takes account of these. The micrometers are vital. On the cheeks of the R.A.S. regulator, they look like an afterthought to me. The cheeks have to be remachined very many times over a long period. No system of measurement and recording is equal to a fixed register point. The fixture on the lathe, shown in Figure 13, was devised by Bill Laycock. I could not have designed it, and I could not have made the cheeks without it. To reduce the radius, a cheek is moved back and the tool advanced to meet it until the surface has been cleaned up to the new radius. It is necessary to be very careful: remove only five-thousandths of an inch at a time, put it back on the clock, re-regulate the clock, and then perform the tests. This process must be repeated until the correct radius has been found. To achieve the desired result for the first time can take years.[59]

Figure 13. Fixture developed and constructed by the author from a design made by William Laycock about 1975. This device enables minute changes to be made on the lathe to the radius of the cheeks. Courtesy of the author.

58. Harrison states that the spring is 'best to be made of Gold, properly allayed with Copper, and to be well hammered before it be brought to its Thinness, (as being then more elastic, than as if, or when allayed with Silver)'. *Ibid.*, p. 46.

59. I cannot be certain, but my experiments appear to show that the cheeks of the R.A.S. regulator had been finished by the time of Harrison's death, because they had been cut to a radius that could be used.

60. Harrison, *A Description Concerning Such Mechanism* (*op. cit.*, note 7), p. 45.

61. Harrison states, in reference to adjusting the thickness of his balance spring, that 'it must (with care) be rubbed to its thickness or strength as in its figure and that with a convenient bit of brass and oil-stone powder as resting upon a bit of wood prepared for the purpose'. Harrison, 'An Explanation of my Watch' (*op. cit.*, note 6).

62. 'If the balance-spring is too strong, it must be made weaker by rubbing it away a little; but, if it be too weak, it must be changed for a stronger [spring]'. [John Harrison], *The Principles of Mr. Harrison's Time-keeper, with Plates of the Same* (London, 1767), p. xi; facsimile reprint by the British Horological Institute in *Principles and Explanations of Timekeepers by Harrison, Arnold and Earnshaw* ([Upton, Notts., England], 1984).

63. The free-pendulum clock, designed by W. H. Shortt (1882-1971) about 1920, is described in F. Hope-Jones's *Electrical Timekeeping* (London: N.A.G. Press, 1940), pp. 162-70. Fedchenko's clocks, dating from the 1950s, were described by Myron Pleasure in the last part of his three-part article, 'Precision Pendulum Clocks', *Horological Journal*, vol. 113, no. 4 (October 1970), pp. 6-10, and by the same author in 'The Fedchenko Clock', *Horological Journal,* vol. 116, no. 3 (September 1973), pp. 3-7, 55, as well as in 'Heavy and Light Pendulums', my response to a reader in the Letters section of *Horological Journal*, vol. 136, no. 11 (May 1994), pp. 367-8.

I built the suspension units for both of my clocks three times, because at first I did not understand the vital importance of insulating the unit from the wall. Harrison tells us that the suspension spring must be 'thin to the Purpose'.[60] I wasted years because I did not understand that he meant that it must be the right thickness. Because the spring does at least 50 per cent of the work, its thickness must be found by repeated experiment, and it has to be brought to much better than a tenth of a thousandth of an inch. To do this, you have to rub it down by hand.[61] We need to remember that Harrison brought H.4 and H.5 to time by rubbing down the balance springs.[62]

Rendering a pendulum isochronous is important for another reason. A swinging pendulum is an energy dissipater. Energy is given to it and energy goes out from it. An exact balance must be established at all times. A pendulum swinging a circular arc moves slower in the longer arcs and faster in the shorter arcs. An increase in air resistance will cause the pendulum to move through the air more slowly, but it will also cause it to swing in a smaller arc. At the smaller arc, the circular deviation causes the pendulum to increase its rate. It follows that there is not a sharply defined arc that will exactly satisfy the energy input and output balance. There is, however, a fuzzy area in which the pendulum can take up its position for any given set of environmental conditions. Such a pendulum will have at least two rates and two arcs that will satisfy those conditions. If it is swinging at its maximum arc and there is no change in the conditions at all, then a stray vibration may shift the arc to a very slightly smaller one. The increase of rate at that smaller arc will abstract more energy. The arc will continue to decline until an energy balance is again achieved. A normal pendulum swinging in air will have arcs and, therefore, rates very close together—so close, in fact, that the error is swamped by other factors. But if we streamline the pendulum or, worse still, put it in a partial vacuum, the fuzzy area gets wider. That is why the Shortt clock, which is housed in a vacuum, jumps its rate, while Dr. Feodosii M. Fedchenko's clock, also in a vacuum but with an isochronous suspension, does not.[63]

Figure 14 shows what the spring and cheeks do. The train is latched and the remontoire spring unwinds. The pendulum moves slower, because it is not being pushed. Then the pallets disengage, and as the arc continues to decline, its rate increases. It does so in a straight line. The gradient represents

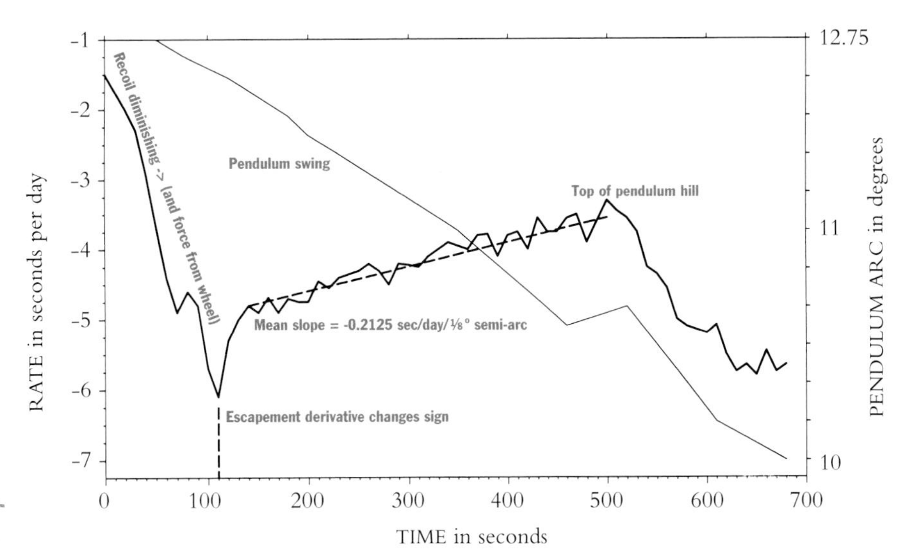


**Figure 14. Graph illustrating the effect that the suspension spring and cheeks have upon the period of oscillation when an interruption in the power to the escapement causes the pendulum's arc to decay.**

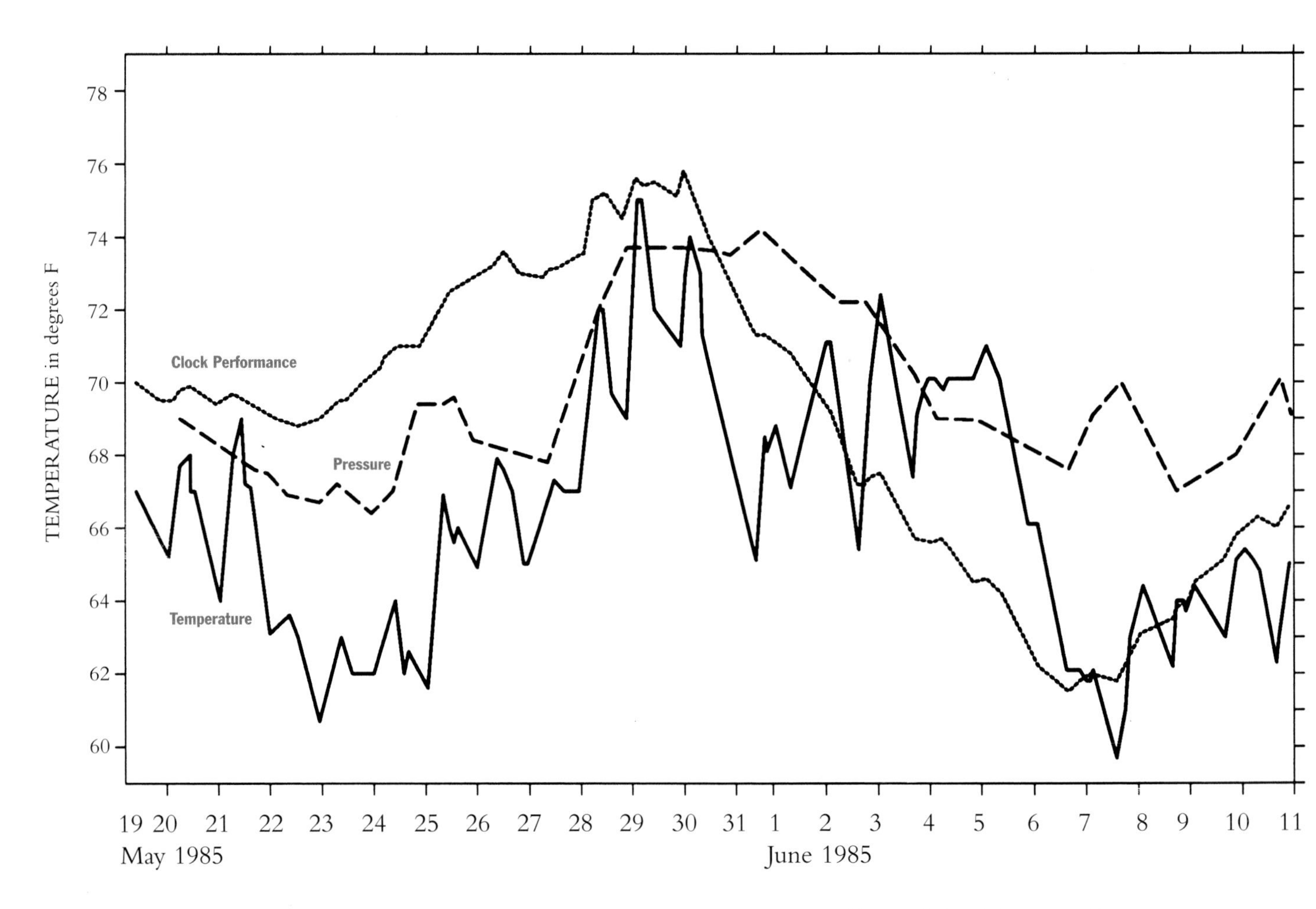


Figure 15. Graph illustrating the accuracy of the Gurney clock over a period of 46 days. During this trial, the clock's maximum variation was never more than one second fast or slow. Its rate followed the general path of pressure and temperature changes but averaged out their sudden fluctuations. The graph width represents two seconds.

the circular deviation required to counter the energy input. At 11° the rate starts to decline again. The cheeks and the spring have stopped doing anything: the arc is too small. That is why cheeks were abandoned before Harrison was born, but you won't find that in any modern book on horology because the writers were ignorant of the reason.

In 1773, John Arnold (1736-1799) installed two clocks with cheeks in the Greenwich Observatory.[64] The success of Harrison's timekeepers was well known, and, because of this, Arnold no doubt suspected that cheeks were a 'good thing'. He could not have tested them properly. With the small arc of swing that the dead-beat escapements of these clocks required, the cheeks can only have been a confidence trick, but Arnold could boast that his regulator had cheeks just like Harrison's. He may have harmed Harrison's reputation by doing that. For his pendulum work, Arnold now has my total contempt. Contrast this totally unscientific approach by someone who was regarded as one of the great makers of the day with Harrison's painstaking analysis of every detail. Harrison tested everything, and he always published his mistakes.

I made my full share of mistakes as well. My regulator is balanced for energy input and arc only. There is still some temperature and barometric error. In addition, my Invar pendulum rod does not carry enough air with it to rub sufficiently against the air through which it is moving. As a result, the arc changes too much with increased energy input and not enough with changed pressure. Figure 15 shows a test of 46 days, during which the clock's maximum error was less than plus or minus one second.[65] Simply from one stellar observation, Harrison would not have seen any of these errors. No

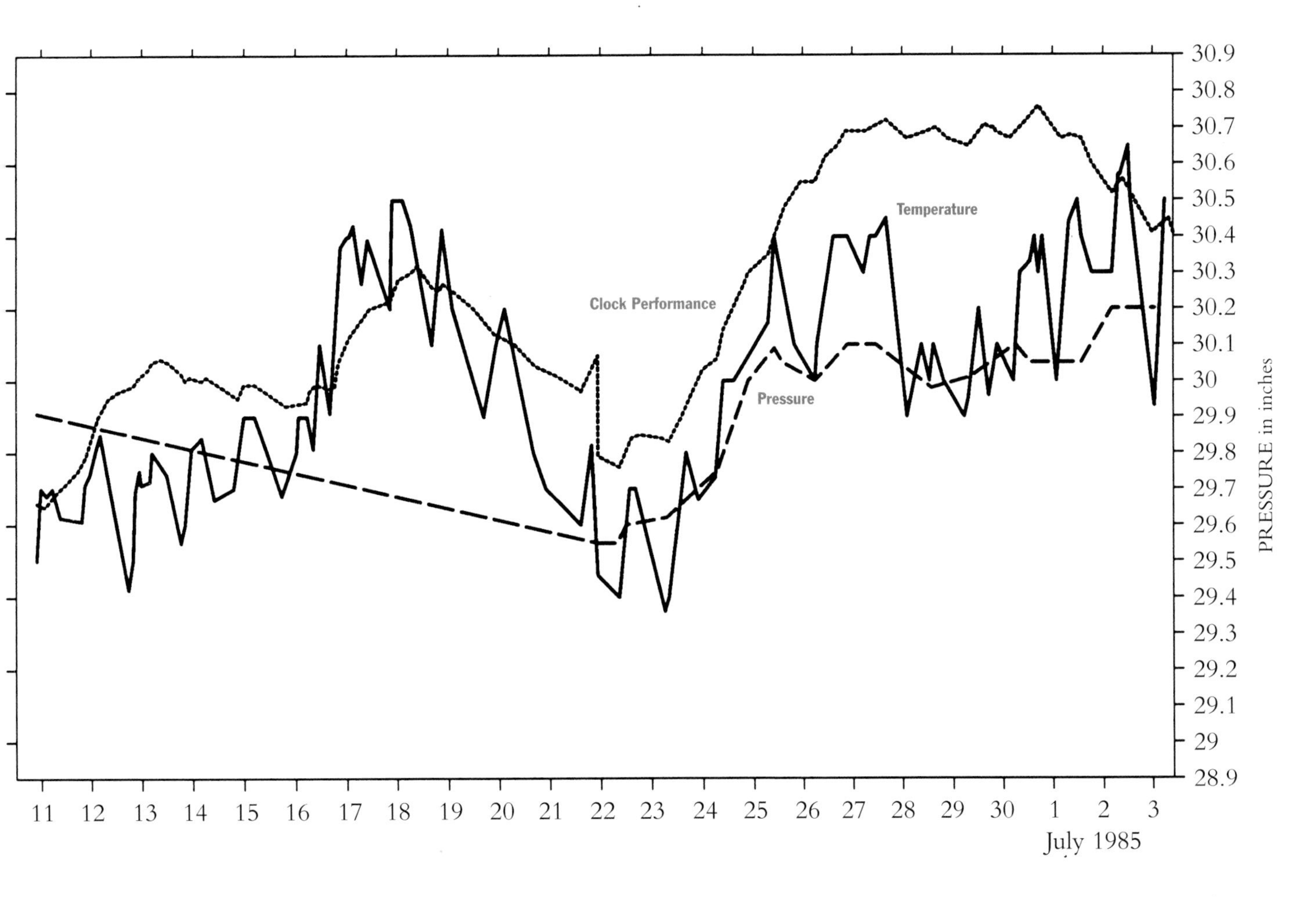


rate has been applied to the clock. In my view, to apply a rate shows either that the clockmaker is too lazy to regulate properly or that the clock won't stand regulation.

As mentioned earlier, the remontoire of the R.A.S. regulator, had it been finished by Harrison, would undoubtedly have been just like the remontoire on H.3.[66] It can't be improved. My remontoire, shown in Figure 16, is of the same design, but the fly makes one turn per unlock. The torque of my spring remains constant with variations in temperature. There is a shift of a few microseconds between being wound and about to be wound, but I used

64. Apart from the cheeks, both clocks were also fitted with gridiron pendulums and dead-beat escapements with ruby pallets. In 1836, the cheeks and original suspension springs of both clocks were removed by Edward J. Dent. See Derek Howse, *Greenwich Observatory: The Royal Observatory at Greenwich and Herstmonceux, 1675-1975,* vol. 3, *The Buildings and Instruments* (London: Taylor and Francis, 1975), pp. 130-2.

65. This test was made in the autumn of 1986, shortly before the clock's installation in Norwich in early 1987. During the first two weeks of its trial in its new location, it gained 0.6 second but lost most of this in the next two weeks and ended up after one month 0.04 second fast. Unfortunately, a serious problem with the winding mechanism interrupted the subsequent rating of the clock. The clock is presently under test at the British Horological Institute's headquarters at Upton Hall, where it was moved in 1993.

66. H.3's remontoire is described in Peter Hastings's article, 'John Harrison—Spacecraft Engineer?', *Horological Journal*, vol. 136, no. 6 (December 1993), pp. 193-7.

**Figure 16. Detail of the grasshopper escapement and remontoire of the Gurney clock. Courtesy of the author.**

cycloidal wheel teeth in the train, which give a constant winding torque. Harrison's train fluctuates because he used straight-sided teeth and chordal pitch to get large rollers in the pinions. So he needed to make the energy input from the remontoire totally constant. I should have used an inertia fly so that a change of air density would not change the speed of winding. It matters! Harrison even used diamond endstones for his escape wheel arbor.

It is important to isolate the escape wheel from the train, especially in recoil. Wheel and spring act like a balance and spring and take part in the oscillating function. The remontoire also allows an exact and repeatable change of energy input for the testing. Harrison did that with little cams at the outer ends of the springs (Figure 17). These cams have five positions; setting the cam in the centre position allows for two adjustments to increase the running arc of the pendulum and two to decrease it. I only have three positions. My spring was not soft enough. There was too little air friction. There is no excuse for this at all. I learned as a schoolboy that three points don't make a proper graph. I saw those cams on H.3 every time I visited Greenwich. I knew what they were for. So why did I not count the positions and act accordingly?

There is another reason for a good remontoire. To study the environmental changes, each must be isolated from all the others. To do that, we need to know that the others have all remained the same. Later in life, Harrison would have had a good thermometer and he could change the temperature, especially in winter. He would also have had a good stick barometer, and, though he could not change the pressure, he could see when nature did. With some effort we might change it. It would save a lot of time. For

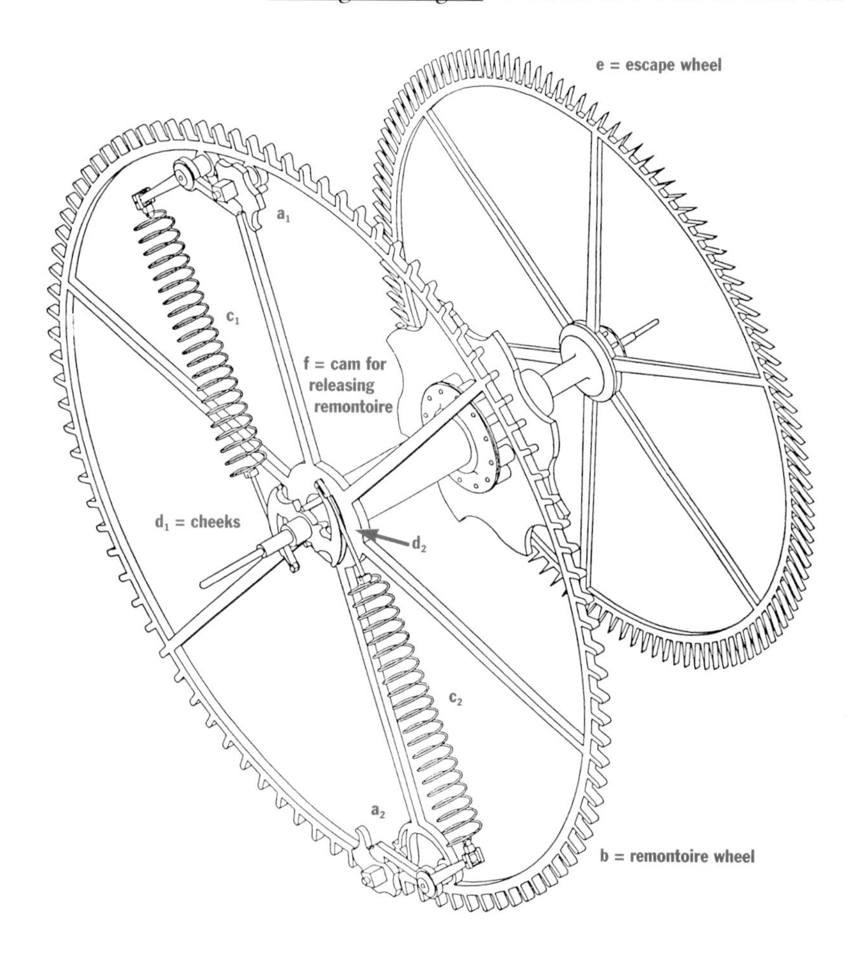


**Figure 17. Drawing of the remontoire used on H.3, showing the cams ($a_1$, $a_2$) on the perimeter of the remontoire wheel (b), which allow five positions for the adjustment in the power delivered by the remontoire springs ($c_1$, $c_2$). The curved 'cheeks' ($d_1$, $d_2$), by equalising the variation in the tension of the springs when they are wound, ensure that a constant driving torque is delivered to the escape wheel (e). The remontoire is released by a mechanism (not shown) activated by the cam (f) mounted on the escape wheel arbor. Drawing by Peter Hastings.**

these changes we have to be sure that the energy input only changes when we change it deliberately. A good remontoire is vital.

Harrison could only obtain absolute time from the stars. Until the advent of the atomic clock and the new time standard that it established in 1967, everyone had to look at the stars to get absolute time, and it was from these observations that all clocks were regulated.[67] To regulate his clocks at his home in Barrow upon Humber, Harrison had used the west side of one of his window frames and the east edge of his neighbour's chimney, which was situated across the road some 25 yards to the south of his house.[68] With this primitive but effective transit instrument, we know that he observed the passing of the same star more than once, for he refers to not 'one of my window frames' but to 'the West side of an exact place of some [the emphasis is mine] one of the upright parts of my Own Window Frames'.[69] While he observed the exact moment at which the star vanished behind his neighbour's chimney, he had another person call out the seconds shown by the clock. It is known from his will that he had a more sophisticated optical transit instrument in his house in Red Lion Square in London. By knowing the precise time difference between the transits of certain stars, several consecutive observations would be used to iron out any human error in these observations. On the following night, the transits of the same stars would occur 3 minutes 56.4 seconds sooner, and, thereby, he could determine exactly how his clock had performed.

This was satisfactory for determining the time when the sky was clear. For the development and adjusting of other timekeepers, however, a reliable regulator of known accuracy was needed so that the time could be obtained at any time of day or night. To achieve this he used two clocks placed in two different rooms in such a way that when their hoods were removed, he could stand in the doorway and hear, as he said, 'the beats of both the pendulums' or 'see the Seconds of one Clock, whilst another Person count the Seconds of the other: by which means I can have the difference of the Clocks to a Small part of a Second.'[70] By leaving one room cool and heating the other with a fire, he could also test the performance of his clocks at different temperatures. Because he had no way of making artificial cold, this work had to be done during the winter.

Temperature error is caused not only by changes in the length of the pendulum, but also by the thinning of the air in heat. This thinning causes an increased arc in heat, but it is not accompanied by much flotation error, like the change in the height of a barometer. In order to bring the temperature compensation into final adjustment, Harrison had to over-compensate to take in the change that would have happened if there were flotation error. Temperature error is a non-linear function: temperature plotted against rate does not produce a straight-line graph. It follows that there will only be a small range of temperature that can be compensated for. So for the true and final temperature testing, what is required is a long run in the temperature conditions that the clock will actually experience: a whole winter followed by a whole summer and then a small adjustment if required, followed by another whole winter followed by a whole summer, and so on until no temperature error can be seen.

As an example of the degree of accuracy that was expected from the finest observatory timekeepers toward the end of the eighteenth century, it

67. For regulating clocks today, we are more fortunate, because the atomic time signal is readily available. The device I use, called a 'Radiocheckrate', was designed by Douglas Bateman, a British government scientist working on the development of precision instruments at a research establishment at Farnborough. The Radiocheckrate compares the atomic time signal with a photoelectric signal sent by the passing of the pendulum. With a readout to a tenth of a millisecond, the exact shift of the clock can be detected any time of day or night.

68. With this arrangement, he would only have been able to get the time to the nearest second if he was lucky, because one second of time in this situation is about one-sixteenth of an inch on the chimney while on the window frame it is about the thickness of a coat of paint. He might have edged the observation surfaces with brass to get a more exact reading.

69. Harrison, 1730 manuscript (*op. cit.*, note 2), p. 11, sect. 19.

70. *Ibid.*, p. 11, sect. 20.

is interesting to mention a regulator that was made by Thomas Earnshaw (1749-1829) for the observatory of the Archbishop of Armagh in Ireland.[71] Nevil Maskelyne, who had recommended that Earnshaw should make the clock, expected the error to be no greater than half a second per day. The regulator, completed in 1792, was tested by Maskelyne at Greenwich. After a year Earnshaw made an adjustment to the temperature compensation and wanted the clock to remain at Greenwich for another year. Without another accurate regulator in Earnshaw's workshop, it would have been hard to adjust the clock there. However, it only stayed at Greenwich another six months. The compensation was still not right, but the clock was dispatched to Ireland, and Earnshaw installed it in Armagh Observatory in 1794. The clock was a success and was so highly esteemed, apparently, that the astronomer the Reverend Doctor Thomas R. Robinson (1792-1882) described it in 1831 as 'The excellent transit clock'.[72] How could it not be when all that was expected was a rate of half a second per day! Robinson used the clock in trials to correct for barometric error. One does not know whether to laugh or weep! Harrison's wooden regulator of 1728 had already solved most of the problem 100 years earlier. In his house in London, it was keeping time to a second per month. It was still there when Earnshaw built his regulator, and it would have had a closer rate than anything that Earnshaw or Arnold managed to build.[73]

Why was it that Harrison's regulator science so quickly lapsed into neglect after his death? One major reason, certainly, was the lack of interest on the part of the Establishment, caused no doubt in part by the bitterness that had arisen over the trials of H.4 and the payment of the £20,000 longitude prize. By insulting the Astronomer Royal, Nevil Maskelyne, verbally and in print for many years, John Harrison and his son William (1728-1815) had really shot themselves in the foot. It is highly unlikely, therefore, that Maskelyne wanted anything further to do with them, and Harrison, not surprisingly, returned this feeling: he stated in his book that he had 'once thought of giving a Clock to the Observatory at Greenwich, but my bad Usage proved too tedious for that'.[74] Thus, the possibility of any cooperation with the Greenwich Observatory that might have ensured the preservation of the regulator technology was effectively negated. Moreover, according to Maskelyne's instruction to Earnshaw in 1792, the need for a better time standard was evidently not considered too pressing at that time.[75]

Another problem was that Harrison worked independently and did not train anyone through whom the details of his very specialised approach would be preserved. He did, however, describe his technology in his book *Concerning Such Mechanism*, published when he was 82, a few months before his death. In this work, he admitted that there was still no 'sufficient Instruction being as yet communicated to the World; and as without which, although as it were my natural Road, would still prove a very tedious Matter to others, it being (as it were) so very much out of their beaten Path'.[76] The last part of this remark emphasises the fact that the nature of his work was very different from the traditional practice of precision horology, and, because of this, Harrison's technology did not lend itself well to words—indeed, words for many of the scientific concepts he attempted to describe did not exist in the eighteenth century. On top of that, he experienced great difficulty expressing himself clearly in writing.[77] The review that appeared in

71. This clock is described by Thomas Earnshaw in *Longitude. An Appeal to the Public:...* (London, 1808); facsimile reprint by the British Horological Institute in *Earnshaw's Appeal* ([Upton, Notts., England], 1986), pp. 39-49. It is interesting to note that Earnshaw, although a recognised chronometer maker, had never made a clock before.

72. T. R. Robinson, 'On the Dependence of a Clock's Rate on the Height of the Barometer', *Memoirs of the Royal Astronomical Society*, vol. 5 (1831), p. 125; quoted in John Griffiths's introduction to the reprint of Earnshaw's *Longitude: An Appeal* (*ibid.*).

73. In 1809, Maskelyne commissioned William Hardy (fl. 1800-1832) to make a regulator for the Royal Observatory. One of Hardy's regulators with his new spring-pallet escapement had been tested by Maskelyne in 1807 and, during the whole two months of its trial, had shown no variation. Hardy's escapement, however, suffered from metal fatigue, and his regulators were incapable of maintaining this rate over a long period. Modern research has shown that the pallets have to be absolutely clean, but because they require oil and oil attracts dust, they do not remain clean for long. Harrison had solved the problem of lubrication by 1726. The Greenwich clock, which Hardy constructed at great expense, had its escapement replaced in 1830. For further information see Charles Allix, 'William Hardy and his Spring-Pallet Regulators', *Antiquarian Horology*, vol. 18, no. 6 (Summer 1990), pp. 607-29, and Philip Woodward, 'The Performance of a 19th Century Regulator by William Hardy', *Horological Journal*, vol. 135, no. 9 (March 1993), pp. 306-12.

74. Harrison, *A Description Concerning Such Mechanism* (*op. cit.*, note 7), p. 52.

75. This situation would, of course, change during the nineteenth century, when significant improvements in astronomical instruments permitted more precise observations to be made.

76. Harrison, *A Description Concerning Such Mechanism* (*op. cit.*, note 7), p. 23.

77. In the past, friends such as James Short (1710-1768) and Taylor White (1701-1771) had assisted him in this regard, but by this time they had died.

1775 stated: 'Any one who reads but a single page of this pamphlet will be convinced that Mr. H. is utterly unqualified to explain, by writing his own notions, or to give a tolerable idea of his inventions'.[78] To a casual reader, this is certainly true. But to those who have a full working knowledge of mechanical horology, a familiarity with Harrison's clocks, and a fundamental knowledge of mechanics and oscillator physics, this work represents a gold mine of information.

Yet although this was Harrison's only published work about his technology, he had not been secretive about this aspect of his work: 'For in particular, I took some pains with Mr. Shepherd, (viz. when he was my friend) but could make nothing of him (viz. any farther than that one wheel turned another) although it was his Desire: Very unfit gentlemen to be my masters'.[79] Somewhat surprisingly, Harrison was never asked to speak about his work at any meeting of the Royal Society, although in 1749 the Society had bestowed on him their highest honour, the Copley Gold Medal. In his book, he offered to show them 'the Draught of the Clock which I have in great Part made, and not only the Draught of the Pallats, as in particular, but also the Pallats themselves, in order that they may see at least some Reason for what I found'.[80]

This offer notwithstanding, Harrison was bitter about the manner in which he had been treated and, elsewhere in the book, stated his refusal to do anything further to instruct others in his technology until he was 'the more freely, or the more genteelly rewarded than what I have as hitherto speakingly been'. And recalling all the expense, time, trouble, and 'scurvy' work to which he had been subjected, he continued, 'for not one Stroke as farther will I take...but as being paid short,...I will also be short'.[81] These are the words of a disgruntled old man who knows he has done the experiments and has found the answers but has not had sufficient encouragement. The damning review of his book that appeared in *The Monthly Review* in October 1775 brought out further bitter resentment of those whose social or academic standing gave them the authority to sit in judgement of his work. Recognising their ignorance of technology and their jealousy of the rewards he had received, he concluded that it was no longer worth responding to the 'Nonsense, Spite and Poison (scandalously scurvy, dirty Work indeed) as runs throughout the Whole of their maliciously groundless Objections, as objecting against Things which are really true and done'.[82]

It is interesting to note the similarity in the way that Harrison and one of his strongest supporters, James Short (1710-1768), left their affairs when they died. Short, who was elected a Fellow of the Royal Society when he was 26, was an optician who gained an international reputation for the perfection of the glass and metal specula he made for his telescopes. Yet when he died, he left no information on the formulae and techniques that he had employed for the making of his specula metal mirrors. With no apprentices to continue his work, the art of the craft that he had perfected during his lifetime died with him.[83] The contents of Short's workshop were auctioned the year following his death, but the tools alone were useless without the hands of knowledge and experience. Harrison followed a similar path, leaving, in addition to his tools and equipment, his writings and drawings. But his writings were extremely difficult to understand without the knowledge and experience that only someone working with him could have gained.

78. *The Monthly Review; or, Literary Journal: From July 1775 to January 1776* (London), vol. 53, Art. VIII (October 1775), p. 320.

79. Harrison, *A Description Concerning Such Mechanism* (*op. cit.*, note 7), p. 42. The 'Mr. Shepherd' referred to here was Anthony Shepherd (fl. 1763-1772), who, by virtue of his appointment as Plumian Professor of Astronomy at Cambridge University, was one of the members of the Board of Longitude. He became a Fellow of the Royal Society in 1763 and published two works relating to the lunar-distance method in 1772. In the preface of one of these works, he also refers to himself as 'Master of Mechanics to his Majesty'.

80. *Ibid.*, p. 54. The fact that this clock was 'in great Part made' suggests that it is the same piece that, two pages earlier, he calls his 'next or second' clock. That the pallets were one part of the clock that had been finished provides further evidence that this clock is the R.A.S. regulator.

81. *Ibid.*, pp. 28-9.

82. *Ibid.*, second printing with appendix, p. 114. (The editor wishes to thank Charles Allix for supplying a copy of the appendix, so that the wording of this quote could be checked and details of this very rare second printing could be added to the bibliography.)

83. Gerard L'E. Turner's paper, 'James Short, F.R.S., and his Contribution to the Construction of Reflecting Telescopes', *Notes and Records of the Royal Society of London*, vol. 24, no. 1 (June 1969), pp. 91-108, provides an interesting insight into the work of James Short. Turner states that 'it is very doubtful whether Short had any secret other than his skill, dexterity, and patience. A cookery book and a well-equipped kitchen are not the only, or even the main, attributes of a good cook'. A good cook invents his own recipes and knows how to blend the ingredients; these were the secrets that Short took with him when he died.

Thus, Harrison's regulator technology remained unappreciated and was forgotten soon after his death.

John Harrison's art was unique, and its loss delayed improvements in the accuracy of precision timekeepers for more than a century. Had there been cooperation to encourage the completion of the R.A.S. regulator, the Royal Observatory might have had the regulator that Harrison hoped would keep time to within a second in a 100 days. Everyone could have benefited. But as it was, it took over 100 years for the Observatory to acquire a timekeeper capable of such accuracy,[84] and it has taken 200 years for the importance of Harrison's regulator technology to be recognised. The point is that much can be learned from the past: Harrison's work, having been neglected for so long, has in the last few years influenced important developments both in oscillator physics and in mechanical engineering.[85] If it is inspiring scientists and craftsmen today, how much more could it have done for us 100 years ago? What is particularly unfortunate is that the neglect of Harrison's work was not due to it being too difficult, nor because it took too long to do, nor because it cost too much. It was neglected and then forgotten because of a gross failure of human relationships.

Research work without proper publication is lost work. The author would like to thank all of the people who have assisted with the publication of this paper:

Pride of place must go to Will Andrewes. It was essential to have a really good, open-minded horologist at the other end who both understands the technology that has been uncovered and has a strong grasp of the horological history of the times under discussion. I should also like to thank Jonathan Betts and Andrew King for checking the work and providing historical and technical insights and references not available to me.

Of course, as I am not a mathematician or an engineer or an historian and I have never had any formal horological training, the work could not have been done at all without the constant support of the Harrison Research Group. For my presentation at the Symposium, I was really the mouthpiece of the team. Mervyn Hobden must be singled out for especial thanks. He not only formed the team in the first place but also brought to the regulator work his massive experience of the testing and adjustment of precision timers and his wide knowledge of the history of science.

Finally, I would like to thank Peggy Liversidge for her many useful suggestions to clarify the text, Ana Dujmovic and Peter Pinch for the transcription of the graph data, Geraldine Gardner for creation of the computer-generated graphs, and Marty Richardson for her help in coordinating these tasks.

I think John Harrison's regulator claims have been vindicated now that they are understood, but the thought that he has had to wait 200 years is not a comfortable one.

84. Fitted with an Invar pendulum rod and isolated from external effects in an evacuated cylinder, the precision regulators of Siegmund Riefler (1847-1912) of Munich are reputed to have attained an accuracy of 0.01 second per day. Although Riefler's clocks were available in the 1890s, the Royal Observatory at Greenwich did not acquire one of this design (made by the Englishman Edwin T. Cottingham [1869-1940]) until 1922. This was replaced as the sidereal time standard within three years by the Shortt clock No. 3, a free-pendulum clock that is reputed to have had an accuracy of one second per year. See Derek Howse, *Greenwich time and the discovery of the longitude* (Oxford: Oxford University Press, 1980), p. 216, and Howse, *Greenwich Observatory* (*op. cit.*, note 64), pp. 142-3.

85. Two members of the Harrison Research Group, Mervyn Hobden and David Harrison, have made contributions to the field of oscillator physics that have been influenced by John Harrison's work. David Harrison's research conducted while he was in the Department of Electrical and Electronic Engineering at the University of Leeds is contained in his Ph.D. thesis 'The Influence of Non-linearities on Oscillator Noise Performance', *ca.* 1987, a copy of which has been deposited at the British Library in London. Another member, Peter Hastings, has used the stopwork designed for Harrison's second marine timekeeper in an application to protect delicate optical components in an infrared spectrograph (see William Andrewes's paper in this volume for further details).

Perfecting the Marine
Timekeeper

See Penney Figure 18 for caption for this and previous page.

# Ferdinand Berthoud and Pierre Le Roy: Judgement in the Twentieth Century of a Quarrel Dating from the Eighteenth Century

*Catherine Cardinal*

Catherine Cardinal is Scientific Director of l'Institut l'Homme et le Temps and Curator of the Musée International d'Horlogerie in La Chaux-de-Fonds, Switzerland. Prior to commencing these appointments in 1987, she was responsible for the collection of clocks, watches, and automata at the Conservatoire National des Arts et Métiers in Paris. She received her doctorate in the history of art from the University of the Sorbonne in 1979.

Cardinal has published several books relating to the history of timekeeping. Her publications, some of which have been translated into English, include *L'horlogerie dans l'histoire, les arts et les sciences* (1983), *Ferdinand Berthoud, 1727-1807* (1985), *La montre des origines au XIXe siècle* (1985), *Le mètre et la seconde: Charles-Edouard Guillaume, prix Nobel de physique* (1988), *La révolution dans la mesure du temps: calendrier républicain, heure décimale, 1793-1805* (1989), *L'homme et le temps en Suisse, 1291-1991* (1991), *Les horloges marines de M. Berthoud* (1994), and *L'orologio e la moda* (1994). Three of these have been written, under her direction, with other authors. She is currently preparing the second volume of the catalogue of watches and table-clocks in the Louvre; the first volume was published in 1984.

# Ferdinand Berthoud and Pierre Le Roy: Judgement in the Twentieth Century of a Quarrel Dating from the Eighteenth Century

*Catherine Cardinal*

In the history of watchmaking, disputes were as frequent in the past as they are today. Let us recall, for example, the difficulties that Christiaan Huygens (1629-1695) encountered in Paris and London around 1675, with opponents such as Isaac Thuret (d. 1706), the Abbé de Hautefeuille (1647-1724), and Robert Hooke (1635-1702/3).[1] The eighteenth century witnessed several quarrels. One of the more significant ones, between the elder Pierre Le Roy (1687-1762)—not the son but the brother of Julien Le Roy (1686-1759)—and Godefroy (fl. mid-18th century), watchmaker to the late Duc d'Orleans, concerned the use of the cylinder escapement invented by George Graham (*ca.* 1674-1751). Another dispute occurred when Pierre-Augustin Caron de Beaumarchais (1732-1799), then known as 'Caron the Son', presented a dead-beat escapement called *à double virgule,* a discovery that both Jean-André Lepaute (1720-1789) and Jean Romilly (1714-1796) claimed for themselves.[2]

The most famous and bitter quarrel of this century was, however, that between Pierre Le Roy (1717-1785)—son of Julien Le Roy and nephew of the above-mentioned Pierre Le Roy—and Ferdinand Berthoud (1727-1807). It revolved around one of the most significant developments of the period, the perfection of the marine chronometer. In 1759, John Harrison had completed his prize-winning watch, H.4, which had been successfully tried on voyages in 1761-2 and 1764. Much work, however, remained to be done in order to make this instrument practicable and commercially viable. The details of the dispute between Le Roy and Berthoud are by now well known and go to prove how important the chronometer was to the clockmakers in terms of financial security, celebrity, and influence. The quarrel began in 1768. But who were these protagonists?

1. Léopold Defossez, *Les savants du 17^e^ siècle et la mesure du temps* (Lausanne, Switzerland, 1946), pp. 192-222.

2. See Catherine Cardinal, *The Watch from its Origin to the XIXth Century* (Secaucus, New Jersey: Wellfleet Press, 1989), pp. 79, 82.

3. See Paul Ditisheim, Roger Lallier, Léopold Reverchon, and le Commandant Vivielle, *Pierre Le Roy et la chronométrie* (Paris: Tardy, 1940) and Jean Le Bot, 'Pierre Le Roy et les horloges marines', in Catherine Cardinal and Jean-Claude Sabrier, eds., *La dynastie des Le Roy, horlogers du Roi* (Tours, France: Musée de Beaux-Arts, 1987), pp. 43-50.

## THE PROTAGONISTS

Born in Paris, Pierre Le Roy was ten years older than his rival.[3] By 1768, when he was 49 years old, Le Roy had held the title of *Horloger du Roi* (Watchmaker to the King) for fourteen years. He was in charge of the well-known Paris workshop founded by his father Julien, the greatest French

Figure 1. General view of Pierre Le Roy's famous *montre marine A*, which was presented to King Louis XV of France on 5 August 1766. This remarkable spring-driven timekeeper, the first to incorporate the crucial elements of the modern marine chronometer (detached [pivoted-detent] escapement, temperature compensation contained in the balance itself, and an isochronous balance spring), was tested on voyages in 1768 and in 1771-2. Length: 35 cm. (13.8 in.); width: 25 cm. (9.8 in.); height: 25 cm. (9.8 in.). Courtesy of the Musée National des Techniques, Conservatoire National des Arts et Métiers, Paris. Inv. no. 1395.

horologist of the first part of the eighteenth century; he was also the inventor of such important devices as the detached escapement and a balance that compensated for variations in temperature. Around the middle of the century, he had decided to build a sea clock. In 1754, he deposited with the Académie Royale des Sciences a sealed description (not opened until 20 June 1763) of his first marine timekeeper. This timekeeper, which was of only six-hour duration, was made in 1756 but never tried at sea. In 1763, he presented to the Académie a marine timekeeper that stood three feet high, but this also was never sent on a trial. Another timekeeper, half the size, was presented in the following year and tried by the astronomer Pierre-Charles Le

Figure 2. Detail of the dial of Le Roy's *montre marine*. Courtesy of the Musée National des Techniques, Conservatoire National des Arts et Métiers, Paris.

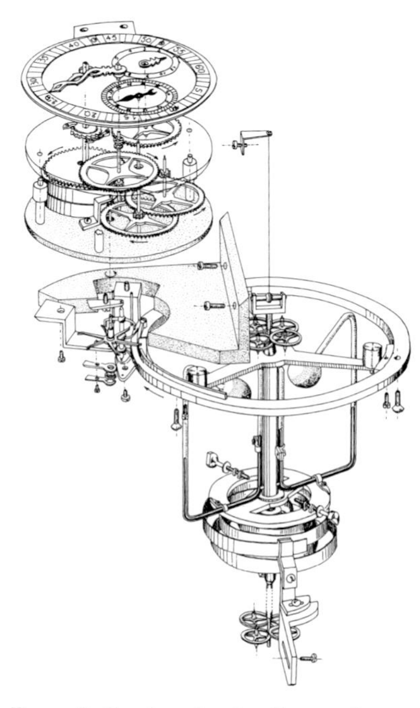

Figure 3. Drawing showing the construction of Le Roy's *montre marine*. Courtesy of Jean Le Bot.

**Figure 4. Detail of the movement and temperature-compensated balance of Le Roy's *montre marine.* The balance is composed of two thermometers, each consisting of a tube and a bulb. The bulb and the upper portions of the tube are filled with alcohol, while the lower parts are filled with mercury. With this arrangement, a rise in temperature, which causes a slight expansion in the balance and loss of strength in the balance spring, is counteracted by the expansion of the mercury toward the axis of the balance. Courtesy of the Musée National des Techniques, Conservatoire National des Arts et Métiers, Paris.**

**Figure 5. Movement of Le Roy's *montre marine.* The detached form of pivoted-detent escapement, invented by Le Roy, provides impulse to the balance rim, not to the balance arbor. Courtesy of the Musée National des Techniques, Conservatoire National des Arts et Métiers, Paris.**

Monnier (1715-1799), but the results are not known. On 5 August 1766, Le Roy presented to the King a timekeeper that he called his third and that was later referred to as *montre A (Ancienne)* (Figures 1-5). This was tried between May and August 1767 at Le Havre by François-César Courtanvaux (1718-1781), Alexandre-Guy Pingré (1711-1796), and Charles Messier (1730-1817). The record of this trial indicates that on 5 July, Le Roy showed the astronomers on board another almost identical timekeeper that is referred to as *montre S (Seconde).*

In 1768, Ferdinand Berthoud, a native of a little Swiss village near Neuchâtel, was 39 years old.[4] Despite his modest beginnings as an unknown young Swiss, he had been building up a great reputation for himself and had managed to become a master watchmaker in France—thanks to an exceptional decision taken by the King's Council on 4 December 1753. In the years leading up to 1768, he had proceeded from one success to another. He had set up a workshop in the rue de Harlay, the same street in which Le Roy also practised his art. He had published some important books, such as *L'Art de conduire et de régler les pendules et les montres*[5] and *Essai sur l'Horlogerie.*[6] A contributor to Diderot's *Encyclopédie,*[7] he presented his research about watches and clocks to the Académie, and, because he counted among his sponsors both French and English scholars, he was elected a member of the Royal Society in 1764.

Like Pierre Le Roy, Berthoud had begun to devote himself in 1754 to work on marine clocks. At the end of that year, he gave the Académie a sealed paper describing a marine timekeeper. In 1760, he constructed his *Horloge Marine* No. 1, the large sea clock that is preserved in the Musée National des Techniques, Conservatoire National des Arts et Métiers, in Paris. It was approved by the Académie in 1764.[8]

He became sufficiently well known to be noticed by the government, and, on two occasions, in 1763 and 1766, he was sent on official visits to

4. For his complete biography, see Catherine Cardinal, ed., *Ferdinand Berthoud, 1727-1807: Horloger mécanicien du Roi et de la Marine* (La Chaux-de-Fonds, Switzerland: Musée International d'Horlogerie, 1984).

5. Ferdinand Berthoud, *L'Art de conduire et de régler les pendules et les montres:...* (Paris, 1759).

6. Ferdinand Berthoud, *Essai sur l'Horlogerie; dans lequel on traite de cet Art rélativement à l'usage civil, à l'Astronomie et à la Navigation,...* (Paris, 1763). This two-volume work has almost 1,000 pages.

7. Denis Diderot and Jean Le Rond d'Alembert, *Encyclopédie, ou Dictionnaire Raisonné des Sciences, des Arts, et des Métiers,* 28 vols. (Paris, 1751-72); articles of interest: 'Equation', vol. 5 (1755); 'Fendre (machine à)', vol. 6 (1756); 'Fusée', vol. 7 (1757); 'Horloge', 'Horloger', 'Horlogerie', vol. 8 (1765); 'Pendule', vol. 12 (1765); 'Répétition', vol. 14 (1765).

8. Cardinal, *Ferdinand Berthoud* (*op. cit.,* note 4), pp. 26, 191.

London to examine Harrison's fourth marine timekeeper, H.4. In 1766, he obtained from Louis XV a special order for two marine clocks. These clocks, numbered '6' and '8' (Figures 6–8) in his series of *horloges marines*, performed successfully in tests and permitted Ferdinand Berthoud to obtain, four years later, the title of *Horloger Mécanicien du Roi et de la Marine* (Watchmaker and Engineer to the King and the Navy).

As we can see, by the year 1768 both watchmakers were very well established in French scientific and technical circles. In fact, they were the favourites of the government as well as of the Académie Royale des Sciences.

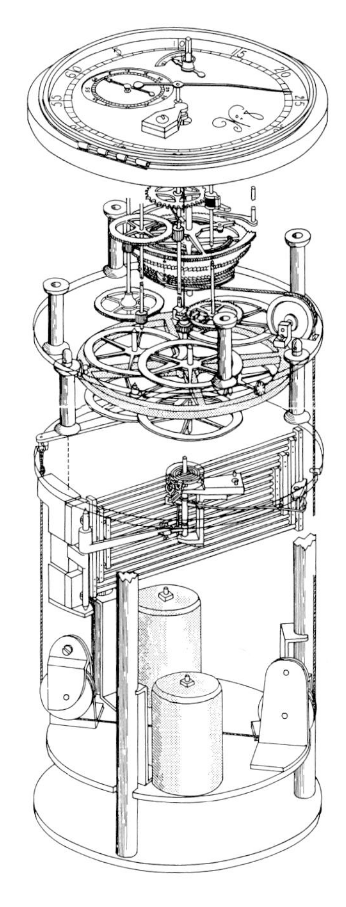

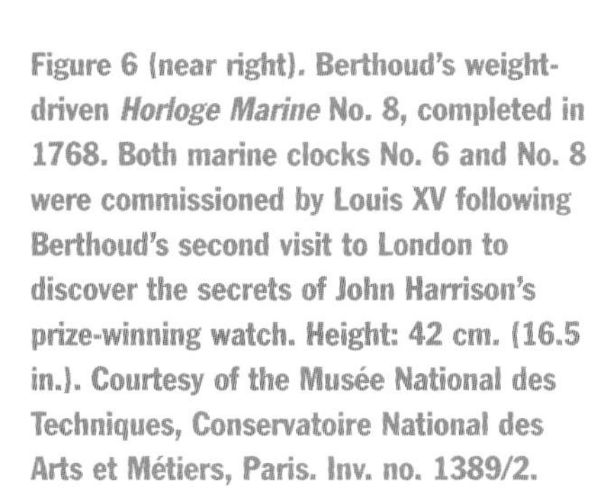

**Figure 6 (near right). Berthoud's weight-driven *Horloge Marine* No. 8, completed in 1768. Both marine clocks No. 6 and No. 8 were commissioned by Louis XV following Berthoud's second visit to London to discover the secrets of John Harrison's prize-winning watch. Height: 42 cm. (16.5 in.). Courtesy of the Musée National des Techniques, Conservatoire National des Arts et Métiers, Paris. Inv. no. 1389/2.**

**Figure 7 (far right). Drawing showing the construction of Berthoud's *Horloge Marine* No. 8. Courtesy of Jean Le Bot.**

**Figure 8 (below). Detail of the dial of Berthoud's *Horloge Marine* No. 8. Diameter of dial: 23.5 cm. (9.25 in.). Courtesy of the Musée National des Techniques, Conservatoire National des Arts et Métiers, Paris.**

Written texts were important in the development of the dispute over the invention of chronometric devices. Equally talented writers, Pierre Le Roy and Ferdinand Berthoud used their pens like swords. Let us mention six publications in turn, all but one of which are by the above authors.

The first one, *Exposé succint des travaux de MM. Harrison et Le Roy,* published in 1768, was written by Pierre Le Roy[9] as an answer to *The Principles of Mr. Harrison's Time-keeper,* published the previous year.[10] Pierre Le Roy's intention was to illustrate the originality of his own work in comparison with Harrison's. While recognising the merit of Harrison's work, he considered his own to be superior.

The condescension expressed in this book provoked a violently critical response, entitled *Examen critique d'un mémoire publié par M. Le Roy,* which was published anonymously the same year. The author proclaimed to be a sailor who, being responsible for testing Berthoud's clocks, had become an admirer of his work. So harsh was the style of this response that the book was destroyed by the order of the *Ministre de la Marine* (Minister of the Navy). Today, only three copies survive.[11] Among the critical observations we may include the following:

> Mr. Le Roy ends his memoir by assuring us that he has reaped the most flattering reward he could ever wish for: *the intimate conviction of having produced a work forever useful to his country and to humanity.* (Page 35 of l'Exp. suc.) We may believe, together with Mr. Le Roy, that he is indeed deeply convinced of the merit of his own work; but in order to convince the scholars of this fact, his work would have to be more useful than it really is; let us even admit that it would be dangerous to employ such defective means when searching for the longitude at sea; this would mean to expose the wealth of the State, the glory of the Prince, and the life of his subjects to very shaky probabilities.[12]

In 1773, Ferdinand Berthoud published his famous work entitled *Traité des Horloges Marines,*[13] which meticulously described his marine timekeepers. The book was dedicated to King Louis XV, whose portrait appears on the frontispiece.

Pierre Le Roy was very upset that he went entirely unmentioned in this book and answered speedily by publishing his *Précis des recherches,*[14] a summary of the development of marine chronometers in France. In it, he claimed: 'I have sacrificed for this important discovery a great part of my fortune;...in spite of my unstable health, I have left my business to go to sea to check the working of my watches'.[15] He was desirous to prove that 'The discoveries on which French marine chronometers rest are the results of my late nights'[16] and added that 'The construction of the marine chronometer that I presented to His Majesty & to the Académie in 1766 is the best hitherto produced'.[17]

Ferdinand Berthoud, in turn, was quick to fight back. A book entitled *Éclaircissemens sur l'invention...des nouvelles machines proposées en France, pour la détermination des longitudes en mer* argued against Le Roy:

> He accuses me of being *his imitator,* he pretends to have invented everything, to have done everything first; he assures us modestly that

9. Pierre Le Roy, *Exposé succint des travaux de MM. Harrison et Le Roy, dans la recherche des Longitudes en mer,...* (Paris, 1768).

10. [John Harrison], *The Principles of Mr. Harrison's Time-keeper, with Plates of the Same* (London, 1767); facsimile reprint by the British Horological Institute in *Principles and Explanations of Timekeepers by Harrison, Arnold and Earnshaw* ([Upton, Notts., England], 1984).

11. According to G. H. Baillie in *Clocks and Watches: An Historical Bibliography* (London: N.A.G. Press, 1951; reprinted London: Holland Press, 1978), p. 277, one of the three surviving copies is inscribed with the words 'par M. de Fleurieu garde de la Marine; cet exemplaire est unique, l'ouvrage n'a pas été publié'. Baillie also mentions that with this copy is a letter from Charles Pierre Claret D'Eveux de Fleurieu (1738-1810), concerning the prohibition of this publication.

12. 'M. le Roy termine son Mémoire, en nous assurant qu'il possède la récompense la plus flatteuse qu'il put désirer: *la conviction intime d'avoir produit un ouvrage à jamais utile à sa patrie & à l'humanité.* (Pag. 35 de l'Exp. suc.) Nous serons persuadés, comme M. le Roy, qu'il est intimement convaincu du mérite de son ouvrage; mais, pour en convaincre les savans, il falloit que l'ouvrage fut plus utile à la patrie, qu'il ne l'est en effet: osons même dire qu'il seroit dangereux d'emploïer des moïens aussi défectueux dans la recherche des longitudes en mer: ce seroit exposer, sur des probabilités trop mal fondées, les richesses de l'État, la gloire du Prince, & la vie des sujects'. [Charles Pierre Claret D'Eveux de Fleurieu], *Examen critique d'un mémoire publié par M. Le Roy, horloger du Roi, sur l'épreuve des horloges propres à déterminer les longitudes en mer et sur les principes de leur construction* (London, 1768), p. 64.

13. Ferdinand Berthoud, *Traité des Horloges Marines, contenant La Théorie, la Construction, la Main-d'oeuvre de ces machines,...* (Paris, 1773).

14. Pierre Le Roy, *Précis des recherches faites en France depuis l'année 1730, pour la détermination des Longitudes en mer, par la mesure artificielle du tems* (Amsterdam, 1773).

15. '...j'ai fait à cette importante découverte le sacrifice d'une grande partie de ma fortune;... j'ai abandonné le soin de mes affaires pour aller suivre sur les mers, malgré une santé chancelante, la marche de mes montres'. *Ibid.,* p. 3.

16. 'Les découvertes sur lesquelles sont établies les montres marines Françoises, sont le fruit de mes veilles'. *Ibid.,* p. 11.

17. 'La construction de la montre marine que j'ai présentée à Sa Majesté & à l'Académie en 1766 est la meilleure qui ait encore paru'. *Ibid.,* p. 29.

*every marine chronometer that will be used in the future will be made very closely on the model of his watch*....I do not possess the presumption to believe that one day better marine chronometers than mine will not be made; neither shall I venture to assert that *my watches will be used in future time as models;* it is to the public, to posterity to pass judgement on their merit.[18]

In 1774, Pierre Le Roy struck back with his last defence, *Suite du Précis sur les Montres Marines de France.*[19] But of course, this quarrel was not carried out on paper alone. The timepieces at stake obviously also had a role to play. By 1768, Pierre Le Roy and Ferdinand Berthoud had built the clocks that decided their respective futures: the 'A' and 'S' for the former; the marine clocks No. 6 and No. 8 for the latter. During the period from 1768 to 1773, these clocks were tried on several voyages at sea and their quality was judged by the Académie and the government.

In 1768, Pierre Le Roy's timekeepers were tested on the ship *L'Enjouée* under the supervision of a member of the Académie, the astronomer Jean-Dominique Cassini (1748-1845).[20] The following year, Ferdinand Berthoud's marine clocks No. 6 and No. 8 were tried under the control of Charles Pierre Claret D'Eveux de Fleurieu (1738-1810), *enseigne des vaisseux du Roy* (Lieutenant of the Royal Ships).[21] During a voyage from 1771 to 1772, both Pierre Le Roy's and Ferdinand Berthoud's timekeepers were tested on the frigate *La Flore* in the presence of the famous scientists Alexandre Pingré and Jean-Charles Borda (1733-1799).[22]

18. 'Il m'accuse d'être *son Copiste*: il prétend avoir tout inventé, tout fait le premier; il nous assure modestement que *toutes les Horloges Marines dont on fera usage dans l'avenir, serons faites, à très peu près, sur le modele de sa Montre....* Je n'ai point la présomption de croire qu'on ne parviendra pas quelque jour à faire des Horloges Marines meilleures que les miennes; je ne me hasarde pas non plus à assurer qu'*à l'avenir ellés serviront de modele:* c'est au Public, c'est à la Postérité à prononcer sur leur mérite'. Ferdinand Berthoud, *Éclaircissemens sur l'invention, la théorie, la construction, et les épreuves des nouvelles machines proposées en France, pour la détermination des longitudes en mer par la mesure du temps...* (Paris, 1773), p. 3.

19. Pierre Le Roy, *Suite du Précis sur les Montres Marines de France, avec un Supplément au Mémoire sur la meilleure manière de mesurer le tems en mer* (Leiden, 1774).

20. Great-grandson of Jean-Dominque (Giovanni Domenico) Cassini (1625-1712).

21. De Fleurieu had written anonymously a publication in support of Berthoud; see note 11.

22. See Jean Le Bot, *Les Chronomètres de marine français au XVIII^e siècle, Quand l'art de naviguer devenait science* (Grenoble, France, 1983).

**Figure 9. Pierre Le Roy's *La Petite Ronde, ca.* 1774. About 1771, after the completion of his two successful marine timekeepers, Le Roy attempted to produce some small precision marine pocket watches, which, when not being carried, were housed in gimballed boxes. Due to their shape, he called them *'Petites Rondes'.* This example is the only one known to have survived in perfect condition. Diameter of box: 11 cm. (4.33 in.); height: 13.5 cm. (5.31 in.). Diameter of watch case: 5 cm. (1.97 in.). Courtesy of the Clock and Watch Museum Beyer Zurich. Photograph by the Musée International d'Horlogerie, La Chaux-de-Fonds, Switzerland.**

**Figure 10. Movement of *La Petite Ronde*, bearing the signature 'Julien le Roy in.^vt et fecit A PARIS'. Pierre Le Roy continued to use his father's name on many of the pieces that came out of his workshop after his father died in 1759. This watch, which incorporates a bimetallic compensation plate to automatically adjust the balance spring for variations in temperature, employs a form of frictional-rest escapement that Pierre Le Roy adapted from that used by his father's friend, Henry Sully (1680-1728). Diameter of back plate: 4.07 cm. (1.6 in.). Courtesy of the Clock and Watch Museum Beyer Zurich. Photograph by the Musée International d'Horlogerie, La Chaux-de-Fonds, Switzerland.**

As a result of the trials, rewards and honorary titles were given to both watchmakers. In 1769, Pierre Le Roy won a prize from the Académie for the best way of determining longitude at sea. In 1770, Ferdinand Berthoud received the title of *Horloger Mécanicien du Roi et de la Marine*, together with an annual pension. Three years later, Pierre Le Roy again won the prize given by the Académie.

### THE CONCLUSION—1773

When all was said and done, Ferdinand Berthoud received the lion's share of the rewards and Pierre Le Roy the consolation prizes. Le Roy was, however, far from consoled. Disgusted by the turn of events, he gave up his research. On 28 May 1773, Ferdinand Berthoud received a firm order for four clocks a year for the King. This constituted perhaps the last blow in the face of his rival.[23]

The financial imbalance between the two inventors was soon obvious. In 1770, Ferdinand Berthoud had obtained an annual pension of 3,000 livres, which was increased by 1,500 livres in 1773. Pierre Le Roy had to wait until 1776 in order to receive a pension of 1,200 livres. In 1768, he had tried without success to obtain a royal stipend. His written request to the Duc de Praslin (1712-1785), *Ministre de la Marine*, has been preserved. In it, we find him referring to the King's custom of honouring those artists and scholars who had made useful discoveries. As the clockmaker of the King, Le Roy hopes to be counted among those worthy of such honours and begs the Duc de Praslin to intervene with the King on his behalf, so that his labours will not go unrecognised and without royal remuneration. After all, Le Roy laments, he has sacrificed a good portion of his fortune for his research.[24]

### AFTER THE DISPUTE

Pierre Le Roy, in poor health, went to live in Viry-Châtillon, a little town by the Seine just south of Paris, where he died on 25 August 1785. Ferdinand Berthoud, on the other hand, merrily continued a long and acclaimed official career, with only a few setbacks at the beginning of the Revolution. He continued his construction of marine clocks until his death in 1807, supplying timekeepers to such famous sailors as Jean-François de Galaup, Comte de La Pérouse (1741-*ca.* 1788). In 1782, he succeeded in selling his tools, machines, and measuring instruments to King Louis XVI while keeping the right to use them. Around 1795, he obtained free accommodation in the Galleries du Louvre. In addition, he became a member of the Légion d'honneur and of the Institut de France.

Berthoud continued publishing into the nineteenth century. His last important book was his *Histoire de la Mesure du Temps par les Horloges* of 1802.[25] It is interesting to note that there he mentions Pierre Le Roy and himself as 'the authors to whom we owe the discovery and invention of marine chronometers'.[26] He adds:

> Although Harrison's English watch is the first to have met success after a test at sea (around 1763), French artisans have no less a claim to the discovery, because, starting in 1754 (a period in which the work and even the name of the English artisan were unknown in France), two French artisans were engaged in the same study.[27]

23. Cardinal, *Ferdinand Berthoud* (*op. cit.*, note 4), pp. 33-7.

24. 'Aujourd'hui que ce travail a été couronné par le plus grand succès lui sera-t-il permis de suplier Vôtre Grandeur dispensatrice des graces dont Sa Majesté honore les artistes et les Sçavants, qui se sont rendus dignes de ses bontés par l'utilité de leur travaux dans la marine; de vouloir bien interposer ses bons offices pour qu'il obtienne le prix précieux auquel il a aspiré, c'est à dire un témoignage de la satisfaction de Sa Majesté, et une Pension qui en comblant d'honneur le dit sieur LeRoy et consolidant en quelque sorte sa fortune très-altérée par les travaux et les recherches fort dispendieuses où l'ont entrainée sa découverte le mettroit à portée de se livrer avec plus d'efficacité aux soins requis pour l'éxecution d'un nombre suffisant de ces machines'. Pierre Le Roy to Monseigneur le Duc de Praslin, 26 March 1768, archive ref. D294, Institut l'Homme et le Temps, La Chaux-de-Fonds, Switzerland.

25. Ferdinand Berthoud, *Histoire de la Mesure du Temps par les Horloges* (Paris, 1802).

26. 'Tels sont les Auteurs auxquels on doit la découverte et l'invention des horloges et des montres à longitudes'. *Ibid.*, p. 281.

27. 'Quoique la montre anglaise HARRISON soit en effet la première dont une épreuve en mer ait constaté (vers 1763) le succès, les Artistes

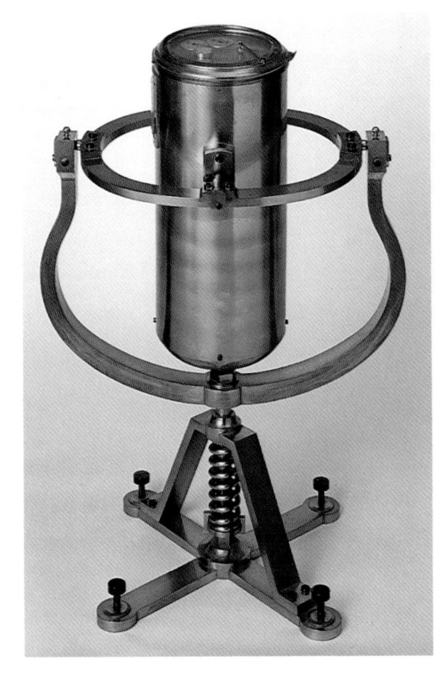

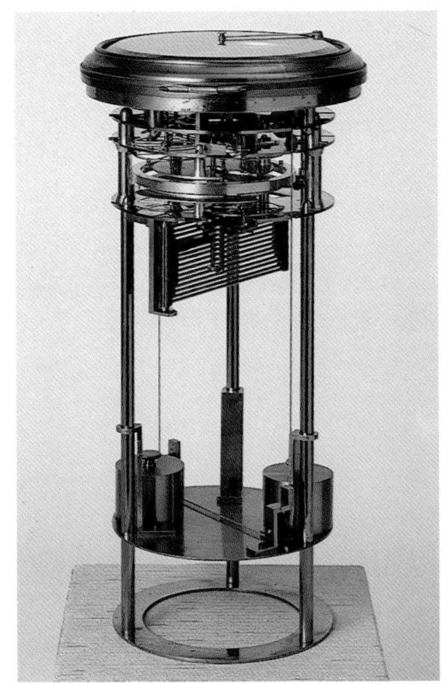

**Figure 11. Berthoud's *Horloge Marine* No. 16, dated 1775. Height: 87 cm. (34.25 in.). Courtesy of the Clock and Watch Museum Beyer Zurich.**

**Figure 12. Movement of *Horloge Marine* No. 16, which is weight-driven and has a pivoted-detent escapement, suspended balance pivoted on antifriction wheels, and gridiron temperature compensation. Outside diameter: 23 cm. (9 in.). Courtesy of the Clock and Watch Museum Beyer Zurich.**

## THE JUDGEMENT OF POSTERITY

Pierre Le Roy died a forgotten man, but he has been allowed to take his revenge in the horological literature that has followed. Many authors, English as well as French, speak very highly of him. They often insist on placing him above Ferdinand Berthoud and even John Harrison. Until the middle of the twentieth century, historians of horology all agreed in recognising him, rather than Ferdinand Berthoud, as the inventor of the marine chronometer in France. In these texts, Berthoud routinely appears as a sycophantic and calculating man.

Even in the nineteenth century, some writers and experienced professional watchmakers had begun to attribute the invention of the marine clock to Pierre Le Roy, drawing special attention to the precision of his timepieces. As an example, we may mention Pierre Dubois (1802-1860), who wrote in 1849:

> After Pierre le Roy, Ferdinand Berthoud worked with much success on machines to measure time at sea; but he was not, as he claims in a number of his writings, the inventor of the marine chronometer....Harrison and especially Pierre Le Roy preceded him in making chronometers that were quite exact in finding longitude at sea.[28]

Toward the end of the century, Alfred Beillard (1850-1939), the founder of a famous watchmaking school in Anet, wrote that Le Roy's qualities 'made him withdraw into a modesty that deprived him of a reputation that his genius would otherwise have made universal'.[29] Another of Le Roy's well-known defenders was Claudius Saunier (1816-1896), director of a watchmaking school, who, in his *Revue Chronométrique,* asserted that Pierre Le Roy's chronometer 'is the real starting point of great progress, of giant steps by the artists who came after him'.[30]

*The Marine Chronometer* by Rupert T. Gould (1890-1948) was an essential step in the recognition of Pierre Le Roy's merits at an international level. Gould's judgement was uncompromising. 'Le Roy's machines', he said, 'are superior to those of Berthoud and Harrison'. He analysed very clearly Le Roy's contribution:

français n'ont pas moins de droit à la découverte, puisque dès 1754 (époque où le travail et le nom même de l'Artiste anglais étoient inconnus en France), deux Artistes français s'occupoient de la même recherche'. *Ibid.*, p. 281.

28. 'Après Pierre Le Roy, Ferdinand Berthoud s'occupa avec beaucoup de succès des machines propres à mesurer le temps en mer; mais il ne fut pas, comme il le dit dans différents écrits, l'inventeur des montres marines, puisque, comme nous venons de le dire, Harrisson *[sic]* et surtout Pierre Le Roy avaient fait, avant lui, des machines fort exactes pour trouver la longitude en mer'. Pierre Dubois, *Histoire de l'horlogerie* (Paris, 1849), p. 217.

29. 'Ses solides qualités, sa noblesse de coeur et sa délicatesse d'esprit, le faisaient se renfermer dans ûne *[sic]* modestie qui l'a privé d'une réputation que son génie eût rendue universelle'. Alfred Beillard, *Recherches sur l'horlogerie, ses inventions et ses célébrités* (Paris, 1895), p. 145.

30. '...il est le véritable point de départ des grands progrès, des pas de géant accomplis par les artistes venus après lui'. Claudius Saunier, 'Notre Dernier Mot sur Pierre Le Roy', *Revue Chronométrique,* vol. 18 (Paris, 1894-5), p. 92.

...the true development of the chronometer, on modern lines, may be traced back to the work of Pierre Le Roy and Ferdinand Berthoud, both of Paris....[I]t may briefly be said that Berthoud was a man of extraordinary talent...while Le Roy was a genius.[31]

In this publication, which has become a sort of Bible for all chronometer lovers, Pierre Le Roy is recognised as the true father of the modern marine chronometer. Gould invoked the following strong image:

The difference [between Harrison and Le Roy] in their machines is fundamental—Harrison built a wonderful house on the sand; but Le Roy dug down to the rock.[32]

The ultimate verdict was pronounced in 1940, in a monograph entirely dedicated to Pierre Le Roy. Hard to find because of its limited edition, it brings together the views of three well-known authors in the field of horology: Paul Ditisheim (1868-1945), Léopold Reverchon (1863-1940), and Roger Lallier (fl. 1940). They jointly praise Pierre Le Roy's work as well as his personality and exclaim: 'Pierre Le Roy is nevertheless the great inventor of chronometry, a man to whom the term "genius" can be justly applied'.[33] Apologetically, they add:

Pierre Le Roy, shy, a little obsequious in his manners, and sometimes even tactless towards great men, nevertheless fills us with the sympathy and admiration one feels for the intellectual creator who does not want to humble himself and bow down in order to give value to his work.[34]

Meanwhile, in spite of this severe criticism of his talents, Ferdinand Berthoud continued to enjoy posthumous glory. In 1907, the 100th anniversary of his death was celebrated sumptuously in the presence of 5,000 people. Among them were many watchmakers from Paris and abroad. On this occasion, a bronze bust was unveiled in Groslay, where the watchmaker was buried. In 1949, after its destruction during the Second World War, the Canton of Neuchâtel offered to have it replaced out of a fund of public donations. In fact, Berthoud's memory has been kept very much alive during the last ten years. In 1982, I organised a special exhibition of Ferdinand Berthoud's clocks and tools in the Conservatoire National des Arts et Métiers in Paris. For this exhibition, a wax figure of Berthoud was made by the Musée Grévin. Two years later, the Musée International d'Horlogerie organised a large exhibition with a catalogue celebrating the life and work of Ferdinand Berthoud.[35] In 1994, finally, a children's book was published that presents his life and his marine clocks.[36] We can, in sum, consider Ferdinand Berthoud a lucky man.

And yet recently the controversy between Le Roy and Berthoud has once again attracted the attention of historians. We think here particularly of the contributions by Jean Le Bot, Jean-Claude Sabrier, and David Landes. Their judgement is as clear as that of their predecessors: Pierre Le Roy's inventions are seen as superior to Ferdinand Berthoud's talented constructions. Their judgement can be taken to be reliable, because nowadays it is no longer necessary to rescue and elevate Pierre Le Roy's memory.

31. Rupert T. Gould, *The Marine Chronometer: Its History and Development* (London: J. D. Potter, 1923), p. 83.

32. *Ibid.*, p. 91.

33. Ditisheim *et al.*, *Pierre Le Roy* (*op. cit.*, note 3), p. 37.

34. *Ibid.*, p. 39.

35. Cardinal, *Ferdinand Berthoud* (*op. cit.*, note 4).

36. Catherine Cardinal, *Les horloges marines de M. Berthoud* (Paris: F. Nathan, 1994).

37. David S. Landes, *Revolution in Time* (Cambridge: Harvard University Press, 1983), p. 169; see also chapter entitled 'The French Connection', pp. 158-70.

**Figure 13. Berthoud's *Montre Marine* No. 6, dated 1777. This is one of the first timekeepers to be housed in the type of box that became the standard used for all marine chronometers. Berthoud's design was produced in an effort to make good chronometers available for merchant shipping. Width of box: 32.8 cm. (12.9 in.). Courtesy of The Time Museum, Rockford, Illinois. Inv. no. 310.**

**Figure 14. Movement of Berthoud's *Montre Marine* No. 6. This spring-driven timekeeper, one of a series of seven *montres marines* made between 1775 and 1778, is of 36-hour duration and incorporates antifriction wheels, a form of pivoted-detent escapement, and gridiron temperature compensation. Diameter of upper plate (back plate): 14.3 cm. (5.6 in.). Courtesy of The Time Museum, Rockford, Illinois.**

## AN ANALYSIS OF THE PERSONALITIES AND WORKS OF THE TWO RIVALS

Despite this somewhat negative assessment of Berthoud's work relative to Le Roy's, however, we should note that Ferdinand Berthoud was probably not quite the egocentric and sycophantic man that several authors have portrayed. David Landes has justly remarked that

> History has not been kind to Ferdinand Berthoud. Le Roy's contemptuous strictures have left their mark, and subsequent generations of horologists have tended to dismiss Berthoud's work as bumbling, uninspired, and irrelevant.[37]

Berthoud possessed a kind of generosity that pushed him to communicate his knowledge through his many literary works. These were richly illustrated and thereby of invaluable help to watchmakers. On the other hand, he obviously also knew how to defend and further his interests, using a mixture of authority and courtesy. The archival record shows us his effectiveness in interpersonal relations.

By the same token, Pierre Le Roy does not seem to have been the modest and reserved man depicted by many authors. One can see that he had disputes with other watchmakers, for example with Lepaute, watchmaker to King Louis XV. Often put out, he acted too abrasively and thereby alienated potential supporters. His character did not help to foster public appreciation of the originality of his work, and he had to die before his work could be properly assessed.

**Figure 15. Berthoud's *Montre à Longitude* No. 65, dated 1796. This timekeeper was described by Berthoud in 1797 in *Suite du Traité des Montres à Longitudes* (Paris), chapter 5, under the title 'Petite horloge horizontale sans rouleau' ('Small horizontal clock without antifriction rollers'). Diameter: 6 cm. (2.36 in.); height: 15 cm. (5.9 in.). Courtesy of the Musée International d'Horlogerie, La Chaux-de-Fonds, Switzerland. Inv. no. 96.**

**Figure 16. Portrait of Ferdinand Berthoud, after 1804, when he was elected *Chevalier de la Légion d'honneur*. Drawing signed 'Morel'. Pencil and chalk on paper. 22.9 x 30.1 cm. (9.02 x 11.85 in.). Courtesy of The Time Museum, Rockford, Illinois. Inv. no. 2724.**

Indeed, Pierre Le Roy's originality was difficult for his contemporaries to appreciate. With the benefit of hindsight, however, it is comparatively easy to understand and appreciate the nature of his contribution. Several clockmakers and historians (such as Jean Le Bot and Jean-Claude Sabrier)[38] have analysed the innovative features of his machines in great detail, and I will therefore limit myself to a few concluding remarks.

Pierre Le Roy established the main principles of the modern chronometer, to wit, a detached escapement, temperature compensation in the balance, and an isochronous balance spring. But he did not succeed in making a machine of which multiple copies could easily be reproduced and which could survive the many dangers of a voyage at sea unscathed. The marine timekeeper that is preserved in the Musée National des Techniques, Conservatoire National des Arts et Métiers, seems to be an experimental model. In fact, it is of such fragility that it cannot be moved without difficulty, and lending it for exhibitions elsewhere represents a major challenge.

By contrast, Ferdinand Berthoud successfully overcame certain problems with which Pierre Le Roy never concerned himself. As he himself repeats time and again in his writings, Berthoud's priorities were strength, simplicity in manufacture and manipulation, and convenience in repair.

His 70 marine timekeepers were used in many voyages of exploration. In 1792, he could proudly exclaim: 'Eighty sea travels have been undertaken with my clocks. Fifty clocks or watches have been used in these campaigns'.[39] In conclusion, we may therefore state that Ferdinand Berthoud was an excellent supplier whose works pleased their purchasers, but that clockmakers preferred his rival.

The author is most grateful to David Landes and Christoph Lüthy, who have assisted in the translation of this text. The illustrations have been selected and described by Will Andrewes and Jean-Claude Sabrier; the latter has also contributed information about Pierre Le Roy's timekeepers.

38. Le Bot, 'Pierre Le Roy' (*op. cit.*, note 3) and Jean-Claude Sabrier, 'La contribution de Ferdinand Berthoud aux progrès de l'horlogerie de marine', in Cardinal, *Ferdinand Berthoud* (*op. cit.*, note 4), pp. 165-84.

39. 'On a fait vingt-quatre voyages en mer avec mes horloges. On a employé cinquante horloges ou montres pour ces campagnes'. Ferdinand Berthoud, *Mémoire sur le travail des horloges et des montres à longitudes inventées par M. Ferdinand Berthoud* (Paris, 1792).

# Thomas Mudge and the Longitude: A Reason to Excel

## *David Penney*

David M. Penney is a free-lance illustrator specializing in the history of technology, particularly horology. His illustrations have been commissioned by the British Museum, the Prescot Museum, The Time Museum, Patek Philippe, Omega, Chopard, George Daniels, and many publishers. He was commissioned by the British Post Office to illustrate the set of four stamps, released in February 1993, to commemorate the tercentenary of the birth of John Harrison.

Since 1975, Penney has been a part-time lecturer and tutor at several art colleges, most recently being appointed Course Leader of "Information and Technical Illustration," a postgraduate course at the Royal College of Art, London. He served as Vice-Chairman, then Chairman, for the Association of Illustrators, London (1983-4). From 1986 to 1992, he served as Editor of *Antiquarian Horology,* the Journal of the Antiquarian Horological Society, in which he co-authored, with Valerie Finch, "William Howells, Watch-maker" as an ancillary part of his research on the life of Thomas Mudge, whose work he has studied for the past twenty years.

# Thomas Mudge and the Longitude: A Reason to Excel

*David Penney*

Though 1993 is the 300th anniversary of John Harrison's birth and much has been published that confirms his principal position in the history of precision timekeeping, there are others who played their part: Thomas Mudge (1715?-1794) is one of them. The aim of this short paper is to give some idea of how important Mudge's contribution was, as was that of his master, George Graham (*ca.* 1674-1751),[1] especially during the 1730s and 1740s.

Thomas Mudge, son of the Reverend Zachariah Mudge (1694-1769), the second son in a family of five children, was apprenticed to George Graham, London's most celebrated clockmaker, in 1730. Thomas, as was then normal for a boy starting his seven-year apprenticeship, was fourteen years of age. This began on 4 May 1730 and was the first in a series of events that was to shape Mudge's and subsequently our lives. To the best of our knowledge, that same year saw a not-so-young man of 37 years of age also make a visit to Graham's workshop.[2] The man was John Harrison (1693-1776), and this was the first of what must have been many visits.

Armed with the utmost belief in his own capabilities and a newly written document detailing his findings,[3] Harrison was able to convince Graham that he, although untrained as a clockmaker, had some interesting ideas and was worth taking seriously—seriously enough for Graham to offer both encouragement and financial support for the proposed 'Sea Clock'. Harrison's handwritten document is dated 10 June 1730, just one month after the start of young Thomas's apprenticeship.

Graham was asked to believe in the extreme accuracy of a series of long-case regulators, largely made of wood, with strange recoiling escapements. Recoil was something that Graham, by 1726, had taken steps to eliminate in favour of dead-beat escapements, both in clocks and watches. The document also contains a description of how Harrison had designed and built efficient compensating pendulums, a fact that may be significant in view of Graham's subsequent encouragement.

A lesser man than Graham could have dismissed Harrison as a charlatan or have felt his position threatened enough to ridicule Harrison's ideas. To

1. Fellow of the Royal Society (1721), Master of the Clockmakers' Company (1722).

2. The exact date of this first meeting is not known, but the eagerness of Harrison's nature at this time precludes a long delay between writing the paper (see note 3) and taking it to London.

3. John Harrison, untitled manuscript signed 'John Harrison, Clockmaker at Barrow; Near Barton upon Humber; Lincolnshire. June 10. 1730', MS. 6026/1, Library of the Worshipful Company of Clockmakers, Guildhall Library, London.

4. For information concerning this timekeeper, see William Andrewes's paper in this volume.

Figure 1. Portrait of Thomas Mudge, *ca.* 1772, by Nathaniel Dance, R.A. (1735-1811). Entered for sale at Sotheby's, London, by descendants of the family, it was purchased by the Science Museum, London, in 1985. A contemporary copy is in the collection of the Worshipful Company of Clockmakers, London. Oil on canvas. 74 x 61 cm. (29.1 x 24 in.). Courtesy of the Science Museum, London. Inv. no. 1985-1362. Photograph by kind permission of Sotheby's, London.

Figure 2. Portrait of George Graham, *ca.* 1745, by Thomas Hudson (1701-1779). At one time this painting was in the possession of the astronomer and mathematician George Parker, 2nd Earl of Macclesfield (*ca.* 1697-1764), President of the Royal Society from 1752 until his death. Note the representation of Graham's mercury-compensated pendulum in the background. Oil on canvas. 120 x 96 cm. (47.2 x 37.8 in.). Courtesy of the Science Museum/ Science and Society Picture Library, London. Inv. no. 1868-249.

his great credit, Graham not only showed real interest in Harrison's work but also gave him support and encouragement for the next twenty years, some of which were quite unproductive. So important was Graham's support of Harrison that it may not be unfair to say that without it Harrison would have had little chance of success and would certainly not have produced a machine like his fourth marine timekeeper, commonly known as H.4.

It is impossible to assess the full effect of this and subsequent meetings of Graham and Harrison on young Thomas at this time, but, as an eager fourteen-year-old with ambitions of his own, the effect was likely to have been great. The small workshop and intimate surroundings, combined with the interest of the workmen, would have ensured that this meeting was the talk not only of Graham's workshop but also of London's horological and scientific communities. No doubt Graham's position and contacts with fellow members of the Royal Society were to prove as important to Harrison as they were to Mudge.

Within five years of this meeting Harrison had completed his first timekeeper, a machine that has come to be known as H.1.[4] Such was the interest in this machine that Harrison's celebrity was ensured; accounts of the timekeeper even appeared in the national newspapers. This was not the case with young Mudge, let alone Graham, during this period. Very little information concerning Graham has survived, this despite Parliament's decision

**Figure 3. Artist's view of Fleet Street, on a postcard printed by T. Nelson and Son in the late nineteenth century. Though drawn over a century after Mudge lived there, this view, showing Ludgate bridge with St. Martin's Church, Ludgate, in front of St. Paul's Cathedral, gives some impression of Fleet Street in the eighteenth century. Mudge's and Graham's shops would be near the viewer, on the left. Private collection.**

to recognise Graham's importance to London life by allowing his burial at Westminster Abbey. He was laid to rest beside his old master, Thomas Tompion (1639-1713), in 1751.

For some of the very few contemporary references that do exist, we are indebted to John Byrom (1692-1763),[5] one of Graham's contemporaries in the Royal Society Club, for keeping a diary.[6] Byrom's diaries, published in the nineteenth century, contain passing mention of Graham and others of the Royal Society during this interesting and horologically important period; most future developments had their roots in Graham's workshop at this time. Graham is mentioned most frequently in the context of meetings of the Royal Society Club.[7] The following three extracts concerning Graham give a tantalisingly brief but significant insight into London and the Royal Society at this period. The first quote comes from an entry dated 15 December 1725. Byrom writes:

> ...thence [we went] with Mildmay to Mr. Graham's about the Jesuit Hildegard's[8] *[sic]* book[9] about a new clock, which he had desired Mildmay to present to the Royal Society. We had some talk with Graham about the longitude.[10]

The eleven years since the passing of the Act of Queen Anne[11] had not, it seems, dulled interest in the subject.

A second entry dates from 4 April 1737:

> ...I met Mr. Woolaston [12] *[sic]* by Will's coffehouse going to Mr. Graham's, and I went with him, and Dr. Hoadly[13] met them there; Mr. Graham showed us a pretty deceptio vis, of two balls moving upwards, which had decieved Sir. H. Sloan[14] *[sic]*, Monsr. Fontenelle,[15] etc.; thence they went and I with them to Dr. Hoadly's, who showed us his orrery and our library telescope.[16]

The last entry is dated 17 May 1748, just three years before Graham's death:

> Lord Baltimore [17] brought a machine that he and some lords fancied to be a great improvement, and there was old Mr. Graham,

5. John Byrom was particularly known as a promotor and teacher of shorthand. He was a Fellow of Trinity College, Cambridge, and was elected a Fellow of the Royal Society in 1723.

6. *The Private Journal and Literary Remains of John Byrom*, ed. Richard Parkinson (Manchester, England, 1854-7). I am indebted to Timothy Underhill, presently writing a Ph.D. thesis on Byrom, for bringing this information to my attention.

7. The Royal Society Club was a discussion group for Fellows of the Society and others, presumably, by invitation. This Club should not be confused with the Royal Society itself.

8. Thomas Hildeyard (1690-1747), who worked in Hereford and Liège.

9. This must relate to the Latin edition published in Liège about 1725. The 1727 English edition is entitled *Chronometrum Mirabile Leodiense: Being A Most Curious Clock, Lately Invented By Thomas Hildeyard, Professor of Mathematicks in the English College of Liege* (London). The clock in question, which is dated 1724, is in the Palacio de la Zarzuela, Spain. It has been described and illustrated as cat. no. 3 in Ramón Colón de Carvajal, *Catálogo de Relojes del Patrimonio Nacional* (Madrid: Editorial Patrimonio Nacional, 1987).

10. Byrom, *Private Journal* (*op. cit.*, note 6).

11. Act 12 Anne *cap.* 15 (1714): *An Act for Providing a Publick Reward for such Person or Persons as shall Discover the Longitude at Sea.'*

12. Francis Wollaston (1694-1774), Fellow of the Royal Society (1723).

13. Benjamin Hoadly, M.D. (1706-1757), Fellow of the Royal Society (1726), physician to George II, 1742, and author of the comedy *The Suspicious Husband*, performed at Covent Garden in 1747.

14. Sir Hans Sloane (1660-1753), Fellow of the Royal Society (1685). At this time, he was physician to George II and President of the Royal Society. Like the others mentioned, he seems to have been as interested in mechanical paradoxes as many people still are today.

15. Bernard le Bouyer de Fontenelle (1657-1757), Fellow of the Royal Society (1732).

16. Byrom, *Private Journal* (*op. cit.*, note 6).

17. Frederick Calvert, sixth Baron Baltimore (1731-1771). Described in the *Dictionary of National Biography* as 'a rake' who 'lived much abroad'.

18. Byrom, *Private Journal* (*op. cit.*, note 6).

19. Letter to the Board of Longitude dated 16 January 1741/2, in Jane Squire, *A Proposal To Determine our Longitude*, 2nd ed. (London, 1743), p. 54.

20. Undated letter, *ca.* 1782, in Thomas Mudge Jr., *A Description, with plates, of The Time-Keeper invented by the late Mr. Thomas Mudge....* (London, 1799; facsimile reprint London: Turner and Devereux, 1977), p. 150.

21. Courtenay A. Ilbert was an important collector of clocks and watches. The major part of his clock and watch collection, which was due to be auctioned in 1958, was saved for the nation, chiefly through the generosity of Mr. Gilbert Edgar, C.B.E., and is on view at the British Museum.

watchmaker, and several of our Society at dinner, and Lord Balt. was satisfied that the engine would not answer the expectation, which he could not see why it would not till Mr. Graham explained the matter to all our satisfactions.[18]

From such references to Graham and the various groupings noted in Byrom's diaries, it is obvious that the young Thomas Mudge could not have had a kinder, more influential master and John Harrison a more well-connected supporter.

Graham's work in precision horology is well known, yet only one scant historical reference exists to his interest in building a longitude timekeeper.[19] In particular, we know of Graham's early interest in temperature compensation, and, thanks to the published remarks of Mudge,[20] we know that Harrison spoke to Graham about placing compensation within the balance of his timekeepers. Graham was against it, and, says Mudge, Harrison came to agree with him. Changes to the position of H.1's gridiron compensation would appear to confirm Mudge's statement. The questions then arise: What investigations did Graham make and when did he stop, if in fact he did, leaving it to John Harrison to proceed?

Apart from the contemporary reference by Jane Squire, one much-altered, some would say suspicious, item has survived and may well prove to be of significance when properly researched. Bought by Courtenay Ilbert (1888-1956)[21] in 1938, it is now at the British Museum. Signed 'Geo. Graham, London', the timepiece, shown in Figures 4 and 5, is of box chronometer size and has an unusual and clear dial layout. The engraving of the dial, especially the signature, is exactly as you would expect from Graham's workshop at this period. The movement has subsequently been decoratively engraved, with an added 'Geo Graham' signature, prior to fire gilding. Though done after Graham's death, deception does not appear to be the reason for these changes.

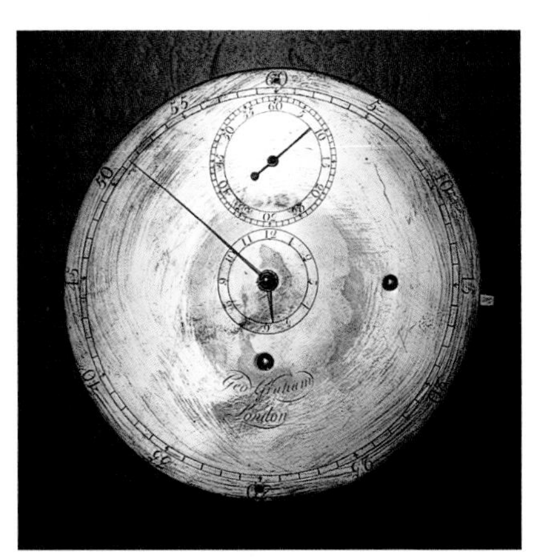


**Figure 4. Dial of the Graham timepiece, *ca.* 1730. Note the unusual layout of hours and minutes achieved with very different-length hands. The second winding square is now used to lock the balance. Originally it seems to have operated a form of 'bolt and shutter' maintaining power. Diameter of dial: 13.97 cm. (5.5 in.). By permission of the Trustees of the British Museum. Inv. no. CAI-2068.**

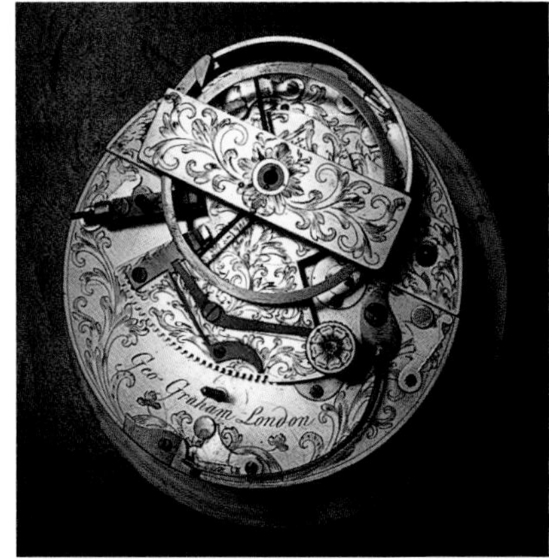


**Figure 5. Movement of the Graham timepiece with regulation disc removed. The decorative engraving was done at a later date. Note the large balance running under a simple balance bridge and the elaborate balance spring clamps. Numerous filled holes in the balance bridge may be signs of experimentation. Diameter of upper plate (back plate): 11.4 cm. (4.5 in.). By permission of the Trustees of the British Museum.**

The plain brass balance is 63 mm. in diameter. The train has pinions of 16, 12, and 12, with an escape pinion of 16, and beats half seconds. The large, 16-tooth, brass cylinder escape wheel is of typical Graham proportions, and the train, to the barrel, runs in hard brass screwed-in settings. The balance holes are jewelled with diamond endstones at both ends.[22]

Because of the dubious nature of certain parts of the timepiece, it has not been on public view for many years, but there remains enough to provide strong evidence that it is not the fake that some believe. In particular, the engraving of the dial in combination with that of the large regulation disc, which match perfectly, is a strong indication that it did originate from Graham's workshop sometime after 1726, the date of Graham's decision to make only his newly developed form of dead-beat watch escapement, the cylinder escapement, instead of the recoiling verge.

The intended use of this timepiece is uncertain, but with both maintaining power and an unusual and elaborate regulation slide, it only needs to have had some form of temperature compensation to convince me that it was an early attempt at producing a longitude timekeeper. At the very least, the size and proportions of this machine seem to have influenced all who came after it: Harrison, Mudge, Arnold, etc.

Whether or not Graham was investigating the possibility of constructing a longitude timekeeper, it is certain that Harrison, encouraged by Graham and others and assisted by his brother James (1704-1766), produced his first sea clock within five years of his first visit to London. Further encouragement came with H.1's sea trial to Lisbon.[23] Spurred on by his success, Harrison immediately set about making a new, improved timekeeper. The second timekeeper, H.2, was completed in 1739, a period that also marked the end of the two Harrison brothers working together, with James, who had moved with him to London, returning home to Barrow. This left John needing to employ workmen to continue. What better place—London—and what better contact—Graham—could Harrison have had in finding such tradesmen, the very tradesmen who would be of great help in making new, all-metal machines?

Exactly who and how many people Harrison employed are not known. What is certain is that some twenty to forty separate trades would have been utilised in making a well-finished watch at this period. It was exactly this division of labour that enabled exquisite items such as H.4 to be produced. Presumably some of the same workmen were employed in making H.3.

One problem of employing workmen was the difficulty Harrison may have had in keeping his pioneering work secret. With regard to what Harrison felt about the problem, we have to thank another diarist, the Reverend William Stukeley (1687-1765).[24] Writing in 1740, he tells us that H.1 was on show at Mr. Graham's, this despite the fact that Harrison was now residing in London.[25] The secrets of Harrison's first machine, at least, were certainly not hidden from the public or the trade—a trade that by then included Thomas Mudge, who, in 1738, had become a journeyman after gaining his freedom of the Clockmakers' Company.

Working in the trade and, if qualified, being called a 'journeyman', meant being employed to make or help make items for other people to name and sell. The opportunity to market one's own creations did not come without effort or connections, or both. Thomas Mudge, unknown and unheard

22. Jewelling, introduced in London in 1704, was little used in the early years. George Graham was the first watchmaker to make general use of jewels in the balance holes and was probably the first maker to use diamond for the endstone.

23. H.1, on board the *Centurion*, left Spithead on 19 May 1736, arriving at Lisbon on 26 May.

24. Fellow of the Royal Society (1718).

25. Stukeley manuscript diaries, Shelfmark MSS England e 125, vol. 5 (of 20), p. 7, 3 July 1740, Bodleian Library, Oxford University.

26. Poor and Church Rates for the parish of St. Dunstan's in the West, MS. 3008/2, Guildhall Library, London. Thomas Mudge began paying rates for the property on the east corner of Bolt Court and Fleet Street, now numbered 151, in the September to December quarter of 1750.

of by anyone outside the trade, waited some twelve years after becoming a journeyman before taking steps to change his position.

A year before his old master's death, Mudge took a lease on workshop premises at 151 Fleet Street, only a few doors away from Graham's, at number 135.[26] George Graham died on Saturday, 16 November 1751, and on Monday, 18 November, the first working day after, Mudge's advertisement appeared. Prior to this, it must be remembered that Mudge operated as a journeyman throughout the 1740s, a period that could well have seen his close contact with, even employment by, Harrison.

One workman who we know for certain worked for Harrison was John Jefferys (fl. 1726-*ca.* 1753).[27] Jefferys was even allowed to sign and date two Harrison-inspired items: a longcase regulator in 1748 and Harrison's own pocket watch. Dated 1753, this watch, known as the 'Jefferys watch', is of fundamental importance to Harrison's later work. It combined ideas that were to be developed into the prize-winning H.4,[28] ultimately leading to the abandonment of his first three marine timekeepers.

Apart from his involvement in making these two Harrison items, Jefferys is little known; there are very few surviving examples of his work that bear his own name. A verge movement numbered '48' exists, as does a complete cased watch numbered '108'. Hallmarked 1743, the watch is of good quality and contains a typical, fairly early, Graham cylinder escapement. The design of the movement shows strong similarities with Graham's output at this period. Indeed, who more likely than Graham would Harrison have turned to in his need for competent workmen?

Jefferys worked as a repeating-motion maker,[29] an exacting branch of the horological trade. It was Jefferys's apprentice, Larcum Kendall (1719-1790), who was to figure so strongly in connection with Harrison and the Board of Longitude in the 1760s and 1770s. Even Kendall has left us with very few pieces, other than his chronometers K.1, K.2, and K.3, that bear his own name. There are two watches with pivoted-detent escapements (one hallmarked 1776, the other 1786-7), one large cylinder watch with a top plate echoing the design of H.4,[30] and a cylinder watch movement that, bar Kendall's signature, is indistinguishable from those made and sold by the firm of Mudge and Dutton in the 1780s.

The first positive proof of an early connection between Kendall and Graham appears in Kendall's obituary notice, which was published in the *Gentleman's Magazine* in 1790. Describing Kendall's early life, the anonymous biographer writes:

> Mr. Kendall was brought up a Quaker, and bound apprentice to a repeating motion maker; both of which he quitted almost as soon as he became his own master, and was, for several years employed by the late Mr. Graham in making horizontal[31] escapements, which at that time, was reckoned a difficult piece of business.[32]

Though H.2 was completed by 1739 and proved significantly better than its predecessor, Harrison immediately started on a new, much more complicated machine. It was during these difficult years that Harrison leaned heavily on the goodwill of the Royal Society and its new President, Martin Folkes (1690-1754). Folkes, in turn, appears to have looked to Graham for guidance in matters horological. Sadly, Graham did not live to see Harrison suc-

27. John (Joseph) Jefferys was apprenticed to Edward Jagger in November 1717 and became free of the Clockmakers' Company in January 1726.

28. See Anthony Randall's paper in this volume.

29. The maker of repeater mechanisms, by which the watch may be made to indicate the hours, quarters, and sometimes minutes by a series of blows on a bell.

30. Privately owned; now lost and believed to have been stolen.

31. A contemporary term for Graham's cylinder escapement.

32. *Gentleman's Magazine*, vol. 60 (1790), part 2, p. 113.

ceed, and it may be no coincidence that, just one year after his death, another Fellow of the Royal Society, John Ellicott (1706-1772),[33] presented a paper to the Society.[34] Of particular importance is a note that appears at the end of the paper. The note refers to temperature compensation being fitted to a pocket watch, a model of which was shown to the Society in 1748, and records that 'at a meeting of a Council of this Society, on the fifteenth of February last' (1751) a watch and a model were shown. The exact words Ellicott uses in describing the watch, 'I produced a watch (which I had made for a gentleman) with this contrivance', are particularly significant. The words 'had made' imply that Ellicott commissioned someone else to make the watch, and we know that Mudge, through a fortuitous accident, was employed by Ellicott to make his most complicated watches.[35] No further mention of this watch is to be found in the records of the Royal Society but subsequent events, as I hope to show, may help to explain this.

Harrison, having lost his most influential supporter and fearing the effects of this paper within the Society, complained of Ellicott stealing his ideas.[36] Why? Harrison could not legitimately claim sole rights to the idea of temperature compensation, nor to the different ways of achieving it. As for compensation being employed in a watch, Ellicott's was some two years at least before the Jefferys watch. One reason why Harrison was so upset may have been Ellicott's use of the trade to make his watch. Was this the same trade employed by Harrison? What better conduit of ideas and information could there be (though such a conduit can, of course, work in both directions)?

The discovery that Mudge was working for Ellicott was made by the Irish watchmaker Michael Smith (d. 1766) on his travels to London in 1751. Smith, holding the position of Court Watchmaker to the King of Spain, was able to report on Mudge's work and, on returning to Madrid, obtain for Mudge an open commission to make anything that was wonderful for the Spanish Court. In a letter dated July 1752, Smith says of the commission, 'I no *[sic]* Ellicott will be greatly vexed for my recommending you'.[37] A truer statement is unlikely to be found in this paper.

Given the connections between Ellicott and Mudge in the late 1740s, it is highly likely that Mudge made the watch with compensation that Ellicott showed the Royal Society in 1751. With the Spanish Royal Commission of 1752 coming so soon after his decision to seek retail business, Mudge was now in a position to pursue his own ideas, in his own name, and sever all ties with those who were now his competitors. The wording of Mudge's first advertisement following Graham's death states:

> Thomas Mudge, watchmaker, late apprentice to Mr. Graham deceased, carries on Business in the same manner Mr. Graham did, at the Dial and one Crown, opposite Bolt and Tun, Fleet Street.[38]

Significantly, later advertisements add that he 'employs (as he always has done) the same workmen Mr. Graham did',[39] offering yet more evidence of the close connections within the trade.

Recognition and the monetary rewards that accompanied it allowed Mudge a most productive and stimulating decade, one, however, that saw his heath decline as his business increased. The first piece Mudge supplied to the King of Spain was a large, gold-cased, quarter-repeating clock-watch with a date aperture in the enamel dial (Figure 6). Numbered '212', it is hallmarked

33. Fellow of the Royal Society (1738).

34. Dated 4 June 1752, the title of Ellicott's paper is 'A Description of Two Methods, by which the Irregularity of the Motion of a Clock, arising from the Influence of Heat and Cold upon the Rod of the Pendulum, may be prevented'. Published in *Philosophical Transactions* of the Royal Society, London, vol. 47, no. 81 (1752), pp. 479-94.

35. Reported by Mudge's anonymous biographer in 'Memoirs of the Life and Mechanical Labours of the late Mr. Thomas Mudge:...', *The Universal Magazine of Knowledge and Pleasure* (London), vol. 97 (July 1795), p. 42.

36. Letters by Short and Harrison, reproduced in a reprint (London, 1753) of Ellicott's 'A Description of Two Methods' (*op. cit.*, note 34). On the title page of the reprint, the above title (with minor alterations) is given, followed by 'Read at the Royal Society, June 4, 1752. To which are added A Collection of Papers, Relating to the same Subject, Most of which were read at several Meetings of the Royal Society'.

37. A. J. Turner and A. C. H. Crisford, 'Documents Illustrative of the History of English Horology, 1: Two Letters Addressed to Thomas Mudge', *Antiquarian Horology*, vol. 10, no. 5 (Winter 1977), p. 580. The horological world is indebted to Martin Bishop for rescuing these two documents during a house clearance in Surrey. The house had been occupied by descendants of the Imray family of scientific instrument makers.

38. *London Daily Advertiser*, 18 November 1751.

39. J. B. Penfold, 'The London Background of George Graham', *Antiquarian Horology*, vol. 14, no. 3 (September 1983), p. 277.

40. Sold by Sotheby's, London, 19 December 1991, lot 220.

41. Turner and Crisford, 'Documents Illustrative' (*op. cit.*, note 37), p. 581.

42. Part of the Ilbert Collection of the British Museum, this is said to have been Ilbert's favourite clock, a sentiment with which many agree.

43. This may point to the clock being an intended longitude contender, considering the work then being done on the production of accurate lunar tables as an alternative method of finding the longitude.

44. It has been said that Mudge did not fully appreciate the importance of his invention. This is not true. In a letter to Count Brühl dated 9 August 1766 (Mudge Jr., *A Description, with plates* [*op. cit.*, note 20], p. 100), Mudge states: 'it has great merit, and will, in a pocket watch particularly, answer the purpose of time-keeping better than any other at present known'. It could be added, 'or since invented'.

Figure 6. Movement and pair cases of the first watch Mudge supplied to the King of Spain, No. 212. Hallmarked 1753, the watch strikes the hours and is a quarter repeater. The workmanship, as you would expect of Mudge, is superb. The design feature of a shaped signature 'cutout' on the potence plate precedes Harrison's use of it in H.4. Overall diameter: 6.6 cm. (2.6 in.). Private collection. Photograph courtesy of Sotheby's, London.

1753.[40] Referring to this watch in a letter dated 1757, Michael Smith says:

> ...ye King wares & allows [this watch] to be the best watch for performance that he ever had, & to my Knowledge the minit hand has not been set to any time these 45 months and in that time never was none [known] to make about a minit difference wh the King by altering the Kirbs Directly brings it to the true time without altering ye hands.[41]

By 1754, maybe earlier, Mudge had produced a spring clock (Figures 7 and 8) that was no doubt his first attempt at producing a portable precision timekeeper.[42] It even has a most accurate lunar indication.[43] The clock has a train remontoire, temperature compensation applied to one of the two balance springs, and the world's first detached-lever escapement, the most common form of escapement used in mechanical watches to this day.[44] The influence of H.3 on this clock is obvious, though Mudge, by including features of his own design, shows himself to be no mere copier of others' ideas. Despite the time and effort that such a clock would have taken to construct, it was never entered for trial, nor does Harrison make any reference to having seen it or heard of its existence. The clock was never sold by Mudge and passed into the Brunel family through Mudge's daughter-in-law, Sophia Kingdom, who married Sir Marc Isambard Brunel (1769-1849), the father of the great engineer Isambard Kingdom Brunel (1806-1859).

In 1755, Mudge supplied the King of Spain with a most wonderful centre-seconds watch of standard size (Figures 9 and 10). Owing, perhaps, something to his earlier work for Ellicott, it contains a train remontoire (a *tour de force* for a watch of this size), a spiral compensation curb acting on one of two contra-pinned balance springs, and an unusual verge escapement (Figure 11) that allows much greater freedom of arc of the balance than does the normal verge. Mudge's verge escapement was occasionally used by other makers, though much later.[45] The train is jewelled to the fusee, with endstones to the third. The workmanship throughout the watch is superb and set a standard that has rarely been equalled. Numbered '260', this watch by Mudge survives as a movement only.[46] It has been said that this watch mimics H.4, just as his spring clock mimics H.3. This cannot be the case, however, as this watch was finished and delivered to the King of Spain some four or five years before H.4 was ready for trial. Although Mudge may have been inspired by knowledge of the Jefferys watch, this watch, in turn, may have had a great influence on Harrison and his future work on H.4. The fact that Mudge made no claims for, nor hardly mentions, this watch is typical of a man who, when taken to task by his patron, Count Brühl (1736-1809), concerning his

45. A watch by Thomas Wright (d. 1792), No. 1591, hallmarked 1780, has this escapement. Given the connection between Wright and Thomas Earnshaw (1749-1829), the escapement may have been fitted by Earnshaw to this and certain other watches described as having 'cylindrical verge' escapements. Earnshaw's watch No. 1522, which was lost aboard HMS *Providence* in 1797, was described as having a 'cylindrical verge' escapement, though this may just refer to a dead-beat verge escapement.

46. This watch movement was donated to the Worshipful Company of Clockmakers by John Grant in 1850. It was restored by George Daniels in 1967 and is now housed in a brass box.

**Figure 7. Ebonised spring clock with detached-lever escapement by Thomas Mudge. Calendar and lunar indications are shown in the top three dials. Overall height: 30.5 cm. (12 in.). By permission of the Trustees of the British Museum. Inv. no. CAI-2118.**

**Figure 8. Movement of Mudge's spring clock showing the large, vertically mounted, circular balance running on antifriction 'rolls'. Mounted on the back plate are the compensation curbs acting on one of the two contra-pinned balance springs. Width of back plate: 18 cm. (7.12 in.). By permission of the Trustees of the British Museum.**

**Figure 9. Movement of Mudge watch No. 260. Note the unpierced balance bridge, what looks like a second bridge, and a fixed regulation disc (with moving hand). The latter was often used by Mudge in his special watches. Diameter of upper plate (back plate): 4.6 cm. (1.81 in.). Courtesy of the Worshipful Company of Clockmakers, London/Bridgeman Art Library, London. Cat. no. 189.**

**Figure 10. Mudge No. 260 with balance bridge and slide plate removed. Note the coiled compensation curb operating on one of the two contra-pinned balance springs. The elaborate use of jewelling predates that of H.4, and the watch may well be earlier than 1755, the date of its sale to the King of Spain. Courtesy of the Worshipful Company of Clockmakers, London/Bridgeman Art Library, London.**

neglect in claiming the invention of the detached-lever escapement, writes: 'as to the honour of the invention, I must confess, I am not at all solicitious about it: whoever would rob me of it does me honour'.[47]

One other event happened during this period that may have had a direct effect on what was to eventually become Mudge's final design for his marine timekeeper. In 1754, a young Swiss astronomer named Johann Jakob Huber (1733-1798),[48] who was looking for someone capable of putting his ideas for a longitude timekeeper into practice, was directed to Mudge's workshop.[49] On 31 October 1755, Huber left London with a ticking but unfinished machine, in a rush to take up his new job as Professor of Astronomy to Frederick the Great. It is highly probable that Mudge's work was influenced by his association with Huber. However, despite the fact that Huber returned to London later in life and no doubt was well informed of all attempts at the longitude prize, he was never to dispute any part of Mudge's machines, nor claim Mudge was stealing his ideas.

All this happened while Harrison was still working on H.3 in an attempt to finally get it ready for trial. It was not until July 1760 that Harrison said it was ready, with a further request that both H.3 and H.4 should be tried in 1761. Events of that year changed so rapidly that by November 1761, Harrison requested that only H.4 should be sent. The period starting with H.4's journey to Jamaica[50] and ending with George III's intervention that led to Harrison getting the full amount of the £20,000 prize in 1773 was a busy one for John Harrison, who was now being helped by his son William (1728-1815). During this period Mudge was busy with commerce, including more items for the King of Spain, and did not pursue his interest in making a longitude timekeeper (his 'gimcrack', as he called it). We have a clue

47. Letter dated 9 August 1776, in Mudge Jr., *A Description, with plates* (*op. cit.,* note 20), p. 100.

48. Fellow of the Royal Society (1752).

49. A very good account by T. P. C. Cuss of Mudge's work for Huber, entitled 'The Huber-Mudge Timepiece with Constant Force Escapement', was published in *Pioneers of Precision Timekeeping,* pp. 93-115, Antiquarian Horological Society Monograph No. 3 ([Ramsgate, England], [1965]). This is required reading for a better understanding of Mudge's work.

50. H.4, on board HMS *Deptford,* left Plymouth on 28 November 1761, arriving at Madeira on 9 December.

as to why in the printed dedication of the book written by Mudge's son, Thomas Jr. (1760-1843), that was published in 1799.

This important book, titled *A Description, with plates...*, is a full, though edited, account of the manufacture and trials of his father's three timekeepers and of his own attempts at setting up a manufactory to produce them commercially in the 1790s.[51] Of his father's lack of longitude activity during the 1750s and 1760s, Thomas Jr. writes:

> Some time before Mr Harrison obtained his reward, my father had formed in his mind the plan of his time-keeper. Several years afterwards, when he had carried it into effect, and the excellence of it was manifested by its performance, he was asked why he had not made it before the reward was granted to Mr Harrison, and publicly disputed the prize with the gentleman. His answer was: that he thought Mr Harrison a great and deserving character and that after having spent almost the whole of a long life in the laborious pursuit of an object, for which his genius so well qualified him, he could not prevail upon himself to attempt the production of anything by which Mr Harrison might be deprived of that reward, to which he was so well entitled.[52]

Mudge's health deteriorated during the 1760s; no doubt a sedentary working life and the inclinations of a perfectionist were beginning to take

51. This book (*op. cit.*, note 20) is fundamental reading for anyone interested in the history of finding longitude at sea.

52. *Ibid.*, dedication.

53. Awarded in 1777 for his work on the grinding of telescope specula.

54. Letter dated 20 April 1773, in Mudge Jr., *A Description, with plates* (*op. cit.*, note 20), p. 47.

55. Obtained on 10 December 1776, the pension provided £150 per annum till 1783, £120 per annum thereafter.

56. Harrison and the Board of Longitude remained in dispute despite the Act of 1773 awarding Harrison the outstanding sum of £8,750. The 1774 Act (Act 14 George III *cap.* 66 [1774]: *An Act for the repeal of all former Acts concerning the Longitude at Sea*) repealed all former Acts and restated the criteria for new awards of £5,000 (1° error), £7,500 (⅔°), and £10,000 (½°).

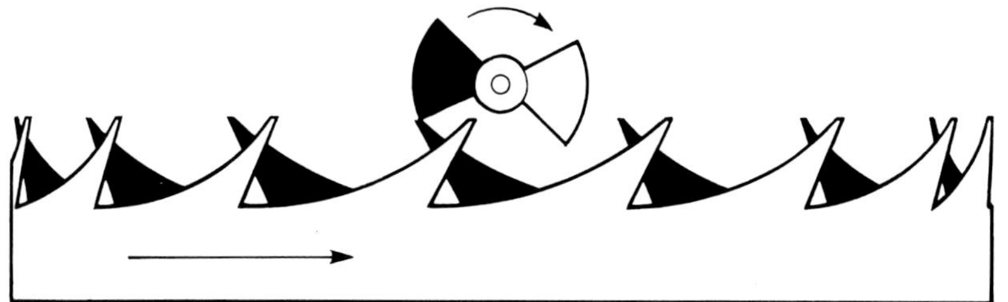

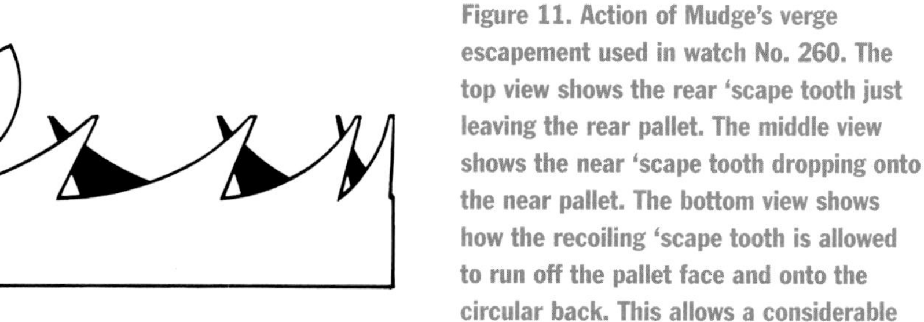

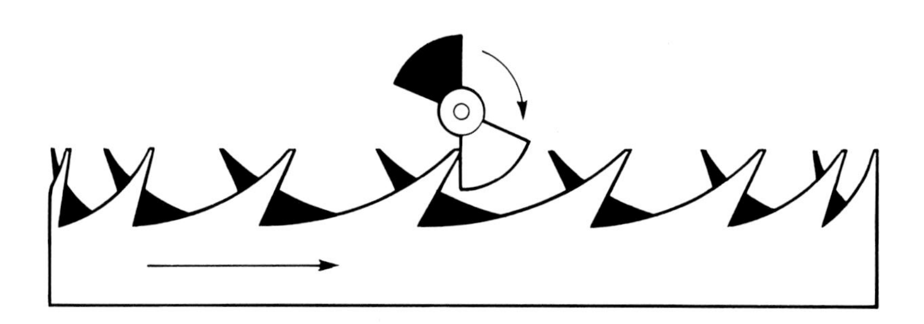

**Figure 11. Action of Mudge's verge escapement used in watch No. 260. The top view shows the rear 'scape tooth just leaving the rear pallet. The middle view shows the near 'scape tooth dropping onto the near pallet. The bottom view shows how the recoiling 'scape tooth is allowed to run off the pallet face and onto the circular back. This allows a considerable increase in balance arc prior to banking. Drawing by the author.**

their toll. Mudge suffered increasingly severe headaches during this period and by the end of the decade had decided to leave London and join his brother John (1721-1793), a doctor of some repute, in his home town of Plymouth. Dr. John Mudge is known for his invention of the inhaler. Not so well known is the fact that he, like Harrison, was awarded the Royal Society's Copley Medal.[53]

It is evident in the published letters of Mudge and especially the one by his brother that he did not want to leave London at this time—a time when he would need the skills of the London trades that earlier had enticed Harrison to make the move to London. In particular, the letter from John Mudge contains a very candid account of Mudge's state of health in the 1770s. It is worth remembering that Mudge was now in his fifties. John's letter was written in reply to Count Brühl, who was at this time, 1773, trying to get Mudge to move back to London. John writes:

> My brother has a full habit, is constitutionally strong, and his want of health is the entire consequence of a want of exercise. He is but ill made for walking and indeed it is too great a fatigue to him; riding on horseback he is totally disqualified for, and hiring a chaise now and then would not suit his finances.[54]

From these and other remarks it is clear that Mudge did not retire to Plymouth to build his first marine timekeeper; rather this was built despite the move. However, there is no doubt that the removal from business would have also freed Mudge from the accompanying stresses and allowed him to concentrate his failing powers. Brühl's court connection did allow him to obtain a small pension for Mudge as watchmaker to the King during his time at Plymouth.[55] This no doubt helped with the hiring of the odd chaise.

The successful second trial of H.4 and part payment to Harrison of £10,000 in 1765 had cleared the way, Mudge felt, for him to make his own machine. This, and the fact that the Board of Longitude prize still stood,[56] provided the opportunity and the stimulation for Mudge to proceed. As

**Figure 12 (left). Mudge's marine timekeeper No. 1. Finished in 1774, the timepiece has elaborate outer and inner wooden cases; the latter is shown here. The four enamel dials indicate hours, minutes, and seconds, with the hour and minute dials also showing degrees of arc (24 hours being equivalent to 360°). The sector dial at the top shows when the timekeeper needs winding. Width of base across plate: 13.2 cm. (5.2 in.). By permission of the Trustees of the British Museum. Inv. no. CAI-2119.**

**Figure 13 (right). Movement of Mudge No. 1. Note the geometrically pierced balance bridge, the twin compensation curbs, and the worm mounted on the top plate. The latter is for setting up Mudge's elaborate barrel assembly, which contains two mainsprings, one above the other. Diameter of upper plate (back plate): 10.7 cm. (4.2 in.). By permission of the Trustees of the British Museum.**

**Figure 14. Action of Mudge's constant-force escapement, which he used in all three of his marine timekeepers. The first view shows the 'top' pallet (shown in line) about to be unlocked by a pin mounted on the balance staff. The second view shows that after unlocking, the 'scape wheel turns and locks on the 'bottom' pallet (shown in solid black). The third view shows the balance being impulsed via the pin. The pallet is being driven by its spiral spring (not shown). The fourth view shows the bottom pallet about to be unlocked. The last view shows the 'scape wheel turning and locking on the top pallet, thus recocking the pallet's impulse spring. This happens while impulse is given to the balance via the bottom pallet. Drawing by the author.**

**Figure 15. Dial and case of Mudge 'Blue', which is now in a later case of the style used by Thomas Mudge Jr. in his chronometer enterprise. It is worth remembering that Mudge was over 60 years of age, with failing health and eyesight, when he embarked on making these two further timekeepers. Purchased by the Mathematisch-Physikalischer Salon, Dresden, in 1925 from a Swedish engineer named Ericson, who is reported to have found it in St. Petersburg, Russia, in 1910. Overall diameter of case: 12.2 cm. (4.8 in.). Courtesy of the Mathematisch-Physikalischer Salon, Dresden, Germany. Inv. no. D IV b11.**

**Figure 16. Upper plate (back plate) of Mudge 'Blue'. Diameter: 10 cm. (3.9 in.). Courtesy of the Mathematisch-Physikalischer Salon, Dresden, Germany.**

stated in a letter to Brühl, he was sure that his own timekeeper would perform as well as, if not better than, Harrison's and, over longer periods, would out perform it.[57]

By 1774 his first machine (Figures 12 and 13) was finished. Mudge was then 58 years of age and was beginning to despair of his ability to adjust the machine fully. As if his bad health was not enough of a problem, the Board of Longitude, in 1774, amended the Longitude Act.[58] This required at least two chronometers to be tested, with more stringent requirements needed from both machines. Despite the problems, two further machines, known as 'Blue' (Figures 15 and 16) and 'Green' (Figures 17 and 18), were completed by 1779. As well as obvious differences from the first machine in dial layout, the two later machines were made to go for one day only. This allowed Mudge to use a smaller, less powerful mainspring and thus do away with the elaborate twin-mainspring assembly used in his first, eight-day timekeeper; spring breakages in this machine had forced Mudge to provide a two-day stop on its going. It is interesting to note that two-day duration, together with the provision of a state-of-wind indication[59] (both part of Mudge's first machine), became standard features of marine chronometers during the nineteenth century. All three of Mudge's marine timekeepers included his constant-force escapement, the action of which is shown in Figure 14.

The testing of the three machines, with an ever-more-vehement debate about the application of 'rates'[60] to all those undergoing trials, is a long story, too long to relate here. Suffice it to say that Mudge's feeling that the Board of Longitude, in particular the Astronomer Royal, Nevil Maskelyne (1732-1811),[61] was intent on minimising the chances for success of a mechanical

57. Undated letter, *ca.* 1774, in Mudge Jr., *A Description, with plates* (*op. cit.*, note 20), p. 71.

58. See note 56.

59. Also called an 'up-and-down' indicator.

60. No chronometer is expected to keep exactly to mean time. It is considered accurate if it either gains or loses the same amount in the same period of time, *e.g.*, one second in 24 hours. By applying the established losing or gaining 'rate' to the chronometer's indicated time, the correct time can be determined. The problem of accelerating changes of rate was known but not then understood. The various different methods of ascertaining the correct or changing rate was one of the main disputes that both the Harrisons and the Mudges had with the Board of Longitude.

61. Nevil Maskelyne, elected a Fellow of the Royal Society in 1758, established the *Nautical Almanac*, first published in 1766, and was awarded the Copley Medal in 1775 for his paper 'Observations on the Attraction of Mountains'.

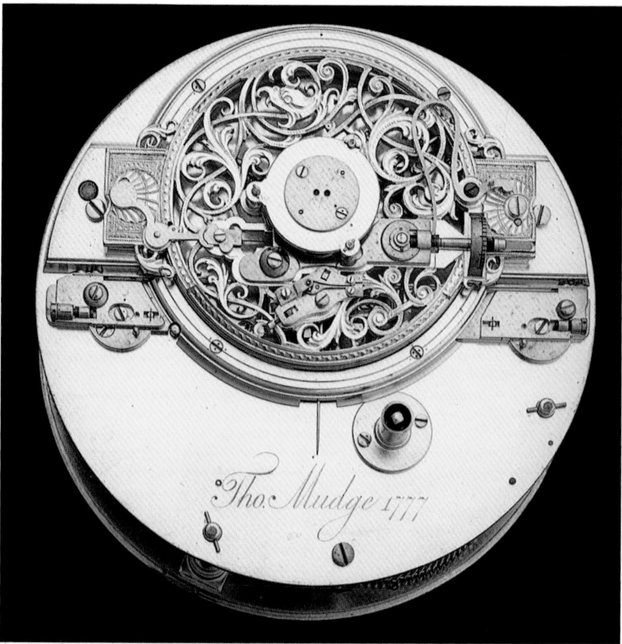


**Figure 17. Dial and case of Mudge 'Green'. It was given this name by Mudge because of the colour of its shagreen-covered case, which was stained green in order to distinguish it from its identical twin, Mudge 'Blue'. The latter presumably had a blue-stained case. Being of one-day duration, 'Green' and 'Blue' are somewhat simpler than his first marine timekeeper, having one large enamel dial with concentric hour and minute hands and having no need for a state-of-wind indicator. 'Green' was sold by Christie's, Geneva, on 8 November 1976, lot 240, for the then-world-record price for a mechanical timekeeper of £60,000. Overall diameter: 11.48 cm. (4.49 in.). Courtesy of The Time Museum, Rockford, Illinois. Inv. no. 1303.**

**Figure 18. Upper plate (back plate) of Mudge 'Green', dated 1777 though not finished until 1779. Note the elaborately pierced and chased balance bridge and the lack of both a compensation curb scale and mainspring set-up worm that were fitted to his first timekeeper. Diameter: 10 cm. (3.94 in.). Courtesy of The Time Museum, Rockford, Illinois.**

solution, to a great extent echoed those of the Harrisons a decade earlier. Despite an advance of £500 from the Board of Longitude in 1777, it was not until 1793 that Mudge was to receive any further reward for his labours. It took a report from a Select Committee of the House of Commons[62] before an award of £3,000, in recognition of his endeavours, was made. Unlike Mudge's actions regarding Harrison, two watchmakers, one of whom made a complete fool of himself, allowed themselves to be used in an attempt to block any award.[63]

The award of £3,000 and an abiding feeling of his father's merits led Thomas Jr., a practising solicitor, to set up a manufactory in order to produce timekeepers on his father's plan. This is described in his book. Some 27 or so machines were produced, none of which, for various reasons, lived up to the accuracy of the originals.[64] The venture was a failure and no wonder: such expensive, complicated machines were not what was wanted by the English Royal and merchant navies. Neither, of course, were Harrison's. By the 1780s, the work of John Arnold (1736-1799), followed by Thomas Earnshaw (1749-1829)—both building on what had been done before—provided the sort of practical machines that navigators wanted.[65] The time for Mudge's and Harrison's timekeepers was past.

The importance of the pioneering work by Harrison, Mudge, Arnold, and others was researched by Rupert Gould (1890-1948)[66] in the early part of the present century. His findings were published in 1923 in his book *The*

62. The *Report from the Select Committee of the House of Commons,...to whom the Petition of Thomas Mudge, Watchmaker, was referred;...* (London, 1793).

63. The watchmakers were John Arnold (1736-1799) and Josiah Emery (1725?-1797), who appeared on behalf of the Board of Longitude to oppose Mudge's petition, as did Sir Joseph Banks (1743-1820), John Arnold's staunchest supporter, and Nevil Maskelyne. Emery, in appearing before the Board to promote the going of one of his watches in comparison to Mudge 'Green' and 'Blue', had to admit that the escapement in his watch was of Mudge's invention. This detached-lever escapement had been communicated to Emery, via Count Brühl, by a model expressly made for that purpose by Mudge.

64. A list of surviving examples of Thomas Mudge Jr.'s chronometer enterprise can be found in Appendix 1 at the end of this paper.

65. See Jonathan Betts's paper on Arnold and Earnshaw in this volume.

66. An obituary 'appreciation' for Lt.-Comdr. Rupert Thomas Gould by H. Alan Lloyd and others was published in the *Horological Journal*, vol. 90, no. 1082 (November 1948), pp. 655-6.

*Marine Chronometer*, the enlarged second edition of which was sadly never published. Despite this, the book still ranks as one of the best horological books ever written.[67] In order to properly compare the performance of the earliest marine timekeepers, Gould applied the modern rating formula[68] to the historic trial records of all the early timekeepers. In all cases, he chose a period giving a continuous trial of 29 weeks. The lower the number, the better. The results, published as Appendix I in his book, do not give the total picture, but, as Gould says, this 'may at least serve to some extent as an indication of how much chronometer-making progressed during the period 1766-1914'.[69]

Gould's table shows the marked development of Harrison's four timekeepers, from a trial number of 1,139 for H.1 to 124.4 for H.4. It shows the continued improvement by the simpler machines of Arnold and Earnshaw, with Arnold No. 36 achieving a trial number of 44.91 and Earnshaw No. 1 a figure of 25.57. It also shows that in this select company all three of Mudge's timekeepers acquit themselves well, with his 'Green' achieving a figure of 58.12 and 'Blue' a figure of 39.39. In particular, Mudge's first machine achieves a trial number that is outstanding. Using the same formula as a standard of comparison, the figure for Mudge No. 1 of 11.73 was not bettered until 1873, some 97 years later, and then by only 0.63.[70]

For all those clockmakers who attempted a solution to the longitude problem, the requirements of the Act[71] demanded a mechanical genius combined with the highest skills of eye and hand. In this respect, Mudge truly did excel.

67. It was as a tribute to the work of Gould and Col. Humphrey Quill (1897-1987) that this paper was presented at the Longitude Symposium. An obituary for Humphrey Quill by Charles Allix and others was published in *Antiquarian Horology*, vol. 17, no. 3 (Spring 1988), pp. 245-7. Quill was author of *John Harrison, the Man who found Longitude* (London: John Baker, 1966).

68. The 'trial number' is obtained by taking the weekly sums of the daily rates and applying the formula a + 2b, where 'a' is the difference between the greatest and least weekly sums and 'b' is the greatest difference between the sums of two consecutive weeks.

69. Rupert T. Gould, *The Marine Chronometer: Its History and Development* (London: J. D. Potter, 1923), Appendix I, p. 258.

70. Weichert's two-day chronometer, No. 2300, with a Kullberg 'flat-rim' balance.

71. In 1774, the year of the completion of his first timekeeper, Mudge was disagreeably surprised to hear of a new Act of Parliament that replaced the 1714 Act of Queen Anne, which had, in effect, been won by John Harrison. Although this new Act (see note 56) offered a further £10,000 prize, it did so under conditions so exacting that it was never won, and it was repealed in 1828.

## APPENDIX 1

### SURVIVING EXAMPLES OF THOMAS MUDGE JR.'S CHRONOMETER ENTERPRISE

| *No.* | *Signature & date* | *Going* | *Provenance/Present whereabouts* |
|---|---|---|---|
| 1 | Howells & Pennington | 8-day | Duke of Saxe Gotha? Now at the British Museum, London. |
| 2 | Howells & Pennington | 1-day | The Royal Navy. Present whereabouts not known. |
| 3 | Howells & Pennington, 1795 | 1-day | The Royal Navy. Sold by Christie's, 5 February 1963, lot 170. Private collection. |
| 4 | Howells & Pennington | 8-day | The Royal Navy. Now at the National Maritime Museum, Greenwich. |
| 5 | Howells & Pennington, 1795 | 1-day | The Royal Navy. Now at the Victory Museum, Portsmouth, England. |
| 6 | Not known | Not known | Danish Navy? Present whereabouts not known. |
| 7 | Howells & Pennington, 1795 | 1-day | Danish Navy? Purchased by the Clockmakers' Company, London, 1881. |
| 8 | Howells & Pennington, 1795 | 1-day | The Royal Navy. Now at the National Maritime Museum, Greenwich. |
| 9 | Not known | Not known | Captain Bowen? Present whereabouts not known. |
| 10 | Pennington, Pendleton & Others, 1796 | 8-day | Duke of Marlborough. Now at the Soane Museum, London. |
| 11 | Pennington, Pendleton & Others, 1796 | 1-day | Spanish Navy? Sold by Sotheby's, 3 December 1976, lot 261. Private collection. |
| 12 | Pennington, Pendleton & Others, 1796 | 2-day | The Royal Navy. Now at the National Maritime Museum, Greenwich. |
| 13 | Not known | Not known | Present whereabouts not known. |
| 14 | Not known | Not known | Alexander Aubert. Sold by Leigh & Sotheby, 21 July 1806, lot 328. Present whereabouts not known. |
| 15 | Not known | Not known | Present whereabouts not known. |
| 16 | Pennington, Pendleton & Others, 1796 | 2-day | The Royal Navy. Now at the Royal Greenwich Observatory Archives, Cambridge University Library. |
| 17 | Pennington, Pendleton & Others, 1797 | 1-day | Now at the Liverpool Museum, England. |
| 18 | Pennington, Pendleton & Others, 1796 | 1-day | Count Brühl. Now at the Mathematisch-Physikalischer Salon, Dresden, Germany. |
| 19 | Not known | Not known | Present whereabouts not known. |
| 20 | Not known | Not known | Present whereabouts not known. |
| 21 | Not known | 2-day | Much altered and renamed 'Barraud No. 325'. Private collection. |
| 22 | Pennington, Pendleton & Others, 1796 | 1-day | Sold by Sotheby's, 22 April 1963, lot 3. Private collection. |
| 23 | Not known | Not known | Present whereabouts not known. |
| 24 | Pennington, Pendleton & Others, 1796 | 1-day | Sold in the auction rooms of Puttick & Simpson, London, November 1964. Private collection. |
| 25 | Pennington, Pendleton & Others, 1797 | 1-day | Sold by Sotheby's, New York, 15 June 1992, lot 502. Private collection. |
| 26 | Pennington, Pendleton & Others, 1798 | 8-day | Marquis of Ripon? Sold by Sotheby's, 27 May 1954, lot 118. Formerly in collection of The Time Museum, Rockford, Illinois; now in private collection. |
| 27 | Pennington, Pendleton & Others, 1797 | 8-day | Much altered and renamed 'Barraud No. 399'. Now at the Mariners' Museum, Newport News, Virginia. |

# Arnold and Earnshaw: The Practicable Solution

## *Jonathan Betts*

**Jonathan Betts, Horology Conservator at the National Maritime Museum since 1979 and Curator of Horology since 1990, is also Horological Advisor to the National Trust of Great Britain, the Wallace Collection, the Harris Collection at Belmont in Kent, and a number of other notable clock collections. He served on the Crafts Council's Conservation Committee for three years.**

**Betts organized the publication and wrote the introduction to the facsimile reprint of *The Principles of Mr. Harrison's Time-keeper* (1767) in 1983. He wrote the *National Trust Pocket Guide to Clocks* (1985) and a small Harrison biographical booklet for the National Maritime Museum (1993). He has written a number of articles and reviews in the horological press and given numerous lectures on horological subjects. At present, his specific research subjects are Thomas Earnshaw (1749-1829) and Lt.-Commander Rupert T. Gould (1890-1948).**

**From 1972 to 1975, Betts studied horology at Hackney College. He then spent a year in London restoring clocks and music boxes and the following five years as a free-lance clock restorer in Ipswich. By his own account, horology has been an incurable affliction in which he has rejoiced ever since.**

# Arnold and Earnshaw: The Practicable Solution

*Jonathan Betts*

## THE PROFOUND INFLUENCE OF JOHN HARRISON (1693-1776)

The year 1767 saw the publication of *The Principles of Mr. Harrison's Time-keeper,*[1] and the horological world was intensely interested to see details of the prizewinner. With H.4's successful trials just a few years before, one would expect 1767 to be the year in which the marine timekeeper successfully solved the longitude problem. In a sense it was, but it was also the beginning of a whole new chapter in the story of portable timekeepers, a period when intuitive and entrepreneurial craftsmen grasped the initiative from the staid, scientific stereotypes of John Harrison, George Graham (*ca.* 1674-1751), John Ellicott (1706-1772), and their kind.

Just as surely as those engaged in the horological world were keen to read *The Principles of Mr. Harrison's Time-keeper*, they were disappointed with what they learned from it. Harrison's disclosure of the mechanism of H.4 had been published by order of the Commissioners of Longitude under the direction of Nevil Maskelyne (1732-1811), who had stated in the preface that it was intended it to be complete instructions and working drawings for 'artists who may be desirous to construct other watches after the model of Mr. Harrison's'.[2] For most ordinary watch- and clockmakers, however, it would have proved either unintelligible or unbelievable. The views of most practitioners must have been echoed by the natural philosopher and author William Emerson (1701-1782), who remarked a couple of years later about finding the longitude that it was 'still a secret, and likely to continue so, for tho many thousands of pounds have been paid for the pretended discovery thereof; we remain just as wise as we were before the discovery, except the ill success of it happens to teach us so much wit as to take better care of our money for the future'.[3]

It was also realised that even if the design did turn out to be a resounding success, the solution was still not practicable because it was such a complicated, and therefore expensive, thing to make. The Board of Longitude was unhappy too. It is recorded that Dr. Anthony Shepherd (fl. 1763-1772), one of the Commissioners, once announced that it had never been the

1. [John Harrison], *The Principles of Mr. Harrison's Time-keeper, with Plates of the Same* (London, 1767); facsimile reprint by the British Horological Institute in *Principles and Explanations of Timekeepers by Harrison, Arnold and Earnshaw* ([Upton, Notts., England], 1984). This work was based on a written description and testimony that Harrison had provided to a special panel appointed by the Commissioners of the Board of Longitude.

2. *Ibid.*, p. vii.

3. William Emerson, *The Mathematical Principles of Geography, Navigation & Dialling* (London, 1770), p. 172.

Board's intention that a timekeeper should qualify,[4] and it is clear that even many years later, the Establishment, in the shape of the Board and the Admiralty, was still not convinced that the solution was entirely trustworthy.

However, although they may not have realised it at the time, *The Principles of Mr. Harrison's Time-keeper* did have a profound effect, in that it demonstrated the small scale on which a successful longitude timekeeper should be made, and, for those prepared to read the text carefully, the work contained a number of fundamental truths, essential in timekeeper design.[5] Now watchmakers, rather than clockmakers, began to take an interest, and the period immediately following the publication of *Principles* is, arguably, one of the most interesting in the whole story of precision timekeeping. In just 30 years, the English marine chronometer was to evolve into its modern form, proven in efficacy and as an article of commerce.

## JOHN ARNOLD (1736-1799) AND THOMAS EARNSHAW (1749-1829)

In the year that *The Principles of Mr. Harrison's Time-keeper* was published, a resourceful and ambitious young watchmaker called John Arnold, a 31-year-old Cornishman then working in London, evidently did read Harrison's *Principles* carefully and at that time embarked on a notable and courageous new direction in his career. He decided to try to improve on Harrison's timekeeper, and he ended up by contributing more, in detail, to the developed chronometer than any other maker.

The Lancastrian Thomas Earnshaw, thirteen years his junior, was just a lad of eighteen in 1767 and was still bound apprentice, almost certainly to the watchmaker William Hughes in High Holborn.[6] It would be some twelve difficult years before Earnshaw was to enter the stage as a maker of precision timekeepers. Starkly different in character, the argumentative and bitter Earnshaw was also to contribute a great deal to the story, but, in so doing, he set virtually the whole of his profession, and much of the

4. Shepherd's comment was recorded by John Arnold and quoted in Thomas Mudge Jr., *A Reply to the Answer of the Rev. Dr. Maskelyne, Astronomer Royal, to a Narrative of Facts, relating to some Time-keepers, constructed by Mr. Thomas Mudge, for the Discovery of the Longitude at Sea,...* (London, 1792), p. 21.

5. Jonathan Betts, introduction to *Principles and Explanations of Timekeepers by Harrison, Arnold and Earnshaw* (*op. cit.*, note 1).

6. In *Longitude. A Full Answer To The advertisement concerning Mr. Earnshaw's Timekeeper In The Morning Chronicle, 4th Feb. and Times 13th Feb.* (London, 1806), p. 72, Alexander Dalrymple (1737–1808) refers to 'Hughes' as being Earnshaw's master.

**Figure 1. Portrait of John Arnold, with his wife Margaret and son John Roger, by Robert Davy (1735?-1793), *ca.* 1787, the year Arnold took his son into business with him. Arnold holds a typical marine chronometer movement of his from this date, with an 'OZ' balance. Oil on canvas. 106 x 127 cm. (41.7 x 50 in.). Courtesy of the Science Museum/Science and Society Picture Library, London. Inv. no. 1868-248.**

Establishment, against him. We will return to Thomas Earnshaw, and to his controversy with John Arnold, in due course.

## JOHN ARNOLD'S EARLY WORK

We know little of John Arnold's early years as a watchmaker. He was certainly a very clever and ambitious young man, and, having arrived in London, he soon made himself known in the highest social circles by presenting the young King George III, in June 1764, with a watch mounted in a ring.[7] We get a glimpse at the character of the determined Cornishman from a description of a brief encounter with him when a young Shetlander, Thomas Mouat, visited the South in 1775. According to Mouat's diary, he and a church minister, one Charles Stewart, met Arnold while the three were travelling together

> ...in the stage fly for London. It holds four inside passengers. Our companion happened to be the famous Arnold, who made a watch in the stone of the King's ring; his conversation was animated, blustering and much adorned with oaths, which were too frequent for Mr. Stewart's ease....[8]

This interesting reference suggests a rather different character from the shy and gentle man that Arnold's biographer, Vaudrey Mercer (1905-1993), believed Arnold to be.[9]

By early 1767 it seems that Arnold had also managed an introduction to the Astronomer Royal, Nevil Maskelyne, as on 27 April Maskelyne presented him with a copy of *Principles,* which had just left the presses; this copy, now in a private collection, is inscribed as a gift from Maskelyne to Arnold. It is also very possible that, through Maskelyne, Arnold was able to inspect H.4 itself at this time. A few months later, in late 1767, the King is said to have encouraged Arnold to improve marine timekeepers by granting him £100.[10]

We believe we know what Arnold's first attempt at a marine timekeeper was like, but alas, its whereabouts is now unknown. However, if one imagines the timekeeper in Figure 2 having a centre seconds hand, this would give a good idea of its general appearance. This first, and vitally important, machine of Arnold's is described by the late, great Lt.-Commander Rupert T. Gould (1890-1948) as the 'detached cylinder timekeeper' and dated at *ca.* 1768. Unfortunately, this vital link in horological history disappeared in the United States during World War II.[11] Gould, who wrote about it prior to its disappearance, does at least provide us with a reasonable description, and this sheds some fascinating light on Arnold's early thinking on timekeeper design. The timekeeper was not signed, but Gould declared it could only be an early Arnold. He described it as follows: 'The balance was almost a facsimile of that used on Harrison's No. 4—so was the compensation curb—except that instead

**Figure 2. One of Arnold's early marine timekeepers (known as 'No number'), *ca.* 1772, made for the Royal Society to issue to James Cook for his second voyage of discovery and taken on board the *Adventure.* Arnold's first marine timekeeper, now missing but believed to be in the United States, looks very like this but is unsigned and has a centre seconds hand rather than subsidiary seconds. Width and depth of box: 14.5 cm. (5.7 in.). By permission of the President and Council of the Royal Society, London. Inv. no. RS 37. Photograph by permission of the National Maritime Museum, Greenwich.**

**Figure 3. Movement of H.4 with the balance bridge removed. The balance, three-turn balance spring, 'cycloidal' adjuster, and bimetallic compensation can all be seen. The engraved bridge covers the third wheel. Diameter of upper plate (back plate): 9.6 cm. (3.75 in.). Courtesy of the National Maritime Museum, Greenwich. Inv. no. Ch.38.**

**Figure 4. Movement of 'No number' marine timekeeper by Arnold. Its technical specification is remarkably similar to H.4 and appears to have been inspired by *The Principles of Mr. Harrison's Time-keeper*, which had been published in 1767. Diameter of upper plate (back plate): 9.8 cm. (3.85 in.). By permission of the President and Council of the Royal Society, London. Inv. no. RS 37. Photograph by permission of the National Maritime Museum, Greenwich.**

of carrying the curb pins, its free end worked in a pivoted, forked lever whose other end carried the pins'.[12]

The earliest timekeepers Arnold actually completed for official trial were from the series he made in the early 1770s. He had first approached the Board of Longitude in May 1770, and, in March 1771, he was before them again with a machine he estimated could be produced for just 60 guineas. This compared very favourably with the cost of Harrison's design: Larcum Kendall (1719-1790) was paid £500 for K.1, the copy of H.4, £200 for K.2, a simplified version without Harrison's remontoire, and £100 for K.3, another simplified version with a 'co-axial crown wheel' dead-beat escapement, a type of frictional-rest escapement.[13] The Board granted Arnold £200 to continue his experiments. This was the first of a number of sums that he was to receive.

The first of Arnold's machines to go to sea were a pair sent in 1771 with Admiral Sir Robert Harland (1715?-1784), who was most impressed with their performance,[14] and three sent with James Cook (1728-1779) on his second voyage to the South Seas in the same year.[15] Two of the early machines tested by Cook survive (in the Royal Society collections), and like the earlier detached cylinder timekeeper, their construction shows how much they were influenced by Harrison's published description of H.4. Harrison shows, both by description and engravings, the scale and the detail of his timekeeper, and Arnold's early machines are virtually identical to it (compare Figures 3 and 4): the diameter of H.4's upper plate is shown in *Principles* as 3 3/4 in. (9.6 cm.), although Harrison says it should be 1/16 in. (0.16 cm.) more, and that of Arnold's machines are just over 3.85 in. (9.8 cm.). Maskelyne states in *Principles* that the balance is 2.2 in. (5.59 cm.) but should be a little larger,[16] and Arnold's balance is 2.35 in. (5.97 cm.) in diameter. H.4 has a three-turn spiral balance spring with a cycloidal correction device, and so do Arnold's early timekeepers. Harrison describes the bimetal compensator, and Arnold employed an almost identical one, although he uses it to move separate curb pins about the balance. Maskelyne also mentions Harrison's statement that the movement of H.4 would be better in a wooden

7. Col. Humphrey Quill (1897-1987) tells us in his biography of Harrison (*John Harrison, the Man who found Longitude* [London: John Baker, 1966], p. 190) that the King had been following the trials of H.4 with close interest and that it is probable that John Harrison was introduced to George III after the conclusion of the second trial of H.4 in July 1764, soon after Arnold had made his presentation of the ring watch. One wonders if Harrison and Arnold ever met.

8. Francis Bamford, 'A Shetlander in St Martin's Lane, 1775', *Furniture History*, vol. 11 (1975), p. 108.

9. Vaudrey Mercer, *John Arnold & Son* (London: Antiquarian Horological Society, 1972), p. 144.

10. *Ibid.*, p. 15.

11. The chances are that this timekeeper is still somewhere in the United States, and I would like to make an appeal for help in tracking it down.

12. Notes on this early Arnold timekeeper are in manuscript in one of Gould's annotated copies of his *The Marine Chronometer: Its History and Development* (London: J. D. Potter, 1923), now in private hands.

13. Unlike a detached escapement, which is in contact with the balance for a minimum amount of time, a frictional-rest escapement is in constant engagement with the balance.

14. John Arnold, *Certificates and Circumstances relative to the going of Mr Arnold's Chronometers* (London, 1791), p. 1.

15. Derek Howse and Beresford Hutchinson, *The Clocks and Watches of Captain James Cook, 1769-1969* (London: Antiquarian Horological Society, 1969).

16. Nevil Maskelyne, 'Notes Taken at the Discovery of Mr. Harrison's Time-keeper', published in [Harrison], *The Principles of Mr. Harrison's Time-keeper* (*op. cit.*, note 1), p. xiv.

box, without its silver outer case, and this is how Arnold chose to case his timekeepers.

Indeed, the only fundamental difference between the two designs is in the escapement. Gould termed Arnold's design the 'see-saw detent escapement'. The detent, which is disposed across the top of the wheel, has the action of a see-saw and locks the wheel on vertical pins standing up from its face. This design incorporated both a 'passing piece' for the mute vibration of the balance and a 'safety action', and, although rather crude, the escapement is highly ingenious for a first design.

As a detached escapement, the arrangement ensures that the mechanism is only in contact with the balance for the minimum amount of time. Even here I would suggest that Arnold's idea might have come from Harrison. In *Principles* Harrison states quite unequivocally: 'and it must be allowed, the less the Wheels have to do with the Balance, the better'.[17] Of course, Harrison's concept of detachment was different from Arnold's more literal interpretation: Harrison was proposing a slight but virtually continuous impulse, and his concept of detachment also embraced the fact that a balance that stores a lot of energy is less affected by variations in impulse. It is therefore possible that Arnold conceived the idea of detachment entirely independently, but it is to my mind just as likely that he first considered the concept on reading Harrison's statement.

## PIERRE LE ROY (1717-1785)

It is sometimes believed that Arnold's design for detached escapements was, in fact, based on the work of Pierre Le Roy in France. However, Le Roy did not publish his *Memoire*[18] describing his detached escapement until 1770, and I cannot find a single piece of evidence to suggest that Arnold knew of Le Roy's work at this time. In his *Memoire*, Le Roy depicts his truly innovative design for a compensation balance with bimetallic rims (inspired, Le Roy tells us, by Harrison's bimetal, probably seen in the French edition of *Principles,* which was also published in 1767),[19] and I feel sure that had Arnold seen the *Memoire* in the early 1770s, he would have adopted this form of balance then and not spent the next twenty years working toward it.

I must say of Le Roy's contribution in general that although, in his *Memoire,* he describes three principles of the developed chronometer—detached escapement, isochronal balance spring, and compensation balance—his actual machines were far from satisfactory in design. Each with a heavy, seconds-beating, five-inch-diameter balance, mounted on a very vulnerable suspension and unlocking and impulsed at the rim, they bear little resemblance to a developed chronometer. In his *Memoire,* Le Roy goes on to make the ill-advised statement that detached escapements are not suitable for chronometers with balances smaller than his machines and recommends the frictional-rest escapement of Henry Sully (1680-1728) for smaller marine timekeepers, similar to his own design in his celebrated watch, *La Petite Ronde.*[20]

For his three statements of principle, one should have great respect for Le Roy, and, for these insights, he was undoubtedly one of the important pioneers. But I must say that I have always been rather surprised that Gould, among other authorities, has placed Le Roy in quite such an honoured position and poured so much adulation upon him.[21]

17. [Harrison], *The Principles of Mr. Harrison's Time-keeper* (*op. cit.*, note 1), p. 20.

18. Pierre Le Roy, *Memoire sur la Meilleure Maniere de Mesurer le Tems en* Mer,... (Paris, 1770). Translated by T. S. Evans, F.L.S., of the Royal Military Academy, Woolwich, under the title 'A Memoir on the best Method of measuring Time at Sea,...', *The Philosophical Magazine* (London), vol. 26 (October 1806-January 1807), pp. 40-68.

19. Père Esprit Pezenas, *Principes de la Montre de Mr. Harrison, avec les planches relatives à la même montre,...* (Paris, 1767).

20. *La Petite Ronde* is shown in Figures 9 and 10 in Catherine Cardinal's paper in this volume.

Le Roy was not the only pioneer who was working on frictional-rest escapements for chronometers. In the early 1770s, the small number of other makers experimenting on precision timekeepers were mostly working with forms of the frictional-rest escapement (these operated without recoil and are therefore also known as 'dead-beat' escapements). After all, the best watches of the day were still reckoned to be those with a dead-beat escapement, usually the type known as a horizontal, or cylinder, escapement. Thus Kendall's design for the timekeeper K.3, designed in 1772 but not completed until 1774, also incorporates a dead-beat escapement, a co-axial crown wheel type commonly known as 'Kendall's escapement'. In 1772, Josiah Emery (1725?-1794) also made a timekeeper (No. 615) with this escapement. The extensively jewelled movement of this important timekeeper, which is clearly of experimental design, is housed in a silver case hallmarked London, 1772/3.[22]

## ARNOLD'S PIVOTED-DETENT WATCHES

Arnold, however, had different ideas, although few people in the London watchmaking trade knew at the time what he was up to. In the early 1770s, while his see-saw escapement timekeepers were away on trial with Cook, Arnold was, in fact, busy designing his first pocket watches with pivoted-detent escapements (a type of detached escapement). These watches formed the second of several different series Arnold was to make, each with a different numbering system and each with a new movement of improved design. Indeed, throughout his life John Arnold was constantly thinking of improvements, regularly updating earlier instruments, and it can be extremely difficult today sorting out which series a given Arnold chronometer comes from.

Unfortunately, Cook's trials of Arnold's see-saw detent timekeepers had not gone well,[23] especially compared with the superb performance achieved by K.1, but the pivoted-detent chronometers that followed were of a much more sophisticated design and were properly to establish Arnold's reputation. In this design the detent pivots in the same plane as the escape wheel and is kept in the path of the wheel under the influence of a return spring. Both the passing arrangement and the safety action have been much improved, and the whole escapement is a thoroughly good design, except perhaps for the fact that the detent still has considerable inertia and that the pivots still require oil.

One of these pivoted-detent watches of Arnold's was used by Captain the Hon. Constantine John Phipps (1744-1792) on his voyage toward the North Pole in 1773, along with a see-saw type marine timekeeper and Kendall's K.2. The watch proved to be the best timekeeper, by far. Nonetheless, although instructions to the expedition's astronomer, Israel Lyons (1739-1775), dictated that longitudes should be found both by lunars and by the timekeepers, there is no doubt that at this early date, lunars would take precedence should any question arise.[24]

As a result of Arnold's work, a few other London watchmakers began to experiment with pivoted detents, and we find Kendall moving on to this detached escapement in his version, with rather crude passing arrangement, in a watch (No. B+m) of 1776.[25] Another example by Kendall (No. B+y), is hallmarked 1786-7.[26] Josiah Emery also advanced to the pivoted detent at this period, and one of his watches (No. 781), dated 1778, with a pivoted-detent escapement, is very similar in principle to Arnold's.[27]

21. Gould, *The Marine Chronometer* (*op. cit.*, note 12), pp. 83-94.

22. This watch is now at the Musée des Arts et Métiers, Paris (ref. no. 19429). Further information on Josiah Emery may to found in Part 1 of an article by the author entitled 'Josiah Emery, Watchmaker of Charing Cross', in *Antiquarian Horology*, vol. 22, no. 5 (Spring 1996), pp. 394-401. A discussion of this watch and its early trials conducted by Count Frederik Sigismund van Byland (1749-1828) appears in Part 2 of the article in *Antiquarian Horology* (Summer 1996).

23. Andrew David, ed., *The Charts and Coastal Views of Captain Cook's Voyages*, 2 vols., Hakluyt Society Publications, extra series no. 43 (London, 1988 and 1992), p. xlvii.

24. Ann Savours, 'A Very Interesting Point in Geography: The 1773 Phipps Expedition towards the North Pole', *Arctic*, vol. 37, no. 4 (December 1984), pp. 402-28.

25. George Daniels, *English and American Watches* (London: Abelard-Schuman, 1967), p. 90, figs. 13a and 13b. Formerly in the Maryatt Collection; now in a private collection.

26. Cecil Clutton and George Daniels, *Clocks and Watches: The collection of the Worshipful Company of Clockmakers* (London: Sotheby Parke Bernet, 1975), p. 57. Now in the collection of the Worshipful Company of Clockmakers, London (cat. no. 422).

27. Anthony G. Randall and Richard Good, *Catalogue of Watches in the British Museum* (London: British Museum, 1990), p. 37. Now in the collection of the British Museum, London (ref. no. 1969, 3-3,1).

**Figure 5. Portrait of Alexander Dalrymple, drawn by George Dance (1741-1825) in 1794. Dalrymple was a great supporter of John Arnold and an early advocate of the use of the marine chronometer (a term he was one of the first to use) as the practical method for finding the longitude at sea. Pencil on paper. 25.4 x 18.7 cm. (10 x 7.4 in.). Courtesy of the National Library of Australia, Canberra.**

In 1775, Arnold took out his first patent, perhaps one of the most notable in horology because it included his unique invention, the cylindrical (helical) balance spring, with the spring formed into a helix rather than a flat spiral, and a design for a compensation balance. Again, it is possible that Arnold's inspiration for the compensation balance was Harrison. Some years later, Arnold himself pointed out that 'the correction piece of my timekeeper is in the balance itself, which Mr Harrison says in his book would be the greatest possible perfection'.[28] Arnold was referring to Harrison's *A Description Concerning Such Mechanism,*[29] published just a few months before Arnold was granted his patent for the compensation balance.

## ALEXANDER DALRYMPLE (1737-1808) AND THE EAST INDIA COMPANY

In 1775, during a voyage to Madras in the East India Company ship *Grenville*, which had on board an Arnold timekeeper (coincidentally, as the property of the V.I.P. passenger George, Lord Pigot [1719-1777]), Mr. Alexander Dalrymple, later hydrographer of that Company, realised the great potential of such an instrument for finding the longitude. Back in London in 1778, he soon made himself known to Arnold, and there began a relationship between the two men that was to ensure both Arnold's continuing commercial success and the establishment of the marine timekeeper in East India Company service as a practically proven method for finding longitude at sea. Dr. Andrew Cook has discussed the importance of this close cooperation between the two, resulting, from 1779, in many pamphlets published by Dalrymple providing instructions and notes on the use of chronometers by Arnold.[30] In some cases, it seems that Dalrymple had a hand in writing Arnold's own pamphlets for him. Cook has also corrected the common misunderstanding that the East India Company bought timekeepers in large numbers.[31] In fact, just as with Royal Navy captains, East India captains usually had to buy their own instruments and only rarely did the Company itself buy timekeepers.

One of Dalrymple's early pamphlets, undated but ascribed by Cook to late 1779 or 1780), is entitled *Some Notes useful to those who have Chronometers at Sea.*[32] This pamphlet marks the first established use of the term 'chronometer' to describe precision portable timekeepers.[33] Dalrymple remarks significantly about the title: 'The Machine used for measuring Time at SEA is here named CHRONOMETER, my friend Mr. Banks agreeing with me in thinking so valuable machine deserves to be known by a *Name,* instead of a Definition. The name *Time-Keeper* is only proper to a *perfect Chronometer'*.[34] Mr. Banks was, of course, Joseph Banks (1743-1820), later Sir Joseph, President of the Royal Society, member of the Board of Longitude, and another staunch supporter of Arnold in the years to come. Following the publication of this pamphlet, the term 'chronometer' was increasingly used to describe marine timekeepers.

**Figure 6. Portrait of Sir Joseph Banks, drawn in 1788, by John Russell (1745-1806). Another great supporter of Arnold, Banks was President of the Royal Society and a long-standing member of the Board of Longitude. In later years he was to prove an implacable enemy of Thomas Earnshaw, opposing all of Earnshaw's claims for reward. Pastel crayon on paper. 61 x 45.7 cm. (24 x 18 in.). Courtesy of The Knatchbull Portrait Collection, England.**

This enlightened attitude of Dalrymple and the East India Company to the use of chronometers contrasts with the Establishment view, held up until the early nineteenth century, that the chronometer would always remain a back-up, a second best to lunars for finding longitudes. Chronometers were, of course, sent on all of the British expeditions of discovery at this period, but the whole concept of a practical marine timekeeper was still regarded as decidedly unproven. Nevil Maskelyne, the Astronomer Royal himself, stated in 1792 of chronometers:

28. John Arnold, in *Report from the Select Committee of the House of Commons,...to whom the Petition of Thomas Mudge, Watchmaker, was referred;...* (London, 1793), p. 80.

29. John Harrison, *A Description Concerning Such Mechanism as will Afford a Nice, or True Mensuration of Time;...* (London, 1775), p. 103.

30. Andrew S. Cook, 'Alexander Dalrymple and John Arnold: Chronometers and the representation of Longitude on East India Company charts', *Vistas in Astronomy,* vol. 28, parts 1/2 (1985), pp. 189-95.

31. *Ibid.,* p. 189.

32. Alexander Dalrymple, *Some Notes useful to those who have Chronometers at Sea* (London, [late 1779 or 1780]).

33. The term 'chronometer' had been used in print before by Jeremy Thacker in *The Longitudes Examin'd...* (London, 1714), by J. T. Desagulier in *A Course in Experimental Philosophy,* 2 vols. (London, 1734), vol. 1, pp. 375-6, when describing a clock by George Graham, and by Benjamin Martin in *Mathematical Institutions* (London, 1764) about a similar clock.

34. Dalrymple, *Some Notes* (*op. cit.,* note 32), p. 1.

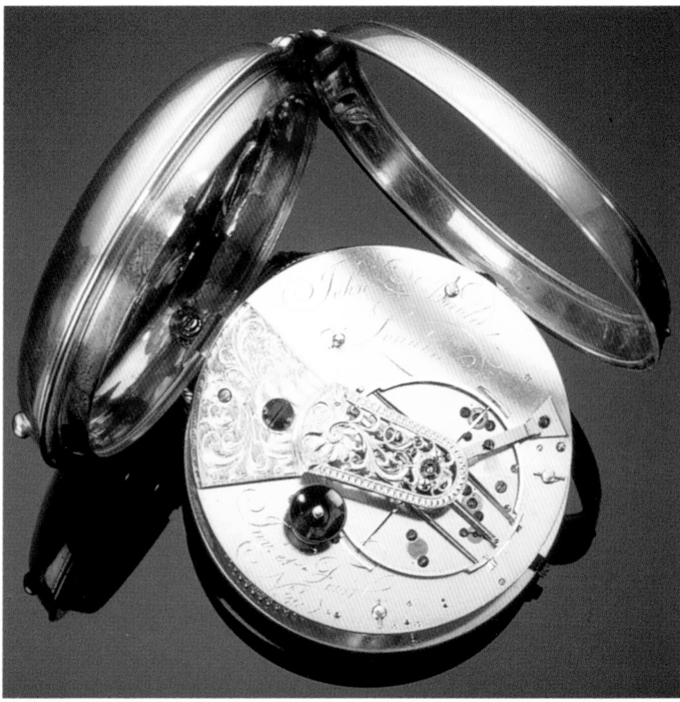

**Figure 7. Arnold pocket chronometer No. 1/36 of 1778. In a 22 ct. gold case, this was from the series Arnold called 'of the best kind'. Its performance when on trial at Greenwich astonished the horological world. About 1780, after Dalrymple had referred to this particular piece as a 'chronometer', this term was increasingly used to describe these precision timekeepers. Diameter of outer case: 7 cm. (2.8 in.). Courtesy of the National Maritime Museum, Greenwich. Inv. no. Ch.646. Photograph by permission of Christie's, London.**

**Figure 8. Movement of No. 1/36. This was the first pocket chronometer to have a compensation balance, as Harrison had recommended, and it was extensively jewelled, as H.4 had been. The movement is of gilt brass and has Arnold's pivoted-detent escapement. Diameter of upper plate (back plate): 5.9 cm. (2.3 in.). Courtesy of the National Maritime Museum, Greenwich. Photograph by permission of Christie's, London.**

> But I should prefer correspondent observations of an eclipse of a bright fixt star by the moon, made by two astronomers furnished with proper instruments, at places not very remote from each other; and a number of correspondent observations of the transits of the moon over the meridian, compared with those of fixt stars, made by two astronomers at two remote places, to *any timekeeper whatsoever*, for determining the relative situation of the two places.[35]

Perhaps the most famous of all the Arnold chronometers from the pivoted-detent series, and some may say the most famous of all Arnold's chronometers, was his No. 1/36 (Figures 7-8).[36] Made in 1778, this is almost certainly the first watch to incorporate a compensation balance. It was initially made for a Major Wood on the recommendation of Dalrymple and was sent to Greenwich for trial in 1779. It performed incredibly well, astonishing Nevil Maskelyne and his assistants and inducing Arnold, in 1780, to publish its trial results in a pamphlet.[37]

The record shows that throughout the whole thirteen-month period covered, the watch's rate hovered around minus 0.75 seconds per day, and that having established a rate, in the first month, of minus 0.31 seconds per day, at the end of the thirteen months the watch was only 124 seconds in error. It has to be said that no register was kept of temperature, but the watch was moved daily from the assistant's pocket to a hook in the assistant's room,

35. Nevil Maskelyne, *An Answer to a Pamphlet entitled "A Narrative of Facts," lately published by Mr. Thomas Mudge, Junior, relating to some Timekeepers, constructed by his Father Mr. Thomas Mudge:...* (London, 1792), p. ix.

36. Now in the collection of the National Maritime Museum, Greenwich (inv. no. Ch.646).

37. John Arnold, *An Account kept during Thirteen Months in the Royal Observatory at Greenwich of The Going of a Pocket Chronometer, Made on a New Construction by John Arnold,...* (London, 1780). In this pamphlet, the use of the term 'chronometer' is applied to a specific timekeeper for the first time.

38. Mercer, *John Arnold* (*op. cit.*, note 9), p. 79.

where it hung at night, so the watch certainly experienced frequent temperature changes and changes in position. Arnold's design for pivoted-detent chronometers, from the series he called 'of the best kind',[38] was thus proving extremely successful.

I can find no evidence at all to show that at this stage Arnold was in any way unhappy with his escapement, but it was at this time that the spring detent made its appearance, leading to one of the most famous controversies in horology—and one in which, in my view, John Arnold is too often seen as the innocent and blameless party.

## THOMAS EARNSHAW

Enter stage the vehement antagonist Thomas Earnshaw, watchmaker of St. Pancras. Earnshaw was 32 in 1781 and, having been out of his apprenticeship for eleven years, had established an excellent reputation within the trade as a watch-finisher. But if he was such a clever and dedicated maker of timekeepers, why hadn't he made his name outside the trade by this time, as Arnold had?

Besides the fact that he was thirteen years younger than Arnold, Earnshaw was, by all accounts, not nearly as resourceful and ambitious, and

**Figure 9. Portrait of Thomas Earnshaw in 1798, by Sir Martin Archer Shee (1769-1850), exhibited in that year at the Royal Academy. Earnshaw is seated next to a typical example of his marine chronometer. A later version of this portrait, showing Earnshaw with quill pen in hand, was painted by Shee, probably in or after 1808, when Earnshaw's *Appeal* was published. Oil on canvas. 92 x 71 cm. (36.2 x 27.9 in.). Courtesy of the Science Museum/ Science and Society Picture Library, London. Inv. no. 1962-374.**

he doesn't appear to have had any interest in social advancement. Arnold had ensured royal and noble patronage and financial help by loans from wealthy friends as support for his development work. With the additional support of his canny and businesslike wife, and with only one son to look after, it is not surprising that Arnold's career had advanced smoothly. In contrast, Earnshaw's chief problem, by far, was money. He had married in 1769, even before finishing his apprenticeship, and within four years had three sons to support as well. By 1774, debtors' prison loomed and, in that year, Earnshaw escaped his creditors by absconding to Dublin. At that time, the best watchmaker in Dublin was John Crosthwaite (1745-1829), and it seems most likely this is where Earnshaw sought shelter and temporary employment.

In the *London Gazette* for December 1774, we find that Earnshaw eventually surrendered himself to the Fleet Prison in order that his debts might be resolved. For the benefit of his creditors, his presence in prison had to be announced three times in the press and, on an appointed day soon after, creditors were then entitled to visit relevant debtors, in this case at the Fleet Prison. The prisoners were then ignominiously paraded in a line, each holding a list of their goods and chattels, in order to come to some arrangement with their creditors. Earnshaw's imprisonment just might explain the curious remark made some years later by the watchmaker Paul P. Barraud (1752-1820), that Earnshaw 'had forfeited his life to the laws of his country'.[39] If this is what Barraud meant it would be ironic, as Barraud himself had faced bankruptcy in 1788.

Over the next six years, Earnshaw evidently worked his way out of debt, and, in 1780, he began to make significant improvements to timekeepers. Late in that year, he turned his attention to improving the pivoted-detent form of detached escapement, and, by the following year, he claimed to have devised the spring-detent escapement, which was to become one of the central features of the developed chronometer. When Arnold claimed a version of this invention as his own by including it in his patent of 1782, there arose one of the most famous controversies in horology, one that lasted into the early years of the nineteenth century, more than a decade beyond Arnold's death. Indeed, it has continued to some degree to the present day.

## THE SPRING-DETENT ESCAPEMENT CONTROVERSY

Figures 10 and 11 show the patent drawings for both Arnold's and Earnshaw's designs for a spring-detent escapement. This escapement has a definite advantage over the pivoted detent in that there are no pivots to oil. In addition, the spring detent can be made more slender, thereby reducing its inertia. Briefly, the facts about the controversy over this invention are as follows: Arnold's version appears in his second patent of May 1782; Earnshaw's form was patented for him by the watchmaker Thomas Wright (d. 1792) in February 1783.[40] But although Arnold was the first to patent it, Earnshaw always claimed that the spring detent had been his original invention in 1781 and that Arnold had first heard of the idea from the watchmakers John (1747?-1808?) and Miles (1755?-1821) Brockbanks, to whom Earnshaw had foolishly described the invention in the same year.

Unfortunately, no chronometers with this escapement by either maker that are early enough to be conclusive are known to exist, but as earlier

39. Gould, *The Marine Chronometer* (*op. cit.*, note 12), p. 126.

40. Ferdinand Berthoud (1727-1807) also implied that he was the inventor of this escapement; see Ferdinand Berthoud, *De La Mesure du Temps, Ou Supplément Au Traité Des Horloges Marines, Et A L'Essai Sur L'Horlogerie;...* (Paris, 1787), pp. 47-57. However, Anthony Randall has shown conclusively that this was a case of 'retrospective wishful thinking' (to put it politely) and that Berthoud's machine No. 24 of 1782, cited by Berthoud, has never had anything but a pivoted-detent escapement; see Anthony Randall, 'L'Oeuvre Chronometrique de Ferdinand Berthoud de 1760-1787. Analyse du Traité des Horloges Marine et du Supplément au Traité des Horloges Marines', *ANCAHA*, vol. 30 (1981), pp. 23-35. Also mentioned by Randall in his lectures on Berthoud given to the Antiquarian Horological Society in London in December 1985 and April 1986.

authors have usually given Arnold the benefit of the doubt, I would like to propose the following as an equally likely scenario for the invention of the spring-detent escapement:[41] I suggest that the story as described by Earnshaw in his published *Appeal* of 1808,[42] although written in an aggressive and polemic style, is basically correct. In 1780, Arnold was satisfied with his pivoted-detent escapement's stunning results and was working on improving his compensation balances and experimenting with terminal curves on his helical balance spring. On hearing from the Brockbanks brothers about a detent on a spring, however, he immediately saw the advantage in it. He was perhaps indignant that this relative newcomer to the field might well have hit upon the ultimate evolution of what he himself had been painstakingly developing for over ten years. He tried to imagine how Earnshaw had arranged the spring detent. We know he tried to get a look at it in Wright's shop, but Wright would not have the watch opened. Surely, Arnold reasoned, the detent couldn't lock in compression; that would not be a sound principle. It must therefore lock in tension and therefore unlock <u>into</u> the wheel. What then to do with the escape wheel teeth, which would have to lock at the base? He decided to resurrect an idea that he had discarded some years before, that of making a <u>cycloidally curved</u> impulse face in an attempt to produce a rolling action at the pallet. He hastily drew a rough sketch for this (and rough the sketch, shown in Figure 10, certainly is) and patented the idea along with the new forms of balance and the truly brilliant terminal curves for isochronism, which he actually <u>had</u> invented.

41. Jonathan Betts, 'The Spring Detent', *Clocks*, vol. 7, no. 7 (January 1985), p. 70.

42. Thomas Earnshaw, *Longitude. An Appeal to the Public:...* (London, 1808); facsimile reprint by the British Horological Institute in *Earnshaw's Appeal* ([Upton, Notts., England], 1986).

Let us look briefly at this peculiar escapement of Arnold's as, to me at least, it remains a mystery. The single main advantage of an epicycloidal wheel-tooth form, be it on an escape wheel or an ordinary train wheel, is that it transmits motion to that pallet or pinion at constant relative speeds,

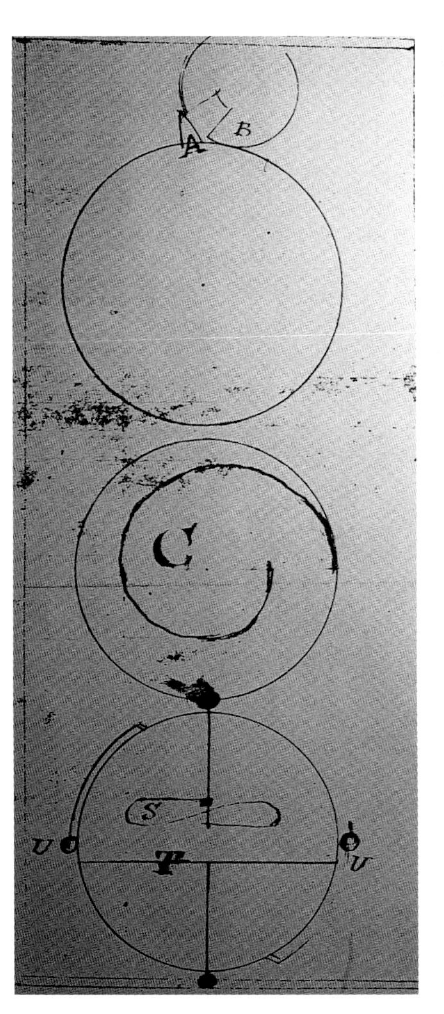


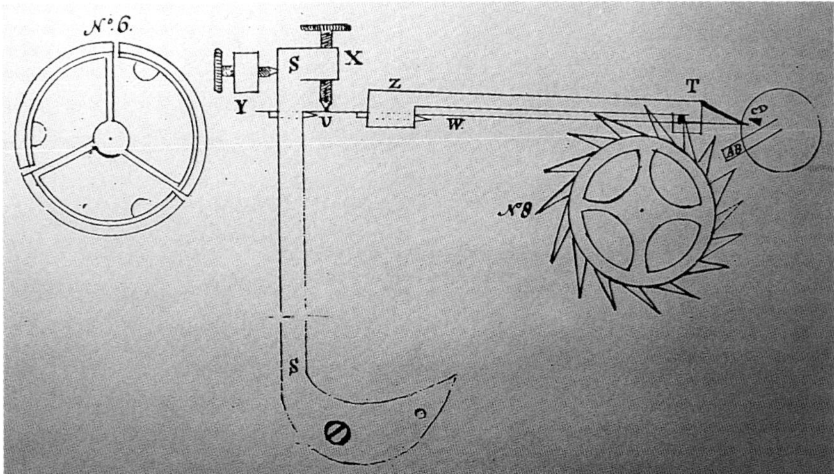


**Figure 10 (left). Part of John Arnold's original patent drawing (no. 1328) of 1782, showing his very rough sketch for the cycloidal tooth and pallet of his spring-detent escapement. The patent also included designs for three types of compensation balance and his terminal curves for isochronous balance springs. Courtesy of the Patent Office, London.**

**Figure 11 (above). Part of Earnshaw's drawing for Thomas Wright's original patent (no. 1354) of 1783. It shows a relatively well-thought-out design for a complete detached spring-detent escapement, though no safety action is shown. Also shown (marked no. 6) is one of Earnshaw's designs for a compensation balance. Courtesy of the Patent Office, London.**

known as 'uniform lead'. However, this advantage only occurs with contact after the line of centres; in other words, cycloidal tooth forms are suited best for gear-wheel transmission, where there are high numbers of teeth following in continual succession. Such a form is not ideal for use in an escapement, as action before the centre line would cause considerable engaging friction. A common misconception about cycloidal tooth forms is that they can also produce a rolling contact between the wheel tooth and the pallet or pinion leaf, but, unfortunately, this is not the case.

We do not know for certain what John Arnold saw as the advantage in his cycloidal tooth forms, but it seems he did believe his escapement would produce rolling impulse. He states of his escapement, in a promotional pamphlet of 1791, that 'friction is utterly excluded from it',[43] and, in 1793, he claims that his escapement 'has no friction whatever',[44] which certainly suggests a rolling action. Further conflicting evidence is found in the description of the escapement published in 1805 by Arnold's son, John Roger Arnold (1769-1843),[45] where he describes, incorrectly, how to generate the tooth form using a rolling circle twice the correct diameter and then, in his drawing, shows the tooth operating before the line of centres and on the edge of the impulse pallet throughout.

The plot thickens. Study of many Arnold chronometers over the last few years, including careful measurement and drawing up of the escapements, has shown that the escapement itself was not then usually constructed correctly, as all of the escapements so far inspected have had the cycloidal tooth face only half engaged with the pallet, ruining any sensible action and invariably causing wear on the wheel teeth. I confess that it is very difficult to guess at what Arnold was up to, but I suggest that the validity of the whole escapement is certainly in question. Like Earnshaw, Arnold was an intuitive, instinctive designer, not a scientific one, and he probably never understood his escapement fully.

Earnshaw maintained throughout his life that the original idea of the spring detent had been his. It has to be said that Arnold never actually claimed the original invention, and, on the several occasions when confronted with the allegation of copying Earnshaw's idea, he always hedged and never actually denied it. Early examples of Earnshaw's spring-detent timekeepers are few and far between, and, because for some years he could not afford to produce watches under his own name, they are all signed by other makers; Earnshaw did not normally sign his own work until the early 1790s. We learn from Earnshaw's *Appeal* that some were sold by these makers under licence from Wright and Earnshaw, and were stamped with the mark 'Wright's Patent'. Three early watches are well known: a pocket chronometer signed by Josiah Jessop (fl. 1780-1794) of London,[46] a pocket chronometer movement signed by Robert Tomlin of London,[47] and a pocket chronometer of 1784 signed by Earnshaw's 'sponsor', Thomas Wright.[48] The earliest of all the extant watches with early Earnshaw escapements is a watch with an astronomical dial, signed by George Margetts (1748-1804).[49] This large watch, numbered '341' and now in a later marine-type box, was clearly made for the Chinese market and dates from *ca.* 1783. The escapement is identical to the patent drawing, with no safety roller and with a detent spring running through the blade and doubling as a passing spring; it even has the adjusting screws on the detent block and the detent spring, as in the patent.

43. Arnold, *Certificates and Circumstances* (*op. cit.*, note 14), p. vii.

44. Mudge Jr., *A Reply* (*op. cit.*, note 4), p. 95.

45. John Roger Arnold, 'Explanation of Time-keepers, constructed by Mr. Arnold. Delivered to the Board of Longitude by Mr. Arnold, March 7th, 1805', included in *Explanations of Time-keepers, constructed by Mr. Thomas Earnshaw and the late Mr. John Arnold* (London, 1806), p. 53; facsimile reprint by the British Horological Institute in *Principles and Explanations* (*op. cit.*, note 1).

46. This watch bears the patent mark and probably dates from 1783 but is in a later case. It is now at The Time Museum, Rockford, Illinois (inv. no. 4127) and is described in Anthony Randall, *The Time Museum Catalogue of Chronometers* (Rockford, Illinois: The Time Museum, 1992), p. 150.

47. This watch, also dating from *ca.* 1783, is stamped with 'TW' and '34' and is now in the British Museum's collection (ref. no. CAI-1752). It was described by Anthony Randall in 'An Early Pocket Chronometer by Thomas Earnshaw, signed Robert Tomlin', *Antiquarian Horology*, vol. 14 , no. 6 (June 1984), pp. 609-15.

48. Naturally, this watch does not have a stamp mark. It is now in private hands. It was described by Andrew Crisford in 'Thomas Wright in the Poultry, London, No. 2228', *Antiquarian Horology*, vol. 9, no. 7 (June 1976), pp. 785-8.

49. On loan to the National Maritime Museum, Greenwich (inv. no. Ch.27).

The movement is not stamped anywhere with Wright's patent mark.

For ordinary watchmakers in the mid-1780s, however, timekeepers with detached escapements were still very much an unknown, and to some extent an unproven, quantity. An interesting experimental pocket timekeeper I discovered some years ago illustrates this point. The watch, which is in private hands, is dated 1786 and is described in a contemporary manuscript, extant with the watch. This important timekeeper is clearly based on Harrison's work, as it has a 21,600 train,[50] as recommended by Harrison for watches smaller than H.4, is jewelled throughout, up to the fusee, and has a Harrison-type bimetallic compensation. In the manuscript, the maker tells us that the watch was designed to have two interchangeable escapements, a cylinder (dead-beat) escapement or a detached escapement, in order to compare each escapement's relative performance, as he says, 'all else remaining the same'. The watch will be the subject of an in-depth study.

At this time, I believe, a typical Arnold marine chronometer would have been made entirely in London. As is well known, rough movements for watches were often supplied from Lancashire (Arnold himself used many Lancastrian craftsmen), but movements didn't always come from there; the London makers were perfectly capable of making them, and did so. Such rough movements were usually bought from Lancashire for convenience, when a number of identical watch movements were required. For marine chronometers, I believe it was, in fact, Earnshaw who first established the supply of standard rough movements from his home county of Lancashire.

In stark contrast to chronometers by the Arnolds, which were being constantly improved and changed in design, one has to look at only one machine to describe an Earnshaw box timekeeper. The vast majority were produced on a remarkably standardised plan, like so many peas in a pod. Typically, they have a round 'porthole' lid with sliding cover (examples—apparently original—have been seen with covers sliding to the front, to the back, or even to the side), silvered brass dial with plain blued-steel poker hands, and movement and bowl in plain gimbals just as seen behind Earnshaw in his portrait (Figure 9). (This depiction suggests that Earnshaw chronometers may usually have had silvered gimbal-rings.) In his marine timekeepers Earnshaw invariably used Arnold's helical balance spring with terminal curves, a fact he naturally avoided mentioning! Earnshaw's design for the compensation balance, with brass fused to the steel wheel and cut to shape on a dividing engine, is, in practice, far better than Arnold's and was eventually adopted by all chronometer balance makers. Another interesting feature of Earnshaw's marine timekeepers is that he used a 'quick train', with the balance beating 130 times per minute.[51] It has been suggested that Earnshaw made such chronometers to specific order, to be used for intercomparison with a regulator: by using the vernier principle, this allowed the error of the chronometer to be judged with great precision and accuracy.[52] I believe that this theory is partially correct; using this technique, it is easy to judge the error of such a chronometer to one-tenth second. However, at present I am of the view that the design was as much for Earnshaw's own use as for the owner's. The evidence suggests that these chronometers were not just an occasional product of Earnshaw's: the 'quick train' appears to have been a standard feature of his marine timekeepers. It is probable, therefore, that Earnshaw used the feature as a means of rapidly checking the error of his marine timekeepers

50. A 21,600 train, meaning the speed of the escapement in the timekeeper, refers to the number of vibrations of the balance in one hour.

51. Because each beat represents two vibrations of the balance, these balances oscillate 15,600 times per hour. Therefore, with a 15,600 train, Earnshaw's marine chronometers were indeed 'quick' compared to the 14,400 train that became the standard used for marine chronometers. They might be considered slow, however, compared to Harrison's 21,600 train (see note 50), or even to H.4, which has an 18,000 train.

52. Christopher Wood, 'The Function of the Quick Train Chronometer', *Antiquarian Horology*, vol. 9, no. 3 (June 1975), pp. 331-6.

during initial testing. He tells, for example, of a chronometer he made for the successful trial against those of Mudge and Brockbanks in 1796, which he brought to mean time in just seventeen hours.[53]

In the course of studying Earnshaw's work, one is often struck by his ability to pare down an idea to its bare essentials. One of his greatest achievements was the design for his marine timekeeper. But it is important to understand that features such as the quick train and, indeed, the very simplicity of the whole design, were born of sheer necessity—the mother of invention. Having no other financial support, Earnshaw had to manufacture reliable and accurate timekeepers quickly. As with so many other great works of art and industry, this creation was thus, in my judgement, a product more of inspired need than of enlightened consideration.

In 1789, Earnshaw was introduced to the Astronomer Royal for the first time, and in due course five of Earnshaw's timekeepers were tried successfully at Greenwich. From this time Earnshaw's business also appears to be on a much sounder footing and more of his products are signed by him, including increasing numbers of box chronometers. Earnshaw's bank account for the early part of the nineteenth century[54] shows that he was then providing himself with a very satisfactory income and, interestingly, making regular payments to Robert Roskell (fl. 1798-1830) of Liverpool. Roskell may well have been an intermediary, supplying rough movements for both marine and pocket chronometers. The trade would evidently have been two way because Earnshaw also appears to have finished pocket and box chronometers for Roskell, which would be sold under Roskell's name. A number of such instruments are known.

From the early 1790s, Earnshaw began his petitioning for reward from the Board of Longitude, and it was from this time that the controversy between Arnold and son and Earnshaw really took off. The whole affair was most unpleasant, but, it must be said, where would horological historians be without controversy? The vast majority of really interesting information comes out of arguments and investigations such as this one.

In 1799, at just 63 years of age, John Arnold died; with his passing, the world of chronometry was vastly the poorer. It would be fair to say that no one has contributed as much in detail to the developed chronometer as he. In 1783, a little tribute to him had been paid by the great mathematician William Ludlam (1717-1788) in a letter to Nevil Maskelyne, which nicely sums him up:

> Now you mention Mr Arnold, I am glad of an opportunity of doing him justice. I confess, from the accounts I had, I once thought he was one of those who talk much and do little [further corroboration that he was not the shy, modest man that Dr. Mercer supposed]....But since the account of the going of his watch, when in your keeping, has been fairly and honestly published, and since I have seen what he has actually done, I have far different thoughts. His contrivances are far simpler and easier than those of Mr Harrison's, or any I have seen, and full as likely as any to answer the end proposed.[55]

And so it was Arnold's son who now bore the brunt of Earnshaw's allegations. In general, the Board of Longitude tried to keep an open mind,

53. Earnshaw, *Longitude. An Appeal* (*op. cit.*, note 42), p. 70.

54. Made available to me by the kind help of David Penney and Valerie Finch.

55. The letter, dated 23 October 1783, is quoted by Maskelyne in the *Report from the Select Committee* (*op. cit.*, note 28), p. 96.

56. Now preserved at the National Maritime Museum, Greenwich (inv. no. Ch.145).

although Nevil Maskelyne does seem to have somewhat favoured Earnshaw's claims, while Sir Joseph Banks, a staunch friend of the Arnold family, joined Alexander Dalrymple in strongly opposing Earnshaw and in recommending that John Roger Arnold should instead be considered for reward, on behalf of his father.

In 1803, formal interviews with fellow watchmakers began to be conducted by the Board, to try to discover who had invented what. Some of these were in Earnshaw's presence and, given Earnshaw's temperament, were probably a pretty uncomfortable experience for some of the interviewees. Ever since the start of the spring-detent escapement controversy in 1782, Earnshaw had managed to set himself against virtually the whole horological community, and now, as he was to discover, that community, whether justified or not, was very ready to condemn his claims for reward.

The Board next decided to request descriptions and drawings of their marine chronometers from both makers, with the intention of circulating these among the other watchmakers for annotated comments. A model of each escapement was also commissioned, but only Earnshaw's survives.[56] When the manuscript of the descriptions arrived, however, the Board made the fatal mistake of sending them straight to the printers without reading them. Upon delivery of the 500 copies of each, they found to their dismay that Earnshaw had included so much spite and invective that the edition had to be destroyed. One or two copies of this first printing have fortunately survived,[57] and it makes lively reading! For example:

> Mr Arnold was a pompous man, and because he made large machines it must be right and the fools followed him in his blunders....[58]

and:

> When the watchmakers who oppos'd me at the Board of Longitude were ask'd the Question which was best—box or pocket timekeepers—their reply was they could not say which had the advantage. Was not that reply a full declaration of their ignorance...?[59]

and:

> ...I think it is plain that their answers were founded in total Ignorance and all the other answers given to the Board were totally false. For instance the very first question put to them: Do you make time keepers? Answer: yes! I call on them to prove, Pennington excepted, whether they ever made one in their lives....[60]

After expurgating many of these vitriolic remarks, both descriptions were reprinted, and this time they were circulated for comment.[61] In their replies, we find many of Earnshaw's great contemporaries in watchmaking firmly putting the knife in. Of the many sarcastic and damning remarks made, one by William Hardy (fl. 1800-1832) is typical. The point concerned how much brass should be used in a compensation balance, and Earnshaw gives a strange analogy of a soldered compensation balance being like two men being pushed about by a third, stronger man. Hardy says of this, that such a description 'merely satisfies his capricious and malicious disposition', and, he goes on, 'the whole of this metaphorical battle, this extraordinary and incomprehen-

57. Thomas Earnshaw, *Explanation of Timekeepers constructed by Thomas Earnshaw* (London, 1805). One of the surviving copies is preserved at the National Maritime Museum, Greenwich (ref. no. C1354). The revised version was published in 1806 in *Explanations of Time-keepers, constructed by Mr. Thomas Earnshaw and the late Mr. John Arnold* (London, 1806); facsimile reprint by the British Horological Institute in *Principles and Explanations* (*op. cit.*, note 1).

58. Earnshaw, *Explanation of Timekeepers* (*ibid.*), p. 9.

59. *Ibid.*, p. 8.

60. *Ibid.*, p. 10.

61. These descriptions were printed on the left half of the page, leaving a wide margin on the right for readers to comment on the text.

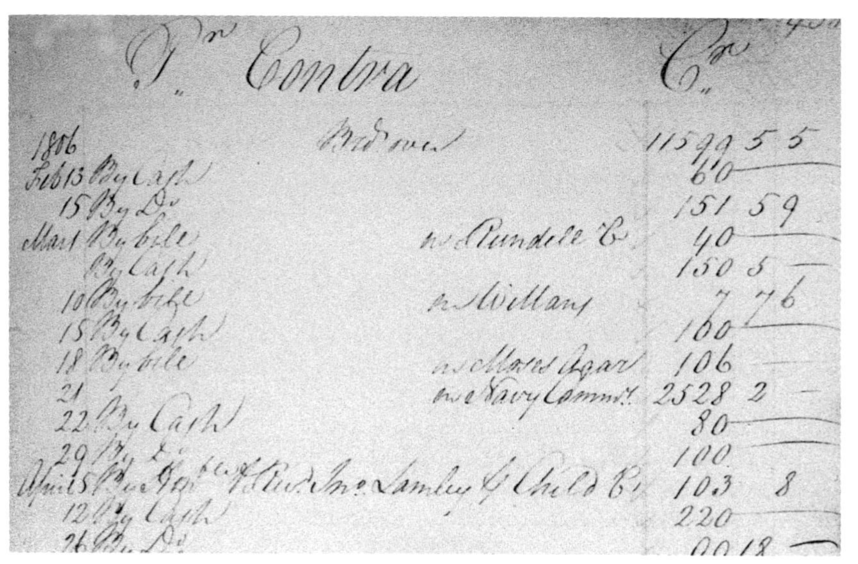

Dr Contra Cr
1806 Brd over 11599 5 5
Feb 13 By Cash 60
15 By Do 151 5 9
Mar 1 By bill on Rundell & Co 40
By Cash 150 5
10 By bill on Williams 7 7 6
15 By Cash 160
18 By bill on Moses Agar 106
21 on Navy Commrs 2528 2 —
22 By Cash 80
29 By Do 100
April 5 By ... Child & Co 103 8
12 By Cash 220

**Figure 12. Earnshaw's account book entry for 21 March 1806, showing under 'Cr' (Credit) the sum of £2528.2.0 (two thousand five hundred and twenty-eight pounds, two shillings, and no pence), paid by the 'Navy Comm.rs' (Navy Commissioners). This amount, with earlier payments, made up Earnshaw's £3,000 Board of Longitude prize. Courtesy of Hoare's Bank, London.**

sible explanation surely requires no uncommon share of *brass*'.[62]

The watchmaker William Howells (1753-1814) complains, jealously, that both makers 'by [their] patents have obtain'd property enough to enable them to ride in their carriages',[63] and, of course, in spite of protests from Banks and Dalrymple, history relates that eventually both Earnshaw and John Roger Arnold were officially rewarded for their trouble, getting an equal £3,000 from the Board of Longitude—a result that, although resented by Earnshaw, was probably very fair, given the contribution made by Arnold's father. Earnshaw's reward is still to be seen today as an entry in his bank account under 21 March 1806, paid by banker's draft from the 'Navy Comm.rs'.[64] The Board's decision did not, however, resolve one of the essential questions of the controversy—who invented the spring-detent escapement—and perhaps that will never be finally settled.

Earnshaw continued to complain bitterly, announcing in a letter to the Board of Longitude, in true paranoid style, the names of his seven 'declared enemies'—John Arnold, John Brockbanks, Paul Barraud, Robert Pennington, Charles Haley, Peter Grimaldi, and James Peto.[65] It is a very sad fact, however, that of all his declared enemies, the greatest enemy Earnshaw ever had was himself.

★ ★ ★ ★ ★

In summary, Harrison could be said to have laid down the foundations for the successful marine timekeeper and, indeed, for all precision watches. Arnold and Le Roy, entirely independently, simplified the main concepts, and Arnold then put those concepts into a practicable, physical form. It then remained for Earnshaw to perfect the escapement and, virtually at one final stroke, create the standard marine chronometer that was to serve the nations of the world so well for the following century and a half.

The author and editors wish to thank Dr. Andrew S. Cook for his help in advising on the portraits of Sir Joseph Banks and Alexander Dalrymple.

62. MS. RGO 14/27, p. 102, Royal Greenwich Observatory Archives, Cambridge University Library.

63. *Ibid.*, p. 146.

64. Manuscript Accounts, Hoare's Bank, London.

65. These were John Brockbanks (1747?-1808?), Paul Barraud (1752-1820), Robert Pennington (1751?-1813?), Charles Haley (*ca.* 1748-1825?), Peter Grimaldi (fl. 1805-1810), and James Peto (fl. 1780-1808). Letter dated 3 July 1805, MS. RGO 14/25, p. 275, Royal Greenwich Observatory Archives, Cambridge University Library.

# James Arthur Lecture

See Cheney Figure 1 for caption
for this and previous page.

# James Arthur: Pioneer Collector and Benefactor

*Robert C. Cheney*

Robert Cheney, who lives in Brimfield, Massachusetts, is a third-generation clock-maker as well as a dealer and consultant specializing in seventeenth-, eighteenth-, and nineteenth-century clocks. He provides services to collectors nationwide and to more than 30 institutions, including Old Sturbridge Village, where he serves as Conservator of Clocks. Cheney has presented lectures on New England clocks to forums throughout the United States and Canada and has contributed articles and book reviews to numerous professional journals. He has served as Assistant Chairman of the Longitude Symposium and as President of the Boston Clock Club and of the Massachusetts Chapter of the National Association of Watch and Clock Collectors (NAWCC). He is a graduate of the University of Massachusetts, a Fellow of the National Association of Watch and Clock Collectors, and co-author of the book *Clock Making in New England, 1725-1825* (1992).

# James Arthur: Pioneer Collector and Benefactor

*Robert C. Cheney*

When James Arthur (1842-1930) left Glasgow and arrived in New York City in 1871, formal collections of antique clocks and watches in America were virtually nonexistent. Textbooks on the subject of timekeepers were predominantly of European authorship and focus and were basically technical treatises with some historical information. Late nineteenth-century American and European collectors were blazing new trails in the acquisition and study of objects with few road maps to guide them. Concrete information on the subject was primarily the spoken word, and it was only as good as the source.

Before leaving Glasgow, Arthur had trained under Professor A. S. Herschel, grandson of the renowned astronomer Sir William Herschel (1738-1822), in mechanics, metal- and woodworking, and the design of machinery. It was this experience that no doubt sparked his interest in timekeepers. As a young man, he began collecting, restoring, and constructing clocks, a passion that followed him for the next 40 years. His successful business life, through the founding of the Arthur Machine Works in New York in 1885, complemented his horological interests and united his vocation and avocation.

Like many successful industrialists of the period, Arthur traveled extensively and completed a world tour in 1897. Abroad, he found others sharing his passion for horology. On these trips, he added to his already growing collection. While the artifacts he collected provided enjoyment of ownership and many workshop projects to restore missing parts, they also served a practical purpose as primary sources of information for learning about horology.

James Arthur's clock-making activities drew heavily on his early training in metal- and woodworking and in gearing theory. It was natural that a man trained in mechanics and engaged in the business of manufacturing machinery and patent models would enjoy the challenges of making clocks and the cases that house them. His notebooks record thirteen plans for longcase clocks, and he constructed 21 cases based on these designs. He made cases of mahogany and oak and decorated some with marquetry inlay of his own design. Although not always artistically conceived or executed, his work does stand as a testament to a craftsman exploring a trade to broaden his knowledge of the larger subject of horology and case making.

**Photograph of James Arthur, 1922. From D. W. Hering, *The Lure of the Clock* (New York: New York University Press, 1932).**

His work as a clockmaker included the manufacture of complete clocks and the re-creation of missing components for items in his collection. While his "original" mechanical designs contributed little to the field of horology, they were complicated and difficult to execute. Most of his work was in brass and wood, but he also utilized his theoretical interests in the craft by experimenting with aluminum as a raw material for clocks. This new material had entered manufacturing in the late nineteenth century, and in 1897 Arthur constructed a large ten-day-duration, floor-standing, four-dial "factory" clock with hands, four sets of dial gearing, and escape wheel made of aluminum. But he concluded that aluminum had a "somewhat sticky friction" and soon abandoned its use for clock wheels. On the positive side, he proposed that aluminum had another more suitable application in temperature-compensated pendulums and constructed examples of his own design.

Throughout a twenty-year period, his intellectual pursuit of horology included numerous accounts of horological inventions published as early as 1901 in technical journals such as *Machinery, Popular Mechanics Magazine,* and the *Jewelers Circular.* In 1909, he wrote *Time and Its Measurement,* which discussed various technical, historical, and philosophical concepts of timekeeping and related topics. His academic legacy was continued after his death by a series of lectures entitled "Time and its Mysteries," which was sponsored by the James Arthur Foundation. Twelve of these were published under the same title in three volumes by New York University Press in 1936, 1940, and 1949. The reintroduction of the James Arthur Lectures by the National Association of Watch and Clock Collectors (NAWCC) in 1984 ensured the continued publication in the *NAWCC Bulletin* of contributions by leading authorities in the field of horology.

When Mr. Arthur's age required him to move to smaller quarters, he began discussions about the disposal of his sizeable collection. The importance of keeping the collection together as a tool for further study finally outweighed the option of a public sale dispersal. At the suggestion of his son-in-law, Irving H. Berg, Chaplain of New York University and later Dean of University College of Arts and Sciences, James Arthur donated his collection to New York University on December 1, 1925. In addition to the 214 clocks, 1,092 watches, and other related materials, Arthur also left a liberal bequest to "care for, enlarge and perpetuate" the collection.

The James Arthur Collection was described by its first curator, Professor D. W. Hering, as a collection of "clocks and watches by distinguished makers from the sixteenth century to the present time." The inventory included continental table and chamber clocks, English lantern and longcase clocks, French skeleton, mantle, and Buhl longcase clocks, examples from Japan, Holland, and America, some contemporary mass-produced pieces, and a large variety of watches. Arthur's clock- and case-making work was also included in the bequest. Unfortunately, despite some fine antique gems, many of the clocks were missing parts of the movement or dial, or had been significantly altered or restored. In addition, some had been collected without their cases. It is probable that Arthur considered incomplete pieces as a personal challenge of resurrection: to restore these clocks to their former glory. Indeed, for James Arthur the mechanic and engineer, the importance of the mechanical idiosyncracies far overshadowed their decorative appeal: more than half of the watches were uncased or otherwise incomplete.

While discerning collectors today sometimes grimace at a horological accumulation of this type, such collections do provide an enormous resource for those interested in technological developments in the history of timekeeping. In fact, it was generally agreed at the time of the bequest that the educational value of a gift of this type to a university was, as stated by Professor Hering, a "great asset to the study of mechanics, engineering, and the fine arts." And, in an ideal world, free of changing educational priorities, soaring institutional budgets, overworked administrators, and space considerations, Arthur's gift would have been the act of "generosity and farsightedness" described by Professor Hering.

The care and responsibility of a sizeable collection, however, is a monumental chore. After dismantling the clocks for delivery to the university, it was soon realized that the new home of the James Arthur Collection did not have enough working space to unpack and assemble many of the clocks. Six years after the bequest, many items were still packed away, awaiting reassembly. The watches, although stored with care, never were properly accessioned or systematically arranged. Despite a generous endowment, New York University soon learned the burdens of owning such a large resource. Therefore, in 1964, it transferred most of the collection, which by that time had grown to over 1,700 clocks and watches, to the Smithsonian Institution, where it remained on loan for almost twenty years.

In December 1982, under the burden of growing economic pressure, New York University announced their decision to dispose of the James Arthur Collection. The *New York Law Journal* (December 1982) announced the terms of the proposal: the university would donate specific items to the Smithsonian Institution, sell others to The Time Museum in Rockford, Illinois, and dispose of the balance of the collection at a public sale. The National Association of Watch and Clock Collectors, however, contended that the collection should remain in a public museum setting. The members of this organization were no strangers to the James Arthur Collection. Brooks Palmer (1900-1974), the third President of the Association from 1951 to 1953, had presented the James Arthur Lecture, "The Early American Clock Making Industry," in 1951. He also served as curator of the collection for ten years and was a visiting lecturer of the James Arthur Foundation. Following legal intervention, the NAWCC was finally named as the permanent custodian, receiving the remainder of the Arthur Collection in 1984 for public display and study at their museum in Columbia, Pennsylvania. The Arthur endowment and the funds generated from the sale of items to The Time Museum presumably were used by New York University for their own purposes.

In 1980, Ward Francillon, then First Vice President of the NAWCC, and other interested horological academics organized and hosted a seminar in King of Prussia, Pennsylvania, to further horological education. In 1981, in his capacity as the eighteenth President of the NAWCC, Francillon continued to emphasize the importance of education, in particular by establishing these seminars as an annual event. Therefore, when the portion of the Arthur Collection was acquired in 1984, the seminars were already established as an ideal vehicle for continuing the James Arthur Lectures, which had formerly been sponsored by New York University. The first of this new series of James Arthur Lectures was held in conjunction with the NAWCC Seminar in Hartford, Connecticut, in 1984, with Dana J. Blackwell, NAWCC Star Fellow and Vice President, presenting a lecture entitled "Horology and the Whole Man." Since that time, various distinguished individuals have been invited to deliver this annual lecture.

At the fourteenth annual NAWCC Seminar, held in association with the Longitude Symposium, the celebrated watchmaker and author George Daniels was invited to be the James Arthur lecturer. His accomplishments in horological design and the creation of unique timepieces were reason enough for this invitation, but his involvement in the restoration of some highly significant marine timekeepers made him a particularly appropriate speaker for this occasion. His main interest at present, however, lies not in the history of timekeeping, but in the survival and growth of the art of watchmaking in the future. James Arthur would no doubt have been pleased that his legacy provided the opportunity for a lecture by such an eminent master of his craft.

The author has drawn heavily on D. W. Hering, *The Lure of the Clock* (New York: New York University Press, 1932; revised and enlarged by Crown Publishers, 1963), and the reader is referred to this source for additional information on James Arthur and his collection.

## APPENDIX 1

## JAMES ARTHUR LECTURES ON "TIME AND ITS MYSTERIES," SPONSORED BY NEW YORK UNIVERSITY, 1932-84

1932 (April 29):
"Time"
*Robert Andrews Millikan*

1933 (May 4):
"Time and Change in History"
*John Campbell Merriam*

1934 (February 6):
"On the Life-time of a Galaxy"
*Harlow Shapley*

1935 (May 16):
"The Beginnings of Time Measurement and the Origins of Our Calendar"
*James Henry Breasted*

1936 (April 2):
"The Time Concept and Time Sense Among Cultured and Uncultured Peoples"
*Daniel Webster Hering*

1937 (April 9):
"What is Time?"
*William Francis Gray Swann*

1938 (April 21):
"Time and Individuality"
*John Dewey*

1939 (April 26):
"Time and the Growth of Physics"
*Arthur H. Compton*

1940 (May 14):
"The Astronomical Scale"
*Henry Norris Russell*

1941 (April 16):
"The Geologic Records of Time"
*Adolph Knopf*

1946 (April 16):
"Time and Historical Perspective"
*James T. Shotwell*

1949 (April 7):
"Developments in Portable Timepieces"
*George P. Luckey*

1951 (October 17):
"The Early American Clock Making Industry"
*Brooks Palmer*

1953 (May 15):
"From Hours to Microseconds: Three Centuries of Timekeeping Progress"
*Arthur L. Rawlings*

1969:
"The Hypothesis of Environmental Timing of the Clock"
*Frank A. Brown Jr.*

"The Cellular-Biochemical Clock Hypothesis"
*J. Woodland Hastings*

1972:
"Physics at the Origin of Time"
*R. Omnès* and *Steven Frautschi*

1975:
"Time and the Atom: Precise Measurement of Time with Atomic Clocks"
"Molecular Beam Spectroscopy with Molecules, Atoms and Neutrons"
*Norman F. Ramsey*

1978:
"Time Without End: Physics and Biology in an Open Universe"
*Freeman J. Dyson*

1980:
"Reality, Illusion and Time"
"Time and Light"
"Beyond the End of Time"
*John Archibald Wheeler*

1984:
"Symmetry Principles in Physics"
"Time as a Dynamical Variable"
"Discrete Theory of General Relativity"
*Tsung Dao Lee*

## JAMES ARTHUR LECTURES AS THE KEYNOTE ADDRESS OF THE NAWCC SEMINARS, 1984-94

1984 (October 25), Hartford, Connecticut:
"Horology and the Whole Man"
*Dana J. Blackwell*

1985 (October 17), Portland, Oregon:
"Paradigms and Clockmaking"
*Douglas H. Shaffer*

1986 (October 23), Dearborn, Michigan:
"Mark Leavenworth, Clockmaker, 1811-1835"
*Snowden Taylor*

1987 (October 22), Rochester, New York:
"The Time of Our Lives"
*David Landes*

1988 (October 20), Santa Monica, California:
"The Importance of Horology in Our Lives"
*Seth Atwood*

1989 (October 26), Washington, D.C.:
"The History of British Public Timekeeping"
*Beresford Hutchinson*

1990 (October 25), Houston, Texas:
"The History of the Watch"
*Henry B. Fried*

1991 (October 24), Ft. Mitchell, Kentucky:
"Horologists Oiled the Industrial Revolution"
*Theodore R. Crom*

1992 (October 22), Cleveland, Ohio:
"Uses of the Atomic Clock"
*Norman F. Ramsey*

1993 (November 4), Cambridge, Massachusetts:
"The Mechanical Watch in the Twenty-First Century: The Renaissance of the Mechanic"
*George Daniels*

1994 (October 28), Toronto, Ontario:
"Horological Ephemera, Its Variety, Availability, and Importance"
*David Penney*

1995 (October 26), Harrisburg, Pennsylvania:
"Clockmaking or Timekeeping"
*Douglas H. Shaffer*

# Watchmaking in the Twenty-First Century: The Renaissance of the Mechanic

*George Daniels*

George Daniels is a watchmaker, author, and horological consultant. His watches, especially designed to improve long-term timekeeping, are among the most highly prized by connoisseurs of the art of precision watchmaking. For his contributions to horology, Daniels was awarded the Victor Kullberg Medal by the Stockholm Watchmakers' Guild (1977), the Tompion Gold Medal of the Worshipful Company of Clockmakers (1981), the Gold Medal of the British Horological Institute (1981), the M.B.E. by Queen Elizabeth II (1982), the Gold Medal of the City and Guilds Institute of London (1990), and an Honorary Degree of Doctor of Science by the City of London University (1994). He has served as President of the British Horological Institute (1980) and Master of the Worshipful Company of Clockmakers (1980).

Daniels started his professional career in horology after leaving the army in 1947. In 1956, he began to specialize in the restoration of historical watches. He produced the first watch of his own design in 1969. His published works include *Watches* (with Cecil Clutton, 1965), *English and American Watches* (1967), *The Art of Breguet* (1975), *Clocks and Watches of the Worshipful Company of Clockmakers* (with Cecil Clutton, 1975), *Watches and Clocks in the Sir David Salomons Collection* (with Ohannes Markarian, 1980), and *Watchmaking* (1981). When not engaged in horological pursuits, he enjoys vintage cars, fast motorcycles, opera, and Scotch whisky.

# Watchmaking in the Twenty-First Century: The Renaissance of the Mechanic

*George Daniels*

I am honoured to find myself today among so many distinguished historians, who will undoubtedly offer much-anticipated new and exciting information on one whom we have come to understand is the greatest horological figure in the history of the timekeeper.

Alas, by comparison, I can only hope to entertain you briefly and at best arouse your curiosity in one particular aspect of my horological interests—but one that, in the past twenty years, has found many new supporters who will carry it enthusiastically into the twenty-first century. The watch has occupied my mind from my very earliest recollections, and the present state of the art is more interesting than ever before in my lifetime.

The twenty-first century will soon be with us. Each in his own profession will, no doubt, see it as a new threshold to an inspiring start for what will undoubtedly be the most momentous, and perhaps traumatic, century in the known history of the world we live in.

Indeed, most modern scientific progress seems to make people unnecessary, so that they are unemployable and become discontented. As a consequence, the buzzword is leisure. Leisure activities are the panacea and will be ever more popular in the twenty-first century. The problem seems to be what to take up: languages—but anyone worth speaking to speaks English; carpentry—out of the question—uses up trees; dressmaking—too transitory—off with the old, on with the new; home mechanics—the washing machine is disposable and the motor car will be old hat by the middle of the century. But wait! The best watches are mechanical and the most resourceful people are watchmakers. Remember Maskelyne's oath of contempt for the watchmakers: 'By God, this will give the mechanics something to crack their teeth on'. But the mechanics cracked him. And 'mechanicking' can crack the anticipated twitches of the twenty-first century. Of course, no one will need a watch in the twenty-first century because, there being nothing needing doing, there will be no need for timetables. But everybody likes a watch. As one maker puts it, it 'tells you something about yourself'. And it is a wonderfully tranquil intellectual and manipulative occupation that will bring great pleasure to its twenty-first-century practitioners.

Yes, I would advise watchmaking to carry you through the next century. The art is already four centuries old, and there is still some way to go so that it remains competitive, but in a friendly way.

Watchmaking in England started a little before 1600 and by 1700 had established a strong and progressive market for good-quality watches. By 1800, the principles of the British precision watch were universally acknowledged, and names such as Harrison, Mudge, Arnold, and Earnshaw, all schoolboy heroes to me, dominated the horological world. By 1900, the industry was in serious decline as a consequence of exploitation by quantity production methods of English inventions and designs. The British artist-craftsmen could not adapt to the new methods, and by 1930, this beautiful and unique art had disappeared. Only in Switzerland, up to about 1950, could a hand-finished watch of a type desired by a connoisseur be purchased. Although beautifully made, such watches lacked the individuality that had characterised the earlier periods, and collectors and connoisseurs turned again to older, more individually styled watches.

The introduction in the 1950s of the electric watch had already led me to take a closer interest in antique watches, and by 1962, I was established as a restorer and specialised in the work of Breguet. In the mid-1960s came the euphoria and extravagant claims for the electronic watch that was to sweep away the old-fashioned mechanical watch and replace it with the amazing quartz timekeeper. I felt great indignation that my lifelong passion, the mechanical watch, was to be so casually set aside for what was, in those days, an inadequate, unproven device marketed by people who had never had any interest in the watch except as a money spinner and who could comprehend

**Figure 1. Dial of a minute-repeating pocket watch, completed in 1987, with subsidiary indications for the phase of the Moon, the temperature (in centigrade), and a perpetual calendar showing the day of the week, the date, the month, the leap year cycle, and the phase of the Moon.**

**Figure 2. Movement of the pocket watch, showing the upper plate (back plate) with a dial showing the days and length of each month and sectors indicating the equation of time (the difference between Sun time and mean time) and the reserve of winding (the number of hours that the watch will continue to run before it needs to be rewound). In addition, this watch incorporates a tourbillon with a co-axial escapement.**

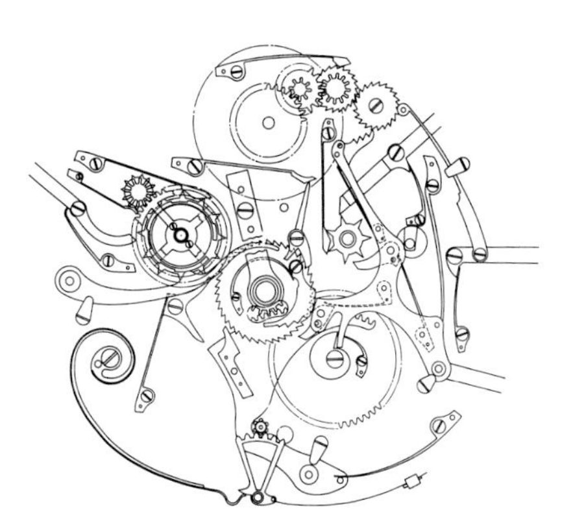

**Figure 3. Drawing for the perpetual calendar mechanism used in this watch. Unlike other perpetual calendar mechanisms that can take up to three hours to change, this arrangement, which was designed about 1985, allows all long and short months and leap year to change automatically at midnight. Drawing by George Daniels.**

nothing of the romance of the history of the watch or of its technical, intellectual, aesthetic, amusing, and useful qualities.

There was not the slightest doubt in my mind that nothing could ever sweep away the mechanical watch. But it was obviously going to have a hard time in the factories, and a large part of the Swiss industry, which had never made any attempt to improve the basic mechanism of the watch, was going to be put out of business.

Every comment about the amazing quartz watch inevitably pointed to comparison with the mechanical watch, so that it was receiving attention that otherwise it would not have had. Some of this interest was of a nostalgic variety and set the minds of collectors thinking again of the mechanical watch and missed opportunities to purchase unique pieces.

So convinced was I by now that this was the moment to restart the artist-craftsman horological industry in London that I turned down the offer of proprietorship of the Breguet Company in Paris simply because, in my more light-headed moments, I thought that Daniels-London would sound better than Breguet-Paris. Twenty-five years have passed since then, and I have never regretted my decision.

The book *Watchmaking,*[1] written to encourage a younger generation to take up the art, has, in fact, done so. I now have correspondents in almost every country in the civilised world, all of whom wish to establish themselves as watchmakers.

All have found the tourbillon watch with detent escapement irresistible, just as I found it irresistible when starting my first watch in 1967. The tourbillon is said to be very difficult to make and to be the ultimate portable timekeeper, giving a great thrill to its maker. But it is not more difficult; it offers the same difficulties as any other watch but merely takes longer to make. Now that the new generation of artist-craftsmen is blossoming, it is not good for everyone to make the same thing.

Whenever possible, I repeat this advice to the Swiss makers who are making serious efforts to set the watchmaking industry into another decline by producing in quantity, with very sophisticated equipment, those pieces that were once esteemed as the province of the artist-craftsman. These over-complex pieces and *grande complications*, increasingly referred to by cynics as 'Fruit-Salads', will soon be flooding the markets so that, being very expensive, they will become unsaleable. And, being constructed on conventional lines, they will need frequent and expensive servicing by those who can afford to buy them and who will naturally be cross about this when they find out. What is required is not more complexity and complication to show how clever the maker is, but more imagination and innovation to attract the intelligent purchaser.

Considering that the twenty-first-century watchmaker will need to make every component, including the case, dial, and hands, in his own workshop without outside assistance, it may be thought too much that his work is expected also to have some attractive quality that makes it desirable to own. But this is what he must achieve. No one needs an expensive hand-made watch—but most people would like to own an original and improving concept made by an artist at his craft. This was the strength of the handful of artists who developed the watch during the past 400 years.

I am reminded of the occasion of a feast at Sydney Sussex College in

1. George Daniels, *Watchmaking* (London: Sotheby Publications, 1981).

Cambridge, given to welcome an international gathering of astronomers. I was flattered to be admitted to this distinguished and extravagant party. I knew that during the course of the evening we would discuss the contribution of the timekeeper to astronomy and I would be expected to say something intelligent about my work, which, earlier that week, had received an award. Suddenly, I was challenged by a young astronomer, who earlier had appeared to be friendly. 'Why do you make mechanical watches', he demanded, 'when quartz watches are better?'

At that time, the quartz watch was not in its present state of advanced development, and I was able to explain why mechanical watches constructed to my designs had certain advantages. At the same time, without making any claims, I managed to leave him with the impression that I had certain other aces up my sleeve for future development. Thus, the honour of my passion was saved for the moment. The incident, nevertheless, made me ponder on the future of the mechanical watch. True, it was great fun making them and people enjoyed owning them, but that was not enough. The intellectual and aesthetic charm of the mechanical watch had been overlooked by a mere scientist. Yet, from his standpoint, the implied criticism was valid and the task should be to develop the mechanical watch, so that it could hold its own in terms of its *raison d'être* and at the same time offer a charm that mere quartz timekeeping could not attain. Exactly what this charm is may depend upon the individual. What is certain is that the Swiss industry, recognising the intrinsic appeal of the mechanical watch, now takes it seriously and has recently reinvented it (I'm sure they won't credit me with giving them this advice some ten years ago), and that great quartz invention, the Swatch, has suddenly turned into a mechanical, automatically wound watch. Thus the wheel has turned full circle, and the mechanical watch is back.

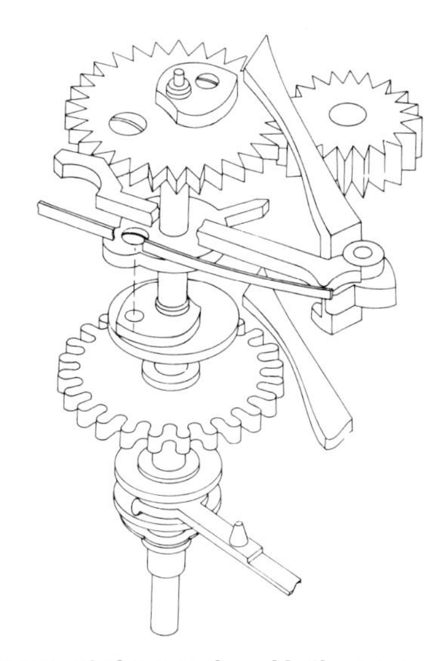

**Figure 4. Wrist-watch, completed in 1990. The sector near the top of the dial indicates the reserve of winding. The seconds ring on the left serves as a chronograph, showing minutes and seconds.**

**Figure 5. Upper plate (back plate) of the wrist-watch, showing the tourbillon with the co-axial escapement and part of the chronograph mechanism.**

**Figure 6. In order to make the watch as thin as possible, the chronograph mechanism of this watch, instead of being spread over the upper plate (back plate), was designed without centre seconds to fit into an unwanted space alongside the movement. The mechanism includes a device for stopping and starting the seconds hand and has a separate operation for zeroing the seconds- and the minute-recording hands. Drawing by George Daniels.**

This resurgence notwithstanding, the problems of precision and reliability have remained a challenge to makers of mechanical watches. During the past 25 years, while the quartz watch attained a high degree of precision and reliability, developments made by the watchmaking industry to improve the accuracy of mechanical watches were limited to their performance in the short term, not in the long term. By attention to mainsprings (wound automatically through the movements of the wearer), wheel trains, and escapements, the amplitude of the vibration of the oscillator has been raised throughout the daily run. This has reduced the isochronal errors of the escapement and thus the short-term variations in daily rate. Improvements in temperature correction and in the orientation of the fixing of the balance spring have further improved the short-term daily rate.

As the result of these developments, a modern, high-grade mechanical watch can now maintain a rate within a few seconds per day, depending upon the use to which it is put. Freshly cleaned and oiled, it can maintain a rate of within one second per day in the short term under observatory conditions of test. The period will depend partly on the size of the watch: the smaller the watch, the shorter the running time between overhauls. A watch of some 30 mm. diameter will need cleaning at one- to one-and-a half-year intervals in order to prevent a variance from this rate, while a watch of, say, 15 mm. diameter and smaller will need more frequent attention. Nevertheless, because of the problems of lubrication, the mechanical watch, even after 400 years of development, has been unable to compete with the long-term reliability of the quartz watch.

The escapement universally used in the mechanical watch, the detached-lever escapement, was invented in 1754 by Thomas Mudge (1715?-1794). Its single fault is the necessity for lubrication at the teeth of the escape wheel: the viscosity of the oil varies with temperature and age to produce a constantly changing rate. Yet this fault lies at the heart of the problem, and the search for a more consistent oil has occupied the minds of horologists since the birth of the precision timekeeper. Improvements have been marginal, however, and have never eradicated the problem. Thus it occurred to me that it would be better to eliminate the necessity for the oil by the introduction of a new escapement. The decision to begin experimenting to produce an oil-free escapement was casually taken considering that the worldwide industry had not even contemplated the possibility. Working alone, I had only myself to please, and if I failed no one would know! From the mid-1970s, therefore, I used only escapements of my own design. The principal feature was the elimination of friction so that oil was not needed. Conventional materials were used, as in all other escapements, because these have been tried and tested throughout the history of the watch and have proven their stability and durability.

It should be noted that escapements such as the spring-detent or so-called chronometer escapement, invented in the eighteenth century, are capable of a very exact performance because they require no lubrication. But such escapements are delicate and give impulse to the oscillator only at alternate vibrations, so they must be used with care to prevent damage and setting. An escapement for modern use must, like the lever escapement, give impulse at each vibration of the oscillator and be sufficiently robust to withstand the often violent agitation of modern use.

Figure 7. Space Traveller's watch, completed in 1982, made to commemorate the American landing on the Moon. The dial indicates sidereal time on the left and mean solar time on the right. Four sectors provide additional information: the equation of time is shown at the top; the phase of the Moon can be seen in the sidereal time chapter ring, and its age is indicated in the sidereal seconds ring; the sector in the mean solar seconds ring shows the date. This watch is also fitted with a centre-seconds chronograph that will indicate either sidereal or mean solar time.

Figure 8. Movement of the Space Traveller's watch, showing the chronograph mechanism and the escapement. The solar-to-sidereal gearing ratio is 1 : 1.002737924. This represents an annual gain of only 0.4 seconds.

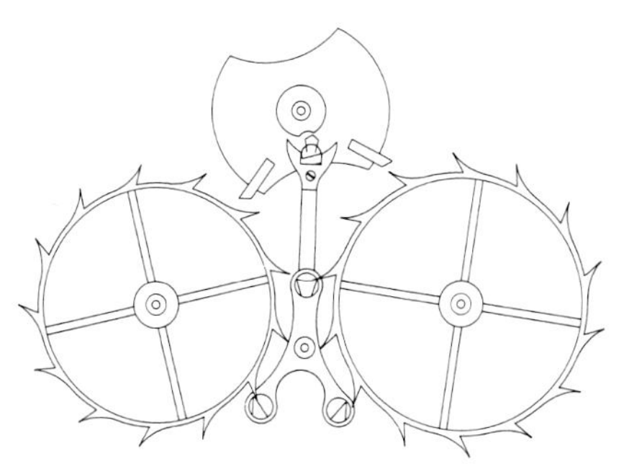

Figure 9. Drawing for the independent double-wheel escapement as used in the Space Traveller's watch. This escapement, conceived at the time of the first lunar landing in 1969 and designed for this watch in 1982, incorporates two independently driven wheels engaging a single oscillator. Drawing by George Daniels.

My first experimental escapement was the independent double-wheel, so called because each escape wheel is separately driven by its own train of wheels terminating in a single oscillator. Tested in daily use during a 30-day trip to Tokyo and back, it arrived home a little less than one second in error. The escapement is complex, expensive to produce, and unsuitable for commercial development, but it had shown the merit of a fresh approach to the design of the mechanical watch. The final watch completed with this escapement is shown in Figures 7, 8, and 9. Named 'The Space Traveller', this watch encapsulates the eighteenth-century watchmakers' wish for a precision mean-time chronometer that could be regulated by the stars without the necessity for tedious arithmetical calculations. It was constructed as an intellectual exercise to include the traditional units of time used in navigation and commemorates the great U.S. voyage of exploration to the Moon.

The success of this escapement was an encouragement to find ways of simplifying and reducing the mechanism to fit into a wrist-watch. Ideally the solution would be an escapement with only one wheel, but this was not possible. When two pivoted components are in peripheral engagement, as with the chronometer escapement, they are turning in opposite directions. Such escapements do not need lubrication, but the impulse can be given to the oscillator only when the direction of its rotation is opposite to the rotation of the escape wheel. The escape wheel can rotate only in one direction, but a lever engaging the balance, as in the lever escapement, will oscillate in unison with and in the opposite direction to the balance. Therefore, when the direction of the balance and the escape wheel are the same, the second impulse must be given via the lever. But the impulse must be radial, as with the escape wheel to balance impulse, in order to avoid sliding friction.

With this in mind, the co-axial escapement shown in Figure 10 was devised. The two wheels on the same axis are necessary to ensure tangential locking of the wheels after the impulse is delivered. The large wheel impulses the balance directly, as with the chronometer escapement; the small wheel

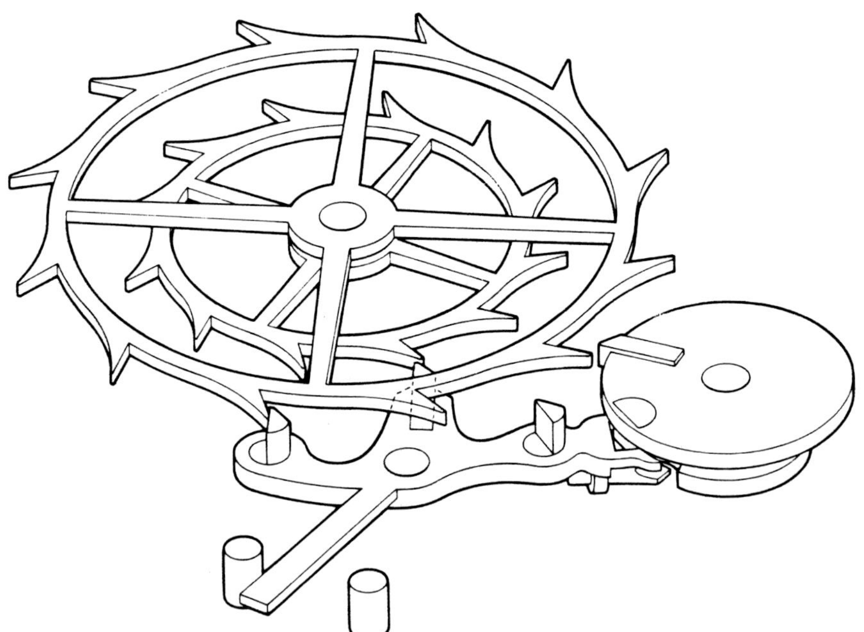

**Figure 10. The first version of the co-axial escapement, designed about 1975. Impulse is given to the balance both directly and via the lever at alternate vibrations without necessity for lubrication. Drawing by George Daniels.**

impulses in the same manner via the fork and roller action, as with the conventional lever escapement. This escapement was fitted to an Omega chronograph wrist-watch. It was found that the balance amplitude was low as a consequence of the large diameter of the escape wheels. The inertia of these components will increase as the square of their radii. This could have been anticipated, but there are many lessons to be learned from a new construction that only practical experiment can reveal. The original total arc of 280° with an escaping angle of 54° gave a ratio of 5.18 : 1. The new, reduced, arc of 200° with an escaping angle of only 36° gives a more favourable ratio of 5.55 : 1. Thus the losing error of the escaping angle is reduced and, with an increase in total arc to 280°, would be further, relatively, reduced to produce a more isochronous escapement.

**Figure 11. Daniel's extra-flat co-axial escapement, designed for modern ultra-thin wrist-watches, was first made in 1982 especially to fit into a 2 mm.-thick Patek Philippe movement. This escapement was used both in the chronograph wrist-watch (Figures 5 and 6) and in the double-sided wrist-watch (Figure 13). Drawing by George Daniels.**

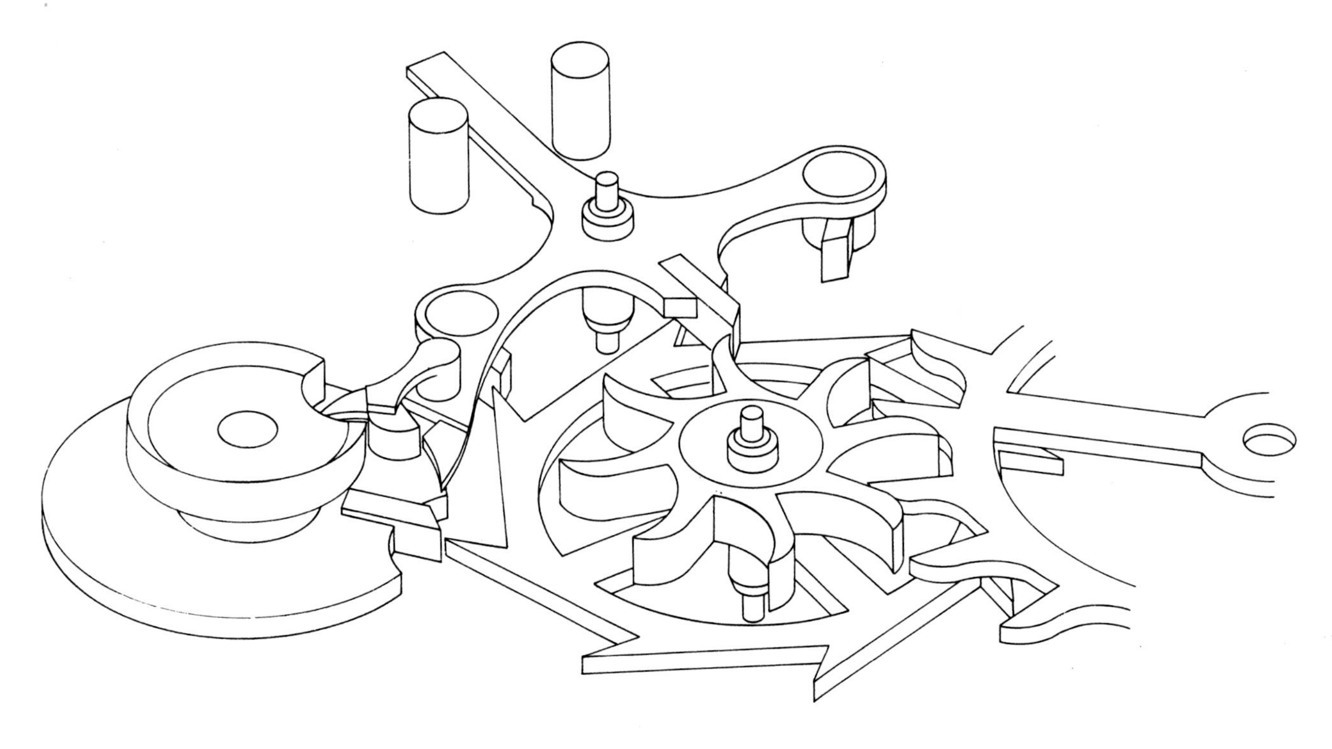

**Figure 12. Double-sided wrist-watch, made in 1991. The front dial includes a subsidiary dial indicating the reserve of winding. The movement casing is hinged in its outer ring, in order to view the tourbillon and the dials showing the day of the week and the date.**

The escapement was redesigned by reducing the number of teeth of the escape wheel from twelve to eight. This reduced the inertia and gave greater intersection of the teeth with the impulse pallets to increase the overall efficiency. At the same time, the escape wheel pinion was modified to serve both a pinion and small escape wheel. By thus dispensing with the conventional escape wheel pinion, the height of the components was reduced so that the escapement could be fitted into a modern thin movement. The arrangement is shown in Figure 11, where the point of the pinion leaf is engaging the lever pallet while the flank of the leaf is engaged to rotate by the train wheel, the latter with half-ogive teeth to reduce inertia. In this form, the escapement has fulfilled all my expectations, with prototypes that have now been running for fifteen years without attention.

All of my own watches manufactured during this period have been fitted with this escapement, which has demonstrated its adaptability, reliability, and constancy of balance amplitude during many years without attention. Now at last the industry is taking a closer look at its potential for use in series production. Prototype escapements employing this system have been fitted by me to a variety of conventional watches, including Omega, Zenith, Patek Philippe, and Rolex. I am optimistic for its future and the improved long-term reliability of the mechanical watch.

It is now 26 years since I completed by first watch. Many have been made since then. All of the components, including the cases, dials, hands, wheels, pinions, escapements with their jewels, springs, screws, and subsidiary mechanisms, have been made in my workshop without assistance—which is to say also without interference. It has been a most enjoyable pastime, if at times an anxious one. But for relaxation one can turn one's attention to such mechanisms as the subsidiary chronograph (Figures 4 and 5), or the perpetual calendar (Figure 3), which gives instantaneous change for all functions at midnight, or the complex minute-repeating watch (Figure 1).

These are pleasant diversions, and I would commend such exercises to the young twenty-first-century mechanic who is building his workshop and

Figure 13. George Daniels, wearing his double-sided wrist-watch. This watch is shown in Figure 12.

developing his ideas. But while enjoying his occupation, he must bear in mind the duty of every professional to endeavour to make a contribution to his art. He will be expected not only to fulfill the historic, technical, intellectual, aesthetic, amusing, and useful qualities of yesterday's watch, but also to beat its performance. No other artist-craftsman's industry is so demanding. As to how he should set about this task is a matter for the individual to decide, but my own methods may offer suggestions, or at least amuse you.

All illustrations courtesy of George Daniels.

# Appendices

See Appendix C Figure 8 for caption for this and previous page.

APPENDIX A

# Acknowledgements

# I. Subscribers to this volume

S. J. Allen
John and Pol Andrewes
Comdr. Richard and Sue Andrewes
*The Antiquarian Scientist*
Adam Jared Apt
Margarida Archinard
Jim Arnfield
S. Pearson Auerbach, M.D.
James F. Austin, M.D.

Dr. William Back
Tom Bales
Charles Bascom
Douglas A. Bateman
Frank H. Beberdick
Rear-Admiral François Bellec
Steven and Marsha Berger
J. L. Berggren
Theodore R. Beyer
Arthur Bjornestad
Dana J. Blackwell
Harold E. Brandmaier
Margaret Brandon
Comdr. William R. Bricker
William B. Bromell
Stephen A. Brownell
Peter H. Buckley
Louis Buda
Max A. Buehlmann
Steve Burgess
Frederick Byl

Arthur E. Capstaff Jr.
Roland P. Carreker
Harold Cherry
David A. Christianson
Jim Cipra
Dewey Clark
Clockroom, British Museum
Martin L. Cohen
James P. Connor
Dr. Andrew S. Cook
Dr. Edward Werner Cook
Lewis Anderson Cook
Len Corwin
William V. Cox
Carlton Cranor
Ted Crom
David J. Cuff

Marshall Damerell
Dr. Alun C. Davies
James H. Davis
Capt. Ignacio de Isusi
Robert J. Deroski
Frank A. Dickof
Frederick A. Dieter
James M. Dowling
Nan Doyle
Julien Dubuc
Prof. Kermit Duckett

Michael Edidin
Evan Edwards
Robert S. Edwards
Christopher and Gill St. J. Ellis
Moritz Elsaesser
André M. Englebert
Jim and Mildred Espy

Dr. Stephen D. Fantone
H. W. Farrer
Werner Faust
Dr. George Feinstein
Timothy F. Flower
Robert Forsberg
Gary Fox
Ward and Mary Francillon
J.T. Fraser
Norman Friedman

Eliza N. Garfield
D. J. I. Garstin
Giovanni Gasparini
Alan J. Genteman
Willard H. Gilmore
Robert M. Girdler
Dr. Richard Gliedman
Dr. David S. Goodman
Ian Graham
William M. Graham
John Grass
Roger P. Grimshaw
Marybess Hudgins Grisham
Dr. Robert D. Gross
Richard F. Gutow

Paul E. Hackamack
Elton W. Hall
Jon Hanson
John S. Heiden
Alan W. Heldman Sr.
Stephen B. Helfant
John Hiew Watch & Clock Repair Centre
Jeffrey R. Hill
B. J. Hobart
Donald R. Hoke
James M. Holland, M.D.
The Horolovar Company
W. M. Huegel
C. J. Hurrion

Peter Ifland
Dr. Fred H. Ingram

Richard L. Jacobs
Andrew James
Prof. and Mrs. Malcolm I. V. Jayson
Graham H. Jeffries
Chester Henry Johnson
Daniel W. Jones Jr.
Everett R. Jones
Johann M. Jorgo

Bertram Kalisher
Richard L. Ketchen
Roger L. Kimmel
J. A. Knobbout

Ronald F. Krasovec
Joseph P. Kuechle

William C. Lance
Karl J. Langer
Jerry Laux
M. C. Leach
Col. George E. Lear
Jud Leech
Frederick R. Levy
Dr. Samuel Licata
Dave Livingston
Erich A. Lorenz
John M. Luykx
Robert T. Lynch

Luis Marden
Ernest Martt
Otsenre E. Matos, M.D.
Nancy L. Maull
John E. McBarron
Robert W. McGillion
Thomas W. McIntyre
W. F. McLean
John W. Meisner
James Metcalf
Paul Middents
Phil and Stella Middleton
Viscount Midleton
M. J. Mintz
Arne B. Molander
Michelle Monto
Vernon Alan Morris
W. F. J. Mörzer Bruyns
James Moss Clockmakers, Inc.
Stanley A. Mueller Jr.
Fortunat F. Mueller-Maerki
John and Susan Murdoch
Brian D. Murphy
Louise Muse

NAWCC, Inc.
David and Carol Newsom
Roderick and Joan Nordell

Jim O'Boyle
Albert L. Odmark
George E. Orr

Keith Palmer
Linda Parsonage
Michael Payne
Charles R. Peck
Robert R. Perron
Christian Pfeiffer-Belli
Dia Philippides and Willem Bakker
Peter D. Pinch
Gunnar Pipping
Stephen and Susan Pitkin
Louise D. Pittaway
Irving Polk
Bill and Kathy Pritchard
Charles F. Prosser

Robert D. Reymond
Charles V. Reynolds Jr.
Robert F. Richards
Elise T. Richardson
Maurice Richesson
Jill Robinson-Terry, M.D.
Edward Rodley
Dan Rootham and Pippa Andrewes
M. Eugene Rudd
Kathy and Dr. Martin D. Ruddock, M.D.
G. Price Russ III
William M. Ryan

Jean-Claude Sabrier
Philip Michael Sadler
Wilf Saver
Harvey Schmidt
Charles H. Schwartz
Peter Schweitzer
David B. Searles II
Lawrence A. Seymour
Douglas H. Shaffer
John and Ann Shallcross
Sid Shapiro
Allan F. Shaw, D.D.S.
Dr. E. P. S. Shellard
William K. Sieber, M.D.
Leonard and Lynn Simon
Antoine Simonin
Clyde Simpson
L. E. Sizemore
Raymond Slagle
Michael Smith
Thomas and Sonya Spittler
Richard Stenning
James J. Stewart
Jim and Marge Stewart
Robert and Christine Stewart
Bernhard Stoeber
Frank S. Streeter
Gisela Striker
Martin Swetsky
Geoffrey H. Sykes
Valerie M. Sykes

Bernie Tekippe
Peter von Thüna
The Time Museum
Patrick and Peggy Timlin
W. David Todd
William R. Topham
Timothy Treffry
Donald L. Treworgy
Norton Tyzack

USS Constitution Museum

John and Dorothy (Mastricola) Verstynen
Clare Vincent
William A. Vint, M.D.
George A. Violin
James R. Voelkel

Edward F. Wale
Edward P. Wallner
Ray Watson
Roderick and Marjorie Webster
Stephen F. Weinstock
Leonard Weiss
Marvin E. Whitney
Eric Whittle
Philip Whyte
James M. Wideman
Frederick Widmer
Donald B. Wing
Robert C. Wing
Eric W. Wolf
Robert Wolf
Dr. C. L. Woodbridge
Eugene B. Woodbury
Thomas Woodbury
Richard M. Wright
Gerard Wynen

# II. Contributors of information and materials

Adler Planetarium and Astronomy Museum, Chicago:
Kate Desulis, Bruce Stephenson, Roderick and Marjorie Webster
American Historical Association, Washington, D.C.
Pippa Andrewes
Art Resource, New York: Alison Gallup
Ashmolean Museum, Oxford: Anne Steinberg

Jonathan Betts
Bibliothèque Mazarine, Paris
Bibliothèque Nationale de France, Paris: Catherine Hofmann,
Monique Pelletier
Boston Public Library: Eugene Zepp, Roberta Zonghi
Lord Brabourne
Bridgeman Art Library, London: Katrien Demoor, Rachel Dunk,
Eleanor Harris, Jemma Wallace, Emma Williams
British Horological Institute, Upton, Notts., England: Eliot Isaacs
The British Museum, London: John Leopold, David Thompson
Giuseppe Brusa

Cambridge University Library, Royal Greenwich Observatory Archives:
Adam Perkins
Bruce Chandler
Christie's, London
Civica Biblioteca Aprosiana, Ventimiglia, Italy
Clements Library, University of Michigan, Ann Arbor: David Bosse
Clock and Watch Museum Beyer Zurich: Theodore Beyer
Collection of Historical Scientific Instruments, Harvard University:
Christopher Cannon, Ana Dujmovic, Maame Ewusi-Mensah,
Robert Forsberg, Lewis Law, Peter Pinch, Martha Richardson,
Amita Shukla, Hans Stöhrer
Andrew Cook
Courtauld Institute of Art, London: Barbara Hilton-Smith,
Sarah Wimbush

Fondren Library, Rice University, Houston
Franciscan University of Steubenville, Ohio: Michael Skinner

Geraldine Gardner
Gemäldegalerie, Berlin (Dahlem): Rainald Grosshans
Gemeentemuseum, The Hague: Peter Couvée
Germanisches Nationalmuseum, Nuremberg: Rainer Schoch
J. Paul Getty Museum, Malibu, California: Jacklyn Burns, Tracey Schuster
George Gibson

Owen Gingerich
Bernard R. Goldstein
Stephen F. Grohe, Inc.

Haags Historisch Museum, The Hague: Robert van Lit
Harvard Law School Library: David Ferris, David Jenkins
Harvard Map Collection, Harvard University: David Cobb, Joseph Garver
Harvard University Art Museums: Mary Clare Altenhofen, Marjorie Cohn, Sandra Grindlay, Clare Rogan
Peter Hastings
Hereford Cathedral, England: Lindsey Mundy, Joan Williams
David Hirst
Historic Urban Plans, Ithaca, New York: Sara Ekholm, Julie Johnson
Hoare's Bank, London
Houghton Library, Harvard University: Anne Anninger, Mariana Oller, Roger Stoddard
Derek Howse
The Huntington, San Marino, California: Lisa Ann Libby

Istituto e Museo di Storia della Scienza, Florence: Mara Miniati, Franca Principe

John Carter Brown Library, Providence, Rhode Island: Susan Danforth; Richard Hurley (photography)

Andrew L. King
Knatchbull Portrait Collection, England: Barbara Tolhurst; Arthur Nash (photography)

David S. Landes
Jean Le Bot

The Mariners' Museum, Newport News, Virginia: Benjamin Trask
Mathematisch-Physikalischer Salon, Dresden: J. Schardin
Barbara McCorkle
Andrew McKinney
Willem Mörzer Bruyns
Andrea Murschel
Musée Boucher de Perthes, Abbeville, France
Musée International d'Horlogerie, La Chaux-de-Fonds, Switzerland: Catherine Cardinal, Minnie Chatelain
Musée National des Arts et Métiers, Conservatoire National des Arts et Métiers, Paris: Frédérique Desvergnes
Museo Nacional de Ciencia y Tecnologia, Madrid: Amparo Sebastián
Museum Boerhaave, Leiden: Peter de Clercq

National Gallery, Washington, D.C.: Missy Beck, John Hand
National Library of Australia, Canberra: Sandra Martens
National Maritime Museum, Old Royal Observatory, Greenwich: Jonathan Betts, Chris Gray, Lindsey Macfarlane, Alan Stimson

National Portrait Gallery, London: Paul Cox, Tim Moreton,
Jill Springall, Kai Kin Yung
National Trust, England
New York Public Library: Miriam Mandelbaum

Observatoire de Paris
Öffentliche Kunstsammlung Basel, Kunstmuseum: Charlotte Gutzwiller;
Martin Bühler (photography)

Palacio Real, Madrid: Ramón Colón de Carvajal
Patent Office, London
David Penney

Rand McNally: Karen Horcher, Edward C. McNally
Rijksmuseum-Stichting, Amsterdam
Royal Astronomical Society, London
Royal Society, London: Sandra Cumming, Sheila Edwards,
Mary Sampson
Royal Swedish Academy of Sciences, Stockholm: Olov Amelin

Jean-Claude Sabrier
Scala Istituto Fotografico Editoriale, Florence: Valentina Bandelloni
Science Museum, London: Neil Brown, Peter Fitzgerald,
Wendy Sheridan
Science Museum/Science and Society Picture Library, London:
Sarah Connolly, Ivor Gwilliams, Cathy Houghton, Venita Paul
Dava Sobel
Société d'Ingénierie et de Microfilmage, Neuilly-sur-Marne, France:
F. X. Labrador
Sotheby's, London: Tina Millar
Gisela Striker
Noel Swerdlow

The Time Museum, Rockford, Illinois: Karon Anderson,
Patricia Atwood, John Shallcross, Dorothy Mastricola Verstynen
Timothy Treffry
Trinity House, Hull, England

Uffizi Gallery, Florence

Clare Vincent

William Andrews Clark Memorial Library, University of California,
Los Angeles: John Bidwell
David Woodward
Worshipful Company of Clockmakers, London: Sir George White
Wuppertaler Uhren Museum, Wuppertal, Germany: Jürgen Abeler

Earl of Yarborough

APPENDIX B

# The Longitude Symposium

# I. Background to the conference

The idea for the Longitude Symposium had its roots in the bicentennial exhibition on the life and work of John Harrison held at Greenwich in 1976, the same year in which the United States celebrated 200 years of independence from Great Britain. Exhibitions commemorating both events were planned at the National Maritime Museum at Greenwich. Those involved in the Harrison exhibition[1] gathered together all of the then-known longcase clocks and marine timekeepers that Harrison had made, apart from one clock, the Brocklesby Park turret clock, that was too large to move. Being the first time that all these pieces had been in the same location at the same time, it was an important occasion for those interested in Harrison's work. Such an opportunity for comparing so many of his clocks in one place at one time had not arisen before.

The main purpose of the exhibition was to bring Harrison's contributions to the attention of the general public, a wonderful opportunity, one would have thought, for the National Maritime Museum. Alas, it was not to be the event that some of us had hoped for. Politics intervened. No promotion of any other exhibition was allowed to take place during the period of the American bicentennial exhibition; not even directional signs to the Harrison exhibition were permitted until it was announced that the former Prime Minister, Harold Wilson, was to come on an official visit with the Clockmakers' Company. The next appropriate occasion to celebrate Harrison's life and work—the tercentenary of his birth—was seventeen years away, which seemed like a long time to wait.[2] Valerie Finch, David Penney, and I made preparations for a book, but, as happens with many promising ideas, the work through which we earned our livings quickly began to absorb all of our time and energy, leaving only folders of information that we hoped one day would serve a useful purpose.

Opportunities at The Time Museum in Rockford, Illinois, soon took me away from England. The Time Museum's remarkable collection had been established about 1970 by Seth Atwood, to whom I had been introduced by Derek Howse. In addition to building a comprehensive collection illustrating the history of timekeeping devices, Atwood's vision included a complete

1. The staff from the Old Royal Observatory involved in the exhibition included Derek Howse (at the time Director of Astronomy and Navigation and hence my immediate superior), his assistant Valerie Finch, and me. Also assisting were Col. Humphrey Quill in an advisory capacity and David Penney, who provided illustrations.

2. A conference on Harrison's work, in fact, took place in Norwich, England, on August 27-28, 1988. This concerned the findings of the Harrison Research Group and Martin Burgess's work in making the Gurney clock, a clock based on Harrison's principles. For further information on this work, see Martin Burgess's paper in this volume.

Four years earlier, in 1984, to celebrate the centenary of the establishment of the international prime meridian at Greenwich, the National Maritime Museum had organized a conference called the "Longitude Zero" Symposium. Although the Longitude Symposium obviously covered some of the same material as this earlier conference, "Longitude Zero" differed in its emphasis as well as in the period it covered. Furthermore, whereas all the speakers at the Longitude Symposium were asked to contribute to a planned program on specific subjects that related to their field of expertise, "Longitude Zero" involved a general call for papers. The proceedings of this conference were published under the title *Longitude Zero 1884-1984* in a special issue of *Vistas in Astronomy*, vol. 28, parts 1/2 (1985), Stuart R. Malin *et al.*, eds. (Oxford, England: Pergamon, 1985).

catalogue of his collection that, had it ever been realized in full, would have been an outstanding encyclopedia of the history of time measurement. In any specialized field, meeting one respected individual brings one quickly into contact with others. As curator charged with the task—indeed, the privilege—of putting Atwood's ambitious plan into effect, I soon came into contact with collectors, dealers, and scholars from all over the world who, after traveling thousands of miles to reach a rather remote corner of northern Illinois, stayed in the hotel in which the museum had been located. This experience of meeting and getting to know so many specialists in various branches of horology was of immense value during the organization of the Longitude Symposium, because through the lasting friendships established at The Time Museum, it took only a few days of telephone calls to arrange the majority of the program. In 1987, I left The Time Museum and took up my present appointment at Harvard University. Working with the Collection of Historical Scientific Instruments, a resource that owes its existence to a remarkable collector named David Pingree Wheatland (Harvard Class of 1922),[3] broadened my horizons from the history of time measurement to the vast spectrum of the history of science.

The idea for holding the Longitude Symposium came about as a result of the seminars organized by the National Association of Watch and Clock Collectors (NAWCC). These meetings, the brainchild of Ward Francillon, provide the members of this large organization with the opportunity of meeting on an annual basis in different areas of the United States to learn about some specific period or aspect of the history of time measurement. Therefore, at the Washington meeting in October 1989 when Francillon requested a volunteer to undertake the responsibilities of the seminar in 1993 (the tercentenary of John Harrison's birth), I offered to organize a conference relating to the history of the marine chronometer.

By holding the seminar at Harvard, I hoped to place John Harrison's achievement in its broad historical context. Having already given much thought to the history of finding longitude at sea, it took only a week or two to establish the basic program and invite the principal speakers, whom from past experience I already knew well. The event that I had in mind, however, was much larger than anything ever planned before for these seminars and required a considerable amount of financial support to bring the idea to reality. Among the enthusiastic and dedicated members of the NAWCC who provided encouragement during periods of doubt and helped to raise funds, there are two friends who deserve special mention: Robert Cheney (the assistant chairman of the symposium) and Jim Cipra, both of whom served as invaluable liaisons with the NAWCC's seminar committee. To encourage donations, the symposium's organizing committee established three categories of support: sponsors, patrons, and benefactors. Cipra immediately went to work with his dedicated group of horologists in California and came up with the first donation. This was a positive beginning, but to succeed in convincing others of the conference's potential, a major gift from outside the horological world was needed.

I turned therefore to a colleague at Harvard, James Baker, who, as a member of the board of trustees of the Perkin Fund, had helped to secure support for the new gallery designed for the Collection of Historical Scientific Instruments in 1989. Upon Baker's recommendation, the Perkin

3. In the section below on the Longitude Symposium exhibitions, further information on Wheatland's contributions to Harvard is provided. A more detailed description of Wheatland's collecting activities and his involvement in establishing the Collection of Historical Scientific Instruments is contained in my article, "The Life and Work of David Pingree Wheatland (1898-1993)," published in the *Journal of the History of Collections* (Oxford University Press), vol. 7, no. 2 (1995), pp. 261-8.

Fund generously donated the sum that I requested. When, with the support of Jim Cipra and Norman Friedman (another member of the symposium's committee), Cheney transmitted this news to the NAWCC at their national convention in 1992, the NAWCC's governing board kindly gave matching funds. Subsequently Bruce Chandler, the third member of the Longitude Symposium's executive committee, approached the directors of the American Section of the Antiquarian Horological Society, who also agreed to become sponsors of what was now appearing to be a major international horological event. In addition, Frank and Linda Vitale, friends in the horological trade, generously supported the occasion. Thus it was that initial financial support was secured, providing a solid base for the fund-raising structure that had been established. A list of all the contributors to this event, which includes many individual chapters of the NAWCC, collectors, and other members of the horological trade, is given in the description of the program that follows.

Harvard University has been mentioned in connection with the Longitude Symposium's first major gift, but I have not expressed the importance of having this institution as the site for the conference. Its significance is impossible to overestimate, because an academic institution of this caliber provides support in so many hidden ways. As the organizational details rapidly began to absorb all of my spare time, I was able to devote part of my time in the museum to the growing needs of the conference. Without this opportunity, many of the exhibitions and other added dimensions would have had to be excluded. Furthermore, Harvard's outstanding resources and reputation provided the most solid foundation that any conference organizer could hope for.

Although the majority of the Longitude Symposium's committee members came from the local chapter of the NAWCC, the Collection of Historical Scientific Instruments was soon established as the base for all the organizing activities. Keeping track of all the finances and projected expenditures quickly became an overwhelming responsibility, so I asked one of the Collection's volunteers, Lewis Law, if he could help. Not in a month of Sundays could I have hoped for anyone more efficient and adept at this onerous task. In addition to keeping me informed on a regular basis of the symposium's financial predicament, Law, who before retirement had worked at Harvard as Director of Computer Services in the Science Center, became a constant and trusted companion. But even with this encouraging help, a full-time assistant was needed for the event to be a success on the day. One of David Wheatland's final gestures of generosity toward his alma mater was a donation that allowed me to hire a remarkable lady upon whom we all came to rely: Martha Richardson, as many symposium participants noticed, ensured that every detail was attended to, before, during, and after the conference.

While the basic program had been established by this time, the major international symposium that we had planned still needed an internationally known banquet speaker. The person had to be well known and highly respected on both sides of the Atlantic, with a deep appreciation for history and an interest in the longitude problem. On Easter Day, 1992, while I was gazing over rows of books at my parents' house in Dorset, England, my eyes lit upon a book called *America*. Alistair Cooke! I wrote to him as soon as I returned. Fortunately, longitude was a subject that united two of our heroes: Harrison for me, and La Salle for Cooke. And it just so happened that La

Salle was exactly 50 years older than Harrison; thus, 1993 was an appropriate occasion to celebrate the lives of both men. Cooke's great generosity in accepting my invitation was of inestimable importance in gaining recognition for the Longitude Symposium.

At this point, the program had been divided into four sessions, with speakers engaged to relate different aspects of the longitude story within the context of their own area of expertise. In this way, the fields of economic history, astronomy, navigation, and horology were well represented. However, during a discussion with Willem Mörzer Bruyns at the International Scientific Instrument Symposium in September 1992, an idea evolved to appoint a chair for each session to provide an introductory talk to describe the impact of the longitude problem upon other disciplines. As a result, the longitude story could also be placed within the context of mathematics, cartography, and the history of science. This important additional interdisciplinary dimension greatly increased the diversity of the audience.

One small but important concern was the fact that the $175 conference admission fee, although reasonable in terms of a three-day symposium, would clearly make it impossible for younger students—perhaps the most important audience of all—to attend. A number of complimentary passes helped to solve the problem for those who lived in the vicinity of Cambridge. Nevertheless, we felt that it was important to provide opportunities for students from other countries, particularly those that were formerly behind the Iron Curtain, who, due to the new economic crisis, were greatly in need of the encouragement that would come from meeting colleagues in their field from other countries. Christoph Lüthy, a former student and good friend, identified a few students involved in the history of science from Russia and the Czech Republic who were anxious to attend. Letters requesting funds to cover their travel and accommodation expenses were sent to many corporations and foundations, but, because of the depressed economic climate, no donations were forthcoming. At the last minute, two of David Wheatland's sisters, Anna Ordway and Martha Ingraham, came to the rescue. Their generosity, combined with a donation from David and Dorothy Harris, allowed four students from Moscow and Prague to attend the conference.

The plans for the exhibitions that accompanied the Longitude Symposium began with an idea for one exhibition illustrating the most important pieces in the history of time measurement. This would have involved borrowing the Harrison timekeepers from Greenwich, the Le Roy and Berthoud timekeepers from Paris, as well as many other artifacts from other museums and institutions in different parts of the world. The idea was enticing, but, even with the help of many good friends in the museum world, the virtual impossibility of doing this within a relatively short period of time soon became apparent. The idea had to be simplified by limiting the exhibits to artifacts that belonged to Harvard.

The artifacts available within Harvard's museums and libraries for the study of the history of astronomy, cartography, science, time, and navigation—five subjects that share a common root in the history of finding longitude at sea—were divided into three groups, each representing a separate exhibition: one showing the clocks, watches, and other timekeeping devices at the Fogg Art Museum; another with the books and artifacts relating to the history of navigation at the Houghton Library; and a third displaying some of the scientific apparatus of the period at the Collection of Historical Scientific Instruments. The situation, however, was to change. In May 1993, David Pingree Wheatland, benefactor and founder of the Collection of Historical Scientific Instruments and generous supporter of the Houghton Library, died. Roger Stoddard, Curator of Rare Books in the Houghton Library, and I decided that we should organize an exhibition at the Widener Library in Wheatland's honor. Then, with four exhibitions already under way at the same time, an excellent opportunity arose for an exhibition of psychological instruments, which, through the history of reaction timing, have an interesting connection to the longitude problem.[4] These instruments, which had never been properly displayed before, were to be placed in a long-term exhibition adjacent to one of the laboratories in the Psychology Department in William James Hall.

To undertake the mounting of all five exhibitions at the same time, in addition to the responsibilities of organizing the Longitude Symposium, was not something that could be done single-handed. The names of all those who contributed their time and energy to these projects are listed below in the section describing the exhibitions. There are, however, five whom I would like to mention here: Sandra Grindlay of the Harvard University Art Museums gave unremitting support and attention to many of the details involved in "The Art of Time." Richard Ketchen devoted countless hours to helping in many ways with this exhibition, as well as attending to most of the restoration of the clocks and the instruments in all the exhibitions. Ed Haack of the Museum of Comparative Zoology came to the rescue at the last minute to undertake the preparation of all the labels for the Wheatland exhibition at the Widener Library. Ed Rodley from the Museum of Science in Boston generously undertook most of the responsibilities involved in the navigation exhibition at the Houghton Library. And Dean Gallant assumed responsibility for most of the chores involved in mounting the exhibition of psychological instruments. Without the diligent support of these and other friends and colleagues, the mounting of all five exhibitions at virtually the same time would have been impossible.

After the Longitude Symposium, I received many gratifying telephone calls and letters. Although, as chairman, these remarks were directed to me, any congratulation must be shared with everybody who was involved in making the event a success. Through the lists of the organizing committee, the contributors, the speakers, and the exhibitions that follow, I have tried to mention every name, but inevitably there are some who have been omitted because they supported the event in some inconspicuous way. To them, I apologize; but they too may share in what for me was the greatest reward: the symposium was a success, not because over 500 people from seventeen countries attended the event, nor because it made some money toward the production of this volume. It was a success because it struck a chord that united the passions and interests of people from many different backgrounds and fields of endeavor. As David Landes mentioned after the event, those of us who were there will long remember the time we spent together.

William J. H. Andrewes
*Chairman of the Longitude Symposium*

4. The reaction times of different individuals, which became an involved area of experimental psychology in the late nineteenth century, had been of concern to astronomers since the end of the eighteenth century, when timing the transit of celestial objects. In 1795, Nevil Maskelyne noticed that his assistant, David Kinnebrook, was reading the times of stellar transits about half a second slower than he was; this problem subsequently resulted in Kinnebrook's dismissal. During the late nineteenth century, instruments were developed so that psychologists could study the problem of varying reaction times. See E. C. Sanford, "The Personal Equation," in the *American Journal of Psychology,* vol. 2 (1888-9), pp. 3-38, 271-98, 403-30; another discussion, largely derived from Sanford, is in E. G. Boring, *A History of Experimental Psychology,* 2nd ed. (New York: Appleton Century Crofts, 1950), pp. 134-53. This development will be described in detail in Rand Evans's forthcoming catalogue of the psychological instruments in the Collection of Historical Scientific Instruments.

# II. Conference information

The Longitude Symposium
Harvard University, Cambridge, Massachusetts
November 4 - 6, 1993
*Organized in association with the National Association of Watch and Clock Collectors by the Collection of Historical Scientific Instruments, Harvard University*

## ORGANIZERS

**Executive Committee:**
William J. H. Andrewes
*Chairman*
Robert C. Cheney
*Assistant Chairman*
Bruce Chandler
*General Editor*

**Articles:**
Herbert A. Gold
Paul Middents

**Coordinator of events:**
Martha Richardson

**Design:**
Pamela Geismar

**Editing:**
William J. H. Andrewes
Bruce Chandler
Susan M. Rossi-Wilcox

**Fund-raising:**
William J. H. Andrewes
Robert C. Cheney
Jim Cipra
Jim Espy
Bill Givens
Frederick L. Koved
Christoph Lüthy (*Russian and Eastern European Scholarship Program*)
David B. Searles II

**Hotel and food arrangements:**
Norman Friedman
Linn W. Hobbs

**Publicity:**
Linn W. Hobbs
Anabela Afonso

**Registration:**
Lewis A. Law
Martha R. Richardson
Jackie Wagner

**Speaker liaison:**
Richard Ketchen

**Treasurer:**
Lewis A. Law

## CONTRIBUTORS

### Sponsors

The Perkin Fund

The National Association of Watch and Clock Collectors

The American Section of the Antiquarian Horological Society

Vitale & Vitale Ltd.

Martha W. Ingraham and Anna W. Ordway *(Russian and Eastern European Scholarship Program)*

### Patrons

Jon Hanson

William L. Scolnik

David B. Searles II

Philip Whyte

The following chapters of the National Association of Watch and Clock Collectors:

- Los Angeles Chapter 56, California
- San Fernando Valley Chapter 75, California
- Western Electrics Chapter 133, California

### Benefactors

Alan W. Heldman

Mr. and Mrs. David F. Harris Jr. *(Russian and Eastern European Scholarship Program)*

The following chapters of the National Association of Watch and Clock Collectors:

- New England Chapter 8
- Philadelphia Chapter 1, Pennsylvania

### Other contributors

R. O. Schmitt Fine Arts

Jonathan Snellenburg

The following chapters of the National Association of Watch and Clock Collectors:

- Chicagoland Chapter 3, Illinois
- Old Dominion Chapter 34, Virginia
- Orange County Chapter 69, California
- San Jacinto Chapter 139 Inc., Texas

Sotheby's and Christie's *(Symposium promotion)*

*The organizers would also like to thank American Airlines for providing the air services.*

## PROGRAM

### THURSDAY, NOVEMBER 4

Charles Hotel, Harvard Square

**19:15** Welcome and introduction
*Robert Cheney, Assistant Chairman, Longitude Symposium*

**19:20** History and background of the James Arthur Lecture
*Ward Francillon, Chairman of the NAWCC Seminar Committee, U.S.A.*

**19:30** James Arthur Lecture
"The Mechanical Watch in the 21st Century: The Renaissance of the Mechanic"
*George Daniels, watchmaker and author, U.K.*

### FRIDAY, NOVEMBER 5

Charles Hotel, Harvard Square

**08:15** Welcome
*Jeremy Knowles, F.R.S., Dean of the Faculty of Arts and Sciences, Harvard University, U.S.A.*

**08:20** Introductory remarks
*William J. H. Andrewes, Chairman of the Longitude Symposium, Harvard University, U.S.A.*

**08:30** "The Point in Vast Emptiness: The Importance of Finding One's Way in an Age of Ocean Sail"
*David Landes, Coolidge Professor of History and Professor of Economics, Harvard University, U.S.A.*

Figure 1. The speakers at the Longitude Symposium. Front row (left to right): Owen Gingerich, David Landes, John Leopold, Anthony Turner, Catherine Cardinal, William Andrewes, and Bruce Chandler; second row (left to right): Willem Mörzer Bruyns, Albert Van Helden, Alan Stimson, Michael Mahoney, Derek Howse, Norman Thrower, Jonathan Betts, and David Penney; back row on the staircase (left to right): Andrew King, Anthony Randall, and Martin Burgess.

**Session I: Early History of Navigation at Sea: Theories and Practice of Finding Longitude**

*Chair: Bruce Chandler, Professor of Mathematics, College of Staten Island, City University of New York, U.S.A.*

**09:15** Longitude in the context of astronomy and mathematics
*Bruce Chandler (Chair)*

**09:30** "The Longitude Problem: The Navigator's Story"
*Alan Stimson, Head of Development and former Curator of Navigation, National Maritime Museum, Greenwich, U.K.*

**10:15** Break and refreshments

**10:45** "Longitude and Jupiter's Moons"
*Albert Van Helden, Lynette S. Autrey Professor of History, Rice University, U.S.A.*

**11:30** "Christiaan Huygens and His Longitude Timekeepers"
*John Leopold, Assistant Keeper, British Museum, U.K.*

**12:15** Panel discussion

**Session II: Early Attempts to Find Longitude**

*Chair: Willem Mörzer Bruyns, Curator of Navigation, Scheepvaartmuseum, Amsterdam, The Netherlands*

**14:00** Longitude in the context of navigation
*Willem Mörzer Bruyns (Chair)*

**14:15** "In the Wake of the Act and Even Before"
*Anthony Turner, historian of scientific instruments, France*

**15:00** "Cranks and Opportunists: Nutty Solutions to the Longitude Problem"
*Owen Gingerich, Professor, Harvard-Smithsonian Center for Astrophysics, U.S.A.*

**15:45** Break and refreshments

**16:15** "The Lunar-Distance Method for Measuring Longitude"
*Derek Howse, author and former Curator in the Department of Navigation and Astronomy, National Maritime Museum, U.K.*

**17:00** Panel discussion

**20:30–22:30** **Reception and exhibition at the Fogg Art Museum**

**SATURDAY, NOVEMBER 6**

Lecture Hall B, Science Center, Harvard University

**Session III: John Harrison**

*Chair: Norman Thrower, Professor of Geography, University of California, Los Angeles, U.S.A.*

**08:30** Longitude in the context of cartography
*Norman Thrower (Chair)*

**08:45** "John Harrison: The Early Years"
*Andrew King, clockmaker, U.K.*

**09:30** "Even Newton Could Be Wrong: The Story of Harrison's First Three Sea Clocks"
*William Andrewes, David P. Wheatland Curator, Collection of Historical Scientific Instruments, Harvard University, U.S.A.*

**10:15** Break and refreshments

**10:45** "The Timekeeper that Won the Longitude Prize"
*Anthony Randall, watchmaker and author, U.K.*

**11:30** "The Scandalous Neglect of Harrison's Regulator Science"
*Martin Burgess, clockmaker, U.K.*

**12:15** Panel discussion

**Session IV: Perfecting the Marine Timekeeper**

*Chair: Michael Mahoney, Professor of History, Princeton University, U.S.A.*

Figure 2. Alistair Cooke, who delivered the Longitude Symposium's concluding address after the banquet.

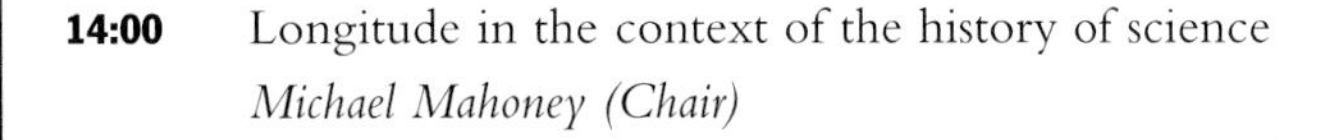

**14:00** Longitude in the context of the history of science
*Michael Mahoney (Chair)*

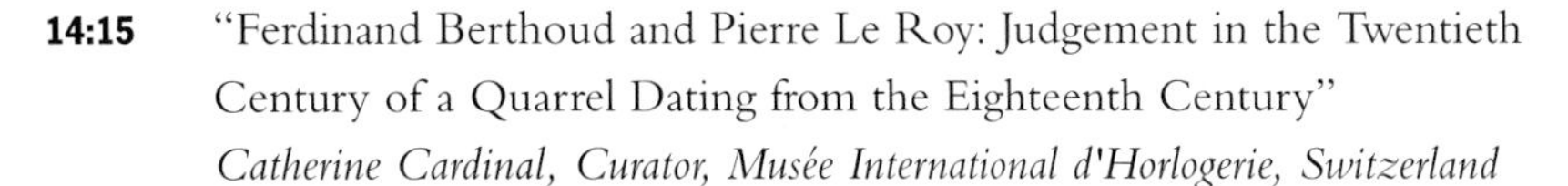

**14:15** "Ferdinand Berthoud and Pierre Le Roy: Judgement in the Twentieth Century of a Quarrel Dating from the Eighteenth Century"
*Catherine Cardinal, Curator, Musée International d'Horlogerie, Switzerland*

**15:00** "Thomas Mudge and the Longitude Prize: A Reason to Excel"
*David Penney, horological illustrator and former editor,* Antiquarian Horology*, U.K.*

**15:45** Break and refreshments

**16:15** "Arnold and Earnshaw: The Practicable Solution"
*Jonathan Betts, Curator of Horology at the National Maritime Museum and Horological Advisor to the National Trust, U.K.*

**17:00** Panel discussion

**17:15** **Presentation of certificates of appreciation, celebrating 50 years of the NAWCC**
James Coulson, President, NAWCC, U.S.A.

**19:30** **Hors d'oeuvres and drinks (cash bar) at Memorial Hall, Harvard University**

**20:00** **Banquet at Memorial Hall, Harvard University**
(Black tie optional)
Speaker and Guest of Honor: Alistair Cooke

# III. Special exhibitions

Five special exhibitions were organized to coincide with the symposium:

**Instruments of Science:**
**An Exhibition of Instruments used for Teaching and Research at Harvard University during the Eighteenth and Nineteenth Centuries**
*Collection of Historical Scientific Instruments,*
*Science Center (Lower Level), 1 Oxford Street*

During the past 400 years, the advancement of human knowledge of nature and the physical world has been greatly enhanced by instrumentation. A detailed examination of the materials and construction of an instrument reveals much about the period in which it was made: the social and economic demands that fostered its creation, the state of scientific and technical knowledge, and the difficulties and limitations of its use.

Figure 3. Apparatus used for teaching and research at Harvard during the eighteenth century. Among the instruments shown is a large telescope made by the famous London maker, James Short (1710-1768). This was commissioned with the assistance of Benjamin Franklin and purchased for 100 guineas in 1767.

Harvard University has been purchasing scientific instruments on a continuous basis for teaching and research since 1765. With the introduction of new technologies and the renovation of laboratories after World War II, some instruments were in danger of being discarded. Recognizing their historical value, David P. Wheatland, a Harvard alumnus of the Class of 1922, undertook to preserve them. With the support of key members of the faculty, the Collection of Historical Scientific Instruments was established in 1947 and held its first public exhibition in February 1949. With the addition of instruments from various departments and private benefactors, the Collection now contains over 15,000 artifacts dating from about 1450 to the present. Significant apparatus, made obsolete by new technologies, continue to be incorporated. A broad range of disciplines, including astronomy, navigation, horology, surveying, geology, calculating, physics, biology, medicine, psychology, electricity, and communication, are represented. Many documents detailing the purchase and use of the instruments have been preserved. The Collection's library comprises over 9,000 books, pamphlets, and catalogues relating to the history of instruments.

The Collection of Historical Scientific Instruments was associated with the Harvard Libraries until 1987, when it became affiliated with the History of Science Department. Like many of the other Harvard collections, its primary purpose is teaching and research, providing students with the opportunity of examining and working with artifacts that have made the progress of science possible.

## The Art of Time:
## Clocks, Watches, and Other Timekeeping Devices from the Harvard Collections

*Fogg Art Museum, 32 Quincy Street*

Among the many diverse collections acquired by Harvard University are a large number of instruments associated with the measurement of time. These come from three main sources: those that were acquired for general use; those that were made specifically for precision timekeeping, particularly for use at the Harvard College Observatory; and those that have been donated. Most of the timekeeping devices are preserved in the Harvard University Art Museums and the Collection of Historical Scientific Instruments. The former houses the magnificent collections bequeathed in 1943 by Grenville L. Winthrop, which include a large number of seventeenth- and eighteenth-century clocks. The latter contains most of the precision clocks acquired by the University and over 700 sundials and other time-finding devices collected by David P. Wheatland.

This exhibition included some of the finest and most historically interesting examples of timekeeping devices acquired by Harvard University during the last 300 years, illustrating some of the developments that have made the precise measurement of time possible. When Harvard College was founded in 1636, mechanical clocks could keep time to about fifteen minutes per day. Twenty years later, the first major breakthrough in accuracy came with the invention of the pendulum clock. Instruments of time embody the elements of science in their design, technology in their construction, and art in their appearance: in the hands of great craftsmen, these are combined into the art of time.

**Figure 4. View near the entrance to the exhibition with a longcase clock by John Knibb (1650-1722), London, *ca.* 1685, in the foreground on the left. The movement of each clock was displayed on a stand next to its case and dial, so that the operation of every mechanism could be closely examined.**

This exhibition was organized by William Andrewes in collaboration with Marjorie Cohn (Harvard University Art Museums) and with the assistance of Carolann Barrett, Nancy Buschini, Anne Driesse, Robert French, Ivan Gaskell, Sandra Grindlay, Danielle Hanrahan, Henry Lie, Timothy Lloyd, Jane Montgomery, Peter Schilling, Anthony Sigel, and Lisa Young (Harvard University Art Museums staff), Robert C. Cheney, Lissa Coolidge, Douglas R. Currie, and Richard L. Ketchen.

**Figure 5. View of the exhibition overlooking the courtyard from the gallery above.**

## From an Art to a Science: Navigation in an Age of Exploration and Discovery

*Houghton Library, Harvard Yard*

With the advent of oceanic navigation, the art of the navigator changed. Practice, instincts, and inherited knowledge alone were no longer sufficient to guide the pilot. As ocean travel increased with the expansion of exploration, trade, and colonization in the Americas, the human and material cost of prolonged voyages and shipwrecks revealed severe weaknesses in the traditional art of navigation. The solution to this problem came from advancements in knowledge and the application of scientific principles. The instruments developed for this purpose made it possible for the navigator to establish the latitude of his ship north or south of the equator and its longitude east or west of a given meridian. With the perfection of these instruments by the end of the eighteenth century, the navigator's art was transformed into a science.

**Figure 6. Reflecting circle (no. 186) by Henri Gambey (1787-1847), France, *ca.* 1820. Courtesy of the Collection of Historical Scientific Instruments, Harvard University. Inv. no. 5312.**

"From an Art to a Science" traced the history of navigation from the fifteenth to the nineteenth century with maps, illustrations from books, and navigational instruments from collections at Harvard: the Harvard Map Collection, the Houghton Library, and the Collection of Historical Scientific Instruments.

This exhibition was organized by Edward J. Rodley (then Research Associate and Registrar at the Museum of Science, Boston) in collaboration with Roger Stoddard (Houghton Library) and William Andrewes and with the assistance of David Cobb (Harvard Map Collection) and Roxana Breckner, Dennis Marnon, and Mariana Oller (Houghton Library).

**Figure 7. World map from Claudius Ptolemy's *Geographia* (Ulm, 1482). By permission of the Houghton Library, Harvard University. Ref. no. Typ Inc 2556.**

**A Natural Philosopher:**
**An Exhibition to Honor the Life and Work of David Pingree Wheatland**
*Widener Library, Harvard Yard*

This exhibition was held in memory of David Pingree Wheatland (Class of 1922), whose determination and perseverance made the Collection of Historical Scientific Instruments possible. After graduating from Harvard, Wheatland entered his family's forestry business in Maine. His principal interest, however, was in physics, and after about six years he returned to work for Professor Leon Chaffee as a research associate at the Cruft Laboratory at Harvard. Toward the end of World War II, he became the first operator of the mammoth computer, the IBM Mark I.

Wheatland developed an interest in collecting books on the early history of electricity and magnetism soon after graduating from Harvard. When he returned to Harvard to work for Professor Chaffee in the development of radio and communications, this fascination continued to grow, and within ten years he had collected over 5,000 titles relating to this subject published before 1820. Between 1941 and 1991, he donated 4,600 rare books to the Houghton Library and established a reference library in the Collection of Historical Scientific Instruments of over 9,000 titles.

Figure 8. David P. Wheatland in 1977 with some of the old Harvard instruments that he preserved. Courtesy of Stephen F. Grohe.

Before World War II, Harvard's old scientific instruments were stored in the attics and basements of the laboratories in which they had been used. Wheatland was able to recognize some of these devices from illustrations in the books he had collected. Realizing that these old instruments were sometimes cannibalized for parts and would be thrown away when the space they occupied was needed, he decided that something had to be done to preserve them. After storing as many as possible in his own office, he inquired about the possibility of finding additional space. This simple request was to change the course of his life. When the Collection of Historical Scientific Instruments was established in 1947, Wheatland was appointed its first curator; he also became its principal benefactor.

His activities in acquiring instruments for his own collection began first with sundials, largely because the compass needles of portable sundials represent one of the earliest uses of magnetism. Through his involvement in preserving Harvard's scientific apparatus, however, Wheatland purchased—over a period of about 50 years—several thousand instruments covering a broad range of scientific disciplines; these artifacts were acquired with the intention that one day they would be used to fill in gaps in the Collection of Historical Scientific Instruments, which at that time concentrated on preserving only those instruments associated with Harvard. About 1984, when Harvard showed no interest in adding Wheatland's instruments to its holdings, some of these were sold. Fortunately, however, a change in policy in

1988 allowed the majority of his instruments to be donated to the Collection of Historical Scientific Instruments.

The instruments and books shown in this exhibition represented a few examples of the remarkable collection David Wheatland assembled during a period when there was a minimal interest in collecting in the fields of science and technology.

This exhibition was organized by William Andrewes and Roger Stoddard (Houghton Library) in collaboration with James Lewis and Pamela Matz (Widener Library) and with the assistance of Robert Forsberg (Collection of Historical Scientific Instruments), Edward Haack (Museum of Comparative Zoology), and Mariana Oller (Houghton Library).

## "...these new prism, pendulum, and chronograph-philosophers."

*William James Hall, 8th Floor, 33 Kirkland Street*

"Prism, pendulum, and chronograph-philosophers" was William James's appellation in his *Principles of Psychology* for the new generation of quantitative, experimental psychologists who made use of laboratory procedures and instruments. Although James founded the first psychological laboratory at Harvard around 1875, his heart remained with the qualitative, empirical method in the "grand style" of philosophical psychology. Despite his reservations, James continued to develop the laboratory and support its expansion. As early as 1895, the laboratory was referred to as "the most unique, the richest, and the most complete in any country."

The instruments assembled in this exhibition were selected from the surprisingly large collection of psychological instruments to have survived the many moves of the laboratory. They are representative examples of the types of instruments used in the early days of experimental psychology and are, in themselves, historical documents of the first order. Here we return to a period when time intervals were produced by the swinging of a pendulum (one-second pendulum chronoscope) or the falling of a weight (Wundt's control hammer); when tone stimuli were produced by sounding pipes (Koenig's organ pipes, Stern's variators), striking tuning forks (tuning fork and resonator box), sounding reeds (Appunn's tonometer, Ellis Harmonium), striking steel

**Figure 9. General view of the exhibition of psychological instruments. Courtesy of the Collection of Historical Scientific Instruments, Harvard University.**

cylinders (Koenig's tuned cylinders), or plucking a string (differential sonometer). It was a time when reaction times were measured by the arc of a pendulum's fall (pendulum reaction-time apparatus) or the spin of a clockwork dial (Münsterberg's chronoscope); when visual instruments did not assume the electric light (Hering's color-blindness apparatus, Titchener's telestereoscope). Through these instruments, we can recall when prior entry and the complication experiment were hot issues in perception (Wundt's complication apparatus), when the horopter was considered an essential psychological concept (horopter apparatus), and when memory drums (Müller's Memory Drum) and learning curves were new innovations.

This semipermanent exhibition has been mounted at the entrance to one of the most modern of psychological laboratories in order to illustrate the long tradition of experimental psychology at Harvard. These instruments were selected from a collection of over 250 early psychological instruments preserved in the Collection of Historical Scientific Instruments.

This exhibition was organized by Rand Evans (East Carolina University), Dean Gallant (William James Hall), and William Andrewes in collaboration with Rob Olson (Robert Olson and Associates, Architects) and with the assistance of Richard Herrnstein, Jerome Kagan, Richard L. Ketchen, Jeff Mifflin, Daniel Schacter, and Sheldon White.

APPENDIX C

# Translations of the Earliest Documents Describing the Principal Methods Used to Find Longitude at Sea

# Translations of the Earliest Documents Describing the Principal Methods Used to Find Longitude at Sea

*Translations revised by Andrea Murschel*
*Introductions by William Andrewes*

Two of the earliest, and ultimately the most practicable, suggestions for finding longitude were published within 40 years of Columbus's epic voyage. Although straightforward in theory, both of these—the lunar-distance method and the timekeeper method—were extremely complicated and difficult to realize in practice, involving knowledge and technique that would take more than two centuries to develop. Interestingly, both methods, which were proposed within sixteen years of each other during the first half of the sixteenth century, became viable for use at sea at about the same time, in the sixth decade of the eighteenth century.

As far as we know, the originators of these methods—Johann Werner (1468-1522), who proposed the lunar-distance method, and Gemma Frisius (1508-1555), who proposed the timekeeper method—never left dry land and therefore had no experience of the difficulties of finding longitude at sea. In fact, both authors refer only to the general theory of finding longitude, without any reference to the practical problems involved, either on land or at sea; neither mentions the significance of the problem to the expansion of exploration, colonization, and trade among seafaring nations.

Both Werner and Gemma were respected scholars in the works of Ptolemy (fl. A.D. 150) and became well known for their publications covering the fields of mathematics, astronomy, and geography.[1] The actual passages in which they advanced the theories that ultimately led to the two most practical solutions to the problem of finding longitude at sea occupy less than a page of the volumes in which they appeared; yet these seemingly insignificant paragraphs embodied the ideas that were to become among their most widely recognized and far-reaching contributions.

This is not the first time in this century that the original texts of Werner's and Gemma's proposals have been described, reproduced, or translated from the Latin. The superb study of the life and work of Gemma Frisius, published by Fernand van Ortroy (1856-1934) in 1920, includes a free translation into French of the passage concerning Gemma's timekeeper method along with a transcription of the text from the fifth (revised) edition of 1553.[2] In 1935, Alexander Pogo published an account of Gemma's method with a

1. A good summary of the life and works of Johann Werner is given in Menso Folkerts's article in the *Dictionary of Scientific Biography*, ed. Charles Coulston Gillispie and Frederic L. Holmes, 18 vols. (New York: Charles Scribner's Sons, 1970-90), vol. 14, pp. 272-7. Additional information may be found in Ernst Zinner's *Die Fränkische Sternkunde im 11. bis 16. Jahrhundert* (Bamberg, Germany, 1934), pp. 111-13 (with a portrait in fig. 4); an earlier biography of Werner is given in Johann Gabriel Doppelmayr's *Historische Nachricht von den Nürnbergischen Mathematicis und Künstlern* (Nuremberg, 1730), pp. 31-5. For the life and works of Gemma Frisius, readers are referred to Fernand van Ortroy's classic study, "Bio-bibliographie de Gemma Frisius," in *Académie Royale de Belgique. Classe des lettres et des sciences morales et politiques: Mémoires* (Brussels), vol. 11, 2nd series (December 1920), 2nd part, pp. 9-358.

2. Ortroy (*ibid.*). The translation is on pp. 66-8, and the transcription of the text from the 1553 edition is in note 1 on p. 67.

3. Alexander Pogo, "Gemma Frisius, his method of determining differences of longitude by transporting timepieces (1530), and his treatise on triangulation (1533)," *Isis*, vol. 22 (2), no. 64 (February 1935), pp. 469-85. The facsimile of the passage from the 1530 edition is on p. 472.

4. Derek Howse, *Greenwich time and the discovery of the longitude* (Oxford: Oxford University Press, 1980), pp. 7-10.

5. Andrea Murschel is a graduate student of Noel Swerdlow at the University of Chicago and has been involved with the collection of astronomical instruments at the Adler Planetarium in Chicago.

facsimile of the text from the first edition of 1530, but with a summary of its content, not a translation.[3] A translation of both Werner's and Gemma's proposals was made by Philip Kay in 1977; part of his translation of Werner's method and all of his translation of Gemma's method from the 1530 edition, with the additional passage from the 1553 edition, were published in 1980 by Derek Howse in *Greenwich time*.[4] This very useful book is now, alas, out of print and hard to find. As a result, it was decided that revised translations of both Johann Werner's and Gemma Frisius's proposals, being of prime importance in the story of finding longitude, should be provided alongside facsimiles of the original texts. Derek Howse kindly supplied copies of Kay's work to serve as a guide for the revised translations below, which were made by Andrea Murschel.[5]

## THE LUNAR-DISTANCE METHOD

Johann Werner was a priest who lived most of his life in Nuremberg, serving during his last fourteen years at St. Johannis Church. He enrolled in the University of Ingolstadt in 1484 and studied in Rome from 1493 to 1497. Since his pastoral duties proved not to be too demanding, he was able to devote much of his time to scientific study, including perhaps the design and construction of some instruments.[6] He had been keenly interested in mathematics from an early age. His description of the lunar-distance method was published in 1514 in a collection of works on mathematical geography; the contents of this book, which included a new translation and commentary on the first book of Ptolemy's *Geography*, were listed on the title page under the heading *In hoc opere haec cõtinentur....*[7] The description of the lunar-distance method appears in note 8 of chapter 4, following a discussion of the principles of finding longitude by lunar eclipses and an explanation of the use of the cross-staff, the instrument he employed to make the required angular-distance measurements.

Werner's proposition was doubtless influenced by the writings of the great astronomer and mathematician Regiomontanus (1436-1476),[8] who had lived for a few years toward the end of his life in Nuremberg. It was not Regiomontanus's method of finding longitude, however, but his techniques for tracking a comet that laid an important part of the foundation for the lunar-distance method. Regiomontanus had used lunar eclipses to determine the longitudes of places.[9] But although these events can be predicted, they occur so infrequently that they do not represent a viable method for finding longitude at sea and are of limited value even on land.[10] In the passage "Concerning finding longitude by lunar eclipses" (see below), Werner clearly illustrates the existing inaccuracy of this

**Figure 1. Portrait of Johann Werner, artist unknown, *ca.* 1600; etching probably based on an earlier portrait. No positively identified portrait of Werner made during his lifetime is known to exist. There has, however, been some speculation that the portrait of an unknown ecclesiastic, painted by Albrecht Dürer (1471-1528) in 1516 (now in the National Gallery in Washington, D.C. [1952.2.17]), is in fact a portrait of Johann Werner, who was a member of the artist's close circle of friends (Dürer is known to have painted a series of private portraits of his friends during the period 1516-18). 8 x 4.8 cm. (3.1 x 1.9 in.). Courtesy of the Germanisches Nationalmuseum, Nuremberg. Inv. no. P 3330, Kapsel 919.**

6. In his article on Werner, Menso Folkerts (*op. cit.*, note 1), p. 273, mentions the following instruments that may have been designed and/or made by Werner: an astrolabe (now in the Germanisches Nationalmuseum, Nuremberg), the clock on the south side of the parish church in Herzogenaurach, and two sundials in the choir of the church at Rosstal. More detailed information is provided in Ernst Zinner, *Astronomische Instrumente des 11. bis 18. Jahrhunderts* (Munich: C. H. Beck, 1956), p. 584.

7. Johann Werner, *In hoc opere haec cõtinentur...* (Nuremberg, 1514).

8. Regiomontanus's real name was Johann Müller. He was born in Königsberg, which literally means "King's Mountain." In 1450, he was enrolled in the University of Vienna as "Johannes Molitoris de Künigsperg" and later sometimes Latinized the place as well as his name to "Johannes de Regio Monte." From this derived the name Regiomontanus.

9. The idea was not new. Using lunar eclipses for finding the longitudes of places had been proposed by Hipparchus (*ca.* 180-125 B.C.) and subsequently by Ptolemy (fl. A.D. 150).

10. In chapter 17—"On finding the longitudes of places"—translated below, Gemma Frisius described some of the problems of using lunar eclipses to find longitude and stated that, as a result, the longitudes of many regions, especially of those discovered by the Spanish, were uncertain or completely unknown.

method by noting Regiomontanus's errors in determining the difference in longitude between Nuremberg and Rome. Werner's own calculations were, however, not much better.

Two of the fundamental requirements for the astronomical solution to the longitude problem were frequency and predictability; Werner recognized that, in principle, the Moon's motion satisfied both of these. Nevertheless, the required precise knowledge of the Moon's true and mean motions for predicting its position in one location and the ability to accurately measure its position in another represented practical problems that would take over 200 years to resolve. The techniques that Werner suggested for tracking the Moon's progress through the heavens had been described by Regiomontanus for tracking the path of the comet of 1472: the first involved measuring its height and azimuth, the second, its latitude and longitude in the meridian, and the third, its angular distance from two stars.[11]

The instrument that both men employed for making the required angle measurements was the cross-staff, a device first described about 1330 by Levi ben Gerson (1288-1344) and later developed in various specialized models for use in astronomy, surveying, and navigation. In order to obtain more precise observations, Levi had marked his instrument with a diagonal scale, a development that is often credited to the Danish astronomer Tycho Brahe (1546-1601); in the design, Levi also took into account the eccentricity of the eye, determining that rays of light focus not on the surface of the eye but "in the crystalline lens, and that its center is the center of vision."[12] Calling his instrument "The Revealer of Profundities," he described it in two poems, the second of which mentions a passage in Genesis (30:37) that refers to Jacob.[13] Perhaps from this derived the term "Jacob staff," another name used for the cross-staff. In Latin, this instrument was often referred to simply as *baculus* (the staff) or *radius astronomicus* (the astronomical pointing rod); the most common of its vernacular names was the Portuguese *balhestilha*.[14] The version that Werner described for making lunar-distance measurements was six to seven feet long and had eight crosspieces with eight corresponding scales that provided a direct reading in degrees.[15] A translation of his description of the use of this instrument, with the original text and illustration, is given below.

Another earlier, but possibly spurious, reference to a method of finding longitude by observations of the Moon's position had been made by the Florentine explorer Amerigo Vespucci (1451-1512), whose first name was given to the newly discovered continent. On August 23, 1499, he determined the local time of the conjunction of the Moon with Mars, and, by comparing his observations with those listed in Regiomontanus's almanac, he deduced that he was 82.5° west of Cadiz. Daniel Boorstin, in *The Discoverers*, gives the following translation of the passage in which Vespucci describes his observations and calculations:

> As to Longitude, I declare that I found so much difficulty in determining it that I was put to great pains to ascertain the east-west distance I had covered. The final result of my labors was that I had nothing better to do than to watch for and take observations at night of the conjunction of one planet with another, and especially of the conjunction of the moon with the other planets, because the moon

11. These methods are described in detail by Ernst Zinner in *Regiomontanus: His Life and Work*, trans. Ezra Brown (Amsterdam: North Holland, 1990), pp. 130-1. An even earlier reference to the lunar-distance method is discussed by W. G. L. Randles in *Portuguese and Spanish Attempts to Measure Longitude in the 16th Century* (Coimbra: Instituto de Investigação Científica Tropical, 1985), pp. 8–9. This monograph gives an English translation of a relevant passage in *Theoretica Planetarum*, a late twelfth-century work (attributed to Gerard of Cremona [*ca.* 1114–1187]) printed in 1478.

12. Levi ben Gerson, *Sefer ha-tekunah* ("The Book of Astronomy"), edited in the Hebrew and translated into English by Bernard R. Goldstein in *The Astronomy of Levi ben Gerson (1288-1344)* (New York: Springer-Verlag, 1985), p. 52.

13. *Ibid.*, p. 72.

14. For further information, see Goldstein (*ibid.*), chaps. 6-11, pp. 51-81, with commentary on pp. 143-62; John Roche, "The Radius Astronomicus in England," *Annals of Science*, vol. 38, no. 1 (1981), pp. 1-32; and Bernard R. Goldstein, "Remarks on Gemma Frisius's *De Radio Astronomico et Geometrico*," in J. L. Berggren and B. R. Goldstein, eds., *From Ancient Omens to Statistical Mechanics: Essays on the Exact Sciences Presented to Asger Aaboe*, vol. 39 of Acta Historica Scientiarum Naturalium et Medicinalium (Copenhagen: University Library, 1987), pp. 167-80. I appreciate the assistance of Bernard Goldstein in bringing these references to my attention. A history of the development of the cross-staff, with a list of surviving examples (currently less than 100 are known), is given by Willem Mörzer Bruyns in *The Cross-staff. History and Development of a Navigational Instrument* (Zutphen, the Netherlands: Walburg Instituut, 1994). Illustrations showing the use of the cross-staff appear in Figure 5 of Alan Stimson's paper and Figure 2 of Derek Howse's paper in this volume.

15. *In hoc opere* (*op. cit.*, note 7), *cap.* 4, note 5, sig. diiii$^v$, with illustration on sig. dv$^r$. For a summary and illustration showing the parts of Werner's cross-staff, see Roche, "The Radius Astronomicus" *(ibid.)*, pp. 12-13. The pagination of Werner's original text given by Roche differs from that in the Houghton Library copy and may be in error.

16. Daniel J. Boorstin, *The Discoverers*, 2 vols. (New York: Harry N. Abrams, 1983), vol. 1, p. 357. The reference occurs in a letter written by Vespucci on July 18, 1500. A summary of the arguments against the originality of this document is given by Ernst Zinner in *Regiomontanus* (*op. cit.*, note 11), pp. 123-5.

is swifter than any other planet....After I had made experiments many nights, one night, the twenty-third of August, 1499, there was a conjunction of the moon with Mars, which according to the almanac [for the city of Ferrara], was to occur at midnight or a half hour before. I found that when the moon rose an hour and a half after sunset, the planet had passed that position in the east.[16]

Vespucci's version of the lunar-distance method, however, was more complex than Werner's, because it involved establishing the Moon's position in relation to a planet—another moving body—rather than to the fixed stars.

Despite the fact that others had described the main elements of the lunar-distance method, Werner has received credit as its originator, because he alone combined those elements into an intelligible and, at least in theory, practicable solution to the longitude problem. No other edition of his work was printed, and evidence that it was not widely distributed is shown by the fact that there were a sufficient number of copies remaining two decades later to warrant a fellow German, Peter Apian (1495-1552), to bind them into his *Introductio Geographica.*[17] Apian, who was one of the most prolific scientific writers of the sixteenth century, included on the title page of this work an illustration showing astronomers making lunar-distance observations (see Figure 2 in Derek Howse's paper in this volume). This, however, was not the first time that Apian had mentioned the method. By the time he reprinted Werner's original description, the method was already well known through the publication of his *Cosmographicus Liber*, a work that was to become one of the most popular scientific textbooks of the sixteenth century. In this volume, first published in 1524,[18] Apian provides a description and an illustration of the lunar-distance method, but without mention of Werner. After being revised and edited by Gemma Frisius in 1529, *Cosmographicus Liber* was reprinted many times and translated into three languages. Fernand van Ortroy was able to find 30 editions that were printed before 1600.[19]

**Figure 2. Portrait of Peter Apian, whose popular work *Cosmographicus Liber* made the lunar-distance method well known throughout Europe. From an engraving by Johann Theodor de Bry (1561-1623) published in Jean-Jacques Boissard, *Bibliotheca sive Thesaurus Virtutis et Gloriae* (Frankfurt, 1628), p. 270.**

Thus, as the need for a solution to the longitude problem continued to grow with the increase in oceanic navigation during the sixteenth century, the lunar-distance method became widely recognized as a potential solution. Attempts to make it practicable by improving the accuracy of celestial maps and knowledge of the Moon's motion resulted in the founding of the Greenwich Observatory in 1675. Nevertheless, the difficulties involved in predicting the Moon's motion continued through the first few decades of the

17. Peter Apian, *Introductio Geographica Petri Apiani in Doctissimas Verneri Annotationes,...* (Ingolstadt, 1533). In the Houghton Library copy at Harvard and in the copy in The Time Museum, Rockford, Illinois, signatures B through K of Werner's text have the same type, the same watermarks on the paper, and the same pagination as the original text of 1514. In Fernand van Ortroy's "Bibliographie de l'Oeuvres de Pierre Apian," published as an article in *Le Bibliographie Moderne,* vol. 5 (March-June and July-October 1901), pp. 7-156 and 284-333, *Introductio Geographica* is not described because he was unable to locate a copy. There appears to be only the one edition of 1533, which is sometimes bound in with other books by Apian (as with the Houghton Library copy), but van Ortroy provides a reference on p. 91 to a possible second edition of 1537, mentioned in 1882 by Sigismond Günther in his article on Peter and Philipp Apian.

18. There are at least three states of the first edition of this book. The copy recorded by the New York Public Library as belonging to Grenville Kane may be the earliest, because it is the only one showing the pillar dial on the lower left of the title page in its complete state. The other copies I have seen show only the base of this sundial, which suggests that the upper portion broke off during printing. Kane's copy bears the title *Cosmographicus Liber a Petro Apiano Mathematico studiose collectus.* This same title is found on the example in the Boston Public Library (which bears the 1524 date on the colophon), and also on one listed by van Ortroy (*ibid.,* p. 117) in the Royal Library in Munich. This title without the complete pillar dial may be the second state produced. The third state bears the title found most frequently, and the one used for later editions of this work: *Cosmographicus Liber Petri Apiani Mathematici studiose collectus.* All of the first edition copies with this title appear to indicate the place of publication (Landshut), but not all bear the date. The pagination and watermarks in the copies I examined are the same: the description of the lunar-distance method is found on fol. 30 (Diii$^{v}$) and fol. 31 (Div$^{r}$), and is illustrated on fol. 32 (Div$^{v}$). In the first edition of *Cosmographicus Liber,* this illustration portrays a rather plain and unimpressive figure demonstrating the use of the cross-staff; in the 1529 and all later editions, both the pose and appearance of the observer have been transformed to represent a more distinguished individual; see Figure 5 of Alan Stimson's paper in this volume.

19. Ortroy, "Bibliographie" (*op. cit.,* note 17). The title was abbreviated to *Cosmographia...* from 1539 onward. The original Latin text was translated into French, Spanish, and Flemish.

eighteenth century to baffle every mathematician, including Isaac Newton (1642-1727), who believed that the lunar-distance method was the only method of finding longitude at sea with any chance of success.[20] The problem of predicting the Moon's motion with sufficient accuracy was eventually solved through the collaboration of a mathematician and an astronomer —Leonhard Euler (1707-1783) and Tobias Mayer (1723-1762)[21]—both of whom, like Johann Werner, had no experience of navigation at sea.

In order to place the lunar-distance method in the context of the period in which it was first suggested, the original text and translation of Werner's lunar-distance proposal are accompanied by two preceding passages, one about finding longitude by lunar eclipses and the other concerning the use of the cross-staff.

20. For Newton's opinion on the methods of finding longitude, see William Andrewes's paper (note 5 and related text) in this volume.

21. For further information on Euler and Mayer, see Bruce Chandler's and Derek Howse's papers in this volume.

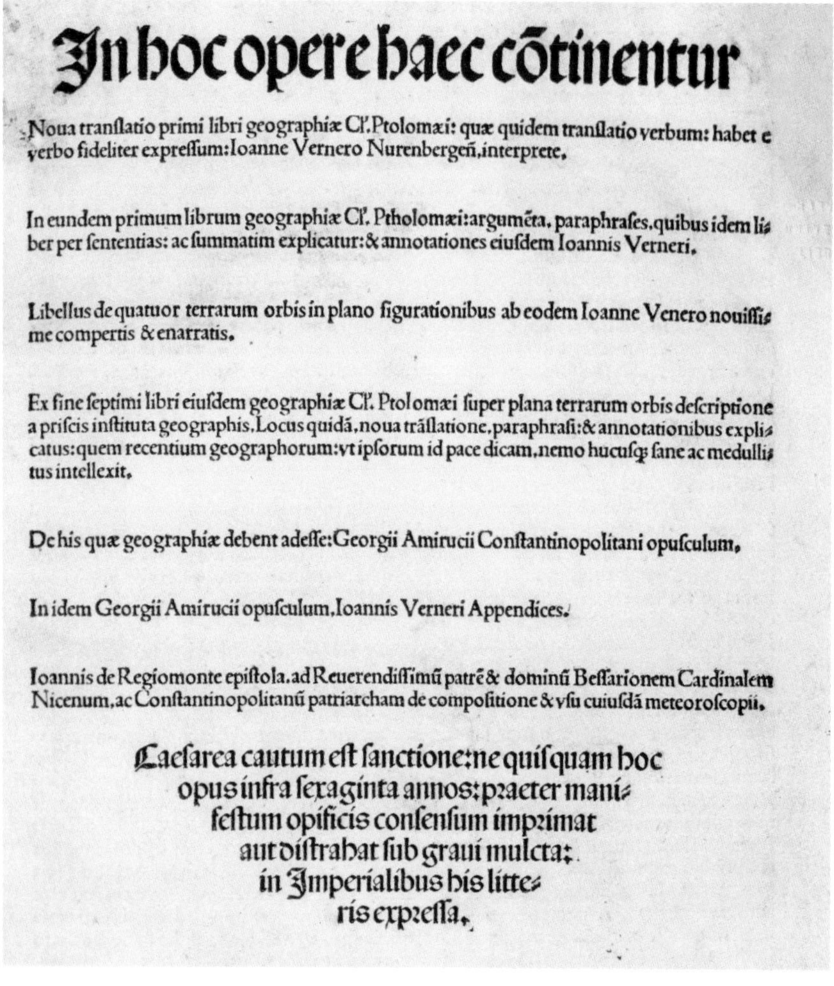

In hoc opere haec cõtinentur

Noua tranflatio primi libri geographiæ Cl'. Ptolomæi: quæ quidem tranflatio verbum: habet e verbo fideliter expreffum: Ioanne Vernero Nurenbergeñ. interprete.

In eundem primum librum geographiæ Cl'. Ptholomæi: arguméta, paraphrafes, quibus idem liber per fententias: ac fummatim explicatur: & annotationes eiufdem Ioannis Verneri.

Libellus de quatuor terrarum orbis in plano figurationibus ab eodem Ioanne Venero nouiffime compertis & enarratis.

Ex fine feptimi libri eiufdem geographiæ Cl'. Ptolomæi fuper plana terrarum orbis defcriptione a prifcis inftituta geographis. Locus quidã, noua trãflatione, paraphrafi: & annotationibus explicatus: quem recentium geographorum: vt ipforum id pace dicam, nemo hucufq; fane ac medullitus intellexit.

De his quæ geographiæ debent adeffe: Georgii Amirucii Conftantinopolitani opufculum.

In idem Georgii Amirucii opufculum, Ioannis Verneri Appendices.

Ioannis de Regiomonte epiftola. ad Reuerendiffimũ patrẽ & dominũ Beffarionem Cardinalem Nicenum, ac Conftantinopolitanũ patriarcham de compofitione & vfu cuiufdã meteorofcopii.

Caefarea cautum eft fanctione: ne quifquam hoc opus infra fexaginta annos: praeter manifeftum opificis confenfum imprimat aut diftrahat fub graui mulcta: in Imperialibus his litteris expreffa.

Figure 3. Title page of Johann Werner's *In hoc opere haec cõtinentur...* (Nuremberg, 1514). By permission of the Houghton Library, Harvard University. Ref. no. Gp 120.21F*.

**Figure 4. Note 2 of *In hoc opere haec cõtinentur...*, sig. cv[v]. By permission of the Houghton Library, Harvard University.**

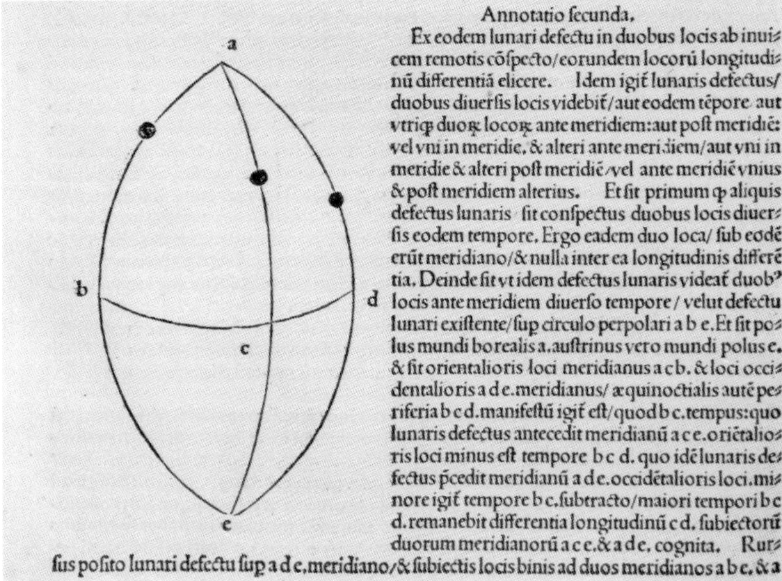


Annotatio secunda.

Ex eodem lunari defectu in duobus locis ab inui- cem remotis cõspecto/eorundem locorũ longitudi- nũ differentiã elicere. Idem igiť lunaris defectus/ duobus diuersis locis videbiť/aut eodem tẽpore aut vtriqꝫ duorꝫ locorꝫ ante meridiem: aut post meridiẽ: vel vni in meridie. & alteri ante meridiem/aut vni in meridie & alteri post meridiẽ/vel ante meridiẽ vnius & post meridiem alterius. Et sit primum qꝫ aliquis defectus lunaris sit conspectus duobus locis diuer- sis eodem tempore. Ergo eadem duo loca/ sub eodẽ erũt meridiano/& nulla inter ea longitudinis differẽ tia. Deinde sit vt idem defectus lunaris videať duob⁹ locis ante meridiem diuerso tempore / velut defectu lunari existente/sup circulo perpolari a b e. Et sit po- lus mundi borealis a. austrinus vero mundi polus e. & sit orientalioris loci meridianus a c b. & loci occi- dentalioris a d e. meridianus/ æquinoctialis autẽ pe- riferia b c d. manifestũ igiť est/quod b c. tempus: quo lunaris defectus antecedit meridianũ a c e. oriẽtalio- ris loci minus est tempore b c d. quo idẽ lunaris de- fectus pcedit meridianũ a d e. occidẽtalioris loci. mi- nore igiť tempore b c. subtracto/maiori tempori b c d. remanebit differentia longitudinũ c d. subiectorũ duorum meridianorũ a c e. & a d e. cognita. Rur- sus posito lunari defectu sup a d e. meridiano/& subiectis locis binis ad duos meridianos a b e. & a

## I) CONCERNING FINDING LONGITUDE BY LUNAR ECLIPSES: NOTE 2[22]

From the same lunar eclipse observed at two places distant from each other, to determine the difference in longitude between these places.

This lunar eclipse will therefore be seen at the two different locations: either at the same time, or before the meridian[23] at each of the two places, or after the meridian, or at the meridian at one and before the meridian at the other, or at the meridian at one and after the meridian at the other, or else before the meridian at one and after the meridian at the other. First, let it be that any lunar eclipse is observed at two different locations at the same time, then these two places are beneath the same meridian, and there is no difference in longitude between them.

Now, let the same lunar eclipse be seen from two locations at different times before the meridian, as if the lunar eclipse were on the meridian *abe*. Let the northern pole of the world be *a* and the southern pole of the world be *e*. Also let *ac[e]* be the meridian for the eastern location and *ade* the meridian for the western location. The equator is *bcd*. It is therefore clear that *bc*, the time by which the lunar eclipse precedes *ace*, the meridian for the eastern position, is less than the *bcd*, the time by which the same lunar eclipse precedes *ade*, the meridian for the western position. Therefore, when the smaller time *bc* is subtracted from the greater time *bcd*, there results *cd*, the difference in longitude determined between *ace* and *ade*, the two meridians under consideration.

On the other hand, let the lunar eclipse be on meridian *ade* and let the two locations investigated be at the two meridians *abe* and *ace*.

22. Werner, *In hoc opere* (*op. cit.*, note 7), *cap.* 4, note 2, sigs. cv[v]–di[r].

23. Werner uses "meridiem" here, but a translation using "noon" is not appropriate, since a lunar eclipse cannot be observed in daylight. In these instances the term will be translated as "meridian" since it fits the meaning that Werner seems to convey with his description of the appearances from different locations. Note, however, that Werner also uses the term "meridianus." In reviewing the intended meaning of this passage, we appreciate the help of Derek Howse, who brought to our attention the comments made to him by the late Donald Sadler, former superintendent of the Nautical Almanac Office.

ce. Igitur defectus lunaris super a d e. circulo perpolari siue meridiano/ vtriq; locoꝝ ad duos meridianos a b e. & a c e. videbiť post meridiem/ minus igitur tempus c d. quo defectus idem lunaris
cõspiciť post meridianũ occidentaliorẽ a c e. subtractũ ex b c d. maiore tempore relinquiť tempus
b c. differẽtia existens longitudinũ duorũ meridianoꝝ super datoꝝ duoꝝ locoꝝ vertices euntiũ.
Deinde datus lunaris defectus sit super meridiano a b e. vnius subiectorũ locorum duorum: &
alter locus sit ad meridianũ a c e. igitur manifestũ est b c. tempus/ seu segmentũ æquatoris esse differentiã longitudinũ duorũ locorũ ad duos meridianos a b c. & a c e. iacentium/ par est ferme ra-
tio/ si lunæ defectus extiterit in meridie vnius duorũ locorũ & post meridiem alteri⁹/ velut si lunæ
defectus cõtingeret in occidentalioris loci meridiano a d e/ videreť ergo ipse post orientalioris lo
ci meridianũ a c e. perspicuũ igiť erit c d. tempus/ quo idem defectus lunaris comitať meridianum
a c e. orientalioris loci/ esse longitudinũ differentiã/ subiectorũ duoꝝ locorum. Esto nunc vt luna
ris defectus cõstitutus ad meridianũ a c e. cõspiciať ante meridiem occidentalioris loci ad meridianum a d e. positi/ & videať post meridiem oriẽtalioris loci/ cuius meridianus sit a b e/ igiť tempus
c d/ quo idem defectus videť ante meridiem occidentalis loci/ iunctum tempori b c. quo ipse lunaris defectus conspiciť post meridianũ a b e. orientalioris loci cõflabit differentiã lõgitudinũ b c d.
inter duos meridianos a b e. & a d e. subiectorũ duorum locorum/ quod oportuit declarare.
Corolarium. Inde etiam liquet/ ꝙ duo loca quibus idem lunaris defectus/ eodem cõspiciť tempore sub eodem existunt meridiano. Et locorũ duorum quibus idem lunæ defectus ante meridiẽ
diuerso videbiť tempore/ orientaliorẽ esse eum cuius tempus defectus ante meridianũ maius exti
terit. Et duorũ locorum quibus idem lunæ defectus post meridiem: diuerso videbiť tempore/
eum rursus esse oriẽtaliorẽ/ cuius tempus post meridianũ quo luna deficit maius fuerit. Et si lunæ
deliquiũ vni duorum locorũ cõtigerit in meridie/ alteri vero ante meridiem videri/ ille locus erit
orientalior/ cui deliquiũ lunæ in meridie videbiť. At vbi lunæ defectus/ vni duorum locoꝝ in me
ridie/ alteri post meridiẽ cõspicieť: ergo ille locus orientalis existit/ cui lunæ deliquiũ post meridiẽ
videtur. Similiter lunæ deliquio cõspecto/ apud vnum duorũ locorum post meridiem/ apud alte-
rum vero ante meridiem ille rursus orientalis existit/ qui lunarem defectum a meridie intuetur.
Obiter quoq; notandũ est q̃d in obseruatione deliquiorũ lunariũ in diuersis locis cõspectorũ/
tempora vel initii vel medii vel finis adinuicem cõparari debent. Quemadmodũ ego Hromæ cõ-
spexi lunæ deliquiũ/ quod fuit anno dñi. 1497. post diem. xviii. Ianuarii/ sub noctis principium.
Eiusdem itaq; lunaris defectus principiũ Hromæ fuit a me visum: post diem decimũ octauũ Ianua
rii horis quinq; minutis. xxiv. Et Nurenbergæ eiusdem deliquii lunaris initiũ cõspectum fuit post
meridiem horis. iv. m̃. lij. fere quemadmodũ computus habuit ephemeridis Ioannis de regiomõ-
te. duratio deniq; eiusdem eclipsis lunaris/ vtrobiq; tãta fuit visa quantã eiusdem Ioannis cõputus
indicauit. Inde patuit/ differentiã longitudinũ vrbis Hromæ & Nurenbergæ prope esse minutoꝝ
xxxii. siue graduũ octo. Et per iam positũ patet corolariũ Hromam esse orientaliorẽ Nurenber-
ga. octo gradibus æquinoctialis. Quamq̃ eiusdẽ vrbis Hromæ & Nurẽbergæ Germaniæ op-
pidi longitudinũ differentiã: Ioannes de Regiomonte/ in problemate. xlv. canonis tabularũ primi mobilis scripsit esse graduum. ix. Sed ego libentius credo illam esse octo tantũ graduũ: quo-
niam iuxta meam obseruationem/ latitudo vrbis Hromæ est/ gra. xlj. m̃. l. Latitudo vero Nurenbergæ gra. xlix. m̃. xxiv. Itinerariũ vero interuallum inter Hromã & Nurenbergam/ germanico-
rum milliariorũ fere. cl. his si propter obliquitatem viarum decima pars detrahatur remanẽt mil-
liaria germanica. cxxxv. quibus/ de maximo circulo siue de ambitu terræ gradus respondent ix.
his ita suppositis per librum de sphæricis triangulis: differentia longitudinũ vrbis Hromæ & Nurenbergæ reperietur etiam gra. viii. quanta iam pridem per lunare deliquium fuerat comperta.
In vetustissimo deniq; libro latitudines quarundam ciuitatum a modernioribus mathematicis
obseruatas cõtinente reperi: eandem vrbis Hromæ latitudinem graduum. xlj. m̃. l. q̃d & meam in-
uentionem esse veram/ plurimum nititur affirmare.

**Figure 5.** Note 2 of *In hoc opere haec cõtinentur...*, sig. di[r]. By permission of the Houghton Library, Harvard University.

Then the lunar eclipse on polar circle or meridian *ade* will be seen after the meridian at each of the locations under meridians *abe* and *ace*. Therefore, time *cd*, at which this lunar eclipse is seen from the western meridian *ace*, subtracted from the greater time *bcd* results in time *bc*, which is the difference in longitude between the two meridians which pass through the poles of the two given locations.

Now let the given lunar eclipse, as well as one of the two places under investigation, be on meridian *abe*, and let the other place be on meridian *ace*. Therefore, it is clear that time *bc*, which is an arc of the equator, is the difference in longitude between the two places lying on the two meridians *ab[e]* and *ace*.

The method is the same if the lunar eclipse occurs on the meridian for one of the two places and after the meridian for the other, just as if the lunar eclipse occurred on meridian *ade*, which is the western location. Then it would be seen after meridian *ace*, which is the eastern location. It will therefore be clear that *cd*, the time by which that lunar eclipse follows meridian *ace* of the eastern place, is the difference in longitude between the two positions under investigation.

Now let the lunar eclipse located at meridian *ace* be seen before the meridian of the western position, which is located at meridian *ade*, and let it be seen after the meridian of the eastern place, the meridian for which is *abe*. Therefore, *cd*, the time by which that eclipse is seen before the meridian of the western location, added to *bc*, the time by which this same lunar eclipse is seen after the meridian *abe* for the eastern position, will be *bcd,* the difference in longitude between the two meridians *abe* and *ade* for the two positions being considered. QED.

Corollary. Accordingly, it is also clear that the two positions at which one lunar eclipse may be seen at the same time are on the same meridian. Moreover, for two places at which the same lunar eclipse is seen before the meridian but at different times, the [western]most is the one for which the time of the eclipse before the meridian is the greatest. Also, for two places at which the same lunar eclipse is seen after the meridian but at different times, in this case the easternmost is the one for which time after the meridian at which the Moon is eclipsed is the greatest. And if the lunar eclipse occurs at the meridian for one of the two locations and before the meridian at the other, then that place will be toward the east, that is, the one at which the lunar eclipse is seen at the meridian. But when the lunar eclipse is seen at the meridian at one of the two locations and after the meridian at the other, then that place will be to the east, that is, the one at which the lunar eclipse appears after the meridian. Similarly, when the lunar eclipse is seen after the meridian at one of the two locations and before the meridian at the other, then that one will be toward the east, which sees the lunar eclipse after the meridian.

Incidentally, it must also be noted that in the observation of lunar eclipses seen from different locations, the times of first contact or of mid-eclipse or of last contact should be compared with one another. For example, I myself saw a lunar eclipse in Rome on the evening of 18 January in the year 1497. First contact for this lunar eclipse appeared to me, in Rome, to be at 5:24 p.m. on 18 January. But in Nuremberg first contact for this same lunar eclipse was seen at approximately 4:52 p.m., as computed from the ephemerides of Johannes of Regiomontanus. Therefore, the duration of this lunar eclipse from beginning to end is as great as the amount computed by Regiomontanus. Hence, it is obvious that the difference in longitude between the city of Rome and of Nuremberg is nearly 32 minutes or 8°. And so, by means of the corollary already set out, it is clear that Rome is east of Nuremberg by 8° of the equator.[24] In Problem XLV of the book of the *Tabulae Primi Mobilis*, Regiomontanus wrote that the difference in longitude between the city of Rome and Nuremberg in Germany is 9°. But I more readily believe that it is only as large as 8°, because according to my own observation the latitude of the city of Rome is 41° 50', whereas the latitude of Nuremberg is 49° 24'. Furthermore, the distance for traveling between Rome and Nuremberg is roughly 150 German miles. If a tenth of this figure is subtracted to account for the indirect route, there remains 135 German miles, to which corresponds 9° of a great circle or the circumference of the Earth. So, with these values having been calculated from a book of spherical triangles, the difference in longitude between the city of Rome and Nuremberg is indeed 8°, just as was already determined from the lunar eclipse.

Finally, the observed latitudes of certain cities in that very old book [Regiomontanus's *Tabulae Primi Mobilis*] has been approximately determined by more recent mathematicians; furthermore, they have made an effort especially to confirm that the latitude of the city of Rome is 41° 50' and that my [value] is accurate.[25]

24. Obviously, the time of one or the other or both of these observations is incorrect. Rome lies only about 1° 30' east of Nuremberg.

25. The current value given for the latitude of Rome is 41° 53' and that of Nuremberg is 49° 27', a variation of 3' greater in both cases than those given by Werner.

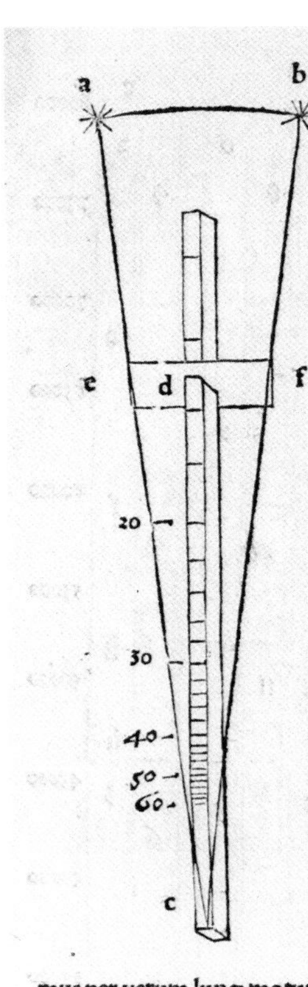


Annotatio ſexta. Duarum ſtellarũ intercapedinẽ officio præ-
miſſi radii viſorii depræhendere. Sint igiẽ propoſitæ ſtellæ duæ a b.
quarũ interſtitiũ velut ſegmentũ a b. maximi circuli per eaſdem ſtel
las deſcripti: intẽtio ſit perſpicuũ facere: Igitur radius obſeruatori⁹
c d. habens pinacidiũ verſatile e f. cũ termino eius c. oculo applicetẽ:
deinde volubile pinacidiũ e f. paulatim admoueatẽ oculo donec per
extremitates e f. eiuſdem pinacidii mobilis duæ datæ ſtellæ a b. examuſſim cõſpiciantẽ: ergo longitudo c d. inter oculũ & pinacidiũ e f.
in ſpacio ad idem pinacidiũ ꝑtinente: docebit quot gra. & m̃. ſit a b.
ſegmentũ: qđ oportuit efficere. Annotatio ſeptima.

Duorũ locorũ/ quos ex lõginquo geographus proſpicit/ angu-
lum quem apud locũ geographi conſiderantis efficiunt obſeruare.
Sint igitur tria loca a b c. & angulũ a c b. quem duo loca a b. ad c. locum cõſtituunt/ intẽtio noſtra ſit obſeruare. igitur geographus ra-
dium viſoriũ hunc cum vna eius extremitate c. oculo applicet/ volu
bile deinde pinacidium e f. oculo ſuo admoueat: vt ꝑ terminos e f.
eiuſdem pinacidii intueatur duo loca a b. rurſus ergo c d. longitudo
inter pinacidiũ e f. & c. oculũ cõſiderãtis exiſtens in ſpacio radii/ qđ
eidem inſeruit pinacidio/ angulũ a c b. in gradibus & minutiis pate-
facit. quod oportuit efficere. Annotatio octaua.

Duorum locorũ qui abinnicem plurimũ diſtant. Longitudinũ
differentiã per motum verum lunæ/ atq; per aliquam ſtellam fixam
quæ vltra quinq; gradus latitudinẽ ab æcliptica non habeat inuenire. In hoc problemate ſupponendũ eſt/ tabulas medii & veri motus lunæ/ ad alterũ ꝓpoſitorũ locorꝝ examuſſim eſſe cõpoſitas/ atq;
iuſtiſſime verificatas: præterea/ ſiderum fixorum quæ huius adhi-
bentur problematis vtilitati/ motus tam in longitudine q̃ in latitudi
ne veraciter innoteſcant. His itaq; ſubiectis. Sint duo loca quæ plurimũ elongentẽ abinuicem/ & intentio ſit eorꝝ differentiã longitudinũ inuenire: igitur geographus accedat ad vnum datorũ locorum/
& in eo per radiũ hunc obſeruatoriũ/ cõſideret ad aliqđ momentũ
cognitũ/ diſtantiã lunæ vniuſq; dictorꝝ fixorum ſiderum/ quæ pa-
rum aut nihil ab ecliptica recedant/ quã quidem diſtantiã ſi diuiſeri-
mus per verum lunæ motum in vna hora/ exibit tempus/ quo luna cum eodem ſidere fixo cõiũge
tur/ ſi talis eorum coniunctio adhuc exiſtit futura/ aut tempus patebit/ quo eadem lunæ & inſpecti
ſideris cõiunctio præteriuerat. Deinde ꝓ meridiano loci alterius abſentis/ eandem lunæ fixiq;
ſideris cõiunctionẽ ex tabulis medii veriq; motus lunæ ꝓ eodem loco abſente verificatis geographus computet. deniq; hæc duo tempora ꝓ meridianis eorundẽ duorum locorũ/ velut de eclipſi
lunæ ſuperius traditũ fuit comparando: inueniet eorundẽ duorũ locorꝝ differentiã longitudinũ.
quam oportuit reperire/ neq; lunæ diuerſitas aſpectus in longitudine/ quæ modica exiſtit geogra
phum perturbet. Et ſi ſcrupuli huius dubio angaẽ. ergo quintũ librum magnæ cõpoſitionis claudii Pto. cõſulat/ ſtatimq; reperiet ex viſa illa lunæ eiuſdẽq; fixi ſideris diſtãtia/ verã eorꝝ elõgationẽ

Argumentum. cap. v.

Figure 6. Notes 6, 7, and 8 of *In hoc opere haec cõtinentur....*, sig. dv$^v$. By permission of the Houghton Library, Harvard University.

## II) CONCERNING THE USE OF THE CROSS-STAFF: NOTE 6[26]

To find the interval [angular distance] between two stars by means of the aforementioned observing staff.

First let *a* and *b* be the two suggested stars, the interval between which is segment *ab* of a great circle drawn through these stars. It is our purpose to find this interval. So the observing staff *cd*, which has a moveable crosspiece *ef*, is placed with the end *c* near the eye. Then the moveable crosspiece *ef* is adjusted little by little until the two given stars *a* and *b* may be seen exactly at the ends *e* and *f* of this moveable crosspiece. Therefore, the distance *cd* between the eye and the crosspiece *ef* [measured on the scale] along the length [of the staff] extending to this same crosspiece, will be known, and it is equal to segment *ab* in degrees and minutes.

26. Werner, *In hoc opere* (*op. cit.*, note 7), *cap.* 4; notes 6, 7 and 8 in this chapter all appear on the same page, sig. dv$^v$.

## NOTE 7

To observe the angle which lies between two places that the geographer sees at a distance and which they form with respect to the location of the observing geographer.

First let *a*, *b*, and *c* be three positions and let *acb* be the angle which contains positions *a* and *b* with respect to position *c*. It is our intention to observe [this angle]. Therefore, let the geographer hold this observing staff with one of its ends *c* near his eye. Let him then adjust the moveable crosspiece *ef* with respect to his eye so that he sees the two positions *a* and *b* at ends *e* and *f* of this crosspiece. So again, measuring [on the scale] along the staff to which the crosspiece is attached, the distance *cd* between the crosspiece *ef* and *c*, which is located at the eye of the observer, clearly shows angle *acb* in degrees and minutes. QED.

### III) CONCERNING FINDING LONGITUDE BY LUNAR DISTANCES: NOTE 8

To find the difference in longitude between two places which are very far from one another by means of the true motion of the Moon and by means of a fixed star which is no more than 5° in latitude from the ecliptic.

In this problem it must be established that the tables for the mean and true motions of the Moon have been precisely computed for one of the two given locations and that they have been accurately verified. Moreover, the [positions] of the fixed stars which are used in solving this problem should be accurately known in longitude as well as in latitude.

With these matters having been established, let there be two places which are very far from one another. It is our purpose to find the difference in longitude between them. Then, let the geographer go to one of the given places and from there observe, with this observing staff at a known time, the distance between the Moon and one of the aforementioned fixed stars which diverges little or not at all from the ecliptic. If we divide this distance by the Moon's true motion for one hour, there will result either the time at which the Moon is in conjunction with this fixed star if such a conjunction of these bodies is yet to occur in the future, or else the time at which that conjunction of the Moon and the observed star did occur.

Next, let the geographer calculate this conjunction of the Moon and the fixed star for the meridian of the other, distant location by means of accurate tables of the Moon's true and mean motions for that distant location. Finally, by comparing these two times for the meridians of those two places, just as was explained above for lunar eclipses, he will find the difference in longitude between these two places. QED.

The Moon's parallax in longitude, which is insignificant, should not worry the geographer. But if he is bothered by doubts about this detail, then let him consult the fifth book of Ptolemy's great work [the *Almagest*] and he will immediately find the true elongation of these bodies from the apparent distance of the Moon from this fixed star.

The idea of using a portable mechanical clock to find longitude was first proposed in 1530 by Gemma Frisius[27] in *...de Principiis Astronomiae & Cosmographiae,...*.[28] An experiment using a portable timekeeper to carry the time from one place to another had, in fact, been made about 100 years earlier by Michael Savonarola (1384-1464), but its purpose was not to determine a difference in longitude.[29] Unlike Werner's book, *...de Principiis Astronomiae & Cosmographiae...* was well received almost immediately, going through eleven editions between 1530 and 1582;[30] in addition, the passage relating to the method of finding longitude was translated into English in 1555 and was subsequently described in English in 1559.[31] The proposal for finding longitude by means of a timekeeper, which appears as chapter 18 in the 1530 edition, makes no mention of the value of the method for navigation at sea. In the fifth edition (1553), in which this section becomes chap-

**Figure 7. Portrait of Gemma Frisius in 1555, the year of his death. In the foreground on the right is a universal ring dial, an instrument he invented about 1533 for finding local time at sea. Copper plate engraving dated 1557 by Jan van Stalburch (active in Louvain 1555-1562). 30.1 x 23.6 cm. (11.8 x 9.3 in.). Courtesy of the Ashmolean Museum, Oxford.**

27. In official documents cited by van Ortroy ("Bio-bibliographie" [*op. cit.*, note 1], pp. 11-12), he was referred to as "Gemma Reyneri de Gruninghâ" (University of Louvain, 1528) and "Gemma Regneri Frisius medicus" (Charles V, 1544). Based on this evidence, van Ortroy asserted that Gemma was his first name and Regnier, his father's name. The name Frisius—coming from Friesland, his homeland—had been adopted as his last name by 1529; in his revised edition of Peter Apian's *Cosmographicus Liber* (*op. cit.*, note 18), he was referred to as Gemma Frisius: *Cosmographicus Liber Petri Apiani Mathematici, studiose correctus, ac erroribus vindicatus per Gemmam Phrysium* (Antwerp, 1529).

28. *Gemma Phrysius de Principiis Astronomiae & Cosmographiae, Deque usu Globi ab eodem editi* (Louvain and Antwerp, 1530).

29. Lynn Thorndike (1882-1965), in *A History of Magic and Experimental Science,* 8 vols. (New York: Columbia University Press, 1923-58), vol. 4, p. 203, cites a dispute between Savonarola and Francesco Bussone (*ca.* 1390-1432), the count of Carmagnola, as to whether the water in the baths of St. Helena was warmer than that in the baths of Abano, a few miles away. To resolve the matter, they dispatched a messenger to Abano with a phial to fill with water and a *"portantem orologium"* (a portable timekeeping device, which was perhaps more likely to have been a sandglass than a spring-driven mechanical clock) to ensure that the sample of water in Abano was taken at the same time as the sample in St. Helena. In this way, when brought together for comparison, both samples would have had the same amount of time to cool. Although neither man was proved right because the waters were found to differ little in heat, the incident is worthy of note as being one of the earliest references of a portable instrument of time being used in an experiment.

30. The following list of editions was compiled from Fernand van Ortroy, "Bio-bibliographie" (*op. cit.*, note 1), with additional reference to *The National Union Catalog Pre-1956 Imprints* and *The British Library General Catalogue of Printed Books to 1975* (all works listed are in Latin unless otherwise stated): Louvain and Antwerp, 1530; Antwerp, 1544; Paris, 1547 (on title page, but dated 1548 at the end); Antwerp, 1548 (with two tracts by Johannes Schöner bound in); Antwerp, 1553 (the enlarged fifth edition with a slightly different title [see note 39]); Paris, 1556 (two editions, one in French and one in Latin); Paris, 1557 (two editions, one in French and one in Latin); Cologne, 1578; Paris, 1582 (French). A further edition, mentioned only in *The National Union Catalog*, appeared in Paris, 1619 (French).

31. The first English translation of Gemma's new method of finding longitude was published by Richard Eden (1521?-1576) in *The Decades of the newe worlde or west India,...wrytten in the Latine tounge by Peter Martyr of Angleria* (London, 1555), pp. 361-2, under the title "A newe maner of fyndynge the Longitudes of regions" in the section "The maner of fyndynge the Longitude of regions by divers wayes after the description of Gemma Phrysius." The second mention of the timekeeper method in English—described in the form of a dialogue—was made by William Cuningham (b. 1531) in *The Cosmographical Glasse, conteinyng the pleasant Principles of Cosmographie, Geographie,*

ter 19, the original passage concerning finding longitude is repeated verbatim, apart from a few typographical changes, but added to it is a short tract recommending the use of water clocks or sandglasses as time-measuring devices to regulate clocks for determining longitude on long journeys, "especially at sea." The very suggestion of using such time-measuring devices is evidence of how poor the early portable mechanical timekeepers were: obviously, by 1553, Gemma had become aware of their impracticability because of their short duration and unreliability. Water clocks and sandglasses at that time were perhaps more dependable for measuring the period of a day, but neither, of course, could have made his scheme practicable.[32]

Although portable mechanical timekeepers were becoming more available at this time,[33] their accuracy and reliability left much to be desired; even on land, no timekeeper was capable of maintaining anything like the rate required for keeping the time of a place of known longitude. Because of these shortcomings, mechanical clocks of the sixteenth and early seventeenth centuries were designed to emphasize other features, either in the design and decoration of their cases, or through the inclusion of some complication, such as astronomical or calendrical dials, or automata that provided some hourly entertainment. Unaware of the practical problems of timekeeping, Gemma stressed, in the first edition of his book, only that care should be taken to set the clock and to ensure that it did not stop on the journey. In the 1553 and later editions, he recognized that changes in the weather affected the performance of clocks, but was under the impression that this could be overcome by making the mechanism with extreme care.[34]

Gemma was only 22 when he published his proposal for finding longitude. Educated in Groningen and at the University of Louvain, he received

**Figure 8. Small portable mechanical timekeeper of the type in use at the time that Gemma suggested his method of finding longitude. Detail from the portrait of the merchant Georg Gisze (b. 1497), painted by Hans Holbein (1497/8-1543) in 1532. Oil on oak panel. Dimensions of portrait: 96.3 x 85.7 cm. (37.9 x 33.7 in.). Courtesy of the Gemäldegalerie, Berlin (Dahlem). Inv. no. 586.**

**Figure 9. View of the case and movement of a small drum-shaped watch of the type shown in the portrait of Georg Gisze and available at the time that Gemma proposed his method. This example, marked "Beauvais," is French and was probably made around 1540. Comparing it with Harrison's fourth marine timekeeper (see Anthony Randall's paper, Figures 8-11 and 18-19) illustrates clearly how much technology had to improve before a mechanical timekeeper could perform well enough to make Gemma's idea practicable. Case diameter: 5.2 cm. (2.05 in.). Courtesy of The Time Museum, Rockford, Illinois. Inv. no. 1070.**

*Hydrographie, or Navigation* (London, 1559), fol. 109. This passage is reproduced in Derek Howse's *Greenwich time* (*op. cit.*, note 4), fig. 4.

32. Water clocks are among the oldest instruments of time measurement. They were superseded by the mechanical clock in the fourteenth century, but, with the surge of interest in the works of the ancient Greeks and Romans—in this case, particularly Vitruvius (d. *ca.* 25 B.C.)—during the fifteenth and sixteenth centuries, the water clock became an instrument of interest in the academic world. Although some of the ingenious designs proposed were made, none would have been practical on board ship or provided a consistent measure of time over an extended period. In contrast, by the beginning of the sixteenth century sandglasses were already being used on board ship for measuring specific intervals of time, such as the four-hour watch on deck and the time traveled on a particular course. During the 1590s, they also came into use, in conjunction with the log and line, for measuring the speed of the ship. Attempts to make them practical for use in finding longitude at sea were still being made, without success, in the first half of the eighteenth century. For further general information on these devices, see David W. Waters, *The Art of Navigation in England in Elizabethan and Early Stuart Times* (London: Hollis & Carter, 1958), and A. J. Turner, *The Time Museum*, vol. 1, *Time Measuring Instruments* (Rockford, Illinois: The Time Museum, 1984), part 3, "Water-clocks, Sand-glasses, Fire-clocks."

33. References to and illustrations of spring-driven timekeepers occur in the fifteenth century, but evidence that these timekeepers were really portable—in other words, small enough to be carried in a pouch or worn as a pendant—does not appear until the early sixteenth century. During the second quarter of the sixteenth century, it is clear from the literature and paintings of the period that small drum- and spherical-shaped watches were in vogue. Nevertheless, very few mechanical timekeepers of any kind made before 1550 have survived; because these devices are seldom dated, the iconography of the period remains an important source of information for determining their period of manufacture. The design of portable timekeepers in the shape of what today we would consider a pocket watch evolved during the second half of the sixteenth century. For further information, see Klaus Maurice, *Die deutsche Räderuhr* (Munich: C. H. Beck, 1976), vol. 1, pp. 87-93, and related illustrations in vol. 2, figs. 474-99, and Hugh Tait, *Catalogue of Watches in the British Museum*, vol. 1, *The Stackfreed* (London: British Museum Publications, 1987), pp. 10-12.

34. Overcoming the effects of changing weather conditions—in particular temperature and barometric pressure—was to become one of the greatest challenges for makers of precision timekeepers. Although Gemma would have been completely unaware of the exact causes of this problem (200 years would pass before they were understood), his observation may be the earliest mention of this in print.

his master of arts degree in 1528 and doctor of medicine degree in 1541, becoming a practicing physician and later a professor of medicine at the University of Louvain. His interests included mathematics and cosmography—he was the teacher of Gerardus Mercator (1512-1594) and of Juan de Rojas Sarmiento (fl. 1544-1551)—and he established a workshop for the commercial production of mathematical and astronomical instruments in which Mercator and, later, Gemma's nephew Gualterus (Walter) Arsenius (fl. 1554-1579) were employed. Although Gemma had a great talent for designing instruments, there is no evidence that he ever made any of his own.

The following is a translation from the first edition (1530) of *Gemma Phrysius de Principiis Astronomiae & Cosmographiae, Deque usu Globi ab eodem editi*. The description of the timekeeper method of finding longitude is preceded by the chapter describing other methods of finding the longitude of places; this has been included because it provides an interesting comparison to Gemma's new method.

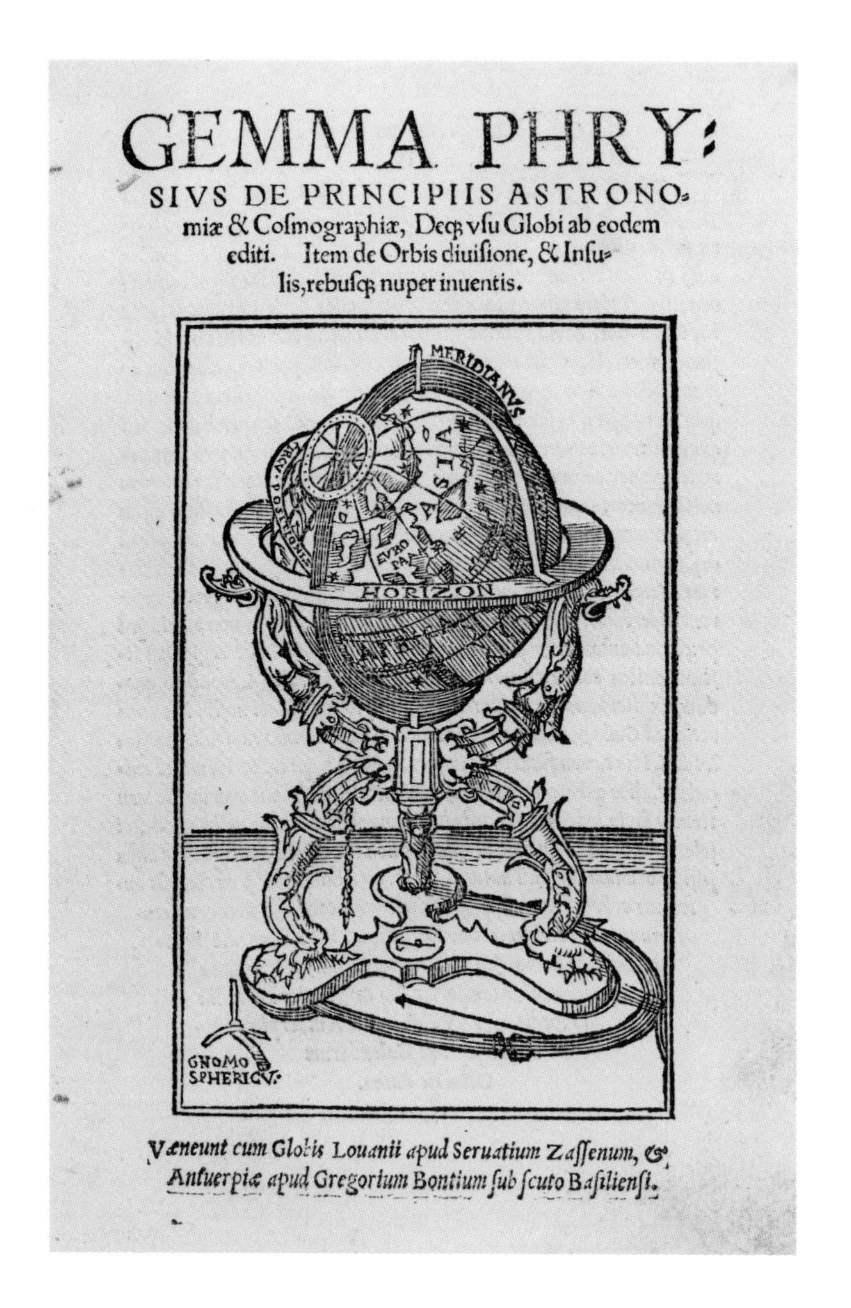
GEMMA PHRY-
SIVS DE PRINCIPIIS ASTRONO-
miæ & Cosmographiæ, Deq; vsu Globi ab eodem
editi. Item de Orbis diuisione, & Insu-
lis, rebusq; nuper inuentis.



Væneunt cum Globis Louanii apud Seruatium Zassenum, &
Antuerpiæ apud Gregorium Bontium sub scuto Basiliensi.

**Figure 10. Title page of *Gemma Phrysius de Principiis Astronomiae & Cosmographiae, Deque usu Globi ab eodem editi* (Louvain and Antwerp, 1530). Courtesy of the John Carter Brown Library at Brown University, Providence, Rhode Island. Ref. no. F530 G323d.**

Figure 11. Chapter 17 of *...de Principiis Astronomiae & Cosmographiae...*, sig. D2r and D2v. Courtesy of the John Carter Brown Library at Brown University.

Caput Decimumſeptimum, De longitudi
tudine regionum inuenienda.

INuento loco Lunæ, conſideranda eſt hora qua eadem talem locum occupet. Deinde ex Ephemeridibus, aut tabulis Alphonſi quota hora Luna in alia regione, aut oppido alio,

D 2 cuius

cuius longitudo nota eſt, ipſa luna eundem locum ſigniferi ingrediatur: Reductis deinde horis ad.24.minor numerus horarum de maiori ſubducendus. Quod vero relinquitur ex horis & minutis ad gradus reducendū eſt, hac lege. Multiplica horas per.15.minuta vero horaria ptire per.4.prouenient gradus æquatoris inter duos meridianos intercepti. Minuta ſi q̄ poſt partitionem ſuperſint, per.15.multiplica, prouenient minuta graduum. Hanc vero differentiā longitudinis inuētam, adde ad longitudinem regionis cognitam, ſi eius loci horæ fuerint plures: aut deme ex eadem longitudine ſi minores fuerint, colliges tandem lōgitudinem loci ignoti, ab inſulis fortunatis. Facilius quidem & promptius idem per globum inueſtigatur geometrice, hac arte. Locum cuius nota eſt longitudo in globo inuentum ſub meridiano collocato, indicem qui circum polum mouetur ad horam qua luna locum præfinitū in tali regione occupat dirigito, deinde globum conuertito donec index horarius ad horam peruenerit, qua quidem tu locum lunæ ignotum quęſieras: gradus æquatoris a meridiano mobili intercisi longitudinem regionis quæſitam aperient. Certiſſima tamē via qua longitudinū differentiæ inueniuntur ea eſt, quæ per idem aliquod quod in omnibus regionibus eodem momento apparet, vtpote per eclipſes lunæ procedit. Cognitis enim diuerſis horis quibus hæc in diuerſis regionibus contingit, non alia ratione longitudo aut arithmetice aut geometrice, q̄ per præcedentem proxime canonem inuenitur. Verum cum hęc neq; ſemper neq; omnibus appareat, deinde præcedens modus ſic ſatis difficilis ſit, neq; idē ſemper promptus propter lunæ coniunctiones, atq; adeo interdū nonnihil a rei veritate deficiat, propter diuerſitatem aſpectus, & latitudinem lunæ: fit vt longitudines multarum regionum præcipue earum quas nuper Hiſpani adiuere, incertas aut penitus nullas habeamus: neq; enim ex itinerum dimenſionibus flexuoſis certi quicq̄ colligi poteſt, confirmante id Ptolomeo libro prim. Coſmographiæ. Subnectam igitur aliquid domi noſtræ natū, quo facili via & quouis tempore longitudines regionum inter proficiſcendum inueniuntur.

*Aliter per globum*

*Certiſſima ratio per Eclypſes.*

35. *Gemma Phrysius de Principiis Astronomiae* (*op. cit.*, note 28), part 2, entitled "De Usu Globi," *cap.* 17, D2r-D2v.

36. The first six lines of the 1530 edition are replaced in the 1553 edition with a more complete passage that translates as follows: "After the position of the Moon has been found by observation as we discussed, at the same time the very moment when the Moon occupies such a position must be very carefully observed by means of a globe. Then from exactly calculated ephemerides, either the Alfonsine Tables or else others from some learned mathematician, the time at which the Moon ought to occupy such a position, in accordance with the observation which was made, must be computed by very meticulous calculation for any precisely known location, for which the longitude is established. Moreover, it is necessary to calculate the time of the day from noon, following astronomical usage, up to 24 hours. Then let the smaller number of hours be subtracted from the larger, or else the earlier time from the later." The rest of the chapter remains the same.

## CHAPTER 17[35]

### ON FINDING THE LONGITUDE OF PLACES.

To find the position of the Moon, the time at which it is located at that position must be observed. Then [it must be determined], from ephemerides or from the Alfonsine Tables, at what hour the Moon is at one position or another, *[End of first page of Gemma's original text]* the longitude of which is known precisely, [exactly as] the Moon itself is entering the same position in the zodiac. Then after reducing the hours to 24, the smaller number of hours must be subtracted from the larger one.[36] The result must be converted from hours and minutes into degrees, like this: Multiply the time by 15', and then divide by 4. Degrees of the equator contained between the two meridians will result. If any minutes remain after the division, multiply them by 15, and minutes of a degree will result. Add this computed difference in longitude to the known longitude of the place if the time for this location was greater, or subtract it from that longitude if it was smaller. You will then find the longitude of the unknown place with respect to the Fortunate Isles.[37]

The same thing may be found more quickly and easily using a globe with this geometrical method: Turn the pointer which revolves about the pole to the time at which the Moon occupies the prescribed location in that region, the longitude of which is known to be beneath a particular meridian on the globe,[38] and then rotate the globe until the hour-pointer shows the time for which you wanted the unknown position of the Moon. The degrees of the equator may be determined from the moveable meridian intersecting the desired longitude of the location.

Nevertheless, the most precise method of finding the difference in longitude is by using something which appears at the same time in every region, just like what happens with lunar eclipses. For when the different times at which this [phenomenon] appears in different regions is known, the longitude is determined by no other geometrical or arithmetic method than by means of the rule just described. But since they do not always appear, nor to everyone, the preceding method is in fact rather difficult, and it is not always at hand because of the [infrequency of] conjunctions of the Moon. Moreover, occasionally it differs somewhat from the true state of affairs because of parallax and the latitude of the Moon. Thus, the longitudes of many regions, especially of those which the Spanish have discovered, are uncertain or completely unknown to us. For nothing certain can be determined from the winding paths of these voyages, as confirmed by Ptolemy in the first book of his *Cosmographia.*[39] Therefore, let me attach something I thought of myself, by which the longitudes of regions can be found with an easy method and at any time while traveling.

37. The Fortunate Isles, the "Isles of the Blest" in Greek mythology, served as the prime meridian used by Ptolemy and correspond with the Canary Islands.

38. It is interesting to note that this is not a Latin grammatical construction; the phrase appears to have been taken from a literal translation of an Arabic text.

39. Ptolemy's *Cosmographia* is, in fact, the same work as *Geographia (The Geography).* Gemma must have had an early edition of this book, because after 1490, the word "cosmographia" in the title, by which this book was first known, was changed to "geographia." The Latin translation of this work from the original Greek was begun in the fifteenth century by the Byzantine scholar Emanuel Chrysoloras (d. 1415) of Constantinople and completed by his student Jacopo Angeli da Scarperia (also known as Jacobus Angelus Florentinus) between 1406 and 1409. Jacopo used the word "cosmographia" in the translation of the title, and this was retained in the first five printed editions (Vicenza, 1475; Rome, 1478; Bologna, dated 1462 but probably a misprint for 1482; Ulm, 1482; Ulm, 1486). The next edition, printed in Rome in 1490, mentioned "geographia" in the title, and in the following 1507 edition—and from then on—*Geographia,* a title more descriptive of its content, was adopted. For further information, see N. M. Swerdlow's chapter "The Recovery of the Exact Sciences of Antiquity: Mathematics, Astronomy, Geography," in Anthony Grafton, ed., *Rome Reborn: The Vatican Library and Renaissance Culture* (New Haven: Yale University Press, 1993), pp. 125–67.

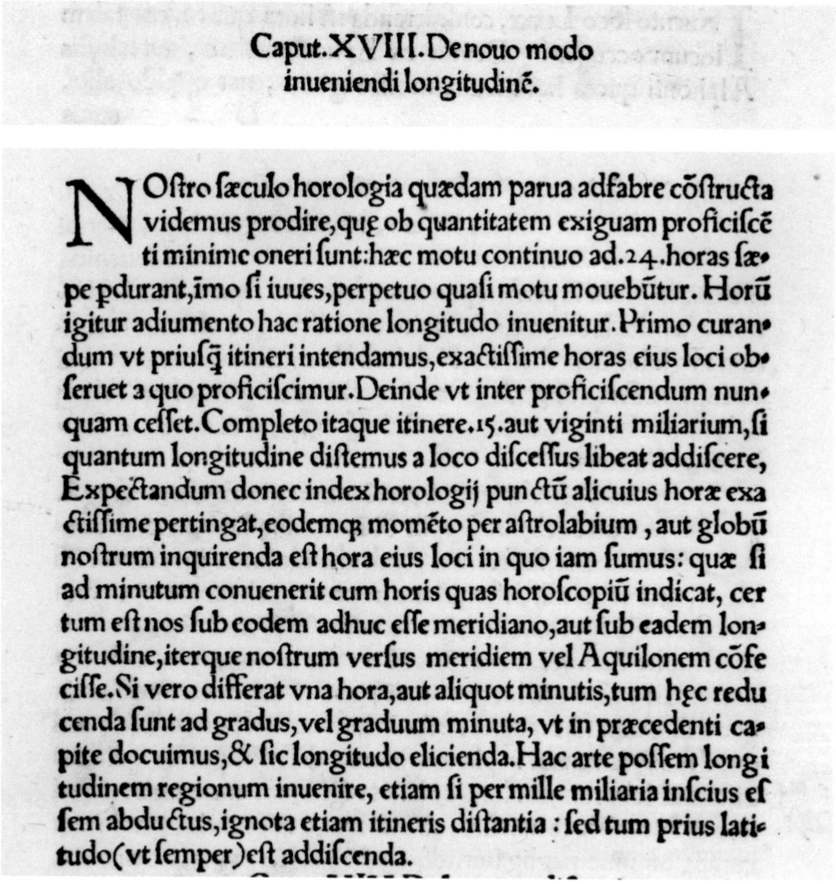

Figure 12. Chapter 18 of ...*de Principiis Astronomiae & Cosmographiae...*, sig. D2v and D3r. Courtesy of the John Carter Brown Library at Brown University.

**Caput.XVIII.De nouo modo inueniendi longitudinē.**

NOſtro ſæculo horologia quædam parua adfabre cōſtructa videmus prodire,quę ob quantitatem exiguam proficiſcē ti minime oneri ſunt:hæc motu continuo ad.24.horas ſæpe ꝑdurant,īmo ſi iuues,perpetuo quaſi motu mouebūtur. Horū igitur adiumento hac ratione longitudo inuenitur.Primo curandum vt priuſq̄ itineri intendamus,exactiſſime horas eius loci obſeruet a quo proficiſcimur.Deinde vt inter proficiſcendum nunquam ceſſet.Completo itaque itinere.15.aut viginti miliarium,ſi quantum longitudine diſtemus a loco diſceſſus libeat addiſcere, Expectandum donec index horologij punctū alicuius horæ exactiſſime pertingat,eodemq; momēto per aſtrolabium , aut globū noſtrum inquirenda eſt hora eius loci in quo iam ſumus: quæ ſi ad minutum conuenerit cum horis quas horoſcopiū indicat, certum eſt nos ſub eodem adhuc eſſe meridiano,aut ſub eadem longitudine,iterque noſtrum verſus meridiem vel Aquilonem cōfeciſſe.Si vero differat vna hora,aut aliquot minutis,tum hęc reducenda ſunt ad gradus,vel graduum minuta, vt in præcedenti capite docuimus,& ſic longitudo elicienda.Hac arte poſſem longitudinem regionum inuenire, etiam ſi per mille miliaria inſcius eſſem abductus,ignota etiam itineris diſtantia : ſed tum prius latitudo(vt ſemper)eſt addiſcenda.

## CHAPTER 18[40]

### CONCERNING A NEW METHOD OF FINDING LONGITUDE.

40. *Gemma Phrysius de Principiis Astronomiae* (*op. cit.*, note 28), part 2, *cap.* 18, fols. D2v-D3r.

We are aware that in our time some small clocks of ingenious construction are being produced which are little burden to a traveler on account of their small size. They often run with a continuous motion for up to 24 hours, and if you wish, they will run with an almost perpetual motion. Longitude is then found with the aid of these clocks according to this method: First, we must take care that before we set out on the journey, the clock is accurately set for the time of the location from which we are traveling. Next, it must never stop on the journey. After 15 or 20 miles have gone by, if we want to learn the distance we have traveled in longitude from the place of departure, we must then wait until the hand of the clock exactly touches the point of any hour and at the same moment find the time for our current location by means of an astrolabe or our globe. If it agrees to the minute with the time the clock shows, it is certain that we are still beneath the same meridian or at the same longitude, and our journey has been made toward the south or the north. But if it differs by an hour or by any number of minutes, then they must be converted into degrees or minutes of a degree, just as we explained in the previous chapter, and in this way the longitude is determined.

By this method I would be able to find the longitude of places, even if I were unknowingly taken 1,000 miles away, and even if the distance of the journey were not known. But first of all (as always) the latitude must be determined.

*Additional passage included in the 1553 and later editions:*[41]

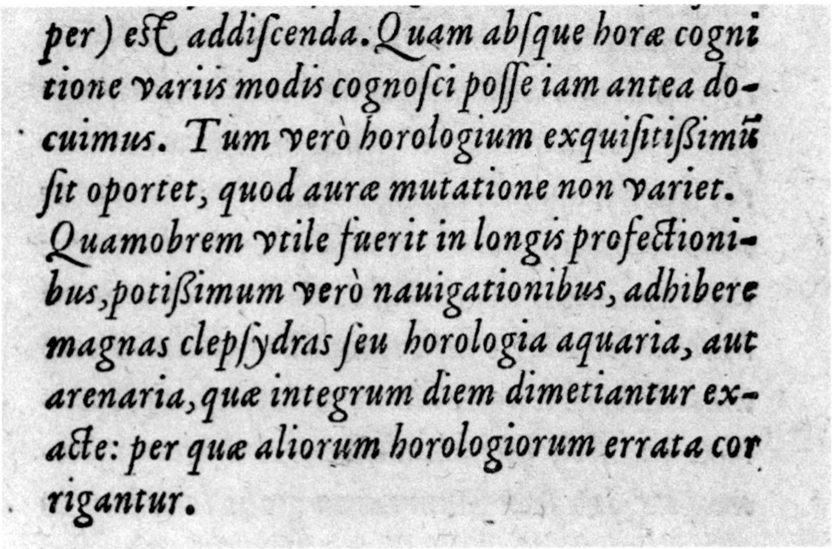

per) eſt addiſcenda. Quam abſque horæ cogni
tione variis modis cognoſci poſſe iam antea docuimus. Tum verò horologium exquiſitiſſimũ
ſit oportet, quod auræ mutatione non variet.
Quamobrem vtile fuerit in longis profectioni-
bus, potiſſimum verò nauigationibus, adhibere
magnas clepſydras ſeu horologia aquaria, aut
arenaria, quæ integrum diem dimetiantur ex-
acte: per quæ aliorum horologiorum errata cor
rigantur.

**Figure 13. Additional paragraph included in the 1553 edition of ...*de Principiis Astronomiae & Cosmographiae...* (chap. 19, p. 65). By permission of the Houghton Library, Harvard University. Ref. no. *NC5.G2848.530de.**

We have already explained above that it can be learned by various methods without knowing the time. Then, more than ever, it is necessary that the clock have been constructed with extreme care, so that it not vary with a change of the weather. Therefore, on long journeys, especially by sea, it is advantageous to use large clepsydras or water clocks or sandglasses, which exactly measure a whole day and through which the errors of the other clocks may be corrected.

41. *Gemmae Phrysii Medici ac Mathematici De Principiis Astronomiae & Cosmographiae, Deque usu Globi ab eodem editi* (Antwerp, 1553), part 2, *cap.* 19 ("De novo modo inveniendi longitudinem"), pp. 64-5.

APPENDIX D

# Finding Local Time at Sea, and the Instruments Employed

# Finding Local Time at Sea, and the Instruments Employed

*William J. H. Andrewes*

> Mr. *Harrison's* Clocks, undoubtedly bid fairest for measuring Time truly at Sea, of all that ever were invented for that Purpose; and his unwearied Pains to bring his curious Machines to still greater Degrees of Exactness, will in Process of Time (we hope) have the desired Effect. But granting we had a just Measure of Time at Sea, there is still requisite a good Method of finding the apparent Time, in that Place where the Ship is; as it is from comparing the true and apparent Time that the Difference of Longitude must be had; and I must frankly acknowledge, that I know of no Method by which to determine the apparent Time at Sea, with less Error than that of some Minutes, which if we take to be only five, this Error, when reduced to its proportional Part of the Equator, will amount to no less than 75 Sea-Miles:....[1]

The problem of finding longitude at sea related not only to finding the time at the reference point from which longitude was to be measured, but also to determining precisely the local, or "apparent," time. As has been stated elsewhere in this book, longitude and time are related in that the Earth rotates 15° every hour. Therefore, if the time at one location can be compared with the time at another, the difference between those times can be converted into the difference in their longitude.[2] It was, however, not until the late 1750s, when both the lunar-distance and the timekeeper methods were developed into viable and practicable solutions to finding the time at the reference point from which longitude was to be measured, that altitude-measuring instruments of the required accuracy became available. Finding local time at sea precisely was an integral part not only of calculating the longitude of a ship, whether by lunars or by timekeepers, but also of checking the going of a timekeeper. The need for finding local time at sea was therefore directly linked to the longitude problem. Yet, with all the attention given to the story of the struggle to win the longitude prize, this aspect of the problem has been overlooked by most historians. When the question was raised by an attendee of the Longitude Symposium, I realized that this publication should include some explanation. This essay, therefore, will exam-

1. William Maitland, *An Essay Towards the Improvement of Navigation, Chiefly with Respect to the Instruments used at Sea* (London, [1750]), pp. 46-7. By "true" time, Maitland means the time of the place from which the longitude of the ship is to be measured, and by "apparent" time, the local time of the ship. Maitland had sailed under the command of Sir Peter Warren (1703-1752) and, therefore, was clearly aware of the practical, as well as the theoretical, problems of navigation. Interestingly, while discussing different methods for finding longitude at sea, he makes no mention of lunar distances, an omission that suggests that this method was not at that time considered a viable alternative by navigators, although it would soon become so.

A "Sea-Mile," commonly known today as a nautical or geographic mile, is the distance of one-sixtieth of a degree of a great circle of the Earth; thus on the equator, it is the distance that the Earth turns every four seconds. Because the Earth is not a perfect sphere, however, estimates of the length of a nautical mile have varied. In the mid-eighteenth century, it was reckoned as about 6,120 feet, but the standard later officially adopted by the British Admiralty was 6,080 feet. The nautical mile is thus about 15 per cent longer than the statute mile of 5,280 feet, which had its origins in Roman times.

2. If the difference between the local times of two places is two hours, then the longitude difference would be 30°. As a more detailed example, the local time at Harvard University is approximately $4^h\ 44^m\ 28^s$ behind that at the Old Royal Observatory at Greenwich; Harvard is therefore 71° 7' 8" west of Greenwich. Because small increments of time difference between places proved highly inconvenient for civil purposes with the growth of public transport by rail and communication by telegraph, the present system of international time zones was introduced in 1884.

ine—albeit briefly—the methods and the evolution of the instruments employed for this purpose.

By the end of the eighteenth century, there were four methods recommended in the principal navigation manuals for determining local time precisely.[3] The first required finding the exact moment of noon, which occurs when the Sun crosses the observer's meridian; this method, known as "equal altitudes," was recommended by Christiaan Huygens (1629-1695) and used during the trial of his marine timekeepers in the 1660s. Its use continued throughout the eighteenth century, and it was the method employed in the early 1760s during both trials of Harrison's prize-winning timekeeper, H.4. The second method, which involved measuring the exact altitude of the Sun by day, and the third, which involved measuring the exact altitude of a star when the horizon was visible, required some prior knowledge of the latitude and the longitude of the ship; these methods were not introduced until the 1760s when precise angle-measuring instruments and improved astronomical tables had been introduced.[4] A fourth method, less well known but also proposed after accurate angle-measuring instruments had been introduced, involved measuring the meridian altitude of the Sun or a star and using spherical trigonometry and logarithms to find the true time from noon when the altitude was taken.[5]

The imaginary line of the meridian marks the point at which the Sun reaches its highest altitude during the day and thus, in addition to the marking the moment of noon, provided the observer with information for determining latitude.[6] To calculate latitude by the Sun, the navigator made frequent observations with an altitude-measuring instrument just before noon to establish when the Sun reached its maximum altitude above the horizon. By subtracting the result from 90° and applying the Sun's declination (found in tables), the latitude of the ship could be calculated.[7]

Because finding latitude and local time both depended on celestial observations, it is natural that the same instruments were employed for both purposes. However, whereas finding latitude required establishing the Sun's altitude at noon (which varies a very small amount around that time), observations for finding local time depended on the far more critical measurements of determining not only the Sun's height above the horizon in the early morning and late afternoon when its altitude is changing rapidly, but also the exact time of these observations. A precise altitude-measuring instrument and an accurate timekeeper were therefore essential. On dry land, the required observations could be made without much difficulty, but on the rocking and pitching deck of a ship moving through the open sea, specialized navigational instruments were needed.

Attempts to find local time at sea date back to the first half of the sixteenth century, when the lunar-distance method and the timekeeper methods for finding longitude were first suggested.[8] Gemma Frisius (1508-1555), who proposed the timekeeper method, devised an ingenious portable sundial, known as a universal ring dial, which indicated local time directly on an horary scale.[9] Despite the fact that this instrument was one of the best sundials of its day, it was still limited in accuracy and was somewhat difficult to use, particularly on the rocking and pitching deck of a ship. Although examples made in the eighteenth century were divided to indicate the time to the nearest minute (15' of arc), such accuracy was impossible to obtain in

3. The navigation manuals used as a general reference for this article were John Robertson's *The Elements of Navigation;...*, 6th ed., revised by William Wales (London, 1796) and John Hamilton Moore's *The New Practical Navigator;...*, 10th ed. (London, 1798). That the former went through seven editions between 1754 and 1805 and the latter nineteen editions between 1772 and 1814 is evidence of the importance of these manuals as navigation changed during the second half of the eighteenth century from a largely experience-based art into a more instrument-dependent science. John Robertson (1712-1776) was appointed first master of the Royal Academy at Portsmouth in 1755 and thereby was the person assigned to set Harrison's fourth marine timekeeper, H.4, to local time before its first trial to Jamaica in 1761.

4. I am grateful to Derek Howse for drawing my attention to the significance of these two methods in finding longitude by lunar distances.

5. A description and examples of this method may be found in John Hamilton Moore's *The New Practical Navigator* (*op. cit.*, note 3), pp. 232-4.

6. A description and brief history of the methods of finding latitude at sea are included in Alan Stimson's paper in this volume.

7. For example, if the Sun's meridian altitude is observed to be 53° 32' N., its zenith distance (found by subtraction from 90°) is 36° 28'. If the date is March 21 (the equinox), the Sun's declination would be 0° and the latitude of the ship would therefore be 36° 28' S., the same as the zenith distance. If, however, the same observation was made on July 29, when the Sun's declination (shown in the tables) is 18° 47' N., then the ship's latitude would be 17° 41' S. Conversely, on November 9 when the Sun's declination is 16° 48' S., the ship's latitude given by this observation would be 53° 16' S. Corrections for the dip of the horizon, refraction, and the semi-diameter of the Sun would have to be applied to the initial observation. A good description and other examples are given in John Robertson's *The Elements of Navigation;...* (*op. cit.*, note 3), vol. 2, pp. 261-5.

8. See Appendix C.

9. Sundials are, of course, a type of altitude- or azimuth-measuring instrument, but they give a reading directly in time, rather than in degrees. Gemma's universal sundial, which also served as a direction-finding instrument, may be seen in his portrait, which is shown in Figure 7 of Appendix C. Gemma described it in *Usus Annuli Astronomici* probably as early as 1533 (see Fernand van Ortroy, "Bio-bibliographie de Gemma Frisius," in *Académie Royale de Belgique. Classe des lettres et des sciences morales et politiques: Mémoires* [Brussels], vol. 11, 2nd series [December 1920], 2nd part, p. 65).

practice. A better method was required.

Given his attempts at producing a reliable marine timekeeper, it is not surprising that Christiaan Huygens addressed this problem. In the description of his marine timekeepers that he published in 1665,[10] he explains two methods to find local time at sea, both of which required the use of a clock or watch. The first required the navigator to establish, by his marine clock, the precise times of sunrise and sunset (Huygens suggested observing the moment when the Sun was divided in half by the horizon); then, by adding half the difference of the time between sunrise and sunset to the time of sunrise, the navigator could calculate the exact time shown by the clock when the Sun had crossed the meridian. Alternatively, he could use the period of sunset to sunrise and, by adding half the difference in time to the time of sunset, determine where the ship was at midnight. There were several problems with this method, however, not least of which were the mists and other atmospheric conditions that would limit the frequency and accuracy of observations made at these times, and the distance that the ship would move in the time between the observations.[11]

Huygens may have recognized these difficulties, because he mentions another method, one that had been employed at sea for improving the determination of latitude.[12] It was essentially the same as his sunrise-sunset method, but, to avoid the difficulties of making observations at these times and to reduce the length of time between observations, an instrument was needed to make precise measurements of the Sun's altitude a few hours before and after noon. With his timekeeper, the navigator would note the time of a particular altitude measurement of the Sun in its ascent before noon, and then determine the exact time at which the Sun reached the same, or equal, altitude in its descent after noon. By adding half the time elapsed between these observations to the time of the first observation, the exact moment when the Sun's meridian passage had occurred could be accurately determined. This method, which became known as "equal altitudes," was commonly adopted for on-shore observations in the eighteenth century. Obviously, both the reliability of the timekeeper and the accuracy of the altitude-measuring instrument—not to mention the skill of the user—were critical, but once the altitudes were accurately measured and timed, the accuracy of a timekeeper could also be established. The following example was given by John Robertson (1712-1776) in his *Elements of Navigation*:[13]

> May, 20, 1772, at $8^h$ $40^m$ forenoon, and at $3^h$ $16^m$ afternoon, by my watch, the Sun had equal altitudes; required the going of the watch?

| | | |
|---|---|---|
| Add together | | $12^h$ $0$ $^m$ |
| | + | 8 40 |
| | + | 3 16 |
| Take the half sum | | 23 56 |
| Rem.[remove] noon by the watch | = | 11 58 |
| Apparent noon | = | 12 0 |
| Watch too slow | | 2 min. |
| May 20th ap. noon | | 11 58 |
| Equal [Equation of] time subtr. | | 4 |
| Mean noon by watch | | 11 54 |

10. [Christiaan Huygens], *Kort Onderwys Aengaende het gebruyck Der Horlogien Tot het vinden der Lenghten van Oost en West* ([The Hague], [1665]); English translation entitled "Instructions Concerning the Use of Pendulum-Watches, for finding the Longitude at Sea," *Philosophical Transactions* of the Royal Society, London *(Phil. Trans.)*, vol. 4, no. 47 (1669), pp. 937-53.

11. Distance traveled was ascertained by measuring the speed of the ship with a log attached to the end of a line. Beginning 60 feet from the log, knots were tied in the line at regular intervals of 51 feet. Then, when the log was cast overboard and floated away from the ship, the number of knots counted in a period of 30 seconds (timed by a sandglass) would indicate the speed of the ship. Hence, the term "knot" was adopted as the nautical measure of speed. With 30 seconds being 1/120th of an hour, the knot interval of 51 feet represented 1/120th of a nautical mile, reckoned at that time as 6,120 feet (see note 1). For a more detailed explanation, see Archibald Patoun's article "Navigation" in the *Encyclopaedia Britannica* (London, 1771), vol. 3, pp. 365-93.

12. In *The Art of Navigation in England in Elizabethan and Early Stuart Times* (London: Hollis and Carter, 1958), p. 71, David W. Waters relates that in 1537 the Portuguese cosmographer Pedro Nuñez (1502-1578) proposed using "double altitudes" for determining latitude. In the early eighteenth century, the Swiss mathematician Nicholas Facio de Duillier (1664-1753), who settled in London and from 1703 devoted much of his time to improvements in navigation, suggested a similar method known as "double altitude and elapsed time." (Facio's greatest contribution, however, was his invention, patented in 1704 with the watchmakers Peter and Jacob Debaufre, of a method of piercing and working rubies and other hard precious stones for use in watches. When highly polished, jewels serve as a very effective bearing surface for watch pivots. Jewelling remained a closely guarded secret in the English watchmaking trade for about 70 years and became a standard feature of marine chronometers.) For further information on methods of finding latitude, see E. G. R. Taylor, *The Mathematical Practitioners of Hanoverian England, 1714-1840* (Cambridge: Cambridge University Press, 1966).

13. Robertson, *Elements of Navigation* (*op. cit.*, note 3), vol. 2, p. 259.

In explaining the equal-altitude method for finding local time at sea, Robertson recommended that the ship be "lying-by," in other words, anchored or moving as little as possible, and that observations be made three to five hours before and after noon, when changes in the Sun's altitude are most pronounced. To overcome the problem of clouds obscuring an observation, he suggested that several of these equal-altitude measurements be made and that, to reduce errors, the results be averaged.[14] William Wales (1734-1798), who had sailed as astronomer aboard the *Resolution* on Cook's second voyage (1772-5) and served as editor of *Elements of Navigation* after Robertson's death, recommended that, to avoid errors, one instrument should be used for each pair of observations. Wales would set the degree scale of his sextant at a point at which he could gradually observe the Sun rising to that altitude, so that his assistant could note the exact time of the observation with a timekeeper. That sextant would then be placed carefully away in its box until the afternoon, when the exact time that the Sun reached the same altitude in its descent would be noted. Other sextants, "as many of them as are at hand," would be used to make additional observations, and, at the end of the day, the results obtained from all the observations could be averaged to find when, by the timekeeper, noon occurred and, thereby, establish if the timekeeper was gaining or losing.[15] To avoid the problems of refraction and other difficulties involved in observing the horizon at sea, Wales recommended using an instrument known as an artificial horizon.[16]

Because this method employed the Sun to find the time and a mechanical clock to keep the time, the difference between Sun time and mean (clock) time had to be taken into account, as shown in the example above. Whereas the mechanical clock provides an even division of time, the observed motion of the Sun during the course of the seasons is not constant, due to the tilt of the Earth's axis relative to its path and the ellipticity of the Earth's orbit about the Sun. As a result, apparent solar time (the time shown by the sundial) and mean solar time (the time shown by the clock) agree only four times each year (on or about April 16, June 14, September 1, and December 25, varying slightly with the leap-year cycle) and diverge at other times by as much as fifteen minutes. The difference between these times, which became known as the equation of time, was of no significance outside the field of astronomy until the introduction of the pendulum and the balance spring greatly increased the accuracy of clocks and watches. In order to use a sundial to set a clock correctly to mean time, the equation of time for that particular day has to be applied to apparent time, the time shown on the sundial. Recognizing the need to account for this variation in the calculations for finding longitude with a timekeeper, Huygens published tables showing the equation of time in *Kort Onderwys*, the work mentioned above in which he described his marine timekeepers and the methods of finding local time at sea.[17]

It is worth noting that while the timekeeper method of finding longitude used mean solar time and therefore required the equation of time to be taken into account when determining local time by the Sun, lunar-distance observations could be made according to either mean solar time or apparent solar time, so long as the time used for the observations was the same as that used in the tables. Derek Howse informs me that the first 66 issues of the *Nautical Almanac* were calculated according to apparent solar time, and that

14. *Ibid.*, pp. 258-61.

15. William Wales, *The Method of Finding Longitude by Timekeepers* (London, 1794), pp. 70-9. In these calculations, an adjustment for the difference between solar and mean time (the equation of time) would have to be applied to the time of apparent noon.

16. *Ibid.*, pp. 70-1. This instrument provided a horizontal reflecting surface, which the observer used instead of the horizon. When Sun's image on this surface is sighted through the telescope of a sextant and brought into coincidence with the Sun's image reflected in the sextant mirrors, the angular measurement will be exactly twice that of the Sun above the horizon. Various designs were developed for use on land and at sea. Wales's preferred artificial horizon for use on land was an oblong trough filled with mercury with a protective roof formed of two glass plates.

17. [Huygens], *Kort Onderwys* (*op. cit.*, note 10), pp. 6-7, shown in Figure 6 of John Leopold's paper. These were the first printed equation tables calculated with a pendulum clock. In the English translation of this work published in 1669 in *Philosophical Transactions* (*op. cit.*, note 10), the equation tables, shown on pp. 940-1, have been adjusted for the Julian calendar still in use in England at that time. The earliest printed tables noting the difference between apparent and mean solar time are probably those contained in the 1492 edition of the Alfonsine Tables. Equation tables in manuscript were provided by Claudius Ptolemy (fl. A.D. 150) in his "Handy Tables," in which the equation of time is included in the table of right ascensions. I am grateful to Noel Swerdlow, who drew my attention to these last two references several years ago.

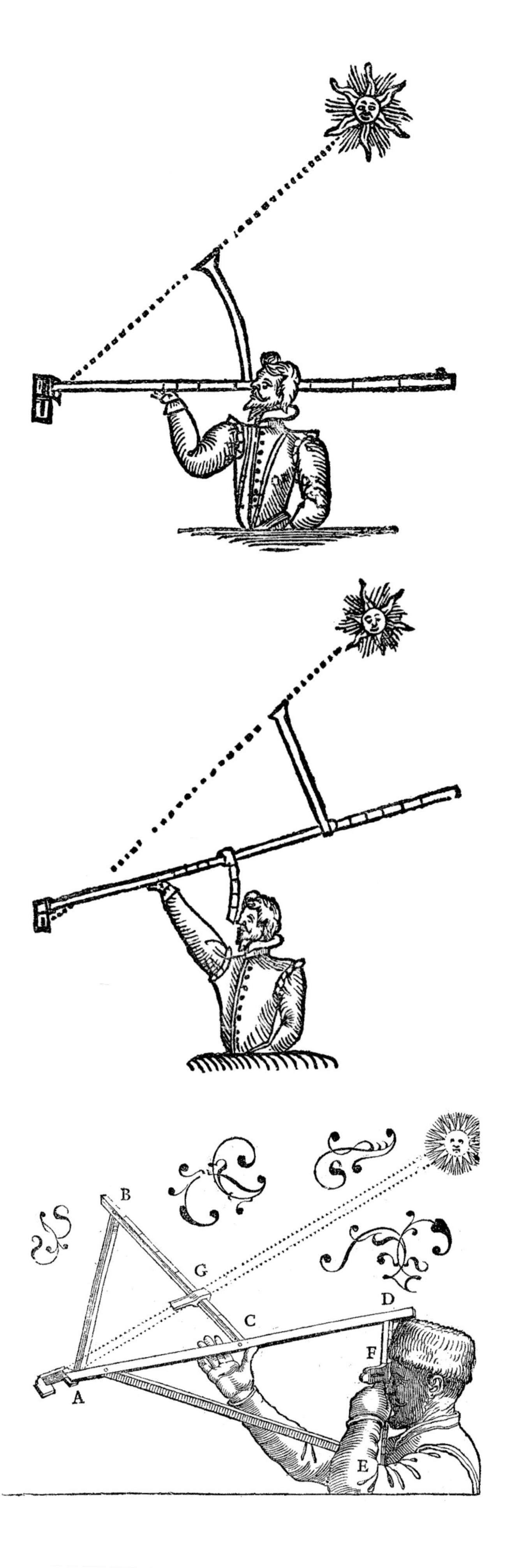


Figure 1. An improved version of the cross-staff, probably from a design by Thomas Har[r]iot (*ca.* 1560-1621), that gave the Sun's altitude by the shadow of the curved vane cast upon the horizon sighting vane and allowed the user to have his back to the Sun when making the observation. This instrument, first described by John Davis in 1595, was limited to observations of up to 45°. From John Davis's *The Seamans Secrets,...* (London, 1607).

Figure 2. John Davis's own design, also first described in 1595, for an improved observing staff, which combined the advantages of the earlier design in sighting the horizon and the Sun's altitude simultaneously, but also permitted observations of up to 90°. From John Davis, *The Seamans Secrets,...* (London, 1607).

Figure 3. Backstaff, or Davis's quadrant (as it was previously known), in the form to which John Davis's design had evolved by the first quarter of the seventeenth century. From William Johnson Blaeu's *The Sea-Mirrour...* (Amsterdam, 1625), first part, sig. CV$^{v}$, p. 30. Courtesy of The Huntington Library, San Marino, California. Ref. no. RB 98695.

it was not until 1834 that the change was made to mean time.[18]

The methods of determining local time by measuring the altitude of either the Sun by day or a star at night appear to have been introduced as part of the solution to the problem of finding longitude at sea by lunar-distance observations, and were made possible by the significant refinements in the computation tables and in altitude-measuring instruments that were developed for this purpose. These had an advantage over the equal-altitude method in that, because there was no need for a delay of several hours between the observations, the ship did not have to be "lying by" when the observations were made. Nevertheless, as stated earlier, these methods did require some account of latitude and longitude of the ship. To find local time by this method, the Sun's altitude above the horizon had to be accurately measured and, once the zenith distance, the co-latitude, and the polar distance had been determined, the horary angle could be computed. An example of using the Sun's altitude to determine local time is included in Figure 7 of Derek Howse's paper, which shows an actual lunar-distance computation.[19] Describing both methods in *The New Practical Navigator...*, John Hamilton Moore (1738-1807) recommended the method using the Sun, because of the difficulties often experienced in seeing the horizon at sea clearly at night.[20]

Although the solution to finding local time at sea required a reliable timekeeper to establish the times of altitude measurements of the Sun, the accuracy of the altitude-measuring instruments was also of primary importance. The significance of these devices has been described in other papers in this volume, but readers may find a brief illustrated history of their development useful.

When the first edition of Robertson's *Elements of Navigation* was published in 1754, there were three types of angle-measuring instruments in use at sea. In the section entitled *Of Instruments for taking Altitudes*,[21] he describes in detail the construction and use of the two that were "best adapted for use at sea": Davis's quadrant (the backstaff) and Hadley's quadrant (the octant). The cross-staff, which had been developed as a navigational instrument in the early sixteenth century, is also briefly mentioned, but as Robertson states, it had been largely superseded by these more precise instruments. Even so, probably because it was inexpensive and simple, its use continued well into the nineteenth century.[22]

In 1595, Captain John Davis (1552-1605), an experienced navigator who had sailed in search of the Northwest Passage in 1587, published a book entitled *The Seamans Secrets*, in which he illustrated two designs for a new altitude-measuring instrument for determining latitude at sea.[23] These are shown in Figures 1 and 2. In addition to increasing the accuracy of solar altitude measurements, these designs overcame some of the inherent problems of the cross-staff in that they allowed the navigator to observe the altitude of the Sun above the horizon without facing the blinding light of the Sun or running the risk of damaging his eye with the end of the staff when making observations on the rolling deck of a ship. With the improvements that Davis introduced during the next ten years, the design became standardized and the instrument was received with great approbation by mariners, who termed it "Davis's quadrant" (Figure 3); today, it is usually referred to as a backstaff.[24]

18. Letter from Derek Howse dated May 18, 1996.

19. A detailed description and other examples of finding local time at sea by the Sun's altitude and by a star's altitude may be found in Robertson, *Elements of Navigation* (*op. cit.*, note 3), vol. 2, pp. 259-61, and in Moore, *The New Practical Navigator* (*op. cit.*, note 3), pp. 226-32.

20. Moore (*ibid.*), p. 232.

21. Robertson, *Elements of Navigation* (*op. cit.*, note 3), vol. 2, book 9 ("Days Works"), sect. 7 ("Of Instruments for taking Altitudes"), articles 36-54, pp. 245-55.

22. Details of and references to the origins of the cross-staff are given in note 12 of Appendix C and its related passage in the text.

23. John Davis, *The Seamans Secrets, Devided into 2. partes, wherein is taught the three kindes of Sayling, Horizontall, Peradoxall, and sayling upon a great Circle:...* (London, 1595). The illustrations of the instruments appear in the second part on signatures L4$^{v}$ and M1$^{r}$.

24. The term "backstaff" was, in fact, also used in the seventeenth century. On p. xi of his excellent historical introduction to John Robertson's *Elements of Navigation* (*op. cit.*, note 3), James Wilson (d. *ca.* 1770) cites that "Captain Charles Saltonstall [fl. 1627-1665], in his book *Navigation*, describes it under the name of a Back-Staff," and that in the account of Capt. Thomas James's famous voyage in search of the Northwest Passage, begun in 1631, two of "Mr. Davis's Back-staves" are mentioned. A useful account of the early development of the backstaff is given by David W. Waters in *The Art of Navigation* (*op. cit.*, note 12), pp. 590-1 and plate 71.

The backstaff was designed specifically for measuring the altitude of the Sun, since only the Sun had sufficient light to cast the shadow of the shadow vane upon the horizon sight. Thus, unlike the cross-staff, it could not be used at night. The large number of surviving examples from the eighteenth century indicate that, even after more accurate instruments (the octant and the sextant) had been invented, the less expensive backstaff, like its predecessor the cross-staff, continued to be used throughout the century.

In 1750, William Maitland (fl. 1740-1757), a teacher of navigation aboard ship, assessed the accuracy of the backstaff, which he was able to do because of the superior instruments that were by then available:

> ...to me it is evident from great numbers of Observations I have made, and been present when made by others with it, that even when there is a good Horizon, a bright Sun, the Ship not rolling much, and everything, in short, favourable for the Observation, the Latitude cannot be determined by this Instrument, nearer than to ten or a dozen Minutes.[25]

The backstaff can in turn be compared for accuracy with its predecessor, the cross-staff, thanks to the work of Christopher St. John Hume Daniel, who used a modern reproduction of the latter during a voyage from Plymouth, England, to San Francisco in 1974-5 aboard a replica of Drake's famous ship, the *Golden Hind*. On this voyage, he recorded that, when using the cross-staff with the back observation method,[26] his readings on two occasions erred less than 20' of arc from those made with modern instruments. Summarizing the accuracy of the instrument, he estimated that it could be relied upon to give "results within a latitude range of twenty nautical miles."[27] Taken together, Maitland's and Daniel's assessments would suggest that the backstaff was about twice as accurate as the cross-staff. Determinations such as these are, however,

25. Maitland, *An Essay* (*op. cit.*, note 1), pp. 35-6. An error of 10' to 12' of arc would represent ten to twelve nautical miles on the equator (see note 1).

26. The cross-staff can, in fact, be used for making back observations, although it was not originally intended to do so. The method of doing this is described by Alan Neale Stimson and Christopher St. John Hume Daniel in *The Cross Staff: Historical Development and Use* (London: Harriet Wynter Ltd., 1977), pp. 6-7, 16-19.

27. *Ibid.*, p. 21. Like Maitland's assessment of the accuracy of the backstaff, Daniel's analysis of the cross-staff was made possible by the fact that he had a more precise technology available to check his observations.

**Figure 4 (left). Robert Hooke's design, which he described in 1666, for a reflecting quadrant for improving the accuracy of lunar-distance observations. From *The Posthumous Works of Robert Hooke,...,* ed. Richard Waller (London, 1705), table 11, fig. 2.**

**Figure 5 (right). Isaac Newton's design for a reflecting quadrant for making lunar-distance measurements. Concealed among Edmond Halley's papers, it was not published until 1742. From *Philosophical Transactions* of the Royal Society, London, vol. 42, no. 465 (1742), with article on pp. 155-6.**

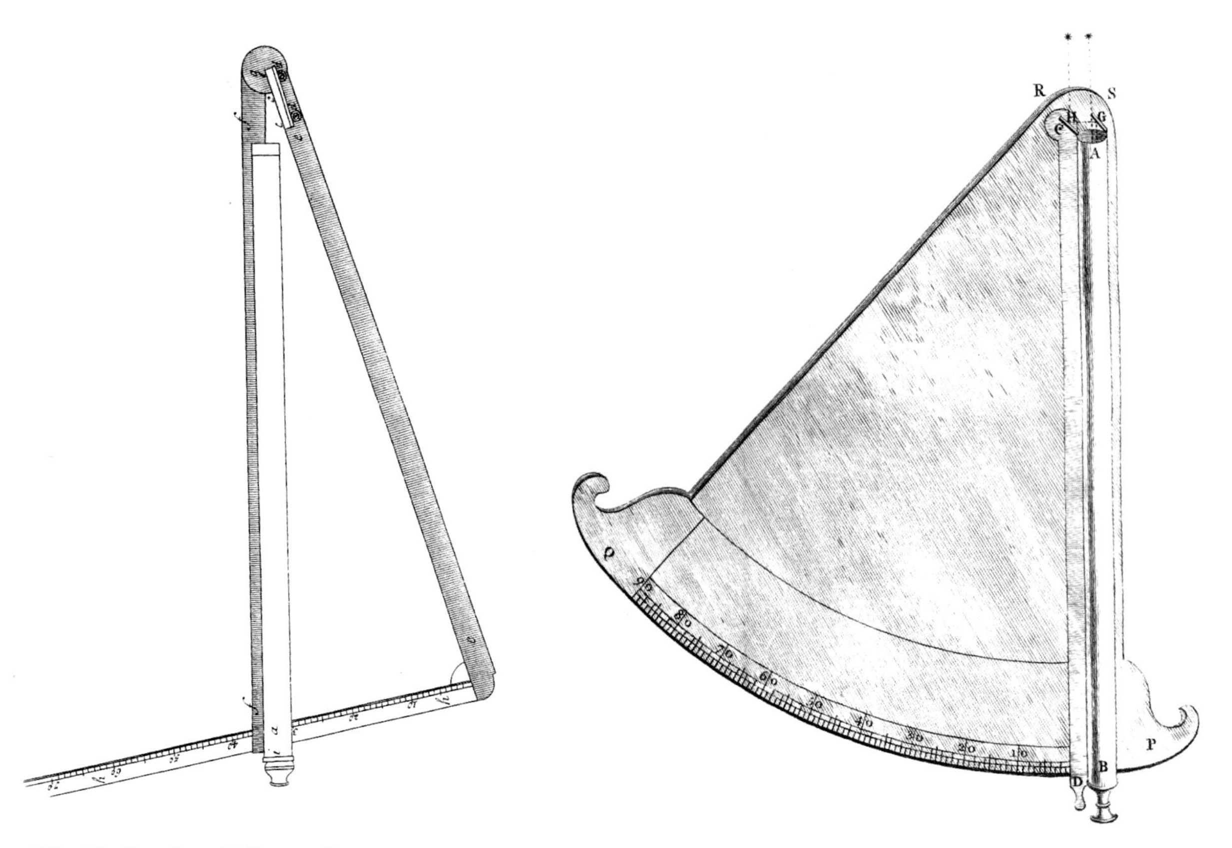

28. Thomas Birch, *The History of the Royal Society of London...*, 4 vols. (London, 1756-7), vol. 2, p. 114.

29. *The Posthumous Works of Robert Hooke,...*, ed. Richard Waller (London, 1705), p. 503. The instrument is shown in table 11, fig. 2.

30. "A true Copy of a Paper found, in the Hand Writing of Sir Isaac Newton, among the Papers of the late Dr. Halley, containing a Description of an Instrument for observing the Moon's Distance from the Fixt Stars at Sea," *Phil. Trans.*, vol. 42, no. 465 (1742), pp. 155-6.

dependent not only upon the skill of the observer, but also upon the skill of the individual who made the instrument and divided its scales.

It was not until 1730 that there appeared a new instrument that, because of its greater accuracy, revolutionized nautical astronomy. This was the reflecting quadrant, so called because it employed mirrors to observe the altitude of celestial bodies. It is clear from all the early accounts that this device was invented not in an attempt to improve observations for latitude or local time, but specifically for making the precise measurements required for finding longitude by the lunar-distance method. The concept of this device was not new. Such an instrument, described as a "new perspective for taking angles by reflexion," had been shown to the Royal Society on September 12, 1666, by Robert Hooke (1635-1702/3), the Society's celebrated curator of experiments.[28] Hooke did not leave any explanation of this instrument other than that contained on a loose sheet of paper, which was discovered after his death by the Secretary of the Royal Society, Richard Waller (*ca.* 1646-1714), who published an illustration with a very brief description (Figure 4).[29] There is also evidence that Isaac Newton (1642-1727) designed a similar instrument (Figure 5), but the description of his "Instrument for observing the Moon's Distance from the Fixt Stars at Sea" remained concealed in the papers of Edmond Halley (*ca.* 1656-1742) and was not discovered until after Halley's death.[30] Neither of these descriptions, however, aroused much interest, and there is no evidence that such an instrument was tried in practice before 1730.

During the summer of that year, Thomas Godfrey (1704-1749), a 26-year-old Philadelphia window glazier, plumber, and self-taught amateur of science, converted a backstaff into a reflecting altitude-measuring instrument (Figure 6). Following successful trials at sea—one voyage to the West Indies in December 1730 and another to Newfoundland in the following February—Godfrey showed his instrument to a well-known Philadelphia

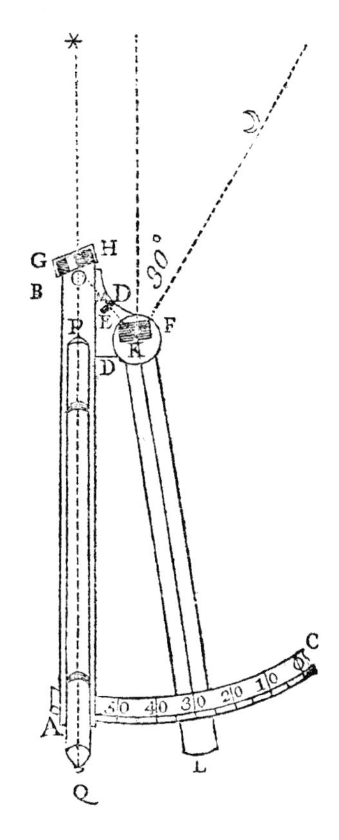


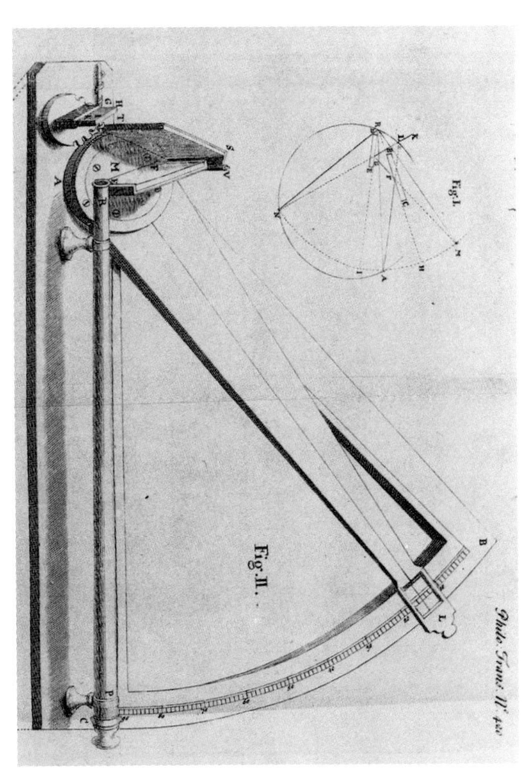


**Figure 6 (this page, left). The earliest known illustration of Thomas Godfrey's reflecting 'quadrant', designed for improving lunar-distance observations at sea. The length (AB) of the main arm of the frame, upon which the telescope is mounted, is described as being about 40 to 45 inches. Observations are limited to 50°, unless a special adjustment is made to the mirror EF. From *The American Magazine and Monthly Chronicle for the British Colonies* (Philadelphia), vol. 1, no. 10 (July 1758), p. 478.**

**Figure 7 (this page, right). Reflecting quadrant designed by John Hadley for making lunar-distance observations. It was his second version, designed for use at sea, that was quickly adopted by mariners with great approbation as a latitude-finding instrument and called Hadley's quadrant (see Figure 6 in Alan Stimson's paper). From *Philosophical Transactions* of the Royal Society, London, vol. 37, no. 420 (1731), plate IV.**

statesman, James Logan (1674-1751), who brought it to the attention of the Astronomer Royal, Edmond Halley, in 1732.[31] From Logan's letter dated May 25, 1732, it is clear that the instrument was designed as a solution to the problem of making lunar-distance observations at sea, and, as a result, he requested at the end of the letter that Godfrey should be considered as a contestant for at least part of the longitude prize:

> ...and therefore, if the longitude could ever be expected to be determined by the motions of the moon, (to which end J. Flamstead's *[sic]* and thy more assiduous labours in observing her, have I suppose been principally levelled) and this instrument be duly made to answer what is proposed, as it may be framed light and easily manageable, thou wilt then with thy accurate tables, have obtained the great desideratum and all that can in this way be had from our satellites. And if the method of discovering the longitude by the moon is to meet with a reward, and this instrument, which for all that I have ever read or heard of, is an invention altogether new, be made use of, in that case I would recommend the inventor to thy justice and notice.[32]

By an interesting coincidence, a very similar device had been described at a meeting of the Royal Society on May 13 of the previous year and was by this time recorded in an article in the Society's *Philosophical Transactions*.[33] The author was a mathematician and scientific mechanist, John Hadley (1682-1744), who had been elected a Fellow of the Royal Society in 1717 and had been appointed its vice-president in 1728. Hadley described two

**Figure 8. Portrait of John Hadley made between 1728, when he was first appointed vice-president of the Royal Society, and his death in 1744. Lithograph from a portrait, artist unknown. 12.7 x 11.43 cm. (5.0 x 4.5 in.). Courtesy of the Science Museum/Science and Society Picture Library, London. Inv. no. 1903-142.**

31. A history, description, and illustration of Thomas Godfrey's reflecting quadrant was published in an anonymous article entitled "On the Invention of the Quadrant, commonly called Hadley's" in *The American Magazine and Monthly Chronicle for the British Colonies* (Philadelphia), vol. 1, no. 10 (July 1758), pp. 475-80, and in vol. 1, no. 11 (August 1758), pp. 527-34. An excellent account of Godfrey's work in the invention of this instrument is given by Silvio Bedini in *At the Sign of the Compass and Quadrant: The Life and Times of Anthony Lamb,* Transactions of the American Philosophical Society, vol. 74, part 1 (Philadelphia, 1984), pp. 37-44. Bedini's most recent research on Godfrey is due to be published in the "History Corner" of *The Professional Surveyor* (May, June, July, and August 1996) under the title "Thomas Godfrey and the Invention of the Octant."

32. "On the Invention of the Quadrant" (*ibid.*), p. 480. This letter to Halley, which contains the illustration of the reflecting quadrant shown in Figure 6, was written on May 25, 1732, but it was not published until 1752. The instrument devised by Godfrey that Logan described and illustrated in a letter dated June 28, 1734, to the Royal Society makes only passing mention to Godfrey's reflecting instrument and concerns instead a device more closely related to the backstaff; it was published under the title "An Account of Mr. Thomas Godfrey's Improvement of Davis's Quadrant, transferred to the Mariner's-Bow,..." in *Phil. Trans.*, vol. 38, no. 435 (1734), pp. 441-50.

33. John Hadley, "The Description of a new Instrument for taking Angles. By John Hadley, Esq; Vice-Pr. R. S. communicated to the Society on May 13. 1731," *Phil. Trans.*, vol. 37, no. 420 (1731), pp. 147-57.

34. A wooden octant of this period is shown in Figure 7 of Alan Stimson's paper in this volume. The origin of the diagonal scale in the fourteenth century is mentioned in the discussion of the cross-staff in Appendix C (note 12 and related text). In its application to the backstaff, this scale was commonly made to read to within 1' of arc. These fine divisions were made possible by the large radius of the observing degree scale.

**Figure 9. An English navigator using an octant. The instrument shown has none of the refinements needed for lunar-distance observations and was probably intended purely for measurements of latitude and local time. The illustration comes from a trade card, *ca.* 1781, of the instrument maker Richard Rust (apprenticed 1744, d. 1785); the device attached to the instrument's frame is an artificial horizon, probably of Rust's invention. Dimensions of card: 13.97 x 17.78 cm. (5.5 x 7 in.). Courtesy of the Science Museum/Science and Society Picture Library, London. Ref. no. 1934-111.**

reflecting quadrants. The first, which can be seen in Figure 7, shows an instrument designed especially for improving the accuracy of lunar-distance measurements on land; it is similar to Godfrey's in that its telescope is mounted on the side frame of the octant, but, whereas Godfrey's instrument required special adjustment for observations beyond 50°, Hadley's could, without adjustment, measure angles up to 90°. Hadley's second design, described and illustrated (Figure 6) in Alan Stimson's paper, was intended more specifically for measuring the altitude of the Sun, the Moon, or a star above the horizon at sea. When the advantages of its greater accuracy and ease of use over both the cross-staff and the backstaff became apparent, it was quickly adopted by mariners as an instrument for finding latitude, and was also used for finding local time by the equal-altitude method. Although today it is usually referred to as an octant, because its frame occupies one-eighth of a circle, in its day it was usually called "Hadley's quadrant." As an instrument for improving the accuracy of lunar-distance measurements for finding longitude, however, both Godfrey's and Hadley's reflecting quadrants were about 25 years ahead of their time, because the Moon's motion could not then be predicted with sufficient accuracy for practicable use.

During the next twenty years, the octant underwent a period of basic refinement as its popularity increased. The surviving examples from this time have frames made of wood, and the degree arc is commonly divided with a diagonal scale.[34] While the 90° scale on Hadley's design was satisfactory for determining latitude, however, it was too limited in practice for making the wider angular measurements required for lunar-distance observations, and its wooden frame was not rigid enough for very precise observations. Furthermore, the diagonal scale, which was also commonly used on the backstaff, was of insufficient accuracy. Thus in 1757, after the lunar-distance method had become a viable option as a result of the work of Leonhard Euler (1707-1783) and Tobias Mayer (1723-1762), John Bird (1709-1776) was commissioned to design an improved version especially for making precise lunar-distance measurements. The instrument he produced was made of brass, fitted with a telescope, and accommodated a larger scale with a vernier, which enabled readings to be made accurately to 1' of arc.[35] This new instrument, with its scale of 120°, had a frame occupying one-sixth of a circle and therefore became known as a sextant. Three sextants, all made by Bird, were tried at sea on a voyage in 1758 and were shown to agree to within 1' of arc of each other.[36] With the advent of the sextant and the accurate tables required for lunar-distance measurements came the development of new

**Figure 10. Portrait of the instrument maker John Bird, *ca.* 1770. His book *The Method of Constructing Mural Quadrants*, published in 1768, is shown on the table. Mezzotint by Valentine Green (1739-1813) published in December 1776, from a portrait by Lewis. 35.5 x 25.4 cm. (14 x 10 in.). Courtesy of the Science Museum/ Science and Society Picture Library, London. Inv. 1857-305.**

The actual accuracy of the divisions, however, no doubt varied. Diagonal scales on octants were usually made to read to 2' of arc; although these instruments often have a large radius, their design places the 90° scale on an arc occupying only 45° and therefore halves the amount of space for the divisions, limiting the accuracy of the readings accordingly. Only the vernier permitted more precise readings.

35. John Bird's first sextant, fitted with a vernier, is shown in Figure 9 of Alan Stimson's paper. The vernier, invented by Pierre Vernier (1584-1638) about 1630, was not commonly applied to instruments until the sextant was developed. This was probably because it required a high degree of precision in the division of the scale (more perhaps than many makers were capable of) and gave a more precise reading than was needed for most applications before the middle of the eighteenth century. A small microscope was later fitted to some sextants to aid the reading of the vernier.

36. Anonymous letter to the editor of *Gentleman's Magazine* in June 1758, published by Nevil Maskelyne as Appendix 12 in his publication of Tobias Mayer's *Tabulae Motuum Solis et Lunae Novae et Correctae* (London, 1770), p. cxxix. On the previous page of this work, it is stated that Abbé de Lacaille (1713-1762) had reported less favorable results in France with similar altitude-measuring instruments: the observations of two astronomers had differed by as much as 8' of arc. This variance may have been due to the inferior quality of the instruments.

**Figure 11. Portrait of Jesse Ramsden about 1789, by Robert Home (1752-1834). Ramsden's right arm rests on one of his early dividing engines, possibly the second, for which he was awarded £615 by the Board of Longitude in October 1774; in the background is his five-foot altitude and azimuth circle, completed for the Palermo Observatory in August 1789. Oil on canvas. 127 x 100.3 cm. (50 x 39.5 in.). By permission of the President and Council of the Royal Society, London.**

methods of finding local time at sea by a single observation of the altitude of the Sun or a star.

Following the introduction of the sextant, the most critical issue was to find a way to maintain a standard of high quality in their production. An instrument, of course, was only as good as its maker, and no one was as skilled as John Bird at dividing the scales by hand. The invention of a machine that automatically divided circular scales on instruments was therefore received with much acclaim by members of the Board of Longitude, who awarded £615 to its ingenious maker, Jesse Ramsden (1735-1800), in October 1774.[37] Ramsden's circle dividing engine increased the speed of production (thus reducing the cost), enabled sextants to be made in a more manageable size (with a radius of twelve inches rather than the more common eighteen or twenty) without compromising accuracy, and provided greater uniformity in the quality of instruments. In so doing, it ushered in a new era of instrument making, during which the trades became more specialized and machine production began gradually to replace the skills of eye and hand.

37. As part of the agreement with the Board of Longitude for the award, Jesse Ramsden published an account of his invention in *Description of an Engine for Dividing Mathematical Instruments* (London, 1777). Subsequently, Ramsden invented a straight-line dividing engine for which the Board of Longitude awarded him £400 in March 1778. Derek Howse is currently researching the history of the awards paid by the Board of Longitude, which will be published in due course.

I am grateful to Mariana Oller for her assistance in obtaining some of the references for this article and to Derek Howse for his insightful and useful comments.

# Bibliography of Published and Manuscript Sources

# Bibliography of Published and Manuscript Sources

USAGES AND ABBREVIATIONS

*Use of brackets and asterisks:*

[ ] (in the author, date, or place-of-publication position): Information known but not so designated on title page.

*[ ] (in the author position): Relating to, but not by, this person.

*Parenthetical references to authors of this book:*

At the end of each entry, the author or authors who have cited this work are given in parentheses, so that readers may work back from the bibliography to the text to see where the material has been used and how it has been interpreted. The following abbreviations have been used:

app.: appendix
fig./figs.: figure/figures
fn./fns.: footnote/footnotes
gen. ref.: general reference

GENERAL BIOGRAPHICAL AND TECHNICAL REFERENCES


*Allgemeine Deutsche Biographie.* 1875-1912. 56 vols. Leipzig: Duncker.

Baillie, G. H. 1951. *Clocks and Watches: An Historical Bibliography.* Vol. 1 (vol. 2 has not been published). London: N.A.G. Press. Reprinted London: Holland Press, 1978.

———. 1951. *Watchmakers and Clockmakers of the World.* 3rd ed. London: N.A.G. Press.

*Biographie Universelle (Michaud) Ancienne et Modern.* 1843-65. Nouv. éd. 45 vols. Paris: Delagrave.

Clifton, Gloria. 1995. *Directory of British Scientific Instrument Makers 1550-1851.* London: Zwemmer (with the National Maritime Museum).

*The Dictionary of National Biography.* 1973 ed. Edited by Sir Leslie Stephen and Sir Sidney Lee. 22 vols. London and Oxford: Oxford University Press.

*Dictionary of Scientific Biography.* 1970-90. Edited by Charles Coulston Gillispie (vols. 1-16) and Frederic L. Holmes (vols. 17-18). 18 vols. New York: Charles Scribner's Sons.

*Dictionnaire de Biographie Française.* 1933-92. Edited by J. Balteau, M. Barroux, and M. Prevost. 18 vols (in progress). Paris: Librairie Letouzey et Ané.

*Dizionario Biografico degli Italiani.* 1960-94. Edited by Alberto M. Ghisalberti. 44 vols. (in progress). Rome: Istituto della Enciclopedia Italiana.

*Encyclopaedia Britannica.* 1910-11. 11th ed. Cambridge: Cambridge University Press.

*English Maritime Books Printed Before 1801 Relating to Ships, Their Construction and Their Operation at Sea....* 1995. Compiled by Thomas R. Adams and David W. Waters. Providence Rhode Island: The John Carter Brown Library, and Greenwich: The National Maritime Museum.

Harley, J. B., and Woodward, David. 1987 and 1992. *The History of Cartography.* 2 vols. (series in progress). Chicago: University of Chicago Press.

Macey, Samuel L., ed. 1994. *Encyclopedia of Time.* New York: Garland.

*Nouvelle Biographie Générale Depuis les Temps les Plus Reculés Jusqu'a nos Jours,....* 1862-66. Edited by M. le Dr. J. C. F. Hoefer. 46 vols. Paris: Firmin Didot Frères.

Randall, Anthony G. 1992. *The Time Museum Catalogue of Chronometers.* Rockford, Illinois: The Time Museum. (Contains a useful, illustrated glossary of technical terms.)

Smith, Alan, ed. 1979. *The Country Life International Dictionary of Clocks.* New York: G. P. Putnam.

Sommervogel, Carlos. 1890-1932. *Bibliothèque de la Compagnie de Jésus.* Nouv. éd. 11 vols. Paris: Alphonse Picard.

Taylor, E. G. R. 1954. *The Mathematical Practitioners of Tudor and Stuart England [1485-1714].* Cambridge: Cambridge University Press.

———. 1966. *The Mathematical Practitioners of Hanoverian England, 1714-1840.* Cambridge: Cambridge University Press.

Thieme, Ulrich, and Becker, Felix. 1907-50. *Allgemeines Lexikon der Bildenden Künstler von der Antike bis zur Gegenwart.* 37 vols. Leipzig: Wilhelm Engelmann (1907-10); E. A. Seemann (1911-1950).

Waterhouse, Ellis. 1981. *The Dictionary of British 18th Century Painters in oils and crayons.* Woodbridge, England: Antique Collectors' Club.

Waters, David W. 1958. *The Art of Navigation in England in Elizabethan and Early Stuart Times.* London: Hollis & Carter.

Williamson, George C., ed. 1903-05. *Bryan's Dictionary of Painters and Engravers.* 5 vols. New York: Macmillan.

CITED REFERENCES


Abeler, Jürgen. 1983. *Die Longitudo zur See.* Wuppertal, Germany: Wuppertaler Uhrenmuseum. (Andrewes fn. 150)

———. 1984. "Die Nachbildung von Harrisons Seechronometer »H1«." *Alte Uhren*, vol. 1/1984 (January), pp. 11-22. (Andrewes fn. 150)

*Acta Eruditorum*. 1682-. Leipzig. (Leopold fn. 90)

Acts of Parliament:

Act 12 Anne *cap.* 15. 1714. *An Act for Providing a Publick Reward for such Person or Persons as shall Discover the Longitude at Sea.* (Andrewes fns. 2, 34; Howse fn. 5; Penney fn. 11; Randall fn. 31; Stimson fn. 23)

Act 3 George III *cap.* 14. 1763. *An Act for the Encouragement of John Harrison, to publish and make known his Invention of a Machine or Watch, for the Discovery of the Longitude at Sea.* (Andrewes fn. 124)

Act 5 George III *cap.* 20. 1765. *An Act for explaining and rendering more effectual Two Acts, One made in the Twelfth Year of the Reign of Queen Anne,...and the other in the Twenty sixth Year of the Reign of King George the Second,....* (Howse fn. 12; Randall fn. 32)

Act 14 George III *cap.* 66. 1774. *An Act for the repeal of all former Acts concerning the Longitude at Sea.* (Penney fns. 56, 71; Randall fn. 50)

Airy, George. 1844-8. "Correspondence with clockmakers." MS. RGO 6/586, Royal Greenwich Observatory Archives, Cambridge University Library. (Andrewes fn. 142)

Aked, Charles K. 1970. "William Derham and the 'Artificial Clockmaker,' " part I. *Antiquarian Horology*, vol. 6, no. 6 (March), pp. 362-72. (Turner fn. 68)

———. 1976. "John Harrison's Paradox." *Horological Journal*, vol. 119, no. 1 (July), pp. 3-5. (Andrewes fn. 110)

Albuquerque, Luis de. 1991. "Portuguese Navigation: Its Historical Development." In Jay A. Levenson, ed., *Circa 1492: Art in the Age of Exploration*, pp. 35-9. New Haven: Yale University Press. (Landes fns. 5, 7, 11-12)

Alimari, Dorotheo. 1714. *The New Method Propos'd by...to discover the Longitude. To which are added, Proper Figures of some Instruments which he hath Invented to that Purpose: With a plain Description of Them....* 2 eds. London. (Turner app. 3)

———. 1715. *Longitudinis uut terra aut mari Investigandae Methodus. Adjectis insuper Demonstrationibus aut Instrumentoru Iconismis.* London. (Turner app. 3)

Allix, Charles. 1990. "William Hardy and his Spring-Pallet Regulators: Historical Notes and List of Known Examples, with reference to how Hardy's Regulators inhibited sales of similar Clocks by Thomas Reid." *Antiquarian Horology*, vol. 18, no. 6 (Summer), pp. 607-29. (Burgess fn. 73)

———, *et al.* 1988. "Obituary: Colonel R. H. Quill, C.B.E., D.S.O., M.V.O." *Antiquarian Horology*, vol. 17, no. 3 (Spring), pp. 245-7. (Penney fn. 67)

Andrewes, William. 1979. "John Harrison: A Study of his Early Work." *Horological Dialogues* (journal of the American Section of the Antiquarian Horological Society), vol. 1, pp. 11-38. (Andrewes fns. 31, 41, 45; Burgess fn. 3; King fns. 10, 30)

———. 1995. "The Life and Work of David Pingree Wheatland (1898-1993)." *Journal of the History of Collections* (Oxford University Press), vol. 7, no. 2, pp. 261-8. (App. B fn. 3)

Anon. 1688. *Curious Enquiries. Being Six Brief Discourses, viz. I. Of the Longitude. II. The Tricks of Astrological Quacks. III. Of the Depth of the Sea. IV. Of Tobacco. V. Of Europes being too full of People. VI. The various Opinions concerning the Time of Keeping the Sabbath.* London. (Gingerich fn. 3)

Anon. 1726. *A Sailor's Proposal for Finding his Longitude by the Moon....* 2nd ed. London. (The first edition, also published in London, has no date of publication.) (Turner app. 3)

Anon. 1772. Form for lunar-distance computation. MS. RGO 14/67, fol. 46, Royal Greenwich Observatory Archives, Cambridge University Library. (Howse fn. 19, fig. 7)

Apianus, Petrus. 1524. *Cosmographicus Liber Petri Apiani Mathematici studiose collectus.* Landshut. (There are two states, or perhaps two different issues, of the first edition of this book. The title of the one that is most frequently found is shown above. All copies of this appear to indicate the place of publication [Landshut], but not all bear the date. Another, less known, state or issue carries a slightly different [perhaps earlier?] title: *Cosmographicus Liber a Petro Apiano Mathematico studiose collectus.*) (Howse fn. 2; Stimson fig. 5; App. C fns. 18, 27)

———. 1529. *Cosmographicus Liber Petri Apiani Mathematici, studiose correctus, ac erroribus vindicatus per Gemmam Phrysium.* Rev. by Gemma Frisius. Antwerp. (Stimson fig. 5; App. C fn. 27)

———. 1533. *Introductio Geographica Petri Apiani in Doctissimas Verneri Annotationes, cõtinens plenum intellectum & iudicium omnis operationis, quae per sinus & chordas in Géographia confici potest, adjuncto Radio astronomico cum quadrante novo Meteoroscopii loco longe utilissimo.* Ingolstadt. (Howse fig. 2; App. C fn. 17)

Apollonius of Perga. 1696. *Apollonii Pergaei Conicorum Libri Quatuor, Serenissimo Principi Joanni Gastoni ab Etruria Dicati una cum Lemmatibus Pappi Alexandrini et Commentariis Eutocii Ascalonitae: Quae olim primus vulgavit omnia Federicus Commandinus Urbinas, e Graeco a se Conversa, Expurgata mendis, & commentariis illustrata: Nuperrimè autem in lucem prodeunt, Ab aliis etiam erratis longè plurimis, Quae, ut primùm edita sunt, identidem irrepserunt, Vindicata.* Pistoia. (Chandler fig. 7)

Applebaum, Wilbur. 1986. "A Descriptive Catalogue of the Manuscripts of Nicolas Mercator, F.R.S. (1620-1687) in Sheffield University Library." *Notes and Records of the Royal Society of London*, vol. 41, no. 1 (October) pp. 27-37. (Turner fn. 25)

Arnold, John. 1780. *An Account kept during Thirteen Months in the Royal Observatory at Greenwich of The Going of a Pocket Chronometer, Made on a New Construction by John Arnold, Having his new-invented Ballance Spring, And A Compensation for the Effects of Heat and Cold In the Ballance. Published by Permission of the Board of Longitude.* London. (Betts fn. 37)

———. 1791. *Certificates and Circumstances relative to the going of Mr Arnold's Chronometers.* London. (Betts fns. 14, 43)

Arnold, John Roger. 1806. "Explanation of Time-keepers, constructed by Mr. Arnold. Delivered to the Board of Longitude by Mr. Arnold, March 7th, 1805." In *Explanations of Time-keepers, constructed by Mr. Thomas Earnshaw and the late Mr. John Arnold. Published by Order of the Commissioners of Longitude.* London. Facsimile reprint by the British Horological Institute in *Principles and Explanations of Timekeepers by Harrison, Arnold and Earnshaw.* [Upton, Notts., England], 1984. (Betts fn. 45)

Arthur, James. 1909. *Time and Its Measurement.* Chicago: H. W. Windsor. (Cheney gen. ref.)

Aubrey, John. 1696. *Miscellanies.* London. (Turner fn. 36)

———. 1847. *The Natural History of Wiltshire, edited...by John Britton.* London. Reprinted 1969. (Turner fn. 43)

———. 1972. *John Aubrey: Three Prose Works. Miscellanies, Remaines of Gentilisme and Judaisme, Observations.* Edited by John Buchanan-Brown. Fontwell, England. (Turner fns. 33, 36)

*[Axe, Thomas]. References concerning the life of Thomas Axe. Aubrey MS. 10, fols. 29$^{r}$, 38; MS. 12, fols. 11$^{r}$-14$^{v}$, 15-16, Bodleian Library, Oxford University. (Turner fns. 34, 37, 39)

B., R. [Robert Burleigh?]. 1714. *Longitude To be found out with A new Invented Instrument, Both By Sea And Land. Also, Some Reasons for finding it thereby, extracted from the Three Years Observations made at Islington by Dr. Edmund Halley, Savilian Professor of Geometry in Oxford, for knowing the true Place of the Moon, and which now are inserted in Mr. Street's Caroline Tables. With A better Method for discovering Longitude, than that lately propos'd by Mr. Whiston and Mr. Ditton.* London. (Gingerich fns. 36-7, fig. 5; Turner app. 3)

Baillie, G. H. 1950. "The 1730 Harrison M.S." *Horological Journal*, vol. 92, no. 1102 (July), pp. 448-50, and vol. 92, no. 1103 (August), pp. 504-6. (Andrewes fn. 29)

———. 1951. *Clocks and Watches: An Historical Bibliography.* Vol. 1 (vol. 2 has not been published). London: N.A.G. Press. Reprinted London: Holland Press, 1978. (Cardinal fn. 11; Turner fn. 45)

Baily, Francis. 1835. *An Account of the Revd. John Flamsteed, the first Astronomer Royal....* London. Reprinted 1960. (Turner fn. 29)

Bamford, Francis. 1975. "A Shetlander in St Martin's Lane, 1775." *Furniture History,* vol. 11, pp. 108-11. (Betts fn. 8)

Barrow-Green, June. 1996. *Poincaré and the Three Body Problem.* Providence, Rhode Island: American Mathematical Society. (Chandler fn. 24)

Baugh, Daniel. 1978. "The Sea-trial of John Harrison's Chronometer, 1736." *Mariner's Mirror,* vol. 64, no. 3 (August), pp. 235-40. (Andrewes fn. 76)

Baxondall, B. 1924. "The Circular Dividing Engine of Edward Troughton, 1793." *Transactions of the Optical Society* (London), vol. 25.3, pp. 136-8. (Stimson fn. 30)

Beaglehole, J. C. 1961. *The Journals of Captain James Cook on his Voyages of Discovery.* 4 vols. and a portfolio of charts and views. Vol. 2, *The Voyage of the* Resolution *and* Adventure, *1772-1775.* Hakluyt Society Publications, extra series no. 35. Cambridge: Cambridge University Press. (Randall fns. 48-9)

———. 1974. *The Life of Captain James Cook R.N., F.R.S.* Hakluyt Society Publications, extra series, vol. 37. London. (Thrower fn. 28)

Bedini, Silvio A. 1984. *At the Sign of the Compass and Quadrant: The Life and Times of Anthony Lamb.* Transactions of the American Philosophical Society, vol. 74, part 1. Philadelphia. (Stimson fn. 24; App. D fn. 31)

———. 1991. *The Pulse of Time: Galileo Galilei, the determination of longitude, and the pendulum clock.* Biblioteca di Nuncius, Studi e Testi, vol. 3. Florence: Leo S. Olschki. (Landes fn. 19; Leopold fn. 8; Mahoney fn. 2; Turner fn. 8; Van Helden fn. 18)

———. 1996. "Thomas Godfrey and the Invention of the Octant." *The Professional Surveyor* (May, June, July, and August), in "History Corner." (App. D fn. 31)

———. In preparation. *Uncommon Genius. The Campani Brothers of Rome.* (Andrewes fn. 66)

Bedwell, William. Concerning Bedwell's suggestion to use water-clocks to find longitude. State Papers Domestic 12/153/27, Public Record Office, London. (Turner fn. 14)

Beeckman, Isaac. 1939-53. *Journal tenu par Isaac Beeckman de 1604 à 1634.* Edited by C. de Waard. 4 vols. The Hague: Martinus Nijhoff. (Leopold fn. 8)

Beillard, Alfred. 1895. *Recherches sur l'horlogerie, ses inventions et ses célébrités.* Paris. (Cardinal fn. 29)

Bennett, J. A. 1985. "The Longitude and the New Science." *Vistas in Astronomy* (special issue for the Longitude Zero Symposium 1984), vol. 28, parts 1/2 (1985), pp. 219-25. (Turner fns. 12, 27)

Bernoulli, Jean. 1771. *Lettres Astronomiques où l'on donne une idée de l'état actuel de l'astronomie pratique dans plusieurs villes de l'Europe.* Berlin. (Andrewes fn. 138)

Berthoud, Ferdinand. 1759. *L'Art de conduire et de régler les pendules et les montres: A l'usage de ceux qui n'ont aucune connaissance d'Horlogerie.* Paris. (Cardinal fn. 5)

———. 1763. *Essai sur l'Horlogerie; dans lequel on traite de cet Art rélativement à l'usage civil, à l'Astronomie et à la Navigation, en établissant des Principes confirmés par l'experience.* Paris. (Cardinal fn. 6)

———. 1773. *Traité des Horloges Marines, contenant La Théorie, la Construction, la Main-d'oeuvre de ces machines, et la manière de les éprouver, pour parvenir, par leur moyen, à la rectification des Cartes Marines, et la détermination des Longitudes en mer.* Paris. (Cardinal fn. 13; Randall figs. 12-13)

———. 1773. *Éclaircissemens sur l'invention, la théorie, la construction, et les épreuves des nouvelles machines proposées en France, pour la détermination des longitudes en mer par la mesure du temps. Servant de Suite à* l'Essai sur l'Horlogerie *&* *au* Traité des Horloges Marines, *et de réponse à un écrit qui a pour titre: Précis des recherches faites en France pour la détermination des longitudes en mer par la mesure artificielle du temps.* Paris. (Cardinal fn. 18)

———. 1787. *De La Mesure du Temps, Ou Supplément Au Traité Des Horloges Marines, Et A L'Essai Sur L'Horlogerie; Contenant les principes de construction, d'exécution & d'épreuves des petites Horloges à Longitude. Et l'application des mêmes principes de construction, &c. aux Montres de poche, ainsi que plusiers constructions d'Horloges Astronomiques, &c.* Paris. (Betts fn. 40)

———. 1792. *Mémoire sur le travail des horloges et des montres à longitude inventées par M. Ferdinand Berthoud.* Paris. (Cardinal fn. 39)

———. 1797. *Suite du Traité des Montres à Longitudes.* Paris. (Cardinal fig. 15)

———. 1802. *Histoire de la Mesure du Temps par les Horloges.* Paris. (Cardinal fns. 25-7)

———. 1807. *Supplément au traité des montres à longitudes avec appendice contenant la notice ou indication des principales recherches ou des travaux faites par F. B. sur diverses parties des machines qui mesurent le temps depuis 1752 à 1807 accompagné d'observations et d'éclaircissement.* Paris. (Randall fn. 40)

Betts, Jonathan. 1984. Introduction to *Principles and Explanations of Timekeepers by Harrison, Arnold and Earnshaw.* British Horological Institute facsimile reprint. [Upton, Notts., England]. (Betts fn. 5)

———. 1985. "The Spring Detent." *Clocks,* vol. 7, no. 7 (January), p. 70. (Betts fn. 41)

———. 1993. *John Harrison.* National Maritime Museum exhibition catalogue. Greenwich, England. (Andrewes fn. 89)

———. 1996. "Josiah Emery, Watchmaker of Charing Cross," parts 1 and 2. *Antiquarian Horology,* vol. 22, no. 5 (Spring 1996), pp. 394-401, and vol. 22, no. 6 (Summer 1996). (Betts fn. 22)

Bigelow, J. E. 1992. "Barometric Pressure Changes and Pendulum Clock Error." *Horological Journal,* vol. 135, no. 2 (August), pp. 62-4. (Burgess fn. 26)

Billingsley, Case. 1714. *A Letter to the Commissioners appointed by the Parliament of Great Britain, for discovering the Longitude at Sea..., Further to Explain his late Proposal for that Discovery, by the Sun, Moon and Stars; and a true Time-keeper.* London. (Turner app. 3)

———. 1714. *The Longitude At Sea, Not to be found by Firing Guns, nor by the Most Curious Spring-Clocks or Watches. But The only True Method for Discovering that Valuable Secret by the Sun, Moon or Stars, and an Exact Time-keeper, with such Necessary Improvements, as have not yet been Describ'd by any other Person; and (with respect to the Term of any Ordinary Voyage) may properly be call'd a Perpetual Motion.* London. (Gingerich fns. 31-5; Turner app. 3)

Birch, T. 1756-7. *The History of the Royal Society of London....* 4 vols. London. (Leopold fn. 37; App. D fn. 28)

Blaeu, William Johnson. 1625. *The Sea-Mirrour Containing, a Briefe Instruction In the Art of Navigation; and a Description of the Seas and Coasts of the Easterne, Northerne, and Westerne Navigation; Collected and Compiled Together Out of the Discoveries of many Skilfull and expert Sea-men, by William Iohnson Blaevw; and Translated out of Dutch into English, by Richard Hynmers.* Amsterdam. (App. D fig. 3)

Board of Longitude Confirmed Minutes. 1737-1824. MS. RGO 14/5, Royal Greenwich Observatory Archives, Cambridge University Library. (Andrewes fns. 87, 106, 111, 114, 116; Randall fns. 8, 16, 27, 33-4, 36; Stimson fn. 28)

Boissard, Jean-Jacques. 1628. *Bibliotheca sive Thesaurus Virtutis et Gloriae.* Frankfurt. (App. C fig. 2)

Bond, Henry. 1676. *The Longitude Found: Or, A Treatise Shewing An Easie and Speedy way, as well by Night as by Day, to find the Longitude, having but the Latitude of the Place, and the Inclination of the Magnetical Inclinatorie Needle.* London. (Thrower fn. 21)

Boorstin, Daniel J. 1983. *The Discoverers.* 2 vols. New York: Harry N. Abrams. (App. C fn. 16)

Boring, E. G. 1950. *A History of Experimental Psychology.* 2nd ed. New York: Appleton Century Crofts. (App. B fn. 4)

Bos, H. J. M. 1980. "Mathematics and rational mechanics." In G. S. Rousseau and Roy Porter, eds., *The Ferment of Knowledge: Studies in the Historiography of Eighteenth-Century Science,* pp. 327-55. Cambridge: Cambridge University Press. (Chandler fn. 12)

Bosscha, J. 1907. "Simon Marius, réhabilitation d'un Astronome Calomnié." *Archives Néerlandaises des Sciences Exactes et Naturelles,* ser. 2, vol. 12, pp. 258-307, 490-527. (Van Helden fn. 11)

Bourne, William. 1574. *A Regiment for the Sea.* London. (Stimson fns. 13-14)

Boyle, Robert. 1660. *New Experiments Physico-Mechanicall Touching the Spring of the Air and its Effects....* London. (Turner fn. 68)

Bradley, James. 1726. "The Longitude of Lisbon, and the Fort of New York, from Wansted and London, determin'd by Eclipses of the First Satellite of Jupiter. By the Reverend Mr. James Bradley, M. A., Astron. Prof. Savil. R.S.S." *Philosophical Transactions* (Royal Society, London), vol. 34, no. 394 (published 1728), pp. 85-90. Reprinted in *Miscellaneous Works and Correspondence of James Bradley*, pp. 58-61. Edited by Stephen Peter Rigaud. Oxford, England, 1832. (Andrewes fn. 75; Van Helden fn. 71)

———. 1832. *Miscellaneous Works and Correspondence of James Bradley*. Edited by Stephen Peter Rigaud. Oxford, England. Reprinted New York: Johnson Reprint Corp., 1972. (Van Helden fns. 70-1)

Bradley, Thomas. 1840. "Five drawings of Harrison's three chronometers (furnished to the Board of the Admiralty by Messrs Arnold & Dent)." MS. RGO 6/586 f213r-f217r, Royal Greenwich Observatory Archives, Cambridge University Library. (Apart from f216r, which shows some of the component parts of H.3, all of these drawings are included in this volume.) (Andrewes fn. 142, figs. 11, 15, 21-2)

Braudel, Fernand. 1979. *Civilisation matérielle, économie et capitalisme.* Vol. 1, *Les structures du quotidien*. Paris: Armand Colin. (Landes fns. 6, 8, 10, 27)

British Horological Institute. 1984. *Principles and Explanations of Timekeepers by Harrison, Arnold and Earnshaw.* British Horological Institute facsimile reprint. [Upton, Notts., England]. (Betts fn. 5)

*The British Library General Catalogue of Printed Books to 1975.* 1979-87. London: Clive Bingley/K. G. Saur. (App. C fn. 30)

Britten, F. J. 1911. *Old Clocks and Watches & their Makers*. 3rd ed. London, 1911. (Andrewes fn. 69)

Brown, Lloyd A. 1941. *Jean Domenique Cassini and his World Map.* Ann Arbor: University of Michigan Press. (Thrower fn. 8)

———. 1949. *The Story of Maps.* Boston: Little, Brown. (Leopold fn. 46; Thrower app. 1)

Brown, Peter. 1987. *The Noel Terry Collection of Furniture and Clocks*. York, England. (Turner fn. 74)

Browne, Robert. 1714. *Methods, Propositions, and Problems, for finding the Latitude; with the Degree and Minute of the equator upon the Meridian. And the Longitude at Sea, by Coelestial Observations only. And also by Watches, Clocks, &c. and to correct them and to know their Alterations.* London. (Turner app. 3)

Bruce, Alexander, Earl of Kincardine. Concerning Bruce's dealings with Christiaan Huygens. Kincardine papers, fols. 162r-v, 170v, 172r-v, 174r-v, 176r-7r, 178r-9r, 181v, 184r, 188v, 195r, Royal Society, London. (The originals of these papers are in private ownership; a copy is at the Royal Society, London.) (Leopold fns. 4, 14, 17-18, 23-4, 26-7, 31, 35, 43)

Bryden, D. J. 1993. "Magnetic Inclinatory Needles: approved by the Royal Society." *Notes and Records of the Royal Society of London*, vol. 47, no. 1 (January), pp. 17-31. (Turner fn. 27)

Burgess, Martin. 1971. "The Grasshopper Escapement: Its Geometry and its Properties." *Antiquarian Horology*, vol. 7, no. 5 (December), pp. 416-22. (Burgess fns. 4, 45)

———. 1994. "Heavy and Light Pendulums" (Letters section). *Horological Journal*, vol. 136, no. 11 (May), pp. 367-8. (Burgess fn. 63)

Bush, Herman. 1885. "Re Harrison's Invention of the Marine Chronometer" (letter to the editor). *Horological Journal*, vol. 28, no. 327 (November), p. 45. (Andrewes fn. 62)

Byrom, John. 1854-7. *The Private Journal and Literary Remains of John Byrom*. Edited by Richard Parkinson. Manchester, England. (Penney fns. 6, 10, 16, 18)

Carboni, Jean-Baptiste. 1726. "Observationes Astronomicae habitae Ulyssipone, Anno 1725, & sub init. 1726, à Rev. P. Johanne Baptista Carbone, Soc. Jes. Communicante Isaaco Sequeyra Samuda, M.D., R.S.S. Coll. Med. Lond. Lic." *Philosophical Transactions* (Royal Society, London), vol. 34, no. 394 (published 1728), pp. 90-101. (Andrewes fn. 75)

Cardinal, Catherine. 1989. *The Watch from its Origin to the XIXth Century.* Secaucus, New Jersey: Wellfleet Press. (Cardinal fn. 2)

———. 1994. *Les horloges marines de M. Berthoud*. Paris: F. Nathan. (Cardinal fn. 36)

———, ed. 1984. *Ferdinand Berthoud, 1727-1807: Horloger mécanicien du Roi et de la Marine*. La Chaux-de-Fonds, Switzerland: Musée International d'Horlogerie. (Cardinal fns. 4, 8, 23, 35, 38)

Carvajal, Ramón Colón de. 1987. *Catálogo de Relojes del Patrimonio Nacional*. Madrid: Editorial Patrimonio Nacional. (Penney fn. 9)

Cassini, Giovanni Domenico (Jean-Dominique). 1668. *Ephemerides Bononienses Mediceorum Syderum ex Hypothesibus, et Tabulis Io: Dominici Cassini*. Bologna. (Van Helden fn. 38, fig. 8)

———. 1676. "An Extract of a Letter written by Signor Cassini to the Author of the Journal des Scavans, containing some Advertisements to Astronomers about the Configurations, by him given of the Satellites of Jupiter, for the years 1676, and 1677, for the verification of their Hypotheses." *Philosophical Transactions* (Royal Society, London), vol. 11, no. 128 (1676), pp. 681-3. (Van Helden fn. 44)

———. 1684. *Les Elemens de l'Astronomie Verifiez par Monsieur Cassini par le rapport de ses Tables aux Observations de M. Richer faites en l'Isle de Caïenne. Avec les Observations de MM. Varin, des Hayes, et de Glos faites en Afrique & en Amerique*. Paris. Published as the fifth work (pp. 1-74) in *Recueil d'Observations Faites en Plusieurs Voyages par Ordre de Sa Majesté, pour Perfectionner l'Astronomie et la Geographie. Avec divers Traitez Astronomiques. Par Messieurs de l'Académie Royale des Sciences.* Paris, 1693. Later published without the appendix concerning the observations of Varin *et al.* as *Les Elemens de l'Astronomie Verifiez par Monsieur Cassini par rapport de ses Tables aux Observations de M. Richer faites en l'Isle de Cayenne* in *Memoires de l'Académie Royale des Sciences. Depuis 1666. jusqu'à 1699*, vol. 8, pp. 55-79. Paris, 1730. (Van Helden fns. 42, 47-50)

———. 1692. *Éclipses du premier satellite de Jupiter pendant l'année.* Paris. (Thrower fn. 9)

———. 1693. *Les Hypotheses et les Tables des Satellites de Jupiter Reformées sur de Nouvelles Observations. Par Monsieur Cassini de l'Académie Royale des Sciences.* Published as the eighth (last) work (pp. 1-52, tables pp. 1-106) in *Recueil d'Observations Faites en Plusieurs Voyages par Ordre de Sa Majesté, pour Perfectionner l'Astronomie et la Geographie. Avec divers Traitez Astronomiques. Par Messieurs de l'Académie Royale des Sciences.* Paris, 1693. Later published in *Memoires de l'Académie Royale des Sciences. Depuis 1666. jusqu'à 1699*, vol. 8, pp. 317-505. Paris, 1730. (Van Helden fns. 45, 51)

Chapin, Seymour L. 1978. "Lalande and the Longitude: A Little Known London Voyage of 1763." *Notes and Records of the Royal Society of London*, vol. 32, no. 2 (March), pp. 165-80. (Randall fn. 17)

Chapman, Allan. 1993. "Scientific Instruments and Industrial Innovation: The Achievement of Jesse Ramsden." In R. G. W. Anderson, J. A. Bennett, and W. F. Ryan, eds., *Making Instruments Count, Essays on Historic Scientific Instruments presented to Gerard L'Estrange Turner,* pp. 418-23. Aldershot, England: Variorum. (Stimson fn. 29)

Clarke, James. 1714. *An Essay wherein a Method is Humbly propos'd for Measuring Equal Time with the utmost Exactness; Without the Necessity of being confin'd to Clocks, Watches or any other Horological Movements; in order to Discover the Longitude at Sea.* London. (Turner app. 3)

———. 1715. *The Mercurial Chronometer improv'd: or a Supplement to a Book entitled An Essay... In which all objections that are in the least bit rational are remov'd, and the Method confirm'd....* London. (Turner app. 3)

Clutton, Cecil, and Daniels, George. 1975. *Clocks and Watches: The collection of the Worshipful Company of Clockmakers.* London: Sotheby Parke Bernet. (Betts fn. 26; Turner fn. 74)

*Connoissance des Temps, Pour l'Année Commune 1773. Publiée Par l'ordre de l'Académie Royale des Sciences, et calculée Par M. de la Lande, de la même Académie.* 1771. Paris. (Between 1762 and 1767, the name of this publication was changed to *Connoissance des Mouvemens Célestes.* Jérôme Lalande served as editor between 1759 and 1775.) (Van Helden fns. 53-55, 68)

Cook, Andrew S. 1985. "Alexander Dalrymple and John Arnold: Chronometers and the representation of Longitude on East India Company charts." *Vistas in Astronomy* (special issue for the Longitude Zero Symposium 1984), vol. 28, parts 1/2 (1985), pp. 189-95. (Betts fns. 30-1)

Cortés, Martín. 1551. *Breve compendio de la sphera y de la arte de navegar con nuevos instrumentos y reglas exemplificado con muy subtiles demonstraciones.* Seville. Translated by Richard Eden, under the title *The Arte of Navigation, conteynyng a compendious description of the Sphere, with the makyng of certen Instrumentes and Rules for Navigations: and exemplified by manye Demonstrations. Wrytten in the Spanyshe tongue by Martin Curtes, And directed to the Emperour Charles the fyfte. Translated out of Spanyshe into Englyshe by Richard Eden.* London, 1561. (Stimson fn. 11)

Cotter, Charles H. 1968. *A History of Nautical Astronomy.* London: Hollis & Carter. (Howse app. 1)

———. 1983. *A history of the navigator's sextant.* Glasgow: Brown, Son & Ferguson. (Mörzer Bruyns fn. 11)

Cottingham, E. T. 1910. "A Description of the Royal Astronomical Society's Harrison Clock, with a brief account of the maker." *Horological Journal*, vol. 52 (May), pp. 137-41. (Burgess fn. 16)

Crisford, Andrew. 1976. "Thomas Wright in the Poultry, London, No. 2228." *Antiquarian Horology*, vol. 9, no. 7 (June), pp. 785-8. (Betts fn. 48)

Cumming, Alexander. 1766. *The Elements of Clock and Watch-work.* London. (Andrewes fn. 128)

Cuningham, William. 1559. *The Cosmographical Glasse, conteinyng the pleasant Principles of Cosmographie, Geographie, Hydrographie, or Navigation.* London. (App. C fn. 31)

Cuss, T. P. C. [1965]. "The Huber-Mudge Timepiece with Constant Force Escapement." In *Pioneers of Precision Timekeeping*, pp. 93-115. Antiquarian Horological Society Monograph No. 3. [Ramsgate, England]. (Penney fn. 49)

Dalrymple, Alexander. [1779 or 1780]. *Some Notes useful to those who have Chronometers at Sea.* London. (Betts fns. 32, 34)

———. 1806. *Longitude. A Full Answer To The advertisement concerning Mr. Earnshaw's Timekeeper In The Morning Chronicle, 4th Feb. and Times 13th Feb. 1806.* London. (Betts fn. 6)

Daniels, George. 1967. *English and American Watches.* London: Abelard-Schuman. (Betts fn. 25)

———. 1981. *Watchmaking.* London: Sotheby Publications. (Daniels fn. 1)

d'Arcons, César. 1655. *Le secret du flux et reflux de la mer et des longitudes, delivré a la sapience eternelle.* Rouen. (Turner fn. 23)

David, Andrew, ed. 1988 and 1992. *The Charts and Coastal Views of Captain Cook's Voyages.* 2 vols. Hakluyt Society Publications, extra series no. 43. London. (Betts fn. 23; Thrower app. 1)

Davids, C. A. 1980. "De zeevaartkunde en enkele maatschappelijke veranderingen in Nederland tussen 1850 en 1914." *Mededelingen van de Nederlandse Vereniging voor Zeegeschiedenis*, no. 40/ 41, pp. 51-83. (Mörzer Bruyns fn. 15)

———. [1985]. *Zeewezen en Wetenschap, de Wetenschap en de ontwikkeling van de Navigatietechniek in Nederland tussen 1585 en 1815.* Amsterdam/Dieren: De Bataafsche Leeuw. (Leopold fns. 44, 67, 73)

———. 1990. "Finding longitude at sea by magnetic declination on Dutch East-Indiamen 1596-1795." *The American Neptune*, vol. 50, pp. 281-90. (Mörzer Bruyns fn. 10)

Davis, John. 1595. *The Seamans Secrets, Devided into 2. partes, wherein is taught the three kindes of Sayling, Horizontall, Peradoxall, and sayling upon a great Circle: also an Horizontall Tyde Table for the easie finding of the ebbing and flowing of the Tydes, with a Regiment newly calculated for the finding of the Declination of the Sunne, and many other most necessary rules and Instruments, not heeretofore set foorth by any.* London. (Stimson fn. 17; App. D fn. 23, figs. 1-2)

Dawson, Percy G.; Drover, C. B.; and Parkes, D. W. 1982. *Early English Clocks.* Woodbridge, England: Antique Collectors' Club. (Turner fn. 75)

Débarbat, Suzanne, and Wilson, Curtis. 1989. "The Galilean Satellites of Jupiter from Galileo to Cassini, Rømer and Bradley." In René Taton and Curtis Wilson, eds., *Planetary Astronomy from the Renaissance to the Rise of Astrophysics*, pp. 144-57. Vol. 2A (of 4) in *The General History of Astronomy*, M. A. Hoskin, ed. Cambridge: Cambridge University Press, 1984-. (Van Helden fns. 12, 36, 43, 52)

Defossez, Léopold. 1946. *Les savants du 17ᵉ siècle et la mesure du temps.* Lausanne, Switzerland. (Cardinal fn. 1)

Derham, William. 1704. "Experiments about the Motion of Pendulums in Vacuo." *Philosophical Transactions* (Royal Society, London), vol. 24, no. 294 (published 1706), pp. 1785-9. (Andrewes fn. 66; Turner fn. 68)

———. 1714. "Observations concerning the motion of Chronometers," dated August 4, 1714, read November 4, 1714. Ref. no. Cl.P.III(2)10, Royal Society Records, London. (Andrewes fn. 13)

Desagulier, J. T. 1734. *A Course in Experimental Philosophy.* 2 vols. London. (Andrewes fn. 13; Betts fn. 33)

*Dictionary of Scientific Biography.* 1970-90. Edited by Charles Coulston Gillispie (vols. 1-16) and Frederic L. Holmes (vols. 17-18). 18 vols. New York: Charles Scribner's Sons. (Chandler fn. 7; App. C fns. 1, 6)

Diderot, Denis, and Alembert, Jean Le Rond d'. 1751-72. *Encyclopédie, ou Dictionnaire Raisonné des Sciences, des Arts, et des Métiers.* 28 vols. Paris. (Andrewes fn. 13; Cardinal fn. 7)

Digges, Dudley. 1761. Captain's and master's logs of the *Deptford* concerning spoiled rations on voyage to conduct sea trials of H.4, December 7, 1761, and December 9, 1761. MSS Adm.51/241 and Adm.52/828, Public Record Office, London. (Randall fn. 10)

Digges, Thomas. 1576. "Errors in the Arte of Navigation commonly practized." Essay in "The Addition," published in Thomas Digges's corrected and augmented edition of Leonard Digges, *A Prognostication everlastinge of righte good effecte,...contayning plaine, briefe, pleasaūt, chosen rules to judge the weather by the Sunne, Moone, Starres, Comets, Rainebow, Thunder, Cloudes,....* London. (Stimson fn. 18)

Dilke, O. A. W. 1985. *Greek and Roman Maps.* Ithaca, New York: Cornell University Press. (Thrower fns. 5-6)

Ditisheim, Paul; Lallier, Roger; Reverchon, Léopold; and Vivielle (le Commandant). 1940. *Pierre Le Roy et la chronométrie.* Paris: Tardy. (Cardinal fns. 3, 33-4)

Doppelmayr, Johann Gabriel. 1730. *Historische Nachricht von den Nürnbergischen Mathematicis und Künstlern.* Nuremberg. (App. C fn. 1)

Drake, Stillman. 1978. *Galileo at Work: His Scientific Biography.* Chicago: University of Chicago Press. (Van Helden fn. 19)

Dubois, Pierre. 1849. *Histoire de l'horlogerie.* Paris. (Cardinal fn. 28)

du Halde, Jean Baptiste. 1735. *Description geographique, historique, chronologique, politique et physique de l'empire de la Chine et de la Tartarie chinoise.* 4 vols. Paris. The Hague, 1736. English translation: London, 1738 and 1741. German translation: 1748. (Landes app. 1)

Duliris, Leonard. 1647. *La Theorie des longitudes redvite en pratique sur le globe celeste, extraordinairment appareillé, pour coignoistre facilement en Mer, combien l'on est eslogné de toutes les terres du monde....* Paris. (Turner fn. 20)

———. 1655. *Ephemeride maritime dressé pour observer en mer la longitude et latitude selon l'invention du Pere Leonard Duliris Recollet. Avec un nouveau moyen de perpeuer l'ephemeride du soleil, pour avoir tousiours exactement sa declinaison....* Paris. (Turner fns. 21-2)

Dunkin, E. [1879]. *The Midnight Sky.* 2nd ed. London. (Howse fig. 9)

Durand, Robert. 1992. *Histoire du Portugal.* Paris: Hatier. (Landes fn. 21)

Earnshaw, Thomas. 1805. *Explanation of Timekeepers constructed by Thomas Earnshaw.* London. (One of the few surviving copies of this is preserved at the National Maritime Museum, Greenwich [ref. no. C1354]). (Betts fns. 57-60)

———. 1805. Proof copy, annotated by William Hardy, of Earnshaw's explanations of his timekeepers, which were circulated to watchmakers for their remarks. MS. RGO 14/27, Royal Greenwich Observatory Archives, Cambridge University Library. (Betts fns. 62-3)

———. 1805. Letter to Board of Longitude listing names of seven "declared enemies," July 3, 1805. MS. RGO 14/25, p. 275, Royal Greenwich Observatory Archives, Cambridge University Library. (Betts fn. 65)

———. 1806. "Explanation of Time-keepers, constructed by Mr. T. Earnshaw. Three of Them Having Been Tried Under the Present Act of Parliament. Delivered to the Board of Longitude, March 7th, 1805." In *Explanations of Time-keepers, constructed by Mr. Thomas Earnshaw and the late Mr. John Arnold. Published by Order of the Commissioners of Longitude.* London. Facsimile reprint by the British Horological Institute in *Principles and Explanations of Timekeepers by Harrison, Arnold and Earnshaw.* [Upton, Notts., England], 1984. (Betts fn. 57)

———. 1808. *Longitude. An Appeal to the Public: stating Mr. Thomas Earnshaw's Claim to the Original Invention of the Improvements in his Timekeepers their Superior Going in Numerous Voyages, and also as tried by the Astronomer Royal by Orders of the Commissioners of Longitude, and his Consequent Right to National Reward.* London. Facsimile reprint by the British Horological Institute in *Earnshaw's Appeal.* [Upton, Notts., England], 1986. (Betts fns. 42, 53; Burgess fns. 71-2)

*[———]. Thomas Earnshaw's bank records. Manuscript Accounts, Hoare's Bank, London. (Betts fn. 64)

Easton, C. 1928. "Klimatologische Studies." *Tijdschrift van het Kon. Ned. Aardrijkskundig Genootschap*, 2e ser., vol. 45, pp. 248ff. (Leopold fn. 69)

Eden, Richard. 1555. *The Decades of the newe worlde or west India,...wrytten in the Latine tounge by Peter Martyr of Angleria.* London. (App. C fn. 31)

———. 1561. *The Arte of Navigation,....* See under Cortés, Martín. (Stimson fn. 11)

Ellicott, John. 1752. "A Description of Two Methods, by which the Irregularity of the Motion of a Clock, arising from the Influence of Heat and Cold upon the Rod of the Pendulum, may be prevented. Read at the Royal Society, June 4, 1752." *Philosophical Transactions* (Royal Society, London), vol. 47, no. 81 (published 1753), pp. 479-94. (King fn. 45; Penney fn. 34; Randall fn. 4)

———. 1753. *A Description of Two Methods, by which The Irregularities in the Motion of a Clock, arising from the Influence of Heat and Cold upon the Rod of the Pendulum, may be prevented. Read at the Royal Society, June 4, 1752. To which are added A Collection of Papers, Relating to the same Subject, Most of which were read at several Meetings of the Royal Society.* London. (Penney fn. 36)

Emerson, William. 1770. *The Mathematical Principles of Geography, Navigation & Dialling.* London. (Betts fn. 3)

*Encyclopaedia Britannica.* 1910-11. "Guaiacum." 11th ed., vol. 12, pp. 646-7. Cambridge: Cambridge University Press. (Andrewes fn. 46)

Favaro, Antonio. 1891. *Nuove Studi Galileiani.* Venice: Tipografia Antonelli. (Van Helden fn. 19)

———. 1891. "Documenti inediti per la storia dei negoziati con la Spagna per la determinazione delle longitudini in mare." *Memorie del Reale Istituto Veneto di Scienze, Lettere ed Arti,* vol. 25, pp. 101-48. (Van Helden fn. 19)

———. 1891. "Documenti inediti per la Storia dei Negoziati con gli Stati Generali d'Olanda per la Determinazione dell Longitudini." *Memorie del Reale Istituto Veneto di Scienze, Lettere ed Arti,* vol. 25, pp. 289-338. (Van Helden fn. 28)

*[Flamsteed, John]. 1675. Concerning Charles II's appointment of Flamsteed as first Astronomer Royal. State Papers Domestic 29/368, fol. 299, and State Papers Domestic 44, p. 10, Public Record Office, London. (Howse fn. 4)

———. 1683. "An account of the Eclipses or ingresses of Jupiters Satellits into his shadow and such Emersions of them from it as will be visible at the Observatory at Greenwich in the three last Months of this year 1683. Sent in a Letter to the Publisher from J. F. Astron. Reg." *Philosophical Transactions* (Royal Society, London), vol. 13, no. 151 (published 1683), pp. 322-3. (Van Helden fn. 59)

———. 1683. "A Letter from Mr. Flamsteed concerning the Eclipses of Saturns [misprint; should be Jupiter's] Satellit's for the year following 1684, with a Catalogue of them, and informations concerning its use." *Philosophical Transactions* (Royal Society, London), vol. 13, no. 154 (published 1683), pp. 404-15. (Van Helden fns. 60-1, fig. 11)

———. 1684. "A Letter from the learned Mr John Flamsteed, Astron. Reg. concerning the Eclipses of Jupiters Satellit's for the Year following 1685. with a Catalogue of them, and informations concerning its use." *Philosophical Transactions* (Royal Society, London), vol. 14, no. 165 (published 1684), pp. 760-5. (Van Helden fn. 62)

———. 1685. "An Abstract of a Letter from Mr. J. Flamsteed, Math. Reg. & F. of the R. S. giving an account of the Eclipses of ♃s [Jupiter's] Satellits, anno 1686; and containing a Table of the Parallaxes of ♃s [Jupiter's] Orb, and an Ephemeris of ♃s [Jupiter's] Geocentric Places for the same year." *Philosophical Transactions* (Royal Society, London), vol. 15, no. 177 (published 1686), pp. 1215-25. (Van Helden fn. 63)

———. 1686. "An Extract of a Letter from Mr. J. Flamsteed Astr. Reg. and Reg. Soc. S. giving his calculation of the Eclipses of Jupiters Satellites for the Year 1687. Together with a Table of the Parallaxes of the Orb, and the Ephemeris of Jupiters Geocentric Place for the same Year; to which is added an Observation of the Eclipse of the Moon, Novemb. 30, 1695. [misprint; should be 1685] made at Lisbon, and Mr. Flamsteed's own Observation of the Eclipse of Jupiter by the Moon on March 31th. past." *Philosophical Transactions* (Royal Society, London), vol. 16, no. 184 (published 1688), pp. 196-206. (Van Helden fns. 63-4)

———. 1725. *Historia Coelestis Britannica.* 3 vols. London. (Howse fn. 8)

Fletcher, D. W. 1951. "Temperature Compensation of Harrison's First 'Marine Timekeeper.' " *Horological Journal*, vol. 93, no. 1113 (June), pp. 377-9. (Andrewes fns. 64-5, 149)

———. 1952. "Restoration of John Harrison's First Marine Timekeeper." *Horological Journal*, vol. 94, no. 1125 (June), pp. 366-9, with editorial introduction on p. 365. (Andrewes fns. 27, 44, 48, 63, 68, 149, fig. 6)

[Fleurieu, Charles Pierre Claret D'Eveux de]. 1768. *Examen critique d'un mémoire publié par M. Le Roy, horloger du Roi, sur l'épreuve des horloges propres à déterminer les longitudes en mer et sur les principes de leur construction.* London. (Cardinal fn. 12)

Folkerts, Menso. 1976. Article on Johann Werner. In Charles Coulston Gillispie, ed., *Dictionary of Scientific Biography,* vol. 14, pp. 272-7. New York: Charles Scribner's Sons. (App. C fns. 1, 6)

Folkes, Martin. 1749. Address delivered at presentation of Copley Gold Medal to John Harrison, November 30, 1749. Ref. no. JBC.XX.181-96, Minutes of the Royal Society, London. (Andrewes fns. 20, 60, 70-1, 78, 88, 92, 107, 109, 113; Burgess fn. 25)

Forbes, Eric G. 1971. *The Euler-Mayer Correspondence (1751-1755): A New Perspective on Eighteenth-Century Advances in the Lunar Theory.* New York: American Elsevier. (Chandler fns. 16-20)

———. 1975. *Greenwich Observatory, The Royal Observatory at Greenwich and Herstmonceux, 1675-1975.* Vol. 1, *Origins and Early History (1675-1835).* London: Taylor and Francis. (Howse fn. 10; Stimson fn. 21; Turner fn. 28)

Force, J. E. 1985. *William Whiston, Honest Newtonian.* Cambridge: Cambridge University Press. (Turner fn. 2)

Ford, Henry. 1922. *My Life and Work.* Garden City, New York: Doubleday. (Mahoney fn. 14)

Fournier, Georges. 1643. *Hydrographie Contenant la Theorie et la Practique de Toutes les Parties de la Navigation.* Paris. (Turner fn. 24)

[Franklin, Benjamin]. 1931. "Account Book of Benjamin Franklin kept by him during his First Mission to England as Provincial Agent 1757-1762." *The Pennsylvania Magazine of History and Biography*, vol. 55, no. 2, pp. 97-133. (Andrewes fn. 83)

———. 1959-93. *The Papers of Benjamin Franklin.* Edited by Leonard W. Labaree *et al.* 30 vols. New Haven: Yale University Press. (Andrewes fns. 124, 128-9)

French, John. 1715. *A Perfect Discovery Of The Longitude at Sea; In Compliance with what's propos'd In a late Act of Parliament. Being the Product of Nine Years Study, and frequent Amendments of a Mathematician, who has us'd the Sea upwards of Twenty five Years, and has had the Experience thereof at Sea. And his Projection hath been View'd and Approv'd on by most of the Mathematicians about London.* London. (Gingerich fns. 14-15, fig. 3; Turner app. 3)

Friedrich von Hessen-Kassel, Crown Prince. 1692-4. Concerning stay in The Hague. MSS 4a 57.24, 25 and 4a 72.2, Hessisches Staatsarchiv Marburg, Germany. (Leopold fn. 93)

Friis, Herman, ed. 1967. *The Pacific Basin: A History of its Geographical Exploration.* New York: American Geographical Society. (Thrower app. 1)

Fulton, John F. 1961. *A Bibliography of the Honourable Robert Boyle....* 2nd ed. Oxford: Oxford University Press. (Turner fn. 68)

Furnivall, J. S. 1939. *Netherlands India: A Study of Plural Economy.* Cambridge: Cambridge University Press. Reprinted 1967. (Landes fns. 2, 9)

F[yler], S[amuel]. 1699. *Longitudinis Inventae Explicatio Non Longa, Or, Fixing the Volatilis'd, And Taking Time On Tiptoe, Briefly Explain'd; By which Rules are given to find the Longitude at Sea by, as truly and exactly as the Latitude is found by the Star in the Tayle of Ursa Minor, call'd the Pole-Star.* London. (Gingerich fns. 4-11, fig. 1; Turner fn. 51)

Galilei, Galileo. 1610. *Sidereus Nuncius.* In Antonio Favaro, ed., *Le Opere di Galileo Galilei,* vol. 3. Edizione Nazionale. 20 vols. Florence, 1890-1909. Translated by Albert Van Helden, under the title *Sidereus Nuncius or the Sidereal Messenger.* Chicago: University of Chicago Press, 1989. (Van Helden fn. 6)

———. 1890-1909. *Le Opere di Galileo Galilei.* Edited by Antonio Favaro. Edizione Nazionale. 20 vols. Florence. Reprinted 1929-39, 1964-6. (Van Helden fns. 6, 10, 16-17, 19-26)

Gardiner, Robert Barlow. 1889. *The Registers of Wadham College, Oxford. (Part 1.) From 1613 to 1719.* London: George Bell and Sons. (Turner fn. 32)

Gassendi, Pierre. 1657. *The mirrour of true nobility and gentility: being the life of the renowned Nicolaus Claudius Fabricius, Lord of Peiresk, Senator of the Parliament at Aix.* London. (Van Helden fn. 14)

Gellibrand, Henry. 1635. *A Discourse Mathematical On The Variation Of The Magneticall Needle. Together with Its admirable Diminution lately discovered.* London. (Thrower fn. 18)

Gemma Frisius. 1530. *Gemma Phrysius de Principiis Astronomiae & Cosmographiae, Deque usu Globi ab eodem editi.* Louvain and Antwerp. (The passage concerning finding longitude at sea with a clock is in part 2, *cap.* 18 ["De novo modo inveniendi longitudinē"], fols. D2v-D3r; see Appendix C for translation.) (Mörzer Bruyns fn. 1; Stimson fn. 20; App. C fns. 28, 35, 40, figs. 10-12)

———. 1553. *Gemmae Phrysii Medici ac Mathematici De Principiis Astronomiae & Cosmographiae, Deque usu Globi ab eodem editi.* Antwerp. (In this edition, the passage concerning finding longitude at sea with a clock is in part 2, *cap.* 19 ["De novo modo inveniendi longitudinem"], pp. 64-5.) (App. C fn. 41, fig. 13)

Gingerich, Owen. 1992. "Astronomy in the Age of Copernicus." *Scientific American,* vol. 267, no. 5 (November), pp. 100-5. (Gingerich fn. 1)

Gingerich, Owen, and Welther, Barbara L. 1983. *Planetary, Lunar, and Solar Positions, New and Full Moons, A.D. 1650-1805.* American Philosophical Society *Memoirs* series, vol. 59S. Philadelphia. (Gingerich fn. 38)

*[Godfrey, Thomas]. 1758. "On the Invention of the Quadrant, commonly called Hadley's." *The American Magazine and Monthly Chronicle for the British Colonies* (Philadelphia), vol. 1, no. 10 (July), pp. 475-80. (App. D fns. 31-2, fig. 6)

Goldstein, Bernard R. 1969. "Preliminary Remarks on Levi Ben Gerson's Contributions to Astronomy." *The Israel Academy of Sciences and Humanities, Proceedings III.* (Stimson fn. 12)

———. 1985. *The Astronomy of Levi ben Gerson (1288-1344).* New York: Springer. (App. C fns. 12-14)

———. 1987. "Remarks on Gemma Frisius's *De Radio Astronomico et Geometrico."* In J. L. Berggren and B. R. Goldstein, eds., *From Ancient Omens to Statistical Mechanics: Essays on the Exact Sciences Presented to Asger Aaboe,* pp. 167-80. Vol. 39 of Acta Historica Scientiarum Naturalium et Medicinalium. Copenhagen: University Library. (App. C fn. 14)

Gordon, George. 1724. *A Compleat Discovery of a Method of Observing the Longitude at Sea.* London. (Turner app. 3; Van Helden fn. 72)

Gould, Rupert T. 1923. *The Marine Chronometer: Its History and Development.* London: J. D. Potter. Reprinted London: Holland Press, 1960. Revised ed.: Woodbridge, England: Antique Collectors' Club, 1990. (Andrewes fns. 11, 16, 27, 47, 52, 84, 89, 137; Betts fns. 12, 21, 39; Burgess fns. 40, 44, 48; Cardinal fns. 31-2; Gingerich fn. 2; Penney fn. 69; Randall fn. 52; Turner fns. 8, 45)

———. 1932. "The Restoration of John Harrison's Third Timekeeper." Published in five parts in the *Horological Journal,* vol. 74 (September 1931-August 1932): March, pp. 105-8; April, pp. 120-7; May, pp. 151-3, 148; June, pp. 166-7; July, pp. 178-81. Also published as a British Horological Institute monograph (n.d). (Andrewes fns. 114, 143)

———. 1933. "The Reconstruction of Harrison's First Timekeeper." *The Observatory, A Monthly Review of Astronomy,* vol. 56, no. 709 (June), pp. 193-6. (Andrewes fns. 69, 144-5, 147-8)

———. 1935. "John Harrison and his Time-keepers." *The Mariner's Mirror,* vol. 21, no. 2 (April), pp. 115-39. Subsequently issued by the National Maritime Museum in a pamphlet that has been reprinted many times since 1958. (Andrewes fn. 146)

Graham, George. 1722. "Mr Graham's acct of Coll Moleworth's Hourglass," June 7, 1722. Ref. no. Cl.P.III(2)15, Royal Society Records, London. (Andrewes fn. 8)

———. 1726. "A Contrivance to avoid the Irregularities in a Clock's Motion, occasion'd by the Action of Heat and Cold upon the Rod of the Pendulum." *Philosophical Transactions* (Royal Society, London), vol. 34, no. 392 (published 1728), pp. 40-4. (Burgess fn. 36; King fn. 44)

Gray, Jeremy. 1992. "Poincaré, topological dynamics, and the stability of the solar system." In P. M. Harman and Alan E. Shapiro, eds., *The investigation of difficult things: Essays on Newton and the history of the exact sciences in honour of D. T. Whiteside,* pp. 503-24. Cambridge: Cambridge University Press. (Chandler fn. 24)

Gray, John M. 1894. *James and William Tassie: A Biographical and Critical Sketch with a Catalogue of Their Portrait Medallions of Modern Personages.* Edinburgh. (Harrison family tree)

Gunther, R. T. 1935. *Early Science in Oxford.* Vol. 10, *The Life and Work of Robert Hooke,* part 4. Oxford, England. (Turner fn. 50)

Guyot, Edmond. 1955. *Histoire de la détermination des longitudes.* La Chaux-de-Fonds, Switzerland. (Chandler fn. 2)

Hadley, John. 1723. "An Account of a Catadioptrick Telescope, made by John Hadley, Esq; F.R.S. With the Description of a Machine contriv'd by him for the applying it to use." *Philosophical Transactions* (Royal Society, London), vol. 32, no. 376 (published 1724), pp. 303-12, with illustration on pp. 280-1. (Van Helden fig. 12)

———. 1731. "The Description of a new Instrument for taking Angles. By John Hadley, Esq; Vice-Pr. R. S. communicated to the Society on May 13. 1731." *Philosophical Transactions* (Royal Society, London), vol. 37, no. 420 (published 1733), pp. 147-57, with illustration opposite the contents page of the August-September 1731 issue. (Stimson fn. 25, fig. 6; App. D fn. 33, fig. 7)

Hakluyt, R. 1599. *Principal Navigations.* London. (Stimson fn. 15)

Haldanby, Francis. 1714. *An Attempt To Discover the Longitude At Sea, Pursuant To what is Proposed in a Late Act of Parliament.* London. (Gingerich fig. 2; Turner app. 3)

Hall, John James. 1930. "An Unrecorded Episode in the History of the Marine Chronometer." *Horological Journal,* vol. 73, no. 868 (December), pp. 65-9. (Turner fn. 45)

Hall, William. 1714. *A New and True Method to Find the Longitude, much more Exacter than that of Latitude by Quadrant also a new Method for the Latitude....* London. (Turner app. 3)

Halley, Edmond. 1682/3. "A Correction of the Theory of the Motion of the Satelite of Saturn, by that Ingenious Astronomer Mr. Edmund Hally." *Philosophical Transactions* (Royal Society, London), vol. 13, no. 145 (published 1683), pp. 82-8. (Van Helden fn. 67)

———. 1683. "A Theory of the Variation of the Magnetical Compass." *Philosophical Transactions* (Royal Society, London), vol. 13, no. 148 (published 1683), pp. 208-21. (Thrower fns. 14, 16)

———. 1686. "An Historical Account of the Trade Winds, and Monsoons, observable in the Seas between and near the Tropicks, with an attempt to assign the Phisical cause of the said Winds." *Philosophical Transactions* (Royal Society, London), vol. 16, no. 183 (published 1688), pp. 153-68, with foldout map of the trade winds opposite p. 151. (Thrower fig. 8)

———. 1687. "Catalogus Eclipsium omnium Satellitum Jovialium Anno 1688 per universam Terram Visibilium; momenta Occultationum eorum in Jovis Umbrâ, ac ex eâdem Egressuum sub Meridiano Londinensi exhibens. Supputante E. H." *Philosophical Transactions* (Royal Society, London), vol. 16, no. 191 (published 1688), pp. 435-9. (Van Helden fn. 63)

———. 1694. "Monsieur Cassini his New and Exact Tables for the Eclipses of the First Satellite of Jupiter, reduced to the Julian Stile, and Meridian of London." *Philosophical Transactions* (Royal Society, London), vol. 18, no. 214 (published 1695), pp. 237-56. (Van Helden fn. 69)

———. 1702. "The Description and Uses of a New and Correct Sea Chart of the whole World, shewing the Variations of the Compass" (explication appearing beneath Halley's world chart of magnetic variation). A copy of this map may be found in the Royal Geographical Society, London. (Thrower fns. 11, 14, 19)

———. 1710. "Observationes Syderum & imprimis Lunae, In Suburbio Londinensi apud Islington. Annis 1682, 1683 & 1684...." Appendix to Thomas Streete, *Astronomia Carolina: A New Theory of the Cœlestial Motions....* 2nd ed. corrected. London. (Turner fn. 52)

———. 1731. "A Proposal of a Method for finding the Longitude at Sea within a Degree, or twenty Leagues. By Dr. Edmund Halley, Astr. Reg. Vice President of the Royal Society. With an Account of the Progress he hath made therein, by a continued Series of accurate Observations of the Moon, taken by himself at the Royal Observatory at Greenwich." *Philosophical Transactions* (Royal Society, London), vol. 37, no. 421 (published 1733), pp. 185-95. (Thrower fn. 25)

Hanna, John. 1725. *An Astronomical Creed, In Ten Articles, With Natural and Moral Reflections from them; To which is subjoyned, the new Method of discovering the Longitude, by the appulse of the Moon to the fixed Stars: Improved By Sir Isaac Newton's Theory of the Moon and Mr. Flamsteed's Geometrical construction of the paralax.* Dublin. (Gingerich fn. 39; Turner app. 3)

Harley, J. B., and Woodward, David. 1987 and 1992. *The History of Cartography.* 2 vols. (series in progress). Chicago: University of Chicago Press. (Thrower fn. 2)

Harris, Joseph. 1783. *The Description and Use of the Globes and the Orrery....* 12th ed. London. (Includes catalogue of John Troughton's instruments.) (Howse fn. 13)

Harrison, David Mark. n.d. (*ca.* 1987). "The Influence of Non-linearities on Oscillator Noise Performance." Ph.D. diss., University of Leeds, England; copy at the British Library, London. (Burgess fn. 85)

Harrison, Edward. 1696. *Idea Longitudinis: Being, a brief Definition Of the best known Axioms For finding the Longitude. A more Rational Discovery thereof, than hath been heretofore Published.* London. (Turner fns. 30, 48)

*[Harrison, James]. 1736. Concerning fine imposed for allowing swine to trespass into a cornfield, May 24, 1736. Barrow Court Rolls, 1730-36, LR3/36/6,7, Public Record Office, London. (King fn. 25)

Harrison, James. 1828. "Observations on Winn's Improvements in Church and Turret Clocks." *Mechanics' Magazine,* vol. 10, no. 260, pp. 2-5. (Andrewes fn. 100)

———. 1829. Letter. *Mechanics' Magazine,* vol. 11, no. 304, p. 264. (King fn. 14)

*[Harrison, John]. 1720-36. Concerning Harrison's activities as juror in the Court of the Lord of the Manor. Barrow Court Rolls, 1720-36, LR3/36/2,3,4,5,6,7, Public Record Office, London. (King fn. 2)

———. 1730. Untitled manuscript signed "John Harrison, Clockmaker at Barrow; Near Barton upon Humber; Lincolnshire. June 10. 1730." MS. 6026/1, Library of the Worshipful Company of Clockmakers, Guildhall Library, London. Transcribed by G. H. Baillie in "The 1730 Harrison M.S." *Horological Journal,* vol. 92, no. 1102 (July 1950), pp. 448-50, and vol. 92, no. 1103 (August 1950), pp. 504-6. (Andrewes fns. 19, 25, 29-31, 34, 49-54, 56-9, 61, 67, 70; Burgess fns. 2, 5, 34, 69-70; King fns. 32, 38, 48, 50-3, figs. 18-19; Penney fn. 3)

———. n.d. (*ca.* 1740-1). "Some account of two of the most particular Circumstances whereby I was induced to Set about to make a third Machine." MS. 3972/2, pp. 5-7, Library of the Worshipful Company of Clockmakers, Guildhall Library, London. Transcribed by Andrew King in "A Manuscript of John Harrison, Circa 1740." *Horological Journal,* vol. 119, no. 2 (August 1976), pp. 4-6. (This is a similar but slightly longer version of another manuscript of the same period entitled "That the Ballances of my Second Machine are, from their Figure or Construction unfit for their intended purpose,....") (Andrewes fn. 99)

———. n.d. (*ca.* 1740-1). "That the Ballances of my Second Machine are, from their Figure or Construction unfit for their intended purpose, altho' not from my first Machine descern'd so to be before hand, viz. by any of my Friends the ablest Virtuosos, yet that they are so [as if, tho' too late discover'd] may be perceiv'd as followeth." MS. 3972/2, pp. 1-3, Library of the Worshipful Company of Clockmakers, Guildhall Library, London. (This is a similar but slightly shorter version of another manuscript of the same period entitled "Some account of two of the most particular Circumstances....") (Andrewes fn. 99, fig. 17)

———. 1752. Letter to James Short concerning use of term "gridiron pendulum," December 12, 1752. MS. 3972/2, pp. 9-10, Library of the Worshipful Company of Clockmakers, Guildhall Library, London. (Andrewes fn. 60)

———. n.d. (*ca.* 1760). "Of the Nature or Phaenomenon of Ballances, or the Solution of a seemingly Paradox therein." MS. 3972/2, pp. 11-16, Library of the Worshipful Company of Clockmakers, Guildhall Library, London. (This is a shorter version of another manuscript of the same period entitled "A Description of the Nature or Phenomenon of Ballances,....") (Andrewes fn. 110)

*[———]. 1761-6. Private "Journal" maintained for John and William Harrison by Walter Williams. The original and a later undated copy are both in private ownership in England; a second copy, transcribed by P. Atkinson and dated 1817 ("An account of the Chronometer, invented by John Harrison, English Horologist, for measuring longitude at sea," MSS Box 110/9, H 17809), is in the State Library of Victoria, Melbourne, Australia. (Andrewes fns. 72, 134; Randall fns. 13, 23-4, 35, 38)

*[———]. 1763. *An Account of the Proceedings, in order to the Discovery of Longitude: in a Letter to the Right Honourable *****, Member of Parliament.* London. (There are two editions of this work, both published in 1763: the first has 46 pages; the second, which includes the speech made by Martin Folkes to the Royal Society on November 30, 1749, has 98 pages. There is evidence that this pamphlet, signed "A MEMBER of the Royal Society," was written for John Harrison by James Short and Taylor White; see Andrewes fn. 28.) (Andrewes fns. 20, 22-3, 28, 32, 73-4, 87, 96-8, 105; Randall fn. 14)

———. 1763. "A Description of the Nature or Phenomenon of Ballances, as I found from Experience in my 3rd. Machine: Or the Solution of a seemingly Paradox therein." MS. 3972/1, pp. 103-19, Library of the Worshipful Company of Clockmakers, Guildhall Library, London. (This is a longer version of another manuscript of the same period entitled "Of the Nature or Phaenomenon of Ballances,....") (Andrewes fn. 110)

———. 1763. "An Explanation of my Watch or Timekeeper for the Longitude and as with a view of other Timekeepers, viz of such as have hitherto been produced in the World and as farther with some Historical Account Coincident to my Procedure," April 7, 1763. MS. 3972/1, pp. 1-102, Library of the Worshipful Company of Clockmakers, Guildhall Library, London. (Andrewes fns. 92, 94; Burgess fns. 6, 52, 55, 61; King fns. 6, 8; Randall fns. 41, 52)

[———]. [1765]. *A Narrative of the Proceedings relative to The Discovery of the Longitude at Sea; by Mr. John Harrison's Time-keeper; Subsequent to those published in the Year 1763.* [London]. (Randall fn. 25)

*[———]. 1765. "Projects." In *The Annual Register*, pp. 113-33. London. (In this section is an article giving a brief history of finding longitude at sea, with an account of Harrison's contributions.) (Andrewes, King, Randall gen. ref.)

[———]. *ca.* 1766; reissued 1770, 1773. *The Case of Mr. John Harrison.* Copies of this rare publication can be found in the Library of the Worshipful Company of Clockmakers, Guildhall Library, London, and in the British Museum, London: *ca.* 1766 edition in WCC MSS 6026/2 and 3973 (No. 18) and BM MS. 214.i.1(102); 1770 edition in WCC MS. 6026/2; 1773 edition in BM MS. 215.1.3(103). (Andrewes fn. 21; King fn. 34)

*[———]. 1767. *The Principles of Mr. Harrison's Time-keeper, with Plates of the Same. Published by order of The Commissioners of Longitude.* London. Preface and introductory "Notes Taken at the Discovery of Mr. Harrison's Time-keeper" by Nevil Maskelyne. French translation with additional material by Père Esprit Pezenas, under the title *Principes de la Montre de Mr. Harrison, avec les planches relatives à la même montre, imprimés à Londres en 1767 par ordre de Mrs. les Commissaires des Longitudes. A Avignon, Chez La Veuve Girard & François Seguin....Se vend A Paris.* Paris, 1767. Facsimile reprint by the British Horological Institute in *Principles and Explanations of Timekeepers by Harrison, Arnold and Earnshaw.* [Upton, Notts., England], 1984. (Betts fns. 1-2, 17; Burgess fn. 62; Cardinal fn. 10; King fn. 3; Randall fn. 44)

———. 1767. *Remarks on a Pamphlet Lately published by the Rev. Mr. Maskelyne Under the Authority of the Board of Longitude.* London. (Andrewes fn. 41; Randall fn. 43)

*[———]. 1773. Recommendations of a special Parliamentary Finance Committee concerning the final award to John Harrison, accepted by the Commons on June 19, 1773. *Journals of the House of Commons*, vol. 34, p. 383. (Randall fn. 46)

———. 1775. *A Description Concerning Such Mechanism as will Afford a Nice, or True Mensuration of Time; Together with Some Account of the Attempts for the Discovery of the Longitude by the Moon; as also an Account of the Discovery of the Scale of Musick.* London. A second printing with an appendix (pp. 109-14) was published later in the same year. (Andrewes fns. 33, 36, 38-9, 52, 62, 115, 121-2, app. 1; Betts fn. 29; Burgess fns. 7, 20, 23-4, 29, 31-2, 34, 43, 52, 55, 57-8, 60, 74, 76, 79-82; King fns. 22, 40-3, 46-7)

*[———]. 1775. Review of *A Description Concerning Such Mechanism....* In *The Monthly Review; or, Literary Journal: From July 1775 to January 1776* (London), vol. 53 (October), Art. VIII, pp. 320-9. (Burgess fn. 78)

———. n.d. Working drawing of the type of grasshopper escapement used in the R.A.S. regulator. MS. 3972/3, p. 5, Library of the Worshipful Company of Clockmakers, Guildhall Library, London. (Burgess fn. 51, fig. 9)

*[———]. 1777. "An Account of the Life of the late Mr. Harrison" (under "Characters"). In *The Annual Register*, pp. 24-6. London. (Andrewes fn. 37; King fn. 39)

*[———]. 1778. "Memoirs of the late Mr. Harrison." *The Universal Magazine of Knowledge and Pleasure* (London), vol. 63 (November), pp. 266-7. (Andrewes fn. 35)

*[———]. 1835. *The Gallery of Portraits: with Memoirs.* 7 vols. (1833-7). London: Charles Knight. (Memoir entitled "Harrison" on pp. 153-5 in vol. 5, with engraved portrait of John Harrison by William Holl II opposite p. 153.) (Andrewes fig. 28)

*[———]. 1889-90. *The Barrow Monthly Monitor* (November-May). Barrow upon Humber, England. (These issues, edited by Canon J. E. Sampson, contain important references to John Harrison.) (King fn. 28)

Harrison, William. 1762. Letter to Robert Atkinson concerning success of sea trial of H.4, April 10, 1762. MS. 6206, Library of the Worshipful Company of Clockmakers, Guildhall Library, London. (Randall fns. 11-12)

Hastings, Peter. 1993. "John Harrison—Spacecraft Engineer?" *Horological Journal*, vol. 136, no. 6 (December), pp. 193-7. (Andrewes fn. 109; Burgess fn. 66)

———. 1993. "A Look at the Grasshopper Escapement." *Horological Journal*, vol. 136, no. 2 (August), pp. 48-53. (Burgess fn. 46)

Hatton, Thomas. 1773. *An Introduction to the Mechanical Part of Clock and Watch Work.* London. Facsimile reprint in *Movements in Time.* Toronto: Turner and Devereux, 1978. (Andrewes fns. 52, 92-3, 132)

Hawkins, Isaac. 1714. *An Essay For The Discovery Of The Longitude at Sea, By several New Methods fully and Particularly laid before the Publick.* London. (Gingerich fn. 19; Turner app. 3)

*[Hearne, Thomas]. 1874. *Letters addressed to Thomas Hearne M.A....* London. (Turner fn. 49)

Heath, Thomas. 1921. *A History of Greek Mathematics.* 2 vols. Oxford: Oxford University Press. (Chandler fn. 25)

Henderson, E. 1867. *The Life of James Ferguson, F.R.S.* Edinburgh. (Andrewes fn. 118)

Hérigone, Pierre. 1634-42. *Cursus mathematicus nova, brevi et clara methodo demonstratus.* 6 vols. Paris. (Van Helden fn. 32)

Hering, D. W. 1932. *The Lure of the Clock.* New York: New York University Press. Revised and enlarged by Crown Publishers, 1963. (Cheney gen. ref.)

Hevelius, Johannes. 1647. *Selenographia.* Gdansk. (Van Helden fn. 33)

———. 1671. "Occultatio Primi Jovialium ab umbra Jovis." *Philosophical Transactions* (Royal Society, London), vol. 6, no. 78, pp. 3029-30. (Van Helden fn. 41)

Hildeyard, Thomas. 1727. *Chronometrum Mirabile Leodiense: Being A Most Curious Clock, Lately Invented By Thomas Hildeyard, Professor of Mathematicks in the English College of Liege.* London. Latin edition: Liège, n.d. (*ca.* 1725). (Penney fn. 9)

Hobbs, William. 1714. *A New Discovery For Finding the Longitude. Humbly Submitted to the Approbation of the Right Honourable the Lords Spiritual and Temporal,....* London. (Turner app. 3)

Hogarth, William. 1753. *Analysis of Beauty.* London. (Andrewes fn. 82)

Holmes, Robert. 1664/5. "A Narrative concerning the success of Pendulum-Watches at Sea for the Longitudes." *Philosophical Transactions* (Royal Society, London), vol. 1, no. 1, pp. 13-15. (Leopold fn. 37)

Hooke, Robert. 1705. *The Posthumous Works of Robert Hooke, M.D., S.R.S., Geom. Prof. Gresh. &c. Containing his Cutlerian Lectures, and other Discourses, Read at the Meetings of the Illustrious Royal Society.* Edited by Richard Waller. London. (App. D fn. 29, fig. 4)

Hope-Jones, F. 1940. *Electrical Timekeeping.* London: N.A.G. Press. (Burgess fn. 63)

Horrins, Johan [anagram of John Harrison (1761-1842), son of William Harrison]. 1835. *Memoirs of a Trait in the Character of George III. of these United Kingdoms; Authenticated by Official Papers and Private Letters in Possession of the Author:....* London. (Randall fns. 6, 45)

Howard, Edward. 1705. *Copernicans Of all Sorts, Convicted: By Proving, that the Earth hath no Diurnal or Annual Motion, as is suppos'd by Copernicans, from the beginning of the World, to this day. As Also That their Hypothesis is Astronomically, Philosophically, and Sensibly False, to all Impartial Apprehensions. To which is annex'd A Treatise of the Magnet: As also how to find the Annual Variation of the Compass, at Land and Sea, Mathematically demonstrated, by a Process Unknown before, for the Improvement of Navigation.* London. (Gingerich fns. 10-11)

Howse, Derek. 1970-1. "The Tompion Clocks at Greenwich and the Dead-beat Escapement." *Antiquarian Horology*, vol. 7, no. 1 (December), pp. 18-34, and vol. 7, no. 2 (March 1971), pp. 114-33. With appendix by Beresford Hutchinson. Reprinted as a booklet by the Antiquarian Horological Society, London. (Burgess fns. 35, 42)

———. 1975. *Greenwich Observatory: The Royal Observatory at Greenwich and Herstmonceux, 1675-1975.* Vol. 3, *The Buildings and Instruments.* London: Taylor and Francis. (Burgess fns. 64, 84)

———. 1980. *Greenwich time and the discovery of the longitude.* Oxford: Oxford University Press. (Burgess fn. 84; Howse app. 1; Landes fn. 24; Mörzer Bruyns fns. 3-4, 9, 15; Turner fns. 2, 5-6, 28; App. C fns. 4, 31)

———. 1989. *Nevil Maskelyne: The Seaman's Astronomer.* Cambridge: Cambridge University Press. (Chandler fns. 15-16; Howse app. 1; Stimson fn. 29; Thrower fn. 27; Van Helden fn. 73)

———, ed. 1990. *Background to Discovery: Pacific Exploration from Dampier to Cook.* Berkeley/Los Angeles: University of California Press. (Thrower app. 1)

Howse, Derek, and Hutchinson, Beresford. 1969. *The Clocks and Watches of Captain James Cook, 1769-1969.* London: Antiquarian Horological Society. (Betts fn. 15)

Howse, Derek, and Sanderson, Michael. 1973. *The Sea Chart.* Newton Abbot, England: David and Charles. (Thrower fn. 27)

Humbert, Pierre. 1933. *Un amateur: Peiresc, 1580-1637.* Paris: Desclée de Brouwer. (Van Helden fns. 9, 15, 27)

———. 1948. "Joseph Gaultier de la Valette, astronome provençale (1564-1647)." *Revue d'Histoire des Sciences et de leurs Applications,* vol. 1, pp. 314-22. (Van Helden fn. 9)

Humble, Richard. 1979. *The Seafarers: The Explorers.* Alexandria, Virginia: Time-Life Books. (Landes fn. 17)

Hunter, Michael. 1975. *John Aubrey and the World of Learning.* London: Duckworth. (Turner fn. 35)

———. 1982. *The Royal Society and its Fellows, 1660-1700: The Morphology of an Early Scientific Institution.* British Society for the History of Science, monograph 4. Chalfont St. Giles, England. (Turner fns. 31, 40)

Huygens, Christiaan. 1658. *Horologium.* The Hague. French translation in *Oeuvres Complètes de Christiaan Huygens,* vol. 17. pp. 41-73. English translation by Ernest L. Edwardes, in *The Story of the Pendulum Clock,* pp. 60-97. Altrincham, England: J. Sherratt, 1977. (Chandler fn. 5; Leopold fns. 6-7)

[———]. [1665]. *Kort Onderwys Aengaende het gebruyck Der Horologien Tot het vinden der Lenghten van Oost en West.* [The Hague]. Reprinted, with French translation, in *Oeuvres Complètes de Christiaan Huygens,* vol. 17, pp. 199-235. Translated into English as "Instructions Concerning the Use of Pendulum-Watches, for finding the Longitude at Sea" (1669). *Philosophical Transactions* (Royal Society, London), vol. 4, no. 47 (published 1670), pp. 937-53. Translated into Latin in *Opera Varia* (Leiden, 1724) and *Opera Mechanica* (Leiden, 1751). (Leopold fns. 38-9, 59; App. D fns. 10, 17)

———. 1673. *Horologium Oscillatorium Sive De Motu Pendulorum Ad Horologia Aptato Demonstrationes Geometricae.* Paris. French translation in *Oeuvres Complètes de Christiaan Huygens,* vol. 18, pp. 69-368. German translation by A. Heckscher and A. von Oettingen, eds., under the title *Die Pendeluhr.* Ostwald's Klassiker der Exakten Wissenschaften, no. 192. Leipzig: Wilhelm Engelmann, 1913. English translation with notes by R. J. Blackwell (introduction by H. J. M. Bos), under the title *Christiaan Huygens' The Pendulum Clock or Geometrical Demonstrations Concerning the Motion of Pendula as Applied to Clocks.* The Iowa State University Press Series in the History of Technology and Science. Ames, Iowa, 1986. (Chandler fns. 5, 7, fig. 1; Leopold fns. 20, 40, 44, 52, 60, 70, 90, 98)

*[———]. 1682. Concerning funding of Huygens's work by the Dutch East India Company, December 31, 1682. VOC 241, General Archives, The Hague. (Leopold fn. 67)

———. 1724. *Opera Varia.* 2 vols. Leiden. (Leopold fn. 101)

———. 1751. *Opera Mechanica.* 2 vols. Leiden. (Leopold fn. 101)

———. 1888-1950. *Oeuvres Complètes de Christiaan Huygens.* 22 vols. The Hague: Martinus Nijhoff. (Andrewes fns. 12, 43, 55, 93; Leopold fns. 1, 4-7, 9-11, 13-23, 25-41, 43-63, 65-95, 97-100, 104-7, figs. 1-5)

Huygens, Constantyn. 1876-88. *Journaal van C. Huygens, den zoon 1670-78, 1688-96.* Werken uitgegeven door het historisch gezelschap te Utrecht, Nieuwe Serie nos. 23, 25, 32, 46. Utrecht. (Leopold fn. 93)

Jones, Dennis. 1991. "E. T. Cottingham, F.R.A.S." *Antiquarian Horology,* vol. 19, no. 6 (Winter), pp. 593-605. (King fn. 12)

Jones, Vincent. 1978. *Sail the Indian Sea.* London/New York: Gordon and Cremonesi. (Landes fns. 12, 20)

*Journal des Sçavans.* 1665-. Paris. (Leopold fn. 63)

*Journals of the House of Commons.* 1711-14. Vol. 17. (Andrewes fn. 5; Turner fns. 1, 3-4, 77)

*[Kendall, John]. 1790. Kendall's obituary notice. *Gentleman's Magazine,* vol. 60, part 2, p. 113. (Penney fn. 32)

Kepler, Johannes. 1611. *Narratio de observatis a se quatuor Iovis satellitibus erronibus.* In *Johannes Kepler Gesammelte Werke.* Vol. 4, *Kleinere Schriften 1602-1611* and *Dioptrice,* pp. 315-25. Edited by Max Caspar and Frans Hammer. Munich: C. H. Beck, 1941. (Van Helden fn. 7)

———. 1627. *Tabulae Rudolphinae, Quibus Astronomicae Scientiae, Temporum longinquitate collapsae Restauratio continetur; A Phoenice illo Astronomorum Tychone,....* Ulm. (The world map by Philippus Eckebrecht of Nuremberg, dated 1630, has been inserted between the index [contents] and p. 1.) (Thrower fn. 7)

———. 1951. *Johannes Kepler Gesammelte Werke.* Vol. 15, *Briefe 1604-1607.* Edited by Max Caspar. Munich: C. H. Beck. (Mahoney fn. 1)

King, Andrew. 1976. "A Manuscript of John Harrison, Circa 1740." *Horological Journal,* vol. 119, no. 2 (August), pp. 4-6. (Andrewes fn. 99)

[———]. 1993. *From a Peal of Bells: John Harrison, 1693-1776.* Usher Gallery exhibition catalogue. [Lincoln, England]: Lincolnshire County Council. (Andrewes fns. 24, 101; King fns. 4, 26)

Kirby, J. L. 1846-7. Log for the *Owen Glendower,* with sketches of taking lunar-distance measurements, August 27, 1846, and February 8, 1847. MS. log M1, National Maritime Museum, Greenwich. (Howse fns. 21-2, figs. 10-11)

Klug, Joseph. 1906. "Simon Marius aus Gunzenhausen und Galileo Galilei." *Abhandelungen der II. Klasse der Königlichen Akademie der Wissenschaften,* vol. 22, pp. 385-526. (Van Helden fn. 11)

Konvitz, Josef. 1987. *Cartography in France, 1660-1846: Science, Engineering, and Statecraft.* Chicago: University of Chicago Press. (Thrower fn. 17)

Lacaille, Nicolas-Louis de. 1757. *Astronomiae Fundamenta.* Paris. (Howse fn. 9)

Lalande, Jérôme de. 1763. "Voyage d'Angleterre." MS. 4345 (3322), Bibliothèque Mazarine, Paris. Transcribed with an introduction by Hélène Monod-Cassidy as *Jérôme Lalande: Journal d'un Voyage en Angleterre.* Oxford, England: Voltaire Foundation at the Taylor Institution, 1980. (Andrewes fns. 28, 85, 117, 126; Randall fn. 17)

———. 1771-81. *Astronomie.* 2nd ed. 4 vols. Paris. (Van Helden fn. 68)

———. 1803. *Bibliographie Astronomique: avec l'histoire de l'astronomie depuis 1781 jusqu'à 1802.* Paris. (Andrewes fn. 28)

Landes, David S. 1983. *Revolution in Time: Clocks and the Making of the Modern World.* Cambridge: Harvard University Press. French translation, with additions, entitled *L'Heure qu'il est: Les horloges, la mesure du temps et la formation du monde moderne.* Paris: Éditions Gallimard, 1987. (Andrewes fn. 36; Cardinal fn. 37; Gingerich fn. 2)

Laplace, Pierre-Simon de. 1802. *Traité de mécanique céleste.* Vol. 3. Paris. (Chandler fig. 4)

Lapthorne, Richard. 1691. Letter to Richard Coffin concerning Thomas Axe's will, September 13, 1691. *Appendix 5th Report,* p. 382a, Historical Manuscripts Commission, London. (Turner fn. 44)

———. 1928. *The Portledge Papers being extracts from the Letters of Richard Lapthorne...to Richard Coffin...from December 10th 1687-August 7th 1697.* Edited by Russell J. Kerr and Ida Coffin Duncan. London. (Turner fn. 44)

Laycock, William S. 1976. "John Harrison, the man who mastered the pendulum." *Horological Journal,* vol. 118, no. 8 (February), pp. 5-13. (Burgess fn. 9)

———. 1976. *The Lost Science of John "Longitude" Harrison.* Ashford, England: Brant Wright Associates. (Burgess fns. 8, 33, 41; Leopold fn. 103)

Le Bot, Jean. 1983. *Les Chronomètres de marine français au XVIII^e siècle, Quand l'art de naviguer devenait science.* Grenoble, France. (Cardinal fn. 22)

———. 1987. "Pierre Le Roy et les horloges marines." In Catherine Cardinal and Jean-Claude Sabrier, eds., *La dynastie des Le Roy, horlogers du Roi,* pp. 43-50. Tours, France: Musée de Beaux-Arts. (Cardinal fns. 3, 38)

Lecky, Squire T. S. 1881. *"Wrinkles" in Practical Navigation.* London. 2nd ed., 1884. (Howse fn. 25; Stimson fn. 31)

Lee, Ronald A. 1969. *The First Twelve Years of the English Pendulum Clock, or the Fromanteel Family and their Contemporaries, 1658-1670.* London. (Turner fn. 75)

Leopold, J. H. 1980. "Christiaan Huygens and his instrument makers." In H. J. M. Bos *et al.*, eds., *Studies on Christiaan Huygens*, pp. 221-33. Lisse, the Netherlands: Swets. (Leopold fn. 2; Mahoney fn. 10)

———. 1981. "L'Invention par Christiaan Huygens du Ressort Spiral réglant pour les Montres." In *Huygens et la France,* pp. 153-7. Paris: Vrin. Shorter version published in *ANCAHA* (journal of the Association Nationale des Collectionneurs et Amateurs d'Horlogerie Ancienne, Paris), vol. 25 (1979), pp. 9-11. (Leopold fn. 64)

———. 1993. "Christiaan Huygens, the Royal Society and Horology." *Antiquarian Horology*, vol. 21, no. 1 (Autumn), pp. 37-42. (Leopold fn. 12)

Le Roy, Pierre. 1763. "Memoire Sur une nouvelle Montre Marine," December 7, 1763. MS. 104, Archives de l'Académie de Sciences de l'Institut de France, Paris. (Andrewes fn. 84)

———. 1768. *Exposé succint des travaux de MM. Harrison et Le Roy, dans la recherche des Longitudes en mer, et des épreuves faites de leurs Ouvrages.* Paris. (Cardinal fn. 9)

———. 1768. Letter to Monseigneur le Duc de Praslin seeking reward for his research, March 26, 1768. Archive ref. D294, Institut l'Homme et le Temps, La Chaux-de-Fonds, Switzerland. (Cardinal fn. 24)

———. 1770. *Memoire sur la Meilleure Maniere de Mesurer le Tems en Mer, Qui a remporté le prix double au jugement de l'Académie Royale des Sciences. Contenant la description de la montre à longitudes, présentée à Sa Majesté le 5 Août 1766. Par M. Le Roy l'aîné, Horloger du Roi.* Paris. Translated by T. S. Evans, under the title "A Memoir on the best Method of measuring Time at Sea, which obtained the double Prize adjudged by the Royal Academy of Sciences; containing the Description of the Longitude Watch presented to His Majesty the 5th of August 1766. By M. Le Roy, Clock-Maker to the King." *The Philosophical Magazine* (London), vol. 26 (October 1806-January 1807), pp. 40-68. (Betts fn. 18)

———. 1773. *Précis des recherches faites en France depuis l'année 1730, pour la détermination des Longitudes en mer, par la mesure artificielle du tems.* Amsterdam. (Cardinal fns. 14-17)

———. 1774. *Suite du Précis sur les Montres Marines de France; avec un Supplément au Mémoire sur la meilleure manière de mesurer le tems en mer.* Leiden. (Cardinal fn. 19)

Leur, J. C. van. 1955. *Indonesian Trade and Society: Essays in Asian Social and Economic History.* The Hague/Bandung: W. van Hoeve. (Landes fn. 21)

Levenson, Jay A., ed. 1991. *Circa 1492: Art in the Age of Exploration.* New Haven: Yale University Press. (Landes fns. 5, 15)

Lloyd, H. Alan, *et al.* 1948. "Rupert Thomas Gould, An Appreciation." *Horological Journal,* vol. 90, no. 1082 (November), pp. 655-6. (Penney fn. 66)

Logan, James. 1734. "An Account of Mr. Thomas Godfrey's Improvement of Davis's Quadrant, transferred to the Mariner's-Bow,...." *Philosophical Transactions* (Royal Society, London), vol. 38, no. 435 (published 1735), pp. 441-50. (App. D fn. 32)

Lombard, Jean. 1612. Observations of the formations of Jupiter's satellites made by Lombard in Marseilles, Malta, Cyprus, and Tripoli in Lebanon and sent to Peiresc. MS. 1803, fols. 251r-277r, Bibliothèque Inguimbertine, Carpentras, France. (Van Helden fn. 13)

Loria, Gino. 1902. *Spezielle Algebraische und Transscendente Ebene Kurven. Theory und Geschichte.* German translation by Fritz Schütte. Leipzig: B. G. Teubner. (Chandler fn. 7)

Maddison, Francis. 1969. *Medieval Scientific Instruments and the Development of Navigational Instruments in the XVth and XVIth Centuries.* Agrupamento de Estudos de Cartografia Antiga, no. 30. Coimbra, Portugal. (Landes fn. 13)

———. 1991. "Tradition and Innovation: Columbus' First Voyage and Portuguese Navigation in the Fifteenth Century." In Jay A. Levenson, ed., *Circa 1492: Art in the Age of Exploration,* pp. 89-94. New Haven: Yale University Press. (Landes fn. 15)

Mahoney, Michael S. 1980. "Christiaan Huygens: The measurement of time and longitude at sea." In H. J. M. Bos *et al.*, eds., *Studies on Christiaan Huygens*, pp. 234-70. Lisse, the Netherlands: Swets. (Leopold fn. 3; Mahoney fn. 3)

Maitland, William. [1750]. *An Essay Towards the Improvement of Navigation, Chiefly with Respect to the Instruments used at Sea.* London. (App. D fns. 1, 25)

Malin, Stuart R., *et al.*, eds. 1985. *Longitude Zero 1884-1984* (special issue of *Vistas of Astronomy*, vol. 28, parts 1/2, comprising the proceedings of the "Longitude Zero" Symposium held at Greenwich in 1984). Oxford, England: Pergamon. (App. B fn. 2)

Marcus, G. J. 1980. *The Conquest of the North Atlantic.* Ipswich, England: Boydell. (Landes app. 1)

Marguet, F. 1917. *Histoire de la longitude à la mer au XVIIIe siècle en France.* Paris. (Howse fn. 3)

———. 1931. *Histoire générale de la Navigation du XVe au XXe Siècle.* Paris. (Howse fn. 17)

Marius, Simon. 1614. *Mundus Iovialis.* Nuremberg. German translation in Joachim Schlör, ed., *Die Welt des Jupiter.* Gunzenhausen, Germany: Schrenk-Verlag, 1988. (Van Helden fns. 11, 29)

Martin, Benjamin. 1764. *Mathematical Institutions.* London. (Betts fn. 33)

Maskelyne, Nevil. 1763. *The British Mariner's Guide containing Complete and Easy Instructions for the Discovery of the Longitude at Sea and Land, within a Degree, by Observations of the Distance of the Moon from the Sun and Stars, taken with Hadley's Quadrant.* London. (Howse fn. 11; Randall fn. 22)

———. [1765]. "Notes taken at the Discovery of Mr. Harrison's Time-keeper." MS. RGO 257, Royal Greenwich Observatory Archives, Cambridge University Library. This manuscript was published following the preface in *The Principles of Mr. Harrison's Time-keeper, with Plates of the Same.* London, 1767. (Betts fn. 16; King fns. 3, 7)

———. 1767. *An Account of the Going of Mr. John Harrison's Watch, at the Royal Observatory, from May 6th, 1766, to March 4th, 1767. Together With The Original Observations and Calculations of the same.* London. (Andrewes fn. 131; Randall fn. 42)

———. 1792. *An Answer to a Pamphlet entitled "A Narrative of Facts," lately published by Mr. Thomas Mudge, Junior, relating to some Time-keepers, constructed by his Father Mr. Thomas Mudge: wherein is given an account of the trial of his first time-keeper, and of the three trials of his two other time-keepers, between the years of 1774 and 1790, by Order of the Board of Longitude, at the Royal Observatory: And also the Conduct of the Astronomer Royal, and the Resolutions of the Board of Longitude, respecting them, are vindicated from Mr. Mudge's misrepresentations.* London. (Betts fn. 35)

Maskelyne, Nevil, ed. 1767-1810. *The Nautical Almanac and Astronomical Ephemeris.* London: Commissioners of Longitude. (Howse fn. 15, figs. 1, 8)

———. *Tables Requisite to be used with the Astronomical and Nautical Ephemeris.* London: Commissioners of Longitude. (Published as a companion volume to *The Nautical Almanac and Astronomical Ephemeris.*) (Howse fn. 15)

Massoteau de St. Vincent. 1742. "Lettre." *Memoires pour l'Histoire des Sciences et des Beaux Arts* (September), pp. 1667-70. Reprinted in the facsimile edition (Paris, 1972) of A. Thiout, *Traité de l'Horlogerie.* 2 vols. Paris, 1741. (Leopold fn. 102)

Maurice, Klaus. 1976. *Die deutsche Räderuhr.* 2 vols. Munich: C. H. Beck. (App. C fn. 33)

———. 1980. "Jost Bürgi, or on Innovation." In Klaus Maurice and Otto Mayr, eds., *The Clockwork Universe,* pp. 87-102. New York: Neale Watson Academic Publications. (Andrewes fn. 93)

May, W. E. 1973. *A History of Marine Navigation.* Henley-on-Thames, England: G. T. Foulis. (Stimson fn. 4)

———. 1976. "How the Chronometer went to Sea." *Antiquarian Horology,* vol. 9, no. 6 (March), pp. 638-63. (Howse fn. 23)

Mayer, Jean. 1975. *Les Européens et les autres: de Cortes à Washington.* Paris: Armand Colin. (Landes fns. 18, 23)

Mayer, Tobias. 1770. *Tabulae Motuum Solis et Lunae Novae et Correctae.* Edited by Nevil Maskelyne. London. (App. D fn. 36)

Medina, Pedro de. 1545. *Arte de Navegar....* Valladolid. (Stimson fig. 3)

Meilink-Roelofsz, M. A. P. 1962. *Asian Trade and European Influence in the Indonesian Archipelago between 1500 and about 1630.* The Hague: M. Nijhoff. (Landes fn. 4)

Mel, C. 1719. "Pantometrum nauticum seu Machina pro invenienda Longitudinae et latitudine locorum in Mari," in *Antiquarius Sacer, quamplurima dubia atque obscuriaria sacræ idicta, ex statu ecclesiastico, politico, militari atque conomoco, Hebraœorum, Romanorum, Graœcorum Illustris & explicans. Cum Mantisia dissertationum....* Frankfurt-am-Main. (Turner app. 3)

Mercer, Vaudrey. 1972. *John Arnold & Son.* London: Antiquarian Horological Society. (Betts fns. 9-10, 38)

———. 1977. *The Life and Letters of Edward John Dent, Chronometer Maker, and some account of his Successors.* London: Antiquarian Horological Society. (Andrewes fns. 139-41)

Meyer, Cornelius. 1696. "Osservationi Delle Comete Che Douranno Seguire, e delle ecclisse del primo satellite di Giove, Et altre propositioni proficue da farsi per commodo ornamento della Città." Published as the third part of Cornelius Meyer, *L'Arte Di Rendere I Fiumi Navigabili In Varij Modi, con altre nuove inventioni, e varij altri segreti. Divisi In Tre Parti.* Rome. (Van Helden fig. 1)

Miller, Leonard F. 1977. "The inventive Gainsborough brothers." *Horological Journal,* vol. 119, no. 11 (May), pp. 5-8. (Andrewes fn. 80)

Miller, Russell. 1980. *The Seafarers: The East Indiamen.* Alexandria, Virginia: Time-Life Books. (Landes fns. 1, 3, 21-2)

Molander, Arne B. 1992. "Columbus and the Method of Lunar Distance." *Terrae Incognitae: The Journal for the History of Discoveries,* vol. 24, pp. 65-78. (Landes fns. 15, 24-5)

Moore, John Hamilton. 1782. *The Practical Navigator.* London. (Stimson fn. 16)

———. 1798. *The New Practical Navigator; being an Epitome of Navigation; containing the different methods of working the Lunar Observations, and all the Requisite Tables used with the Nautical Almanac, in Determining the Latitude and Longitude, and Keeping a Complete Reckoning at Sea.* 10th ed. London. (App. D fns. 3, 5, 19, 20)

Morison, Samuel Eliot. 1974. *The European Discovery of America: The Southern Voyages, 1492-1616.* Oxford: Oxford University Press. (Landes fn. 16)

Morpurgo, Enrico. 1970. *Nederlandse klokken- en horlogemakers vanaf 1300.* Amsterdam: Scheltema en Holkema, N. V. (Turner fn. 50)

Mörzer Bruyns, Willem F. J. 1992. "Navigation on Dutch East India Company Ships Around the 1740s." *The Mariner's Mirror,* vol. 78, no. 2, pp. 143-54. (Mörzer Bruyns fn. 10)

———. 1994. *The Cross-staff. History and Development of a Navigational Instrument.* Zutphen, the Netherlands: Walburg Instituut. (App. C fn. 14)

*[Mudge, Thomas]. 1750. Poor and Church Rates for the parish of St. Dunstan's in the West, recording lease for Thomas Mudge's London workshop. MS. 3008/2, Guildhall Library, London. (Penney fn. 26)

*[———]. 1793. *Report from the Select Committee of the House of Commons, to whom it was referred to consider of the Report which was made from the Committee to whom the Petition of Thomas Mudge, Watchmaker, was referred; and who were directed to examine into the Matter thereof, and also to make Enquiry into the Principles on which Mr Mudge's Timekeepers have been constructed.* London. (Betts fns. 28, 55; Howse fn. 16; Penney fn. 62)

*[———]. 1795. "Memoirs of the Life and Mechanical Labours of the late Mr. Thomas Mudge: With a striking Likeness of that celebrated Artist, engraved by Baker, from an original Painting by Dance." *The Universal Magazine of Knowledge and Pleasure* (London), vol. 97 (July), pp. 41-7. (Penney fn. 35)

Mudge, Thomas Jr. 1792. *A Narrative of Facts relating to some Time-Keepers, constructed by Mr Thomas Mudge, for the Discovery of the Longitude at Sea: together with Observations upon the Conduct of the Astronomer Royal Respecting Them.* London. (Randall fn. 51)

———. 1792. *A Reply to the Answer of the Rev. Dr. Maskelyne, Astronomer Royal, to A Narrative of Facts, relating to some Time-keepers, constructed by Mr. Thomas Mudge, for the Discovery of the Longitude at Sea, &c, to which is added A Short Explanation of the most proper methods of calculating a mean daily rate, With Remarks on some Passages in Dr. Maskelyne's Answer, by his Excellency The Count de Bruhl.* London. (Betts fns. 4, 44)

———. 1799. *A Description, with plates, of The Time-keeper invented by the late Mr. Thomas Mudge. To which is prefixed A Narrative, by Thomas Mudge, his Son, of measures taken to give effect to the invention since the reward bestowed upon it by the House of Commons in the year 1793; a republication of a tract by the late Mr. Mudge on the improvement of time-keepers; and a series of letters written by him to His Excellency Count Bruhl, between the years 1773 and 1787....* London. Facsimile reprint London: Turner and Devereux, 1977. (Andrewes fn. 63; Penney fns. 20, 44, 47, 51-2, 54, 57)

*The National Union Catalog Pre-1956 Imprints.* 1968-80. 685 vols. London. (App. C fn. 30)

Nautonnier, Guillaume de. 1603. *Meycometrie de leymant c'est a dire la maniere de mesvrer les longitudes par le moyen de l'eymant....* Venès. (Turner fn. 16)

"Navigation: Finding the future." 1993. Editorial in Science and Technology section of *The Economist,* vol. 329, no. 7836 (November 6), p. 115. (Landes fn. 29)

Needham, Joseph. 1959. *Science and Civilisation in China.* Vol. 3, *Mathematics and the Sciences of the Heavens and the Earth.* Cambridge: Cambridge University Press. (Stimson fn. 3)

———. 1962. *Science and Civilisation in China.* Vol. 4, *Physics and Physical Technology.* Cambridge: Cambridge University Press. (Stimson fn. 3)

Neugebauer, O. 1975. *A History of Ancient Mathematical Astronomy.* Part 2. New York: Springer-Verlag. (Chandler fn. 28)

Newton, Isaac. 1687. *Philosophiae Naturalis Principia Mathematica.* London. 2nd ed.: Cambridge, 1713. (Chandler fns. 6, 10, fig. 3)

———. 1730. *Opticks, or a Treatise of the Reflections, Refractions, Inflections & Colours of Light.* 4th ed. London. Reprinted New York: Dover Publications, 1952. (Mahoney fns. 7-8)

[———]. 1742. "A true Copy of a Paper found, in the Hand Writing of Sir Isaac Newton, among the Papers of the late Dr. Halley, containing a Description of an Instrument for observing the Moon's Distance from the Fixt Stars at Sea." *Philosophical Transactions* (Royal Society, London), vol. 42, no. 465 (published 1744), pp. 155-6 plus illustration. (App. D fn. 30, fig. 5)

———. 1959-77. *The Correspondence of Isaac Newton* [for the years 1661-1727]. Edited by H. W. Turnbull *et al.* 7 vols. Cambridge: Cambridge University Press. (Andrewes fns. 1, 4-7, 9; Gingerich fn. 30; King fn. 49)

———. 1974. *The Mathematical Works of Isaac Newton.* Edited by D. T. Whiteside. Vol. 6. Cambridge: Cambridge University Press. (Chandler fn. 11)

Norman, Robert. 1521. *The newe Attractive, Containyng a short discourse of the Magnes or Lodestone, and amongest other his vertues, of a newe discouered secret and subtill propertie, concernyng the Declinyng of the Needle, touched therewith vnder the plaine of the Horizon.* London. (Thrower fn. 20)

North, Douglass C. 1968. "Sources of Productivity Change in Ocean Shipping, 1600-1850." *Journal of Political Economy,* vol. 76, no. 5 (September/October), pp. 953-70. (Landes fn. 28)

Nunis, Doyce B. 1982. *The 1769 Transit of Venus: The Baja California Observations of Jean-Baptiste Chappe d'Auteroche, Vicente de Doz, and Joaquín Velázquez Cárdenas de Léon.* Los Angeles: Natural History Museum of Los Angeles County. (Van Helden fn. 5)

Oldenburg, Henry. 1965-86. *The Correspondence of Henry Oldenburg.* Edited by A. R. and M. B. Hall. 13 vols. Madison: University of Wisconsin Press, 1965-73 (vols. 1-9); London: Mansell, 1975-7 (vols. 10-11); London: Taylor and Francis, 1986 (vols. 12-13). (Leopold fns. 35, 44, 51, 59)

Ortelius, Abraham. 1570. *Theatrum Orbis Terrarum.* Antwerp. English edition: London, 1606. (Van Helden fns. 3-4)

Ortroy, Fernand van. 1901. "Bibliographie de l'Oeuvres de Pierre Apian." *Le Bibliographie Moderne,* vol. 5 (March-June and July-October), pp. 7-156, 284-333. (App. C fns. 17, 19)

———. 1920. "Bio-bibliographie de Gemma Frisius." *Académie Royale de Belgique. Classe des lettres et des sciences morales et politiques: Mémoires* (Brussels), vol. 11, 2nd series (December), 2nd part, pp. 9-358. (App. C fns. 1-2, 27, 30; App. D fn. 9)

Oudemans, J. A. C., and Bosscha, J. 1903. "Galilée et Marius." *Archives Néerlandaises des Sciences Exactes et Naturelles,* ser. 2, vol. 8, pp. 115-89. (Van Helden fn. 11)

Palmer, William. 1715. *A great Improvement in Watchwork; which may be of great use at Sea, for discovering the Longitude. Humbly offer'd to the Consideration of the Learned. With some Remarks on another Way of Discovering the Longitude....* York, England. (Turner app. 3)

Pares, Jean. n.d. "Jean-Baptiste Morin (1583-1656) et la quérelle des longitudes 1634-1647." Thèse de 3e cycle, l'Université de Paris. (Turner fns. 10, 13)

Parkinson, Sydney. 1773. *A Journal of a Voyage to the South Seas, in his Majesty's Ship, The Endeavour. Faithfully transcribed from the Papers of the late Sydney Parkinson.* Edited by Stanfield Parkinson. London. (Thrower fig. 11)

Patoun, Archibald. 1771. "Navigation." In *Encyclopaedia Britannica; or, a Dictionary of Arts and Sciences, Compiled upon a New Plan...,* vol. 3, pp. 365-93. London. (Appendix D fn. 11)

Peiresc, Nicolas-Claude Fabri de. 1610-12. "Astronomica," containing telescopic observations by Peiresc, Joseph Gaultier de la Valette, and Jean Lombard, as well as Gaultier's unpublished tables of the motions of Jupiter's satellites. MS. 1803, Bibliothèque Inguimbertine, Carpentras, France. (Van Helden fns. 9, 12)

———. n.d. (pre-1636/7). Record of a timepiece method of finding longitude presented by "sieur Boulenger mathematicien" to the Comte de Soisson. MS. fr. 9531, fol. 116r, Bibliothèque Nationale, Paris. (Turner fn. 19)

Penfold, J. B. 1983. "The London Background of George Graham." *Antiquarian Horology,* vol. 14, no. 3 (September), pp. 272-80. (Penney fn. 39)

Pepys, Samuel. 1935. *The Tangier Papers of Samuel Pepys.* Edited by E. Chappell. London: Navy Record Society. (Stimson fn. 1)

Pezenas, Père Esprit. 1767. *Principes de la Montre de Mr. Harrison, avec les planches relatives à la même montre, imprimés à Londres en 1767 par ordre de Mrs. les Commissaires des Longitudes. A Avignon, Chez La Veuve Girard & François Seguin....Se vend A Paris.* Paris. (Betts fn. 19)

Picard, Jean. 1680. *Voyage d'Uranibourg, ou Observations Astronomiques Faites en Dannemarck par Monsieur Picard de l'Académie des Sciences.* Paris. Published as the first part of the third work (pp. 1-29) in *Recueil d'Observations Faites en Plusieurs Voyages par Ordre de Sa Majesté, pour Perfectionner l'Astronomie et la Geographie. Avec divers Traitez Astronomiques. Par Messieurs de l'Académie Royale des Sciences.* Paris, 1693. Later published in *Memoires de l'Académie Royale des Sciences. Depuis 1666. jusqu'à 1699,* vol. 7, part 1, pp. 193-230. Paris, 1729. (Van Helden fn. 39)

Picard, Jean, and de La Hire, Philippe. 1693. "Pour la Carte de France corrigée sur les Observations de MM. Picard & de la Hire." This passage was first published as part of the third work (pp. 91-2, after the section concerning the surveys of France made by Picard and de La Hire) in *Recueil d'Observations Faites en Plusieurs Voyages par Ordre de Sa Majesté, pour Perfectionner l'Astronomie et la Geographie. Avec divers Traitez Astronomiques. Par Messieurs de l'Académie Royale des Sciences.* Paris. The map, showing the old and new surveys of the French coastline, is mounted on a stub bound into the volume following this passage. The passage was later published in *Memoires de l'Académie Royale des Sciences. Depuis 1666. jusqu'à 1699,* vol. 7, part 1, pp. 429-30, with the map bound in as a foldout plate opposite p. 1. Paris, 1729. (Thrower fn. 10; Van Helden fn. 46, fig. 9)

Pipping, Gunnar. 1977. *The Chamber of Physics: Instruments in the History of Sciences Collections of the Royal Swedish Academy of Sciences, Stockholm.* Stockholm: Almqvist & Wiksell. (Thrower fn. 23)

Pitot, Allain. 1716. *L'Automate De Longitude. Nouveau Systéme d'Hydrométrie; par les Périodes d'un Mouvement Nautique, qui marque à un Cadran, les Lieuës qu'un Navire fait dans sa Route. Présenté à nos Seigneurs les Commissaires de la Grande Bretagne pour l'Examen des Décourvertes sur la Longitude.* London. (Gingerich fns. 21-2; Turner app. 3)

Plank, Stephen. 1714. *An Introduction To The Only Method For Discovering Longitude. Humbly Presented, for the Good of the Publick.* London. (Gingerich fig. 4; Turner fn. 55, app. 3)

———. 1720. *An Introduction To a true Method For the Discovery of Longitude at Sea. Humbly offer'd to the Consideration of Both Houses of Parliament.* London. (Gingerich fns. 16-18; Turner app. 3)

Pleasure, Myron. 1970. "Precision Pendulum Clocks." *Horological Journal,* vol. 113, no. 2 (August), pp. 7-11; vol. 113, no. 3 (September), pp. 12-13; and vol. 113, no. 4 (October), pp. 6-10. (Burgess fn. 63)

———. 1973. "The Fedchenko Clock." *Horological Journal,* vol. 116, no. 3 (September), pp. 3-7, 55. (Burgess fn. 63)

Ploeg, W. 1934. *Constantyn Huygens en de Natuurwetenschappen.* Rotterdam: Nijgh & van Ditmar. (Leopold fns. 8, 96)

Pogo, Alexander. 1935. "Gemma Frisius, his method of determining differences of longitude by transporting timepieces (1530), and his treatise on triangulation (1533)." *Isis,* vol. 22 (2), no. 64 (February), pp. 469-85. (App. C fn. 3)

Pound, James. 1719. "New and accurate Tables for the ready Computing of the Eclipses of the first Satellite of Jupiter, by Addition only. By the Reverend Mr James Pound, R.S.S." *Philosophical Transactions* (Royal Society, London), vol. 30, no. 361 (published 1720), pp. 1021-34. (Van Helden fn. 70)

Pratt, Derek. 1993. "The first replica of John Harrison Sea Clock H1." *Horological Journal,* vol. 136, no. 3 (September), pp. 88-91. (Andrewes fn. 150)

———. 1994. "Len Salzer—the only man to make two replicas of John Harrison's Sea Clock, H1." *Horological Journal,* vol. 136, no. 8 (February), pp. 262-5. (Andrewes fn. 150)

Prickard, A. O. 1916. "The Mundus Jovialis of Simon Marius." *The Observatory,* vol. 39. (Van Helden fn. 11)

Proctor, George. 1736. Concerning Harrison's role in trial of H.1 on voyage to Lisbon, May-June 1736. MS. Adm.1/379, Public Record Office, London. (Andrewes fn. 76)

Ptolemy, Claudius. 1475. *Cosmographia.* Vicenza. (This work was later, and more commonly, known as *Geographia.* The first five printed editions [Vicenza, 1475; Rome, 1478; Bologna, dated 1462 but probably a misprint for 1482; Ulm, 1482; Ulm, 1486] were known as *Cosmographia* since that word was used in the translation of the title. The next edition, printed in Rome in 1490, used the translation "geographia" in the title, and in the following 1507 edition—and from then on—*Geographia,* a title more descriptive of its content, was adopted.) (Van Helden fn. 3; App. B fig. 6; App. C fn 39)

———. 1898. *Claudii Ptolemaei Geographia.* Edited by C. F. A. Nobbe. 3 vols. Leipzig. English translation by Edward Luther Stevenson, in *Claudius Ptolemy: The Geography.* New York: Dover, 1991. (Chandler fn. 1)

Quill, Humphrey. 1954. "A James Harrison Turret Clock at Brocklesby Park, Lincolnshire." *Horological Journal,* vol. 96, no. 1146 (March), pp. 156-9, and vol. 96, no. 1147 (April), pp. 234-6. (King fn. 14)

———. 1966. *John Harrison, the Man who found Longitude.* London: John Baker. (Andrewes fns. 24, 41, 47, 87, 103; Betts fn. 7; Burgess fns. 1, 15, 82; King fns. 1, 10; Penney fn. 67; Randall fns. 7, 29, 37, 53)

———. 1971. "The Grasshopper Escapement." *Antiquarian Horology,* vol. 7, no. 4 (September), pp. 288-96. (Burgess fns. 4, 47, 50)

———. 1976. *John Harrison: Copley Medallist and the £20,000 Longitude Prize.* Monograph No. 11. Ticehurst, England: Antiquarian Horological Society. This short history is a revised and updated version of an earlier article published in *Notes and Records of the Royal Society of London,* vol. 18, no. 2 (December 1963). (Andrewes fn. 45)

Ramsden, Jesse. 1777. *Description of an Engine for Dividing Mathematical Instruments.* London. (App. D fn. 37)

*[———]. 1787. Concerning the price of a sextant by Ramsden. MS. RGO 14/18, fol. 257, Royal Greenwich Observatory Archives, Cambridge University Library. (Howse fn. 14)

Randall, Anthony G. 1981. "L'Oeuvre Chronometrique de Ferdinand Berthoud de 1760-1787. Analyse du Traité des Horloges Marine et du Supplément au Traité des Horloges Marines." *ANCAHA* (journal of the Association Nationale des Collectionneurs et Amateurs d'Horlogerie Ancienne, Paris), vol. 30, pp. 23-35. (Betts fn. 40)

———. 1984. "An Early Pocket Chronometer by Thomas Earnshaw, signed Robert Tomlin." *Antiquarian Horology*, vol. 14, no. 6 (June), pp. 609-15. (Betts fn. 47)

———. 1986. "Ferdinand Berthoud: The influence of his contemporaries. Part 1, and his 'Première Montre Astronomique.' " *Antiquarian Horology,* vol. 16, no. 2 (June), pp. 149-65. (Randall fn. 20)

———. 1989. "The Technology of John Harrison's Portable Timekeepers." *Antiquarian Horology,* vol. 18, no. 2 (Summer), pp. 145-60, and vol. 18, no. 3 (Autumn), pp. 261-77. Also issued as a reprint, with addition and corrections, by *Antiquarian Horology*. (Andrewes fn. 90; Randall fn. 54)

———. 1992. *The Time Museum Catalogue of Chronometers.* Rockford, Illinois: The Time Museum. (Betts fn. 46)

Randall, Anthony G., and Good, Richard. 1990. *Catalogue of Watches in the British Museum.* London: British Museum. (Betts fn. 27)

Randles, W. G. L. 1985. "Portuguese and Spanish Attempts to Measure Longitude in the 16th Century." *Vistas in Astronomy* (special issue for the Longitude Zero Symposium 1984), vol. 28, parts 1/2 (1985), pp. 235-41. Reprinted, with additional appendix, in *Boletim da Biblioteca da Universidade de Coimbra,* vol. 39 (Centro de Estudos de História e Cartografia Antiga) and associated monograph. (Turner fn. 7; App. D fn. 11)

Rees, Abraham. 1819-20. *The Cyclopedia; or Universal Dictionary of Arts, Sciences and Literature.* London. Reprinted in *Rees's Clocks Watches and Chronometers (1819-20).* Rutland, Vermont: Charles E. Tuttle Co., 1970. (Andrewes fn. 52; Burgess fn. 49)

Riccioli, Giovanni Battista. 1651. *Almagestum Novum.* Bologna. (Van Helden fns. 30, 34)

———. 1672. *Geographia et Hydrographia Reformata.* 2nd ed. Venice. (Van Helden fn. 35)

Robertson, John. 1754. *The Elements of Navigation; Containing the Theory and Practice. With all the necessary Tables. To which is added, A Treatise of Marine Fortification. For the Use of the Royal Mathematical School at Christ's Hospital, and the Gentlemen of the Navy.* 2 vols. London. (Stimson fig. 8). 6th ed., edited by William Wales, 1796. (App. D fns. 3, 7, 13, 19, 21, 24)

Robinson, Arthur H., *et al.* 1978. *Elements of Cartography.* 4th ed. New York: John Wiley and Sons. (Thrower app. 1)

Robinson, T. R. 1831. "On the Dependence of a Clock's Rate on the Height of the Barometer." *Memoirs of the Royal Astronomical Society*, vol. 5. (Burgess fn. 72)

Robison, John. 1822. *A System of Mechanical Philosophy.* 4 vols. Edinburgh. (King fns. 18-19)

Roche, John J. 1981. "The Radius Astronomicus in England." *Annals of Science*, vol. 38, no. 1, pp. 1-32. (App. C fns. 14-15)

——— 1982. "Harriot, Galileo, and Jupiter's Satellites." *Archives Internationales d'Histoire des Sciences,* vol. 32, pp. 9-51. (Van Helden fn. 8)

Ronan, Colin A. 1960. "Laurence Rooke (1622-1662)." *Notes and Records of the Royal Society of London,* vol. 15, pp. 113-18. (Van Helden fns. 37, 56)

Rooke, Laurence. 1667. "Discourse Concerning the Observations of the Eclipses of Jupiter's Satellites." In Thomas Sprat, *The History of the Royal Society of London for the Improving of Natural Knowledge,* pp. 183-9. London. (Van Helden fns. 57-8)

Royal Society. 1665-. *Philosophical Transactions. Giving some Account of the Present Undertakings, Studies and Labours of the Ingenious, in many Considerable Parts of the World.* London. (Andrewes fns. 66, 75; Burgess fn. 36; King fns. 31, 44-5; Leopold fns. 37, 59; Penney fns. 34, 36; Randall fn. 4; Stimson fn. 25, fig. 6; Thrower fns. 14, 16, 25, fig. 8; Turner fn. 68; Van Helden fns. 41, 44, 59-64, 67, 69-71, figs. 10-11; App. D fns. 10, 17, 30, 32-3, figs. 5, 7)

Sabrier, Jean-Claude. 1984. "La contribution de Ferdinand Berthoud aux progrès de l'horlogerie de marine." In Catherine Cardinal, ed., *Ferdinand Berthoud, 1727-1807: Horloger mécanicien du Roi et de la Marine,* pp. 165-84. La Chaux-de-Fonds, Switzerland: Musée International d'Horlogerie. (Cardinal fn. 38; Randall fn. 40)

Sadler, D. H. 1968. *Man is not Lost: A record of two hundred years of astronomical navigation with the Nautical Almanac, 1767-1967.* London: Her Majesty's Stationery Office. (Howse app. 1)

Salzer, Len. 1983. "The Making of a Full Scale Model of No. 1." *Horological Journal,* vol. 125, no. 9 (March), pp. 16-18. (Andrewes fn. 150)

Sanford, E. C. 1888-9. "The Personal Equation." *American Journal of Psychology,* vol. 2, pp. 3-38, 271-98, 403-30. (App. B fn. 4)

*[Saunderson, Nicholas]. 1752. "The Life of Dr. Nicholas Saunderson, late Lucasian Professor of the Mathematics in the University of Cambridge. *The Universal Magazine of Knowledge and Pleasure* (London), vol. 10, no. 69 (May), pp. 193ff. (King fig. 1)

Saunier, M. Claudius. 1894-5. "Notre Dernier Mot sur Pierre Le Roy. Ses inventions et publications." *Revue Chronométrique: Journal de L'Horlogerie Française* (Paris), vol. 18, pp. 92-4, 141-3. (Cardinal fn. 30)

Savours, Ann. 1984. "A Very Interesting Point in Geography: The 1773 Phipps Expedition towards the North Pole." *Arctic,* vol. 37, no. 4 (December), pp. 402-28. (Betts fn. 24)

Seed, Patricia. 1994. "Enacting Colonialism: The Politics and Ceremony of European Rule over the New World, 1492-1640." Typescript. (Landes fn. 14)

Seller, John. 1672. *Practical Navigation or An Introduction to that whole Art.* London. (Stimson fig. 4)

Shapin, Steven. 1989. "The Invisible Technician." *American Scientist,* vol. 77, no. 6, pp. 554-63. (Mahoney fn. 11)

Shepherd, Anthony, ed. 1772. Preface to *Tables for Correcting the Apparent Distance of the Moon and a Star from the Effects of Refraction and Parallax. Published by Order of the Commissioners of Longitude.* Cambridge, England. (Stimson fn. 26)

Short, James. 1752. "A Letter of Mr. James Short, F.R.S. to the Royal Society, concerning the Inventor of the Contrivance in the Pendulum of a Clock, to prevent the Irregularities of its Motion by Heat and Cold." *Philosophical Transactions* (Royal Society, London), vol. 47, no. 88 (published 1753), pp. 517-24. (Andrewes fn. 127; King fn. 31)

Smith, Robert. 1749. *Harmonics, or the Philosophy of Musical Sounds.* Cambridge, England. 2nd ed., 1759, and postscript, 1762. Reprinted New York, 1966. (Andrewes fn. 122)

Snyder, John P. 1993. *Flattening the Earth: Two Thousand Years of Map Projections.* Chicago: University of Chicago Press. (Thrower app. 1)

Sorrenson, Richard. 1993. "Scientific Instrument Makers at the Royal Society of London, 1720-1780." Ph.D. diss., Princeton University. (Mahoney fn. 9)

Sotheran, Henry, and Co. 1921. *Bibliotheca Chemico-Mathematica: Catalogue of Works in Many Tongues on Exact and Applied Science.* London: Henry Sotheran and Co. (Items 8955-70, pp. 453-4, in vol. 2 list books and manuscripts connected with John Harrison and illustrate a page from item 8960, a transcript that Harrison made of Professor Saunderson's lectures on mechanics. Vol. 1 of the second supplement [Henry Sotheran, Ltd., 1937], lists other items [4110-17, p. 309], including an undated autograph manuscript entitled "Of the Nature of a Pendulum,...." The present whereabouts of all of these interesting documents is not known.) (Andrewes fn. 47; King fn. 1)

Sprat, Thomas. 1667. *The History of the Royal-Society of London, for the Improving of Natural Knowledge.* London. (Leopold fn. 21; Van Helden fn. 57-8)

Squire, Jane. 1743. *A Proposal To Determine our Longitude.* 2nd ed. London. (Gingerich fns. 40-1; Penney fn. 19)

Stewart, Larry. 1993. *The Rise of Public Science: Rhetoric, Technology, and Natural Philosophy in Newtonian Britain, 1660-1750.* Cambridge: Cambridge University Press. (Turner fn. 2)

Stimson, Alan N. 1985. "Some Board of Longitude Instruments in the Nineteenth Century." In P. R. de Clercq, ed., *Nineteenth-Century Scientific Instruments and their Makers,* pp. 93-115. Amsterdam: Rodopi/Leiden: Museum Boerhaave. (Stimson fn. 27)

———. 1988. *The Mariner's Astrolabe: A survey of known, surviving sea astrolabes.* Utrecht: HES. (Stimson fn. 10)

Stimson, Alan N., and Daniel, Christopher St. John Hume. 1977. *The Cross Staff: Historical Development and Use.* London: Harriet Wynter Ltd. (App. D fns. 26-7)

Streete, Thomas. 1661. *Astronomia Carolina. A New Theorie of the Coelestial Motions. Composed according to the Best Observations and most Rational Grounds of Art. Yet far more Easie, Expedite and Perspicuous than any before Extant. With Exact and most Easie Tables thereunto, and Precepts for the Calculation of Eclipses &c.* London. (Gingerich fn. 38)

———. 1664. *An Appendix to Astronomia Carolina containing a Proposition touching the Discovery of true Longitude, and the observation of three Lunar Eclipses.* London. (Turner fn. 15)

———. 1710. *Astronomia Carolina: A New Theory of the Cœlestial Motions....* 2nd ed. corrected, with appendix by Edmond Halley. London. (Turner fn. 52)

Struik, D. J. 1969. *A Source Book in Mathematics, 1200-1800.* Cambridge: Harvard University Press. (Chandler fn. 7)

Stukeley, William. 1740. Concerning display of H.1 at George Graham's workshop, July 3, 1740. Stukeley manuscript diaries, Shelfmark MSS England e 125, vol. 5, p. 7, Bodleian Library, Oxford University. (Penney fn. 25)

———. 1883. *The Family Memoirs of the Rev. William Stukeley, M.D. and the Antiquarian and other Correspondence of William Stukeley, Roger & Samuel Gale, Etc.* Surtees Society, publication no. 76, vol. 2. Durham, England. (Andrewes fn. 81)

Sully, Henry. 1716. "Montre pour la mer, inventée par M. Sully" (article no. 177 for the year 1716), vol. 3, pp. 93-4 and plate 1, published in M. [Jean-Gaffin] Gallon, *Machines et Inventions approuvées par l'Académie Royale des Sciences, depuis son établissement jusqu'à present; avec leur Description.* 7 vols. Paris, 1735 (vols. 1-6) and 1777 (vol. 7). (Andrewes fn. 16)

———. 1726. *Description Abrégée d'une Horloge d'une nouvelle invention, Pour la juste mesure du Temps sur Mer.* Paris. (There are two parts to this work, both published in 1726: the first, with a preface dated January 31, has 48 pages; the second with a preface dated December 31, has an additional 242 pages. A second revised edition entitled *Description abrégée d'une horloge d'une nouvelle invention, pour l'Usage de la Navigation avec le jugement de l'Académie Royale des Sciences sur cette invention...* appeared in 1728. Apparently only 300 copies were printed, however, and it was virtually unknown until one appeared at auction in Paris in 1994 [Hervé Chayette, Laurence Calmels, *La Bibliothèque Horlogère de Monsieur R. P.,* June 14-15, 1994, lot 468].) (Andrewes fn. 16, fig. 2)

Swerdlow, N. M. 1993. "The Recovery of the Exact Sciences of Antiquity: Mathematics, Astronomy, Geography." In Anthrony Grafton, ed., *Rome Reborn: The Vatican Library and Renaissance Culture,* pp. 125-67. New Haven: Yale University Press. (App. C fn. 39)

Tait, Hugh. 1987. *Catalogue of Watches in the British Museum.* Vol. 1, *The Stackfreed.* London: British Museum Publications. (App. C fn. 33)

"Tardy" [pseud. of Henri Lengellé]. 1972. *Dictionnaire des Horlogers Français.* Paris. (Turner fn. 50)

Taylor, E. G. R. 1954. *The Mathematical Practitioners of Tudor and Stuart England* [1485-1714]. Cambridge: Cambridge University Press. (Gingerich fns. 12, 20, 43; Turner fn. 62)

———. 1966. *The Mathematical Practitioners of Hanoverian England, 1714-1840.* Cambridge: Cambridge University Press. (Gingerich fns. 13, 42; App. D fn. 12)

———. 1971. *The Haven-Finding Art: A History of Navigation from Odysseus to Captain Cook.* New ed. London: Hollis & Carter for the Institute of Navigation. (Landes app. 1; Stimson fns. 5-6)

Tempera, Antimo [pseud. of Matteo Campani degli Alimeni]. 1668. *L'Oriuolo Giusto D'Antimo Tempera Utilissimo a'Naviganti* [The Accurate Clock of Antimo Tempera, Most Useful to Navigators]. Rome. (Andrewes fn. 66)

Thacker, Jeremy. 1714. *The Longitudes Examin'd. Beginning with a short Epistle to the Longitudinarians, and Ending with the Description of a smart, pretty Machine Of my Own, Which I am (almost) sure will do for the Longitude, and procure me The Twenty Thousand Pounds.* London. (Andrewes fns. 13-14, fig. 1; Betts fn. 33; Turner fn. 72, app. 3)

Theaker, Robert. 1665. *A Light to the Longitude or the Use of an Instrument call'd the Seaman's Director speadily resolving all Astronomical Cases and Questions concerning the Sun, Moon, Stars with several propositions whereby sea-men may find at what meridian and Longitude they are at, in parts of the world.* London. (Turner fn. 26)

Thicknesse, Philip. 1788. *A Sketch of the Life and Paintings of Thomas Gainsborough Esq.* London. (Andrewes fn. 80)

Thiout, A. 1741. *Traité de l'Horlogerie.* 2 vols. Paris. Facsimile reprint with extra material: Paris, 1972. (Leopold fn. 102)

Thomas, Ivor. 1957. *Selections Illustrating the History of Greek Mathematics.* 2 vols. Cambridge: Harvard University Press. (Chandler fn. 25)

Thorndike, Lynn. 1923-58. *A History of Magic and Experimental Science.* 8 vols. New York: Columbia University Press. (App. C fn. 29)

Thrower, Norman J. W. 1972. *Maps and Man: An Examination of Cartography in Relation to Culture and Civilization.* Englewood Cliffs, New Jersey: Prentice-Hall. (Thrower app. 1)

———. 1996. *Maps and Civilization: Cartography in Culture and Society.* Chicago: University of Chicago Press. (Thrower app. 1)

Thrower, Norman J. W., ed. 1981. *The Three Voyages of Edmond Halley in the* Paramore, *1698-1701.* 2 vols. Hakluyt Society Publications, 2nd series, no. 156-7. London. (Thrower fns. 12, 17, 19; Van Helden fns. 65-6)

———. 1984. *Sir Francis Drake and the Famous Voyage, 1577-1580. Essays Commemorating the Quadricentennial of Drake's Circum-navigation of the Earth.* Berkeley: University of California Press. (Landes app. 1)

Timber Development Association. n.d. "Lignum vitae." Hardwood: Timber leaflet 67. London. (Andrewes fn. 46)

Tobler, Waldo R. 1966. "Medieval Distortions: The Projections of Ancient Maps." *Annals of the Association of American Geographers,* vol. 56, pp. 351-60. (Thrower fn. 1)

Toynbee, H. 1859. "A few more words on lunars." *Nautical Magazine,* vol. 28 (October). (Howse fn. 24)

[Troughton, John]. 1783. "A Catalogue of mathematical philosophical and optical instruments, made and sold by John Troughton, successor to Benjamin Cole...." In Joseph Harris, *The Description and Use of the Globes and the Orrery....* 12th ed. London. (Howse fn. 13)

Truesdell, Clifford. 1968. "A Program toward Rediscovering the Rational Mechanics of the Age of Reason." In *Essays in the History of Mathematics,* pp. 85-137. New York: Springer-Verlag. (Chandler fn. 13)

Tuck, Paul. 1989. "Fine Examples of Antiquarian Horology seen during the Scottish Tour, 1988." *Antiquarian Horology,* vol. 18, no. 2 (Summer), pp. 181-9. (Turner fn. 74)

Turner, A. J. 1972. "The Introduction of the Dead-beat Escapement: A New Document." *Antiquarian Horology,* vol. 8, no. 1 (December), p. 71. (Burgess fn. 35)

———. 1984. *The Time Museum.* Vol. 1, *Time Measuring Instruments.* Rockford, Illinois: The Time Museum. (App. C fn. 32)

———. 1984. "L'Angleterre, la France et la navigation: le contexte historique de l'oeuvre chronométrique de Ferdinand Berthoud." In Catherine Cardinal, ed., *Ferdinand Berthoud, 1727-1807: Horloger mécanicien du Roi et de la Marine,* pp. 143-64. La Chaux-de-Fonds, Switzerland: Musée International d'Horlogerie. Reprinted in A. J. Turner, *Of Time and Measurement: Studies in the History of Horology and Fine Technology,* chap. 14. Aldershot, England: Variorum, 1993. (Turner fn. 11)

———. 1985. "France, Britain and the Resolution of the Longitude Problem in the 18th Century." *Vistas in Astronomy* (special issue for the Longitude Zero Symposium 1984), vol. 28, parts 1/2 (1985), pp. 315-19. (Chandler fn. 4; Turner fn. 11)

———. 1987. *Early Scientific Instruments: Europe 1400-1800.* London: Sotheby's Publications. (Landes app. 1; Stimson fn. 12)

———. 1994. "Learning & Language in the Somerset Levels, Andrew Paschall at Chedsey." In W. D. Hackmann and A. J. Turner, eds., *Learning, Language & Invention,* pp. 297-308. Paris/Aldershot, England: Variorum. (Turner fn. 38)

———. In preparation. "One of 'Two Brass Semicircles': an unidentified instrument in the Orrery collection identified." (Turner fn. 59)

Turner, A. J., and Crisford, A. C. H. 1977. "Documents Illustrative of the History of English Horology, 1: Two Letters Addressed to Thomas Mudge." *Antiquarian Horology,* vol. 10, no. 5 (Winter), pp. 580-2. (Penney fns. 37, 41)

Turner, Gerard L'E. 1969. "James Short, F.R.S., and his Contribution to the Construction of Reflecting Telescopes." *Notes and Records of the Royal Society of London,* vol. 24, no. 1 (June), pp. 91-108. (Burgess fn. 83)

Van Helden, Albert. 1977. *The Invention of the Telescope.* Transactions of the American Philosophical Society, vol. 67, part 4. Philadelphia. (Van Helden fn. 1)

van Helden, Anne C. 1991. "The Age of the Air-pump." *Tractrix. Yearbook for the History of Science, Medicine, Technology and Mathematics,* vol. 3, pp. 149-72. (Turner fn. 68)

Vanpaemel, G. 1989. "Science Disdained: Galileo and the Problem of Longitude." In C. S. Maffeoli and L. C. Palm, eds., *Italian Scientists in the Low Countries in the XVIIth and XVIIIth Centuries,* pp. 111-29. Amsterdam: Rodopi. (Van Helden fn. 28)

Vernay, Arthur S. n.d. (*ca.* 1940). *Catalogue of the Vernay Collection of Early English Furniture, Porcelains, Old English Silver, Glassware and other Art Objects.* New York. (Andrewes fn. 120, fig. 27)

Wales, William. 1794. *The Method of Finding Longitude by Timekeepers.* London. (App. D fns. 15-16)

Wallis, R. V., and Wallis, P. J. In preparation. *Bibliography of British Mathematics and its Applications.* Newcastle-upon-Tyne, England: Phibb. (Turner fn. 62)

Wallman, Henry. 1992. "Do Variations in Gravity mean that Harrison approached the Limit of Pendulum Accuracy?" *Horological Journal,* vol. 135, no. 1 (July), pp. 24-6. (Burgess fn. 26)

Ward, John. 1695. *A Compendium of Algebra. Consisting of plain, easie and Concise Rules for the Speedy attaining to that Art....* London. 2nd ed., enlarged. London, 1724. (Turner app. 4)

———. 1707. *The Young Mathematician's Guide being a plain and easy Introduction to the Mathematicks in five Parts....* London. Reissue with cancel title page, 1709; 2nd ed. corrected, 1713; 3rd ed. corrected, 1719; 4th ed., 1722; 5th ed. corrected, 1728; 6th ed., 1734; 7th ed., 1740 (the first edition with the additional *History of Logarithms...*); 8th ed., 1747; 9th ed., 1752; 10th ed., 1758; 11th ed., 1762; 12th ed., 1771 (by Samuel Clarke). (Turner fn. 58, fig. 1, app. 4)

———. 1714. *A Practical Method to Discover the Longitude at Sea, By a New Contrived Automaton. Freed from all the Various Effects of Air in different Climates, &c. And not Liable to Disorder by the Irregular Motion of a Ship. The whole Method Rendered Plain and Easy to be Understood by every Mariner....* London. (Turner fns. 56, 58, 60, 64-70, 73, app. 4)

———. 1714. *Clavis usurae or, a Key to Interest both simple and compound....* London. 2nd ed., London, 1740. (Turner app. 4)

———. 1730. *The Posthumous Works of Mr. John Ward...published by a particular Friend of the author's from the original Manuscript and revisd by Mr George Gordon, Mathematician in London.* London. 2nd ed., 1765. (Turner app. 4)

Waters, David W. 1958. *The Art of Navigation in England in Elizabethan and Early Stuart Times.* London: Hollis & Carter. (Stimson fn. 7; Turner fns. 17-18; App. C fn. 32; App. D fns. 12, 24)

———. 1966. *The Sea- or Mariner's Astrolabe.* Agrupamento de Estudos de Cartografia Antiga, no. 15. Coimbra, Portugal. (Stimson fn. 9)

———. 1976. *Science and Technique of Navigation in the Renaissance.* National Maritime Museum Monograph. London. (Stimson fn. 8)

Werner, Johann. 1514. *In hoc opere haec cõtinentur Noua translatio primi libri geographiae Cl'. Ptolomaei:....* Nuremberg. (The passage concerning Werner's description of the lunar-distance method of finding longitude is in *cap.* 4, note 8, sig. dv$^{v}$; see Appendix C for translation.) (Howse fn. 1; Mörzer Bruyns fn. 6; Stimson fn. 19; App. C fns. 7, 15, 22, 26, figs. 3-6)

Whewell, William. 1858. *History of the Inductive Sciences, from the Earliest to the Present Time.* 3rd ed., with additions. 2 vols. New York: D. Appleton. (Chandler fn. 14)

Whiston, William. 1721. *The Longitude And Latitude Found by the Inclinatory Or Dipping Needle; Wherein the Laws of Magnetism are also discover'd. To which is prefix'd, An Historical Preface; and to which is subjoin'd, Mr. Robert Norman's New Attractive, or Account of the first Invention of the Dipping Needle.* London. (Turner app. 3)

Whiston, William, and Ditton, Humphry. 1714. *A New Method For Discovering the Longitude Both At Sea and Land, Humbly Proposed to the Consideration of the Publick.* London. (Gingerich fns. 23-9; Turner fn. 2, app. 3)

———. 1715. *A New Method For Discovering the Longitude Both At Sea and Land, Humbly Proposed to the Consideration of the Publick.* 2nd ed. London. (Turner app. 3)

Wilson, Curtis A. 1980. "Perturbations and Solar Tables from Lacaille to Delambre: the Rapprochement of Observation and Theory, Part I." *Archive for History of Exact Sciences* (New York), vol. 22, no. 1/2, pp. 53-304. (Chandler fns. 21-4)

Wise, M. Norton, ed. 1994. *The Values of Precision.* Princeton, New Jersey: Princeton University Press. (Mahoney fn. 15)

Wood, Christopher. 1975. "The Function of the Quick Train Chronometer." *Antiquarian Horology,* vol. 9, no. 3 (June), pp. 331-6. (Betts fn. 52)

Wood, F. T. 1931. "Notes on London Booksellers and Publishers." *Notes & Queries,* vol. 161. (Turner fn. 42)

Wood, Peter H. 1984. "La Salle: Discovery of a Lost Explorer." *American Historical Review,* vol. 89, pp. 294-323. (Van Helden fn. 2)

Woodward, Philip. 1993. "The Performance of a 19th Century Regulator by William Hardy." *Horological Journal,* vol. 135, no. 9 (March), pp. 306-12. With introduction by John Redfern. (Burgess fn. 73)

Wright, Michael. 1989. "Robert Hooke's Longitude Timekeeper." In Michael Hunter and Simon Schaffer, eds., *Robert Hooke: New Studies,* pp. 63-118. Woodbridge, England: D. S. Brewer. (Turner fn. 71)

Yoder, J. G. 1988. *Unrolling Time: Christiaan Huygens and the mathematization of nature.* Cambridge: Cambridge University Press. (Leopold fns. 3, 90)

Youngson, A. L. 1960. "Alexander Bruce, F.R.S., Second Earl of Kincardine (1629-1681)." In H. Hartley, ed., *The Royal Society, its Origins and Founders,* pp. 251-8. London: Royal Society. (Leopold fns. 12, 14, 23)

Zach, Franz Xaver von. 1799. *Allgemeine Geographische Ephemeriden. Verfasset von einer Gesellschaft Gelehrten und herausgegeben.* Vol. 3. Weimar. (Chandler fig. 6)

Zinner, Ernst. 1934. *Die Fränkische Sternkunde im 11. bis 16. Jahrhundert.* Bamberg, Germany. (App. C fn. 1)

———. 1941. "Zur Ehrenrettung des Simon Marius." *Vierteljahresschrift der Astronomischen Gesellschaft,* vol. 76, pp. 23-75. (Van Helden fn. 11)

———. 1956. *Astronomische Instrumente des 11. bis 18. Jahrhunderts.* Munich: C. H. Beck. (App. C fn. 6)

———. 1990. *Regiomontanus: His Life and Work.* Translated by Ezra Brown. Amsterdam: North Holland. (App. C fns. 11, 16)

# Index

# Index

NOTES:

Lifespan dates are given for all deceased persons for whom this information is known. If the year of both birth and death are not known, the best available information is given; in some cases this is the year in which the person is mentioned in the text.

A page number in plain face (*e.g.,* 66) indicates mention on that page and sometimes also in a related footnote; a page number followed by "n" (*e.g.,* 66n) indicates mention only in a footnote; a page number followed by "c" (*e.g.,* 66c) indicates mention in a caption; a page number in bold (*e.g.,* **66**) indicates an illustration.

## C



## D

## E

## H

**I**



**J**



**K**

## M

**N**

## S



## T

**U**



**V**



**W**



**Y**



**Z**

World map by Pierre Moullard-Sanson, Paris, 1695. The information on the map is purported to allow the calculation of longitude.